Fractional
Quantum Hall Effects
New Developments

Fractional Quantum Hall Effects

New Developments

Editors

Bertrand I Halperin

Harvard University, USA

Jainendra K Jain

Penn State University, USA

World Scientific

NEW JERSEY · LONDON · SINGAPORE · BEIJING · SHANGHAI · HONG KONG · TAIPEI · CHENNAI · TOKYO

Published by

World Scientific Publishing Co. Pte. Ltd.
5 Toh Tuck Link, Singapore 596224
USA office: 27 Warren Street, Suite 401-402, Hackensack, NJ 07601
UK office: 57 Shelton Street, Covent Garden, London WC2H 9HE

Library of Congress Cataloging-in-Publication Data
Names: Halperin, Bertrand I., editor. | Jain, Jainendra K., editor.
Title: Fractional quantum Hall effects : new developments / editors,
    Bertrand I. Halperin, Harvard University, Jainendra K. Jain, Penn State University.
Description: New Jersey : World Scientific, [2020] | Includes bibliographical references and index.
Identifiers: LCCN 2020019145 (print) | LCCN 2020019146 (ebook) |
    ISBN 9789811217487 (hardcover) | ISBN 9789811218224 (paperback) | ISBN 9789811217494 (ebook)
Subjects: LCSH: Quantum Hall effect.
Classification: LCC QC612.H3 F73 2020 (print) | LCC QC612.H3 (ebook) | DDC 537.6/226--dc23
LC record available at https://lccn.loc.gov/2020019145
LC ebook record available at https://lccn.loc.gov/2020019146

British Library Cataloguing-in-Publication Data
A catalogue record for this book is available from the British Library.

For any available supplementary material, please visit
https://www.worldscientific.com/worldscibooks/10.1142/11751#t=suppl

https://doi.org/10.1142/9789811217494_fmatter

Contents

Foreword

I accept with pleasure the invitation to write a foreword for the book *Fractional Quantum Hall Effects: New Developments*. The Editors have asked me to write on my personal views on how the field has progressed in surprising ways, and on the implications of the quantum Hall effect to metrology, FQHE, and to the general field of topological phases, topological particles and topological quantum computation.

The discovery of the quantum Hall effect in 1980 was driven purely by experiment. Up until this time, nobody had realized that the Hall resistivity in two-dimensional systems could be measured directly without any knowledge of device parameters. I was aware of the importance of the Josephson effect for the realization of a voltage standard and fascinated by the fact that a quantized resistance was observed without the then-expected corrections due to localized electrons. I recall that David Thouless was most intrigued by the extreme precision of the quantization, which is probably what motivated him to apply his experience from topology to explain the effect in terms of a Chern number.

It has of course been a most fulfilling and rewarding experience for me to witness the evolution of quantum Hall effect and its branching out into numerous directions. Most of these would have been impossible to foresee at the time of the discovery in 1980.

The metrological implication of quantum Hall effect was mentioned in the discovery paper itself. Experiments have established that the quantized Hall resistances in Si, GaAs and graphene agree to at least 10 significant digits. It is therefore widely believed that the quantized value is identical to the combination of the fundamental constants given by h/e^2, which has now been adopted as the fundamental unit of resistance, called R_K. The quantum Hall effect later played a crucial role in the realization of an electronic kilogram based on a fixed value for the Planck constant.

An unusual feature of the quantum Hall effect has been appreciated since the early days of the field, namely that the bulk is insulating and the boundary of the device contains "metallic stripes". A quarter century later, this led to what are known as topological insulators. We now have a zoo of two- and three-dimensional topological insulators, for which the electrical conduction takes place at the boundary or the surface. Many cousins of the integer quantum Hall effect have also been born, notably the quantum spin Hall effect and quantum anomalous Hall effect.

Among the developments inspired by the integer quantum Hall effect, a special place is reserved for the fractional quantum Hall effect, discovered by Dan Tsui, Horst Stormer and Art Gossard in 1982. In contrast to the phenomena mentioned in the preceding paragraph, the fractional quantum Hall effect arises due to the Coulomb interaction between electrons. The corresponding energy gaps are smaller than for the integer quantum Hall effect and therefore this effect has not had applications in metrology, even though the accuracy of the fractional quantization has been demonstrated with an experimental uncertainty smaller than 0.1 ppm.

The fractional quantum Hall effect has been the birthplace for many concepts that would otherwise have appeared unbelievable. First came fractional charge and fractional statistics for the quasiparticle excitations. Then came composite fermions, whose integer quantum Hall effect explained the fractional quantum Hall effect, with impressive predictions for the spin polarizations of different ground states. Then we learned about the Fermi sea and pairing of composite fermions. The paired state, possibly realized in the 5/2 fractional quantum Hall effect, is believed to be a topological superconductor of composite fermions, whose excitations are predicted to be Majorana particles obeying non-Abelian statistics. An intense search for Majorana particles is currently going on in fractional quantum Hall effect as well as in other systems, both because of their intrinsically exotic nature and because they can potentially be useful for topological quantum computation. This can serve as an example of how ideas from fundamental physics can find applications.

The field of fractional quantum Hall effect continues to be robust and vibrant. Experiments continue to surprise us with unexpected discoveries. I give two recent examples, discussed at length in this book. In the first one, precise measurement of the Fermi wave vector of composite fermions revealed surprising behavior that has forced theorists to better understand the role of particle-hole symmetry. In the second example, the measurement of the thermal Hall effect at 5/2 has produced a value different from theoretical predictions, thus offering new insights into the nature of pairing. Part of the excitement in the field is caused by the new physics observed in bi- and multilayer systems and the ever-growing family of two-dimensional systems of high quality, such as graphene and ZnO. (I had started my research on two-dimensional systems with transport measurements on bulk tellurium, where, depending on the etching process, a two-dimensional hole gas with nice Shubnikov–de Haas oscillations could be created. Almost four decades later, wonderful quantum Hall effect has been seen in tellurene — see G. Qiu *et al.*, *Nano Lett.* **18**, 5760 (2018).) All of these materials bring their own unique fingerprints into the problem of quantum Hall effect, thus propelling the field forward.

While we have made dramatic progress in our understanding, not all is understood. In particular, I would like to stress that a complete understanding of the microscopic picture for transport in the fractional quantum Hall effect, which includes electrical contacts, edge properties, disorder and the flow of the electrical current, still needs more research. As detailed measurements in the integer quantum Hall effect have taught us, the actual edge is much more complex than a theorist's idealized notion of it.

As a field grows, it becomes harder for a newcomer, and sometimes even for an expert, to find relevant information in a quick and convenient fashion. This book captures the recent developments in the fractional quantum Hall effect in a readily accessible format, with chapters on a variety of topics of current interest written by acknowledged experts. It is a welcome addition to the field.

Klaus von Klitzing
Max Planck Institute for Solid State Research, Stuttgart

Preface

The first question for any new book is: What need does it fulfill? At least seven edited books and monographs have already been published in the field of fractional quantum Hall effects (FQHE):

(1) *The Quantum Hall Effect*, Eds. R. E. Prange and S. M. Girvin (Springer-Verlag, 1987)
(2) *The Quantum Hall Effects: Integral and Fractional*, T. Chakraborty (Springer Series in Solid State Sciences, 1995)
(3) *Perspectives in Quantum Hall Effects*, Eds. S. Das Sarma and A. Pinczuk (Wiley-Interscience, 1996)
(4) *Composite Fermions: A Unified View of the Quantum Hall Regime*, Ed. Olle Heinonen (World-Scientific, 1998)
(5) *Quantum Hall Effects: Field Theoretical Approach and Related Topics*, Francis Ezawa (World Scientific, first edition 2001, second edition 2008, third edition 2013)
(6) *The Quantum Hall Effect*, D. Yoshioka (Springer Series in Solid State Sciences, 2002)
(7) *Composite Fermions*, J. K. Jain (Cambridge University Press, 2007)

Why do we subject the physics community to yet another book on FQHE? It is worth spelling out the considerations that prompted us to undertake this project.

The 1982 discovery of FQHE sparked one of the most consequential areas of research in modern condensed matter physics. Quantum liquids had already been known for over seventy years, the best-known examples being superconductors and He-3 and He-4 superfluids. However, all of these previous quantum liquids, while different in detail, have an underlying commonality: They are Bose–Einstein condensates of one kind or another, theoretically described by an order parameter or by the concept of off-diagonal long-range order. The FQHE provides a new paradigm for collective quantum behavior. It does not involve any Bose–Einstein condensation or order parameters, but instead, it has a topological origin. The beauty and the richness of the emergent structures arising in the field of FQHE could hardly be more breathtaking. The FQHE has served either as the genesis or as a realization of dramatic concepts such as fractional quantization of the Hall conductance, charge fractionalization, quasiparticles with fractional statistics, composite fermions, non-Abelian anyons, skyrmions, excitonic superfluidity, chiral Luttinger liquids, non-Fermi liquids, topological superconductivity, Majorana fermions, fractionally quantized thermal Hall conductance, and topological quantum computation.

Against this backdrop, it appears to us that the number of currently available books is actually rather small and fails to do justice to the field. Many of the above-mentioned topics are either inadequately covered or altogether absent in these books. Most importantly, in the latter category are a large number of exciting experimental and theoretical developments that have taken place since the publication, more than a dozen years ago, of the most recent book listed above.

We therefore believe that a collection of articles written by experts focusing on the recent developments will prove timely and useful for the advancement of the field. The choice of topics covered in the book has been guided by the most important experimental discoveries in the recent years, as well as new developments in FQHE theory that are closely tied to experimental observations. We have striven for a book that should be useful for researchers in the field as well as for newcomers and interested outsiders who wish to educate themselves on the current status of the field. While the primary focus is on the relatively recent developments, many chapters contain a pedagogical component to ensure, to the extent allowed by length limitations, self-completeness and broad accessibility. A significant fraction of the book should be understandable to a diligent graduate student, although some of the more advanced topics would doubtless require consultation with the previous literature.

The book consists of a total of ten chapters. We provide here a synopsis of each chapter.

The Chapter 1 by Jainendra Jain contains a pedagogical introduction to composite fermion theory and a summary of the ways in which it has enabled an understanding of the physics of FQHE. This is followed by a review of recent theoretical works that perform detailed quantitative comparisons with experiments for spin polarization transitions and for the phase boundary of the crystal; these calculations treat Landau level mixing by the fixed phase diffusion Monte Carlo method. A density functional approach for the FQHE is described, which is capable of capturing certain topological properties of the state. The chapter ends with a review of the parton paradigm for the FQHE, which produces Abelian as well as non-Abelian states beyond the standard composite-fermion theory, and a discussion of which of these states are possibly realizable in experiments.

In Chapter 2, Bertrand Halperin addresses the physics of the half-filled Landau level in several contexts where significant progress in our understanding has recently been made. The issues of particle-hole symmetry, anisotropy and geometry are discussed particularly in the context of the half-filled lowest Landau level. Effects of strong disorder and theoretical issues concerning the transition between different quantized Hall states in the presence of disorder are also reviewed. Recent puzzles and their possible resolutions regarding the nature of the quantized Hall state observed in the second Landau level at half filling are briefly discussed, along with some remarks on the experimental search for non-Abelian statistics in this system. A final section is devoted to phenomena that can occur in two-component quantum Hall systems, such as a system with two nearby layers, when the filling factor for each component is close to one half.

In Chapter 3, the first of five chapters with an experimental focus, Mansour Shayegan describes experimental investigations of composite fermions near a half-filled Landau level through geometric resonance measurements. This chapter reviews evidence that the Fermi wave vector away from half-filling is determined by the density of the minority carriers, and discusses implications of this result for particle-hole symmetry. It further reports on several other phenomena: The degree to which anisotropy of the electron Fermi sea is transferred to the composite-fermion Fermi sea; evidence for fully spin-polarized composite fermions near $\nu = 5/2$, a necessary condition for the existence of topological p-wave pairing at $\nu = 5/2$; using the composite-fermion Fermi sea to probe the Wigner crystal in a nearby layer; and spontaneous Bloch magnetism for composite fermions at low densities.

Moty Heiblum and Dmitri Feldman offer, in Chapter 4, an account of the insights gained regarding the topological order of various FQHE states by probing their edges. They review the earlier shot-noise measurements of fractional charge for the primary FQHE states, and they discuss experimental manifestations of interference phenomena in systems with constrictions that allow tunneling between edges. They discuss thermal conduction and up-stream neutral modes, and they describe recent measurements of the quantized thermal Hall effect in various FQHE states, including the one at filling factor $5/2$, which raise some very puzzling questions.

In Chapter 5, Gábor Csáthy reports on new physics revealed at ultra-low temperatures and high hydrostatic pressures. It includes a rich interplay between FQHE, integer quantum Hall effect and charge-ordered nematic phases in the second Landau level. Evidence is presented for new odd denominator states, and also for phase transitions at even denominator fractions as a function of hydrostatic pressure. The author reviews systematic studies for the excitation gap of the $5/2$ state as a function of density and disorder.

Chapter 6, authored by Joseph Falson and Jurgen Smet, reviews FQHE in ZnO based heterostructures, which have seen tremendous improvements in mobility during the last decade. ZnO has allowed exploration of parameter space not available in other systems, in particular, through its tunable spin susceptibility. An intricate array of odd and even denominator FQHE and insulating states are reviewed, as well as transitions between them caused by crossings of Landau levels with different orbital or spin quantum numbers.

Over the past decade and a half, graphene and graphene-based structures have opened up a totally new area of fractional (as well as integer) quantum Hall research, with an effort that has been taken up by many research laboratories across the world. These developments are reviewed in Chapter 7, coauthored by Cory Dean, Philip Kim, Jia Li and Andrea Young. After introducing the single particle Landau levels in single and bilayer graphene, the authors discuss transport as well as capacitance measurements probing single and multicomponent FQHE in mono-layer and bilayer graphene, and also drag measurements in double-layer graphene where two monolayers are separated by an insulating boron nitride layer. The authors also discuss the moiré superlattice in orientationally-aligned graphene-boron nitride heterostructures, which arises from the slight mismatch in lattice constants, and has enabled the realization of new regimes in the integer quantized Hall effect, as well as in FQHE.

The book returns to a more theoretical focus with Chapter 8, by Steven Simon. The chapter provides a pedagogical account of many useful concepts, such as the entanglement spectrum, quantum Hall edges, non-Abelian braid statistics, and Hall viscosity, in the context of certain simple FQHE wave functions (such as Laughlin's) that are exact solutions of truncated parent Hamiltonians. This chapter also contains a discussion of relevant ideas from conformal field theory, Jack polynomials and symmetric polynomials.

In Chapter 9, Ady Stern deals with fractional quantum Hall states that support non-Abelian quasiparticles. He also provides a pedagogical introduction to proposals for engineering such states, either by combining FQHE states with superconductivity or by carefully controlling tunnel coupling in bilayer systems. His chapter reviews some of these ideas through the coupled wire construction for FQHE. Possible applicability to quantum computation is also discussed.

The Chapter 10 by Nigel Cooper reviews various possible ways of creating artificial magnetic fields for ultra-cold bosons in harmonic traps and optical lattices, as well as their possible FQHE states. Experimental probes for verifying these states are mentioned. The chapter also contains a discussion of the feasibility of FQHE in other bosonic systems, including spin $1/2$ quantum magnets (where the spins effectively behave as hard core bosons), engineered qubit arrays (which map into quantum magnets) and polariton systems.

The reader may note that several themes of the book recur in multiple chapters. Most notable among these are the topics of non-Abelian statistics and FQHE at fractions with even denominator. Chapters 3–5 discuss these topics in the context of experiments on GaAs heterostructures. Even denominator fractions in ZnO and in graphene are discussed, respectively, in Chapters 6 and 7. Wider discussions of the meaning of non-Abelian statistics and of the many varieties of them that could be realized, in principle, in FQHE states are discussed in Chapters 8 and 9, along with ideas on how to construct them. Theoretical issues related to the nature of the FQHE states at $\nu = 5/2$ and $7/2$, as well as the search for signs of non-Abelian statistics in interference experiments, are discussed in Chapter 2.

While we have tried to capture most of the major recent developments, it is inevitable that we have had to omit discussion of many significant research contributions for lack of space, and doubtless have overlooked others. We extend our apologies to those who feel that their work has not been adequately represented in the book.

The graphic on the cover page is an artist's depiction of fractional quantum Hall states in double layer graphene, where two-component composite fermions are realized that bind both intra-layer and inter-layer flux quanta of an emergent gauge field. It is closely based on a figure made by Jia Li, one of the authors of Chapter 7 devoted to FQHE in graphene, and Michelle Miller, an undergraduate student at Brown University.

We are deeply grateful to Prof. Klaus von Klitzing for kindly agreeing to write the Foreword for this book. His seminal discovery of the quantum Hall effect was the beginning of it all, and the field has benefited tremendously from his constant encouragement and leadership.

Finally, our sincerest thanks are owed to all of the contributors for the hard work they put into producing remarkably lucid and informative chapters. We hope that the readers will find the book a useful addition to the FQHE literature.

Bertrand I. Halperin

Harvard University

Jainendra K. Jain

Pennsylvania State University

Editors

Chapter 1

Thirty Years of Composite Fermions and Beyond

J. K. Jain

Physics Department, 104 Davey Laboratory,
Pennsylvania State University, University Park, Pennsylvania 16802, USA
jkj2@psu.edu

This chapter begins with a primer on composite fermions, and then reviews three directions that have recently been pursued. It reports on theoretical calculations making detailed quantitative predictions for two sets of phenomena, namely spin polarization transitions and the phase diagram of the crystal. This is followed by the Kohn–Sham density functional theory of the fractional quantum Hall effect. The chapter concludes with recent applications of the parton theory of the fractional quantum Hall effect to certain delicate states.

Contents

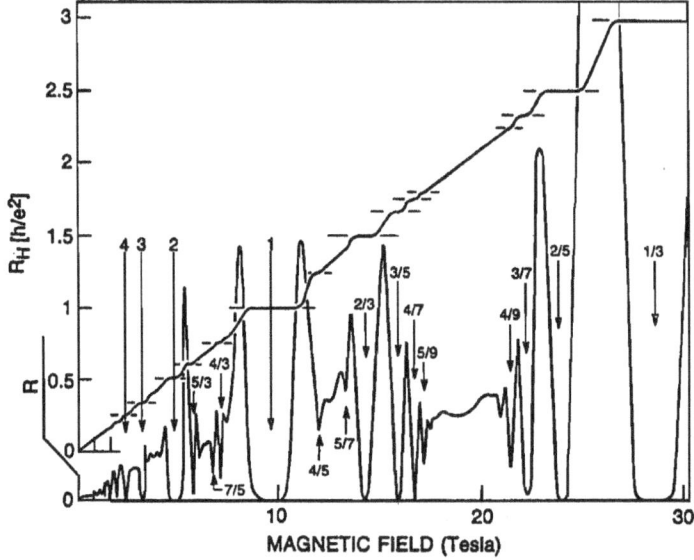

Fig. 1.   The Hall and the longitudinal resistances, R_H and R_L, respectively. The fractions asso-
ciated with the plateaus (or the resistance minima) are indicated. Source: H. L. Stormer and D.
C. Tsui, "Composite fermions in the fractional quantum hall effect," in *Perspectives in Quantum
Hall Effects*, pp. 385–421 (Wiley-VCH Verlag GmbH, 2007).[1]

1. The Mystery of the Fractional Quantum Hall Effect

The fractional quantum Hall effect (FQHE) is among the most stunning manifestations of quantum mechanics at macroscopic scales (Fig. 1). It occurs when electrons
are driven into an extreme quantum corner by confining them to two dimensions,
cooling them down to very low temperatures, and exposing them to a strong magnetic field. The term FQHE does not refer to a single observation but encompasses
a myriad of non-trivial states and phenomena. A fractional quantum Hall (FQH)
state is characterized by a precisely quantized plateau in the Hall resistance at
$R_H = h/fe^2$, where f is a fraction, approximately centered at the Landau level
(LL) filling factor $\nu = f$. (The nominal number of filled LLs, called the filling
factor, is given by $\nu = \rho\phi_0/B$, where ρ is the density, $\phi_0 = hc/e$ is called the
flux quantum, and B is the magnetic field. See Appendix A.1.) The plateau in
R_H is accompanied by a minimum in longitudinal resistance R_L, which vanishes as
$R_L \sim e^{-\Delta/2k_B T}$ as the temperature tends to zero, indicating the presence of a gap
Δ in the excitation spectrum. To date, close to 100 fractions have been observed
in the best quality samples. The number of FQH states is greater than the number of observed fractions because, in general, many distinct FQH states can occur
at a given fraction, differing in their spin polarization, valley polarization or some
other quantum number. Experimentalists have measured the energy gaps, collective modes, spin polarizations, spin wave excitations, transport coefficients, thermal

Hall effect, etc. for many of these FQH states as a function of density, quantum well width, temperature, and the Zeeman energy. Measurements have been performed in two-dimensional and also bilayer systems made of a variety of materials, such as GaAs, AlAs and ZnO quantum wells, heterostructures, and graphene. The FQHE is a data rich field.

To bring out the non-triviality of these observations it is helpful to introduce the "minimal" model Hamiltonian for the FQHE:

$$H = \sum_{j<k=1}^{N} \frac{1}{|\boldsymbol{r}_j - \boldsymbol{r}_k|} \text{ (LLL subspace)}, \tag{1}$$

which describes a two-dimensional system of electrons confined to the lowest LL (LLL). We have used the magnetic length $l = \sqrt{\hbar c/eB}$ as the unit of length and $e^2/\epsilon l$ as the unit of energy (ϵ is the dielectric constant of the background material), and suppressed the term representing interaction with a uniform positively charged background. In writing Eq. (1) we have assumed $\nu < 1$ and the limit of very high magnetic field, $\kappa \equiv (e^2/\epsilon l)/\hbar\omega_c \to 0$, where $\hbar\omega_c = \hbar eB/m_b c$ is the cyclotron energy (m_b is the electron band mass). In this limit the interaction is unable to cause LL mixing and, hence, electrons are strictly confined to the LLL. This Hamiltonian, which is to be solved within the Hilbert space of the LLL states,[a] has been stripped off of all features that are inessential to the FQH physics. In particular, the quantum-well width, LL mixing and disorder have all been set to zero in Eq. (1); these cause quantitative corrections but are not necessary for the phenomenon of the FQHE. For the same reason, the periodic potential due to the lattice has also been neglected, which is justified because the magnetic length, which controls the size of the wave function, is large compared to the lattice constant. The minimal Hamiltonian clarifies, in essence, that the physics of FQHE is governed by the Coulomb interaction alone. It is also noteworthy that the minimal model contains no free parameters, i.e. all sample specific parameters (e.g. the dielectric constant) can be absorbed into the measurement units. The FQHE is actually the most strongly correlated state in the world: the strength of correlations is measured by the ratio of the interaction energy to the kinetic energy, and the latter is absent here.

At the most fundamental level, the puzzle of the FQHE may be stated as follows. In the absence of interaction, all configurations (that is, all Slater determinant basis functions) of electrons in the LLL are degenerate ground states. There are very many of them. Even for a small system, say $N = 100$ electrons at $\nu = 1/3$, the number of degenerate ground states is $\binom{300}{100} \sim 10^{83}$, which is on the order of the number of quarks in the entire Universe. With so many choices, the electrons in the LLL are enormously frustrated. At the same time, the observed phenomenology is telling us that the system is on the verge of a spectacular non-perturbative

---

[a]For states in a different LL, this Hamiltonian needs to be solved within the Hilbert space of that LL. The matrix elements of the Coulomb interaction depend on the LL index and thus produce different behaviors in different LLs.

reorganization as soon as the repulsive Coulomb interaction is turned on. In particular, the observation of FQHE implies that at certain special filling factors nature conspires to eliminate the astronomical degeneracy to yield unique, non-degenerate ground states, which are certain entangled linear superpositions of all of the basis functions. This raises many questions. What is the organizing principle? What is the mechanism of the FQHE? What makes certain filling factors special? What is unique about the ground states at these fractions? What are their wave functions, and what physics do they represent? What are their excitations? What role does the spin degree of freedom play? What is the quantitative theory? How do gaps depend on the filling factor? What are the neutral collective modes and their dispersions? ... Finally, what other surprising phenomena lurk around the corner?

It turns out that we theorists can add to the wealth of FQHE data by performing our own experiments on the computer. A system on the computer is fully defined by two integers[b]: the number of electrons (N) and the number of magnetic flux quanta ($2Q$) to which they are exposed. The dimension of the Hilbert space is finite for a given ($N, 2Q$) system (assuming the LLL constraint), and when it is not too large, a brute force diagonalization can be performed to obtain the exact eigenstates and eigenenergies. This information exists for hundreds of ($N, 2Q$) systems, typically with $N < 18 - 20$ for today's computer, producing tens of thousands of exact eigenstates and eigenenergies. While the laboratory experiments present us with a few correlation and response functions, the computer experiments deliver the complete genomes of miniature FQH systems in the form of long lists of numbers that represent projections of all eigenstates along all directions in the very large Hilbert space. The availability of exact solutions for small systems is a powerful feature of the FQHE, because it allows a detailed and unbiased testing of any candidate theory.

The reader will surely not be surprised to learn that an exact analytical solution of Eq. (1), which gives all eigenfunctions and eigenenergies for all filling factors, does not exist. It is a certain bet that such a solution will never be found.[c] That may not worry a practitioner of condensed matter physics. After all, a satisfactory understanding of certain other systems of interacting electrons has been achieved without an exact solution. There is an important difference from these other systems, however. To illustrate, let us take the example of a weakly-coupled superconductor. Its understanding relies fundamentally on the availability of a "normal sate," namely

[b]This statement refers to the so-called spherical geometry, in which electrons move on the surface of a sphere subject to a radial magnetic field. In the periodic (torus) geometry, the aspect ratio (defined by the modular parameter) and the quasi-periodic boundary conditions are additional variables.
[c]It is possible to construct short range model interactions that produce certain simple FQH wave functions as exact zero-energy ground states.[2] See the Chapter by Steve Simon for examples. It should be noted, however, that these model interactions are constructed for already known wave functions; they are not solvable for excited states; and different model interactions are needed for different wave functions.

the Fermi sea, which is obtained when we switch off the interaction between electrons. This provides a unique and well-defined starting point. The minimal model for superconductivity, due to Bardeen, Cooper and Schrieffer (BCS), considers electrons with a weak attractive interaction (with strength small compared to the Fermi energy), and explains superconductivity as a pairing instability of the Fermi sea as a result of this interaction. This instability involves a rearrangement of electrons only in a narrow sliver near the Fermi energy. In contrast, there is no normal state for the FQHE. Switching off the interaction produces not a unique state but a large number of degenerate ground states. The FQHE cannot be understood as an instability of a known state. The absence of a natural starting point coupled with the fact that the Coulomb interaction is not small compared to any other energy scale makes the FQH problem intractable to the usual perturbative or quasi-perturbative treatments.

How do we proceed, then? As always, the goal of theory is to identify the simple underlying principles that provide a unified explanation of the complex behavior displayed by the interacting system. These principles should provide an intuitive understanding of the qualitative features of the phenomenology, and at the same time guide us toward a quantitative theory that is necessary for a detailed confirmation. Section 2 describes the unfolding of many important experimental facts and theoretical ideas in the 1980s that led to the postulate that nature relieves the frustration, i.e. eliminates the degeneracy of the partially occupied LLL, by creating a new kind of topological particles called composite fermions, which themselves can be taken as weakly interacting for many purposes. (In other words, the non-perturbative role of the repulsive interaction is to produce composite fermions; the rest is perturbative.) Section 2 provides a pedagogical introduction to the foundations of the composite fermion (CF) theory as well as its prominent verifications. Section 3 reports on detailed quantitative comparisons of the experimentally observed phase diagram of the spin polarization and the interplay between the crystal phase and the FQHE with theoretical calculations including the effects of finite quantum well width and LL mixing. Section 4 shows how the Kohn–Sham density functional theory can be formulated for the strongly correlated FQH state by exploiting the CF physics. The chapter concludes in Sec. 5 with the parton theory of the FQHE, which produces states beyond the CF theory, including many non-Abelian states (i.e. states that support quasiparticles obeying non-Abelian braid statistics). This section also gives a brief account of recent work indicating that some of these are plausible candidates for certain delicate states observed in higher GaAs or graphene LLs and in the LLL in wide quantum wells.

2. Composite Fermions: A Primer

This section contains an introduction to the essentials of the CF theory. A newcomer to the field may find it useful for the remainder of this chapter, and, perhaps, also for some other chapters in the book.

2.1. Background

The birth of the field was announced by the discovery of the integer quantum Hall effect (IQHE) by von Klitzing in 1980,[3] which, in hindsight, marked the beginning of the topological revolution in modern condensed matter physics. Von Klitzing observed that the Hall resistance is precisely quantized at $R_H = h/ie^2$, where i is an integer, with the plateau occurring in the vicinity of filling factor $\nu \approx i$. The quantization is exact as far as we now know, and the equality of the resistance on the $i = 1$ plateau in different samples has been established to an extremely high precision (a few parts in ten billion). The longitudinal resistance R_L shows a minimum at $\nu = i$, behaving as $R_L \sim \exp(-\Delta/2k_BT)$ as a function of temperature T. A gap Δ can be extracted from the temperature dependence of the longitudinal resistance. The most remarkable aspect of the IQHE is the universality of the quantization, which is utterly oblivious to the details such as which two-dimensional (2D) system is being used, what is the sample size or geometry, what band structure electrons occupy, what is their effective mass, or the nature or strength of disorder. The IQHE was not predicted, but was almost immediately explained by Laughlin[4] in 1981 as a consequence of the formation of LLs combined with disorder induced Anderson localization of states. Soon thereafter in 1982, Thouless et al.[5] related the Hall conductance to a topological quantity known as the Chern number, and a few years later Haldane[6] showed that bands with non-zero Chern numbers do not require a uniform external magnetic field. These works later served as inspiration for the field of topological insulators.

With the IQHE explained, the story seemed complete, and Tsui, Stormer and Gossard[7] set out to look for the Wigner crystal.[8] These authors' aim was to expose electrons to such high magnetic fields that they are all forced into the LLL. With their kinetic energy thus quenched, it is left entirely to the Coulomb repulsion to determine their state. What else could the electrons do but form a crystal?[9] In 1982, Tsui, Stormer and Gossard discovered instead a Hall plateau quantized at $R_H = h/(1/3)e^2$. This was not anticipated by any theory.

Laughlin again made a quick breakthrough in 1983.[10] He began by noting that a general LLL wave function must have the form $\Psi = F(\{z_j\})\exp(-\sum_i |z_i|^2/4)$, where $z_j = x_j - iy_j$ represents the coordinates of the jth electron as a complex number and $F(\{z_j\})$ is a holomorphic function of z_j's that is antisymmetric under exchange of two particles. (See Appendix A.1.) He then considered a Jastrow form $F(\{z_j\}) = \prod_{j<k} f(z_j - z_k)$, which builds in pairwise correlations and has been found to be useful in the studies of helium superfluidity. Imposing the conditions of antisymmetry under particle exchange and a well defined total angular momentum fixes $f(z_j - z_k) = (z_j - z_k)^m$, where m is an odd integer. That leads to the wave function

$$\Psi_{1/m} = \prod_{1 \le j < k \le N} (z_j - z_k)^m \exp\left[-\frac{1}{4}\sum_i |z_i|^2\right]. \tag{2}$$

This wave function describes a state at $\nu = 1/m$, and has been found to be an excellent representation of the exact ground state at $\nu = 1/3$ obtained in computer studies (results shown below). Laughlin postulated that it represents an incompressible state, i.e. it takes a non-zero energy to create an excitation of this state. With a flux insertion argument, he showed that the elementary excitation of this state has a fractional charge of magnitude e/m relative to the ground state (this argument actually relies only on the incompressibility of the state, not on the microscopic physics of incompressibility). He further wrote an ansatz wave function for the positively charged quasihole located at z_0 as

$$\Psi_{1/m}^{\text{quasihole}} = \prod_{k=1}^{N} (z_k - z_0)\Psi_{1/m}. \tag{3}$$

Laughlin also suggested a wave function for the negatively charged quasiparticle, but a better wave function for it is now available.

At this stage in early 1983 the story again seemed both elegant and complete. It only remained to test the Laughlin wave function, to measure the fractional charge of the excitations, and to look for a plateau quantized at $1/5$. Subsequent exploration showed, however, that the $1/3$ plateau was only the tip of the iceberg. Over the next few years, as experimentalists improved the conditions by removing dirt and thermal fluctuations, a deluge of new fractions revealed a large structure that was not a part of Laughlin's theory.

In a parallel development, the concept of particles obeying fractional braid statistics in two dimensions was being pursued, which subsequently played an important role in the theory of the FQHE. This possibility was introduced by Leinaas and Myrheim,[11] and by Wilczek[12] who christened these particles anyons. These particles are defined by the property that a closed loop of one particle around another has a non-trivial path-independent phase associated with it. (This is referred to as statistics because an exchange of two particles can be viewed as half a loop of one particle around another followed by a rigid translation.) Anyons can be defined only in two dimensions, because here, if one removes particle coincidences (say, by assuming an infinitely strong hard core repulsion), then each particle sees punctures at the positions of all other particles, and a closed path that encloses another particle cannot be continuously deformed into a path that does not. Wilczek[12] modeled anyons as charged bosons or fermions with gauge flux tubes bound to them carrying a flux $\alpha\phi_0$; the statistical phase then arises as the Aharonov–Bohm (AB) phase due to the bound flux. The list of particles in a particle-physics textbook does not contain any anyons, but nothing precludes the possibility that certain emergent particles in a strongly correlated condensed-matter system may behave as anyons. Nature seemed to oblige almost immediately. Halperin[13] proposed that Laughlin's quasiholes are realizations of anyons, which was confirmed by Arovas, Schrieffer and Wilczek in an explicit Berry phase calculation.[14]

In what is known as the hierarchy theory, Haldane[2] and Halperin[13] sought to understand the general FQH states based on the paradigm of the Laughlin sates.

The Laughlin fraction $\nu = 1/m$ serves as the point of departure. As the filling factor is varied away from $\nu = 1/m$, quasiparticles or quasiholes are created. A natural approach, in the spirit of the Landau theory of Fermi liquids, is to view the system in the vicinity of $\nu = 1/m$ in terms of a state of these quasiparticles or quasiholes. The hierarchy approach considers the possibility that these may form their own Laughlin-like states to produce new daughter incompressible states, which would happen provided that the interaction between the quasiparticles or quasiholes is repulsive with the short-distance part dominating. Beginning with the daughter states, their own quasiparticles or quasiholes (which have different charges and braid statistics than those of the $\nu = 1/m$ state) may produce, again provided that their interaction has the appropriate form, grand-daughter FQH states. A continuation of this family tree *ad infinitum* suggests the possibility, in principle, of FQHE at all odd denominator fractions.

Important ideas were proposed to address the question of what makes the Laughlin wave function special. A key property of this wave function is that it has no wasted zeros, that is, when viewed as the function of a single coordinate, say z_1, all of the zeros of the polynomial part of the wave function $\prod_{j<k}(z_j - z_k)^m$ are located on the other particles. This follows from the fundamental theorem of algebra: a simple power counting shows that the wave function, viewed as a function of one coordinate, is a polynomial of degree $m(N-1)$, i.e. has $m(N-1)$ zeros, which are all accounted for by the m zeros on each of the remaining $N-1$ particles.[d] Because of the holomorphic property of the wave function, each zero is actually a vortex, that is, it has a phase 2π associated for any closed loop around it. Building upon this observation and Wilczek's flux attachment idea, Girvin and MacDonald[15] introduced a singular gauge transformation that attaches an odd number (m) of gauge flux quanta to each electron to convert the Laughlin wave function into a bosonic wave function that is everywhere real and non-negative and also has algebraic off-diagonal long-range order. Zhang, Hansson and Kivelson[16] formulated a Chern–Simons (CS) field theory for the $\nu = 1/m$ state in which the singular gauge transformation is implemented through a CS term. In a mean field approximation, the effect of the external magnetic field is canceled by the m flux quanta bound to the bosons, thus producing a system of bosons in a zero effective magnetic field; the FQHE of electrons at $\nu = 1/m$ thus appears as a Bose–Einstein condensation of these bosons.[16]

2.2. *Postulates of the CF theory*

The motivation for the CF theory came from the following observation: If you mentally erase all numbers in Fig. 1, you will notice that it is impossible to tell the FQHE from the IQHE. All plateaus look qualitatively identical. This observa-

---

[d]The property of "no wasted zeros" cannot be satisfied for fractions other than $\nu = 1/m$. For example, an electron in the wave function of the $\nu = 2/5$ state sees, neglecting order one corrections, $5N/2$ zeros, only N of which are located at the other electrons.

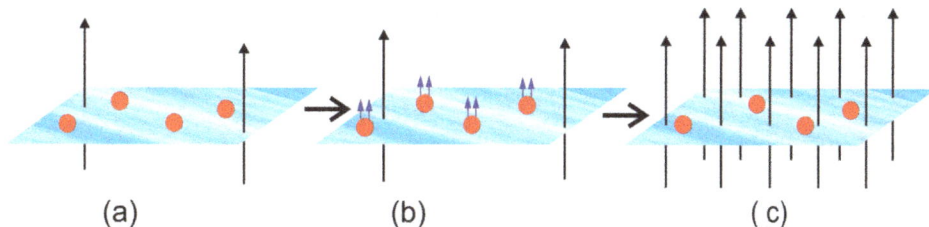

Fig. 2. Deriving FQHE from the IQHE through composite-fermionization. We (a) begin with an integer quantum Hall state at $\nu^* = n$, (b) attach two magnetic flux quanta to each electron to convert it into a composite fermion, and (c) spread out the attached flux to obtain electrons in a higher magnetic field. If the gap does not close during the flux smearing process, it produces a FQH state at $\nu = n/(2n+1)$. More generally, allowing the initial magnetic field to be positive or negative, i.e. $\nu^* = \pm n$, and attaching $2p$ flux quanta produces FQHE at $\nu = n/(2pn \pm 1)$.

tion suggests a deep connection between the FQHE and the IQHE. Can the well understood IQHE serve as the paradigm for understanding the FQHE? This question inspired the proposal that a new kind of fermions are formed, and their IQHE manifests as the FQHE of electrons.[17,18]

The intuitive idea, explained in Fig. 2, is as follows.[17] Let us begin with the integer quantum Hall (IQH) state of non-interacting electrons at $\nu^* = \pm n$ in a magnetic field $B^* = \rho\phi_0/\nu^*$. The sign of B^* indicates whether it is pointing in the positive or negative z direction. Now we attach to each electron an infinitely thin, massless magnetic solenoid carrying $2p$ flux quanta pointing in $+z$ direction. The bound state of an electron and $2p$ flux quanta is called a composite fermion.[e] The flux added in this manner is unobservable. To see this, consider the Feynman path integral calculation of the partition function, which receives contributions from all closed paths in the configuration space for which the initial and the final positions of electrons are identical, although the paths may involve fermion exchanges, which produces an additional sign $(-1)^P$ for P pairwise exchanges. The excess or deficit of an integral number of flux quanta through any closed path changes the phases only by an integer multiple of 2π and thus leaves the phase factors unaltered, and the fermionic nature of particles guarantees that the phase factors of paths involving particle exchanges also remain invariant. The new problem defined in terms of composite fermions is thus identical (or dual) to the original problem of non-interacting electrons at B^*. The middle panel of Fig. 2 thus represents the $\nu^* = \pm n$ integer quantum Hall (IQH) state of composite fermions in magnetic field B^*. (The quantities corresponding to composite fermions are conventionally marked by an asterisk or the superscript CF.)

This exact reformulation prepares the problem for a mean-field approximation that was not available in the original language. Let us adiabatically (i.e. slowly compared to $\hbar/\Delta$, where Δ is the gap) smear the flux attached to each electron until

---

[e]The bound state of an electron and a flux is a model of an anyon.[12] When the flux is an even integer number of flux quanta, the bound state comes a full circle into a fermion.

it becomes a part of the uniform magnetic field. At the end, we obtain particles moving in an enhanced magnetic field

$$B = B^* + 2p\rho\phi_0, \tag{4}$$

which is identified with the real applied magnetic field. This implies

$$\nu = \frac{n}{2pn \pm 1}, \tag{5}$$

where $\pm$ corresponds to the CF filling $\nu^* = \pm n$. If the gap does not close during the flux smearing process, i.e. if there is no phase transition, then we have obtained a candidate incompressible state at a fractional filling factor. To be sure, we know from general considerations that the system must undergo a complex evolution through the flux smearing process. The cyclotron energy gap of the IQHE must somehow evolve into an entirely interaction induced gap, and the wave function of n filled LLs into a LLL wave function. The electron mass, which is not a parameter of the LLL problem, is not simply renormalized but must be altogether eliminated during the above process. A satisfactory quantitative description of the evolution of the interacting ground state as the attached flux is spread from point flux to a uniform magnetic field is not known.

To make further progress, we abandon the idea of theoretically implementing the flux smearing process, but rather use the above physics as an inspiration to make an ansaz directly for the final state. A mean field theory suggests[17]

$$\Psi^{\text{MF}}_{\nu = \frac{n}{2pn \pm 1}} = \prod_{j<k} \left(\frac{z_j - z_k}{|z_j - z_k|} \right)^{2p} \Phi_{\pm n}, \tag{6}$$

where the multiplicative factor is a pure phase factor associated with $2p$ flux quanta bound to electrons. Here $\Phi_{-n} = [\Phi_n]^*$ is the wave function of n filled LLs in a negative magnetic field, and the magnetic length in the gaussian factor of $\Phi_{\pm n}$ is chosen so as to ensure that the wave function Ψ describes a state at the desired filling factor. A little thought shows that this wave function has serious deficiencies: it does not build good correlations, as can be seen from the fact that $|\Psi^{\text{MF}}_{\nu = \frac{n}{2pn \pm 1}}| = |\Phi_{\pm n}|$; it has a large admixture with higher LLs; and for $\nu = 1/(2p + 1)$ it produces the wave function $\Psi_{1/(2p+1)} \sim \prod_{j<k} (z_j - z_k)^{2p+1}/|z_j - z_k|^{2p}$, where we have used $\Phi_1 \sim \prod_{j<k}(z_j - z_k)$ (suppressing the ubiquitous Gaussian factors for notational ease), rather than the Laughlin wave function. Many of these problems are eliminated by dropping the denominator,[17] which does not alter the topological structure. That gives:

$$\Psi^{\text{unproj}}_{\nu = \frac{n}{2pn \pm 1}} = \prod_{j<k} (z_j - z_k)^{2p} \Phi_{\pm n}. \tag{7}$$

This wave function explicitly builds good correlations for repulsive interactions, because the configurations wherein two particles approach close to one another have probability vanishing as r^{4p+2}, where r is the distance between them, and are thus strongly suppressed. For $\nu = 1/(2p + 1)$ we recover the Laughlin wave function $\Psi_{1/(2p+1)} \sim \prod_{j<k}(z_j - z_k)^{2p+1}$, but with the new physical interpretation as the $\nu^* = 1$ IQH state of composite fermions. Going from Ψ^{MF} to Ψ^{unproj} also significantly reduces admixture with higher LLs, producing wave functions that are predominantly in the LLL as measured by their kinetic energy.[19,20] Because strictly LLL wave functions are convenient for many purposes, we project Ψ^{unproj} explicitly into the LLL to obtain

$$\Psi_{\nu=\frac{n}{2pn\pm1}} = \mathcal{P}_{\mathrm{LLL}} \prod_{j<k} (z_j - z_k)^{2p} \, \Phi_{\pm n} \,, \qquad (8)$$

with the hope that the nice correlations in the unprojected wave function will survive LLL projection.

Further generalizing to arbitrary filling factors, we obtain the final expression[f]:

$$\Psi^\alpha_{\nu=\frac{\nu^*}{2p\nu^*\pm1}} = \mathcal{P}_{\mathrm{LLL}} \prod_{j<k} (z_j - z_k)^{2p} \Phi^\alpha_{\pm\nu^*} \qquad (9)$$

where α labels different eigenstates (not to be confused with the statistics parameter), and ν is related to the CF filling ν^* by

$$\nu = \frac{\nu^*}{2p\nu^* \pm 1} \,. \qquad (10)$$

Equation (9) may be taken as the defining postulate of the CF theory. While the line of reasoning leading to it was physically motivated, the wave functions Ψ^α_ν are mathematically rigorously defined and allow us to make detailed predictions that can be tested against experiments. It is also possible, in principle, to unpack these wave functions to obtain explicit expansions of all eigenstates along all basis functions and compare with exact computer results.

Equation (9) encapsulates the remarkable assertion of the CF theory, namely that all low-lying eigenstates at arbitrary filling factors in the LLL can be compactly represented by the single equation, which contains no adjustable parameters, and which, as discussed next, reveals in a transparent fashion the emergence of new topological particles that experience a reduced magnetic field.

<u>Reading the physics from the wave functions in Eq. (9):</u> To see what physics Eq. (9) represents, let us inspect it afresh, pretending ignorance of the physical motivation that led to it. Disregarding the LLL projection for the moment, there are two important ingredients in the wave function: the Jastrow factor $\prod_{j<k}(z_j - z_k)^{2p}$ and the IQH wave function $\Phi_{\pm\nu^*}$. (i) The Jastrow factor attaches $2p$ vortices to electrons. [A particle, say z_1, sees $2p$ vortices at the positions of all other particles, due to the factor $\prod_{j=2}^{N}(z_1 - z_j)^{2p}$.] The bound state of an electron and $2p$ quantized vortices is interpreted as an emergent particle, namely the composite fermion. (ii) Because the vortices are being attached to electrons in the state $\Phi_{\pm\nu^*}$, the right-hand side is naturally interpreted as a state of composite fermions at $\pm\nu^*$. (iii) The relation $\nu = \nu^*/(2p\nu^* \pm 1)$ can be derived from the wave function by determining

---

[f]The wave functions in Eqs. (7)–(9) are sometimes referred to as the Jain states and the fractions in Eq. (5) as the Jain sequences.

the angular momentum of the outermost occupied orbit. (iv) The effective magnetic field for composite fermions arises because the Berry phases induced by the bound vortices partly cancel the AB phases due to the external magnetic field. The Berry phase associated with a closed loop of a composite fermion enclosing an area A is given by the sum $-2\pi BA/\phi_0 + 2\pi 2pN_{enc}$, where the first term is the AB phase of an electron going around the loop, and the second term is the Berry phase of $2p$ vortices going around N_{enc} electrons inside the loop. Interpreting the sum as an effective AB phase $-2\pi B^* A/\phi_0$ produces, with $N_{enc} = \rho A$, the effective magnetic field $B^* = B - 2p\rho\phi_0$. (v) The composite fermions are said to be non-interacting because the only role of the interaction is to bind vortices to electrons through the Jastrow factor $\prod_{j<k}(z_j - z_k)^{2p}$ to create composite fermions, and $\Phi_{\pm\nu^*}$ on the right-hand side of Eq. (9) is the wave function of non-interacting fermions. (vi) We can also see that a composite fermion is a topological particle, because a vortex is a topological object, defined through the property that a closed loop of any electron around it produces a Berry phase of 2π, independent of the shape or the size of the loop. (vi) We finally come to $\mathcal{P}_{LLL}$. The LLL projection renormalizes composite fermions in a very complex manner, producing extremely complicated wave functions. We postulate that the projected wave functions are adiabatically connected to the unprojected ones, and therefore describe the same physics. In other words, we assume that LLL projection does not cause any phase transition. While the physics of vortex binding is no longer evident after LLL projection, it is possible to test many qualitative features of the formation of composite fermions with the LLL theory, e.g. the similarity of the spectrum to that of non-interacting fermions at B^*.

To summarize: Interacting electrons in the LLL capture $2p$ quantized vortices each to turn into composite fermions. Composite fermions experience an effective magnetic field $B^* = B - 2p\rho\phi_0$, because, as they move about, the vortices bound to them produce Berry phases that partly cancel the effect of the external magnetic field. Composite fermions form their own Landau-like levels, called Λ levels (ΛLs), in the reduced magnetic field, and fill ν^* of them. (Recall that all of this physics occurs in the LLL of electrons. The LLL of electrons effectively splits into ΛLs of composite fermions.) The occupation of ΛL orbitals is defined by analogy to the occupation of LL orbitals at ν^*. See Fig. 3 as an example. This physics is described by the electronic wave function in Eq. (9), where the right-hand side is interpreted as the wave function non-interacting composite fermions at filling factor ν^*.

It ought to be noted that no real flux quanta are bound to electrons. The flux quantum in Fig. 2 is to be understood as a model for a quantum vortex. While the model of composite fermions as point fluxes bound to electrons is not to be taken literally, it is topologically correct and widely used due to its pictorial appeal and the fact that it yields the correct B^*. In the same vein, an external magnetometer will always measure the field B, not B^*. The effective field B^* is internal to composite fermions, and composite fermions themselves must be used to measure it.

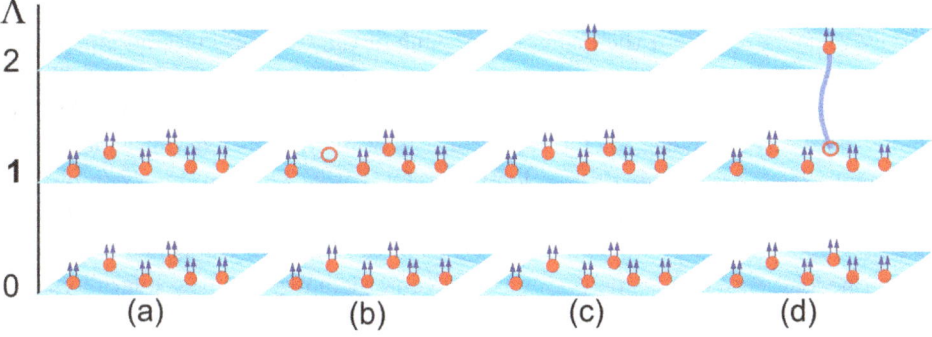

Fig. 3.   Schematic Λ level diagrams for: (a) an incompressible ground state; (b) a quasihole, i.e. a missing composite fermion; (c) a quasiparticle, i.e. an additional composite fermion; and (d) a neutral exciton. The $\nu = 2/5$ FQH state is taken for illustration, which maps into $\nu^* = 2$ of composite fermions.

Construction of CF spectra at arbitrary ν: Suppose we are asked to construct the low-energy spectrum at an arbitrary filling factor ν. We first choose the positive even integer $2p$ in Eq. (10) so as to obtain the largest possible value of $|\nu^*|$. We then construct the basis $\{\Phi_{\pm\nu^*}^{\beta}\}$ of all states, labeled by β, with the lowest kinetic energy. We multiply each basis function by $\prod_{j<k}(z_j - z_k)^{2p}$, project it into the LLL, and postulate that $\{\mathcal{P}_{\mathrm{LLL}}\Phi_{\pm\nu^*}^{\beta}\prod_{j<k}(z_j - z_k)^{2p}\}$ gives us the (in general, non-orthogonal) basis for the lowest band of eigenstates of interacting electrons at ν. For many important cases, this produces unique wave functions with no free parameters. For example, the ground state at $\nu = n/(2pn + 1)$ is related to the ground state at $\nu^* = n$, whose wave function is the Slater determinant:

$$\Phi_n = \mathrm{Det}[\eta_\alpha(\boldsymbol{r}_k)] \,, \tag{11}$$

where $\eta_\alpha(\boldsymbol{r})$ are given in Appendix A.1. Using the projection method of Refs. 21 and 22, the wave function for the ground state at $\nu = n/(2pn+1)$ can be expressed, quite remarkably, as a single Slater determinant:

$$\Psi_{\frac{n}{2pn+1}} = \mathcal{P}_{\mathrm{LLL}}\mathrm{Det}[\eta_\alpha(\boldsymbol{r}_k)] \prod_{j<k}(z_j - z_k)^{2p} \equiv \mathrm{Det}[\eta_\alpha^{\mathrm{CF}}(\boldsymbol{r}_k)]. \tag{12}$$

The elements of this determinant,

$$\eta_\alpha^{\mathrm{CF}}(\boldsymbol{r}_k) \equiv \mathcal{P}_{\mathrm{LLL}}\eta_\alpha(\boldsymbol{r}_k) \prod_{i(i\neq k)}(z_k - z_i)^p, \tag{13}$$

can be evaluated analytically[18] and are interpreted as "single-CF orbitals." The single Slater determinant form for the incompressible states is not only conceptually pleasing but is what enables calculations for systems with 100–200 (or more) particles, for which it would be impossible to store projections on individual Slater determinant basis functions. The wave functions for a single quasiparticle, a single quasihole, and the neutral excitations of the $\nu = n/(2pn \pm 1)$ states, which are images of analogous excitations of the $|\nu^*| = n$ IQH states (see Fig. 3), are also uniquely given by the CF theory, with no adjustable parameters. In these cases, it only remains to obtain the expectation value of the Coulomb interaction, which requires evaluation of a $2N$-dimensional integral, easily performed by the Monte Carlo method. For general fillings, when the topmost partially occupied ΛL has many composite fermions, the CF basis consists of many states, and it is necessary to diagonalize the Coulomb interaction in the CF basis. That can be accomplished numerically by a process called CF diagonalization[23]. (The dimension of the CF basis is exponentially small compared to that of the full LLL Hilbert space.) Basis functions for excited bands can be similarly constructed by composite-fermionizing states in the excited kinetic energy bands at $|\nu^*|$.

The above wave functions are written for electrons in the disk geometry. Other useful geometries are the spherical geometry[2] and the periodic (or the torus) geometry.[24] Wave functions for composite fermions in the spherical geometry were constructed almost three decades ago (see Ref. 18 and references therein), and recently that has been accomplished also for the torus geometry.[25,26] We will not show in this article, for simplicity, the wave functions for the spherical and torus geometries; an interested reader can find them in the literature.

The CF theory naturally gives wave functions. Many other quantities of interest can be obtained from the wave functions, such as energy gaps, dispersions, pair correlation function, static structure factor, entanglement spectrum, charge and braid statistics of the excitations, etc. Efficient numerical methods for LLL projection[21,22] and CF diagonalization[23] have been developed, which allow treatment of large systems. Because all wave functions are confined, by construction, to the LLL, the energy differences depend only on the Coulomb interaction and have no dependence on the electron mass.

Chern-Simons field theory and conformal field theory: A complementary approach for treating composite fermions is through the CS field theory of composite fermions formulated by Lopez and Fradkin,[27] and by Halperin, Lee and Read (HLR)[28] (see Halperin's chapter). It has proved very successful in making detailed contact with experiments, especially for the low-energy long-wave length properties of the compressible state at and in the vicinity of the half filled Landau level. Conformal field theory based approaches are reviewed by Hansson *et al.*[29] and also in the chapter by Simon.

2.3. *Qualitative verifications*

The title of a 1993 article by Kang, Stormer *et al.*[30] posed the question: "How Real Are Composite Fermions?"

It was natural to question composite fermions. After all, they are very complex, non-local objects. Even a single composite fermion is a collective bound state of all electrons, because all electrons participate in the creation of a vortex. One may wonder: Are such bound states really formed? If they are, in what sense do they behave as particles? Do they have the standard traits that we have come to associate with particles, such as charge, spin, statistics, etc.? To what extent is it valid to treat them as weakly interacting? How can they be observed? How can we verify that they see an effective magnetic field and form LL-like ΛLs? These are all important questions, which can ultimately be answered only by putting predictions of the CF theory to the test against experiments and exact computer calculations.

Fortunately, the CF theory leads to many predictions, because weakly interacting fermions exhibit an enormously rich phenomenology. We only need to flip through a standard condensed matter physics textbook to remind ourselves of all of the well studied phenomena and states of electrons, and predict analogous phenomena and states for composite fermions. Let us begin with an account of how the qualitative consequences of composite fermions match up with the experimental phenomenology. Quantitative tests of the CF theory are considered in the next subsection.

The most immediate evidence for the formation of composite fermions can be seen in Fig. 4, due to Stormer.[31] Here the upper panel is plotted as a function of the effective magnetic field B^* seen by composite fermions carrying two vortices, which simply amounts to shifting the upper panel leftward by an amount $\Delta B = 2\rho\phi_0$. A close correspondence between the data in the upper panel and the lower panel is evident. This is a powerful demonstration of emergence of particles in the LLL that behave as fermions in an effective magnetic field $B^* = B - 2\rho\phi_0$, which is the defining property of composite fermions, and of the formation of Landau-like Λ-levels inside the LLL of electrons.

An important corollary of the above correspondence is the explanation of the FQHE as the IQHE of composite fermions. The fractions $\nu = n/(2n+1) = 1/3$, $2/5$, $3/7$, etc. map into the integers $\nu^* = n = 1, 2, 3$, etc. A schematic view of the $2/5$ state is shown in Fig. 3(a). If one attached a mirror image of the lower panel for negative magnetic fields, one would see that the fractions $\nu = n/(2n-1) = 2/3, 3/5, 4/7, \cdots$ align with integers $\nu^* = -2, -3, -4 \cdots$ (in negative magnetic field). The fractions $n/(4n \pm 1)$ in the upper panel map into simpler fractions $n/(2n \pm 1)$ of composite fermions carrying two vortices, but they can also be understood as $\nu^* = \pm n$ IQHE of composite fermions carrying four vortices, as can be confirmed by plotting the upper panel as a function of B^* seen by composite fermions carrying four vortices, which would amount to shifting it leftward by $\Delta B = 4\rho\phi_0$. The fractions $n/(2pn\pm 1)$ and their hole partners $1-n/(2pn\pm 1)$ indeed are the prominently observed fractions in the LLL.[g] There is evidence[31–34] for ten members of the sequences $\nu = n/(2n \pm 1)$ and six members of the sequences $\nu = n/(4n \pm 1)$. The IQHE of composite fermions produces only odd denominator fractions; this can be traced back to the fermionic nature of composite fermions, which requires $2p$ to be an even integer. The CF theory thus provides a natural explanation for the fact that most of the observed fractions have odd denominators. The weak residual

---

[g]The states at $\nu = 1 - n/(2pn \pm 1)$ can be understood by formulating the original problem in terms of holes in the LLL, and then making composite fermions by attaching vortices to holes and placing them in $\nu^* = n$ IQH states.

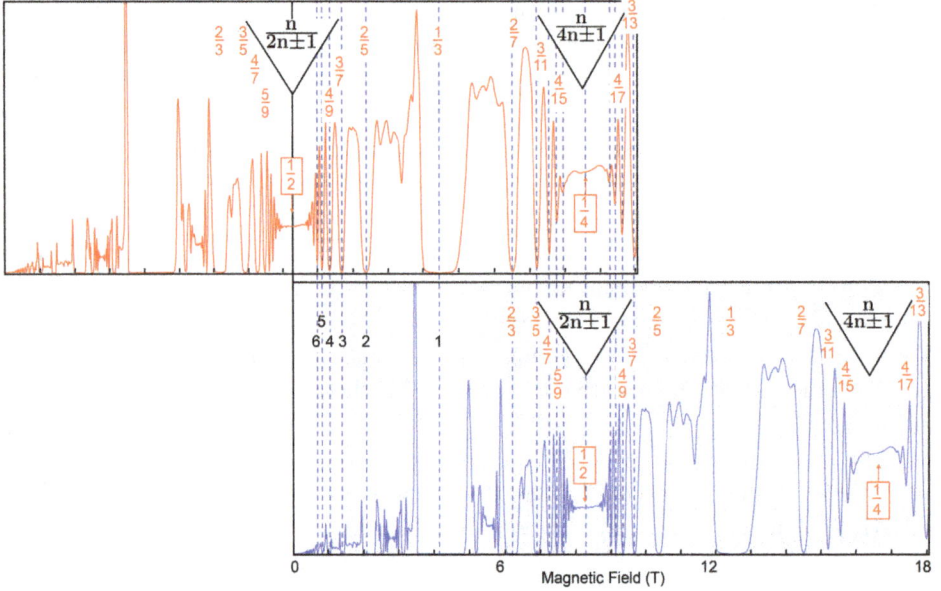

Fig. 4.   In the upper panel, the FQHE trace is plotted as a function of $B^* = B - 2\rho\phi_0$, which is the effective magnetic field seen by composite fermions carrying two vortices. A correspondence of the FQHE around $\nu = 1/2$ can be seen with the IQHE of electrons in the lower panel. The filling factor $\nu = 1/2$ maps into zero magnetic field; the fractional filling factors $n/(2n + 1)$ into inter fillings n; and the fractional filling factors $n/(4n \pm 1)$ around $\nu = 1/4$ map into simpler fractions $n/(2n \pm 1)$. The fractions $n/(4n \pm 1)$ can also be mapped into integers by plotting the top panel as a function of $B^* = B - 4\rho\phi_0$, the magnetic field seen by composite fermions carrying four vortices. Each panel is taken from Pan et al.[32] Source: H. L. Stormer, private communication.[31]

interaction between composite fermions can (and does) produce further fractions, including those with even denominators, but these are expected to be more delicate, just as the FQHE of electrons is weaker than their IQHE.

Notably, the CF theory obtains all fractions of the form $\nu = n/(2pn \pm 1)$ and $\nu = 1 - n/(2pn \pm 1)$ on the same conceptual footing. The earlier dichotomy of "Laughlin states" and "other states" may therefore be dispensed with; drawing such a distinction would be akin to differentiating between the $\nu = 1$ and the other IQH states.

The excitations of all FQH states are simply excited composite fermions. The lowest energy positively or negatively charged excitation of the $\nu = n/(2pn\pm1)$ state is a missing composite fermion in the nth ΛL or an additional composite fermion in the $(n+1)$th ΛL, as shown in Figs. 3(b) and 3(c). These are sometimes referred to as a quasihole or a quasiparticle. The neutral excitation is a particle-hole pair, i.e. an exciton, of composite fermions [Fig. 3(d)]. The activation gap deduced from the temperature dependence of the longitudinal resistance is identified with the energy required to create a far separated pair of quasiparticle and quasihole. As seen in the next subsection, the microscopic CF theory provides an accurate estimate for the energy gaps, but some insight into their qualitative behavior may be obtained by introducing a phenomenological mass for composite fermions and interpreting the gap as the cyclotron energy of composite fermions.[28] The CF cyclotron energy at $\nu = n/(2pn \pm 1)$ is written as $\hbar\omega_c^* = \hbar\frac{eB^*}{m^*c} = \hbar\frac{eB}{(2pn\pm 1)m^*c} \equiv \frac{C}{2pn\pm 1}\frac{e^2}{\epsilon l}$. The last equality follows because all energy gaps in a LLL theory must be determined by the Coulomb energy alone, and implies that the CF mass behaves as $m^* \sim \sqrt{B}$. Direct calculation of gaps along $\nu = n/(2n+1)$ for $n \leq 7$ using the microscopic CF theory[28,35] has found that the gaps, quoted in units of $e^2/\epsilon l$, are approximately proportional to $1/(2n+1)$, with best fit for a system with zero thickness given by $C = 0.33$.[28] This corresponds to a CF mass of $m^* = 0.079\sqrt{B[T]}\ m_e$ for parameters appropriate for GaAs, where $B[T]$ is quoted in Tesla and m_e is the electron mass in vacuum. The experimentally measured activation gaps deduced from the Arrhenius behavior of the longitudinal resistance are found to behave as $\frac{C'}{2pn\pm 1}\frac{e^2}{\epsilon l} - \Gamma$, where Γ is interpreted as a disorder induced broadening of ΛLs.[36,37] The CF mass can be deduced from the slope; not unexpectedly, its value depends somewhat on finite thickness, LL mixing and disorder. Neutral excitons of composite fermions have been investigated extensively in light scattering experiments.[38-46]

So far we have assumed that the magnetic field is so high that all electrons, or composite fermions, are fully spin polarized, i.e. effectively spinless. The spin physics of the FQHE is explained in terms of spinful composite fermions.[47,48] Now the integer filling of composite fermions is given by $\nu^* = n = n_\uparrow + n_\downarrow$, where $n_\uparrow$ and $n_\downarrow$ are the number of filled ΛLs of spin up and spin down composite fermions. This immediately leads to detailed predictions for the allowed spin polarizations for the various FQH states as well as their energy ordering. Transitions between differently spin polarized states can be caused by varying the Zeeman energy, and are understood in terms of crossings of ΛLs with different spins. These considerations also apply to the valley degree of freedom. Spin/valley polarizations of the FQH states have been determined as a function of the spin/valley Zeeman energy, and the ΛL fan diagram for composite fermions has been constructed.[49-54] Section 3.3 is devoted to the phase diagram of spin polarization of the FQH states.

A striking experimental fact is the absence of FQHE at $\nu = 1/2$. As seen in Fig. 4, $\nu = 1/2$ in the upper panel aligns with zero magnetic field of the lower panel. In an influential paper, HLR predicted[28] that the 1/2 state is a Fermi sea of composite fermions in $B^* = 0$. Extensive verifications of the CF Fermi sea (CFFS) and its Fermi wave vector now exist.[30,55-63] The semiclassical cyclotron orbits in the vicinity of $\nu = 1/2$ have been measured by surface acoustic waves,[55] magnetic focusing,[56,57] and commensurability oscillations in periodic potentials.[30,58-61] These are considered direct observations of composite fermions. The measured cyclotron radius is consistent with $R_c^* = \hbar k_F^*/eB^*$ with $k_F^* = \sqrt{4\pi\rho}$, as appropriate for a fully polarized CF Fermi sea. The CF cyclotron radius is much larger than, and thus clearly distinguishable from, the radius of the orbit an electron would execute in the external magnetic field. The temperature dependence of the spin polarization of the

1/2 state measured by NMR experiments is consistent with that of a Fermi sea of non-interacting fermions.[51,64] Shubnikov–de Haas oscillations of composite fermions have been observed and analyzed to yield the CF mass and quantum scattering times.[65,66] The cyclotron resonance of composite fermions has been observed by microwave radiation, with a wave vector defined by surface acoustic waves.[67,68] The CFFS is discussed in further detail in the chapters by Halperin and Shayegan.

In summary, when filtered through the prism of composite fermions, the exponentially large number of choices that were available to electrons disappear, giving way to a host of unambiguous predictions, which have been confirmed by extensive experimental studies. These predictions may appear obvious, even inevitable, once you accept composite fermions, but they are non-trivial from the vantage point of electrons, and would not have been evident without the knowledge of composite fermions.

2.4. Quantitative verifications against computer experiments

Let us next come to the quantitative tests of the CF theory. When originally proposing the wave functions in Eqs. (7)–(9) relating the FQHE to IQHE through composite fermions, the author believed that they were toy models that would describe the correct phase but did not expect them to be accurate representations of the actual Coulomb states. After all, these wave functions are in general enormously complicated after projection into the LLL. Extensive computer calculations in subsequent years proved otherwise.

This subsection presents comparisons of results from two independent calculations. The first is a brute force diagonalization of the Coulomb Hamiltonian within the LLL Hilbert space, which produces exact eigenenergies and eigenfunctions. The second constructs wave functions of the CF theory and obtains their exact energy expectation values.[h] Neither of the calculations contains any adjustable parameters.

A convenient geometry is the spherical geometry[2] where N electrons move on the surface of a sphere subjected to a total flux of $2Q\phi_0$, where $2Q$ is quantized to be an integer. Figures 5–7 show typical comparisons between the CF theory (dots) and exact results (dashes). To gain a better appreciation, we recall certain basic facts about the spherical geometry. An electron in the jth LL ($j = 0, 1, \cdots$, with $j = 0$ labeling the LLL) has an orbital angular momentum $|Q| + j$. The degeneracy of the jth LL is $2(|Q| + j) + 1$, corresponding to the different z components of the angular momentum. For a many-electron system, the total orbital angular momentum L is a good quantum number, used to label the eigenstates. For a non-interacting system, it is straightforward to determine all of the possible L values for a given $(N, 2Q)$ system. To analyze the exact spectra of interacting electrons in terms of composite fermions, we need to make use of the result that the CF theory relates the interacting electrons system $(N, 2Q)$ to the non-interacting CF system $(N, 2Q^*)$

---

[h]This calculation often uses the Monte Carlo method which involves statistical uncertainty, but several significant figures can be obtained exactly with currently available computational resources.

with

$$2Q^* = 2Q - 2p(N-1) \, . \tag{14}$$

This relation follows from the spherical analog of Eq. (9), $\Psi_{2Q} = \mathcal{P}_{\text{LLL}} \Phi_{2Q^*} \Phi_1^{2p}$, by noting that the flux of the product is the sum of fluxes (Φ_1 occurs at $2Q_1 = N-1$), and that the flux remains invariant under LLL projection. Intuitively, the relation between $2Q$ and $2Q^*$ can be understood from the observation that for any given composite fermion, all of the other $N-1$ composite fermions reduce the flux by $2p\phi_0$ each. A corollary of this relation is that the incompressible states do not occur at $2Q = \nu^{-1}N$ but rather at $2Q = \nu^{-1}N - \mathcal{S}$, where $\mathcal{S}$ is called the "shift." For the IQH state at $\nu = n$, the shift is simply $\mathcal{S} = n$, which follows from the fact that the degeneracy of the jth LL is $2|Q| + 2j + 1$. According to the CF theory, the incompressible FQH state at $\nu = \frac{n}{2pn\pm1}$ occurs at shift $\mathcal{S} = 2p \pm n$, because the shift of the product $\Phi_{\pm n}\Phi_1^{2p}$ is the sum of the shifts, which is preserved under LLL projection. The shift is N-independent, and in the thermodynamic limit we recover $\lim_{N\to\infty} N/2Q = \nu$ irrespective of the value of the shift. It is noted that different candidate states for a given filling factor may produce different shifts.

Let us now see what features of the exact spectra are explained by the CF theory by taking some concrete examples. Figure 5 shows exact Coulomb spectra (dashes) for some of the largest systems for which exact diagonalization has been performed. Each dash represents a multiplet of $2L+1$ degenerate eigenstates. The energies (per particle) include the electron-background and background-background interaction. Only the very low energy part of the spectrum is shown. The total number of independent multiplets at each L is shown at the top. Each eigenstate in this figure is thus a linear superposition of ~one hundred thousand to several million independent basis functions. All states would be degenerate in the absence of the Coulomb interaction. The emergence of certain well-defined bands at low energies is a manifestation of non-perturbative physics arising from interaction.

The interacting electron systems $(N, 2Q) = (14, 39)$, $(16, 36)$, and $(18, 37)$ map into CF systems $(N, 2Q^*) = (14, 13)$, $(16, 6)$, and $(18, 3)$. The ground states correspond to 1, 2 and 3 filled ΛLs, and thus have $L = 0$, precisely as seen in the exact spectra. The lowest energy (neutral) excitations for $(N, 2Q^*) = (14, 13)$, $(16, 6)$, and $(18, 3)$ consist of a pair of CF-hole and CF-particle with angular momenta 6.5 and 7.5, 4 and 5, and 3.5 and 4.5, respectively. These produce states at $L = 1, 2, \cdots L_{\text{max}}$ with $L_{\text{max}} = 14$, 9 and 8. The L quantum numbers of the lowest excited branch in the exact spectra agree with this prediction, except that there is no state at $L = 1$. It turns out that when one attempts to construct the wave function for the CF exciton at $L = 1$, the act of LLL projection annihilates it,[69] bringing the CF prediction into full agreement with the quantum numbers seen in the exact spectra.

Going beyond the qualitative explanation of the origin and the structure of the bands, the CF theory gives parameter-free wave functions for the ground states and the lowest energy neutral excitations at all fractions $\nu = n/(2pn \pm 1)$, obtaining by composite-fermionizing the corresponding wave functions at $\nu^* = n$. The dots

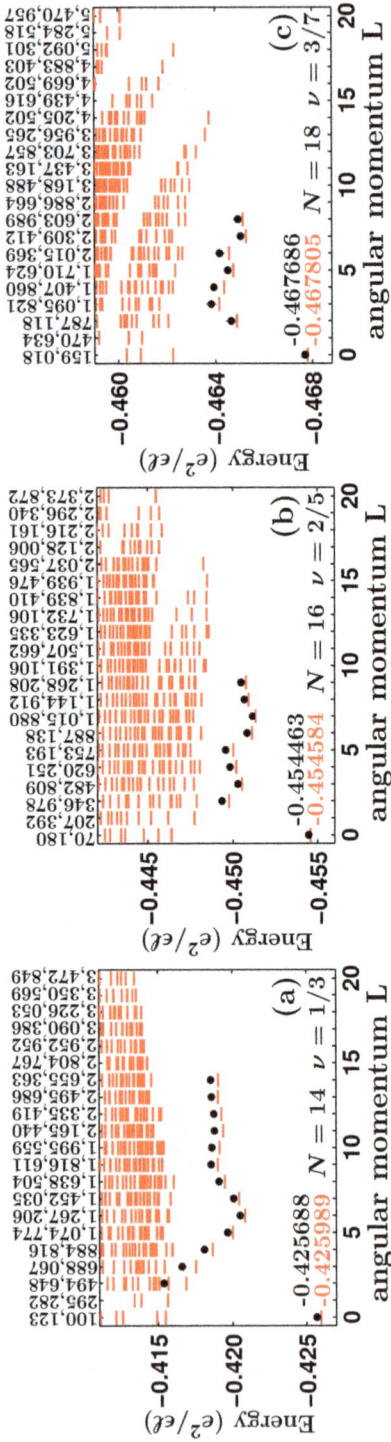

Fig. 5. These figures show a comparison between the energies (per particle) predicted by the CF theory (dots) and the exact Coulomb energies (dashes), both obtained without any adjustable parameters. Panels (a)–(c) show spectra for $(N, 2Q) = (14, 39)$, $(16, 36)$, and $(18, 37)$, which are finite size representations of the $1/3$, $2/5$ and $3/7$ states. The wave function for ground state at $\nu = 1/3$ is the same as the Laughlin wave function. Source: A. C. Balram, A. Wójs, and J. K. Jain, *Phys. Rev. B* **88**, 205312 (2013),[70] and J. K. Jain, *Annu. Rev. Condens. Matter Phys.* **6**, 39–62, (2015).[71]

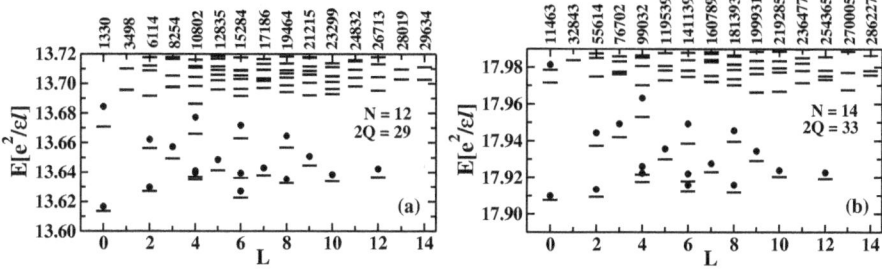

Fig. 6.   Comparison of exact Coulomb spectra (dashes) with the prediction of CF theory (dots) for $(N, 2Q) = (12, 29)$ and $(14, 33)$. The dimensions of the Hilbert space in the individual L sectors are shown at the top. Source: S. Mukherjee, S. S. Mandal, A. Wójs, and J. K. Jain, *Phys. Rev. Lett.* **109**, 256801 (2012).[72]

show the expectation values of the Coulomb interaction for these wave functions. The energies of the ground states agree to within $\sim 0.07\%$., 0.03% and 0.04% for the $1/3$, $2/5$ and $3/7$ systems shown in the figures. Further, the CF theory reproduces the qualitative features of the exact dispersion of the neutral exciton (the wave vector of the neutral exciton is given by $k = L/l\sqrt{Q}$) and predicts its energy (relative to the ground state) with a few % accuracy. The CF theory provides a similarly accurate account of the fractionally charged quasiparticle and quasihole for all fractions $\nu = n/(2n \pm 1)$. These are either an isolated CF particle in an otherwise empty ΛL or an isolated CF hole in an otherwise filled ΛL, as depicted in Figs. 3(b) and 3(c). The CF hole in an otherwise full lowest ΛL reproduces Laughlin's wave function for the quasihole of the $1/(2p + 1)$ state, albeit from a different physical principle.

Figure 6 shows comparisons away from the special fillings.[72] The quantum numbers of the states in the low energy band identifiable in the exact spectra are identical to those for non-interacting fermions at $2Q^*$. As an example, consider the electron system $(N, 2Q) = (12, 29)$ (left panel of Fig. 6) which maps into the CF system $(N, 2Q^*) = (12, 7)$. Here, the lowest energy configurations have filled lowest ΛL (accommodating $2|Q^*| + 1 = 8$ composite fermions), and four composite fermions in the second ΛL, each with angular momentum $|Q^*| + 1 = 9/2$. The predicted total angular momenta L (for fermions) are given by $\frac{9}{2} \otimes \frac{9}{2} \otimes \frac{9}{2} \otimes \frac{9}{2} = 0^2 \oplus 2^2 \oplus 3 \oplus 4^3 \oplus 5 \oplus 6^3 \oplus 7 \oplus 8^2 \oplus 9 \oplus 10 \oplus 12$, which match exactly with the L multiplets seen in the lowest band in the left panel of Fig. 6. A similar calculation successfully predicts the L quantum numbers of the lowest band of $(N, 2Q) = (14, 33)$ (right panel of Fig. 6). Diagonalization of the Coulomb interaction in the reduced CF basis produces the dots in Fig. 6.

The CF Fermi sea at $\nu = 1/2$ is obtained in the $n \to \infty$ limit of the $n/(2n \pm 1)$ fractions in the spherical geometry, or by composite-fermionizing the wave function of the electron fermi sea in the torus geometry.[26,73–79] The left panel of Fig. 7 shows the exact spectrum at $\nu = 1/2$ in the torus geometry, along with the CF energies for the lowest energy states in several momentum sectors.[79]

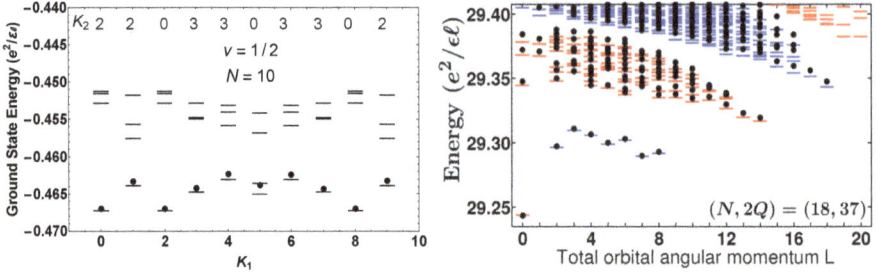

Fig. 7. Left panel: Comparison between the exact spectra (dashes) and the CF spectra (dots) for $N = 10$ particles at $\nu = 1/2$ in the periodic torus geometry. The momentum K_1 is given on the x-axis. For each K_1, the K_2 (shown at the top) is the momentum of the lowest energy state. For the torus geometry, the spectra for K_1 and $K_1 + N$ are identical. Right panel: Comparison of the CF and exact spectra for the lowest four bands at $\nu = 3/7$ in the spherical geometry. The alternating bands in the exact spectra are shown in different colors for contrast. Source: S. Pu, M. Fremling, and J. K. Jain, *Phys. Rev. B* **98**, 075304 (2018);[79] A. C. Balram, A. Wójs, and J. K. Jain, *Phys. Rev. B* **88**, 205312 (2013).[70]

Higher bands are often not clearly identifiable in the exact spectra, presumably because of the broadening induced by the residual interaction between composite fermions. Interestingly, more and more bands become visible as we go to higher CF fillings. For example, four reasonably well defined bands can be seen at $\nu = 3/7$ in Fig. 5. The CF theory gives a good account of the higher bands as CF kinetic energy bands, which involve excitations of one or more composite fermions across one or several ΛLs. Figure 7 shows a comparison between the CF theory and the exact spectrum for four lowest bands. A subtle point is that while the one-to-one correspondence between the FQHE spectra of $(N, 2Q)$ and the IQHE spectra of $(N, 2Q^*)$ is perfect for the lowest band for all LLL spectra studied so far, it is imperfect for higher bands, where the IQHE spectra have a slightly greater number of states. However, when one constructs wave functions by taking the IQH states, multiplying by the Jastrow factor and then performing LLL projection, the last step annihilates many of the states, and, remarkably, the surviving linearly independent states provide a faithful account of the bands seen in the exact FQHE spectra.[70,80,81] (What mathematical structure underlies such elimination of states is not yet understood.) The dots in Fig. 7 are obtained by a diagonalization of the Coulomb interaction in the CF basis derived from all IQH states with energies up to $3\,\hbar\omega_c$. Balram *et al.*[70] have performed an extensive study of the higher bands of many systems, showing that the correct counting for higher bands can be obtained by projecting out states of certain excitons of the $(N, 2Q^*)$ systems.

The CF theory allows, in principle, a systematic improvement of energies by allowing mixing with higher ΛLs. An example can be seen in right panel of Fig. 7, where the ground state and the single exciton energies have improved substantially compared to those in Fig. 5. In practice, the accuracy of the zeroth order CF theory is sufficient for most purposes because corrections due to other effects (e.g. LL mixing or finite width) are larger.

In summary, for all LLL systems studied by exact diagonalization, the CF theory faithfully predicts the structure of the lowest band (i.e. the number of states and their quantum numbers). It never misses any state, nor does it ever predict any false states. Furthermore, it predicts the eigenfunctions and eigenenergies almost exactly.[i] In other words, all low energy wave functions obtained in exact diagonalization studies of electrons in the LLL can be succinctly and accurately synthesized into a single, parameter-free equation, Eq. (9). These studies prove, at the most microscopic level possible, the formation of composite fermions and the relation between the FQHE and the IQHE that they entail.

2.5. *Remarks*

We close the section with some remarks.

Universality of wave functions: As noted above, the wave functions in Eq. (9) contain no free parameters for the ground states as well as the charged and neutral excitations at $\nu = n/(2pn \pm 1)$.[j] How is it then possible that these wave functions so accurately represent the eigenstates of the Coulomb interaction? What if one were to choose some other interaction? Insight into this issue comes from numerical diagonalization studies that demonstrate that the actual eigenfunctions at these fractions are surprisingly insensitive to the detailed form of the interaction so long as it is sufficiently strongly repulsive at short distances. Luckily, the Coulomb interaction in the LLL belongs in that limit. In that sense, the FQHE wave functions in the LLL are universal.[k] The good luck continues in that the CF theory captures precisely this limit. FQHE can also occur when the short range part of the repulsive interaction is not strong, as, for example, is the case for Coulomb interaction in the second LL; the wave functions for many second LL FQH states are more sensitive to the form of the interaction, and the agreement with candidate wave functions is not as decisive as that in the LLL.

Observation of ΛLs: Electrons and their LLs were known prior to the discovery of the IQHE. In contrast, the FQHE was discovered first, and its similarity to the IQHE gave a clue into the existence of composite fermions and their ΛLs. While the LLs can be derived for a single electron, composite fermions and their ΛLs provide a single-particle-like interpretation of the inherently many body wave functions of interacting electrons in the LLL. The formation of ΛLs within the LLL of electrons can be seen in a variety of ways. In computer calculations, the low-energy spectrum of interacting electrons in the LLL at ν splits into bands that have a one-to-one

---

[i]This shows that even though the microscopic wave functions in Eqs. (7)–(9) are motivated by the physics of weakly interacting composite fermions, they incorporate the knowledge of inter-CF interactions.

[j]For $\nu \neq n/(2pn \pm 1)$, the basis functions for the lowest band contain no free parameters, although their mixing and splittings depend on the specific form of the interaction.

[k]This may be contrasted with the Hartree–Fock Fermi-liquid and the BCS wave functions that explicitly depend on the interaction.

correspondence with the kinetic-energy bands of non-interacting electrons at ν^*, and the eigenfunctions of interacting electrons at ν are related to those of non-interacting electrons at ν^* through composite-fermionization. In experiments, the ΛLs appear remarkably similarly as the LLs, for example, through peaks in the longitudinal resistance R_L (see Fig. 1).

Use of higher LLs: One may ask why the path to the FQH wave functions in the LLL should pass through IQH wave functions involving higher LLs. We begin by noting that there is no fundamental reason to insist on strictly LLL wave functions in the first place. While restricting the Hilbert space to the LLL is convenient for computer calculations, it is not a necessary condition for FQHE. LL mixing is always present in experiments, indicating that the phase diagram of the FQHE extends to regions with non-zero LL mixing. The job of theory is to identify a point inside the FQH phase where the physics is the simplest, and approach the physical point perturbatively starting from there. The CF theory demonstrates that allowing a small admixture with higher LLs makes it possible to construct wave functions that reveal the physics of the FQHE in a transparent manner. The LLL projections of these wave functions accurately represent the exact Coulomb solutions, but are extremely complicated and could not have been guessed directly within a LLL theory. Finally, it ought to be stated that the use of higher LLs is not merely a technical matter but is intimately tied to the CF physics and the analogy between the FQHE and the IQHE.

Particle-hole symmetry: When we restrict to the Hilbert space of the LLL, the Hamiltonian with a two-body interaction satisfies an exact symmetry called the particle-hole (PH) symmetry. This refers to the fact that the PH transformation $c_j \rightarrow h_j^\dagger, c_j^\dagger \rightarrow h_j$, which relates the state at ν to a state at $1-\nu$, leaves the interaction Hamiltonian invariant modulo an overall additive term. In other words, the eigenspectra at ν and $1-\nu$ are identical (apart from a constant overall shift) when plotted in units of $e^2/\epsilon l$, and the eigenstates are exactly related by PH transformation. In particular, at $\nu = 1/2$, unless PH symmetry is spontaneously broken, the Fermi sea wave function must be equal to its PH conjugate. PH symmetry cannot be defined in the presence of LL mixing. It should be noted that PH symmetry is not a necessary condition for the observation of the FQHE and the CFFS, given that real experiments always involve some LL mixing, which causes no (measurable) correction to the value of the quantized Hall resistance.

The interplay between the emergence of composite fermions and the PH symmetry of electrons has attracted attention in recent years. It has led, on the one hand, Son to propose an effective theory that views composite fermions as Dirac particles,[82] and, on the other, to improved calculations within the CS field theory of HLR. These developments are discussed in the chapter by Halperin. How about the microscopic theory of composite fermions as defined by the LLL-projected wave functions in Eq. (9)? PH symmetry is neither imposed on these wave functions nor *a priori* evident, but explicit calculations have demonstrated that they satisfy

PH symmetry to an extremely high degree[1] for both the FQH states[47,84,85] and the CFFS.[74,78,79] This is a corollary of the fact that these wave functions are very close to the Coulomb eigenstates, which satisfy the PH symmetry exactly. The wave functions in Eq. (9) are constructed by composite-fermionizing the IQH states and Fermi sea of non-relativistic electrons.

New emergent structures due to inter-CF interaction: As always, explanation of finer and finer features of experimental observations requires increasingly more sophisticated theoretical models and approximations. The model of non-interacting electrons explains the most robust phenomenon, namely the IQHE, but the interaction between electrons causes new structure, namely the FQHE. Analogously, the model of non-interacting composite fermions explains FQHE at $\nu = n/(2pn \pm 1)$ and $\nu = 1 - n/(2pn \pm 1)$, which exhaust a large majority of the observed fractions, but not all. Certain fractions require a consideration of the residual interaction between composite fermions, which is complex but can be determined within the CF theory.[18,86] The FQH states at $\nu = 4/11$ and $\nu = 5/13$ are examples of FQHE of composite fermions.[33,87,88] Another example of new physics arising from the inter-CF interaction is the $5/2$ state, which is believed to occur because of a p-wave pairing instability of the CF Fermi sea[89,90] (see the chapters by Halperin, and Heiblum and Feldman). One may ask how pairing can arise in a model with purely repulsive interaction. It arises because the objects forming pairs are not electrons but composite fermions. The interaction between composite fermions, which is a complex function of the interaction between electrons, is weak, and nothing really forbids it from being attractive. Explicit calculations indicate that at $\nu = 5/2$, the binding of two vortices by electrons over-screens the repulsive Coulomb interaction between electrons to produce a weakly attractive interaction between composite fermions.[91] (In contrast, the inter-CF interaction remains repulsive[91] at $\nu = 1/2$, where the interaction between electrons is more strongly repulsive than that at $\nu = 5/2$.) Certain other paired states of composite fermions are considered in Sec. 5.

FQHE in graphene: In recent years, graphene has produced extensive FQHE. For the Dirac electrons of graphene, LLs occur for positive and negative energies,

---

[1]From the fact that $\nu = n/(2n+1)$ maps into $\nu^* = n$ whereas its hole partner $\nu = 1 - n/(2n+1) = (n+1)/(2n+1)$ into $\nu^* = -(n+1)$, it may appear that the CF theory does not respect PH symmetry. That is not correct. The state obtained from composite-fermionization of $\nu^* = -(n+1)$ is equivalent to the hole partner of the state obtained from composite-fermionization of $\nu^* = n$ in all topological aspects. Their edge physics are identical as are their mean-field gaps (see Supplemental Material of Ref. 83). Furthermore, the explicit wave functions constructed in the two approaches have almost perfect overlap.[47,84] Incidentally, as discussed in Sec. 3.3, the mapping of $\nu = n/(2n + 1)$ and $\nu = (n + 1)/(2n + 1)$ into $\nu^* = n$ and $\nu^* = -(n + 1)$, respectively, is crucial for explaining the qualitatively different spin physics at these filling factors. For example, $\nu = 1/3$, which maps into $\nu^* = 1$, is predicted to be always fully spin polarized, whereas $\nu = 2/3$, which maps into $\nu^* = -2$, is predicted to admit both fully spin polarized and spin singlet states, depending on whether the Zeeman energy is larger or smaller than the CF cyclotron energy. Both spin singlet and fully spin polarized states have been observed at $\nu = 2/3$; the nature of these states and the phase transition between them are quantitatively well explained by the CF theory.

have a spacing proportional to $\sqrt{|n|}$ where n is the LL index, and the $n = 0$ LL is located at zero energy. When one restricts the Hilbert space to a specific LL, the LLs of Dirac electrons differ from those of non-relativistic electrons in two aspects. First, there is additional degeneracy in graphene because of two valleys. The valley degree of freedom can be accommodated into the CF theory in the same manner as the spin. Second, the Coulomb matrix elements are in general different from those in the LLs of non-relativistic electrons. It turns out that for the $n = 0$ LL, the Coulomb matrix elements for Dirac and non-relativistic electrons are identical (for a strictly 2D system). The observed FQHE in the graphene $n = 0$ LL corresponds precisely to what is expected from the CF theory. The Coulomb matrix elements in the $n = 1$ graphene LL are different from those of the $n = 1$ LL of non-relativistic electrons and closer to those of the $n = 0$ LL. Indeed, the FQHE in the $n = 1$ graphene LL is also explained nicely in terms of non-interacting composite fermions. The status of FQHE in graphene is reviewed in the chapter by Dean, Kim, Li and Young.

The role of topology in FQHE: It is useful to ask the question:[92,93] What can we say about the properties of a FQH state without knowing its microscopic origin? Here one assumes a gapped state at a certain filling factor and asks what quantum field theory would produce a non-zero Hall conductance. Electrons, being high energy objects, are not a part of this theory, which, as any effective field theory, deals with the low-energy physics. This line of reasoning naturally leads to CS theories with emergent gauge fields.[92,93] These theories make precise predictions for certain quantities that are of topological origin, i.e. are invariant under continuous changes of the Hamiltonian so long as no phase boundary is breached (which is why their calculation does not require a microscopic understanding). In particular, the CS theories reveal the existence of quasiparticles with fractional charge and fractional braid statistics.[92,93]

The current chapter focuses on the microscopic mechanism of the FQHE. You may recall seeing an animated GIF in a continuous loop, perhaps in a physics department colloquium, showing a coffee mug adiabatically metamorphosing into a doughnut and back, to drive home the fact that the two share the same genus-one topology. The coffee mug and the doughnut are of course different objects, as even a topologist may ascertain by performing the experiment, with care, of biting hard or pouring hot coffee into them. A master chef ready to prepare a doughnut will need to know, aside from its toroidal shape, the various ingredients as well as the recipe for how to put them together. We are similarly concerned in this chapter with the microscopic ingredients of the FQHE (composite fermions) and how they are assembled into various states (IQHE, Fermi sea, crystal, etc.) to produce the phenomenology. We are concerned with microscopic wave functions and calculation of measurable quantities. It turns out, nonetheless, that topology lies at the front and center of the CF theory, for the simple reason that composite fermions themselves are topological particles. The attached vortices endow composite fermions with a U(1) topological character, which, in turn, manifests directly through the effective magnetic field experienced by composite fermions. The effective magnetic field has been measured and is responsible for the explanation or prediction of the vast body of unexpected phenomenology of the FQHE. All of the qualitative phenomenology of composite fermions thus has topological origin. In fact, the FQHE is doubly topological. Recall that IQHE is topological because electrons fill topological bands (LLs) characterized by non-zero Chern numbers. In FQHE, topological particles (composite fermions) fill topological bands (ΛLs). The two topological quantum numbers characterizing a FQH state are $2p$, the CF vorticity, and n, the number of filled ΛLs. It is worth stressing that while all topological properties of the FQHE can be derived starting from the CF theory, the existence of composite fermions and their effective magnetic field, which relate to the microscopic origin of the FQHE, cannot be derived from the purely topological perspective mentioned in the preceding paragraph.

Fractional charge and fractional braid statistics: An attentive reader may have noticed that the above explanations of the FQHE and other related phenomena make no mention of fractional charge and fractional braid statistics. That composite fermions are fermions is beyond question. Their fermionic nature is central to the explanations of the FQHE as the IQHE of composite fermions and of the $1/2$ state as the Fermi sea of composite fermions. Furthermore, computer calculations confirm, beyond doubt, that the quasiparticles and quasiholes are nothing but excited composite fermions or the holes they leave behind, as depicted in Fig. 3. At the same time, the existence of fractional charge and fractional braid statistics for the quasiparticles or quasiholes can be inferred from no more than the assumption of a gap at a fractional filling factor; in fact, the allowed values for them can be derived without an understanding of the microscopic origin of the FQHE[m]. In spite of the appearances, there is no contradiction. The fractional charge and fractional braid statistics can be derived within the CF theory as follows. Consider the state at a filling factor $\nu^* = n$ with two additional composite fermions in the $(n+1)$th ΛL. One may seek an effective formulation of the problem in terms of only two particles by integrating out all composite fermions in the lower filled ΛLs. This must be done with care, however, because the two composite fermions in the $(n+1)$th ΛL are topologically correlated with the composite fermions in lower filled ΛLs as well (i.e. see $2p$ vortices on them). The effect of the lower filled ΛLs is to "screen" both the charge and the braid statistics of the composite fermions in the $(n+1)$th ΛL. There are several ways within the CF theory to derive[18,95] the fractional charge $e^* = e/(2pn \pm 1)$ and braid statistics parameter $\alpha = 2p/(2pn \pm 1)$ for the quasipar-

---

[m]The value of the filling factor puts constraints on the allowed values for the charge and braid statistics of the quasiparticles.[94] Assuming an incompressible state at $\nu = n/(2pn \pm 1)$, adiabatic insertion of a unit flux produces, à la Laughlin,[10] an excitation of charge $en/(2pn \pm 1)$. This in general is a collection of several elementary quasiparticles. Assuming that we have a single type of elementary quasiparticles, the requirement that an integer number of them also produce an electron gives $e^* = e/[k(2pn \pm 1)]$, where k is an arbitrary integer. The simplest choice corresponds to $k = 1$. Braid statistics of the elementary quasiparticles can be deduced analogously from general considerations.[94]

ticles of the $\nu = n/(2pn \pm 1)$ FQH state. These are the simplest values allowed by general considerations, and are also in agreement with those produced previously by the hierarchy theory.[13]

The CF theory goes beyond these quantum numbers and gives a precise microscopic account of the quasiparticles and quasiholes of all $\nu = n/(2pn \pm 1)$ states, which allows us to calculate their density profiles, energies, interactions, dispersions, etc. Most remarkably, the CF theory reveals that the quasiparticles of all $\nu = n/(2pn \pm 1)$ FQH states are, in a deep sense, the same objects, namely composite fermions, which are also the particles that form the ground states. Composite fermions remain sharply defined even when the concept of fractional charge and fractional braid statistics ceases to be meaningful, e.g. at $\nu = 1/2$ (where the state is compressible), or when a ΛL is sufficiently populated that the composite fermions in that ΛL are strongly overlapping.

3. Quantitative Comparison with Laboratory Experiments

Given the accuracy of the CF theory as seen in computer experiments, we can dispense with exact diagonalization and study systems of composite fermions. With the help of convenient numerical methods for LLL projection[21,22] and CF diagonalization,[23] we can go to large systems (with as many as 200 composite fermions or more) to explore phenomena that are not accessible in exact diagonalization studies, and also to obtain thermodynamic limits for various quantities of experimental interest. Numerous observables, such as excitation gaps, dispersions of the neutral CF exciton, dispersions of spin waves, phase diagrams of various states as a function of parameters, have been calculated (see Refs. 18 and 71 for a review). A *priori*, one should expect a few-percent agreement between theory and experiment (which can be systematically further improved if so desired). That indeed would have been the case had we been dealing with a phenomenon in atomic or high energy physics, but the FQH systems, in spite of being among the most pristine and the best characterized of all condensed matter systems, present additional complications. Unlike experiments in atomic or high energy physics, FQH experiments in different laboratories and different samples produce different numbers, because the experimental results are modified by features that were set to zero in computer studies mentioned in the previous section, namely finite quantum well width, LL mixing and disorder. These must be included in the theoretical calculation for a precise quantitative comparison. It is somewhat ironic that we have an extremely accurate quantitative understanding of the nontrivial part of the physics, namely the FQHE, but our understanding of the corrections due to finite width, LL mixing and disorder is less precise. That is the reason why quantitative comparisons with experiments, while decent, do not reflect the full potential of the CF theory.

This section is devoted to recent calculations[96,97] that incorporate the effects of finite width and LL mixing (Secs 3.1 and 3.2) to the best extent currently possible.

Because we do not include disorder, we focus on thermodynamic quantities that are not expected to be very sensitive to disorder, as opposed to quantities such as excitations gaps that are more strongly affected by disorder.

Section 3.3 considers transitions between differently spin polarized FQH states. These are understood, physically, as ΛL crossing transitions as the Zeeman energy is varied relative to the CF cyclotron energy. The critical Zeeman energies at which these transitions are observed are a direct measure of the differences between the Coulomb energies of the competing states. Comparisons with experiments show that after incorporating finite width and LL mixing corrections, the CF theory obtains these energy differences, which are on the order of 1% of the individual energies, with a few-percent accuracy. These calculations also shed light on the dissimilarities observed between the behaviors at $\nu = n/(2n \pm 1)$ and $\nu = 2 - n/(2n \pm 1)$.

Section 3.4 deals with the competition between the liquid and the crystal phases as a function of filling factor and LL mixing. It provides evidence that the crystal phase is not an ordinary, featureless Wigner crystal of electrons but contains a series of crystals of composite fermions with different vorticity. The essential theoretical picture is that as the filling factor is lowered, at some point composite fermions begin to bind fewer than the maximal number of vortices available to them and use the remaining freedom to form a crystal of composite fermions. Given how favorable the CF correlations are, it should not be surprising that nature would exploit them even in the crystal phase to find the lowest energy state. In particular, theoretical calculations show that the crystal of composite fermions with two attached vortices is energetically favored over the FQH state of composite fermions with four attached vortices for a narrow range of filling factors between $\nu = 1/5$ and $\nu = 2/9$, thus explaining the observed insulating phase between the 1/5 and 2/9 FQH liquid states. The CF crystal beats the FQHE here by a mere $\sim$0.0005 $e^2/\epsilon l$ per particle, which is an indication of the theoretical accuracy required to capture the physics of the re-entrant crystal phase. Calculations further show that the enhanced LL mixing in low-density p-doped GaAs quantum wells also stabilizes a crystal in between $\nu = 1/3$ and 2/5, as seen experimentally.

One may ask: Given that the underlying CF physics is already well established, why expend a substantial amount of effort toward calculating numbers very precisely? The reason, from a general perspective, is that progress in physics often relies on a precise quantitative understanding of experiments, which prepares the ground for new discoveries. Significant quantitative deviations between theory and experiment are inevitably found as more accurate tests are performed and as new regimes are explored, pointing to new physics. In the context of the FQHE, an additional motivation for seeking a precise microscopic understanding of experiments is simply that we can. The FQHE is a rare example of a highly nontrivial strongly-correlated state for which it has been possible to achieve a detailed microscopic description in the quantum chemistry sense. Given that an understanding of the role of interactions is a primary goal of modern condensed matter physics, it appears to be of value to push the comparison between theory and experiment in FQHE to its limits.

3.1. *Finite width corrections: Local density approximation*

The nonzero transverse width of GaAs-Al$_x$Ga$_{1-x}$As heterojunctions and quantum wells can be incorporated into theory by using an effective 2D interaction given by:

$$V^{\text{eff}}(r) = \frac{e^2}{\epsilon} \int dz_1 \int dz_2 \frac{|\xi(z_1)|^2 |\xi(z_2)|^2}{[r^2 + (z_1 - z_2)^2]^{1/2}}, \tag{15}$$

where $\xi(z)$ is the transverse wave function, z_1 and z_2 denote the real coordinates perpendicular to the 2D plane (z here is not to be confused with the complex in-plane coordinate introduced previously), and $r = \sqrt{(x_1 - x_2)^2 + (y_1 - y_2)^2}$. The interaction $V^{\text{eff}}(r)$ is less repulsive at short distances than the ideal 2D interaction $e^2/\epsilon r$. We need a model for $\xi(z)$. At zero magnetic field, a realistic $\xi(z)$ for any given density and quantum well width can be obtained by solving the Schrödinger and Poisson equations self-consistently in the density functional theory with the exchange-correlation functional treated in a local density approximation (LDA).[98] (For an earlier model, see Ref. 99.) The resulting $V^{\text{eff}}(r)$ depends on both quantum well width and the electron density. It is customary to assume that $\xi(z)$ remains unaffected by the application of a magnetic field perpendicular to the 2D plane.

3.2. *LL mixing: fixed phase diffusion Monte Carlo method*

The parameter $\kappa = (e^2/\epsilon l)/\hbar\omega_c$, the ratio of the Coulomb interaction to the cyclotron energy, provides a measure of LL mixing. It is related to the standard parameter r_s of electrons (namely the interparticle separation in units of the Bohr radius) through $\kappa = (\nu/2)^{1/2} r_s$. LL mixing is suppressed in the limit $\kappa \to 0$. For small values of κ, the effect of LL mixing can be treated in a perturbative approach[100–108] that modifies the 2D interaction. However, the reliability of the perturbative treatment for typical experiments is unclear, given that $\kappa \sim 0.8 - 2$ for n-doped GaAs and $\kappa \sim 2 - 20$ in p-doped GaAs systems. A lack of quantitative understanding of LL mixing has been an impediment to the goal of an accurate comparison between theory and experiment.

We treat the effect of LL mixing through the nonperturbative method of fixed-phase diffusion Monte Carlo (DMC) calculations.[109–111] This is a generalization of the powerful DMC method[112,113] for obtaining the "exact" ground state energies for certain interacting systems. We give here a brief account of the method; more details can be found in the literature.

Let us assume that the ground state wave function is real and non-negative, as is the case for bosons. The Schrödinger equation for imaginary time ($t \to it$)

$$-\frac{\partial}{\partial t}\Psi(\mathcal{R}, t) = (H - E_T)\Psi(\mathcal{R}, t) \tag{16}$$

can then be viewed as a diffusion equation with the wave function $\Psi(\mathcal{R},t)$ interpreted as the density of the diffusing particles. Here $\mathcal{R}$ collectively represents the coordinates of all the particles and E_T is a conveniently chosen energy offset. Let us now begin with an initial trial function $\Psi(\mathcal{R},t=0)$ which can be expressed in terms of the exact eigenstates Φ_α as $\Psi(\mathcal{R},t=0) = \sum_\alpha C_\alpha \Phi_\alpha$. Its evolution in imaginary time is given by

$$\Psi(\mathcal{R},t) = \sum_\alpha C_\alpha e^{-(E_\alpha - E_T)t}\Phi_\alpha \to C_0 e^{-(E_0 - E_T)t}\Phi_0 \text{ for } t \to \infty . \tag{17}$$

Thus, in the large imaginary time limit the evolution operator projects out the ground state provided it has non-zero overlap with the initial trial wave function. DMC is a stochastic projector method for implementing this scheme through an importance sampling method using a trial or guiding wave function. In the absence of a potential, we have the distribution of random walkers (or diffusing Brownian particles) in the $2N$ dimensional configuration space. In the presence of a potential, the most effective method is through a branching (or a birth/death) algorithm in which either a walker dies with some probability in regions of high potential energy, or new walkers are created in regions of low potential energy, according to certain rules. The probability distribution of the walkers converges to the ground state in the limit $t \to \infty$. The energy offset E_T controls the population of the walkers; one adjusts E_T to keep the walker population at around 100–1000. The energy offset must be adjusted to the ground state energy to obtain a stationary distribution. Alternatively the energy can be obtained from an average of the so-called local energy. One typically keeps the acceptance ratio at around 99%.

The DMC method cannot be applied directly to FQH systems, which, due to the broken time-reversal symmetry, have complex valued eigenfunctions. For such systems, an approximate strategy known as the fixed-phase DMC was introduced by Ortiz, Ceperley and Martin (OCM)[109] which searches for the ground state in a restricted subspace. (The fixed phase DMC is closely related to the fixed node DMC used for real wave functions.[114]) Following OCM, we substitute $\Psi(\mathcal{R}) = \Phi(\mathcal{R})e^{i\varphi(\mathcal{R})}$ where $\Phi(\mathcal{R}) = |\Psi(\mathcal{R})|$ is real and non-negative. The term "phase" in fixed phase DMC is used for the phase $\varphi(\mathcal{R})$ of the wave function, and not for the phase (e.g. liquid, crystal) of the system. The variational energy of the system of interacting electrons in a magnetic field is given by $\langle\Psi(\mathcal{R})|H|\Psi(\mathcal{R})\rangle = \langle\Phi(\mathcal{R})|H_R|\Phi(\mathcal{R})\rangle$ with $H_R = \sum_{j=1}^{N}\left[p_j^2 + [\hbar\nabla_j\varphi(\mathcal{R}) + (e/c)\mathbf{A}(\mathbf{r}_j)]^2\right]/2m + V_{\text{Coulomb}}(\mathcal{R})$. Now, keeping the phase $\varphi(\mathcal{R})$ fixed and varying $\Phi(\mathcal{R})$ gives us the lowest energy within the subspace of wave functions defined by the phase sector $\varphi(\mathcal{R})$. This minimization can be conveniently accomplished by applying the DMC method to the imaginary time Schrödinger equation $-\hbar\frac{\partial}{\partial t}\Phi(\mathcal{R},t) = [H_R(\mathcal{R}) - E_T]\Phi(\mathcal{R},t)$. The essence of the fixed phase DMC is to transform the fermionic problem into a bosonic one at the expense of an additional vector potential in the Hamiltonian that essentially corresponds to a fictitious magnetic field. The fixed phase DMC produces the lowest energy in the chosen phase sector, and hence a variational upper bound for the

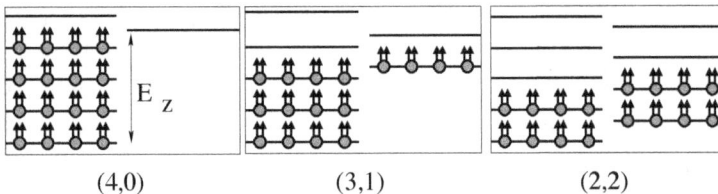

Fig. 8.   Schematic view of the FQH state at $\nu = 4/9$ (or $4/7$), which maps into $\nu^* = 4$ filled Λ levels, as a function of the Zeeman energy, E_Z. The three possible states are $(n_\uparrow, n_\downarrow) = (4, 0)$, $(3, 1)$, and $(2, 2)$, which are fully polarized, partially polarized, and spin singlet, respectively.

exact ground state energy. It would produce the exact ground state energy if we knew the phase of the exact ground state, which we do not.

The accuracy of the energy obtained from fixed phase DMC is critically dependent on the choice of the phase $\varphi(\mathcal{R})$. Güçlü and Umrigar[115] found in exact diagonalization studies of certain small systems (maximum density droplets) that the phase of the wave function is not significantly altered by LL mixing. Following their lead, the calculations shown below use the accurate LLL wave functions of the CF theory as the trial wave functions to fix the phase $\varphi(\mathcal{R})$. In cases where a comparison has been made, fixing the phase with the more accurate LLL wave function (e.g. the CF Fermi sea versus the Pfaffian wave function at $\nu = 1/2$) produces lower energy for up to the largest values of κ considered. This choice has another advantage: it keeps the system in the topological sector defined by the LLL trial wave function. Nonetheless, the results are subject to this assumption regarding the phase, the validity of which can ultimately be justified only by a detailed comparison of the numerical results with experiments. The calculations use the generalization by Melik-Alaveridan, Bonesteel and Ortiz[110,111] of the fixed-phase DMC method to the spherical geometry through a stereographic projection.

3.3. Spin phase transitions

The explanation of FQHE as the IQHE of composite fermions also gives an understanding of the spin physics. The FQHE at $\nu = n/(2pn \pm 1)$ still maps into IQHE of composite fermions at $|\nu^*| = n$ but, in general, we have $n = n_\uparrow + n_\downarrow$, where $n_\uparrow$ and $n_\downarrow$ are the number of occupied up-spin and down-spin Λ levels. These states are labeled $(n_\uparrow, n_\downarrow)$. The allowed spin polarizations are then given by $\gamma = (n_\uparrow - n_\downarrow)/(n_\uparrow + n_\downarrow)$. Figure 8 depicts the situation for $\nu = 4/9$ or $4/7$, where three distinct spin polarizations are possible. In particular, in the limit of zero Zeeman energy, the states at fractions with even numerators are predicted to be spin singlet, whereas those at fractions with odd numerators are predicted to be fully spin polarized for $n = 1$ and partially spin polarized for $n \geq 3$.

Experimentally, transitions between differently spin polarized FQH states can be driven by tuning the Zeeman energy, which can be accomplished either by application of an additional parallel magnetic field (tilted field experiments), or by changing the density. A wealth of experimental information exists for the critical energies where such transitions occur,[40,49,51,54,64,116–123] and the number of transitions seen in experiments is generally in agreement with the prediction from the CF theory. The physical picture is that the transitions are essentially ΛL crossing transitions occurring due to a competition between the CF cyclotron energy and the Zeeman splitting.

To obtain a more quantitative comparison, it is convenient to quote the Zeeman energy in units of the Coulomb energy, which we denote as $\alpha_Z = E_Z/(e^2/\epsilon l)$. The critical Zeeman energy for the transition between two successive states $(n_\uparrow, n_\downarrow)$ and $(n_\uparrow - 1, n_\downarrow + 1)$ is given by

$$\alpha_Z^{\text{crit}} = \frac{E_Z^{\text{crit}}}{e^2/\epsilon l} = (n_\uparrow + n_\downarrow)\left[\frac{E_{(n_\uparrow,n_\downarrow)} - E_{(n_\uparrow-1,n_\downarrow+1)}}{e^2/\epsilon l}\right]. \tag{18}$$

where $E_{(n_\uparrow,n_\downarrow)}$ is the per particle Coulomb energy of the states $(n_\uparrow, n_\downarrow)$. The critical Zeeman energy is thus a direct measure of the difference between the Coulomb energies of the two competing states. These energy differences are on the order of 1% or less of the individual Coulomb energies,[48] and their calculation thus serves as a sensitive test of the quantitative accuracy of the theory.

Let us first consider the ideal system with no LL mixing and no finite width corrections. In this limit, the LLL wave functions for the states $(n_\uparrow, n_\downarrow)$ are accurately given by

$$\Psi_{\frac{n}{2pn\pm1}}^{\text{full}} = A[\Psi_{(n_\uparrow,n_\downarrow)}u_1 \cdots u_{N_\uparrow}d_{N_\uparrow+1} \cdots d_N] \tag{19}$$

with

$$\Psi_{(n_\uparrow,n_\downarrow)} = \mathcal{P}_{\text{LLL}}\Phi_{\pm n_\uparrow}(z_1, \cdots z_{N_\uparrow})\Phi_{\pm n_\downarrow}(z_{N_\uparrow+1} \cdots z_N)\prod_{j<k}(z_j - z_k)^{2p} \tag{20}$$

Here A denotes antisymmetrization, and u_j and d_j are the up and down spinor wave functions. The LLL projection is performed using the method in Refs. 21, 22, 18 and 84. We note that these wave functions automatically satisfy Fock condition, i.e. have eigenstates of S^2 with eigenvalue $S(S+1)$ with $S = S_z$. For the calculation of the Coulomb energy, it is sufficient to work with Eq. (20) rather than Eq. (19).

From these wave functions, the energies of the ground states of various spin polarizations have been calculated for many filling factors.[48,84,124,125] The thermodynamic energies are determined from an extrapolation of finite system results. The predicted critical Zeeman energies are shown in Fig. 9, along with experimental results. [The theoretical critical energies shown in this figure are actually determined from exact diagonalization studies.[125] The CF energies for the states at $n/(2n+1)$ are very accurate with the standard projection method. For the $n/(2n-1)$ states, on the other hand, the standard projection method[21,84] slightly overestimates the probability of spatial coincidence of electrons in the nonfully polarized states, and thereby overestimates their energies. The hard-core projection of Ref. 47 produces very accurate energies, but is not amenable to large scale numerical evaluations.]

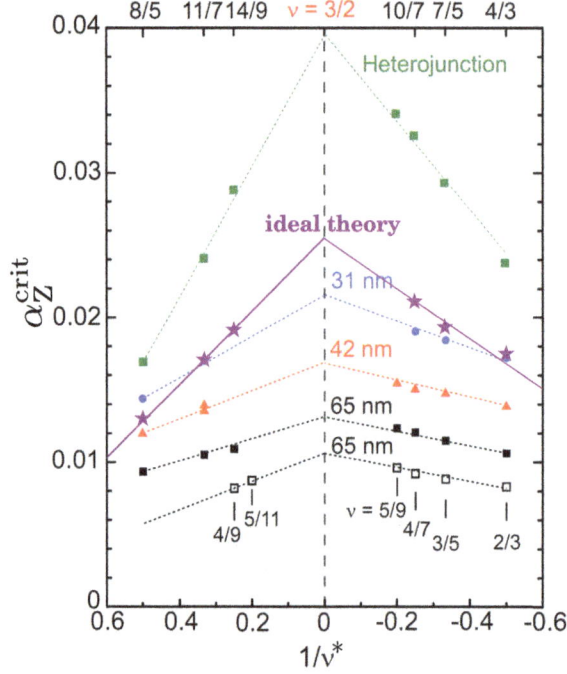

Fig. 9.   Critical Zeeman energies for spin phase transitions. The purple stars marked "ideal theory" show the theoretical prediction for the critical Zeeman energies for transitions from the fully spin polarized state into a partially spin polarized or a spin singlet state for states at $\nu = n/(2n \pm 1)$; these are obtained from exact diagonalization for a system with zero thickness and no LL mixing.[125] All other symbols are from experiments. The top green symbols are taken from Du *et al.*,[49] obtained in the heterojunction geometry. All other results are from Liu *et al.*[123] for experiments in quantum wells of various thicknesses shown on the figure. All experimental results are for filling factors of the form $\nu = 2 - n/(2n \pm 1)$, shown on the top, except for the lowest results, which are for filling factors of the form $\nu = n/(2n \pm 1)$ shown on the figure. Source: Y. Liu, S. Hasdemir, A. Wójs, J. K. Jain, L. N. Pfeiffer, K. W. West, K. W. Baldwin, and M. Shayegan, *Phys. Rev. B* **90**, 085301 (2014).[123]

The theory successfully captures the energy ordering of the differently polarized states, produces critical Zeeman energies that are generally consistent with experiments, and also captures the tent-like behavior of the critical Zeeman energy around $\nu = 1/2$.

Some discrepancy between theory and experiment remains, however, which is not surprising given that the theoretical calculations omit the effects of finite width and LL mixing. Here are the primary deviations: (i) The actual numbers for α_Z^{crit} can be off by up to a factor of 2–3. (ii) The spin phase transitions appear to be strongly affected by the breaking of PH symmetry due to LL mixing. For a system confined to the LLL and interacting by a two-body interaction, there is an exact PH symmetry relating filling factors ν and $2 - \nu$, which implies that α_Z^{crit} for a fraction $\nu = 2 - n/(2n \pm 1)$ is identical to that for $\nu = n/(2n \pm 1)$. That is not the

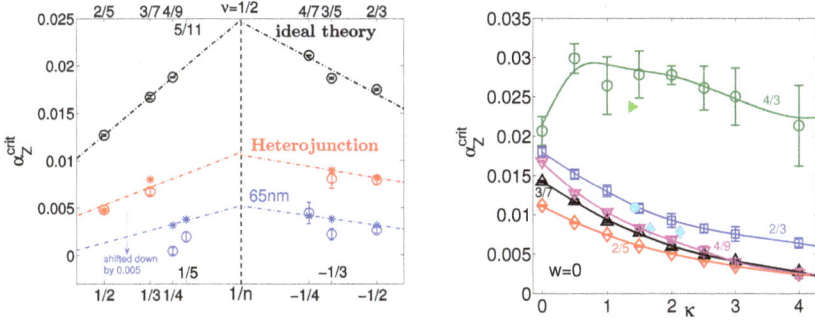

Fig. 10. Comparison between experimental critical Zeeman energies $\alpha_Z^{\mathrm{crit}} = E_Z^{\mathrm{crit}}/(e^2/\epsilon l)$ with theoretical results from fixed phase DMC. Left panel: The blue and red stars show results from experiments on a 65 nm wide quantum well and a heterojunction, taken from Liu *et al.*,[123] Engel *et al.*[118] and Kang *et al.*[119] (For the experiment of Kang *et al.*, we estimate the value of the Landé factor g_0 by assuming that it changes linearly and passes through zero at a pressure of roughly 18 Kbar.[126]) The blue and red circles show the results from fixed phase DMC calculation for corresponding widths and densities. The results for the 65 nm quantum well are shifted down by 0.005 for ease of depiction. The theoretical results without including the effects of LL mixing and finite width are also shown for comparison. The dashed lines are a guide to the eye. Right panel: Theoretical critical Zeeman energies for $w = 0$ as a function of the LL mixing parameter κ obtained from the DMC method for $\nu = 4/3$ (green circle), 2/3 (blue square), 4/9 (magenta downward triangle), 3/7 (black upward triangle), and 2/5 (red diamond). The solid lines are an approximate guide to the eye. The filled symbols indicate the experimental data from heterojunction samples at $\nu = 2/3$ (light blue) and 4/3 (green) taken from Eisenstein *et al.*[117] (circle), Engel *et al.*[118] (diamond), and Du *et al.*[49] (rightward triangle). Source: Y. Zhang, A. Wójs, and J. K. Jain, *Phys. Rev. Lett.* **117**, 116803 (2016).[96]

case in experiments, however. Spin transitions are readily observed for fractions $\nu = 2 - n/(2n \pm 1)$ but not for $\nu = n/(2n \pm 1)$ (even after using densities so that the fractions are seen at the same B). As an example, the $\nu = 8/5$ was the first state where a spin transition was observed,[116] but a transition at 2/5 could be seen only after reducing the Landé g-factor substantially by the application of hydrostatic pressure.[119] (iii) For the heterojunction samples, the measured critical Zeeman energies at $\nu = 2 - n/(2n \pm 1)$ lie above the ideal theoretical values. This is surprising because both finite width and LL mixing reduce the Coulomb energies, and therefore should generally reduce the critical Zeeman energies.

Reference 96 has investigated how the results are modified when we include the effects of finite width and LL mixing, by evaluating the thermodynamic limits of the Coulomb energies of the relevant states as a function of the quantum well widths and densities. Figure 10 (left panel) shows theoretical results for several states of the form $n/(2n \pm 1)$ obtained using the experimental parameters (width, density and LL mixing), along with the experimental results.

How about the breaking of PH symmetry seen in experiments? It is computationally expensive to deal with the states at $\nu = 2 - n/(2n \pm 1)$, and therefore Ref. 96 only compared the spin transitions at $\nu = 2/3$ and $\nu = 4/3$ for a zero width system. To obtain accurate results, the initial trial wave function is chosen as (i) the exact LLL ($\kappa = 0$) Coulomb ground states for the spin singlet states at 2/3 and 4/3; (ii) Eq. (20) for the fully polarized state at $\nu = 2/3$; and (iii) $\Phi_{1\uparrow}\Psi_{1/3\downarrow}$ for the partially polarized state at $\nu = 4/3$. Figure 10 (right panel) shows that the α_z^{crit} for 4/3 is substantially higher than that for 2/3 for the typical experimental value of $\kappa \approx 1-2$. This figure also contains the experimental data from GaAs-Al$_x$Ga$_{1-x}$As heterojunction samples, because these have the smallest effective width. (The comparison with zero width results is meaningful, because at relatively large κ the results are not particularly sensitive to the width.)

The high degree of agreement between theory and experiment seen in Fig. 10 demonstrates that the CF theory predicts the energy difference between the Coulomb energies of differently spin polarized states, which can be as small as $\sim 0.002e^2/\epsilon l$ for the systems studied, to within a few % accuracy. These comparisons also provide an *a posteriori* justification for fixing the phase using the LLL wave functions.

FQHE has also been observed in systems with valley degeneracies, such as AlAs quantum wells,[52,53] graphene,[54,127–131] and H-terminated Si(111) surface.[132] In many of these studies, transitions between differently valley polarized states have been observed. The CF theory can be generalized to treat such systems,[125,133] but a careful treatment of the finite width and LL mixing corrections has not yet been performed.

3.4. *Phase diagram of the CF crystal*

As noted above, Tsui, Stormer and Gossard's motivation for going to higher magnetic fields was to look for the Wigner crystal. While the crystal phase is superseded by the formation of a CF liquid for a range of filling factors, a crystal must ultimately be stabilized as the filling factor is reduced and the electrons behave more and more classically (as the distance between them measured in units of the magnetic length increases). An insulating phase is observed at very low fillings, which is interpreted as a pinned crystal state. Extensive experimental work probing the state in transport and optical experiments[33,134–155] has revealed a rich interplay between the crystal and FQHE. Direct evidence for a periodic lattice at very low filling factors has been obtained through commensurability oscillations in the CF Fermi sea in a nearby layer[153] (see the Chapter by Shayegan).

The experimental facts relevant to our discussion below can be summarized as follows. For n-doped GaAs samples, in the limit of zero temperature, an insulating phase is seen for $\nu < 1/5$, and also for a narrow range of fillings between 1/5 and 2/9. These features have persisted as the sample quality has significantly improved. The fact that an insulating state is flanked by two strongly correlated FQH liquids (1/5 and 2/9) supports the notion that the insulator is a pinned crystal rather than a state with individual carrier freeze-out. The behavior in p-doped GaAs systems is qualitatively different[140,141,147,148] from that in n-doped GaAs systems.

In low-density p-doped GaAs systems, an insulating phase is observed for filling factors below 1/3, and even between 1/3 and 2/5. The FQH states at 1/3 and 2/5 are robust, however. Experiments in ZnO quantum wells[156] also show insulating phases intermingled with the FQH states at 1/3, 2/5, 3/7 etc.

Early theoretical studies[157,158] suggested that the crystal should be stabilized for filling factors below approximately $\nu \approx 1/6.5$.[157,158] These only considered competition between the Laughlin state and the crystal at filling factors of the form $\nu = 1/m$, and thus could not account for the re-entrant crystal phase between 1/5 and 2/9 in the n-doped GaAs systems. Several authors[109,159–161] attributed the difference between n- and p-doped GaAs to the stronger LL mixing in p-doped GaAs quantum wells due to the larger effective mass of holes. (LL mixing is also much larger in ZnO quantum wells.) They showed that LL mixing generally favors the crystal phase by studying the competition between the Laughlin liquid and the crystal state at fractions $\nu = 1/3$, 1/5 and 1/7 through variational,[159–161] diffusion,[109] and path integral Monte Carlo.[162] These studies also considered only the $\nu = 1/m$ FQH states.

More recent calculations addressing these issues have shown that a quantitative explanation of the above experimental facts requires a consideration of composite-fermion crystals[163–166] rather than ordinary electron crystals (i.e. vortices are bound to electrons in the crystal phase as well). There are two types of CF crystals (CFCs):

Type-I CFC: The crystal in which all composite fermions arrange themselves on a lattice is called a type-I CFC, sometimes referred to simply as a CFC. An insulating phase is obtained when a type-I CFC is pinned by disorder.

In the disk geometry, the wave function for a type-I CFC is given by

$$\Psi^{\text{CFC}} = \prod_{j<k} (z_j - z_k)^{2p} \Psi^{\text{EC}}, \tag{21}$$

where $\Psi^{\text{EC}} = \frac{1}{\sqrt{N!}} \sum_P \epsilon_P \prod_{j=1}^{N} \phi_{\boldsymbol{R}_j}(\boldsymbol{r}_{Pj})$ is the Hartree–Fock electron crystal (EC) in which electrons are placed in maximally localized wave packets at $\boldsymbol{R}_j = (X_j, Y_j)$, with $\phi_{\boldsymbol{R}}(\boldsymbol{r}) = \frac{1}{\sqrt{2\pi}} \exp\left(-\frac{1}{4}(\boldsymbol{r} - \boldsymbol{R})^2 + \frac{i}{2}(xY - yX)\right)$. The filling factor ν of the CFC is related to the filling factor ν^* of the EC by the standard relation $\nu = \nu^*/(2p\nu^* + 1)$. The vorticity $2p$ is a non-negative even integer, treated as a variational parameter, and it is assumed that $\nu^* < 1$ (so an electron crystal may be formed within the LLL). Reference 165 tested the CFC wave function against the exact Coulomb ground state wave function at total angular momenta $L = 7N(N-1)/2$ and $L = 9N(N-1)/2$, which correspond to $\nu = 1/7$ and 1/9, for a system of $N = 6$ particles in the disk geometry. (The CFC wave function was projected into the appropriate angular momentum L for this calculation, which in effect produces a rotating crystal.[167]) The lowest energy CFCs were obtained for $2p = 4$ at $\nu = 1/7$ and $2p = 6$ at $\nu = 1/9$. The overlaps of the CFC wave functions with the exact Coulomb ground states were found to be 0.997 at $\nu = 1/7$ and 0.999 at $\nu = 1/9$. These overlaps are significant given that the dimensions of the Hilbert spaces are large (117,788 and 436,140), and are also much better than the overlaps of the exact ground states with the Laughlin wave functions (0.71 and 0.66). The energies of the CFCs are also very close to the exact energies: they are 0.016% (0.006%) higher than the exact energies at $\nu = 1/7$ ($\nu = 1/9$). These calculations establish the validity of the CFC wave functions at low fillings.

Type-II CFC/FQHE: We also need a model for the FQH state as a continuous function of ν. The incompressible states correspond to $\nu^* = n$ filled ΛLs. For non-integer values of ν^* the topmost ΛL is partially occupied. What state these composite fermions will form (some possibilities being Wigner crystal, bubble crystal, stripes, Fermi sea, FQH liquid, paired state) is governed by the interaction between them and is a complex issue in itself. We note, however, that the dominant contribution to the total energy comes from the "kinetic energy" of these composite fermions, and the interaction between them is relatively weak.[n] It is therefore reasonable to expect that any configuration that builds repulsive correlations between composite fermions should be a decent first approximation. We will assume, for simplicity, that the CF-particles or CF-holes in the topmost partially-filled ΛL form a crystal. This crystal rides on a FQH state, and is called a type-II crystal by analogy to the Abrikosov flux lattice in a type-II superconductor.[169] The system exhibits FQHE when a type-II crystal is pinned by disorder. For that reason, the type-II CFC will often be labeled simply as "FQHE" in this subsection. Following our earlier discussion (see Fig. 3), the type-II CFCs are Wigner crystals of fractionally-charged quasiparticles or quasiholes of a FQH state. The idea that the quasiparticles or quasiholes of an incompressible FQH state should form a crystal, provided that their density is small and there is no disorder, has long been a part of the FQHE literature; see Halperin[13] for example. The CF theory, however, provides accurate wave functions that enable reliable estimates of their energies.

To study the interplay between FQHE and CFC states and determine the thermodynamic limits of various energies, it is convenient to employ the spherical geometry. A difficulty here is that a hexagonal lattice cannot be fitted perfectly on the surface of a sphere. We work with the "Thomson crystal" instead, wherein we choose our crystal sites that minimize the Coulomb energy of point charges on a sphere. This is the famous Thomson problem[170] which had been proposed in 1900 as the model of an atom. The positions of electrons in a Thomson crystal have been evaluated numerically and are available in the literature.[171–173] As expected, the Thomson lattice locally has a triangular structure but contains some defects. The correction to energy due to defects is negligible in large systems, and can be eliminated altogether by evaluating thermodynamic limits. A type-I CFC on the sphere is constructed by first forming a Hartree–Fock electron crystal at flux $2Q^* = 2Q - 2p(N-1)$, and then attaching to each electron $2p$ vortices by multiplication by an appropriate Jastrow factor. As before, $2p$ is a non-negative even integer, treated as a variational parameter, and we choose $\nu^* < 1$ (i.e. $2Q^* + 1 > N$).

---

[n]The inter-CF interaction is suppressed[86,168] because the total charge of a CF-particle or a CF-hole, $e^* = e/(2pn \pm 1)$, is small and also spread out.

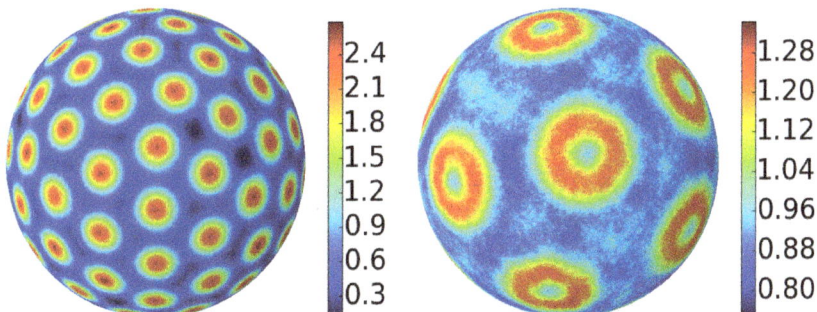

Fig. 11.   Density profiles of two crystals for a total of $N = 96$ electrons at filling factors slightly higher than $1/3$. Left shows a type-I electron crystal for $\nu = 0.394$ ($2Q = 240$). The right panel shows a type-II CF crystal for $\nu = 0.351$ ($2Q = 270$), where the composite fermions in the partially filled second ΛL (i.e. quasiparticles of the $1/3$ state) form a crystal. The density is given in units of the average density. All results are for $\kappa = 0$. Source: J. Zhao, Y. Zhang, and J. K. Jain, *Phys. Rev. Lett.* **121**, 116802 (2018).[97]

The FQH states at $\nu = n/(2pn \pm 1)$ map into $\nu^* = n$ of composite fermions, as discussed earlier. To calculate the energy the FQH state as a continuous function of ν, we assume that at non-integer values of $\nu^* > 1$, the composite fermions in the topmost partially filled ΛL form type-II CFC, again modeled as a Thomson crystal on the sphere. This is expected to be an excellent approximation when the density of composite fermions in the partially filled ΛL is small. It would be more appropriate to consider a crystal of CF-holes when a ΛL is more than half full, but the wave function for that state is technically more complicated to work with. We expect, however, that the CF-particle crystal will continue to produce a reasonable approximation for the energy even when a ΛL is more than half full, given that it properly captures the kinetic energy of composite fermions and that it is guaranteed to produce an accurate total energy in the limit when the ΛL becomes completely full. For $\nu^* < 1$, we assume a crystal of CF-holes in the lowest ΛL, i.e. a crystal of Laughlin quasiholes.

Figure 11 depicts examples of two crystals on the surface of a sphere. Both panels are for a total of 96 particles at filling factors slightly above $1/3$. The left panel shows a type-I electron crystal; the density profile of a type-I CFC is very similar. The right panel depicts a type-II CFC of composite fermions in the second ΛL. The density profile of an isolated composite fermion in the second ΛL resembles a smoke ring (just as that of an electron in the second LL does), producing a smoke-ring crystal when the CF density in the second ΛL is small. More intricate density patterns appear for type-II CFCs in higher ΛLs or when the CF-particles begin to overlap.

Figure 12 displays the energies per particle for various type-I CFC and FQH (i.e. type-II CFC) states as a function of the filling factor.[166] (The calculations

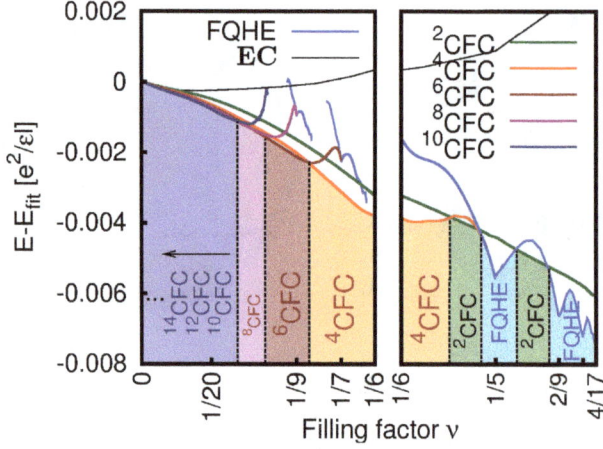

Fig. 12.   Energy per particle as a function of the filling factor for various CF crystal and FQH states. The label ^{2p}CFC refers to a type-I crystal of composite fermions carrying $2p$ vortices, and EC to the type-I electron crystal. The label FQHE refers to a type-II CF crystal (see text) which will show quantized Hall conductance in the presence of pinning by disorder. The energy of the FQH state as a function of ν exhibits downward cusps at the magic filling factors. All energies are quoted relative to the reference energy $E_{\text{fit}} = -0.782133\nu^{1/2} + 0.2623\nu^{3/2} + 0.18\nu^{5/2} - 15.1e^{-2.07/\nu}$. The regions $0 < \nu < 1/6$ and $1/6 < \nu < 4/17$ are shown in separate panels because different filling factor scales are used for them. Source: A. C. Archer, K. Park, and J. K. Jain, *Phys. Rev. Lett.* **111**, 146804 (2013).[166]

presented in Figs. 12 and 13 assume zero width.) The energy of the FQH state has cusps at the special filling factors $\nu = n/(2pn \pm 1)$. The curve for the energy of the type-I crystal of composite fermions carrying two vortices intersects the FQHE curve between 1/5 and 2/9, thus explaining the appearance of a crystal state in between these two filling factors. The CFC beats the FQH liquid in this region by a tiny energy of ∼0.0005 $e^2/\epsilon l$ per particle. (The simple Hartree–Fock crystal of electrons, labeled EC in Fig. 12, has a much higher energy and fails to produce a re-entrant transition here.) As the filling factor is further lowered, a sequence of transitions take place into type-I crystals of composite fermions with increasingly higher vorticity; this persists all the way to $\nu = 0$, although the energy differences between various kinds of crystals become vanishingly small in that limit. References 166, 174 have determined the shear modulus of the various crystals and predicted a discontinuity at the transition points, which should reflect in various observables, for example, in the magneto-phonon energy[154] or the melting temperature of the crystal (although, in practice, disorder will broaden the transitions).

The type-I CFC has the same periodicity and lattice constant as the simple Hartree–Fock electron crystal and the two also have very similar density profiles. Why, then, does a type-I CFC provide a better description at low filling factors? The reason is because, unlike the Hartree–Fock electron crystal, the CF

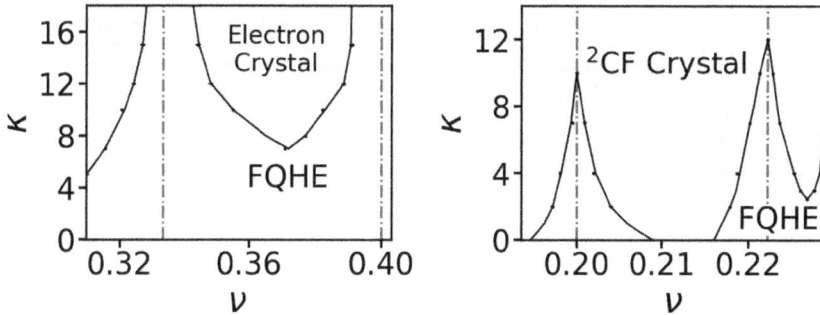

Fig. 13. Left panel: The phase diagram of the electron crystal and the FQHE in a filling factor range including $\nu = 1/3$ and $\nu = 2/5$ as a function of the LL mixing parameter κ. While at $1/3$ and $2/5$ the FQH states are very robust to LL mixing, for intermediate fillings the type-I electron crystal appears for $\kappa \gtrsim 7$. Rght panel: The theoretical phase diagram of the type-I ^{2}CF crystal and FQH state in a filling factor range including $\nu = 1/5$ and $\nu = 2/9$. The type-I electron crystal has substantially higher energy than the type-I ^{2}CF crystal in this filling factor region. Source: J. Zhao, Y. Zhang, and J. K. Jain, *Phys. Rev. Lett.* **121**, 116802 (2018).[97]

crystal also properly accounts, through the Jastrow factor, for correlations between the zero-point fluctuations of neighboring electrons around their equilibrium positions.

To address the difference between the behaviors in p and n doped GaAs systems, Ref. 97 has included the effect of LL mixing using fixed phase DMC, using the above wave functions to fix the phase. The resulting phase diagrams in the ν-κ plane are shown in Fig. 13. The most striking feature they reveal is the strong ν dependence of the phase boundary separating the FQH and the crystal phases. For example, FQHE at $\nu = 1/3$ and $2/5$ survives up to the largest value of κ $(= 18)$ considered, but the electron crystal appears already at $\kappa \gtrsim 7$ for certain ν in between $1/3$ and $2/5$, and at even lower values of κ for $\nu < 1/3$. Another notable feature is that in the vicinity of $\nu = 1/5$ and $2/9$, LL mixing induces a transition into the strongly correlated ^{2}CF crystal rather than an electron crystal.

In n-type GaAs quantum wells, with $\epsilon = 12.5$ and $m_b = 0.067 m_e$, the LL mixing parameter is given by $\kappa \approx 2.6/\sqrt{B[T]} \approx 1.28\sqrt{\nu/(\rho/10^{11}\text{cm}^{-2})}$. For typical densities, we have $\kappa \lesssim 1.0$ in the vicinity of $\nu = 1/3$ and $\nu = 1/5$. For these values we expect a crystal phase only between $1/5$ and $2/9$. The κ for holes in p-doped GaAs is ≈ 5.6 times that for electrons at the same B.[105] Santos *et al.*[141] find that an insulating phase appears between $\nu = 1/3$ and $\nu = 2/5$ at $\rho \approx 7 \times 10^{10}$ cm^{-2}, which corresponds to $\kappa \approx 5$. Given various approximations made in our calculation and our neglect of disorder (disorder should favor a crystal, because a crystal can more readily adjust to it than an incompressible liquid), we regard the level of agreement to be satisfactory. Similar considerations apply to ZnO quantum wells[156] for which κ is ~ 6.4 times larger than that for n-doped GaAs systems.[105]

Why does LL mixing favor the crystal phase? Both the liquid and the crystal states lower their energies by taking advantage of LL mixing, but one can expect that the crystal has more flexibility, because LL mixing allows the wave packet at each site to become more localized. The competition is subtle and complicated, however, and only a detailed calculation can tell if and where a transition into a crystal takes place.

There is additional experimental support for CF nature of the crystal. Jang et al.[154] have measured vibrations of the crystal phase using electron tunneling spectroscopy and found that the stiffening of the resonance is consistent with the shear modulus evaluated in Refs. 166 and 174 for the CF crystal. Evidence for type-II crystals has been seen by Zhu et al.[175] in optical experiments through observation of collective pinning modes in the vicinity of $\nu = 1/3$. In a theoretical work, Shi and Ji[176] have predicted that CF nature of the type-I crystal results in a magnetoroton-like phonon.

4. Kohn–Sham Density Functional Theory of the FQHE

This section is a minimally modified reproduction of an article by Yayun Hu and the author.[177]

The Kohn–Sham (KS) density-functional theory (DFT) uses the electron density to construct a single particle formalism that incorporates the complex effects of many-particle interactions through a universal exchange correlation function.[178] It is an invaluable tool for treating systems of interacting electrons spanning the disciplines of physics, chemistry, materials science and biology, but very little work has been done[179–181] toward applying this method to the FQHE. The reasons are evident. To begin with, even though the KS-DFT is in principle exact, its accuracy, in practice, is dictated by the availability of exchange correlation (xc) potentials, and it works best when the xc contribution is small compared to the kinetic energy. In the FQHE problem, the kinetic energy is altogether absent (at least in the convenient limit of very high magnetic fields) and the physics is governed entirely by the xc energy. A more fundamental impediment is that, by construction, the KS-DFT eventually obtains a single Slater determinant solution, whereas the ground state for the FQHE problem is an extremely complex, filling factor-dependent wave function that is not adiabatically connected to a single Slater determinant. In particular, a mapping into a problem of non-interacting electrons in a KS potential will produce a ground state that locally has integer fillings, whereas nature displays preference for certain fractional fillings. Finally, a mapping into a system of weakly interacting electrons will also fail to capture the topological features of the FQHE, such as fractional charge and fractional braid statistics for the quasiparticles. At a fundamental level, these difficulties can be traced back to the fact that the space of ground states in the LLL is highly degenerate for non-interacting electrons, and the interaction causes a non-perturbative reorganization to produce the FQHE.

4.1. *KS equations for composite fermions*

To make progress, we exploit the fact that the strongly interacting electrons in the FQH regime turn into weakly interacting composite fermions, which suggests using an auxiliary system of non-interacting composite fermions to construct a KS-DFT formulation of the FQHE. This is the approach taken here. A crucial aspect of our KS theory is that it properly incorporates the physics of long range gauge interaction between composite fermions induced by the Berry phases due to the quantum mechanical vortices attached to them, which is responsible for the topological properties of the FQHE, such as fractional charge and statistics.[18,95,182] That effectively amounts to using a non-local exchange-correlation potential. Certain previous DFT formulations of the FQHE[179–181] employ a local exchange-correlation potential and thus do not capture the topological features of the FQHE.

We consider the Hamiltonian for fully spin polarized electrons confined to the LLL:

$$\hat{\mathcal{H}} = \hat{H}_{ee} + \int d\boldsymbol{r} V_{ext}(\boldsymbol{r}) \hat{\rho}(\boldsymbol{r}) . \tag{22}$$

Within the so-called magnetic-field DFT,[183–186] the Hohenberg–Kohn (HK) theorem also applies to interacting electrons in the FQH regime and implies that the ground state density and energy can be obtained by minimizing the energy functional

$$E[\rho] = F[\rho] + \int d\boldsymbol{r} V_{ext}(\boldsymbol{r}) \rho(\boldsymbol{r}), \tag{23}$$

where the HK functional is given by[187,188]

$$F[\rho] = \min_{\Psi_{LLL} \to \rho(\boldsymbol{r})} \langle \Psi_{LLL} | \hat{H}_{ee} | \Psi_{LLL} \rangle \equiv E_{xc}[\rho] + E_H[\rho]. \tag{24}$$

(The B dependence of the energy functional has been suppressed for notational convenience). Here $E_{xc}[\rho]$ and $E_H[\rho]$ are the xc and Hartree energy functionals of electrons and Ψ_{LLL} represents a LLL wave function. The conventional KS mapping into non-interacting electrons is problematic due to the absence of kinetic energy.

We instead map the FQHE into the auxiliary problem of "non-interacting" composite fermions. Even though we use the term non-interacting, the Berry phases associated with the bound vortices induce a long range gauge interaction between composite fermions, as a result of which they experience a density dependent magnetic field $B^*(\boldsymbol{r}) = B - 2\rho(\boldsymbol{r})\phi_0$, where $\phi_0 = hc/e$ is a flux quantum. We therefore write

$$\left[\frac{1}{2m^*} \left(\boldsymbol{p} + \frac{e}{c} \boldsymbol{A}^*(\boldsymbol{r}; [\rho]) \right)^2 + V_{KS}^*(\boldsymbol{r}) \right] \psi_\alpha(\boldsymbol{r}) = \epsilon_\alpha \psi_\alpha(\boldsymbol{r}), \tag{25}$$

where $V_{KS}^*(\boldsymbol{r})$ is the KS potential for composite fermions, m^* is the CF mass (taken to be $m^* = 0.079\sqrt{B[T]}\, m_e$; see Sec. 2.3), and $\nabla \times \boldsymbol{A}^*(\boldsymbol{r}; [\rho]) = B^*(\boldsymbol{r})$. As a result of the gauge interaction, the solution for any given orbital depends, through the $\rho(\boldsymbol{r})$

dependence of the vector potential, on the occupation of all other orbitals. Equation (25) must therefore be solved self-consistently, i.e. the single-CF orbitals $\psi_\alpha(r)$ must satisfy the condition that the ground state density $\rho(r) = \sum_\alpha c_\alpha |\psi_\alpha(r)|^2$, where $c_\alpha = 1$ (0) for the lowest energy occupied (higher energy unoccupied) single-CF orbitals, is equal to the density that appears in the kinetic energy of the Hamiltonian. The energy levels of Eq. (25) are the self-consistent ΛLs. For the special case of a spatially uniform density and constant V_{KS}^*, Eq. (25) reduces to the problem of non-interacting particles in a uniform B^*. Importantly, once a self-consistent solution is found for a given $V_{KS}^*(r)$, for the corresponding density in the Hamiltonian in Eq. (25), the ground state satisfies, by definition, the self-consistency condition and also the variational theorem, and the standard proof for the HK theorem follows.[177] We define the CF kinetic energy functional as

$$T_s^*[\rho] = \min_{\Psi \to \rho} \langle \Psi | \frac{1}{2m^*} \sum_{j=1}^N \left(p_j + \frac{e}{c} A^*(r_j; [\rho]) \right)^2 | \Psi \rangle, \qquad (26)$$

where we perform a constrained search over all single Slater determinant wave functions Ψ that correspond to the density $\rho(r)$, following the strategy of the generalized KS scheme.[177,189]

The next key step is to write $E_{xc}[\rho] = T_s^*[\rho] + E_{xc}^*[\rho]$, or $F[\rho] = T_s^*[\rho] + E_H[\rho] + E_{xc}^*[\rho]$. (Note that $T_s^*[\rho]$ and thus $E_{xc}[\rho]$ is a non-local functional of the density.) Such a partitioning of $F[\rho]$ can, in principle, always be made given our assumptions, but is practically useful only if the $T_s^*[\rho]$ and $E_H[\rho]$ capture the significant part of $F[\rho]$, and the remainder $E_{xc}^*[\rho]$, called the exchange-correlation energy of composite fermions, makes a relatively small contribution. This appears plausible given that the CF kinetic energy term captures the topological aspects of the FQHE, and also because the model of weakly interacting composite fermions has been known to be rather successful in describing a large class of experiments.

Minimization of the energy $E[\rho] = T_s^*[\rho] + E_H[\rho] + E_{xc}^*[\rho] + \int dr V_{ext}(r)\rho(r)$ with respect to $\rho(r) = \sum_\alpha c_\alpha |\psi_\alpha(r)|^2$, subject to the constraint $\int dr \psi_\alpha^*(r)\psi_\beta(r) = \delta_{\alpha\beta}$, yields[177] Eq. (25) with

$$V_{KS}^*[\rho, \{\psi_\alpha\}] = V_H(r) + V_{xc}^*(r) + V_{ext}(r) + V_T^*(r), \qquad (27)$$

where $V_H(r) = \delta E_H/\delta\rho(r)$ and $V_{xc}^*(r) = \delta E_{xc}^*/\delta\rho(r)$ are the Hartree and CF-xc potentials. The non-standard potential $V_T^*(r) = \sum_\alpha c_\alpha \langle \psi_\alpha | \delta T^*/\delta\rho(r) | \psi_\alpha \rangle$ with $T^* = \frac{1}{2m^*} \left(p + \frac{e}{c} A^*(r; [\rho]) \right)^2$ arises due to the density-dependence of the CF kinetic energy. V_T^* describes the change in T_s^* to a local disturbance in density for a fixed choice of the KS orbitals. Because $V_T^*(r)$ depends not only on the density but also on the occupied orbitals, we are actually working with what is known as the "orbital dependent DFT".[190]

Having formulated the CF-DFT equations, we now proceed to obtain solutions for some representative cases. The primary advantage of our approach is evident without any calculations. Take the example of a uniform density FQH state at $\nu = n/(2pn \pm 1)$. It is an enormously complicated state in terms of electrons, but maps into the CF state at filling factor $\nu^* = n$ with a spatially uniform magnetic field, thereby producing the correct density. For non-uniform densities, the state of non-interacting composite fermions will produce configurations where composite fermions *locally* have $\nu^* \approx n$, which corresponds to an electronic state where the local filling factor is $\nu \approx n/(2pn \pm 1)$, which is a reasonable description, and certainly a far superior representation of the reality than any state of non-interacting electrons.

For a more quantitative treatment we need a model for the xc energy. To this end, we assume the LDA form $E^*_{xc}[\rho] = \int d\mathbf{r}\, \epsilon^*_{xc}[\rho(\mathbf{r})]\rho(\mathbf{r})$, where $\epsilon^*_{xc}[\rho]$ is the xc energy per CF. We express all lengths in units of the magnetic length and energies in units of $e^2/\epsilon l$. The density is related to the local filling factor as $\nu(\mathbf{r}) = \rho(\mathbf{r})2\pi l^2$. We take the model $\epsilon^*_{xc}[\rho] = a\nu^{1/2} + (b - f/2)\nu + g$, with $a = -0.78213$, $b = 0.2774$, $f = 0.33$, $g = -0.04981$. The form is chosen empirically so that the sum of ϵ^*_{xc} and the CF kinetic energy accurately reproduces the known electronic xc energies at $\nu = n/(2n + 1)$. (The term $a\nu^{1/2}$ is chosen to match with the known classical value of energy of the Wigner crystal in the limit $\nu \to 0$.[191]) Although optimized for $\nu = n/(2n + 1)$, we shall uncritically assume this form of $\epsilon^*_{xc}(\nu)$ for all ν. Our aim here is to establish the proof-of-principle validity and the applicability of our approach and its ability to capture topological features, which are largely robust against the precise form of the xc energy. The xc potential is given by $V^*_{xc} = \delta E^*_{xc}/\delta \rho(\mathbf{r}) = \frac{3}{2}a\nu^{1/2} + (2b - f)\nu + g$. We note that while the CF xc potential V^*_{xc} is a continuous function of density, the *electron* xc potential V_{xc} has derivative discontinuities at $\nu = n/(2n \pm 1)$, arising from the kinetic energy of the composite fermions.

In our applications below, we will consider N electrons in a potential $V_{ext}(\mathbf{r}) = -\int d^2\mathbf{r}' \frac{\rho_b(\mathbf{r}')}{\sqrt{|\mathbf{r}-\mathbf{r}'|^2+d^2}}$ generated by a two-dimensional uniform background charge density $\rho_b = \nu_0/2\pi l^2$ distributed on a disk of radius R_b satisfying $\pi R_b^2 \rho_b = N$ at a separation of d from the plane of the electron liquid. This produces an electron system at filling factor $\nu = \nu_0$ in the interior of the disk. We use $\nu_0 = 1/3$ and $d/l \to 0$ in our calculations below. For the vector potential, we assume circular symmetry and choose the gauge $\mathbf{A}^*(\mathbf{r}) = \frac{rB(r)}{2}\mathbf{e}_\phi$, with $\mathcal{B}(r) = \frac{1}{\pi r^2}\int_0^r 2\pi r' B^*(r')dr'$. We obtain self-consistent solutions of Eqs. (25) and (27) by an iterative process.

4.2. Density profile of the FQH droplet

As a first application, we consider the density profile of the $\nu_0 = 1/3$ droplet. Figure 14 shows the density profiles calculated from Laughlin's trial wave function as well as that obtained from exact diagonalization at total angular momentum $L = 3N(N - 1)/2$.[192] Also shown are the density profiles obtained from the above KS equations. The density profile from our CF-DFT captures that obtained in exact diagonalization well, especially for $N \geq 10$. Remarkably, it reproduces the characteristic shape near the edge where the density exhibits oscillations and overshoots

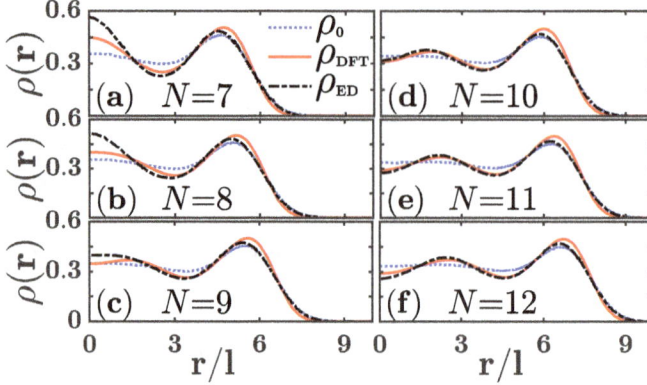

Fig. 14.   Density profile for 1/3 droplets. This figure shows the density of a system of N composite fermions. ρ_0 is the density for Laughlin's 1/3 wave function,[10] and ρ_{ED} is obtained from exact diagonalization (ED) of the Coulomb interaction at total angular momentum $L_{total} = 3N(N-1)/2$.[192] The density ρ_{DFT} is calculated from the solution of the KS equations for composite fermions in an external potential produced by a uniform positively charged disk of radius R so that $\pi R^2 \rho_b = N$. The total angular momentum of the CF state is L^*_{tot}, which is related to the total angular momentum of the electron state by $L_{tot} = L^*_{tot} + N(N-1)$.[193] The CF-DFT solution produces $L^*_{tot} = N(N-1)/2$, which is consistent with $L_{tot} = 3N(N-1)/2$. All densities are quoted in units of $(2\pi l^2)^{-1}$, the density at $\nu = 1$. We take $\rho_b = 1/3$. Source: Y. Hu and J. K. Jain, *Phys. Rev. Lett.* **123**, 176802 (2019).[177]

the bulk value before descending to zero. This qualitative behavior is fairly insensitive to the choice of V^*_{xc}, and is largely a result of the self-consistency requirement in Eq. (25).

4.3. Screening by the FQH state

We next consider screening of an impurity with charge $Q = \pm e$ at a height h directly above the center of the FQH droplet. The strength of its potential

$$V_{imp}(r) = \frac{Q}{\sqrt{|r|^2 + h^2}} \qquad (28)$$

can be tuned by varying h. Panels (a)–(e) in Fig. 15 show the density ρ for certain representative values of h. It is important to note that the CF orbitals in the self-consistent solution form strongly renormalized ΛLs (i.e. include the effect of mixing between the unperturbed ΛLs). Panels (f)–(j) show the occupation of the ΛLs. The presence of the impurity either empties some CF orbitals from the lowest ΛL or fills those in higher ΛLs. Each empty orbital in the lowest ΛL corresponds to a charge 1/3 quasihole, whereas each filled orbital in an excited ΛL to a charge $-1/3$ quasiparticle.[18] The excess charge is defined as $\delta q = \int_{|r|<r_0} d^2 r[\rho_0 - \rho(r)]$ in a circular area of radius $r_0 = 10l$ around the origin. Panel (p) shows how δq and L^*_{tot} change as a function of the potential at the origin $V_{imp}(r=0) = -Q/h$. The excess charge δq is seen to be quantized at an integer multiple of $\pm 1/3$.

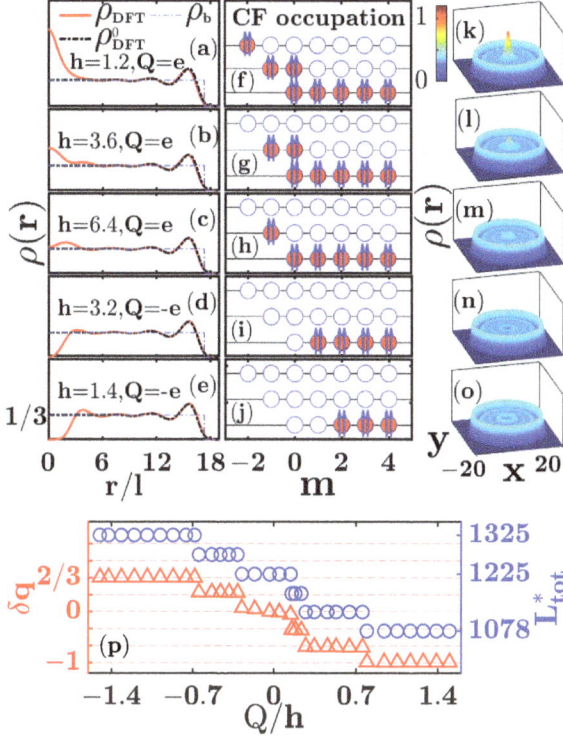

Fig. 15. Screening and fractional charge. This figure shows how the 1/3 state screens a charged impurity of strength $Q = \pm e$ located at a perpendicular distance h from the origin. The panels (a)–(e) and (k)–(o) show the self-consistent density $\rho_{\mathrm{DFT}}(\boldsymbol{r})$. Also shown are $\rho_{\mathrm{DFT}}^{0}(\boldsymbol{r})$, the "unperturbed" density (for $Q = 0$), and ρ_b, which is the density of the positively charged background. Panels (f)–(j) show the occupation of renormalized ΛLs in the vicinity of the origin; each composite fermion is depicted as an electron with two arrows, which represent quantized vortices. (The single particle angular momentum is given by $m = -n, -n+1, \cdots$ in the nth ΛL.) The panel (p) shows the evolution of the excess charge δq and the total CF angular momentum L_{tot}^{*} as a function of the impurity potential strength at the origin $V_{\mathrm{imp}}(r = 0) = Q/h$. Change in the charge at the origin is associated with a change in L_{tot}^{*}. The system contains a total of $N = 50$ composite fermions. For $h = \infty$, we have $L_{\mathrm{tot}}^{*} = 1225$ and $\delta q = 0$. For one and two quasiholes, we have $L_{\mathrm{tot}}^{*} = 1225$ and 1275, whereas for one, two and three quasiparticles we have $L_{\mathrm{tot}}^{*} = 1175$, 1127 and 1078, precisely as expected from the configurations in panels (f)–(j).[193] Source: Y. Hu and J. K. Jain, *Phys. Rev. Lett.* **123**, 176802 (2019).[177]

4.4. *Fractional braid statistics*

We finally come to fractional braid statistics. Particles obeying such statistics, called anyons, are characterized by the property that the phase associated with a closed loop of a particle depends on whether the loop encloses other particles. In particular, for Abelian anyons, each enclosed particle contributes a phase factor of $e^{i2\pi\alpha}$, where α is called the statistics parameter. [For non-interacting bosons (fermions), α is an even (odd) integer.] In the FQHE, the quasiparticles are excited composite fermions and quasiholes are missing composite fermions. Let us consider quasiholes of the 1/3

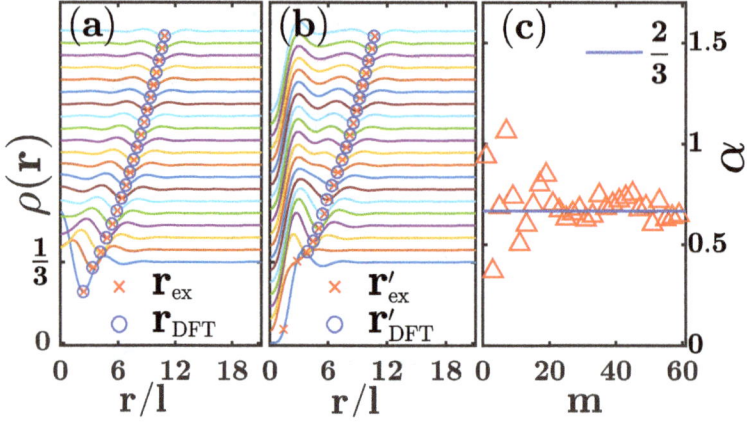

Fig. 16. Fractional braid statistics. Panel **(a)** shows the electron density for a system with a quasihole in angular momentum m orbital, with m changing from 1 to 20 for the curves from the bottom to the top. (Each successive curve has been shifted up vertically for clarity.) Panel **(b)** shows the same in the presence of another quasihole at the origin. For each m, we indicate the expected position of the outer quasihole (red cross) as well as the position obtained from the DFT density determined by locating the local minimum (blue circle). Panel **(c)** shows the calculated statistics parameter $\alpha \equiv (r_{\mathrm{DFT}}^2 - r_{\mathrm{DFT}}'^2)/6l^2$. The calculation has been performed for $N = 200$ composite fermions at $\nu_0 = 1/3$. Source: Y. Hu and J. K. Jain, *Phys. Rev. Lett.* **123**, 176802 (2019).[177]

state for illustration. A convenient way to ascertain the statistics parameter within our KS-DFT is to ask how the location of a quasihole in angular momentum m orbital changes when another quasihole is inserted at the origin in the $m = 0$ orbital. Let us first recall the expected behavior arising from fractional braid statistics. In an effective description, the wave function of a single quasihole in angular momentum m orbital is given by $z^m e^{-|z|^2/4l^{*2}}$ ($z \equiv x - iy$), which is maximally localized at $r_{\mathrm{ex}} = (2m)^{1/2}l^* = (6m)^{1/2}l$, with $l^* = \sqrt{3}l$ (as appropriate for $\nu_0 = 1/3$). When another quasihole is present at the origin, it induces an additional statistical phase factor $e^{i2\pi\alpha}$, where α is the statistics parameter. This changes the wave function of the outer quasihole to $z^{m-\alpha} e^{-|z|^2/4l^{*2}}$, which is now localized at $r_{\mathrm{ex}}' = [6(m-\alpha)]^{1/2}l$. We now determine α from our KS-DFT formalism.

A quasihole can be treated in a constrained DFT[194] wherein we leave a certain angular momentum orbital unoccupied. The panels (a) and (b) of Fig. 16 show the self-consistent KS density profiles of the state with a quasihole in angular momentum m, without and with another quasihole in the $m = 0$ orbital. The locations of the outer quasihole, r_{DFT} and r_{DFT}', are determined from the minimum in the density. These are in reasonable agreement with the expected positions r_{ex} and r_{ex}' (provided $m > 3$). More importantly, the calculated statistics parameter $\alpha \equiv (r_{\mathrm{DFT}}^2 - r_{\mathrm{DFT}}'^2)/6l^2$ is in excellent agreement with the expected fractional value of $\alpha = 2/3$[13,18] provided that the two quasiparticles are not close to one another, indicating that our method properly captures the physics of fractional braid statistics. The small deviation from 2/3 for large m arises from the fact that the density of the unperturbed system itself has slight oscillations due to the finite system size, which causes a slight shift in the position of the local minimum due to an additional quasihole. Correcting for that effect produces a value much closer to $\alpha = 2/3$.[177]

These studies demonstrate that the Kohn–Sham DFT faithfully captures the topological characteristics of the FQH state. This opens a new strategy for exploring a variety of problems of interest.

5. Looking Beyond Composite Fermions: The Parton Paradigm

Soon following the CF theory, a generalization was introduced in 1989 known as the parton construction,[195,196] which further exploits the connection between the FQHE and the IQHE. While composite fermions are the building blocks of the CF theory, IQH states are the building blocks of the parton theory. The parton construction produces candidate FQH states that are products of IQH states. These include all of the states of the CF theory but also states beyond the CF theory. The states in the latter category are interesting in their own right, but also because, as shown by Wen,[197] they include non-Abelian states. All of the states of the parton theory are in principle valid, and one can attempt to construct a model interaction whose ground state is well represented by a given parton state. The important question, however, is whether the new (beyond-CF) states are realized in some known systems. One of the simplest candidates beyond the CF theory, namely the 221 state (defined below), is a non-Abelian state at $\nu = 1/2$. It was considered in early 1990s by the author and his collaborators as a candidate for the 5/2 state[198] (i.e. 1/2 in the second LL) but was not found to be stabilized by the second LL Coulomb interaction. In 1991, a Pfaffian wave function was introduced by Moore and Read,[89] which is also a non-Abelian state at $\nu = 1/2$ (although distinct from the 221 state). The Pfaffian state was seen in numerical diagonalization studies[199] to provide a reasonable description for the 5/2 FQHE. As a result of these developments, interest in the parton construction subsided. However, a recent work by Balram, Barkeshli and Rudner (BBR)[200] has breathed a new life into the parton theory. These authors have demonstrated that a different non-Abelian state from the parton construction, labeled $\bar{2}\bar{2}111$ (see below), does provide a good account of the 5/2 FQHE. This work has inspired further studies that have indicated possible realizations of certain other beyond-CF states as well.

It is a remarkable fact that the physics of strong correlations in the FQHE can be captured by wave functions that are products of Slater determinants of IQH states. One wonders if all experimentally realized FQH states conform to this paradigm. That, in the author's view, would be very satisfying, and also appears to be the case so far, as discussed below.

Section 5.1 outlines the parton construction, and gives a brief account of the topological properties of these states, appearance of non-Abelian statistics, and

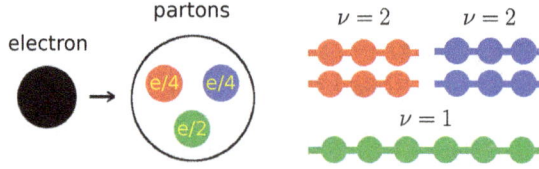

Fig. 17.   Parton construction of the 221 state at $\nu = 1/2$. Figure taken from Y. Wu, T. Shi, and J. K. Jain, *Nano Letters* **17**, 4643–4647 (2017).[201]

connection of some of these states to topological superconductivity of composite fermions. Rest of the section discusses several states that are of possible experimental relevance.

5.1. *The parton construction: Abelian and non-Abelian states*

We begin by asking if it is possible to construct new incompressible states from known incompressible states, such as the IQH states. The parton construction seeks to accomplish this goal in the following manner (Fig. 17). We first decompose each electron into m fictitious particles called partons, which, in the simplest implementation, are taken to be fermionic. We then place each species of partons into an IQH state with filling factor n_λ, where $\lambda = 1, \cdots, m$ labels different parton species. Finally, we glue the partons back together to recover the physical electrons. It is intuitively sensible that the resulting state will be incompressible.

The density of each species of partons must be the same as the density of the physical electrons, which implies, recalling $\rho = \nu eB/hc$, that each species must satisfy $n_\lambda q_\lambda = \nu e$, where q_λ is the charge of the λ-parton. Substituting $q_\lambda = e\nu/n_\lambda$ into $\sum_\lambda q_\lambda = e$ gives the relation $\nu = (\sum_\lambda n_\lambda^{-1})^{-1}$. The wave function for the partons is given by $\prod_{\lambda=1}^m \Phi_{n_\lambda}(\{z_j^\lambda\})$. We identify $z_j^\lambda = z_j$ and project into the LLL to obtain "the $n_1 \cdots n_m$ state"°

$$\Psi_\nu^{n_1\cdots n_m} = \mathcal{P}_{\mathrm{LLL}} \prod_{\lambda=1}^m \Phi_{n_\lambda}(\{z_j\}), \quad \nu = \left(\sum_{\lambda=1}^m \frac{1}{n_\lambda}\right)^{-1}, \quad q_\lambda = \frac{\nu}{n_\lambda}. \tag{29}$$

Negative values of q_λ produce negative filling factors n_λ, which correspond to partons in a negative magnetic field. For notational ease, it is customary to write $-n = \bar{n}$, with $\Phi_{-n} = \Phi_{\bar{n}} = [\Phi_n]^*$. The LLL projection, as before, is not expected to alter the topological character of the unprojected product state. Interestingly, even though the partons are unphysical, they leave their footprints in the physical world: an excitation in the factor Φ_{n_λ} has a charge $q_\lambda = e\nu/n_\lambda$ associated with it. In general, it is expected that the lowest energy quasiparticles correspond to the excitations in the factor Φ_{n_λ} with the largest $|n_\lambda|$, as they have the smallest charge. The $n_1 \cdots n_m$ state occurs at shift $\mathcal{S} = \sum_{\lambda=1}^m n_m$.

°These have been called Jain (parton) states in the literature.

The standard states of Eq. (8) are a part of the parton construction because $\Phi_1 \sim \prod_{j<k}(z_j - z_k)$. Specifically, we have

$$\Psi_{\nu=\frac{n}{2pn+1}}^{n11\cdots} = \mathcal{P}_{LLL}\Phi_n\Phi_1^{2p} = \mathcal{P}_{LLL}\Phi_n\prod_{j<k}(z_j - z_k)^{2p}, \qquad (30)$$

$$\Psi_{\nu=\frac{n}{2pn-1}}^{\bar{n}11\cdots} = \mathcal{P}_{LLL}\Phi_{\bar{n}}\Phi_1^{2p} = \mathcal{P}_{LLL}[\Phi_n]^*\prod_{j<k}(z_j - z_k)^{2p}. \qquad (31)$$

The fact that the states in Eq. (8) can be written as products of IQH states was the original motivation for the parton construction. The standard CF theory corresponds to $n_1\cdots n_m$ states with no more than one integer different from 1.

A field theoretical description[P] of the wave functions in Eq. (29) was developed by Wen and others.[197,200,202–205] Let us outline the derivation of the CS theory for the standard $\nu = n/(2pn \pm 1)$ FQHE starting from the parton construction. We begin by noting that the CS Lagrangian for the $\nu = \pm n$ state of charge q fermions is given by[92,93]

$$\mathcal{L} = \mp\frac{1}{4\pi}\sum_{j=1}^{n} a^j \partial a^j + \frac{1}{2\pi}\sum_{j=1}^{n} t^j A \partial a^j, \qquad (32)$$

where j is the LL index, A is the physical vector potential (treated as a non-dynamical background field), a^j is an emergent gauge field associated with the jth LL, $t = (q, q, \cdots, q)^T$ is the charge vector (same charge in each LL), and we have used the notation $a \partial b = \epsilon^{\mu\nu\delta} a_\mu \partial_\nu b_\delta$ and also set $e = \hbar = c = 1$. The particle current density is given by $J^\mu = \delta S/\delta A_\mu = (1/2\pi)\sum_{j=1}^{n} t^j \epsilon^{\mu\nu\lambda}\partial_\nu a^j_\lambda$.

The CS theory for the state at $n/(2n \pm 1)$ is constructed as follows. Let us represent the electron operator as $c = f_1 f_2 f_3$, where the parton f_1 is in the $\nu = \pm n$ state and the partons f_2 and f_3 are in $\nu = 1$ state. Of course, the partons are unphysical and the final theory must glue the partons into physical electrons. The redundancy of the electron operator implies an internal local gauge symmetry in which the local U(1) transformation $f_1 \to e^{i\theta_1} f_1$, $f_2 \to e^{i\theta_2} f_2$ and $f_3 \to e^{-i\theta_1-i\theta_2} f_3$ leaves the theory invariant. This constraint is imposed by introducing two local gauge fields, denoted b_1 and b_2 below. The Lagrangian for the $\pm n11$ state is given by

$$\mathcal{L} = \left[\mp\frac{1}{4\pi}\sum_{j=1}^{n} a_1^j \partial a_1^j + \frac{1}{2\pi}\sum_{j=1}^{n} t_1^j A \partial a_1^j\right] + \left[-\frac{1}{4\pi}a_2\partial a_2 + \frac{1}{2\pi}t_2 A \partial a_2\right]$$

$$+ \left[-\frac{1}{4\pi}a_3\partial a_3 + \frac{1}{2\pi}t_3 A \partial a_3\right] + \left[b_1\sum_{j=1}^{n}\partial a_1^j + b_2\partial a_2 - (b_1 + b_2)\partial a_3\right]. \qquad (33)$$

Here, $t_1 = (q_1, q_1, \cdots, q_1)^T$, $t_2 = q_2$ and $t_3 = q_3$, where $q_1 = \pm 1/(2n \pm 1)$, $q_2 = q_3 = n/(2n \pm 1)$ are the charges of the three partons in units of e. The terms in the first three square brackets on the right come from the individual factors,

---

[P]The discussion here owes greatly to insights from Ajit Balram and Maissam Barkeshli.

and the last square brackets contain the constraints. The constraints yield $\partial a_2 = \partial a_3 = \sum_{j=1}^{n} \partial a_1^j$. These are equivalent to $a_2 = a_3 + c$ and $a_2 = \sum_{j=1}^{n} a_1^j + d$, with $\epsilon^{\mu\nu\lambda} \partial_\nu c_\lambda = 0$ and $\epsilon^{\mu\nu\lambda} \partial_\nu d_\lambda = 0$. Substituting into Eq. (33) and noting that the terms containing c and d vanish,[205] the final form of the CS Lagrangian is obtained:

$$\mathcal{L} = -\frac{1}{4\pi} \sum_{i,j=1}^{n} a^i K^{ij} \partial a^j + \frac{1}{2\pi} \sum_{j=1}^{n} t^j A \partial a^j, \quad K^{ij} = \pm\delta_{ij} + 2p, \quad t = (1, 1, \cdots, 1)^T,$$

(34)

where we have generalized to $\nu = n/(2pn \pm 1)$.

The CS theory of Abelian FQH states is in general given by a Lagrangian of the type shown in Eq. (34), which is defined by a symmetric integer valued K matrix and a charge vector t. The $n_1 n_2 \cdots n_m$ state where no integer other than 1 is repeated is an Abelian state (as explained below). In this case, there are $\sum_\lambda |n_\lambda|$ gauge fields prior to projection into the physical space, but the $m - 1$ U(1)-constraints gluing the partons reduce the number of physical gauge fields to $\sum_\lambda |n_\lambda| - (m - 1)$, which gives the dimension of the K matrix. The K matrix and the t vector can be determined in the manner outlined above, and encode information about many topological properties of the state.[92,93,204,206,207] The Hall conductance is given by $\sigma_{xy} = \sum_{i,j} t^i (K^{-1})^{ij} t^j$; the charge of the quasihole coupled to the field a^i is $q^i = \sum_j (K^{-1})^{ij} t^j$; the relative braid statistics of the quasiholes are $a^{ij} = (K^{-1})^{ij}$; and the ground state degeneracy on genus-g surface is $|\det K|^g$. The charge and statistics are obtained by adding to the Lagrangian a term $\sum_j l^j a_\mu^j J'^\mu$, where J'^μ is the quasiparticle current and l^j are positive (negative) integers representing the number of quasiparticles (quasiholes) coupled to the gauge field a^j. The K-matrix also contains information about edge states. Assuming that there is no edge reconstruction, the dimension of the matrix gives the number of independent edge modes and the number of positive (negative) eigenvalues gives the number of downstream (upstream) modes. The central charge c is equal to the number of downstream minus the number of upstream modes, which is not affected by edge reconstruction. The central charge can be experimentally ascertained by a measurement of the thermal Hall conductance,[208] which is given by $\kappa_{xy} = c\pi^2 k_B^2 T/3h$.

Wen showed[197] that the parton construction also produces non-Abelian states, which are $n_1 n_2 \cdots n_m$ states where an integer ≥ 2 is repeated. In particular, he considered states of the form $\Psi_{n/m} = [\Phi_n]^m$, i.e. $nn \cdots n$ states, for which all partons have charge $q_\lambda = e/m$. Because all parton species are indistinguishable, only those states are physical that are invariant under a local SU(m) transformation within the parton space. This local SU(m) symmetry is implemented through a non-Abelian SU(m) gauge field coupled to the partons. Integrating out the parton fields yields an SU(m)$_n$ CS theory. The quasiparticles in this theory are particle or hole excitations in Φ_n, dressed by an SU(m)$_n$ CS gauge field. The non-Abelian braid properties of the excitations are determined from the properties of the SU(m)$_n$ CS theory. Wen also considered the edge theory of the $\Psi_{n/m} = [\Phi_n]^m$ state. Before imposing the constraint that combines the unphysical partons into physical electrons, mn chiral edge states arise from n LLs of each of the m partons, producing a central charge of $c = mn$. One must now project the theory into the physical space[q] by eliminating all of the fluctuations that transform non-trivially under SU(m) transformation. The projection can be carried out by using the level-rank duality of Wess–Zumino–Witten models in conformal field theory as follows. Bosonization of mn chiral fermions (assuming the same velocity for them) gives a U(mn)$_1$ algebra, the Hilbert space of which can be represented as a direct product of a U(1) Kac–Moody algebra, an SU(n)$_m$ Kac–Moody algebra, and an SU(m)$_n$ Kac–Moody algebra. The central charges of these three algebras add to mn: $1 + m(n^2 - 1)/(m + n) + n(m^2 - 1)/(m + n) = mn$. Projection is equivalent to removing the Hilbert space of SU(m)$_n$ Kac–Moody algebra, which leaves the central charge $c = 1 + m(n^2 - 1)/(m + n) = n(mn + 1)/(m + n)$. The central charge for the complex conjugate state $[\Psi_{n/m}]^* = [\Phi_n^*]^m = [\Phi_{\bar{n}}]^m$ is given by $c = -n(mn + 1)/(m + n)$.

Let us take some examples. For the Laughlin $1/m$ state, we have $n = 1$, which gives $c = 1$. Furthermore, the SU(m)$_1$ CS theory is abelian, implying Abelian statistics for the excitations (which is why repeated 1's do not yield non-Abelian statistics). For the 22 state Φ_2^2, we get $c = 5/2$ and the fusion rules for quasiparticles correspond to the SU(2)$_2$ Ising topological quantum field theory. The $\bar{2}\bar{2}$ state $[\Phi_2]^{*2}$ has $c = -5/2$.

In addition to $[\Phi_n]^m$, states containing factors of $[\Phi_n]^m$, with $n \geq 2$ and $m \geq 2$, are also non-Abelian. An interesting state is the 221 state $\Phi_2^2 \Phi_1$ at $\nu = 1/2$. At the mean field level (before fusing partons into physical electrons) it has five chiral edge states, i.e. $c = 5$. Gauge constraint must project out SU(2)$_2 \times$U(1), which has central charge $3/2 + 1 = 5/2$, producing the central charge of $c = 5 - 5/2 = 5/2$ for the 221 state. The same remains true for the 2211111 states at $\nu = 1/4$, because both 1 and 111 have central charge $c = 1$. The lowest energy quasiparticles, which are excitations in the factors Φ_2, have Ising fusion rules.

The states containing factors of Φ_2^2 can be interpreted as topological f-wave superconductors of composite fermions.[200] To see this, note that the central charge of an s wave superconductor is zero, whereas the $(p_x \pm ip_y)^l$ superconductor has central charge $c = \pm l/2$. Now consider the wave function $\Psi_l^{\text{paired}} \Phi_1$, where Ψ_l^{paired} is the wave function of a paired state with relative angular momentum l pairing. This wave function has filling factor $\nu = 1$, Ising fusion rules for the quasiparticles (vortices in the superconductor), and central charge $c = 1 + l/2$. Furthermore, there is a unique topological quantum field theory for each central charge satisfying these properties. It therefore follows that $\Psi_{l=3}^{\text{paired}} \Phi_1$ and Φ_2^2 are topologically equivalent, i.e. belong to the same universality class. The 221 wave function $\Phi_2^2 \Phi_1$ at $\nu = 1/2$ and 22111 wave function $\Phi_2^2 \Phi_1^3$ at $\nu = 1/4$ are similarly topologically equivalent to the f-wave superconductor of composite fermions, $\Psi_{l=3}^{\text{paired}} \Phi_1^{2p}$, with $c = 5/2$.

---

[q]This projection, which glues the partons back to produce the physical electrons, is not to be confused with the LLL projection.

Other topological superconductors of composite fermions have been considered in the past. The Pfaffian wave function $\Psi_{l=1}^{\text{paired}}\Phi_1^2$, which is a $p_x + ip_y$ superconductor of composite fermions, has central charge $c = 1 + 1/2 = 3/2$, with the factor Φ_1^2 contributing 1 and the Pfaffian factor $1/2$. Its hole partner, the anti-Pfaffian wave function, has central charge $c = 1 - 3/2 = -1/2$, because it is the Pfaffian of holes (contributing $-3/2$) in the background of $\nu = 1$ state (contributing $+1$). Finally, the so-called PH Pfaffian[82] $\Psi_{l=-1}^{\text{paired}}\Phi_1^2$ has $c = 1/2$. When occurring in the second LL, the central charge for these states has additional contribution of $+2$ from the lowest filled Landau level.

The topological properties can, in principle, be derived directly from the wave functions. For the $\Psi_{n/m} = [\Phi_n]^m$ state with $n \geq 2$ and $m \geq 2$, specifying the positions of the quasiholes, in general, does not fully specify the wave function, because different distributions of the quasiholes in different factors do not necessarily produce identical wave functions. This lies at the root of non-Abelian statistics. The state $\Psi_{1/m} = [\Phi_1]^m$ does not produce non-Abelian statistics because here a hole in Φ_1 at position η corresponds to multiplication by the factor $\prod_j(z_j - \eta)$, and thus the wave function for several quasiholes simply produces an overall multiplicative factor $\prod_{j,\alpha}(z_j - \eta_\alpha)$ independent of which factors of Φ_1 the holes were created in originally, thus defining the wave function uniquely. One can, in principle, obtain the braid statistics of the quasiparticles from the explicit wave functions, and the properties of the edge states and the central charge by studying the entanglement spectra.

All of the states of Eq. (29) are mathematically well defined and presumably occur for some specially designed model interactions. The LLL is known to stabilize composite fermions. Can states beyond the CF theory be realized in experiments? For that one must look to higher LLs, to monolayer or bilayer graphene, or to LLL systems in wide quantum wells, all of which have different Coulomb matrix elements than purely two-dimensional electrons in the LLL. We review recent work that has found certain states of Eq. (29) to be promising candidates for experimentally observed FQH states.

We do not discuss here a further generalization of the parton construction, called the projective construction, where the electron is represented as $c = f_0(f_1 f_2 + \cdots + f_{2k-1} f_{2k})$, where f_i represent fermion species. This can produce many other non-Abelian states, such as the Pfaffian state, and has been used to determine the bulk and edge field theories and other topological properties of these states.[209-213]

5.2. $\bar{2}\bar{2}111$ *at* $\nu = 5/2$

The 221 state $\mathcal{P}_{\text{LLL}}\Phi_2^2\Phi_1$ is the simplest state beyond the standard CF theory.[195,196,201,214] It is also interesting because it occurs at an even denominator fraction and is thought to support non-Abelian quasiparticles. It was considered as a candidate for the 5/2 FQHE[198] but deemed unsatisfactory, because exact diagonalization in the second LL does not produce an incompressible state at the corresponding "shift" on the sphere (see, for example, Ref. 215).

BBR have considered[200] the $\bar{2}\bar{2}111$ state as a candidate for the half filled second LL of the 5/2 FQH state. This state can be constructed conveniently by evaluating the LLL projection as[r]

$$\Psi_{\nu=1/2}^{\bar{2}\bar{2}111} = \mathcal{P}_{LLL}[\Phi_2^*]^2\Phi_1^3 \sim [\mathcal{P}_{LLL}\Phi_2^*\Phi_1^2]^2/\Phi_1 = [\Psi_{2/3}]^2/\Phi_1. \tag{35}$$

BBR showed that this state has a reasonably high overlap with the exact ground state. The $\bar{2}\bar{2}111$ state belongs in the same universality class as the anti-Pfaffian. The two states occur at the same shift $(S = -1)$, have decent overlaps, and produce very similar entanglement spectra.[200] Furthermore they have the same central charge. To see this, we note that $[\Phi_2]^{*2}\Phi_1^3 \sim \Psi_{l=-3}^{\text{paired}}[\Phi_1]^*\Phi_1^3 \sim \Psi_{l=-3}^{\text{paired}}\Phi_1^2$, where $\sim$ refers to topological equivalence and we have made use of the fact that multiplication by $[\Phi_1]^*\Phi_1$ does not alter the topological structure. The $\bar{2}\bar{2}111$ state thus has $c = 1 - 3/2 = -1/2$, which is the same as that for the anti-Pfaffian state.

As seen in the context of $\nu = 1/2$, many different states can be constructed for a given fraction. How does one decide which of these is plausible? The answer to this question must ultimately come from detailed calculations, and will depend on the form of the interaction. (As seen below, the 221 state may also be realized under different conditions.) The rule of thumb is that the more 1's, the better, and the fewer non-1's, the better.

5.3. $\bar{3}\bar{2}111$ at $\nu = 2 + 6/13$

Kumar *et al.*[216] reported the formation of a FQH state at $2 + 6/13$. Many facts suggest that this is unlikely to be analogous to the $6/13$ state in the LLL, which is understood as six filled ΛLs of composite fermions, or, alternatively, as the 611 state. While the path to the $6/13$ state in the LLL passes through 1/3, 2/5, 3/7, 4/9 and 5/11, the last three are not observed in the second LL,[217] and even 2+2/5 is believed to be distinct from the standard 211 state of the LLL.[218–223] Furthermore, 6/13 is close to half filling, where the lowest and the second LLs exhibit qualitatively distinct behaviors. The observation of $2 + 6/13$ thus gives a clue into the different organizing principle in the second LL. (The $\nu = 7/13$ FQH state observed in the $n = 1$ LL of bilayer graphene[224] is likely the hole partner of the $2 + 6/13$ state in GaAs quantum wells.)

Following the BBR insight, Balram *et al.*[205] considered the sequence $\bar{n}\bar{2}111$, which corresponds to the wave functions:

$$\Psi_{\nu=2n/(5n-2)}^{\bar{n}\bar{2}111} \sim \frac{[\mathcal{P}_{LLL}\Phi_{\bar{n}}\Phi_1^2][\mathcal{P}_{LLL}\Phi_{\bar{2}}\Phi_1^2]}{\Phi_1} = \frac{\Psi_{n/(2n-1)}\Psi_{2/3}}{\Phi_1}. \tag{36}$$

The first member of this sequence is $\bar{1}\bar{2}111$ at $\nu = 2/3$, which is essentially identical to the standard $\bar{2}11$ state. The second member is the 1/2 state discussed in the

---

[r]The interaction in the nth LL is fully defined by the Haldane pseudopotentials $V_m^{(n)}$, which are the energies of two electrons in relative angular momentum m. The problem of interacting electrons in the nth LL is formally equivalent to that of electrons in the LLL interacting with an effective interaction $V^{\text{eff}}(r)$ that produces pseudopotentials $V_m^{(n)}$. This mapping allows us to conveniently work within the LLL even for the higher LL states, so long as LL mixing is disallowed.

previous subsection. Encouragingly, the third member $\bar{3}\bar{2}111$ occurs at 6/13. In the second LL, its energy (-0.366) is lower than that of the 611 state (-0.355), in contrast to the lowest LL where the energy of the $\bar{3}\bar{2}111$ and 611 states are -0.438 and -0.453, respectively (all energies are thermodynamic limits, quoted in units of $e^2/\epsilon l$). Additionally, $\bar{3}\bar{2}111$ has a reasonably high overlap of 0.754 with the exact 12 particle state in the second LL. These results make the $\bar{3}\bar{2}111$ state plausible.

The $\bar{3}\bar{2}111$ state has quasiparticles with charges $\mp 3/13$ and $\mp 2/13$ which correspond to particles and holes in the $\Phi_{\bar{3}}$ and $\Phi_{\bar{2}}$ factors; these can be combined to produce an excitation with charge $\pm 1/13$, but that is a composite object. $\bar{3}\bar{2}111$ occurs at a shift $S = -2$ on the sphere. To obtain other topological properties, we consider the low-energy effective theory of the edge. Before we glue the partons to recover electrons, there are a total of eight edge states: three from $\Phi_{\bar{3}}$, two from $\Phi_{\bar{2}}$, and one from each Φ_1. Gluing the partons gives four constraints, reducing the number of independent edge modes to four. Following the methods outlined above, one obtains the K matrix[205]

$$K_{\bar{3}\bar{2}111} = \begin{pmatrix} -2 & -1 & 0 & 1 \\ -1 & -2 & 0 & 1 \\ 0 & 0 & -2 & 1 \\ 1 & 1 & 1 & 1 \end{pmatrix}, \tag{37}$$

and the charge vector $t = (0,0,0,1)^T$. The ground state degeneracy on a manifold with genus g is $|\det(K)|^g = 13^g$. The K matrix above has one positive and three negative eigenvalues, giving central charge $c = -2$. In contrast, the 611 state occurs at shift $S = 8$, and has central charge $c = 6$ with all edge modes moving downstream (assuming the absence of edge reconstruction). The $\bar{3}\bar{2}111$ and 611 states may, in principle, be distinguished by shot noise experiments, which have been used to measure the presence of upstream modes,[225–228] or by a measurement of the thermal Hall conductance.[229,230] (For $\nu = 2 + 6/13$, we must also add $c = 2$ coming from the two edge states of the lowest filled LL.)

Levin and Halperin[231] have proposed to obtain a FQHE at $\nu = 2 + 6/13$ in a hierarchy starting from the anti-Pfaffian. Although this construction does not produce a microscopic wave function, it is possibly topologically equivalent to the $\bar{3}\bar{2}111$ state.

5.4. 221 at $\nu = 1/2$ in single and multi-layer graphene

A FQHE at $\nu = 1/2$ has been seen in the $n = 3$ LL of graphene.[232] Exact diagonalization studies using the interaction pseudopotentials of the $n = 3$ graphene LL do not support any of the known single or two-component candidate incompressible states. However, a slight change of the interaction stabilizes the 221 state,[232] which makes it a plausible candidate for the observed FQHE (given that the actual interaction is modified, for example, due to LL mixing or screening by a nearby conducting layer). A definitive identification will require further investigation.

A model Hamiltonian can be constructed for which the 221 state $\Phi_2^2\Phi_1$ (without the LLL projection) is the exact and unique zero energy ground state. In this model, one takes the lowest three LLs with orbital index $n = 0, 1, 2$ to be degenerate and considers the Trugman–Kivelson interaction[233] $V_{TK} = 4\pi\nabla^2\delta^{(2)}(\mathbf{r})$ between electrons.[s] The kinetic energy is zero because $\Phi_2^2\Phi_1$ involves only the lowest three LLs, and the interaction energy is zero because the wave function vanishes as r^3 when two electrons approach one another. One can further show that $\Phi_2^2\Phi_1$ is the unique state with these properties,[201,214] as also confirmed in exact diagonalization studies in the spherical geometry for $N = 6$ and 8 particles.[201]

The model where the lowest three LLs are degenerate but well separated from other LLs appears unphysical, but it turns out that precisely this situation occurs in multilayer graphene. The low-energy Hamiltonian of Bernel stacked bilayer graphene (BLG) and ABC stacked trilayer graphene (TLG) can be approximately described, for each of the two valleys, by[235,236]

$$H = T_J \begin{bmatrix} 0 & (\pi_x + i\pi_y)^J \\ (\pi_x - i\pi_y)^J & 0 \end{bmatrix}. \tag{38}$$

Here $\boldsymbol{\pi} = \boldsymbol{p} + (e/c)\boldsymbol{A}$ is the canonical momentum operator, $J = 2$ (3) for BLG (TLG) is the chirality, and T_J is a constant depending on microscopic details. The zeroth LL of Eq. (38) contains J-fold degenerate states, the wave functions for which, in the simplest approximation, are the wave functions of the lowest J LLs of non-relativistic fermions. The degeneracy of the LLs is split by various features left out in Eq. (38), and the splitting can be tuned by applying a transverse electric field.[237–240] For a proper choice of parameters, it appears possible to obtain situations where two or three orbital levels are approximately degenerate, producing the ideal condition for the realization of the 221 state.

Wu *et al.*[201] have investigated if the 221 state can be realized in these systems for the Coulomb interaction. Figure 18 shows the overlap of the exact Coulomb ground state for eight particles with the 221 state as a function of the LL splitting ω_c (quoted in units of $e^2/\epsilon l$). Overlaps are also shown for the Pfaffian and the CF Fermi sea states. For a range of splittings near $\omega_c = 0$, the 221 state has a large overlap with the exact Coulomb state for both BLG and TLG. For large positive ω_c the CF Fermi sea is obtained, as expected, in both BLG and TLG. For large negative ω_c in a BLG, ordering of the lowest two LLs is inverted, stabilizing the Pfaffian 1/2 state in the $N = 1$ orbital. These results suggest that the 221 state should occur for both BLG and TLG in the vicinity of $\omega_c = 0$. Should this state be observed, it would be the first example where LL mixing is fundamentally responsible for creating a new FQH state. The 221 state may be relevant to the 1/2 FQH state reported in BLG,[237] and possibly in TLG.[241] It ought to be noted that the actual wave functions for the BLG and TLG graphene LLs are more complicated than the

---

[s]If the lowest *two* LLs with $n = 0, 1$ are taken to be degenerate, this model produces the unprojected 2/5 state $\Phi_2\Phi_1^2$ as the unique zero energy ground state.[196] In this case, the state is seen, in numerical studies,[234] to evolve continuously into the LLL 2/5 state, without any gap closing, as the splitting between the two levels is increased to infinity.

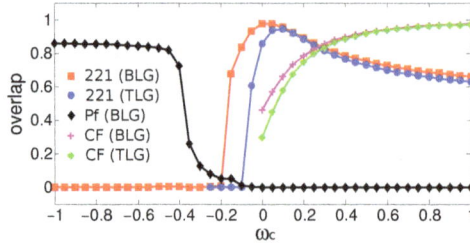

Fig. 18. Overlap between the exact Coulomb ground states of model BLG and TLG Hamiltonians with various trial wave functions: the $\Phi_2^2\Phi_1$ state (221), the CF Fermi liquid state (CF), and the Pfaffian state (Pf). The BLG (TLG) Hamiltonian is defined by a model in which the lowest two (three) LLs are considered, with a variable LL splitting of ω_c. Results are shown for $(N, 2Q) = (8, 11)$ for which the $L = 0$ subspace contains 418 (18212) independent states in BLG (TLG). Y. Wu, T. Shi, and J. K. Jain, *Nano Letters* **17**, 4643–4647 (2017).[201]

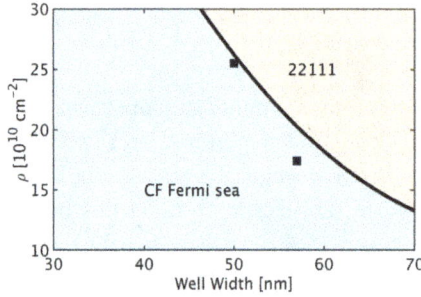

Fig. 19. The calculated phase diagram at $\nu = 1/4$ as a function of the quantum well width and density considering single-component candidate states. Only the CFFS and 22111 states are realized. Black squares, taken from Luhman *et al.*[242] and Shabani *et al.*,[243] indicate the experimental phase boundary. Source: W. N. Faugno, A. C. Balram, M. Barkeshli, and J. K. Jain, *Phys. Rev. Lett.* **123**, 016802 (2019).[244]

model considered above (see the chapter by Dean, Kim, Li and Young), which will need to be incorporated in a more realistic calculation.

5.5. 22111 *at* $\nu = 1/4$ *in wide quantum wells*

There exists evidence for FQHE at filling factor $\nu = 1/4$ in wide quantum wells.[242,243,245,246] This FQHE is induced by a change in the form of the interaction due to finite width, as the $\nu = 1/4$ state is known to be a CFFS for small widths. Faugno *et al.*[244] have determined the variational energies of many single component states: the CF Fermi sea, 22111, $\bar{2}\bar{2}11111$, $\text{Pf}[(z_j - z_k)^{-1}]\Phi_1^4$, and $\text{Pf}[(z_j - z_k)^{-3}]\Phi_1^4$, as a function of the quantum well width and the density, with the finite width effect treated in LDA (see Sec. 3). The CF Fermi sea is seen to become unstable to the 22111 state as the density and/or the quantum well width is increased. The calculated phase diagram shown in Fig. 19 is in good agreement with the onset of the 1/4 FQHE in experiments. Wide quantum wells can behave like bilayer systems and one may ask if the observed state might be an incompressible bilayer state. An energetic comparison between the one- and two-component states is complicated by the fact that the energy separation between the symmetric and antisymmetric subbands (Δ_{SAS}) is known much less precisely than the Coulomb energy differences between the various candidates states. Faugno *et al.*[244] have also considered a large number of two-component candidate states at $\nu = 1/4$ for an ideal bilayer system consisting of two two-dimensional planes, and found that no incompressible state is stabilized for any value of the interlayer separation. This result, combined with the agreement between theory and experiment in Fig. 19, supports the view that the observed $\nu = 1/4$ FQHE in wide quantum wells has a single component origin.

5.6. $\bar{2}\bar{2}\bar{2}1111$ *for* $\nu = 2 + 2/5$

Balram *et al.*[247] have considered states of the form $\bar{2}^k 1^{k+1}$ (in obvious notation) at $\nu = 2/(k+2)$, described by the wave function $\mathcal{P}_{LLL}[\Phi_2^*]^k \Phi_1^{k+1} \sim [\Psi_{2/3}]^k / \Phi_1^{k-1}$. They have shown that these states are in the same universality class as the hole conjugates of the so-called parafermion states[218] at $\nu = k/(k+2)$. There is theoretical evidence that for $k = 3$ this state is relevant for the FQHE at $\nu = 2 + 3/5$ (and, via particle hole symmetry, also for $\nu = 2 + 2/5$).

Acknowledgments

This chapter features results from fruitful collaboration with many wonderful students and colleagues, including Alexander Archer, Ajit Balram, Maissam Barkeshli, William Faugno, Mikael Fremling, Yayun Hu, Manish Jain, Yang Liu, Sudhansu Mandal, Sutirtha Mukherjee, Kwon Park, Loren Pfeiffer, Songyang Pu, Mark Rudner, Diptiman Sen, Mansour Shayegan, Jurgen Smet, G. J. Sreejith, Manisha Thakurathi, Csaba Töke, Arkadiusz Wójs, Yinghai Wu, Yuhe Zhang, and Jianyun Zhao. The author is deeply grateful to all of them. Thanks are also due to Ajit Balram, Yayun Hu, Dwipesh Majumder, Songyang Pu and Yinghai Wu for help with figures, to the US Department of Energy for financial support under Grant no. DE-SC0005042, and to the Infosys Foundation for enabling a visit to the Indian Institute of Science, Bangalore, where part of this article was written. The author expresses gratitude to Bert Halperin for many valuable suggestions on the manuscript.

A.1. Landau levels in the symmetric gauge

This Appendix is reproduced from Ref. 71. The Hamiltonian for a non-relativistic electron moving in two-dimensions in a perpendicular magnetic field is given by

$$H = \frac{1}{2m_b}\left(\boldsymbol{p} + \frac{e}{c}\boldsymbol{A}\right)^2,$$

(A.1)

where m_b is the band mass of the electron. Choosing the symmetric gauge $\boldsymbol{A} = \frac{\boldsymbol{B} \times \boldsymbol{r}}{2} = \frac{B}{2}(-y, x, 0)$, and taking the units of length as the magnetic length $l = \sqrt{\hbar c/eB} = 1$, the Hamiltonian becomes

$$H = \hbar\omega_c \left(a^\dagger a + \frac{1}{2} \right), \tag{A.2}$$

where $\hbar\omega_c = \hbar eB/m_b c$ is the cyclotron energy and the ladder operators are defined as $a^\dagger = \frac{1}{\sqrt{2}} \left(\frac{\bar{z}}{2} - 2\frac{\partial}{\partial z} \right)$ and $a = \frac{1}{\sqrt{2}} \left(\frac{z}{2} + 2\frac{\partial}{\partial \bar{z}} \right)$ in terms of the complex coordinates $z = x - iy = re^{-i\theta}$, $\bar{z} = x + iy = re^{i\theta}$. Further defining $b = \frac{1}{\sqrt{2}} \left(\frac{z}{2} + 2\frac{\partial}{\partial z} \right)$ and $b^\dagger = \frac{1}{\sqrt{2}} \left(\frac{\bar{z}}{2} - 2\frac{\partial}{\partial \bar{z}} \right)$, one can check that $[a, a^\dagger] = 1$, $[b, b^\dagger] = 1$, and all other commutators vanish. The LL index n is the eigenvalue of $a^\dagger a$, and the z component of the angular momentum operator is defined as $L = -i\hbar\frac{\partial}{\partial\theta} = -\hbar(b^\dagger b - a^\dagger a) \equiv -\hbar m$, with $m = -n, -n+1, \cdots 0, 1, \cdots$ in the nth LL. The single particle eigenstates are obtained in the standard manner by successive applications of ladder operators

$$|n, m\rangle = \frac{(b^\dagger)^{m+n}}{\sqrt{(m+n)!}} \frac{(a^\dagger)^n}{\sqrt{n!}} |0, 0\rangle, \tag{A.3}$$

with eigenenergies $E_n = \hbar\omega_c \left(n + \frac{1}{2} \right)$, and the bottom state $\langle r|0, 0\rangle \equiv \eta_{0,0}(r) = \frac{1}{\sqrt{2\pi}} e^{-\frac{1}{4}z\bar{z}}$ is annihilated by a and b. The single-particle states are especially simple in the LLL ($n = 0$):

$$\eta_{0,m} = \langle r|0, m\rangle = \frac{(b^\dagger)^m}{\sqrt{m!}} \eta_{0,0} = \frac{z^m e^{-\frac{1}{4}z\bar{z}}}{\sqrt{2\pi 2^m m!}}. \tag{A.4}$$

Aside from the ubiquitous Gaussian factor, a general single particle state in the lowest LL is an analytic function of z, i.e. it does not involve any $\bar{z}$. A general many-particle wave function confined to the LLL therefore has the form $\Psi = F[\{z_j\}] \exp\left[-\frac{1}{4} \sum_i |z_i|^2 \right]$ where $F[\{z_j\}]$ is an antisymmetric function of the z_j.

The LL degeneracy can be obtained by considering a region of radius R centered at the origin, and asking how many single particle states lie inside it. For the LLL, the eigenstate $|0, m\rangle$ has its weight located at the circle of radius $r = \sqrt{2m} \cdot l$. The largest value of m for which the single particle state falls inside our circular region is given by $M = R^2/2l^2$, which is also the total number of single particle eigenstates in the LLL that fall inside the disk (neglecting order one corrections). Thus, the degeneracy per unit area is $M/\pi R^2 = 1/(2\pi l^2) = B/\phi_0$ which is the number of flux quanta (with a single flux quantum defined as $\phi_0 = hc/e$) penetrating the sample through a unit area. The filling factor, which is the nominal number of filled LLs, is equal to the number of electrons per flux quantum, given by

$$\nu = \frac{\rho}{B/\phi_0} = 2\pi l^2 \rho, \tag{A.5}$$

where ρ is the 2D density of electrons.

The wave function Φ_n of the state with n fully filled LL (in which all states inside a disk of some radius are filled) is precisely known; it is the Slater determinant formed from the occupied single particle orbitals. The wave function of the lowest filled LL, Φ_1, has a particularly simple form (apart from a normalization factor):

$$\Phi_1 = \begin{vmatrix} 1 & 1 & 1 & .. \\ z_1 & z_2 & z_3 & .. \\ z_1^2 & z_2^2 & z_3^2 & .. \\ . & . & . & .. \\ . & . & . & .. \end{vmatrix} \exp\left[-\frac{1}{4}\sum_i |z_i|^2\right] = \prod_{j<k}(z_j - z_k) \exp\left[-\frac{1}{4}\sum_i |z_i|^2\right]. \quad (A.6)$$

References

1. H. L. Stormer and D. C. Tsui, Composite fermions in the fractional quantum Hall effect. In *Perspectives in Quantum Hall Effects*, pp. 385–421. (Wiley-VCH Verlag GmbH, 2007) ISBN 9783527617258 doi: 10.1002/9783527617258.ch10. URL http://dx.doi.org/10.1002/9783527617258.ch10.

2. F. D. M. Haldane, Fractional quantization of the Hall effect: A hierarchy of incompressible quantum fluid states, *Phys. Rev. Lett.* **51**, 605–608 (1983). doi: 10.1103/PhysRevLett.51.605. URL http://link.aps.org/doi/10.1103/PhysRevLett.51.605.

3. K. V. Klitzing, G. Dorda and M. Pepper, New method for high-accuracy determination of the fine-structure constant based on quantized Hall resistance, *Phys. Rev. Lett.* **45**, 494–497 (1980). doi: 10.1103/PhysRevLett.45.494. URL http://link.aps.org/doi/10.1103/PhysRevLett.45.494.

4. R. B. Laughlin, Quantized Hall conductivity in two dimensions, *Phys. Rev. B* **23**, 5632–5633 (1981). doi: 10.1103/PhysRevB.23.5632. URL http://link.aps.org/doi/10.1103/PhysRevB.23.5632.

5. D. J. Thouless, M. Kohmoto, M. P. Nightingale and M. den Nijs, Quantized Hall conductance in a two-dimensional periodic potential, *Phys. Rev. Lett.* **49**, 405–408 (1982). doi: 10.1103/PhysRevLett.49.405. URL http://link.aps.org/doi/10.1103/PhysRevLett.49.405.

6. F. D. M. Haldane, Model for a quantum Hall effect without Landau levels: Condensed-matter realization of the "parity anomaly", *Phys. Rev. Lett.* **61**, 2015–2018 (1988). doi: 10.1103/PhysRevLett.61.2015. URL https://link.aps.org/doi/10.1103/PhysRevLett.61.2015.

7. D. C. Tsui, H. L. Stormer and A. C. Gossard, Two-dimensional magnetotransport in the extreme quantum limit, *Phys. Rev. Lett.* **48**, 1559–1562 (1982). doi: 10.1103/PhysRevLett.48.1559. URL http://link.aps.org/doi/10.1103/PhysRevLett.48.1559.

8. E. Wigner, On the interaction of electrons in metals., *Phys. Rev.* **46**, 1002 (1934).

9. Y. E. Lozovik and V. I. Yudson, Feasibility of superfluidity of paired spatially separated electrons and holes; a new superconductivity mechanism., *JETP Lett.* **22**, 11 (1975).

10. R. B. Laughlin, Anomalous quantum Hall effect: An incompressible quantum fluid with fractionally charged excitations, *Phys. Rev. Lett.* **50**, 1395–1398 (1983). doi: 10.1103/PhysRevLett.50.1395. URL http://link.aps.org/doi/10.1103/PhysRevLett.50.1395.

11. J. Leinaas and J. Myrheim, On the theory of identical particles, *Il Nuovo Cimento B Series 11* **37**(1), 1–23 (1977). ISSN 0369-3554. doi: 10.1007/BF02727953. URL http://dx.doi.org/10.1007/BF02727953.

12. F. Wilczek, Quantum mechanics of fractional-spin particles, *Phys. Rev. Lett.* **49**, 957–959 (1982). doi: 10.1103/PhysRevLett.49.957. URL http://link.aps.org/doi/10.1103/PhysRevLett.49.957.

13. B. I. Halperin, Statistics of quasiparticles and the hierarchy of fractional quantized Hall states, *Phys. Rev. Lett.* **52**, 1583–1586 (1984). doi: 10.1103/PhysRevLett.52.1583. URL http://link.aps.org/doi/10.1103/PhysRevLett.52.1583.

14. D. Arovas, J. R. Schrieffer and F. Wilczek, Fractional statistics and the quantum Hall effect, *Phys. Rev. Lett.* **53**, 722–723 (1984). doi: 10.1103/PhysRevLett.53.722. URL http://link.aps.org/doi/10.1103/PhysRevLett.53.722.

15. S. M. Girvin and A. H. MacDonald, Off-diagonal long-range order, oblique confinement, and the fractional quantum Hall effect, *Phys. Rev. Lett.* **58**, 1252–1255 (1987). doi: 10.1103/PhysRevLett.58.1252. URL http://link.aps.org/doi/10.1103/PhysRevLett.58.1252.

16. S. C. Zhang, T. H. Hansson and S. Kivelson, Effective-field-theory model for the fractional quantum Hall effect, *Phys. Rev. Lett.* **62**, 82–85 (1989). doi: 10.1103/PhysRevLett.62.82. URL http://link.aps.org/doi/10.1103/PhysRevLett.62.82.

17. J. K. Jain, Composite-fermion approach for the fractional quantum Hall effect, *Phys. Rev. Lett.* **63**, 199–202 (1989). doi: 10.1103/PhysRevLett.63.199. URL http://link.aps.org/doi/10.1103/PhysRevLett.63.199.

18. J. K. Jain, *Composite Fermions* (Cambridge University Press, New York, US, 2007).

19. N. Trivedi and J. K. Jain, Numerical study of Jastrow-Slater trial states for the fractional quantum Hall effect, *Mod. Phys. Lett. B* **05**(07), 503–510 (1991). doi: 10.1142/S0217984991000599.

20. R. K. Kamilla and J. K. Jain, Variational study of the vortex structure of composite fermions, *Phys. Rev. B* **55**, 9824–9827 (1997). doi: 10.1103/PhysRevB.55.9824. URL http://link.aps.org/doi/10.1103/PhysRevB.55.9824.

21. J. K. Jain and R. K. Kamilla, Composite fermions in the Hilbert space of the lowest electronic Landau level, *Int. J. Mod. Phys. B* **11**(22), 2621–2660 (1997). doi: 10.1142/S0217979297001301.

22. J. K. Jain and R. K. Kamilla, Quantitative study of large composite-fermion systems, *Phys. Rev. B* **55**, R4895–R4898 (1997). doi: 10.1103/PhysRevB.55.R4895. URL http://link.aps.org/doi/10.1103/PhysRevB.55.R4895.

23. S. S. Mandal and J. K. Jain, Theoretical search for the nested quantum Hall effect of composite fermions, *Phys. Rev. B* **66**, 155302 (2002). doi: 10.1103/PhysRevB.66.155302. URL http://link.aps.org/doi/10.1103/PhysRevB.66.155302.

24. F. D. M. Haldane and E. H. Rezayi, Periodic Laughlin-Jastrow wave functions for the fractional quantized Hall effect, *Phys. Rev. B* **31**, 2529–2531 (1985). doi: 10.1103/PhysRevB.31.2529. URL http://link.aps.org/doi/10.1103/PhysRevB.31.2529.

25. M. Hermanns, Composite fermion states on the torus, *Phys. Rev. B* **87**, 235128 (2013). doi: 10.1103/PhysRevB.87.235128. URL http://link.aps.org/doi/10.1103/PhysRevB.87.235128.

26. S. Pu, Y.-H. Wu and J. K. Jain, Composite fermions on a torus, *Phys. Rev. B* **96**, 195302 (2017). doi: 10.1103/PhysRevB.96.195302. URL https://link.aps.org/doi/10.1103/PhysRevB.96.195302.

27. A. Lopez and E. Fradkin, Fractional quantum Hall effect and Chern-Simons gauge theories, *Phys. Rev. B* **44**, 5246–5262 (1991). doi: 10.1103/PhysRevB.44.5246. URL http://link.aps.org/doi/10.1103/PhysRevB.44.5246.

28. B. I. Halperin, P. A. Lee and N. Read, Theory of the half-filled Landau level, *Phys. Rev. B* **47**, 7312–7343 (1993). doi: 10.1103/PhysRevB.47.7312. URL http://link.aps.org/doi/10.1103/PhysRevB.47.7312.

29. T. H. Hansson, M. Hermanns, S. H. Simon and S. F. Viefers, Quantum Hall physics: Hierarchies and conformal field theory techniques, *Rev. Mod. Phys.* **89**, 025005 (2017). doi: 10.1103/RevModPhys.89.025005. URL https://link.aps.org/doi/10.1103/RevModPhys.89.025005.

30. W. Kang, H. L. Stormer, L. N. Pfeiffer, K. W. Baldwin and K. W. West, How real are composite fermions?, *Phys. Rev. Lett.* **71**, 3850–3853 (1993). doi: 10.1103/PhysRevLett.71.3850. URL http://link.aps.org/doi/10.1103/PhysRevLett.71.3850.

31. H. L. Stormer. Private communication.

32. W. Pan, H. L. Stormer, D. C. Tsui, L. N. Pfeiffer, K. W. Baldwin and K. W. West, Fractional quantum Hall effect of composite fermions, *Phys. Rev. Lett.* **90**, 016801 (2003). doi: 10.1103/PhysRevLett.90.016801. URL http://link.aps.org/doi/10.1103/PhysRevLett.90.016801.

33. W. Pan, H. L. Stormer, D. C. Tsui, L. N. Pfeiffer, K. W. Baldwin and K. W. West, Transition from an electron solid to the sequence of fractional quantum Hall states at very low Landau level filling factor, *Phys. Rev. Lett.* **88**, 176802 (2002). doi: 10.1103/PhysRevLett.88.176802.

34. J. K. Jain, A note contrasting two microscopic theories of the fractional quantum Hall effect, *Indian Journal of Physics.* **88**, 915–929 (2014). ISSN 0974-9845. doi: 10.1007/s12648-014-0491-9. URL http://dx.doi.org/10.1007/s12648-014-0491-9.

35. V. W. Scarola, S.-Y. Lee and J. K. Jain, Excitation gaps of incompressible composite fermion states: Approach to the Fermi sea, *Phys. Rev. B* **66**, 155320 (2002). doi: 10.1103/PhysRevB.66.155320. URL http://link.aps.org/doi/10.1103/PhysRevB.66.155320.

36. R. R. Du, H. L. Stormer, D. C. Tsui, L. N. Pfeiffer and K. W. West, Experimental evidence for new particles in the fractional quantum Hall effect, *Phys. Rev. Lett.* **70**, 2944–2947 (1993). doi: 10.1103/PhysRevLett.70.2944. URL http://link.aps.org/doi/10.1103/PhysRevLett.70.2944.

37. H. C. Manoharan, M. Shayegan and S. J. Klepper, Signatures of a novel Fermi liquid in a two-dimensional composite particle metal, *Phys. Rev. Lett.* **73**, 3270–3273 (1994). doi: 10.1103/PhysRevLett.73.3270. URL http://link.aps.org/doi/10.1103/PhysRevLett.73.3270.

38. A. Pinczuk, B. S. Dennis, L. N. Pfeiffer and K. West, Observation of collective excitations in the fractional quantum Hall effect, *Phys. Rev. Lett.* **70**, 3983–3986 (1993). doi: 10.1103/PhysRevLett.70.3983. URL http://link.aps.org/doi/10.1103/PhysRevLett.70.3983.

39. M. Kang, A. Pinczuk, B. S. Dennis, M. A. Eriksson, L. N. Pfeiffer and K. W. West, Inelastic light scattering by gap excitations of fractional quantum Hall states at $1/3 \geq \nu \leq 2/3$, *Phys. Rev. Lett.* **84**, 546–549 (2000). doi: 10.1103/PhysRevLett.84.546. URL http://link.aps.org/doi/10.1103/PhysRevLett.84.546.

40. I. V. Kukushkin, J. H. Smet, K. von Klitzing and K. Eberl, Optical investigation of spin-wave excitations in fractional quantum Hall states and of interaction between composite fermions, *Phys. Rev. Lett.* **85**, 3688–3691 (2000). doi: 10.1103/PhysRevLett.85.3688. URL http://link.aps.org/doi/10.1103/PhysRevLett.85.3688.

41. M. Kang, A. Pinczuk, B. S. Dennis, L. N. Pfeiffer and K. W. West, Observation of multiple magnetorotons in the fractional quantum Hall effect, *Phys. Rev. Lett.* **86**, 2637–2640 (2001). doi: 10.1103/PhysRevLett.86.2637. URL http://link.aps.org/doi/10.1103/PhysRevLett.86.2637.

42. I. Dujovne, A. Pinczuk, M. Kang, B. S. Dennis, L. N. Pfeiffer and K. W. West, Evidence of Landau levels and interactions in low-lying excitations of composite fermions at $1/3 \leq \nu \leq 2/5$, *Phys. Rev. Lett.* **90**, 036803 (2003).

doi: 10.1103/PhysRevLett.90.036803. URL `http://link.aps.org/doi/10.1103/PhysRevLett.90.036803`.

43. I. Dujovne, A. Pinczuk, M. Kang, B. S. Dennis, L. N. Pfeiffer and K. W. West, Composite-fermion spin excitations as ν approaches 1/2: Interactions in the Fermi sea, *Phys. Rev. Lett.* **95**, 056808 (2005). doi: 10.1103/PhysRevLett.95.056808. URL `http://link.aps.org/doi/10.1103/PhysRevLett.95.056808`.

44. I. V. Kukushkin, J. H. Smet, V. W. Scarola, V. Umansky and K. von Klitzing, Dispersion of the excitations of fractional quantum Hall states, *Science* **324**(5930), 1044–1047 (2009). doi: 10.1126/science.1171472. URL `http://www.sciencemag.org/content/324/5930/1044.abstract`.

45. T. D. Rhone, D. Majumder, B. S. Dennis, C. Hirjibehedin, I. Dujovne, J. G. Groshaus, Y. Gallais, J. K. Jain, S. S. Mandal, A. Pinczuk, L. Pfeiffer and K. West, Higher-energy composite fermion levels in the fractional quantum Hall effect, *Phys. Rev. Lett.* **106**, 096803 (2011). doi: 10.1103/PhysRevLett.106.096803. URL `http://link.aps.org/doi/10.1103/PhysRevLett.106.096803`.

46. U. Wurstbauer, D. Majumder, S. S. Mandal, I. Dujovne, T. D. Rhone, B. S. Dennis, A. F. Rigosi, J. K. Jain, A. Pinczuk, K. W. West and L. N. Pfeiffer, Observation of nonconventional spin waves in composite-fermion ferromagnets, *Phys. Rev. Lett.* **107**, 066804 (2011). doi: 10.1103/PhysRevLett.107.066804. URL `http://link.aps.org/doi/10.1103/PhysRevLett.107.066804`.

47. X. G. Wu, G. Dev and J. K. Jain, Mixed-spin incompressible states in the fractional quantum Hall effect, *Phys. Rev. Lett.* **71**, 153–156 (1993). doi: 10.1103/PhysRevLett.71.153. URL `http://link.aps.org/doi/10.1103/PhysRevLett.71.153`.

48. K. Park and J. K. Jain, Phase diagram of the spin polarization of composite fermions and a new effective mass, *Phys. Rev. Lett.* **80**, 4237–4240 (1998). doi: 10.1103/PhysRevLett.80.4237. URL `http://link.aps.org/doi/10.1103/PhysRevLett.80.4237`.

49. R. R. Du, A. S. Yeh, H. L. Stormer, D. C. Tsui, L. N. Pfeiffer and K. W. West, Fractional quantum Hall effect around $\nu = 3/2$: Composite fermions with a spin, *Phys. Rev. Lett.* **75**, 3926–3929 (1995). doi: 10.1103/PhysRevLett.75.3926. URL `http://link.aps.org/doi/10.1103/PhysRevLett.75.3926`.

50. R. R. Du, A. S. Yeh, H. L. Stormer, D. C. Tsui, L. N. Pfeiffer and K. W. West, g factor of composite fermions around $\nu = 3/2$ from angular-dependent activation-energy measurements, *Phys. Rev. B* **55**, R7351–R7354 (1997). doi: 10.1103/PhysRevB.55.R7351. URL `http://link.aps.org/doi/10.1103/PhysRevB.55.R7351`.

51. I. V. Kukushkin, K. V. Klitzing and K. Eberl, Spin polarization of composite fermions: Measurements of the Fermi energy, *Phys. Rev. Lett.* **82**, 3665–3668 (1999). doi: 10.1103/PhysRevLett.82.3665. URL `http://link.aps.org/doi/10.1103/PhysRevLett.82.3665`.

52. N. C. Bishop, M. Padmanabhan, K. Vakili, Y. P. Shkolnikov, E. P. De Poortere and M. Shayegan, Valley polarization and susceptibility of composite fermions around a filling factor $\nu = 3/2$, *Phys. Rev. Lett.* **98**, 266404 (2007). doi: 10.1103/PhysRevLett.98.266404. URL `http://link.aps.org/doi/10.1103/PhysRevLett.98.266404`.

53. M. Padmanabhan, T. Gokmen and M. Shayegan, Density dependence of valley polarization energy for composite fermions, *Phys. Rev. B* **80**, 035423 (2009). doi: 10.1103/PhysRevB.80.035423. URL `http://link.aps.org/doi/10.1103/PhysRevB.80.035423`.

54. B. E. Feldman, A. J. Levin, B. Krauss, D. A. Abanin, B. I. Halperin, J. H. Smet and A. Yacoby, Fractional quantum Hall phase transitions and four-flux states in graphene, *Phys. Rev. Lett.* **111**, 076802 (2013). doi: 10.1103/Phys-

RevLett.111.076802. URL http://link.aps.org/doi/10.1103/PhysRevLett.111. 076802.

55. R. L. Willett, R. R. Ruel, K. W. West and L. N. Pfeiffer, Experimental demonstration of a Fermi surface at one-half filling of the lowest Landau level, *Phys. Rev. Lett.* **71**, 3846–3849 (1993). doi: 10.1103/PhysRevLett.71.3846. URL http://link.aps.org/ doi/10.1103/PhysRevLett.71.3846.

56. V. J. Goldman, B. Su and J. K. Jain, Detection of composite fermions by magnetic focusing, *Phys. Rev. Lett.* **72**, 2065–2068 (1994). doi: 10.1103/PhysRevLett.72.2065. URL http://link.aps.org/doi/10.1103/PhysRevLett.72.2065.

57. J. H. Smet, D. Weiss, R. H. Blick, G. Lütjering, K. von Klitzing, R. Fleischmann, R. Ketzmerick, T. Geisel and G. Weimann, Magnetic focusing of composite fermions through arrays of cavities, *Phys. Rev. Lett.* **77**, 2272–2275 (1996). doi: 10.1103/PhysRevLett.77.2272. URL http://link.aps.org/doi/10.1103/PhysRevLett.77.2272.

58. J. H. Smet, *Ballistic transport of composite fermions in semiconductor nanostructures*, In *Composite Fermions*, chapter 7, pp. 443–491. (World Scientific Pub Co Inc, 1998). doi: 10.1142/9789812815989_0007. URL http://www.worldscientific.com/ doi/abs/10.1142/9789812815989_0007.

59. R. L. Willett, K. W. West and L. N. Pfeiffer, Geometric resonance of composite fermion cyclotron orbits with a fictitious magnetic field modulation, *Phys. Rev. Lett.* **83**, 2624–2627 (1999). doi: 10.1103/PhysRevLett.83.2624. URL http://link.aps. org/doi/10.1103/PhysRevLett.83.2624.

60. J. H. Smet, S. Jobst, K. von Klitzing, D. Weiss, W. Wegscheider and V. Umansky, Commensurate composite fermions in weak periodic electrostatic potentials: Direct evidence of a periodic effective magnetic field, *Phys. Rev. Lett.* **83**, 2620–2623 (1999). doi: 10.1103/PhysRevLett.83.2620. URL http://link.aps.org/doi/ 10.1103/PhysRevLett.83.2620.

61. D. Kamburov, M. Shayegan, L. N. Pfeiffer, K. W. West and K. W. Baldwin, Commensurability oscillations of hole-flux composite fermions, *Phys. Rev. Lett.* **109**, 236401 (2012). doi: 10.1103/PhysRevLett.109.236401. URL http://link.aps.org/doi/10. 1103/PhysRevLett.109.236401.

62. T. Gokmen, M. Padmanabhan and M. Shayegan, Transference of transport anisotropy to composite fermions, *Nature Physics* **6**, 621–624 (2010).

63. D. Kamburov, Y. Liu, M. Shayegan, L. N. Pfeiffer, K. W. West and K. W. Baldwin, Composite fermions with tunable Fermi contour anisotropy, *Phys. Rev. Lett.* **110**, 206801 (2013). doi: 10.1103/PhysRevLett.110.206801. URL http://link.aps.org/ doi/10.1103/PhysRevLett.110.206801.

64. S. Melinte, N. Freytag, M. Horvatic, C. Berthier, L. P. Lévy, V. Bayot and M. Shayegan, NMR determination of 2D electron spin polarization at $\nu = 1/2$, *Phys. Rev. Lett.* **84**, 354–357 (2000). doi: 10.1103/PhysRevLett.84.354. URL http: //link.aps.org/doi/10.1103/PhysRevLett.84.354.

65. R. Du, H. Stormer, D. Tsui, L. Pfeiffer and K. West, Shubnikov-de Haas oscillations around $\nu = 1/2$ Landau level filling factor, *Solid State Communications* **90**(2), 71–75 (1994). ISSN 0038-1098. doi: http://dx.doi.org/10.1016/0038-1098(94)90934-2. URL http://www.sciencedirect.com/science/article/pii/0038109894909342.

66. D. R. Leadley, R. J. Nicholas, C. T. Foxon and J. J. Harris, Measurements of the effective mass and scattering times of composite fermions from magnetotransport analysis, *Phys. Rev. Lett.* **72**, 1906–1909 (1994). doi: 10.1103/PhysRevLett.72.1906. URL http://link.aps.org/doi/10.1103/PhysRevLett.72.1906.

67. I. V. Kukushkin, J. H. Smet, K. von Klitzing and W. Wegscheider, Cyclotron resonance of composite fermions, *Nature* **415**, 409–412 (2002).

68. I. V. Kukushkin, J. H. Smet, D. Schuh, W. Wegscheider and K. von Klitzing, Dispersion of the composite-fermion cyclotron-resonance mode, *Phys. Rev. Lett.* **98**, 066403 (2007). doi: 10.1103/PhysRevLett.98.066403. URL http://link.aps.org/doi/10.1103/PhysRevLett.98.066403.

69. G. Dev and J. K. Jain, Band structure of the fractional quantum Hall effect, *Phys. Rev. Lett.* **69**, 2843–2846 (1992). doi: 10.1103/PhysRevLett.69.2843. URL http://link.aps.org/doi/10.1103/PhysRevLett.69.2843.

70. A. C. Balram, A. Wójs and J. K. Jain, State counting for excited bands of the fractional quantum Hall effect: Exclusion rules for bound excitons, *Phys. Rev. B* **88**, 205312 (2013). doi: 10.1103/PhysRevB.88.205312. URL http://link.aps.org/doi/10.1103/PhysRevB.88.205312.

71. J. K. Jain, Composite fermion theory of exotic fractional quantum Hall effect, *Annu. Rev. Condens. Matter Phys.* **6**, 39–62 (2015). doi: 10.1146/annurev-conmatphys-031214-014606.

72. S. Mukherjee, S. S. Mandal, A. Wójs and J. K. Jain, Possible anti-Pfaffian pairing of composite fermions at $\nu = 3/8$, *Phys. Rev. Lett.* **109**, 256801 (2012). doi: 10.1103/PhysRevLett.109.256801. URL http://link.aps.org/doi/10.1103/PhysRevLett.109.256801.

73. E. Rezayi and N. Read, Fermi-liquid-like state in a half-filled Landau level, *Phys. Rev. Lett.* **72**, 900–903 (1994). doi: 10.1103/PhysRevLett.72.900. URL http://link.aps.org/doi/10.1103/PhysRevLett.72.900.

74. E. H. Rezayi and F. D. M. Haldane, Incompressible paired Hall state, stripe order, and the composite fermion liquid phase in half-filled Landau levels, *Phys. Rev. Lett.* **84**, 4685–4688 (2000). doi: 10.1103/PhysRevLett.84.4685. URL http://link.aps.org/doi/10.1103/PhysRevLett.84.4685.

75. J. Shao, E.-A. Kim, F. D. M. Haldane and E. H. Rezayi, Entanglement entropy of the $\nu = 1/2$ composite fermion non-Fermi liquid state, *Phys. Rev. Lett.* **114**, 206402 (2015). doi: 10.1103/PhysRevLett.114.206402. URL http://link.aps.org/doi/10.1103/PhysRevLett.114.206402.

76. J. Wang, S. D. Geraedts, E. H. Rezayi and F. D. M. Haldane, Lattice Monte Carlo for quantum Hall states on a torus, *Phys. Rev. B* **99**, 125123 (2019). doi: 10.1103/PhysRevB.99.125123. URL https://link.aps.org/doi/10.1103/PhysRevB.99.125123.

77. S. D. Geraedts, J. Wang, E. H. Rezayi and F. D. M. Haldane, Berry phase and model wave function in the half-filled Landau level, *Phys. Rev. Lett.* **121**, 147202 (2018). doi: 10.1103/PhysRevLett.121.147202. URL https://link.aps.org/doi/10.1103/PhysRevLett.121.147202.

78. M. Fremling, N. Moran, J. K. Slingerland and S. H. Simon, Trial wave functions for a composite Fermi liquid on a torus, *Phys. Rev. B* **97**, 035149 (2018). doi: 10.1103/PhysRevB.97.035149. URL https://link.aps.org/doi/10.1103/PhysRevB.97.035149.

79. S. Pu, M. Fremling and J. K. Jain, Berry phase of the composite-fermion Fermi sea: Effect of Landau-level mixing, *Phys. Rev. B* **98**, 075304 (2018). doi: 10.1103/PhysRevB.98.075304. URL https://link.aps.org/doi/10.1103/PhysRevB.98.075304.

80. X. G. Wu and J. K. Jain, Excitation spectrum and collective modes of composite fermions, *Phys. Rev. B* **51**, 1752–1761 (1995). doi: 10.1103/PhysRevB.51.1752. URL http://link.aps.org/doi/10.1103/PhysRevB.51.1752.

81. M. L. Meyer, O. Liabøtrø and S. Viefers, Linear dependencies between composite fermion states, *Journal of Physics A: Mathematical and Theoretical* **49**(39), 395201 (2016). doi: 10.1088/1751-8113/49/39/395201. URL https://doi.org/10.1088%2F1751-8113%2F49%2F39%2F395201.

82. D. T. Son, Is the composite fermion a Dirac particle?, *Phys. Rev. X* **5**, 031027 (2015). doi: 10.1103/PhysRevX.5.031027. URL http://link.aps.org/doi/10.1103/PhysRevX.5.031027.

83. A. C. Balram, C. Tőke and J. K. Jain, Luttinger theorem for the strongly correlated Fermi liquid of composite fermions, *Phys. Rev. Lett.* **115**, 186805 (2015). doi: 10.1103/PhysRevLett.115.186805. URL http://link.aps.org/doi/10.1103/PhysRevLett.115.186805.

84. S. C. Davenport and S. H. Simon, Spinful composite fermions in a negative effective field, *Phys. Rev. B* **85**, 245303 (2012). doi: 10.1103/PhysRevB.85.245303. URL http://link.aps.org/doi/10.1103/PhysRevB.85.245303.

85. A. C. Balram and J. K. Jain, Nature of composite fermions and the role of particle-hole symmetry: A microscopic account, *Phys. Rev. B* **93**, 235152 (2016). doi: 10.1103/PhysRevB.93.235152. URL http://link.aps.org/doi/10.1103/PhysRevB.93.235152.

86. S.-Y. Lee, V. W. Scarola and J. K. Jain, Structures for interacting composite fermions: Stripes, bubbles, and fractional quantum Hall effect, *Phys. Rev. B* **66**, 085336 (2002). doi: 10.1103/PhysRevB.66.085336. URL http://link.aps.org/doi/10.1103/PhysRevB.66.085336.

87. N. Samkharadze, I. Arnold, L. N. Pfeiffer, K. W. West and G. A. Csáthy, Observation of incompressibility at $\nu = 4/11$ and $\nu = 5/13$, *Phys. Rev. B* **91**, 081109 (2015). doi: 10.1103/PhysRevB.91.081109. URL http://link.aps.org/doi/10.1103/PhysRevB.91.081109.

88. W. Pan, K. W. Baldwin, K. W. West, L. N. Pfeiffer and D. C. Tsui, Fractional quantum Hall effect at Landau level filling $\nu = 4/11$, *Phys. Rev. B* **91**, 041301 (2015). doi: 10.1103/PhysRevB.91.041301. URL http://link.aps.org/doi/10.1103/PhysRevB.91.041301.

89. G. Moore and N. Read, Nonabelions in the fractional quantum Hall effect, *Nucl. Phys. B* **360**, 362–396 (1991). ISSN 0550-3213. doi: 10.1016/0550-3213(91)90407-O. URL http://www.sciencedirect.com/science/article/pii/055032139190407O.

90. N. Read and D. Green, Paired states of fermions in two dimensions with breaking of parity and time-reversal symmetries and the fractional quantum Hall effect, *Phys. Rev. B* **61**, 10267–10297 (2000). doi: 10.1103/PhysRevB.61.10267. URL http://link.aps.org/doi/10.1103/PhysRevB.61.10267.

91. V. W. Scarola, K. Park and J. K. Jain, Rotons of composite fermions: Comparison between theory and experiment, *Phys. Rev. B* **61**, 13064–13072 (2000). doi: 10.1103/PhysRevB.61.13064. URL http://link.aps.org/doi/10.1103/PhysRevB.61.13064.

92. A. Zee, *Quantum Field Theory in a Nutshell* (Cambridge University Press, New York, US, 2010).

93. D. Tong, Lectures on the quantum Hall effect, *arXiv e-prints*. art. arXiv:1606.06687 (2016).

94. W. P. Su, Statistics of the fractionally charged excitations in the quantum Hall effect, *Phys. Rev. B* **34**, 1031–1033 (1986). doi: 10.1103/PhysRevB.34.1031. URL http://link.aps.org/doi/10.1103/PhysRevB.34.1031.

95. G. S. Jeon, K. L. Graham and J. K. Jain, Berry phases for composite fermions: Effective magnetic field and fractional statistics, *Phys. Rev. B* **70**, 125316 (2004). doi: 10.1103/PhysRevB.70.125316. URL http://link.aps.org/doi/10.1103/PhysRevB.70.125316.

96. Y. Zhang, A. Wójs and J. K. Jain, Landau-level mixing and particle-hole symmetry breaking for spin transitions in the fractional quantum Hall effect, *Phys. Rev. Lett.*

117, 116803 (2016). doi: 10.1103/PhysRevLett.117.116803. URL http://link.aps.org/doi/10.1103/PhysRevLett.117.116803.

97. J. Zhao, Y. Zhang and J. K. Jain, Crystallization in the fractional quantum Hall regime induced by Landau-level mixing, *Phys. Rev. Lett.* **121**, 116802 (2018). doi: 10.1103/PhysRevLett.121.116802. URL https://link.aps.org/doi/10.1103/PhysRevLett.121.116802.

98. M. W. Ortalano, S. He and S. Das Sarma, Realistic calculations of correlated incompressible electronic states in GaAs-$Al_xGa_{1-x}As$ heterostructures and quantum wells, *Phys. Rev. B* **55**, 7702–7714 (1997). doi: 10.1103/PhysRevB.55.7702. URL http://link.aps.org/doi/10.1103/PhysRevB.55.7702.

99. F. C. Zhang and S. Das Sarma, Excitation gap in the fractional quantum Hall effect: Finite layer thickness corrections, *Phys. Rev. B* **33**, 2903–2905 (1986). doi: 10.1103/PhysRevB.33.2903. URL https://link.aps.org/doi/10.1103/PhysRevB.33.2903.

100. A. H. MacDonald, Influence of Landau-level mixing on the charge-density-wave state of a two-dimensional electron gas in a strong magnetic field, *Phys. Rev. B* **30**, 4392–4398 (1984). doi: 10.1103/PhysRevB.30.4392. URL http://link.aps.org/doi/10.1103/PhysRevB.30.4392.

101. V. Melik-Alaverdian and N. E. Bonesteel, Composite fermions and Landau-level mixing in the fractional quantum Hall effect, *Phys. Rev. B* **52**, R17032–R17035 (1995). doi: 10.1103/PhysRevB.52.R17032. URL http://link.aps.org/doi/10.1103/PhysRevB.52.R17032.

102. G. Murthy and R. Shankar, Hamiltonian theory of the fractional quantum Hall effect: Effect of Landau level mixing, *Phys. Rev. B* **65**, 245309 (2002). doi: 10.1103/PhysRevB.65.245309. URL http://link.aps.org/doi/10.1103/PhysRevB.65.245309.

103. W. Bishara and C. Nayak, Effect of Landau level mixing on the effective interaction between electrons in the fractional quantum Hall regime, *Phys. Rev. B* **80**, 121302 (2009). doi: 10.1103/PhysRevB.80.121302. URL http://link.aps.org/doi/10.1103/PhysRevB.80.121302.

104. A. Wójs, C. Tőke and J. K. Jain, Landau-level mixing and the emergence of Pfaffian excitations for the $\nu = 5/2$ fractional quantum Hall effect, *Phys. Rev. Lett.* **105**, 096802 (2010). doi: 10.1103/PhysRevLett.105.096802. URL http://link.aps.org/doi/10.1103/PhysRevLett.105.096802.

105. I. Sodemann and A. H. MacDonald, Landau level mixing and the fractional quantum Hall effect, *Phys. Rev. B* **87**, 245425 (2013). doi: 10.1103/PhysRevB.87.245425. URL http://link.aps.org/doi/10.1103/PhysRevB.87.245425.

106. S. H. Simon and E. H. Rezayi, Landau level mixing in the perturbative limit, *Phys. Rev. B* **87**, 155426 (2013). doi: 10.1103/PhysRevB.87.155426. URL http://link.aps.org/doi/10.1103/PhysRevB.87.155426.

107. M. R. Peterson and C. Nayak, More realistic Hamiltonians for the fractional quantum Hall regime in GaAs and graphene, *Phys. Rev. B* **87**, 245129 (2013). doi: 10.1103/PhysRevB.87.245129. URL http://link.aps.org/doi/10.1103/PhysRevB.87.245129.

108. M. R. Peterson and C. Nayak, Effects of Landau level mixing on the fractional quantum Hall effect in monolayer graphene, *Phys. Rev. Lett.* **113**, 086401 (2014). doi: 10.1103/PhysRevLett.113.086401. URL http://link.aps.org/doi/10.1103/PhysRevLett.113.086401.

109. G. Ortiz, D. M. Ceperley and R. M. Martin, New stochastic method for systems with broken time-reversal symmetry: 2D fermions in a magnetic field, *Phys. Rev. Lett.* **71**, 2777–2780 (1993). doi: 10.1103/PhysRevLett.71.2777. URL http://link.aps.org/doi/10.1103/PhysRevLett.71.2777.

110. V. Melik-Alaverdian, N. E. Bonesteel and G. Ortiz, Quantum Hall fluids on the Haldane sphere: A diffusion Monte Carlo study, *Phys. Rev. Lett.* **79**, 5286–5289 (1997). doi: 10.1103/PhysRevLett.79.5286. URL http://link.aps.org/doi/10.1103/PhysRevLett.79.5286.

111. V. Melik-Alaverdian, G. Ortiz and N. Bonesteel, Quantum projector method on curved manifolds, *Journal of Statistical Physics* **104**(1–2), 449–470 (2001). ISSN 0022-4715. doi: 10.1023/A:1010326231389. URL http://dx.doi.org/10.1023/A%3A1010326231389.

112. P. J. Reynolds, D. M. Ceperley, B. J. Alder and W. A. Lester Jr., Fixed-node quantum Monte Carlo for molecules, *J. Chem. Phys.* **77**(11), 5593–5603 (1982). doi: http://dx.doi.org/10.1063/1.443766. URL http://scitation.aip.org/content/aip/journal/jcp/77/11/10.1063/1.443766.

113. W. M. C. Foulkes, L. Mitas, R. J. Needs and G. Rajagopal, Quantum Monte Carlo simulations of solids, *Rev. Mod. Phys.* **73**, 33–83 (2001). doi: 10.1103/RevModPhys.73.33. URL http://link.aps.org/doi/10.1103/RevModPhys.73.33.

114. C. A. Melton and L. Mitas, Quantum Monte Carlo with variable spins: Fixed-phase and fixed-node approximations, *Phys. Rev. E* **96**, 043305 (2017). doi: 10.1103/PhysRevE.96.043305. URL https://link.aps.org/doi/10.1103/PhysRevE.96.043305.

115. A. D. Güçlü and C. J. Umrigar, Maximum-density droplet to lower-density droplet transition in quantum dots, *Phys. Rev. B* **72**, 045309 (2005). doi: 10.1103/PhysRevB.72.045309. URL http://link.aps.org/doi/10.1103/PhysRevB.72.045309.

116. J. P. Eisenstein, H. L. Stormer, L. Pfeiffer and K. W. West, Evidence for a phase transition in the fractional quantum Hall effect, *Phys. Rev. Lett.* **62**, 1540–1543 (1989). doi: 10.1103/PhysRevLett.62.1540. URL http://link.aps.org/doi/10.1103/PhysRevLett.62.1540.

117. J. P. Eisenstein, H. L. Stormer, L. N. Pfeiffer and K. W. West, Evidence for a spin transition in the $\nu = 2/3$ fractional quantum Hall effect, *Phys. Rev. B* **41**, 7910–7913 (1990). doi: 10.1103/PhysRevB.41.7910. URL http://link.aps.org/doi/10.1103/PhysRevB.41.7910.

118. L. W. Engel, S. W. Hwang, T. Sajoto, D. C. Tsui and M. Shayegan, Fractional quantum Hall effect at $\nu=2/3$ and 3/5 in tilted magnetic fields, *Phys. Rev. B* **45**, 3418–3425 (1992). doi: 10.1103/PhysRevB.45.3418. URL http://link.aps.org/doi/10.1103/PhysRevB.45.3418.

119. W. Kang, J. B. Young, S. T. Hannahs, E. Palm, K. L. Campman and A. C. Gossard, Evidence for a spin transition in the $\nu = 2/5$ fractional quantum Hall effect, *Phys. Rev. B* **56**, R12776–R12779 (1997). doi: 10.1103/PhysRevB.56.R12776. URL http://link.aps.org/doi/10.1103/PhysRevB.56.R12776.

120. A. S. Yeh, H. L. Stormer, D. C. Tsui, L. N. Pfeiffer, K. W. Baldwin and K. W. West, Effective mass and g factor of four-flux-quanta composite fermions, *Phys. Rev. Lett.* **82**, 592–595 (1999). doi: 10.1103/PhysRevLett.82.592. URL http://link.aps.org/doi/10.1103/PhysRevLett.82.592.

121. N. Freytag, Y. Tokunaga, M. Horvatić, C. Berthier, M. Shayegan and L. P. Lévy, New phase transition between partially and fully polarized quantum Hall states with charge and spin gaps at $\nu = \frac{2}{3}$, *Phys. Rev. Lett.* **87**, 136801 (2001). doi: 10.1103/PhysRevLett.87.136801. URL http://link.aps.org/doi/10.1103/PhysRevLett.87.136801.

122. L. Tiemann, G. Gamez, N. Kumada and K. Muraki, Unraveling the spin polarization of the $\nu = 5/2$ fractional quantum Hall state, *Science* **335**(6070), 828–831 (2012). doi: 10.1126/science.1216697. URL http://www.sciencemag.org/content/335/6070/828.abstract.

123. Y. Liu, S. Hasdemir, A. Wójs, J. K. Jain, L. N. Pfeiffer, K. W. West, K. W. Baldwin and M. Shayegan, Spin polarization of composite fermions and particle-hole symmetry breaking, *Phys. Rev. B* **90**, 085301 (2014). doi: 10.1103/PhysRevB.90.085301. URL http://link.aps.org/doi/10.1103/PhysRevB.90.085301.

124. K. Park and J. K. Jain, Spontaneous magnetization of composite fermions, *Phys. Rev. Lett.* **83**, 5543–5546 (1999). doi: 10.1103/PhysRevLett.83.5543. URL http://link.aps.org/doi/10.1103/PhysRevLett.83.5543.

125. A. C. Balram, C. Töke, A. Wójs and J. K. Jain, Fractional quantum Hall effect in graphene: Quantitative comparison between theory and experiment, *Phys. Rev. B* **92**, 075410 (2015). doi: 10.1103/PhysRevB.92.075410. URL http://link.aps.org/doi/10.1103/PhysRevB.92.075410.

126. D. R. Leadley, R. J. Nicholas, D. K. Maude, A. N. Utjuzh, J. C. Portal, J. J. Harris and C. T. Foxon, Fractional quantum Hall effect measurements at zero $\sim g$ factor, *Phys. Rev. Lett.* **79**, 4246–4249 (1997). doi: 10.1103/PhysRevLett.79.4246. URL http://link.aps.org/doi/10.1103/PhysRevLett.79.4246.

127. X. Du, I. Skachko, F. Duerr, A. Luican and E. Y. Andrei, Fractional quantum Hall effect and insulating phase of Dirac electrons in graphene, *Nature* **462**, 192–195 (2009).

128. K. Bolotin, F. Ghahari, M. D. Shulman, H. Stormer and P. Kim, Observation of the fractional quantum Hall effect in graphene, *Nature* **462**, 196–199 (2009). doi: 10.1038/nature08582.

129. C. R. Dean, A. F. Young, P. Cadden-Zimansky, L. Wang, H. Ren, K. Watanabe, T. Taniguchi, P. Kim, J. Hone and K. L. Shepard, Multicomponent fractional quantum Hall effect in graphene, *Nature Physics* **7**, 693–696 (2011).

130. B. E. Feldman, B. Krauss, J. H. Smet and A. Yacoby, Unconventional sequence of fractional quantum Hall states in suspended graphene, *Science.* **337**(6099), 1196–1199 (2012). doi: 10.1126/science.1224784. URL http://www.sciencemag.org/content/337/6099/1196.abstract.

131. F. Amet, A. J. Bestwick, J. R. Williams, L. Balicas, K. Watanabe, T. Taniguchi, and D. Goldhaber-Gordon, Composite fermions and broken symmetries in graphene, *Nat. Commun.* **6**, 5838 (2015). doi: 10.1038/ncomms6838. URL http://dx.doi.org/10.1038/ncomms6838.

132. T. M. Kott, B. Hu, S. H. Brown and B. E. Kane, Valley-degenerate two-dimensional electrons in the lowest Landau level, *Phys. Rev. B* **89**, 041107 (2014). doi: 10.1103/PhysRevB.89.041107. URL http://link.aps.org/doi/10.1103/PhysRevB.89.041107.

133. A. C. Balram, C. Töke, A. Wójs and J. K. Jain, Phase diagram of fractional quantum Hall effect of composite fermions in multicomponent systems, *Phys. Rev. B* **91**, 045109 (2015). doi: 10.1103/PhysRevB.91.045109. URL http://link.aps.org/doi/10.1103/PhysRevB.91.045109.

134. M. Shayegan. Case for the magnetic-field-induced two-dimensional Wigner crystal. In *Perspectives in Quantum Hall Effects*, p. 343–384. (Wiley-VCH Verlag GmbH, 2007). ISBN 9783527617258. doi: 10.1002/9783527617258.ch10. URL http://dx.doi.org/10.1002/9783527617258.ch10.

135. H. A. Fertig. Properties of the electron solid. In *Perspectives in Quantum Hall Effects*, p. 71–108. (Wiley-VCH Verlag GmbH, 2007). ISBN 9783527617258. doi: 10.1002/9783527617258.ch10. URL http://dx.doi.org/10.1002/9783527617258.ch10.

136. H. W. Jiang, R. L. Willett, H. L. Stormer, D. C. Tsui, L. N. Pfeiffer and K. W. West, Quantum liquid versus electron solid around $\nu = 1/5$ Landau-level filling, *Phys. Rev. Lett.* **65**, 633–636 (1990). doi: 10.1103/PhysRevLett.65.633.

137. V. J. Goldman, M. Santos, M. Shayegan and J. E. Cunningham, Evidence for two-dimensional quantum Wigner crystal, *Phys. Rev. Lett.* **65**, 2189–2192 (1990). doi: 10.1103/PhysRevLett.65.2189.

138. F. I. B. Williams, P. A. Wright, R. G. Clark, E. Y. Andrei, G. Deville, D. C. Glattli, O. Probst, B. Etienne, C. Dorin, C. T. Foxon and J. J. Harris, Conduction threshold and pinning frequency of magnetically induced Wigner solid, *Phys. Rev. Lett.* **66**, 3285–3288 (1991). doi: 10.1103/PhysRevLett.66.3285. URL https://link.aps.org/doi/10.1103/PhysRevLett.66.3285.

139. M. A. Paalanen, R. L. Willett, R. R. Ruel, P. B. Littlewood, K. W. West and L. N. Pfeiffer, Electrical conductivity and Wigner crystallization, *Phys. Rev. B* **45**, 13784–13787 (1992). doi: 10.1103/PhysRevB.45.13784. URL http://link.aps.org/doi/10.1103/PhysRevB.45.13784.

140. M. B. Santos, Y. W. Suen, M. Shayegan, Y. P. Li, L. W. Engel and D. C. Tsui, Observation of a reentrant insulating phase near the 1/3 fractional quantum Hall liquid in a two-dimensional hole system, *Phys. Rev. Lett.* **68**, 1188–1191 (1992). doi: 10.1103/PhysRevLett.68.1188.

141. M. B. Santos, J. Jo, Y. W. Suen, L. W. Engel and M. Shayegan, Effect of Landau-level mixing on quantum-liquid and solid states of two-dimensional hole systems, *Phys. Rev. B* **46**, 13639–13642 (1992). doi: 10.1103/PhysRevB.46.13639. URL https://link.aps.org/doi/10.1103/PhysRevB.46.13639.

142. H. C. Manoharan and M. Shayegan, Wigner crystal versus Hall insulator, *Phys. Rev. B* **50**, 17662–17665 (1994). doi: 10.1103/PhysRevB.50.17662.

143. L. Engel, C.-C. Li, D. Shahar, D. Tsui and M. Shayegan, Microwave resonances in low-filling insulating phases of two-dimensional electron and hole systems, *Physica E* **1**(14), 111–115 (1997). doi: http://dx.doi.org/10.1016/S1386-9477(97)00025-8.

144. C.-C. Li, J. Yoon, L. W. Engel, D. Shahar, D. C. Tsui and M. Shayegan, Microwave resonance and weak pinning in two-dimensional hole systems at high magnetic fields, *Phys. Rev. B* **61**, 10905–10909 (2000). doi: 10.1103/PhysRevB.61.10905.

145. P. D. Ye, L. W. Engel, D. C. Tsui, R. M. Lewis, L. N. Pfeiffer and K. West, Correlation lengths of the Wigner-crystal order in a two-dimensional electron system at high magnetic fields, *Phys. Rev. Lett.* **89**, 176802 (2002). doi: 10.1103/PhysRevLett.89.176802.

146. Y. P. Chen, R. M. Lewis, L. W. Engel, D. C. Tsui, P. D. Ye, Z. H. Wang, L. N. Pfeiffer and K. W. West, Evidence for two different solid phases of two-dimensional electrons in high magnetic fields, *Phys. Rev. Lett.* **93**, 206805 (2004). doi: 10.1103/PhysRevLett.93.206805.

147. G. A. Csáthy, H. Noh, D. C. Tsui, L. N. Pfeiffer and K. W. West, Magnetic-field-induced insulating phases at large r_s, *Phys. Rev. Lett.* **94**, 226802 (2005). doi: 10.1103/PhysRevLett.94.226802.

148. W. Pan, G. A. Csáthy, D. C. Tsui, L. N. Pfeiffer and K. W. West, Transition from a fractional quantum Hall liquid to an electron solid at Landau level filling $\nu = \frac{1}{3}$ in tilted magnetic fields, *Phys. Rev. B* **71**, 035302 (2005). doi: 10.1103/PhysRevB.71.035302. URL https://link.aps.org/doi/10.1103/PhysRevB.71.035302.

149. G. Sambandamurthy, Z. Wang, R. Lewis, Y. P. Chen, L. Engel, D. Tsui, L. Pfeiffer and K. West, Pinning mode resonances of new phases of 2D electron systems in high magnetic fields, *Solid State Commun.* **140**(2), 100–106 (2006).

150. Y. P. Chen, G. Sambandamurthy, Z. H. Wang, R. M. Lewis, L. W. Engel, D. C. Tsui, P. D. Ye, L. N. Pfeiffer and K. W. West, Melting of a 2D quantum electron solid in a high magnetic field, *Nature Phys.* **2**, 452–455 (2006). doi: 10.1038/nphys322.

151. Y. Liu, D. Kamburov, S. Hasdemir, M. Shayegan, L. N. Pfeiffer, K. W. West,

and K. W. Baldwin, Fractional quantum Hall effect and Wigner crystal of interacting composite fermions, *Phys. Rev. Lett.* **113**, 246803 (2014). doi: 10.1103/PhysRevLett.113.246803. URL http://link.aps.org/doi/10.1103/PhysRevLett.113.246803.

152. C. Zhang, R.-R. Du, M. J. Manfra, L. N. Pfeiffer and K. W. West, Transport of a sliding Wigner crystal in the four flux composite fermion regime, *Phys. Rev. B* **92**, 075434 (2015). doi: 10.1103/PhysRevB.92.075434. URL https://link.aps.org/doi/10.1103/PhysRevB.92.075434.

153. H. Deng, Y. Liu, I. Jo, L. N. Pfeiffer, K. W. West, K. W. Baldwin and M. Shayegan, Commensurability oscillations of composite fermions induced by the periodic potential of a Wigner crystal, *Phys. Rev. Lett.* **117**, 096601 (2016). doi: 10.1103/PhysRevLett.117.096601. URL http://link.aps.org/doi/10.1103/PhysRevLett.117.096601.

154. J. Jang, B. M. Hunt, L. N. Pfeiffer, K. W. West and R. C. Ashoori, Sharp tunnelling resonance from the vibrations of an electronic Wigner crystal, *Nature Physics* **13**(4), 340–344 (2017). ISSN 1745-2473. doi: 10.1038/NPHYS3979.

155. S. Chen, R. Ribeiro-Palau, K. Yang, K. Watanabe, T. Taniguchi, J. Hone, M. O. Goerbig and C. R. Dean, Competing fractional quantum Hall and electron solid phases in graphene, *Phys. Rev. Lett.* **122**, 026802 (2019). doi: 10.1103/PhysRevLett.122.026802. URL https://link.aps.org/doi/10.1103/PhysRevLett.122.026802.

156. D. Maryenko, A. McCollam, J. Falson, Y. Kozuka, J. Bruin, U. Zeitler and M. Kawasaki, Composite fermion liquid to a Wigner solid transition in the lowest Landau level of zinc oxide, *Nature Commun.* **9**, 4356 (2018). doi: 10.1038/s41467-018-06834-6. URL https://www.nature.com/articles/s41467-018-06834-6.

157. P. K. Lam and S. M. Girvin, Liquid-solid transition and the fractional quantum-Hall effect, *Phys. Rev. B* **30**, 473–475 (1984). doi: 10.1103/PhysRevB.30.473.

158. D. Levesque, J. J. Weis and A. H. MacDonald, Crystallization of the incompressible quantum-fluid state of a two-dimensional electron gas in a strong magnetic field, *Phys. Rev. B* **30**, 1056–1058 (1984). doi: 10.1103/PhysRevB.30.1056.

159. X. Zhu and S. G. Louie, Wigner crystallization in the fractional quantum Hall regime: A variational quantum Monte Carlo study, *Phys. Rev. Lett.* **70**, 335–338 (1993). doi: 10.1103/PhysRevLett.70.335.

160. R. Price, P. M. Platzman and S. He, Fractional quantum Hall liquid, Wigner solid phase boundary at finite density and magnetic field, *Phys. Rev. Lett.* **70**, 339–342 (1993). doi: 10.1103/PhysRevLett.70.339.

161. P. M. Platzman and R. Price, Quantum freezing of the fractional quantum Hall liquid, *Phys. Rev. Lett.* **70**, 3487–3489 (1993). doi: 10.1103/PhysRevLett.70.3487.

162. W. J. He, T. Cui, Y. M. Ma, C. B. Chen, Z. M. Liu and G. T. Zou, Phase boundary between the fractional quantum Hall liquid and the Wigner crystal at low filling factors and low temperatures: A path integral Monte Carlo study, *Phys. Rev. B* **72**, 195306 (2005). doi: 10.1103/PhysRevB.72.195306.

163. H. Yi and H. A. Fertig, Laughlin-Jastrow-correlated Wigner crystal in a strong magnetic field, *Phys. Rev. B* **58**, 4019–4027 (1998). doi: 10.1103/PhysRevB.58.4019.

164. R. Narevich, G. Murthy and H. A. Fertig, Hamiltonian theory of the composite-fermion Wigner crystal, *Phys. Rev. B* **64**, 245326 (2001). doi: 10.1103/PhysRevB.64.245326.

165. C.-C. Chang, G. S. Jeon and J. K. Jain, Microscopic verification of topological electron-vortex binding in the lowest Landau-level crystal state, *Phys. Rev. Lett.* **94**, 016809 (2005). doi: 10.1103/PhysRevLett.94.016809.

166. A. C. Archer, K. Park and J. K. Jain, Competing crystal phases in the lowest Landau level, *Phys. Rev. Lett.* **111**, 146804 (2013). doi: 10.1103/PhysRevLett.111.146804.

167. C. Yannouleas and U. Landman, Two-dimensional quantum dots in high magnetic fields: Rotating-electron-molecule versus composite-fermion approach, *Phys. Rev. B* **68**, 035326 (2003). doi: 10.1103/PhysRevB.68.035326. URL http://link.aps.org/doi/10.1103/PhysRevB.68.035326.

168. S.-Y. Lee, V. W. Scarola and J. K. Jain, Stripe formation in the fractional quantum Hall regime, *Phys. Rev. Lett.* **87**, 256803 (2001). doi: 10.1103/PhysRevLett.87.256803. URL http://link.aps.org/doi/10.1103/PhysRevLett.87.256803.

169. A. C. Archer and J. K. Jain, Static and dynamic properties of type-II composite fermion Wigner crystals, *Phys. Rev. B* **84**, 115139 (2011). doi: 10.1103/PhysRevB.84.115139.

170. J. J. Thomson, On the structure of the atom: An investigation of the stability and periods of oscillation of a number of corpuscles arranged at equal intervals around the circumference of a circle; with application of the results to the theory of atomic structure, *Phil. Mag.* **7**, 237 (1904).

171. D. J. Wales and S. Ulker, Structure and dynamics of spherical crystals characterized for the Thomson problem, *Phys. Rev. B* **74**, 212101 (2006). doi: 10.1103/PhysRevB.74.212101. URL https://link.aps.org/doi/10.1103/PhysRevB.74.212101.

172. D. J. Wales, H. McKay and E. L. Altschuler, Defect motifs for spherical topologies, *Phys. Rev. B* **79**, 224115 (2009). doi: 10.1103/PhysRevB.79.224115. URL https://link.aps.org/doi/10.1103/PhysRevB.79.224115.

173. The minimum energy locations can be found at http://thomson.phy.syr.edu/.

174. J.-W. Rhim, J. K. Jain and K. Park, Analytical theory of strongly correlated Wigner crystals in the lowest Landau level, *Phys. Rev. B* **92**, 121103 (2015). doi: 10.1103/PhysRevB.92.121103. URL https://link.aps.org/doi/10.1103/PhysRevB.92.121103.

175. H. Zhu, Y. P. Chen, P. Jiang, L. W. Engel, D. C. Tsui, L. N. Pfeiffer and K. W. West, Observation of a pinning mode in a Wigner solid with $\nu = 1/3$ fractional quantum Hall excitations, *Phys. Rev. Lett.* **105**, 126803 (2010). doi: 10.1103/PhysRevLett.105.126803. URL http://link.aps.org/doi/10.1103/PhysRevLett.105.126803.

176. J. Shi and W. Ji, Dynamics of the Wigner crystal of composite particles, *Phys. Rev. B* **97**, 125133 (2018). doi: 10.1103/PhysRevB.97.125133. URL https://link.aps.org/doi/10.1103/PhysRevB.97.125133.

177. Y. Hu and J. K. Jain, Kohn-Sham Theory of the Fractional Quantum Hall Effect, *Phys. Rev. Lett.* **123**, 176802 (2019). doi: 10.1103/PhysRevLett.123.176802. URL https://link.aps.org/doi/10.1103/PhysRevLett.123.176802.

178. G. Giuliani and G. Vignale, *Quantum Theory of the Electron Liquid*. (Cambridge University Press, The Edinburgh Building, Cambridge CB2 2RU, UK, 2008).

179. M. Ferconi, M. R. Geller and G. Vignale, Edge structure of fractional quantum Hall systems from density-functional theory, *Phys. Rev. B* **52**, 16357–16360 (1995). doi: 10.1103/PhysRevB.52.16357. URL http://link.aps.org/doi/10.1103/PhysRevB.52.16357.

180. O. Heinonen, M. I. Lubin and M. D. Johnson, Ensemble density functional theory of the fractional quantum Hall effect, *Phys. Rev. Lett.* **75**, 4110–4113 (1995). doi: 10.1103/PhysRevLett.75.4110. URL http://link.aps.org/doi/10.1103/PhysRevLett.75.4110.

181. J. Zhao, M. Thakurathi, M. Jain, D. Sen and J. K. Jain, Density-functional theory of the fractional quantum Hall effect, *Phys. Rev. Lett.* **118**, 196802 (2017).

doi: 10.1103/PhysRevLett.118.196802. URL https://link.aps.org/doi/10.1103/PhysRevLett.118.196802.

182. Y. Zhang, G. J. Sreejith, N. D. Gemelke and J. K. Jain, Fractional angular momentum in cold-atom systems, *Phys. Rev. Lett.* **113**, 160404 (2014). doi: 10.1103/PhysRevLett.113.160404. URL http://link.aps.org/doi/10.1103/PhysRevLett.113.160404.

183. C. J. Grayce and R. A. Harris, Magnetic-field density-functional theory, *Physical Review A* **50**(4), 3089 (1994).

184. W. Kohn, A. Savin and C. A. Ullrich, Hohenberg–Kohn theory including spin magnetism and magnetic fields, *Int. J. Quant. Chem.* **100**(1), 20–21 (2004).

185. E. I. Tellgren, S. Kvaal, E. Sagvolden, U. Ekström, A. M. Teale and T. Helgaker, Choice of basic variables in current-density-functional theory, *Phys. Rev. A* **86**, 062506 (2012). doi: 10.1103/PhysRevA.86.062506. URL https://link.aps.org/doi/10.1103/PhysRevA.86.062506.

186. E. I. Tellgren, A. Laestadius, T. Helgaker, S. Kvaal and A. M. Teale, Uniform magnetic fields in density-functional theory, *The J. Chem. Phys.* **148**(2), 024101 (2018).

187. M. Levy, Universal variational functionals of electron densities, first-order density matrices, and natural spin-orbitals and solution of the v-representability problem, *Proc. Nat. Acad. Sci.* **76**(12), 6062–6065 (1979).

188. E. H. Lieb, *Int. J. Quantum Chem.* **24**, 243 (1983).

189. A. Seidl, A. Görling, P. Vogl, J. A. Majewski and M. Levy, Generalized Kohn-Sham schemes and the band-gap problem, *Phys. Rev. B* **53**, 3764–3774 (1996). doi: 10.1103/PhysRevB.53.3764. URL https://link.aps.org/doi/10.1103/PhysRevB.53.3764.

190. S. Kümmel and L. Kronik, Orbital-dependent density functionals: Theory and applications, *Rev. Mod. Phys.* **80**, 3–60 (2008). doi: 10.1103/RevModPhys.80.3. URL https://link.aps.org/doi/10.1103/RevModPhys.80.3.

191. L. Bonsall and A. A. Maradudin, Some static and dynamical properties of a two-dimensional Wigner crystal, *Phys. Rev. B* **15**, 1959–1973 (1977). doi: 10.1103/PhysRevB.15.1959.

192. E. V. Tsiper and V. J. Goldman, Formation of an edge striped phase in the $\nu = \frac{1}{3}$ fractional quantum Hall system, *Phys. Rev. B* **64**, 165311 (2001). doi: 10.1103/PhysRevB.64.165311. URL https://link.aps.org/doi/10.1103/PhysRevB.64.165311.

193. J. K. Jain and T. Kawamura, Composite fermions in quantum dots, *EPL (Europhysics Letters)* **29**(4), 321 (1995). URL http://stacks.iop.org/0295-5075/29/i=4/a=009.

194. B. Kaduk, T. Kowalczyk and T. Van Voorhis, Constrained density functional theory, *Chemical Reviews* **112**(1), 321–370 (2011).

195. J. K. Jain, Incompressible quantum Hall states, *Phys. Rev. B* **40**, 8079–8082 (1989). doi: 10.1103/PhysRevB.40.8079. URL http://link.aps.org/doi/10.1103/PhysRevB.40.8079.

196. J. K. Jain, Theory of the fractional quantum Hall effect, *Phys. Rev. B* **41**, 7653–7665 (1990). doi: 10.1103/PhysRevB.41.7653.

197. X. G. Wen, Non-Abelian statistics in the fractional quantum Hall states, *Phys. Rev. Lett.* **66**, 802–805 (1991). doi: 10.1103/PhysRevLett.66.802. URL http://link.aps.org/doi/10.1103/PhysRevLett.66.802.

198. R. Willett, J. P. Eisenstein, H. L. Störmer, D. C. Tsui, A. C. Gossard and J. H. English, Observation of an even-denominator quantum number in the fractional quantum Hall effect, *Phys. Rev. Lett.* **59**, 1776–1779 (1987). doi: 10.1103/PhysRevLett.59.1776. URL http://link.aps.org/doi/10.1103/PhysRevLett.59.1776.

199. R. H. Morf, Transition from quantum Hall to compressible states in the second Landau level: New light on the $\nu = 5/2$ enigma, *Phys. Rev. Lett.* **80**, 1505–1508 (1998). doi: 10.1103/PhysRevLett.80.1505. URL http://link.aps.org/doi/10.1103/PhysRevLett.80.1505.

200. A. C. Balram, M. Barkeshli and M. S. Rudner, Parton construction of a wave function in the anti-Pfaffian phase, *Phys. Rev. B* **98**, 035127 (2018). doi: 10.1103/PhysRevB.98.035127. URL https://link.aps.org/doi/10.1103/PhysRevB.98.035127.

201. Y. Wu, T. Shi and J. K. Jain, Non-Abelian parton fractional quantum Hall effect in multilayer graphene, *Nano Letters* **17**(8), 4643–4647 (2017). doi: 10.1021/acs.nanolett.7b01080. URL http://dx.doi.org/10.1021/acs.nanolett.7b01080. PMID: 28649831.

202. B. Blok and X. G. Wen, Effective theories of the fractional quantum Hall effect: Hierarchy construction, *Phys. Rev. B* **42**, 8145–8156 (1990). doi: 10.1103/PhysRevB.42.8145. URL http://link.aps.org/doi/10.1103/PhysRevB.42.8145.

203. B. Blok and X. G. Wen, Effective theories of the fractional quantum Hall effect at generic filling fractions, *Phys. Rev. B* **42**, 8133–8144 (1990). doi: 10.1103/PhysRevB.42.8133. URL http://link.aps.org/doi/10.1103/PhysRevB.42.8133.

204. X.-G. Wen, Theory of the edge states in fractional quantum Hall effects, *International Journal of Modern Physics B* **06**(10), 1711–1762 (1992). doi: 10.1142/S0217979292000840. URL http://www.worldscientific.com/doi/abs/10.1142/S0217979292000840.

205. A. C. Balram, S. Mukherjee, K. Park, M. Barkeshli, M. S. Rudner and J. K. Jain, Fractional quantum Hall effect at $\nu = 2 + 6/13$: The parton paradigm for the second Landau level, *Phys. Rev. Lett.* **121**, 186601 (2018). doi: 10.1103/PhysRevLett.121.186601. URL https://link.aps.org/doi/10.1103/PhysRevLett.121.186601.

206. X.-G. Wen, Edge excitations in the fractional quantum Hall states at general filling fractions, *Modern Physics Letters B* **05**(01), 39–46 (1991). doi: 10.1142/S0217984991000058. URL https://www.worldscientific.com/doi/abs/10.1142/S0217984991000058.

207. J. E. Moore and X.-G. Wen, Classification of disordered phases of quantum Hall edge states, *Phys. Rev. B* **57**, 10138–10156 (1998). doi: 10.1103/PhysRevB.57.10138. URL http://link.aps.org/doi/10.1103/PhysRevB.57.10138.

208. C. L. Kane and M. P. A. Fisher, Quantized thermal transport in the fractional quantum Hall effect, *Phys. Rev. B* **55**, 15832–15837 (1997). doi: 10.1103/PhysRevB.55.15832. URL http://link.aps.org/doi/10.1103/PhysRevB.55.15832.

209. X.-G. Wen, Projective construction of non-Abelian quantum Hall liquids, *Phys. Rev. B* **60**, 8827–8838 (1999). doi: 10.1103/PhysRevB.60.8827. URL https://link.aps.org/doi/10.1103/PhysRevB.60.8827.

210. M. Barkeshli and X.-G. Wen, Effective field theory and projective construction for Z_k parafermion fractional quantum Hall states, *Phys. Rev. B* **81**, 155302 (2010). doi: 10.1103/PhysRevB.81.155302. URL http://link.aps.org/doi/10.1103/PhysRevB.81.155302.

211. M. Barkeshli and J. McGreevy, Continuous transition between fractional quantum Hall and superfluid states, *Phys. Rev. B* **89**, 235116 (2014). doi: 10.1103/PhysRevB.89.235116. URL http://link.aps.org/doi/10.1103/PhysRevB.89.235116.

212. C. Repellin, T. Neupert, B. A. Bernevig and N. Regnault, Projective construction of the F_k read-rezayi fractional quantum Hall states and their excitations on the torus geometry, *Phys. Rev. B* **92**, 115128 (2015). doi: 10.1103/PhysRevB.92.115128. URL https://link.aps.org/doi/10.1103/PhysRevB.92.115128.

213. H. Goldman, R. Sohal and E. Fradkin, Landau-Ginzburg theories of non-Abelian quantum Hall states from non-Abelian bosonization, *Phys. Rev. B* **100**, 115111 (2019). doi: 10.1103/PhysRevB.100.115111. URL https://link.aps.org/doi/10.1103/PhysRevB.100.115111.

214. S. Bandyopadhyay, L. Chen, M. T. Ahari, G. Ortiz, Z. Nussinov and A. Seidel, Entangled Pauli principles: The DNA of quantum Hall fluids, *Phys. Rev. B* **98**, 161118 (2018). doi: 10.1103/PhysRevB.98.161118. URL https://link.aps.org/doi/10.1103/PhysRevB.98.161118.

215. A. Wójs, Transition from Abelian to non-Abelian quantum liquids in the second Landau level, *Phys. Rev. B* **80**, 041104 (2009). doi: 10.1103/PhysRevB.80.041104. URL http://link.aps.org/doi/10.1103/PhysRevB.80.041104.

216. A. Kumar, G. A. Csáthy, M. J. Manfra, L. N. Pfeiffer and K. W. West, Nonconventional odd-denominator fractional quantum Hall states in the second Landau level, *Phys. Rev. Lett.* **105**, 246808 (2010). doi: 10.1103/PhysRevLett.105.246808. URL http://link.aps.org/doi/10.1103/PhysRevLett.105.246808.

217. V. Shingla, E. Kleinbaum, A. Kumar, L. N. Pfeiffer, K. W. West and G. A. Csáthy, Finite-temperature behavior in the second Landau level of the two-dimensional electron gas, *Phys. Rev. B* **97**, 241105 (2018). doi: 10.1103/PhysRevB.97.241105. URL https://link.aps.org/doi/10.1103/PhysRevB.97.241105.

218. N. Read and E. Rezayi, Beyond paired quantum Hall states: Parafermions and incompressible states in the first excited Landau level, *Phys. Rev. B* **59**, 8084–8092 (1999). doi: 10.1103/PhysRevB.59.8084. URL http://link.aps.org/doi/10.1103/PhysRevB.59.8084.

219. E. H. Rezayi and N. Read, Non-Abelian quantized Hall states of electrons at filling factors 12/5 and 13/5 in the first excited Landau level, *Phys. Rev. B* **79**, 075306 (2009). doi: 10.1103/PhysRevB.79.075306. URL http://link.aps.org/doi/10.1103/PhysRevB.79.075306.

220. G. J. Sreejith, Y.-H. Wu, A. Wójs and J. K. Jain, Tripartite composite fermion states, *Phys. Rev. B* **87**, 245125 (2013). doi: 10.1103/PhysRevB.87.245125. URL http://link.aps.org/doi/10.1103/PhysRevB.87.245125.

221. W. Zhu, S. S. Gong, F. D. M. Haldane and D. N. Sheng, Fractional quantum Hall states at $\nu = 13/5$ and 12/5 and their non-Abelian nature, *Phys. Rev. Lett.* **115**, 126805 (2015). doi: 10.1103/PhysRevLett.115.126805. URL http://link.aps.org/doi/10.1103/PhysRevLett.115.126805.

222. R. S. K. Mong, M. P. Zaletel, F. Pollmann and Z. Papić, Fibonacci anyons and charge density order in the 12/5 and 13/5 quantum Hall plateaus, *Phys. Rev. B* **95**, 115136 (2017). doi: 10.1103/PhysRevB.95.115136. URL http://link.aps.org/doi/10.1103/PhysRevB.95.115136.

223. K. Pakrouski, M. Troyer, Y.-L. Wu, S. Das Sarma and M. R. Peterson, Enigmatic 12/5 fractional quantum Hall effect, *Phys. Rev. B* **94**, 075108 (2016). doi: 10.1103/PhysRevB.94.075108. URL http://link.aps.org/doi/10.1103/PhysRevB.94.075108.

224. A. A. Zibrov, C. R. Kometter, H. Zhou, E. M. Spanton, T. Taniguchi, K. Watanabe, M. P. Zaletel and A. F. Young, Tunable interacting composite fermion phases in a half-filled bilayer-graphene Landau level, *Nature* **549**, 360–364 (2017). doi: 10.1038/nature23893. URL http://www.nature.com/nature/journal/v549/n7672/full/nature23893.html.

225. A. Bid, N. Ofek, H. Inoue, M. Heiblum, C. L. Kane, V. Umansky and D. Mahalu, Observation of neutral modes in the fractional quantum Hall regime, *Nature* **466** (7306), 585–590 (2010). ISSN 0028-0836. doi: 10.1038/nature09277.

226. M. Dolev, Y. Gross, R. Sabo, I. Gurman, M. Heiblum, V. Umansky and D. Mahalu, Characterizing neutral modes of fractional states in the second Landau level, *Phys. Rev. Lett.* **107**, 036805 (2011). doi: 10.1103/PhysRevLett.107.036805. URL http://link.aps.org/doi/10.1103/PhysRevLett.107.036805.

227. Y. Gross, M. Dolev, M. Heiblum, V. Umansky and D. Mahalu, Upstream neutral modes in the fractional quantum Hall effect regime: Heat waves or coherent dipoles, *Phys. Rev. Lett.* **108**, 226801 (2012). doi: 10.1103/PhysRevLett.108.226801. URL http://link.aps.org/doi/10.1103/PhysRevLett.108.226801.

228. H. Inoue, A. Grivnin, Y. Ronen, M. Heiblum, V. Umansky and D. Mahalu, Proliferation of neutral modes in fractional quantum Hall states, *Nature Communications* **5**, 4067 (2014). ISSN 2041-1723. doi: 10.1038/ncomms5067.

229. M. Banerjee, M. Heiblum, A. Rosenblatt, Y. Oreg, D. E. Feldman, A. Stern and V. Umansky, Observed quantization of anyonic heat flow, *Nature* **545**, 75–79 (2017). ISSN 0028-0836. doi: 10.1038/nature22052.

230. M. Banerjee, M. Heiblum, V. Umansky, D. E. Feldman, Y. Oreg and A. Stern, Observation of half-integer thermal Hall conductance, *Nature* **559**, 205–210 (2018). ISSN 1476-4687. doi: 10.1038/s41586-018-0184-1. URL https://doi.org/10.1038/s41586-018-0184-1.

231. M. Levin and B. I. Halperin, Collective states of non-Abelian quasiparticles in a magnetic field, *Phys. Rev. B* **79**, 205301 (2009). doi: 10.1103/PhysRevB.79.205301. URL http://link.aps.org/doi/10.1103/PhysRevB.79.205301.

232. Y. Kim, A. C. Balram, T. Taniguchi, K. Watanabe, J. K. Jain and J. H. Smet, Even denominator fractional quantum Hall states in higher Landau levels of graphene, *Nature Physics* **15**(2), 154–158 (2019). ISSN 1745-2481. doi: 10.1038/s41567-018-0355-x. URL https://doi.org/10.1038/s41567-018-0355-x.

233. S. A. Trugman and S. Kivelson, Exact results for the fractional quantum Hall effect with general interactions, *Phys. Rev. B* **31**, 5280–5284 (1985). doi: 10.1103/PhysRevB.31.5280. URL http://link.aps.org/doi/10.1103/PhysRevB.31.5280.

234. E. H. Rezayi and A. H. MacDonald, Origin of the ν=2/5 fractional quantum Hall effect, *Phys. Rev. B* **44**, 8395–8398 (1991). doi: 10.1103/PhysRevB.44.8395. URL http://link.aps.org/doi/10.1103/PhysRevB.44.8395.

235. E. McCann and V. I. Fal'ko, Landau-level degeneracy and quantum Hall effect in a graphite bilayer, *Phys. Rev. Lett.* **96**, 086805 (2006). doi: 10.1103/PhysRevLett.96.086805. URL http://link.aps.org/doi/10.1103/PhysRevLett.96.086805.

236. Y. Barlas, K. Yang and A. H. MacDonald, Quantum Hall effects in graphene-based two-dimensional electron systems, *Nanotechnology* **23**(5), 052001 (2012). doi: 10.1088/0957-4484/23/5/052001. URL https://doi.org/10.1088%2F0957-4484%2F23%2F5%2F052001.

237. Y. Kim, D. S. Lee, S. Jung, V. Skákalová, T. Taniguchi, K. Watanabe, J. S. Kim and J. H. Smet, Fractional quantum Hall states in bilayer graphene probed by transconductance fluctuations, *Nano Letters* **15**(11), 7445–7451 (2015). doi: 10.1021/acs.nanolett.5b02876. URL http://dx.doi.org/10.1021/acs.nanolett.5b02876. PMID: 26479836.

238. R. Côté, W. Luo, B. Petrov, Y. Barlas and A. H. MacDonald, Orbital and interlayer skyrmion crystals in bilayer graphene, *Phys. Rev. B* **82**, 245307 (2010). doi: 10.1103/PhysRevB.82.245307.

239. V. M. Apalkov and T. Chakraborty, Stable Pfaffian state in bilayer graphene, *Phys. Rev. Lett.* **107**, 186803 (2011). doi: 10.1103/PhysRevLett.107.186803. URL http://link.aps.org/doi/10.1103/PhysRevLett.107.186803.

240. K. Snizhko, V. Cheianov and S. H. Simon, Importance of interband transitions for the fractional quantum Hall effect in bilayer graphene, *Phys. Rev. B* **85**, 201415 (2012). doi: 10.1103/PhysRevB.85.201415. URL http://link.aps.org/doi/10.1103/PhysRevB.85.201415.

241. W. Bao, Z. Zhao, H. Zhang, G. Liu, P. Kratz, L. Jing, J. Velasco, D. Smirnov, and C. N. Lau, Magnetoconductance oscillations and evidence for fractional quantum Hall states in suspended bilayer and trilayer graphene, *Phys. Rev. Lett.* **105**, 246601 (2010). doi: 10.1103/PhysRevLett.105.246601. URL https://link.aps.org/doi/10.1103/PhysRevLett.105.246601.

242. D. R. Luhman, W. Pan, D. C. Tsui, L. N. Pfeiffer, K. W. Baldwin and K. W. West, Observation of a fractional quantum Hall state at $\nu = 1/4$ in a wide GaAs quantum well, *Phys. Rev. Lett.* **101**, 266804 (2008). doi: 10.1103/PhysRevLett.101.266804. URL https://link.aps.org/doi/10.1103/PhysRevLett.101.266804.

243. J. Shabani, T. Gokmen and M. Shayegan, Correlated states of electrons in wide quantum wells at low fillings: The role of charge distribution symmetry, *Phys. Rev. Lett.* **103**, 046805 (2009). doi: 10.1103/PhysRevLett.103.046805. URL https://link.aps.org/doi/10.1103/PhysRevLett.103.046805.

244. W. N. Faugno, A. C. Balram, M. Barkeshli and J. K. Jain, Prediction of a non-Abelian fractional quantum Hall state with f-wave pairing of composite fermions in wide quantum wells, *Phys. Rev. Lett.* **123**, 016802 (2019). doi: 10.1103/PhysRevLett.123.016802. URL https://link.aps.org/doi/10.1103/PhysRevLett.123.016802.

245. J. Shabani, T. Gokmen, Y. T. Chiu and M. Shayegan, Evidence for developing fractional quantum Hall states at even denominator 1/2 and 1/4 fillings in asymmetric wide quantum wells, *Phys. Rev. Lett.* **103**, 256802 (2009). doi: 10.1103/PhysRevLett.103.256802. URL https://link.aps.org/doi/10.1103/PhysRevLett.103.256802.

246. J. Shabani, Y. Liu, M. Shayegan, L. N. Pfeiffer, K. W. West and K. W. Baldwin, Phase diagrams for the stability of the $\nu = \frac{1}{2}$ fractional quantum Hall effect in electron systems confined to symmetric, wide GaAs quantum wells, *Phys. Rev. B* **88**, 245413 (2013). doi: 10.1103/PhysRevB.88.245413. URL https://link.aps.org/doi/10.1103/PhysRevB.88.245413.

247. A. C. Balram, M. Barkeshli and M. S. Rudner, Parton construction of particle-hole-conjugate Read-Rezayi parafermion fractional quantum Hall states and beyond, *Phys. Rev. B* **99**, 241108 (2019). doi: 10.1103/PhysRevB.99.241108 URL https://link.aps.org/doi/10.1103/PhysRevB.99.241108.

Chapter 2

The Half-Full Landau Level

Bertrand I. Halperin

Physics Department, Harvard University,
17 Oxford Street, Cambridge, MA 02138, USA
halperin@physics.harvard.edu

At even-denominator Landau level filling fractions, such as $\nu = 1/2$, the ground state, in most cases, has no energy gap, and there is no quantized plateau in the Hall conductance. Nevertheless, the states exhibit non-trivial low-energy phenomena. Open questions concerning the proper description of these systems have attracted renewed attention during the last few years. Issues at $\nu = 1/2$ include consequences of particle-hole symmetry, which should be present for a spin-aligned system in the limit where one can neglect mixing between Landau levels. Other issues concern questions of anisotropy and geometry, properties at non-zero temperature, and effects of relatively strong disorder. In cases where one does find a gapped even-denominator quantized Hall state, such as $\nu = 5/2$ in GaAs structures, major questions have arisen about the nature of the quantum state, which will be discussed briefly in this chapter. The chapter will also discuss phenomena that can occur in a two-component system near half filling, i.e. when the total filling factor ν_{tot} is close to 1.

Contents

1. Introduction

The fractional quantized Hall effect was first observed in 1982, in a two-dimenisional electron gas confined in a GaAs heterostructure.[1] Over the next few years, many fractional quantized states were observed in GaAs structures, almost exclusively at fractions with odd denominator. In 1987, Willett et al. reported the existence of a quantized state at even denominator filling fractions $\nu = 5/2$ and $\nu = 7/2$,[2] and in 1992, Suen et al.[3] reported the observation of a plateau at $\nu = 1/2$ in a wide quantum well, with (most likely) several occupied subbands. However, no quantized Hall state was observed at $\nu = 1/2$ in a narrow GaAs quantum well or heterostructure.

In the vicinity of $\nu = 1/2$, the Hall conductance was found to vary smoothly, essentially linearly with the Landau level filling fraction, as in a classical Hall conductor, while the longitudinal resistivity varied continuously, with a magnitude depending on the quality of the sample. Nevertheless, after Willett et al.[4] reported anomalous behavior in the propagation of surface acoustic waves near $\nu = 1/2$ in a high-quality GaAs sample, in 1990, it became clear that something non-trivial was happening in this regime.

An explanation for the surface-acoustic-wave anomaly was provided by the theory of Halperin, Lee, and Read (HLR).[5] In this picture, the state at $\nu = 1/2$ was described as a Fermi sea of composite fermions, interacting with a Chern–Simons gauge field, such that the effective magnetic field felt by the fermions is zero, on average. The HLR theory also made numerous other predictions for the low-energy properties of a quantum Hall system near $\nu = 1/2$, and near other even-denominator fractions, which have been verified in experiments, at varying degrees of accuracy.

Despite its successes, the HLR theory left many questions unanswered. As in the conventional Fermi liquid theory of electrons in zero magnetic field, the HLR theory is a low-energy effective theory, which depends on parameters whose values can only be obtained from independent calculations or observations. Moreover, the HLR theory contains infra-red divergences, whose consequences are only partially understood. Other questions concern issues of compatibility with the requirements of particle-hole symmetry, which should be present in the limit where the bare electron mass is taken to zero and mixing between Landau levels can be ignored, assuming that the spins of electrons in the partially full Landau level are completely aligned by the applied magnetic field.[6]

Issues related to particle-hole symmetry have received considerable attention in the last few years. In particular, in 2015, D. T. Son proposed a new description of the half-filled Landau level, employing a Fermi sea of relativistic Dirac fermions, with a manifest particle-hole symmetry.[7,8] While the Son–Dirac theory has several advantages over the HLR theory, it turns out that the two theories make identical predictions for physically observable quantities in many important cases. (HLR also has some advantages over the Son theory, in that one can more readily see how it may be derived from the microscopic Hamiltonian for electrons in a semiconductor.) Nevertheless, there are certain properties for which it is not currently clear how a correct particle-hole-symmetric prediction can emerge in the HLR theory in the case where the microscopic Hamiltonian is particle-hole symmetric, so there remain open questions about whether the two-theories are fundamentally equivalent. These issues will be a major focus of the present chapter. [See, especially, Secs. 2, 3, and 6.] Several alternate approaches to problems of the half-full Landau level and how these approaches may relate to the HLR and Son–Dirac theories, will be discussed in Sec. 4. Other related issues, including possible effects of anisotropy and geometry, strong disorder, and finite temperatures, will be reviewed in Sec. 5.

Questions related to particle-hole symmetry also play a role in understanding the nature of the fractional quantized Hall state that has been observed at filling fraction 5/2. We shall touch only briefly on these issues, in Sec. 7 of the present chapter, as a more detailed discussion may be found in the chapter by Heiblum and Feldman.

Thus far, we have implicitly assumed that the partially filled Landau level under consideration is completely spin polarized, and that there are no low-lying transverse modes or valley degeneracies to worry about, so we can neglect any degrees of freedom for the electrons other than their orbital motion in the x–y plane. It is only under this assumption that the situation at $\nu = 1/2$ or $\nu = 5/2$ can be described as having a half-filled Landau level equivalent to itself under a particle-hole transformation. By contrast, if the electron spins are not completely polarized, then particle hole-symmetry, in the absence of mixing between Landau levels, will only relate a state at filling factor ν to a state at filling $2 - \nu$. In this case, the condition of half-filling actually occurs at filling factor $\nu = 1$.

The picture of electrons with two "internal" components is also relevant to bilayer systems, when the active electrons are fully spin polarized. Introducing a pseudospin index $\tau = \pm 1$ to differentiate the layers, one finds that τ plays a role analogous to the z-component of spin in a single-layer system. Experimentally, the ability to make measurements with separate contacts to the two layers and the ability to vary the distance between layers relative to the magnetic length have enabled the study of a rich variety of phenomena that could not be studied in single layer systems.

For a spin-polarized bilayer, in the absence of Landau level mixing, the condition of an inert filled Landau level is achieved when the total filling ν_{tot} is equal to 2.

Thus, the half-filled condition is achieved at $\nu_{\text{tot}} = 1$, and there should be at least an approximate particle-hole symmetry about that value of ν_{tot}. States of two-component systems near $\nu_{\text{tot}} = 1$ will be discussed Sec. 8 below.

2. $\nu \approx 1/2$: HLR and Son–Dirac Formulations

2.1. *Definition of the problem*

We start by considering the simplest model for a two-dimensional system of spin-polarized interacting electrons in a strong magnetic field, with a Landau level filling fraction ν that is equal to or close to $\nu = 1/2$. We assume that in the absence of the magnetic field, the electrons can be described by a parabolic dispersion with a bare band mass m, so that the Hamiltonian in the field may be written in the form

$$H_0 = \sum_j \frac{|\mathbf{p}_j + \mathbf{A}(\mathbf{r}_j)|^2}{2m} + V_2, \tag{1}$$

where V_2 is a two-body interaction of the form

$$V_2 = \frac{1}{2} \sum_{i \neq j} v_2(\mathbf{r}_i - \mathbf{r}_j). \tag{2}$$

Here $\mathbf{r}_j$ and $\mathbf{p}_j$ are the position and momentum of electron j, and $\mathbf{A}$ is the vector potential due to a uniform magnetic field B in the z-direction. In the case where v_2 is a long-range potential, the Hamiltonian must include interactions between the electrons and a uniform neutralizing background, which can be taken into account by omitting the zero-wave-vector component of v_2. In the presence of impurities, one must add a one-body potential $V_1(\mathbf{r}_j)$ which depends on position; for the present, however, we shall consider a system without impurities, so we take $V_1 = 0$. Except where otherwise stated, we use units where the electron charge is $e = -1$, and $\hbar = c = 1$. In these units, the quantum of conductance e^2/h equals $1/2\pi$, and the quantum of flux is $h/|e| = 2\pi$.

2.2. *Particle-hole symmetry*

For the system described by (1) and (2), in the limit where $m \to 0$ while the strength of the electron-electron interactions is held constant, there will be no mixing between Landau levels. This is because the energies separating Landau levels become infinite, while the energy differences between many-body states within a Landau level remain finite. As stated above, it can then be shown, for the case of pure two-body interactions, that there is an exact particle-hole symmetry, such that for any eigenstate of the Hamiltonian at a filling factor ν, there is a corresponding eigenstate at filling factor $1 - \nu$.[6] (This assumes that we compare systems at the same magnetic field but different electron densities.) Furthermore, the energy difference between a pair of states at filling ν will be identical to the energy separation between the corresponding states at $1 - \nu$. At $\nu = 1/2$, this means that an energy eigenstate must either be equivalent to itself under a particle-hole transformation, or the state must be degenerate. In addition, particle-hole symmetry imposes relations between response functions and transport properties for systems at conjugate filling factors, and it imposes certain restrictions on these quantities at $\nu = 1/2$, as will be discussed further below. In the following discussions, we shall generally abbreviate particle-hole symmetry as PH symmetry.

In actual experiments, since $m \neq 0$, the PH symmetry should not be exact. However, in most cases the effects of Landau level mixing are small, and it should be a good approximation to neglect them. Since numerical calculations are greatly simplified if one can neglect Landau level mixing, almost all calculations have been carried out under this assumption. Moreover, from a theoretical point of view, the limit of $m \to 0$ is important to understand, and any complete theory should be able to describe properly this limit.

In comparing theory with experiment, one should be aware that in experiments the filling factor ν is most frequently varied by changing the magnetic field at fixed electron density, rather than by varying the density. In this case, the data should not be precisely symmetric between states at ν and $1-\nu$, even if Landau level mixing is neglected. Thus, in order to explore questions of PH symmetry, the experimental data should be corrected to account for the change in magnetic field.

It should be noted that if one includes three-body interactions in addition to the two-body interactions, PH symmetry will be absent even if one neglects Landau level mixing. Similarly, one may understand effects of Landau-level mixing by noting that in the case of small but finite m, three-body interactions will be generated if one uses perturbation theory to eliminate the effects of higher Landau levels.

In the presence of impurities, PH symmetry will only be strictly present if, statistically, there is an equivalence between positive and negative impurities. However, PH symmetry may be a good approximation if the contributions of individual impurities are weak, so that the disorder potential is well approximated by Gaussian fluctuations about its mean.

In this chapter, except where otherwise stated, we shall be concerned with properties of a large flat system in the thermodynamic limit, far from any boundaries. PH symmetry will always be strongly broken at the boundaries of a sample. In addition, PH symmetry must be defined with care on a closed curved surface without boundaries, such as a sphere.

For a system where the kinetic energy and two-body-interactions obey circular symmetry, if one can neglect Landau level mixing, then the two-body interaction at a fixed magnetic field is completely characterized by a discrete set of parameters v_l, the Haldane pseudopotentials, which are the matrix elements of V_2 between two-body states with relative angular momentum l. For spin-aligned electrons, only odd l terms are relevant. For a planar sample, the only difference between electrons in different Landau levels is that the values of v_l will be different, for the same bare interaction V_2.

The fact that the Hamiltonian has PH symmetry at half filling, when restricted to a single Landau level, does not necessarily imply that the ground state will preserve that symmetry in the thermodynamic limit. It could happen that the PH symmetry is spontaneously broken, so that one can find two degenerate ground states, related to each other by PH symmetry, but not symmetric in themselves. As will be discussed below, available evidence strongly suggests that there is no broken PH symmetry at half filling of the lowest Landau level for the Hamiltonian (1) with parameters appropriate to a narrow quantum well in GaAs, *i.e.*, at $\nu = 1/2$ or $3/2$. However, numerical calculations have suggested that PH symmetry should be spontaneously broken in the second Landau level at half filling, *i.e.*, $\nu = 5/2$ or $7/2$, where a gapped FQH state is observed experimentally. This will be discussed, briefly, in Sec. 7 below.

2.3. *The HLR hypothesis*

The fermion-Chern–Simons approach employed in HLR began with an exact unitary transformation, a singular gauge transformation, where the many-body electron wave function is multiplied by a phase factor that depends on the positions of all the electrons, such that the transformed Hamiltonian acquires a Chern–Simons gauge field a_μ, with two negative flux quanta attached to every electron. The transformed problem may be expressed in Lagrangian form by the following Lagrangian density:

$$\mathcal{L}_0 = \bar\psi \left(iD_t - \mu + \frac{\mathbf{D} \cdot \mathbf{D}}{2m} \right) \psi - \frac{ada}{8\pi} + \mathcal{L}_{\text{int}} \tag{3}$$

$$ada \equiv \varepsilon^{\mu\nu\lambda} a_\mu \partial_\nu a_\lambda \tag{4}$$

$$D_\mu \equiv \partial_\mu + i(a_\mu - A_\mu), \tag{5}$$

and μ takes on the values $(0, x, y)$. Taking the variation of the Lagrangian with respect to a_0, we obtain the constraint

$$\nabla \times \mathbf{a} = -4\pi \, \bar\psi\psi = -4\pi \, n_{\text{el}}(\mathbf{r}). \tag{6}$$

In these equations, ψ is the Grassmann field for a set of transformed "composite fermions" (CFs), whose density $\bar\psi\psi$ is identical to the electron density $n_{\text{el}}(\mathbf{r})$.

At this stage, we have merely transformed one insoluble problem to another. However, the transformed problem admits a sensible mean-field approximation, whereas the original problem did not. In particular, if the Landau level is half full, so that there is one electron for each quantum of electromagnetic flux, the mean field problem describes a set of non-interacting fermions in zero magnetic field. To go beyond mean-field theory, one must include the effects of fluctuations in the gauge field and fluctuations in the two-body potential. The central hypothesis of HLR is that, in principle, one could obtain the correct properties of the system by starting from the mean field solution, treating the omitted fluctuation terms via perturbation theory. In other words, one could start with the mean field model of non-interacting fermions in zero field, and one could imagine turning on gradually the Coulomb interaction and the interactions via the gauge field, while simultaneously turning on the external magnetic field in such a manner that the effective magnetic field remains zero. HLR assumes that the interacting ground state can be reached from the mean-field solution by turning on the perturbing terms adiabatically, without encountering any phase transition. Among the consequences of this assumption are that the ground state at $\nu = 1/2$ should be compressible, that there should be no energy gap, and that there should be something like a Fermi surface, with a well-defined Fermi wave vector, $k_F = 4\pi n_{\text{el}}$.[5,9-11] This may be contrasted with the gapped quantized Hall states at the Jain fractions $\nu = p/(2np + 1)$, discussed in earlier work using composite fermions and a Chern–Simons gauge field.[12-15] The HLR analysis was, of course, motivated by this earlier work. (See also Ref. 16.)

In analogy with the Landau theory of a Fermi system in the absence of a magnetic field, one may hope that by integrating out high frequency fluctuations of the fermion and gauge fields, one may obtain a renormalized effective theory, with some resemblance to the original theory, which can be used to calculate response functions and other phenomena at low frequencies. The simplest guess for the low energy theory is given by a Lagrangian of the same form as the bare HLR theory, but with a renormalized effective mass m^* for the composite fermions, which one might then treat in the Random Phase Approximation (RPA). (A renormalized mass is necessary, for example, to obtain the correct specific heat at low temperatures, or to obtain the correct energy gaps for FQH states near to $\nu = 1/2$, since the energy scale is actually determined by the strength of the Coulomb interaction, rather than by the bare electron mass m, when $m \to 0$.) The meaning of the RPA, here, is that response functions are calculated by assuming the composite fermions behave as free fermions of mass m^* in response to the effective electromagnetic field, which is the difference of the real applied electromagnetic field and the Chern–Simons fields produced by the induced variations in fermion density and current. Specifically, by taking variations of the action governed by (3), one finds that a non-zero value of the fermion current density $\mathbf{j}$ leads to a Chern–Simons electric field of the form

$$\mathbf{e} = 4\pi \hat{z} \times \mathbf{j}, \tag{7}$$

while variations in the fermion density lead to variations in the Chern–Simons magnetic field $b = \nabla \times \mathbf{a}$ prescribed by (6). In the HLR formulation, the density and currents of the composite fermions are the same as those of the electrons, at least in the long-wavelength limit, even after renormalization. This leads to relations between the electro- magnetic response functions and free-fermion response functions, at the RPA level, which can be summarized as

$$\hat{\rho}(\mathbf{q}, \omega) = \hat{\rho}^{\text{cf}}(\mathbf{q}, \omega) + \hat{\rho}^{\text{cs}}, \tag{8}$$

where $\hat{\rho}(\mathbf{q}, \omega)$ is the electrical resistivity tensor at wave vector $\mathbf{q}$ and frequency ω, while $\hat{\rho}^{\text{cf}}(\mathbf{q}, \omega)$ is the corresponding tensor for free composite fermions, and

$$\hat{\rho}^{\text{cs}} \equiv 4\pi \hat{\epsilon} \tag{9}$$

is the Chern–Simons resistivity, with $\hat{\epsilon}$ being the unit antisymmetric tensor in two-dimensions.

If one simply uses the RPA with a renormalized effective mass to calculate the long-wavelength response at finite frequencies, one will not obtain correct answers. For example, the response function will violate Kohn's theorem, which says that for the system defined by Eq. (1), the density response function at long wavelengths should be dominated by a single pole at the bare cyclotron frequency, $\omega_c = B/m$, whereas the RPA would predict a pole at a frequency $\omega_c^* = B/m^*$. As in conventional Fermi liquid theory, this problem is corrected if one includes effects of a Landau interaction parameter F_1, which describes the interaction energy associated with a uniform displacement of Fermi surface. An approximation which includes the effects of F_1 together with a renormalized m^* has been called the Modified Random Phase Approximation, or MRPA. A further generalization, designed to take into account effects of the Zeeman and cyclotron energy variations in the case of a non-uniform magnetic field, has been denoted the Magnetized Modified Random Phase Approximation, or MMRPA. However in the case of a uniform magnetic field, or in the case where the effective electron g-factor is equal to two, there is no difference between MMRPA and MRPA.

A more complete theory, capable of describing arbitrary distortions of the composite fermion Fermi surface should include interaction parameters F_l describing distortions of the Fermi surface for an arbitrary circular harmonic l. Moreover, in order to describe response functions at non-zero wave vectors, we expect that the RPA must be further modified to include vertex corrections that depend on wave vector and frequency.

2.4. Son–Dirac

In the 2015 article mentioned above,[7] Son proposed a model in which massless Dirac particles interact with a gauge field by a Lagrangian density of the form

$$\mathcal{L}_D = \psi^\dagger (iD_t - \mu - iv_D \, \mathbf{D} \times \sigma)\psi +$$

$$+ \left[\frac{AdA}{8\pi} + \frac{adA}{4\pi} \right] + \mathcal{L}_{\text{int}}, \tag{10}$$

$$D_\mu \equiv \partial_\mu + i \, a_\mu, \tag{11}$$

where ψ is a two-component Grassmann spinor, $\mathbf{A}$ is the external magnetic field, σ are the Pauli spin matrices, and $\mathcal{L}_{\text{int}}$ is a term which represents the two-body interaction V_2. The velocity v_D is an input parameter, like the effective mass m^* in the HLR theory, which must be taken either from experiment or from an independent microscopic calculation.

In the Son–Dirac formulation, the composite fermions see an effective magnetic field $b(\mathbf{r})$ which is related to the electron density and the applied magnetic field in the same way as in HLR:

$$b = \nabla \times \mathbf{a} = 4\pi n_{\text{el}} - \nabla \times \mathbf{A}. \tag{12}$$

However, the electron density and the composite fermion density are not necessarily identical. Rather, the density of Dirac composite fermions is tied to the (local) value of the magnetic field

$$n_{\mathrm{DF}} =: \psi^\dagger \psi := \frac{1}{4\pi} \nabla \times \mathbf{A}, \tag{13}$$

which is equal to the density of electrons in the ground state at $\nu = 1/2$, but may deviate from the electron density more generally. Similarly, the current of the Dirac fermions is related to the local electric field by

$$\mathbf{j}_{\mathrm{DF}} = -\frac{1}{4\pi} \hat{z} \times \mathbf{E}, \tag{14}$$

while the effective electric field felt by the Dirac fermions is given by

$$\mathbf{e}_{\mathrm{DF}} = -\nabla a_0 - \partial_t \mathbf{a} = \hat{z} \times (4\pi \mathbf{j}_{\mathrm{el}}) - \mathbf{E}. \tag{15}$$

The electrical conductivity tensor, for a long-wavelength electric field is then given by

$$\hat{\sigma} = -\hat{\varepsilon} \, \hat{\rho}^{\mathrm{DF}} \, \hat{\varepsilon} + \hat{\sigma}^{\mathrm{CS}}, \tag{16}$$

where $\hat{\rho}^{\mathrm{DF}} = (\hat{\sigma}^{\mathrm{DF}})^{-1}$ is the resistivity tensor of the Dirac fermions, and

$$\hat{\sigma}^{\mathrm{CS}} = -\frac{1}{4\pi} \hat{\varepsilon}. \tag{17}$$

Because there is no Chern–Simons term of the form ada, and because the density of fermions is unchanged if the density of electrons is varied while the magnetic field is held constant, the Son–Dirac theory is explicitly particle-hole symmetric about $\nu = 1/2$. However, the Son–Dirac Lagrangian in its original form, is not sufficient to describe arbitrary deviations from the $\nu = 1/2$ ground state. As in the HLR approach, one must include the equivalent of Landau interaction parameters to describe arbitrary deviations of the shape of the Fermi surface, and one must include vertex corrections and higher-order interactions in order to describe correctly responses at finite wave vectors or nonlinear behavior.

2.5. *Contrasts*

There are several striking differences between the Son–Dirac and HLR formalisms. Most obvious is the use of two-component massless Dirac fermions in one case and non-relativistic spinless fermions in the second. However, because there is a non-zero fermion density in both cases, and we are concerned with low energy properties of the system, neither the filled negative energy states in the Dirac model, nor the parabolic dispersion in the case of HLR, is of obvious importance. The difference between the two-band structures that would seem to have most effect is the presence in the Dirac case, but not in HLR, of a Berry phase of π, as one moves adiabatically around the Fermi surface.

A second apparent difference between the two theories arises from the absence of a Chern–Simons ada term in the Son formulation, and the related result that the density of fermions will be different in the two theories when the filling factor deviates from 1/2. However, this difference may be more a matter of representation than of physics. The density of fermions in a relativistic theory is normally defined by including only the positive energy states, and ignoring the filled sea of negative energy states. For $\nu \neq 1/2$, the effective magnetic field felt by the fermions will be non-zero, and in the Dirac formulation there will be a finite density of occupied states precisely at zero energy. In the Son–Dirac formulation, one counts in the fermion density precisely half of the fermions in zero energy states. If, instead, one were to omit all zero-energy fermions, the density of fermions would then be equal to that of the electrons, just as in the HLR approach.

The last point can be made clearer if one employs a generalization of the Son approach, which enables one to consider Dirac particles with a non-zero mass.[17] Consider the Lagrangian density

$$\mathcal{L}_D = \psi^\dagger (iD_t - \mu - iv_D \, \mathbf{D} \times \sigma - m_D \, \sigma^z) \psi +$$

$$+ \left[\frac{AdA}{8\pi} + \frac{adA}{4\pi} - \frac{ada}{8\pi} \frac{m_D}{|m_D|} \right] + \mathcal{L}_{\text{int}}, \tag{18}$$

$$D_\mu \equiv \partial_\mu + i \, a_\mu, \tag{19}$$

We are interested in a situation in which the Fermi level is inside the band of positive energy fermion states and the lower Dirac band has been integrated out, which is the origin of the $\frac{1}{8\pi} ada \, \text{sign}(m_D)$ term in (18).

The Lagrangian (18) reduces to (10) if one takes the limit $m_D \to 0$. In this limit, the contribution of the ada term in (18) in any application is precisely canceled by the contribution from the Berry curvature, which is completely concentrated at the bottom of the occupied states in the positive energy Dirac band. (By convention, this contribution, which is ill-defined for $m_D = 0$, is omitted in that case.) Thus, in the limit $m_D \to 0$, the Lagrangian (18) may be replaced by the form (10), in which m_D is precisely zero and the ada term is simply omitted; i.e., there is no longer a Chern–Simons term in the action for the gauge field a_μ. In the following discussion, we confine ourselves to the case $m_D{=}0$, except where otherwise specified.

The question we will return to below is to what extent the differences in formulation between Son–Dirac theory and HLR result in differences in predictions for physical observables, and whether in the end the two theories are fundamentally different.

2.6. Infrared divergences

Discussions of the asymptotic low-frequency behavior at $\nu = 1/2$ are complicated by the occurrence of infrared divergences, predicted by both the HLR and Son–Dirac theories. In the case of Coulomb-like electron-electron interactions, which fall off as $1/r$ at large electron separations r, it was predicted in Ref. 5 that there will be a relatively-innocuous logarithmic divergence in the composite-fermion effective mass. Also the decay rate of a composite fermion near the Fermi energy should be smaller, by a logarithmic factor than the energy. Subsequent calculations using a renormalization group concluded that the mass would only diverge as a square-root of the logarithm of the distance from the Fermi surface.[18] In either case, the composite fermion system may be considered a kind of marginal Fermi liquid, interacting with a gauge field at long wavelengths. Since the infrared divergences are expected to be identical in the Son–Dirac and HLR theories, it seems reasonable to ignore the divergence in m^* when comparing the two theories. Alternatively, the infrared divergences can be eliminated entirely if one wishes to consider a model where the interaction v_2 falls off more slowly than $1/r$.

In the case of short range interaction, the infrared divergences are much stronger, and quasiparticles can no longer be precisely defined. Nonetheless, important features of the Fermi surface are expected to survive, such as the existence of some type of singularity in various response functions at wave vector $q = 2k_F$, at $\nu = 1/2$. Moreover, many of the quantities discussed below, are predicted in both Son–Dirac and HLR to have a behavior independent of the value of m^*, and it is expected that these predictions should remain correct even in the case of short-range interactions. Other quantities, such as the compressibility, which for a non-interacting Fermi system would be proportional to the value of the effective mass, are found to remain finite, essentially because the divergence in m^* is canceled by similar divergence in the various Landau parameters F_l.[9–11]

3. Predictions for Behavior Near $\nu = 1/2$

3.1. *Predictions where HLR and Son–Dirac agree*

Nearby quantized states

As mentioned above, there is a large set of phenomena where predictions of the Son–Dirac theory coincide at the RPA level with the previous predictions of HLR. In both cases, at $\nu = 1/2$ the density of composite fermions is equal to the density of electrons, $n_{\rm el} = B/4\pi$, and the mean-field ground state at $\nu = 1/2$ is a sea of composite fermions in zero effective magnetic field, with a Fermi wave vector given by

$$k_F = [4\pi n_{\rm el}]^{1/2}. \tag{20}$$

Slightly away from $\nu = 1/2$, the fermions see an effective magnetic field

$$\Delta B = B - 4\pi n_{\rm el}. \tag{21}$$

In both theories, a fractional quantized Hall state with an energy gap is expected when the density of composite fermions is such as to fill an integer number of Landau levels in the effective magnetic field ΔB. In HLR, as in Jain's composite fermion theory, the density of composite fermions is equal to $n_{\rm el}$, and the gapped state occurs when

$$2\pi n_{\rm el} = p\Delta B, \tag{22}$$

with p a positive or negative integer. The actual filling fraction is then

$$\nu \equiv \frac{2\pi n_{\text{el}}}{B} = \frac{p}{(2p+1)} = \frac{1}{2} - \frac{1}{2(2p+1)}. \tag{23}$$

In the Son–Dirac theory, due to the Berry phase, one predicts that the Fermi energy is in an energy gap when the density of Dirac fermions n_{DF} satisfies

$$2\pi n_{\text{DF}} = p_{\text{DF}} \Delta B, \tag{24}$$

where p_{DF} is a *half* odd integer. However, $n_{\text{DF}} = n_{\text{el}}/2\nu$, which differs from the density of electrons, when $\nu \neq 1/2$. Using simple algebra, one finds

$$\nu = \frac{1}{2} - \frac{1}{4p_{\text{DF}}}. \tag{25}$$

Consequently, the Son–Dirac theory leads to the *same* values of ν for the gapped states as Jain's theory or HLR [Eq. (23)], with the identification $p_{\text{DF}} = p + 1/2$. The allowed electron densities at the gapped states are symmetric about $\nu = 1/2$, if the magnetic field is held fixed.

Energy gaps

In the mean field theory based on HLR, the energy gap at a quantized Hall state of the form (23) is given by

$$E_g = \frac{|\Delta B|}{m^*}, \tag{26}$$

where m^* is the renormalized effective mass at $\nu = 1/2$. If this effective mass and the energy gaps are calculated at fixed magnetic field while the electron density is varied, then the predicted energy gaps will be symmetric about $\nu = 1/2$, as required by PH symmetry.

In Ref. 5, HLR considered the effects of the predicted infrared divergences in m^* on the size of the energy gaps. For the case of Coulomb interactions, where the divergence is only logarithmic, it was argued that (26) could still be used, provided that m^* was evaluated self-consistently at an energy distant from the Fermi surface by an amount of order E_g. In effect, this leads to an asymptotic behavior, for $\nu \to 1/2$, of the form[11]

$$E_g \sim \frac{\pi}{2} \frac{e^2}{\epsilon l_B} \frac{1}{|2p+1|(C + \ln|2p+1|)}, \tag{27}$$

where ϵ is the background dielectric constant, l_B is the magnetic length, and C is a constant which depends on the short range behavior of the interaction. This result is also consistent with PH symmetry. For the case of short range interactions, where one predicts $m^* \propto |\delta E|^{-1/3}$, one is led to an asymptotic prediction

$$E_g \propto |\Delta B|^{3/2}, \tag{28}$$

but there is no precise prediction for the prefactors.

As remarked earlier, however, Nayak and Wilczek,[18] have argued, using a renormalization group analysis applied to the HLR Lagrangian, that the effective mass divergences should have a somewhat different form than those predicted by HLR in Ref. 5. In the case of Coulomb interactions, they predict that effective mass should diverge as the square-root of the logarithm of the distance from the Fermi surface, which would imply that the energy gaps of the Jain states should vary asymptotically as

$$E_g \propto |\Delta B| / [\ln |\Delta B|]^{1/2}. \tag{29}$$

Although there are no published analyses of the effects of infrared divergences using the Son–Dirac theory, the predictions should be identical to those based on the HLR Lagrangian for all of these quantities.

Morf, d'Ambrumenil, and Das Sarma have compared the predictions of (27) with results for energy gaps at fractions 1/3, 2/5, 3/7, and 4/9 that they obtained by extrapolation of exact diagonalizations on a sphere.[19] They found that (27) fit their data points better than the simple formula $E_g = C/|\Delta B|$, though both formulas contain a single fitting parameter. However, it is impossible to draw firm conclusions from such small systems, and the data would also be compatible with an asymptotic dependence of the form (29).

Conductivity at non-zero wave vector

An important result of HLR was the prediction that the longitudinal wave-vector-dependent electrical conductivity at $\nu = 1/2$, in the limit of vanishing frequency and small finite wave vector, should be given, in the absence of impurities, by

$$\sigma_{xx}(q) = \frac{q}{8\pi k_F}, \tag{30}$$

where we have taken the wave vector $\mathbf{q}$ to lie in the x-direction. Note that this result is independent of m^*, and it is believed that the result should be unaffected by the infrared divergences in the energy spectrum of composite fermions.

In the presence of impurities, (30) should hold for q larger than the inverse of the composite fermion mean free path l_{cf}, which we assume to be long compared to the Fermi wavelength. For $q l_{cf} < 1$, σ_{xx} should become independent of q and should reduce to the macroscopic longitudinal conductivity, $\propto l_{cf}^{-1}$. The wave-vector dependence of the conductivity σ_{xx} is responsible for the shift in the propagation velocity of a surface acoustic wave observed by Willett *et al.* in 1990,[4] which was mentioned in the Introduction as evidence that something peculiar was happening at $\nu = 1/2$.

Within HLR, the result (30) is a consequence of the fact that the transverse wave-vector-dependent conductivity for non-interacting composite fermions, $\sigma_{yy}^{cf}(q)$, actually diverges as q^{-1}, in the absence of impurity scattering. To obtain the electrical conductivity, according to Eq. (8), one should first calculate the CF resistivity tensor, by inverting the CF conductivity tensor. At $\nu = 1/2$, in the absence of impurities, the CF resistivity tensor is diagonal, so $\rho_{yy}^{cf}(q)$ is just the inverse of $\sigma_{yy}^{cf}(q)$. In $\hat{\rho}$, however, following Eq. (8), the off-diagonal part is much larger than the diagonal, and one obtains the result (30) for $\sigma_{xx}(q)$ on inverting $\hat{\rho}$.

The prediction (30) is also obtained in the Son–Dirac theory, using Eq. (16). More generally, the anomalous wave-vector-dependence of the electrical conductivity reflects the fact that composite fermions at $\nu = 1/2$ can travel in a straight line over distances that are very large compared to the microscopic magnetic length.

What happens to these results as one deviates slightly from $\nu = 1/2$? As the composite fermions now see an effective magnetic field ΔB, we should expect them to move in circles, with an effective cyclotron radius

$$R_c^* = k_F/|\Delta B|. \tag{31}$$

The non-local conductivity of the composite fermions will clearly be cut off for distances larger than $2R_c^*$ and $\sigma_{xx}(q)$ will accordingly deviate from its $\nu = 1/2$ value when $|\Delta B|$ is large enough that $qR_c^* < 1$. However, the dependence on $|\Delta B|$ is not generally monotonic. It was predicted in HLR that for a fixed value of q, there should be *oscillations* as a function of $|\Delta B|$ in the region $qR_c^* > 1$, provided that l_{cf} is sufficiently large. Observation of the consequent magneto-oscillations in surface acoustic wave propagation, by Willett *et al.* in 1993,[20] provided an important confirmation of the HLR theory.

The HLR prediction for magneto-oscillations in $\sigma_{xx}(q)$ was based on a semiclassical analysis of the dc conductivity tensor for composite fermions in the presence of the effective magnetic field ΔB and weak impurity scattering. Roughly speaking, the features in $\sigma_{xx}(q)$ could be understood as arising from a commensurability condition between the effective cyclotron diameter $2R_c^*$ and the wave length $2\pi/q$. More precisely, the analysis predicted that for small q and very small disorder, there would be peaks in $\sigma_{xx}(q)$ as a function of ΔB, which would occur when

$$qR_c^* = z_n, \tag{32}$$

where $z_n \approx \pi(n + \frac{1}{4})$ is the nth zero of the J_1 Bessel function. Equivalently, using (31), the peaks should occur at wave vectors given by

$$q_n = z_n|\Delta B|/k_F. \tag{33}$$

Related commensurability oscillations have been predicted for other quantities near $\nu = 1/2$. We shall discuss this subject further below, since possible deviations from the predictions of (33) have played a role in the comparisons between HLR and Son–Dirac.

Hall conductance at $\nu = 1/2$

If mixing between Landau levels is neglected, PH symmetry requires that the Hall conductivity at $\nu = 1/2$, in response to a spatially uniform electric field, should be precisely given by

$$\sigma_{yx} = -\sigma_{xy} = \frac{1}{4\pi}, \tag{34}$$

regardless of the applied frequency. This should be true even in the presence of impurities, provided that the disorder potential V_{imp} is PH symmetric in a statistical sense. (This condition means that if one chooses the uniform background potential such that the average $\langle V_{\text{imp}} \rangle = 0$, then all odd moments of the disorder potential must vanish.)

In 1997, Kivelson *et al.* pointed to a possible difficulty in reconciling the HLR theory with this requirement, in the context of the dc conductivity.[21] Their argument went as follows. Within the HLR approach, the electron resistivity tensor and the resistivity tensor of the composite fermions are related according to Eqs. (8) and (9) above. This means that in order to obtain the PH symmetric result (34) for σ_{yx}, when $\rho_{xx} \neq 0$, it is then necessary that $\sigma_{xy}^{\text{cf}} = -1/4\pi$. However, it was argued that σ_{yx}^{cf} is necessarily equal to zero at $\nu = 1/2$. This seemed evident because, in the absence of impurities, the composite fermions see an average effective magnetic field equal to zero, which is effectively invariant under time reversal. The presence of impurities leads to non-uniformities in the electron density, which lead to local fluctuations in the effective magnetic field $b(\mathbf{r})$ that turn out to be the dominant source of scattering of composite fermions, under conditions where the correlation length for the impurity potential is large compared to the Fermi wave length. If the impurity potential is statistically PH symmetric, then there will be equal probability to have a positive or negative value of b at any point, so that the resulting perturbation to the composite fermions should again be invariant under time reversal in a statistical sense, which would imply that $\sigma_{yx}^{\text{cf}} = 0$.

However, there is a flaw in this argument. As was shown by Wang *et al.* in 2017,[22] when the HLR theory is properly evaluated, it actually gives the correct PH symmetric result for the dc conductivity in this problem, at least to order l_{cf}^{-2}. The error in the earlier reasoning is that while the fluctuations in the electrostatic potential are highly screened at long wavelengths, and their contribution to the CF scattering rate is small compared to the fluctuations in $b(\mathbf{r})$, they must still be taken into account when computing σ_{yx}^{cf}. The residual potential fluctuations are correlated with fluctuations in $b(\mathbf{r})$ in a manner which breaks the statistical time-reversal symmetry present in the b fluctuations alone. For example, the local density of electrons, and therefore of CFs, will be slightly larger in a region with $b > 0$ and smaller in a region with $b < 0$. When these correlated fluctuations are taken into account, it turns out that one recovers precisely the result $\sigma_{yx}^{\text{cf}} = -1/4\pi$ required by PH symmetry. Note that σ_{yx}^{cf} is actually a small correction to the leading diagonal piece, σ_{xx}^{cf}, which diverges as l_{cf} in the limit of weak disorder.

Weiss oscillations

An effect closely related to the above-mentioned oscillations in $\sigma_{xx}(q)$, is the oscillatory behavior of the long-wavelength dc magnetoresistance ("Weiss oscillations") in the presence of an external periodic potential with a period $2\pi/q$.[23–27] Under appropriate conditions, if the wave vector q is fixed and the electron concentration is varied at fixed magnetic field, theory predicts that there should be a series of minima in the longitudinal resistance which should occur, at least approximately, when the conditions satisfy (33).

If one wishes to treat this equation as a precise prediction, however, one must specify precisely the definition of the Fermi wave vector k_F. If one defines k_F as the square-root of 4π times the density of composite fermions, which is equal to the density of electrons in HLR but is determined by the density of flux quanta in Son–Dirac, one would find that the wave vectors q given by (33) would agree in the two theories at first order in $|\Delta B|$ but would disagree at order $|\Delta B|^2$. Furthermore, the HLR result would violate PH symmetry at this order.[27]

However, a more careful evaluation of the HLR theory found the same result for the positions of the resistance minima as the Son–Dirac theory, at least to order $|\Delta B|^2$, so this discrepancy is eliminated.[22,28] The analysis in both cases was carried out at the RPA level, assuming a temperature small compared to the effective Fermi energy $E_F^* = k_F^2/m^*$, but larger than $|\Delta B|/m^*$. Under these conditions, quantum effects due to discreteness of the CF energy spectrum, such as the formation of quantized fractional Hall states, can be ignored. However, commensurability effects related to the diameter of the semiclassical CF cyclotron orbits can persist to higher temperatures. The calculations assume that there is a small amount of background disorder, with a large but finite CF mean-free-path, and the results are only precise in this limit.

We note that, similar to the case of $\sigma_{xx}(q)$, in order to get the correct answer here using the HLR formulation, it is necessary to take into account the variations in the screened electrostatic potential as well as in the effective magnetic field b resulting from the external periodic perturbation. The contribution of the electrostatic potential, though small, leads to a phase shift in the oscillations, which is just sufficient to bring the calculation based on HLR into agreement with that based on Son–Dirac.

Weiss oscillations have been of particular interest because experimental measurements have been able to locate the induced minima in magnetoresistance with great accuracy.[27] (See the discussion in the chapter by M. Shayegan.) Experiments have demonstrated PH symmetry for these positions with an accuracy that is able to rule out an asymmetry of the size that would have been expected from a naïve use of HLR. Of course, this is not a test of HLR versus Son–Dirac, since both theories actually predict the same results for the quantities in question. Interestingly, the measured positions do deviate from the predicted positions by an amount which is PH symmetric but is of a similar magnitude to the difference between Son–Dirac and naive HLR. The sign and magnitude of the observed deviation is such that the data can be well fit by the form (33) with a choice of k_F dictated by the density of electrons for $\nu > 1/2$ and by the density of holes for $\nu < 1/2$. This means that if one wishes to fit the data to (33), one should choose k_F to satisfy $k_F l_B = [\min(2\nu, 2 - 2\nu)]^{1/2}$, rather than the value $k_F = l_B^{-1}$ predicted by the RPA-type calculations of Refs. 22 and 28. The reason for this observed deviation is not known. It may be a result of corrections due to factors such as to the finite density of impurities, but it could be that there would be corrections of the observed magnitude even in the limit of small impurity density. Recent work by Mitra and Mulligan has suggested that the discrepancy may be an effect of gauge fluctuations.[29]

Minima in the magneto-exciton spectrum and maxima in the static structure factor

Another quantity that has been predicted to show commensurability oscillations near $\nu = 1/2$ is the spectrum $\omega(q)$ of neutral excitations in a quantized Hall state of the form $\nu = p/(2p+1)$, for large values of $|p|$. The lowest frequency branch is predicted to have a series of minima, known as magnetoroton minima, which occur at wave vectors given by (33) in the limit of large $|p|$ or small $|\Delta B|$.[30] As in the case of the Weiss oscillations, a naive application of the HLR theory, in which k_F is determined by the density of electrons, would predict a PH asymmetry at order $|\Delta B|^2$. However, a correct evaluation of the formulas in Ref. 30 gives results which are equivalent to setting $k_F = l_B^{-1}$ in (33). Thus one obtains a PH symmetric result, in agreement with the Son–Dirac approach.[22]

Associated with the minima of the magnetoroton spectrum are maxima in the projected static structure factor $\tilde{S}(q)$. Specifically, if $\tilde{\rho}(\mathbf{q})$ is the density operator at wave vector $\mathbf{q}$, projected into the lowest Landau level, then $\tilde{S}$ may be defined as

$$\tilde{S}(q) = \langle \tilde{\rho}(\mathbf{q}) \tilde{\rho}(-\mathbf{q}) \rangle \approx -i \int \frac{d\omega}{2\pi} \chi(\omega + i0^+ \mathrm{sgn}(\omega), q), \qquad (35)$$

where χ is the electron-density response function, and the integral is restricted to frequencies smaller than the bare cyclotron frequency ω_c. The last equality becomes exact in the limit $\omega_c \to \infty$, with interactions held fixed, and the integral is then taken over all finite frequencies.

Nguyen and Son[31] have obtained an analytic formula for $\tilde{S}(q)$ for the Jain states in the limit of large $|p|$, which may be written as

$$\tilde{S}(q) = \frac{1 + |2p+1|}{4}(q l_B)^4 \frac{J_2(z)}{z J_1(z)}, \qquad (36)$$

where $z = |2p+1| q l_B$. This indeed predicts sharp maxima at the positions given by (33), with $k_F = l_B^{-1}$. The quantity $|2p+1|$ is symmetric about $\nu = 1/2$, as it is the denominator of the fraction ν for the Jain state.

Balram *et al.*[32] have calculated the electron pair-correlation functions for a series of Jain states close to $\nu = 1/2$, using CF trial wave functions, for up to 200 particles in both spherical and torus geometries. They extract an effective value of k_F for each filling factor by fitting the results to a functional form with oscillations controlled by k_F, and they plot the results for $k_F l_B$ as a function of $(\nu - 1/2)$. The results exhibit PH symmetry to a very good approximation, and they extrapolate to a value close to the required value $k_F l_B = 1$ at $\nu = 1/2$. On either side of

$\nu = 1/2$, however, the extracted values are close to the values one would obtain by defining k_F by the density of minority carriers, $k_F l_B = [\min(2\nu, 2 - 2\nu)]^{1/2}$, deviating from the predicted value $k_F l_B = 1$ in the same sense as was found in the Weiss oscillation experiments. On the other hand, Nguyen and Son argue that the result (36) should be correct even at the next to leading order in $1/|p|$, at least in the case of short-range electron-electron interaction, which would seem to prevent deviations of this form in that case. It is not known what is the significance of the difference between this prediction and the results of Balram *et al.*

Because the static structure factor and the magnetoroton spectrum can be defined in the absence of impurities, they are more convenient to investigate theoretically than the Weiss oscillations. The exciton modes are predicted to be undamped in the vicinity of the magnetoroton minima, so the positions of these minima can also be precisely defined in principle. However, to the best of our knowledge there are no experiments that have been able to measure either the excitation spectrum or the electron structure factor at small wave vectors close to $\nu = 1/2$ with a precision that would be relevant here.

3.2. *Areas where HLR and Son–Dirac disagree at the RPA level*

Although the positions of maxima in $\tilde{S}(q)$ and minima in the magnetoroton spectra at the Jain states are identical in the HLR and Son–Dirac, at least to order $|\Delta B|^2$, there are deviations at this order between the two theories in their predictions for the frequency spectrum away from the minima, and for $\tilde{S}(q)$ away from its maxima. An important example is in the behavior of $\tilde{S}(q)$ in the limit $q \to 0$.

It is known that $\tilde{S}(q)$ must vanish proportional to q^4, for $q \to 0$, when the system is in a gapped quantized Hall state. Furthermore, we may write

$$\tilde{S}(q) = \tilde{s}_4 \, q^4 l_B^4 + \mathcal{O}(q^6 l_B^6), \tag{37}$$

where $\tilde{s}_4$ is a constant.

Following the notation of Ref. 33, we introduce a parameter $N > 0$ which is equal to our parameter p for $p > 0$, *i.e.*, for $\nu < 1/2$, but $N = |p| - 1$, for $p < 0$ or $\nu > 1/2$. Thus, we may write, in the two cases,

$$\nu = \frac{1}{2} \mp \frac{1}{2(2N + 1)}. \tag{38}$$

Note that $(2N+1) = |2p+1|$, and the Jain states are obtained when N is a positive integer. As follows from Eq. (36), the Son–Dirac formalism is consistent with (37), with the prediction

$$\tilde{s}_4 = \frac{2N + 1}{32}, \tag{39}$$

for both $\nu > 1/2$ and $\nu < 1/2$. This is consistent with particle-symmetry, and in fact coincides with exact results, obtained by other means, for electrons confined to the lowest Landau level. However, as was shown in Ref. 33, if $\tilde{S}(q)$ is evaluated at a Jain state in the HLR formalism at the MRPA level, with the F_1 parameter chosen so that the Kohn mode is at infinite frequency, one obtains the results

$$\tilde{s}_4 = \frac{1}{8} \frac{N^2}{2N+1}, \quad (\nu < 1/2), \tag{40}$$

$$\tilde{s}_4 = \frac{1}{8} \frac{(N+1)^2}{2N+1}, \quad (\nu > 1/2). \tag{41}$$

Although these results agree with (39) to leading order in N, and are PH symmetric to that order, they deviate from the correct results and violate PH symmetry at the next order.

Another property where HLR at the RPA level gives an incorrect answer for electrons confined to the lowest Landau level, is for the q^2 correction of the wave-vector-dependent Hall conductivity at $\nu = 1/2$.[34] Specifically, we define $\sigma_H(q) \equiv -j_x(\mathbf{q})/E_y(\mathbf{q})$, where $\mathbf{j}(\mathbf{q})$, is the electrical current at wave vector $\mathbf{q}$, induced by an electric field in the y-direction, $E_y(\mathbf{q})$, computed in the limit $q \to 0$, $\omega \to 0$, $v_F q/\omega \to 0$. (As a result of Onsager symmetry, this quantity will be independent of the direction of $\mathbf{q}$ in a system with overall rotational symmetry.) The exact result for this quantity, in the limit $m \to 0$, is given by

$$\sigma_H(q) = \frac{1}{4\pi} \left(1 - \frac{q^2 l_B^2}{4} \right) + ..., \tag{42}$$

where the omitted terms vanish faster than q^2. However, the term proportional to q^2 is absent in an analysis based on the HLR theory. On the other hand, the HLR formulation does lead to a correct result for $\sigma_H(q)$ in the gapped Jain states. Following the parametrization (38), one finds

$$\sigma_H(q) = \frac{1}{2\pi} \left(\nu \mp q^2 l_B^2 \frac{N^2}{2N+1} \right). \tag{43}$$

The deviations from $1/4\pi$ are antisymmetric about $\nu = 1/2$, as required.

Geraedts *et al.*[35] have given us an important example where the Son–Dirac theory gives a highly non-trivial prediction that is not evident in the HLR approach, but which, nevertheless, does not seem to be incompatible with HLR. The authors introduce an operator $P(\mathbf{r})$, which is proportional to $n_{el}(\mathbf{r})\nabla^2 n_{el}(\mathbf{r})$ projected to the lowest Landau level, and they study the correlation function for the Fourier transform, $\langle P_{-\mathbf{q}} P_{\mathbf{q}} \rangle$, for q close to $2k_F$. According to the Dirac theory, this correlation function should have no observable singularity at $q = 2k_F$, because $P(\mathbf{r})$ is even under PH inversion, and fluctuations in such quantities should not give rise to backscattering across the Fermi surface at $q = 2k_F$. Geraedts *et al.* studied this correlation function numerically, for electrons confined to the lowest Landau level at half filling, using density-matrix renormalization group (DMRG) methods, and found the $2k_F$ singularity to be missing, as predicted. At the same time, they do observe a singularity at $q = 2k_F$ in the density correlation function $\langle n_{-\mathbf{q}}^{el} n_{\mathbf{q}}^{el} \rangle$, which is consistent with the Son–Dirac theory, because the electron density operator is not even under PH inversion.

There does not seem to be any obvious reason in HLR theory why $\langle P_{-\mathbf{q}} P_{\mathbf{q}} \rangle$ should be immune from a singularity at $q = 2k_F$, even if one imposes the requirement of particle-hole symmetry. In order to actually calculate this response function in the HLR theory, however, one would have to know the correct form of the renormalized vertex that couples $P_{\mathbf{q}}$ to the composite fermions at $q = 2k_F$. In general, the operator $P_{\mathbf{q}}$ should have a portion that couples to the CF density operator at $2k_F$ and a portion that couples to multi-particle excitations. Although correlations of the CF density operator will have a $2k_F$ singularity in the HLR formalism, if it happens that coupling of P to the CF density operator vanishes at $q = 2k_F$ in the limit of $m \to 0$, there will be no $2k_F$ singularity in $\langle P_{-\mathbf{q}} P_{\mathbf{q}} \rangle$ in this case. However, we are not aware of any *a priori* argument that this should happen.

4. Other Approaches to $\nu = 1/2$

There are, of course, many theoretical approaches to the behavior of electrons in a partially filled Landau level other than the HLR and Son–Dirac descriptions. Although in many cases these approaches are not designed to describe precisely the asymptotic low-energy behavior in the limit of $\nu \to 1/2$, they may nonetheless have very good numerical accuracy and they can often help to understand the physics in this limit. We discuss here several of these alternate approaches.

4.1. *"Hamiltonian" formulation and description as a dipolar Fermi liquid*

As composite fermions in the Son–Dirac description do not couple directly to the electromagnetic gauge field $\mathbf{A}(\mathbf{r}, t)$, it is possible to interpret them as neutral objects. Indeed, Son's construction was partly inspired by works in the earlier literature proposing that the actual low-energy quasiparticles, obtained from the bare fermions of the HLR theory after "screening" by relaxation of the high-frequency plasma modes occurring above the bare cyclotron of the electrons, are electrically neutral at $\nu = 1/2$ but carry an electric dipole moment $\mathbf{d}$ determined by the canonical momentum $\mathbf{p}$ of the quasiparticle.[36–41] Specifically, the relation between these quantities is

$$\mathbf{d} = l_B^2 \, \hat{z} \times \mathbf{p} \qquad (44)$$

Thus, the neutral CFs of Son–Dirac theory are, in some sense, closer to the true low-energy quasiparticles than the charged CFs of the HLR picture.

In one notable approach, Murthy and Shankar[38] obtained neutral quasiparticles starting from the unrenormalized HLR Hamiltonian, after a unitary transformation in which the field operators for CFs and Chern–Simons vector potential were transformed to new operators ψ_{MS} and $\mathbf{a}_{\mathrm{MS}}$, which were chosen to have a number of desirable properties. For example, the fermion operators ψ_{MS} are decoupled from the gauge fields $\mathbf{a}_{\mathrm{MS}}$ in the transformed Hamiltonian in the long-wavelength limit.

The gauge fields describe the high-energy magnetoplasma modes, while the transformed fermions, which are important for the low frequency behavior, carry the desired electric dipole moment at $\nu = 1/2$. (A separate benefit of this approach was that by treating the transformed Hamiltonian in a suitable approximation, Murthy and Shankar were able to obtain estimates of the quasiparticle effective mass that were tied to the electron-electron interaction and remain finite in the limit $m \to 0$.) In more recent work, Murthy and Shankar have generalized their Hamiltonian approach to a PH-symmetric description in terms of Dirac fermions whose density is determined by the magnetic field, as in Son's formulation.[42]

As noted by Murthy and Shankar, their formulation of the $\nu = 1/2$ problem has strong analogies to the Bohm–Pines description of a three-dimensional Fermi gas with long-range $(1/r)$ Coulomb interactions.[43] There, too, one employs a unitary transformation to separate low-energy fermions from a high-energy plasma mode. By contrast, the original HLR formulation is more like the Landau–Silin approach to the three-dimensional Coulomb gas, where quasiparticles are renormalized by the short-range portions of the Coulomb interaction, but they retain the original electron charge and still interact with the long-range part of the Coulomb interaction.

In order to properly describe the low energy properties of the electron system at $\nu = 1/2$, however, the Fermi liquid formed by the dipolar quasiparticles must have some peculiar properties. In particular, the $l = 1$ Fermi liquid parameter F_1 must have a critical value such that a uniform displacement of the Fermi surface costs no energy.[44] In fact, the dipolar Fermi liquid will have zero-energy modes even at finite wave vectors, similar to gauge degrees of freedom, which are not observable in any physical measurement. The relation between the peculiar Fermi liquid theory of dipolar quasiparticles and HLR theory was explored in some detail in Ref. 45.

The close relation between Son–Dirac CFs and the actual low energy excitations is less clear away from $\nu = 1/2$. While the Son–Dirac CFs are presumably still neutral, the low energy excitations in a quantized Hall state of form $\nu = p/(2p+1)$ are not neutral but have a charge $1/(2p+1)$. Furthermore, the excitations are no longer fermions but obey fractional statistics, only approaching Fermi statistics in the limit $\nu \to 1/2$.[46]

The effects of F_1 in the dipolar Fermi liquid are quite similar to the effects of the coupling between the CFs and the gauge potential a_μ in the Son–Dirac formulation at $\nu = 1/2$. However, the dipolar Fermi liquid picture developed in the 1990s did not include the Berry phase characteristics featured in Son's formulation.

4.2. Trial wave functions and other microscopic theories

The focus of the discussion in this chapter has, thus far, been on effective theories designed to elucidate the low-energy properties of a quantum Hall system at, or very near to, $\nu = 1/2$. In general, these theories depend on parameters, such as the effective mass or Fermi velocity, which can only be related to parameters of the original Hamiltonian by other means. Methods used to estimate these parameters starting from a microscopic Hamiltonian include fits to results from exact diagonalization of finite sized systems, as well as a variety of approximate calculations.

We have already mentioned the Hamiltonian approach of Ref. 38 as a way to obtain approximate values for energies. The most accurate approaches, however, have made use of "variational" trial wave functions, for which one can numerically calculate expectation values of the Hamiltonian. Originally introduced by Laughlin to describe FQH states of the form $\nu = 1/m$, trial wave functions were the basis for Jain's composite fermion theory, which gave a natural description for FQH states of the form $p/(2p+1)$, and which were the inspiration for the non-relativistic fermion–Chern–Simons theory discussed above. (See the chapter by Jain for more detailed descriptions of these wave functions.) For the case of $\nu = 1/2$, the Jain trial wave function takes the form

$$\Psi = P_{\text{LLL}} \Psi_{\text{FS}} \prod_{j<k} (z_j - z_k)^2 \prod_l e^{-|z_l|^2/4l_B^2}, \tag{45}$$

where P_{LLL} is a projection operator onto the lowest Landau level, Ψ_{FS} is the wave function for a Fermi sea of non-interacting electrons in zero magnetic field, and $z_j = x_j - iy_j$ is the position of electron j in complex notation. Although trial wave functions such as Jain's are entirely within a single Landau level, they are not generally the exact ground state of a Hamiltonian with pure two-body interactions, so they need not obey precisely the requirements of PH symmetry. For example, the two-point correlation function for fluctuations in the electron density obtained from a CF trial wave function at a filling fraction ν will not be precisely the same as that obtained for filling fraction $1 - \nu$ obtained with the same CF prescription. Nevertheless, numerical calculations have found that properly chosen trial wave functions give results which are very nearly PH symmetric, just as they give remarkably good agreement with ground state energies obtained by exact diagonalization of finite systems.[32,47] (See the chapter by Jain for more details.)

Several recent papers have used model wave functions to look for a Berry phase accumulated by the many-body wave function on a torus when a single CF is moved around a closed path on the surface or interior of the CF Fermi sea at $\nu = 1/2$.[47,48] Using a highly efficient scheme for evaluating properties of a model wave functions based on a lattice Monte Carlo technique introduced by Haldane, they have been able to study systems with up to 69 electrons. They find that the many body wave function accumulates a Berry phase of π when a CF is moved around the Fermi surface, analogous to the Berry phase that would be accumulated from moving a non-interacting Dirac fermion around the Fermi surface in the Son–Dirac scheme. A π Berry phase is also obtained when the CF is moved along a closed path inside the Fermi sea which encloses the origin but does not come too close to it. The model wave functions employed, however, are based on the non-relativistic CF formulation of Jain and HLR. Thus, it would not seem that this should be interpreted as an argument favoring Son–Dirac over the HLR formulation of $\nu = 1/2$, or that there is necessarily a fundamental difference between the two theories.

It may be worth noting that a recent paper by Pu *et al.* has studied how the accumulated Berry phase varies when mixing between Landau levels is included.[49] They find that the Berry phase evolves rapidly from the original value of π to a value close to zero, under a relatively small amount of Landau level mixing.

4.3. *Interpretation of Son–Dirac in terms of vortices*

The indirect manner in which the electromagnetic gauge field A_μ is coupled to the composite fermions in Son's formulation (10) has led to the interpretation that the fermions in Son–Dirac should be interpreted to represent "vortices" in the electron gas, rather than the electrons themselves, and that the Son–Dirac theory is a low-energy "dual" description to the original description in terms of electrons.[7,50,51] Wang and Senthil[50] explain this point of view as follows. If we consider the trial wave function (45) for $\nu = 1/2$, then the factor $\prod(z_j - z_k)^2$ may be interpreted as producing a 4π vortex about each original electron. However, this factor will also lead to a suppression of the background particle density in the vicinity of any given electron, such that precisely one electron will be missing from the background at filling factor $1/2$. Therefore, adding in the charge of the chosen electron, the low energy composite fermion will be locally neutral, but it still retains the vorticity.

The Jain trial wave function for a state of the form $\nu = p/(2p+1)$, with $p > 0$, has a form similar to (45), but with Ψ_{FS} replaced by the wave function for an integer quantized Hall state with p filled Landau levels. After projection to the lowest Landau level, this leads to additional zeroes in the wave function, such that the total number of 4π vortices is no longer equal to the number of electrons but is equal to one-half the number of flux quanta. As the number of fermions in the Son–Dirac formulation is equal to one-half the number of magnetic flux quanta, it is therefore equal to the number of vortices. Thus, it is natural to identify the underlying fermions in Son's formalism with vortices.

As originally noted by Read,[36] the electric dipole moment of the CFs at $\nu = 1/2$ can be understood as arising from a spatial separation of the electron and its screening vortex charge. Since the direction of the dipole changes as a CF is moved around the Fermi surface, it may not be surprising that a Berry phase will be associated with this process. At other even denominator fractions of the form $\nu = 1/2m$, where m is an integer, where the trial wave function has $2m$ zeroes for each electron, one can again argue that the low energy CFs are neutral dipoles, arising from a spatial separation of the electron and the screening vortex charge. Wang and Senthil[50] argued that the Berry phase at the Fermi surface should have the value of π/m at $\nu = 1/2m$. This coincides with the Son–Dirac picture at $\nu = 1/2$, but differs for fractions with $m > 1$. Since PH symmetry is no longer present at fractions with $m > 1$, it is not surprising that Son's original model is not applicable to those cases. However, a Berry phase different from π can be represented within the Son formalism if one replaces the massless Dirac fermions in Eq. (10) by Dirac fermions with an appropriate mass. Moreover, if, as in (10), there is no Chern–Simons term in the action for the gauge field a_μ, then a Berry phase of π/m is *necessary*, in order to satisfy the requirement that states corresponding to integer fillings of the effective Landau levels should correspond to the electron filling factors $\nu = p/(2mp+1)$, as obtained in the Jain CF construction.

The transformation of a problem involving electrons into a problem involving vortices has a related precedent in the case of Bose systems. It was proposed long ago, for example, that the phase transition between the superfluid and normal state of a Bose liquid at finite temperatures in three-dimensions, or at zero temperature in (2+1) dimensions, could be interpreted in terms of a proliferation of vortex lines, and that the vortex lines in turn could be interpreted as world lines for a system of bosons interacting with a non-compact U(1) gauge field.[52,53] The transformation from particles to vortex-like variables has even earlier precedents in the case of Ising systems, and is known as a duality transformation. In the last few years, there have been many theoretical developments involving duality transitions in Fermi systems, inspired in large measure by the successes of the Son–Dirac theory. For example, duality transformations have been used to establish connections between the problem of a half-full Landau level and states that could be realized on the surface of a three-dimensional topological insulator.[50,54–58] For further examples, see the discussions of a "network of dualities" in Refs 51, 59 and 60.

5. Other Issues

5.1. *Issues of anisotropy and geometry*

Thus far, we have focused on systems in which the microscopic Hamiltonian is effectively invariant under rotations, and we have assumed that this isotropy is maintained in the final ground state. However, recent experiments have been able to measure distortions of the CF Fermi surface in systems where the electronic Hamiltonian is significantly anisotropic, and have been able to compare anisotropies of the CF Fermi surface with those of the electron Fermi surface, measured or calculated, for zero perpendicular magnetic field. The reader is referred to the chapter by M. Shayegan for a discussion of these experiments and references to theoretical efforts to explain them.

A second issue concerns the possibility for spontaneous development of uniaxial anisotropy, in systems where the microscopic Hamiltonian has little or no anisotropy. Theoretical work carried out in the 1970s, before observations of either the integer or fractional quantized Hall effect, noted that according to the Hartree–Fock approximation, a partially filled Landau level would be unstable to the formation of charge density waves.[61] Moreover at one-half filling, the likely ground state was predicted to be a single charge density wave, or stripe pattern, which would break both translational and orientational symmetry, and would lead to a large anisotropy in the electrical properties. The subsequent discovery of fractional quantized Hall effects showed that the actual ground states, at least in the middle of the lowest Landau level, cannot be properly described by a Hartree–Fock approximation, and the possibility of charge-density waves was therefore largely ignored for more than a decade. However, in 1996, Koulakov *et al.* pointed out that in higher Landau levels, the correlations responsible for formation of fractional quantized Hall states become much less important, and Hartree–Fock should again become reliable.[62,63] In 1998, experiments by the Eisenstein group at Caltech revealed a strongly anisotropic resistivity in high quality GaAs samples at low temperatures, which was consistent with the formation of charge-density-wave order,[64] for states near half filling of the third and higher Landau levels, *i.e.*, for states with ν in the vicinity of 9/2, 11/2, 13/2, etc. Moreover, it was found that in the second Landau level, application of an in-plane magnetic field sufficient to destroy the gapped quantized Hall states at $\nu = 5/2$ and 7/2, also led to a strongly anisotropic resistance. However, continuing experimental and theoretical work suggests that the actual electronic states may be better described, in most cases, as a *nematic*, with uniaxial anisotropy but at most short-range stripe-like translational order.[65–72] Theoretical investigations have included questions about the nature of the transition between the nematic and isotropic states. See also Ref. 73 and references therein.

Considerable theoretical and experimental efforts have been devoted in order to identify small microscopic perturbations, including growth anisotropies or effects of a tilted magnetic field, that would select between one or another of the possible orientations for nematic order. Similarly, there has been considerable effort to study the competition and possible phase transitions between states of nematic order and isotropic states, such as a Fermi liquid or a gapped quantum Hall state near half-filling. This competition is clearly of particular importance in the second Landau level, where different phases have been observed depending on details of the system. Experiments have also produced evidence for spontaneously-developed anisotropy in gapped fractional quantized Hall states at certain filing fractions.[74] Such anisotropies could be found, for example, in transport properties at finite temperature. Mechanisms by which anisotropy might develop in a gapped quantum Hall state have also been explored theoretically. See, e.g. Refs. 75 and 76 and references therein. The reader is referred to the chapter by Csáthy for further discussion and references on all of these points.

There have also been numerous works in the theoretical literature discussing the response of a quantized Hall state to variations of the metric tensor describing the underlying spatial geometry. The "shift parameter", which characterizes changes in the number of flux quanta necessary to realize a quantized Hall state with a given number of electrons on a sphere, is a well-known example of such a geometric effect.[77,78] Closely related to the shift parameter is the Hall viscosity, which describes electrical currents that can be induced by a changing metric tensor in a quantized Hall state.[79] It has been suggested that metric fluctuations produced by an acoustic wave might be used to excite quadrupolar modes of a gapped quantum state.[80] Geometric aspects of states in the lowest Landau level and their response to anisotropy have been emphasized in various investigations.[81–83]

It is not entirely clear whether there should be unique definition of the shift parameter for a gapless state such as at $\nu = 1/2$, and one must be careful about the proper definition of the Hall viscosity in this case. However, the Son–Dirac theory has been explicitly formulated in the presence of a general metric tensor, so it could be used to investigate the response to variations in the geometry.[8] The HLR approach as originally formulated applied only to the case of a planar geometry, but extensions to a general metric have been recently developed.[84,85]

5.2. *Thermal effects*

Among the fundamental questions one may ask about a gapless state in strong magnetic field are the behaviors of thermal properties, such as the thermal conductivity and thermoelectric effects, in the limits of low finite temperatures and small impurity densities. Within either the HLR or Son–Dirac picture, energy current is carried by CFs, and the longitudinal thermal conductivity κ_{xx}, at $\nu = 1/2$, should be proportional to $C_p v_F l_{cf}$, with the specific heat C_p proportional to T, ignoring possible effective-mass divergences. By contrast, the electrical conductivity σ_{xx} will go to zero proportional to $1/l_{cf}$, for large mean-free-path. It is clear, therefore, that the ratio $\kappa_{xx}/T\sigma_{xx}$ will vastly exceed the Wiedemann–Franz value under these conditions.

Thermoelectric effects at $\nu = 1/2$ and $\nu = 3/2$ were discussed two decades ago, by Cooper *et al.*[86] within the HLR picture. More recently, Potter *et al.*[17] calculated both the diagonal and off-diagonal components of Seebeck tensor $\hat{S}$, which relates the generated electric field to the heat current in a geometry where no electrical current is allowed, within the Son–Dirac picture as well as HLR. While they found identical results in the two formalisms for the leading, diagonal, component of $\hat{S}$, they found a difference between predictions using HLR and Son–Dirac for the Hall component, which is smaller than the diagonal component by a factor of order $(k_F l_{cf})^{-1}$ in a clean system.

However, there are subtleties here which require further investigation. On the one hand, the analysis of Ref. 17 did not take into account the effects of fluctuations in the electrostatic potential due to the imperfect screening of impurities, which are important in the calculation of σ_{yx}, as we have seen. In addition, there can be a large contribution to the thermal Hall conductance from regions near the sample boundary, where PH symmetry will be strongly broken. At this point, it is not completely clear whether there is an actual difference between the two theories in their prediction for the Hall contribution. Recent measurements by Liu *et al.*[87] of S_{xy} in a GaAs 2DEG at $\nu = 1/2$ found values that were smaller than the values predicted by Potter *et al.* and that fell off faster with decreasing temperature than the expected linear dependence.

It should also be noted, that predictions for the ratio between the values of S_{xx} at $\nu = 1/2$ and $\nu = 3/2$ in Ref. 86 were not found to be in good agreement with experimental data available at the time. Reference 87 reported measurements of S_{xx}

at $\nu = 1/2$, whose magnitude was consistent with expectations, but measurements were not reported at $\nu = 3/2$, so a comparison cannot be made.

5.3. *Effects of strong disorder*

In Sec 3.1, we discussed the behavior of the the conductivity tensor at $\nu = 1/2$ in the limit of weak, but non-zero, scattering due to disorder. We argued using either the HLR or Son–Dirac picture, that there should be a value of σ_{xx} proportional to the inverse mean-free-path for CFs, l_{cf}^{-1}, which was implicitly assumed to be small, proportional to the impurity concentration for weak disorder. However, this analysis was based on the use of a Boltzmann equation, treating the CFs as non-interacting fermions in a random effective magnetic field (and correlated electrostatic potential), which we do not expect to be strictly correct in the limit of zero temperature. As pointed out in Ref. 5, a diagrammatic analysis of the effects of fluctuations suggests that these should be corrections to σ_{xx} proportional to $\ln(T \, l_{cf}/v_F)$, which, depending on their sign, should drive σ_{xx} towards zero or to a value of the order of $1/4\pi$ in the limit of $T \to 0$. In practice, these corrections may be negligible for experiments on the highest-quality samples, but they are of interest from a theoretical point of view. In any case, in samples of lower quality, one can certainly encounter values of σ_{xx} that are not small compared to $1/4\pi$, and we would like to understand what can happen in that case.

In 1983, Khmel'nitskiï,[88] proposed a theory of phase transitions between integer quantized Hall states in the presence of disorder, based on a renormalization group (RG) involving two parameters, σ_{xx} and σ_{yx}. He argued that the RG should have a periodic dependence on the variable σ_{yx}, with stable fixed points corresponding to the quantized Hall states, with $\sigma_{xx} = 0$ and σ_{yx} equal to an arbitrary integer multiple of $1/2\pi$, and that there should be a set of unstable fixed-points with

$$\sigma_{yx} = \frac{n + 1/2}{4\pi}, \quad \sigma_{xx} = \sigma_c, \tag{46}$$

where n is an arbitrary integer, and the critical conductance σ_c should be independent of n and of order $1/4\pi$. In addition, there should be an unstable fixed point at $\sigma_{xx} = \infty$. According to this analysis, if one starts with any value of σ_{yx} that is not of the form $(n+1/2)/4\pi$, one should flow at large length scales to the nearest integer quantized Hall state, with $\sigma_{xx} = 0$. On the other hand, if one starts with a value of σ_{yx} that is precisely equal to $(n+1/2)/4\pi$, the value of σ_{yx} should remain constant, and one should flow to the associated fixed point with $\sigma_{xx} = \sigma_c$ and the initial value of σ_{yx}. The qualitative flow pattern implied by this analysis is illustrated by the black lines and arrows in Fig. 1. [Note: The RG flows are invariant under the interchange of σ_{xy} and $\sigma_{yx} = -\sigma_{xy}$. We use σ_{yx} in our discussions, because it is positive for electrons with $B > 0$.]

Khmel'nitskiï's picture, and subsequent work by Pruisken,[89] did not take into account the existence of fractional quantized Hall states, which one should find in interacting electron systems of sufficiently high quality. Kivelson, Lee, and Zhang[90] proposed a generalization of the phase diagram, motivated by Jain's CF picture, which included fractional quantized Hall states at arbitrary odd-denominator fractions. RG flow lines associated with some of these states and the allowed transitions between them are indicated by colored curves in Fig. 1. The description of transitions between the various integer quantized Hall states is not affected, as the unstable fixed points (46) are not in the region of the new flow lines. Regions of the phase diagram that flow to fractional quantized Hall states can only be accessed if the disorder is not too strong, and if the starting value of σ_{yx} is not too close to $(n + 1/2)/2\pi$.

The fixed points of the RG flow represent possible values of the conductivity tensor in the limit of an infinite system at $T = 0$. At non-zero temperatures, or in a finite system, the RG flow can stop before one reaches a fixed point, so other values of the conductivity are possible. Because the RG flow rates are slow in some regions of the phase diagram, there will typically be ranges of the magnetic field, in any sample, where the values of σ_{yx} and σ_{xx} vary continuously with filling factor, and are almost independent of temperature, to the lowest attainable temperatures. We may describe such a region as an intermediate *metallic phase*.

According to the flow diagram illustrated in Fig. 1, direct transitions are not allowed between arbitrary pairs of quantized Hall states. In particular, at $T = 0$, it is not possible to have a direct transition between a fractional quantized Hall plateau with $\nu < 1/2$ and one with $\nu > 1/2$, as one varies the electron filling factor. One would have to first pass through the insulating state at $\nu = 0$ and the integer state at $\nu = 1$, which we would describe as *re-entrant* integer states. Of course, for a

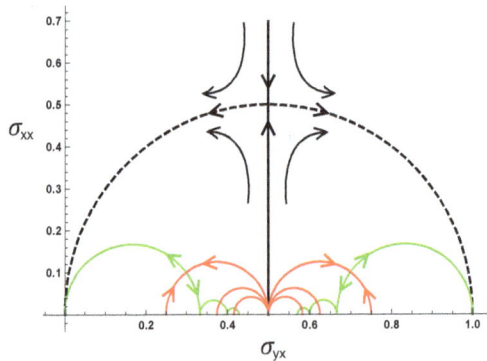

Fig. 1.   Proposed renormalization-group flow in the plane of σ_{yx} and σ_{xx}, in units of e^2/h. The vertical solid black line separates points which flow, at large length scales, into the insulating fixed point, with $\sigma_{xx} = \sigma_{yx} = 0$ and the integer quantized Hall state with $\sigma_{yx} = e^2/h$. Points on the vertical solid black line flow into the unstable fixed point at $\sigma_{xx} = \sigma_{yx} = e^2/2h$. The dashed black curve shows the flow trajectories from the unstable fixed point to the stable fixed points at $\sigma_{xx} = 0$. Red curves separate a region which flows to a fractional quantized fixed point from a region which flows to another fraction or to integer quantized fixed points. Green curves show trajectories flowing from the unstable fixed point on a red curve into a quantized fixed point at $\sigma_{xx} = 0$. Only six fractional state are shown, with $\nu = 1/3, 2/5, 3/7, 4/7, 2/5$ and $2/3$.

finite system or at finite temperatures, the fractional quantized Hall plateaus could be separated by an intermediate metallic state rather than by re-entrant quantized Hall states. [*cf.* Ref. 5.]

Beyond the location of fixed points and phase boundaries, an RG calculation should also make predictions for the *rates* at which physical properties evolve as a function of increasing length scale or decreasing temperature. If one starts with $\sigma_{xx} = \sigma_c$ and σ_{yx} very close to $1/4\pi$, the renormalization group analysis picture says that the deviation $\Delta\sigma_{yx} \equiv \sigma_{yx} - (2n+1)/4\pi$ should grow with the measuring length scale L as $L^{1/\tilde{\nu}}$, where $\tilde{\nu}$ is a critical exponent determined by form of the β-function near the critical point, while σ_{xx} should remain roughly constant as long as $|\Delta\sigma_{yx}|$ remains small. (We denote the exponent as $\tilde{\nu}$ to distinguish it from the filling factor ν.) For an infinite system at non-zero temperature, the conductivity tensor should be determined by stopping the renormalization group flow at a length scale L_T determined by the temperature, following a power law which may be written in the form $L_T \propto T^{-z_T}$. The exponent z_T is supposed to arise from the temperature dependence of the mean-free-path for inelastic scattering.[89] If the filling factor is varied through the value $1/2$, while the temperature is fixed, σ_{yx} should undergo its transition from a value close to zero to a value close to $1/2\pi$ over an interval $\Delta\nu^*$ of filling fraction proportional to T^κ, where, following the RG, $\kappa = (z_T\tilde{\nu})^{-1}$. The precise values of these exponents have been the objects of continuing theoretical and experimental studies.

To obtain a theoretical estimate of the exponent $\tilde{\nu}$, the RG picture can be compared with results for the properties of a non-interacting electron in the lowest Landau level in the presence of a random potential. In this case it has been found that all eigenstates are localized, except at a critical energy E_c, which will sit in the middle of the Landau level in the case of disorder with PH symmetry. The localization length ξ is found to diverge as the energy E approaches E_C, and it is expected that the divergence should have the form of a power law, $\xi \propto |E - E_c|^{-\bar{\nu}}$. It has been argued that short-range electron electron interactions will appear as an irrelevant perturbation at the RG fixed point describing the critical point of the non-interacting system, and consequently, for short-rage interactions, the exponent $\tilde{\nu}$ should be identical to $\bar{\nu}$.[91] On the other hand, Coulomb interactions, which fall off as $1/r$, are found to be relevant at the non-interacting fixed point, so the value of $\tilde{\nu}$ may be different from the non-interacting $\bar{\nu}$ in that case. The theoretical discussions also assume that there are no long-range correlations in the disorder potential or macroscopic inhomogeneities in the system. Long-range-correlated disorder can convert the problem to a classical percolation problem, which certainly has a different value of the critical exponents.

Many of the numerical studies have employed a simplified model, introduced by Chalker and Coddington (CC) in 1988,[92] in which regions with quantized Hall conductances $\nu = 1$ and $\nu = 0$ form a regular checkerboard pattern. Electrons flow along the edges between the squares in a definite direction, such that each vertex has two incoming and two outgoing bonds. Scattering at a node is characterized by an S-matrix, in which there is a probability p to scatter to the right and $1 - p$ to scatter to the left. In the CC model, the probability p is assumed to be the same at every vertex, while the phase accumulated by an electron traveling along a link is assumed to be random. The transition point between states with macroscopic quantized Hall conductance occurs at $p = 1/2$. Although early simulations of the CC model gave results for $\bar{\nu}$ of order 2.35, more recent studies of larger systems give results closer to 2.6. (See, e.g. Refs. 93–95.) More recently, Zirnbauer has argued that the correct asymptotic value for the CC model should be $\bar{\nu} = \infty$, and that apparent values in the range of 2.3 to 2.6 reflect slow flow by a marginal variable towards the asymptotic fixed point.[96] Moreover, Zirnbauer calls into question the entire validity of a two-parameter RG as a description of this problem.

Regardless of the interpretation, it appears that there is a discrepancy between the values of $\bar{\nu}$ obtained from numerical studies fo the CC model and the value of $\tilde{\nu} \approx 2.38 \pm 0.06$ extracted from experiments.[97] One possibility is that the difference is due to effects of electron-electron interactions that are missing from the CC model. It is also possible that long-range correlations in the impurity potential are affecting the experimental results. However, Gruzberg *et al.* have suggested that the problem arises from the regular geometric arrangement of scattering vertices in the CC model. They have studied numerically a set of models in which some fraction of the vertices have scattering probabilities $p = 0$ or $p = 1$, effectively removing them from the network, so that the remaining vertices are connected in an irregular way.[98,99] According to their analysis, the exponent obtained for their model is $\bar{\nu} \approx 2.37$, in good agreement with experiment.

Beyond the value of $\bar{\nu}$, many theoretical investigations have been concerned with the multifractal structure of the electron wave function at the critical point for the non-interacting problem. (See, e.g. Refs. 91 and 96). This structure is probably not accessible in experiments but theory can be compared with numerical simulations.

Since the value of $\tilde{\nu}$ for system with short-range interactions is identical to the value of $\bar{\nu}$ for a non-interacting electron system, the size-dependence of the conductivity tensor for a finite system at $T = 0$ should be similar for the two systems. However, the behavior at finite temperature will generally be different in the two situations. For a non-interacting system, the conductivity tensor at non-zero temperature T will be equal to the conductivity of the same system at $T = 0$, but averaged over Fermi level, weighted by the derivative of the Fermi function at temperature T. Since states of the non-interacting system are localized at all energies except E_c, they do not contribute to σ_{xx}, and the resulting longitudinal conductivity for $T \neq 0$ will be zero, regardless of the chemical potential E_F. If the value of E_F is allowed to vary, σ_{yx} will vary smoothly between a value $n/2\pi$, for $E_F \ll E_c$, and the value $(n+1)/2\pi$, for $E_F \gg E_c$, over a range of filling factors $\Delta\nu^*$ that will be of order T/W, where W is the width of the Landau level due to the disorder. This may be contrasted with the RG prediction for interacting systems that σ_{yx} and σ_{xx} should both be finite and vary smoothly over a range of the form $\Delta\nu^* \propto T^\kappa$, with $\kappa = (z_T\tilde{\nu})^{-1}$. According to Ref. 91, the value of z_T for short-range interactions should be $z_T = 1.23$, suggesting a value of κ in the range 0.314 to 0.346. However, experiments obtain a value $\kappa = 0.42 \pm 0.01$,[91,100] which would imply $z_T \approx 1.0 \pm 0.1$. It is not known theoretically what the value of z_T should be in the case of Coulomb interactions, and it appears beyond current capabilities to obtain a reliable value from computer simulations.

In the paper by Kivelson *et al.*[90] mentioned earlier, it was suggested that at $T \neq 0$, in the vicinity of the transition point between insulator and integer quantized Hall conductor, the Hall resistivity ρ_{yx} should remain constant, at a value equal to 2π, throughout the region where ρ_{xx} changes from a value $\gg 1/2\pi$ to a value $\ll 1/2\pi$. This constancy of ρ_{yx} is equivalent to the statement that the complex conductivity should fall on a semicircle,

$$\sigma_{xx}^2 + (\Delta\sigma_{yx})^2 = (1/4\pi)^2. \tag{47}$$

Invoking PH symmetry, we expect that at a fixed temperature, $\sigma_{xx}(\nu)$ should be symmetric about $\nu = 1/2$, while $\Delta\sigma_{yx}$ should be antisymmetric. This implies that at a fixed non-zero temperature T, the resistivity near the transition, at a distance $\Delta\nu$ from the critical value, should obey the *duality relation*

$$\rho_{xx}(\Delta\nu) = \frac{(2\pi)^2}{\rho_{xx}(-\Delta\nu)} \tag{48}$$

These equations also imply $\sigma_c = 1/4\pi$. The predictions (47) and (48) are well-supported by experiments.[100] (By contrast, the analysis of Zirnbauer[96] predicts $2\pi\sigma_c = 0.6366... .$)

Recently, Kumar, Kim, and Raghu[101] have analyzed the transition between the $\nu = 0$ insulating state and the $\nu = 1$ quantized Hall state by comparing the formulation in terms of electrons with a formulation based on CFs, using both the HLR and Son–Dirac formulations. They find a duality between the two descriptions, which leads to a self-duality at the critical point and, hence, to the prediction $\sigma_c = 1/4\pi$. The analysis of Ref. 101 assumes one can ignore residual interactions between the CFs, though it does not assume validity of a Boltzmann equation. However, it is substantiated by comparisons with an analysis using a generalized Chalker–Coddington-type network model for the CFs, as well as one based on a nonlinear sigma model.

In earlier work, Dolan[102] proposed an analytic form for the flow diagram, based on several assumptions, which implied Eqs. (47) and (48), as well as $\sigma_c = 1/4\pi$. One assumption was that the diagram should be consistent with an assumed symmetry expressed in terms of the action of the modular group on the complex conductivity, $\sigma_{yx} + i\sigma_{xx}$, in the upper-half plane. (This assumption may be understood as a natural extension of the transformations by flux attachment employed by Jain.) A second key assumption was that the β-function for the complex conductivity, which describes the flow of parameters under a change of length scale, should be a meromorphic function of the complex conductivity, (after multiplication by a real function of σ_{xx} and σ_{yx}). This assumption was motivated by an analogy with apparently unrelated work on Yang–Mills gauge theory in four-dimensions, but was not derived from a microscopic theory of quantum Hall systems. The assumption of modular invariance implies that if the flow trajectories from the unstable point $\sigma_{xx} = \sigma_{yx} = 1/4\pi$ to the stable fixed points with $\sigma_{xx} = 0$ and $\sigma_{yx} = 0$ or $1/2\pi$ lies on the semicircle (47), and the boundary between regions which flow to the two stable fixed points is given by the line $\sigma_{xx} = 1/4\pi$, then the trajectories and phase boundaries related to fractional quantized Hall states, represented by the colored curves in Fig. 1 will all be semicircles.

It should be emphasized that a renormalization group analysis cannot, by itself, determine the detailed phase diagram of a real physical system. A β-function based on two parameters will apply only at length scales large compared to any microscopic length, so other variables must be considered at short distances. Some simplification occurs if one assumes that both the disorder potential and electron-electron interaction are much smaller than the cyclotron energy, so that Landau level mixing can be neglected even in the presence of disorder, and one assumes that the disorder is PH symmetric. Then if the Landau level is precisely half-full, the value of σ_{yx} is fixed at $1/4\pi$ at all length scales and temperatures. Then the transition between integer quantized Hall plateaus at $T = 0$ will occur precisely when the filling factor passes through $1/2$. When the disorder potential is strong enough so that Landau level mixing needs to be taken into account, the transition will generally be shifted to larger values of the filling factor. For very large disorder, the transition could be shifted to very high filling factors, so that the system remains localized and is insulating at $T = 0$ for any achievable electron density and magnetic field strength.

6. Open Questions About $\nu \approx 1/2$

Discrepancies between the predictions of HLR and Son–Dirac at the RPA level, such as those discussed above in Sec 3.2, raise the question whether the two theories are fundamentally different, or whether they will become equivalent when corrections beyond RPA are taken into account. If one goes beyond the RPA level, both the HLR theory and the Son–Dirac theory contain an infinite number of parameters, such as vertex corrections, to account for fluctuations at non-zero wave vectors and frequencies. (Vertex corrections mght include corrections to the interaction between CFs and the gauge fields at finite wavevector, as well as corrections to the interactions between CFs and perhaps wavevector dependent corrections to the Lagrangian for the gauge fields themselves. The parameters describing these corrections will generally depend on details of the microscopic Hamiltonian, but we do not know how to calculate their values for any particular model. Based on our investigations, it appears possible to choose values for the vertex corrections in the

HLR theory that will remove the discrepancies cited in Refs. 33 and 34, but it is not clear whether these corrections can be chosen so that they are compatible with Galilean invariance and other microscopic properties of the starting Hamiltonian (1) or (3). This remains an open problem for future investigation.

If it turns out that there is no way to reconcile the HLR theory with PH symmetry, the question will be to understand where HLR goes wrong, when it seems to give the correct answer to so many problems. The basic assumption that correct low-energy theory can be obtained by an adiabatic continuation from the mean-field HLR ground state could turn out to be false, but one might then expect many more discrepancies than we have found.

While the Son–Dirac theory, together with regular vertex corrections allowed by symmetry, together with proper account of the infrared divergences discussed in Sec. 2.6, appears to be a correct theory of the behavior near $\nu = 1/2$ for a system restricted to the lowest Landau level, it is currently not clear how one can derive the theory from a microscopic Hamiltonian. This is also an unsolved problem.

At the *physical* level, we may distinguish between two types of properties. On the one hand, there are properties which should display PH symmetry asymptotically close to $\nu = 1/2$, regardless of whether the microscopic Hamiltonian has PH symmetry or not. Examples that we have discussed include the values of ν associated with quantized fractional Hall states, and the energy gaps at these states, at least to lowest order in $|\Delta B|$. Based on the agreement between HLR and Son–Dirac at the RPA level, it seems most likely that this will also be true for properties such as the peak positions of the Weiss oscillations at order $|\Delta B|^2$, or the value of the Hall conductance at $\nu = 1/2$, to order l_{cf}^{-2}. Calculation of these quantities within HLR did not depend on the value of the Landau parameter F_1, and therefore did not assume restriction to a single Landau level.

Clearly, there must be other properties which do depend on the presence or absence of PH symmetry in the underlying microscopic Hamiltonian. In fact, one would expect that virtually any quantitative property would manifest an absence of PH symmetry at sufficiently high order in ΔB, when there is mixing between Landau levels. We may speculate that the quantities where one finds a discrepancy between HLR and Son–Dirac are of this type, *i.e.*, these quantities depend on microscopic details and should agree with the Son–Dirac prediction only in the case where one can neglect Landau level mixing. If the discrepancy between HLR and Son–Dirac can be repaired by inclusion of appropriate vertex corrections, then these corrections should depend on the microscopic details.

On a broader level, there remain in general, perhaps philosophical, questions about how much accuracy one should expect from a low-energy theory of a system with gauge fields and a Fermi surface.

If it turns out, in the end, that the HLR approach is fundamentally incompatible with PH symmetry, then several possibilities remain. One is that HLR is simply wrong, for reasons that we do not yet understand. A second possibility, which has been suggested,[103] is that for a system restricted to the lowest Landau level, HLR describes correctly the states with $\nu < 1/2$, but not states with $\nu > 1/2$. For the latter states, one would need to first make a PH transformation, apply HLR as a description of the holes, and then transform back to electrons. Exactly at $\nu = 1/2$, this would imply the existence of two inequivalent ground states, distinguishable by an appropriate local measurement. In view of the emergent PH symmetry obtained in HLR at the RPA level for a wide variety of properties, as well as the nearly perfect emergent PH symmetry observed in calculations based on composite fermion trial wave functions, this seems like a rather unlikely possibility, and there is no evidence for such a situation in any numerical calculations that we are aware of.

7. Gapped Quantized Hall States at Even-Denominator Fractions

Although there is no indication of an energy gap or quantized Hall plateau at filling factors 1/2 or 3/2 in narrow-well GaAs samples, quantized Hall plateaus at a half-filled Landau level have been observed in other situations. Best known are the quantized Hall states at $\nu = 5/2$ and $\nu = 7/2$ in GaAs structures, which are discussed in several chapters of this book. (See, e.g., the chapters by Heiblum and Feldman and by Csathy.) An assortment of even-denominator fractional quantized Hall states have also been seen in graphene structures, as discussed in the chapter by Dean, Kim, Li, and Young, and in ZnO structures, as discussed in the chapter by Falson and Smet. The precise natures of these various states are still under debate.

The quantized Hall state at $\nu = 5/2$ in GaAs was first observed by Willett *et al.* in 1987.[2] In 1991, Moore and Read proposed a microscopic wave function, involving a Pfaffian of the variables $(z_i - z_j)^{-1}$, where z_i is the position of electron i in complex notation, which provided a possible explanation for the 5/2 state.[104] From the beginning, it was realized that the Pfaffian state could be understood in a composite fermion picture, if the Fermi surface became gapped by the formation of Cooper pairs, analogous to p-wave superconductivity. The connection to p-wave superconductivity, and its topological implications, was further elucidated, most notably, in the 2001 work by Read and Green.[105] In 2007, it was realized that the PH conjugate of the Pfaffian state, denoted the anti-Pfaffian, which necessarily has the same energy as the Pfaffian in a model restricted to a single Landau level with purely two-body forces, is actually a topologically distinct state.[106,107] Therefore, if either of these states is the correct ground state in a model with PH symmetry at the microscopic level, it means that this symmetry must be spontaneously broken for the ground state at half filling.

Numerical calculations based on exact diagonalizations of finite systems and using the density-matrix renormalization group have strongly supported the idea that exact ground state at half-filling with interaction parameters appropriate to the second Landau level should have good overlap with either the Pfaffian or anti-Pfaffian state and should fall into the corresponding topological class.[108-115] Calculations which take into account PH symmetry breaking effects due to Landau level mixing have tended to favor the anti-Pfaffian over the Pfaffian.[112,114-116] However, as discussed in the chapter by Heiblum and Feldman, recent experimental results have failed to agree with predictions for either the Pfaffian or anti-Pfaffian! Rather, they seem consistent with another proposed state, the PH-Pfaffian.[7]

The various states mentioned above can all be described in terms of BCS pairing of CFs, in either the HLR or Son–Dirac language. Within HLR, the Pfaffian, PH-Pfaffian, and anti-Pfaffian are described, respectively, as arising from pairing in channels with angular momentum l equal to 1, -1, and -3. Within Son–Dirac, the Pfaffian, PH-Pfaffian, and anti-Pfaffian are described, respectively, as arising from pairing in channels with angular momentum l equal to 2, 0, and -2. The PH symmetry between Pfaffian and anti-Pfaffian states is manifest in the Son–Dirac description, but it is not inconsistent with HLR. What is required, there, is that the interaction between CFs should have the same strength in the $l = 1$ and $l = -3$ channels.

At least within the Son–Dirac formalism, it might seem that the PH-Pfaffian should be a more natural candidate to explain the 5/2 state than either the Pfaffian or anti-Pfaffian. However, in numerical calculations, no trial wave function in the same topological class as the PH-Pfaffian has been found to have reasonable overlap with the exact ground state of a finite system, in the parameter regime where one finds a gapped state at half-filling. Moreover, numerical studies in the spherical geometry do not find evidence for an incompressible state at the shift corresponding to the PH Pfaffian. Indeed, it has been proposed that a gapped ground state in this class is impossible for a PH symmetric model restricted to a single Landau level.[117]

As discussed in the chapter by Heiblum and Feldman, there have been suggestions that a state in the class of the PH-Pfaffian could be stabilized by effects of disorder and/or strong Landau level mixing, which have been generally ignored in numerical calculations.[118] Several papers have explored the possibility that such a state could arise in an inhomogeneous system with a mixture of Pfaffian and anti-Pfaffian region, but it appears unlikely that such an inhomogeneous situation can occur in practice.[116]

Recently, experiments on single-layer graphene have demonstrated the existence of gapped quantized Hall states in the $N = 3$ Landau level, at filling factors 21/2, 23/2 25/2, and 27/2.[119] Based on numerical calculations, it was argued that the states could be best explained using a "221 parton construction", originally formulated by Jain, in 1989.[120] In terms of its topological properties, this state is identical to a BCS state of CFs with pairing in the channel $l = 3$, in the language of HLR, or $l = 4$ in the Son–Dirac–formulation. Alternatively, the state might be explained by the PH conjugate of the 221 parton state, which would be described by pairing in the $l = -5$ channel in HLR or $l = -4$ in Son–Dirac.

7.1. Non-Abelian statistics

As shown originally by Moore and Read, a striking property of the Pfaffian state is that the elementary charged excitations from the state should exhibit non-Abelian statistics. This would also occur in the anti-Pfaffian, PH-Pfaffian, and the 221 parton states discussed above, and in various other states that could conceivably be found in a half-filled Landau level, but not in all conceivable states. States with non-Abelian statistics of various types have also been proposed to occur in states at other filling fractions.

In this chapter, we shall only review a few key points related to the experimental search for non-Abelian statistics at $\nu = 5/2$. A more complete discussion of non-Abeilan statistics in various gapped fractional quantized Hall states may be found in several other chapters of this book, including the ones by Simon, by Stern, and by Heiblum and Feldman.

In all of the states that have been proposed for $\nu = 5/2$, the elementary charged excitations are supposed to have charge $\pm 1/4$. For the states supporting non-Abelian statistics, such as the Moore–Read, if there are an even number n such quasiparticles that are each localized in space, say by pinning to charged impurities that are sufficiently well separated from each other, then the ground state of the system will be a Hilbert space of dimension $2^{n/2}$. [The Hilbert space dimension will be $2^{(n+1)/2}$ if n is odd.] In a finite system, the energy levels in this Hilbert space will not be truly degenerate but will have energy splittings that fall off exponentially with the separation between quasiparticles and with their separations from the boundary. If, by varying the Hamiltonian, one moves the quasiparticles around in such a way as to interchange their positions and/or wind them around each other, at a rate which is adiabatically slow on the scale of the energy gap to all other excitations, and if the set of quasiparticle positions at the end is the same as it was at the beginning, then the system will remain within the manifold of ground states and new ground state will be related to the old by a unitary transformation. If, at the same time, the interchange is fast compared to the exponentially small energy splittings between the ground states, the unitary transformation will be independent of all details other than the space-time topology of braiding procedure. In general, the unitary transformation will depend on the order in which quasiparticles are interchanged, so that the set of transformations forms a non-Abelian representation of the braid group.

The non-Abelian properties of this system can be described by saying that there is a localized zero-energy Majorana mode attached to each quasiparticle in the bulk, as well as a set of propagating Majorana modes at the boundary. In the case where the number n of enclosed charge $\pm 1/4$ quasiparticles is odd, one of the boundary Majorana modes will be precisely at zero energy; if n is even, then the lowest boundary mode will have a finite energy, inversely proportional to the length of the boundary.

An alternative to the physical braiding of localized quasiparticles is to carry out a Fabry–Perot-type interferometry experiment, as described in the chapter of Heiblum and Feldman. It is predicted that due to the non-Abelian statistics, if the area of the interferometer is held fixed while the magnetic field is varied, and no quasiparticles are allowed to enter or leave the interferometer region, then the interference pattern would have different periodicities in the cases where the enclosed quasiparticle number is even or odd. In practice, one would expect the quasiparticle number itself to vary as the magnetic field is changed, so one is led to an interference pattern that alternates between two periodicities. In a series of papers dating back to 2007, Willett and coworkers have reported observations of such an even-odd alternation in carefully prepared GaAs samples, at filling factors 5/2 and 7/2.[121] Most recently, they have reported extensive measurements on eleven different samples, with careful analyses of the oscillatory dependences of the observed interferometer resistance on magnetic field and on voltages applied to gates designed to change the area of the interferometer.[122] The results appear to give strong support to the existence of non-Abelian quasiparticles of the expected type.

Nevertheless, there are aspects of these experiments that are not well understood. The interferometer areas needed to fit the observations were of order 0.25 μm^2, while the lithographic structures were squares ranging from 2.5 to 5.7 μm on a side. Also, since interferometer area must presumably connect the openings in the defining gates of the two sides of the interferometer region, the width in the perpendicular direction must be less than 0.1 μm. It is not clear what physical factors could give rise to a stable interferometer area of such a small size in these samples. There may also be questions about the extent to which it is appropriate to talk about the existence of a $\nu = 5/2$ quantized Hall state in a region of these dimensions.

On a more general level, it may be worth emphasizing that while the energy splitting between different ground states for fixed locations of the quasiparticles is supposed to fall off exponentially with the separation between quasiparticles, the expected decay length ξ is not particularly small. One would expect ξ to be at least a few times larger than the magnetic length, which is approximately 1 nm at a field of 5 T.

An experiment that needs to perform topologically controlled braiding by interchanging the positions of two localized quasiparticles would have to keep the quasiparticles sufficiently far away from each other and from all other quasiparticles that the energy splitting is small compared to the inverse of the braiding time. If the pre-exponential factor in the energy splitting is of the order of the energy gap in the 5/2 state (say 200 mK) and if the obtainable braiding time were of order 10 nsec, the minimum separation would have to be greater than 5ξ. However, the requirements for a Fabry–Perot interference experiment may be less severe, and interactions between quasiparticles may actually be helpful in stabilizing the interference pattern.[122,123]

8. Two-Component Systems: Spin-Polarized Bilayers

We consider here a system of two parallel layers separated by an insulator, such that tunneling between layers may be neglected. We assume that the electron spins are fully aligned, so we can neglect the spin degree of freedom. Let us introduce a pseudospin variable $\tau = \pm 1$ to distinguish the layers, and let us introduce a set of three Pauli matrices $\vec{\tau}$ that operate on the pseudospin index. Since the electron number in each layer is separately conserved in our system, the operator τ_z commutes with the Hamiltonian, and the global U(1) symmetry generated by this operator is a symmetry of the problem. Nevertheless, in the presence of Coulomb interactions, this rotational symmetry may be spontaneously broken in the ground state, as we shall see.

In general, the electron-electron interactions will differ for two electrons in the same or opposite layers. Typically, the Coulomb repulsion will become independent of the layer index for in-plane electron separations that are large compared to the distance between layers, while at small in-plane separations the repulsive potential between electrons in different layers will be weaker than for electrons in the same layer.

In the limit of zero separation between the planes, where the interactions are independent of the layer indices, the Hamiltonian becomes identical to that of a single spin-degenerate layer, where the Hamiltonian has an SU(2) spin symmetry. In that case, if mixing between Landau levels can be neglected, the ground state at total filling $\nu_{\text{tot}} = 1$ can be found exactly. The solution is given precisely by the Hartree–Fock approximation and may be described as a ferromagnetic Slater determinant state, with all spins pointing in the same direction of space. As there is precisely one electron per flux quantum, the state is a filled Landau level for electrons in the selected spin state, so it is essentially an integer quantized Hall state. The favored spin direction is arbitrary in the absence of a Zeeman field, but even a weak Zeeman field will be sufficient to fix the direction of polarization. It should be noted, however, that though the ground state is completely spin aligned, low-energy Goldstone modes involving spin fluctuations can be important at finite temperatures.[124,125] Moreover, coupled spin-charge excitations can have dramatic effects on the behavior of a spin-degenerate single-layer system slightly away from $\nu = 1$.[126–128] Related phenomena can also occur in systems where carriers sit in two degenerate values within the two-dimensional Brillouin zone.[129,130]

For the bilayer case we are interested in, Hartree–Fock is no longer exact, but it is still expected to be a good approximation when the separation between the layers is sufficiently small compared to the magnetic length l_B, which is also roughly the separation between neighboring electrons in a single layer. The Hartree–Fock ground state will be a filled Landau level, with all electrons having the same pseudospin polarization, described by a vector $\langle \vec{\tau} \rangle$ of unit magnitude. If the electron concentration is separately specified in the two layers, then the value of τ_z will be fixed. If there are electrons in both layers, the magnitude $|\langle \tau_z \rangle|$ will be less than

1. Then, the component of $\langle \vec{\tau} \rangle$ perpendicular to the axis will be non-zero, and can point in an arbitrary direction within the x–y plane. Since the value of τ_z is conserved by the Hamiltonian in the absence of interlayer tunneling, the Hamiltonian is invariant under rotations about pseudospin z-axis, so a non-zero expectation value of a polarization component in the x–y plane means that the U(1) symmetry has been spontaneously broken. In the Hartree–Fock approximation, the ground state is a single filled Landau level for electrons with pseudospins aligned in the selected direction, so the ground state is often referred to as an *interlayer-coherent integer quantized Hall state*. (The state is also referred to as a quantum Hall ferromagnet).[131–136]

The interlayer coherent state is found to have an energy gap, essentially the Hartree–Fock exchange energy, for any deviation in the total electron density away from $\nu_{\text{tot}} = 1$. However, there is no energy gap for variations in the difference of the densities in the two layers, and this difference can be varied continuously within the coherent state. Elementary excitations associated with variations in the relative density involve dynamically coupled oscillations in the orientation of the in-plane component of $\langle \vec{\tau} \rangle$. They are Goldstone modes resulting from the broken U(1) symmetry, and they have an energy that vanishes proportional to their wave vector q. They may alternatively be described as magnons, in the ferromagnetic analogy.

An alternate way of describing the interlayer coherent state is to start from a state consisting of a filled Landau layer in one layer and an empty Landau layer in the other. One may then add excitons, consisting of an electron in the empty layer bound to a hole in the filled layer. Under appropriate circumstances, when a finite density of excitons is present, the ground state may be described as a Bose condensate of these excitons. Then the broken U(1) symmetry may be described as a broken *phase symmetry* for the Bose condensate. As would be expected for a Bose condensate, the interlayer coherent state behaves like a superfluid for the movement of the excitons. As neutral particles, excitons do not carry a net electric current, but an exciton current will carry a counterflow electrical current, which is equal and opposite in the two layers. The superfluid property means that in a state of constant current, there can be no gradient in the chemical potential for excitons, so any voltage gradients must be identical in the two layers. Since the system is in a quantized Hall state for the net current, currents can flow without dissipation, and any net current will lead to a quantized Hall voltage in both layers.

During the 1990's and early 2000's, properties of the interlayer coherent state were extensively explored, theoretically and experimentally, in GaAs double-well systems. The most dramatic effects were manifest after techniques were developed to make separate electrical contacts to the two layers, In situations where tunneling between layers could be neglected, one could observe directly the phenomenon of counterflow current without a voltage drop and the closely related phenomenon of quantized Hall drag.[135] In particular, when a current I is allowed to pass through one layer, while no current is allowed to flow in the second layer, a quantized Hall voltage of Ih/e^2 is observed in *both* layers, with no longitudinal voltage in either layer.

In the case where there is weak tunneling between the layers, the tunneling operator will break the U(1) symmetry of the Hamiltonian. If the applied magnetic field is perpendicular to the layers, one may choose a gauge in which the tunneling operator is proportional to a particular combination of τ_x and τ_y, independent of position. An arbitrarily small amount of tunneling will then align the pseudospin of the ground state so that its x–y component points in the favored direction. Furthermore, since the value of τ_z is no longer conserved, its value must now be chosen to minimize the total energy. If the Hamiltonian is symmetric with respect to the occupation of the two layers, the lowest energy state will have $\langle \tau_z \rangle = 0$. However, $\langle \tau_z \rangle$ will be different from zero if there is a bias favoring one layer or the other.

If interlayer tunneling is important, it is not possible to conduct a drag experiment in which current flows in the two layers are separately controlled. However, one can apply a voltage difference between the two layers and measure the tunnel current between the layers. In the interlayer coherent state, there will be a sharp increase in the tunneling current at very low voltage difference between the layers, analogous to Josephson tunneling in a superconductor.

In GaAs samples where the layers are close enough for the interlayer coherent state to occur, there will typically be tunneling between layers that cannot be neglected at $\nu_{\text{tot}} = 1$, if the applied magnetic field is perpendicular to the sample. However tunneling effects can be suppressed in a tilted magnetic field. The in-plane field component then gives a momentum boost to electrons tunneling between the layers, which can destroy the coherent Josephson effect.[136]

Very recently, it has been possible to construct Coulomb-coupled bilayer systems using two graphene sheets, separated by an insulating layer of about 2.5 nm of hBN. In this system, tunneling is negligible even in the absence of an in-plane magnetic field, because of the large energy gap of hBN. Results of various experiments on Coulomb-coupled graphene systems are discussed in the chapter by Dean, Kim, Li and Young.

8.1. *Transition to the interlayer coherent state*

On the theoretical side, there is much interest in what happens to the interlayer coherent state in an ideal sample as the separation between layers is increased. The most important variable determining the relative importance of interlayer and intralayer Coulomb interactions is the ratio between the interlayer separation d and the magnetic length l_B. Experimentally, it is difficult to vary d continuously, but one can vary the electron density and hence the value of l_B at a given value of ν_{tot}. Additional issues concern the effects of finite temperature, of imbalance between the layers, of disorder, and/or of weak tunneling between layers.

For two spin-polarized layers with no disorder and $\nu = 1/2$ in each, one might expect that for sufficiently large separations, the ground state would consist of two uncoupled layers, with a Fermi surface and gapless excitations in each layer. However, in analogy with the Cooper instability to superconductivity for a conventional Fermi surface, one should consider the possibility of pairing instabilities of the coupled composite fermion seas for arbitrarily weak interactions. Instabilities have been considered in several channels. The situation is complicated because of the long-range gauge fields and resulting infrared divergences at the Fermi surfaces of the separated layers. Nevertheless, it is generally believed that there should be an energy gap of some kind in the ideal case for arbitrary separations of the layers, though the gap may decrease very rapidly with increasing layer separation and may be completely unmeasurable for large separations. What is not clear is what will be the symmetry of the resulting gapped state. One possibility is that the gapped state at large separations is topologically distinct from the interlayer coherent state, in which case there must be one or more sharp phase transitions between the ideal ground states as the separation is increased, Another possibility, however, is that the states are topologically identical, so that in principle there could be only a crossover, and no sharp transition at any finite separation, as the separation is increased. This possibility is not immediately obvious, as the interlayer coherent state is thought of as a condensate of excitons made from particles and holes with no gauge interactions, while, in the HLR picture, the separated layers are described by two Fermi surfaces of electron-like CFs, with two separate gauge fields.

A recent analysis by Sodemann *et al.*[137] has explained in some detail how this can occur from the point of view of both the Son–Dirac and HLR approaches. Within the HLR approach, Sodemann *et al.* find that one can realize a state that is topologically equivalent to the interlayer coherent integer quantized state by coupling the two electron-like CF systems in a BCS-like pair state with $p_x + ip_y$ symmetry. A trial wave function embodying this type of pairing for CFs was introduced in 2008 by Möller *et al.*, whose energy could be evaluated by Monte Carlo techniques and which could be compared with exact diagonalization studies for small systems.[138] They found very good agreement for intermediate layer separations $(d/l_B \approx 1)$ but poorer agreement at small separations. However, they were able to obtain good overlap with exact calculations at smaller separations by using a slightly more complicated wave function with the same symmetry as the pairing wave function.[139] For large separations, the pairing amplitude becomes very small, and trial wave function cannot be distinguished from two independent CF liquids in a system of limited size. An alternate approach to describing the ground state at various layer separations would be to first make a particle-hole transformation in one of the two layers, and attach Chern–Simons flux to the holes in that layer. One can then create a state topologically equivalent to the interlayer coherent state by pairing the hole-like CFs and the electron-like CFs in a conventional *s*-wave BCS state. So far, however, there have been no numerical calculations using trial wave functions based on this construction.

The question of whether there is actually a smooth crossover between the behavior of two separated layers and the interlayer coherent state, or whether there is some type of phase transition separating different phases has been examined in numerical calculations of finite systems.[138–140] However, there does not seem to be a firm conclusion about this issue.

The situation at balanced layer filling may be contrasted with the situation at $\nu_{tot} = 1$ with unequal filling. For an ideal isolated layer, there should be an infinite number of stable gapped states with filling factors of the Jain form $\nu = p/(2p+1)$. As these states each have a finite energy gap, a pair of such layers with Jain filling factors ν_1 and $\nu_2 = 1 - \nu_1$ would not be sensitive to very weak interactions between the layers, so they would remain as independent gapped states for sufficiently large separations. These separately gapped states could not be connected, without a phase transition, to the interlayer coherent state, which has no energy gap for the difference in densities of the layers.

What happens if the density in each layer is intermediate between two neighboring Jain states? For an isolated layer with short range interactions, it is expected that close to $\nu = 1/2$, there will be a series of first-order phase transitions between quantized states with successive integer values of p, which implies that for intermediate filling factors the system would phase separate into domains of Jain states with different densities. With unscreened Coulomb interactions, one would expect there to be no true phase separation; rather, the system would form an array of large finite domains, in a structure with a broken translational invariance whose period would be sensitive to the precise filling factor of layer. In any case, if the well-separated layers have unequal fillings with $\nu_{tot} = 1$, there must be at least one sharp phase transition as the separation is decreased, if the final state is a translationally-invariant interlayer coherent state.

In experiments on GaAs bilayer systems, it appears that there is a sharp phase transition, even in the case of balanced layer occupation, as one increases the value of d/l_B through a critical value of order 1.8.[141]

In recent work, Eisenstein et al.[142] have studied the tunneling current between layers of a GaAs heterostructure in the regime of d/l_B slightly larger than the critical value for the formation of the interlayer-coherent quantized state at $\nu_{tot} = 1$. Measuring the tunnel current as a function of bias voltage, they found that the Coulomb pseudogap, which generally inhibits interlayer tunneling at low energies, was itself suppressed as the critical value of d/l_B was approached. The results could be interpreted as a reflection of interlayer electron-hole correlations, which became stronger as the critical value of d/l_B is approached. Studies of the dependence of the pseudogap on the density difference between the two layers found a larger suppression effect at filling-factor differences $\Delta\nu = \pm0.08$ than at equal filling. This may reflect the curious fact, previously observed but not yet understood, that the critical value of d/l_B for the formation of the quantized interlayer coherent state is actually a local minimum at $\Delta\nu = 0$.[141] From a larger perspective, these experiments demonstrate the usefulness of tunneling measurements for investigating subtle correlations between Coulomb-coupled layers in the quantum Hall regime.

In general, a quantized Hall conductance is theoretically precise only in the limit of zero temperature. Because charged excitations in a quantized Hall state have finite energy, there will be a non-zero density of such excitations at any finite temperature, and one does not expect to have a sharp phase transition as the temperature is increased. However, in the interlayer coherent state, the excitations which can destroy exciton superfluidity are vortices in the Bose condensate, which have a logarithmically diverging energy at $T = 0$. Thus, one would expect a mathematically well defined phase transition as the temperature is raised, of the Kosterlitz–Thouless type, where counterflow superfluidity is lost.

Experiments to study the temperature-driven transition have been carried out in Coulomb-coupled graphene double layers.[143] Interpretations are complicated by possible effects of disorder and because the observed transition required finite measuring currents. Nevertheless, the experiments have enabled one to track behavior of the transition temperature as one varies the electron density, and hence d/l_B. It was found that the effective transition temperature increases with density for small values of d/l_B, as one would expect for a Bose condensate of tightly bound excitons. However around d/l_B of order 0.8, as the excitons begin to overlap more strongly, the transition temperature starts to decrease with increasing density. This is what one might expect qualitatively if one is entering a regime of BCS-like pairing, as the Fermi level will rise with increasing density, and the ratio of the interlayer interactions to intralayer interactions will decrease.

In the regime of relatively large d/l_B, the counterflow resistance was found to increase sharply near the critical temperature, with a dependence on temperature and measuring current similar to what one would expect for a Kosterlitz–Thouless transition. On the other hand, for smaller values of d/l_B, the temperature dependence of the counterflow resistance was better fit by a simple Arrhenius law, down to the lowest temperatures where it could be observed. Arrhenius behavior has been also observed for GaAs double layers. This has been attributed to the effects of disorder, which might lead to a finite density of vortices even at $T = 0$, which then might be unpinned and contribute to transport at finite temperatures.[135] However, a full explanation for these observations has not been developed.

9. Conclusions

Although problems related to quantum Hall phenomena at, or in the vicinity of a half full Landau level have been actively studied for nearly three decades, the subject remains active and exciting, due to new experimental observations and to new theoretical developments. In this chapter, we have reviewed the current status of the field, with emphasis on new developments and unanswered questions. A major focus of the chapter has been on issues surrounding systems of spin-polarized electrons in the lowest Landau level, as in GaAs narrow quantum wells, where no quantized Hall plateau is observed at filling $\nu = 1/2$. However, we have also discussed, briefly, situations where a quantized Hall plateau is observed, such as at $\nu = 5/2$ and $\nu = 7/2$ in GaAs structures. Finally, we have discussed the phenomena in double-well systems where the half full condition is achieved when the total filing factor is $\nu_{\text{tot}} = 1$.

We find that there remain many unanswered questions in the field.

Acknowledgments

Over the years, I have benefited greatly from discussions with many colleagues, too numerous to name, on the various topics covered in this chapter. However, I would like to thank particularly Chong Wang, Ady Stern, and Jainendra Jain for their helpful comments on preliminary versions of the manuscript.

References

1. D. C. Tsui, H. L. Stormer and A. C. Gossard, Two-dimensional magnetotransport in the extreme quantum limit, *Phys. Rev. Lett.* **48**, 1559–1562 (1982). doi: 10.1103/PhysRevLett.48.1559. URL https://link.aps.org/doi/10.1103/PhysRevLett.48.1559.
2. R. Willett, J. P. Eisenstein, H. L. Störmer, D. C. Tsui, A. C. Gossard and J. H. English, Observation of an even-denominator quantum number in the fractional quantum Hall effect, *Phys. Rev. Lett.* **59**, 1776–1779 (1987). doi: 10.1103/PhysRevLett.59.1776. URL http://link.aps.org/doi/10.1103/PhysRevLett.59.1776.
3. Y. W. Suen, L. W. Engel, M. B. Santos, M. Shayegan and D. C. Tsui, Observation of a ν=1/2 fractional quantum Hall state in a double-layer electron system, *Phys. Rev. Lett.* **68**, 1379–1382 (1992). doi: 10.1103/PhysRevLett.68.1379. URL https://link.aps.org/doi/10.1103/PhysRevLett.68.1379.
4. R. L. Willett, M. A. Paalanen, R. R. Ruel, K. W. West, L. N. Pfeiffer and D. J. Bishop, Anomalous sound propagation at $\nu = 1/2$ in a 2D electron gas: Observation of a spontaneously broken translational symmetry? *Phys. Rev. Lett.* **65**, 112–115 (1990). doi: 10.1103/PhysRevLett.65.112. URL https://link.aps.org/doi/10.1103/PhysRevLett.65.112.
5. B. I. Halperin, P. A. Lee and N. Read, Theory of the half-filled Landau level, *Phys. Rev. B* **47**, 7312–7343 (1993). doi: 10.1103/PhysRevB.47.7312. URL http://link.aps.org/doi/10.1103/PhysRevB.47.7312.
6. S. M. Girvin, Particle-hole symmetry in the anomalous quantum Hall effect, *Phys. Rev. B* **29**, 6012–6014 (1984). doi: 10.1103/PhysRevB.29.6012. URL http://link.aps.org/doi/10.1103/PhysRevB.29.6012.
7. D. T. Son, Is the composite fermion a Dirac particle?, *Phys. Rev. X.* **5**, 031027 (2015). doi: 10.1103/PhysRevX.5.031027. URL http://link.aps.org/doi/10.1103/PhysRevX.5.031027.
8. D. T. Son, The Dirac composite fermion of the fractional quantum Hall effect, *Ann. Rev. Condensed Matter Physics.* **9**(1), 397–411 (2018). doi: 10.1146/annurev-conmatphys-033117-054227. URL https://doi.org/10.1146/annurev-conmatphys-033117-054227.
9. B. L. Altshuler, L. B. Ioffe and A. J. Millis, Low-energy properties of fermions with

singular interactions, *Phys. Rev. B* **50**, 14048–14064 (1994). doi: 10.1103/PhysRevB.50.14048. URL http://link.aps.org/doi/10.1103/PhysRevB.50.14048.

10. Y. B. Kim, A. Furusaki, X.-G. Wen and P. A. Lee, Gauge-invariant response functions of fermions coupled to a gauge field, *Phys. Rev. B* **50**, 17917–17932 (1994). doi: 10.1103/PhysRevB.50.17917. URL http://link.aps.org/doi/10.1103/PhysRevB.50.17917.

11. A. Stern and B. I. Halperin, Singularities in the Fermi-liquid description of a partially filled Landau level and the energy gaps of fractional quantum Hall states, *Phys. Rev. B* **52**, 5890–5906 (1995). doi: 10.1103/PhysRevB.52.5890. URL http://link.aps.org/doi/10.1103/PhysRevB.52.5890.

12. J. K. Jain, Composite-fermion approach for the fractional quantum Hall effect, *Phys. Rev. Lett.* **63**, 199–202 (1989). doi: 10.1103/PhysRevLett.63.199. URL http://link.aps.org/doi/10.1103/PhysRevLett.63.199.

13. A. Lopez and E. Fradkin, Fractional quantum Hall effect and Chern–Simons gauge theories, *Phys. Rev. B* **44**, 5246–5262 (1991). doi: 10.1103/PhysRevB.44.5246. URL http://link.aps.org/doi/10.1103/PhysRevB.44.5246.

14. M. Greiter and F. Wilczek, Exact solutions and the adiabatic heuristic for quantum Hall states, *Nucl. Phys. B* **370**, 577–600 (1992).

15. B. Rejaei and C. W. J. Beenakker, Vector-mean-field theory of the fractional quantum Hall effect, *Phys. Rev. B* **46**, 15566–15569 (1992). doi: 10.1103/PhysRevB.46.15566. URL http://link.aps.org/doi/10.1103/PhysRevB.46.15566.

16. V. Kalmeyer and S.-C. Zhang, Metallic phase of the quantum Hall system at even-denominator filling fractions, *Phys. Rev. B* **46**, 9889–9892 (1992). doi: 10.1103/PhysRevB.46.9889. URL http://link.aps.org/doi/10.1103/PhysRevB.46.9889.

17. A. C. Potter, M. Serbyn and A. Vishwanath, Thermoelectric transport signatures of Dirac composite fermions in the half-filled Landau level, *Phys. Rev. X.* 6 (3): 031026 (2016). doi: 10.1103/PhysRevX.6.031026.

18. C. Nayak and F. Wilczek, Non-Fermi liquid fixed point in 2 + 1 dimensions, *Nucl. Phys. B* **417**(3), 359–373 (1994). ISSN 0550-3213. doi: https://doi.org/10.1016/0550-3213(94)90477-4. URL http://www.sciencedirect.com/science/article/pii/0550321394904774.

19. R. H. Morf, N. d'Ambrumenil and S. Das Sarma, Excitation gaps in fractional quantum Hall states: An exact diagonalization study, *Phys. Rev. B* **66**, 075408 (2002). doi: 10.1103/PhysRevB.66.075408. URL http://link.aps.org/doi/10.1103/PhysRevB.66.075408.

20. R. L. Willett, R. R. Ruel, K. W. West and L. N. Pfeiffer, Experimental demonstration of a Fermi surface at one-half filling of the lowest Landau level, *Phys. Rev. Lett.* **71**, 3846–3849 (1993). doi: 10.1103/PhysRevLett.71.3846. URL http://link.aps.org/doi/10.1103/PhysRevLett.71.3846.

21. S. A. Kivelson, D.-H. Lee, Y. Krotov and J. Gan, Composite-fermion Hall conductance at $\nu = 1/2$, *Phys. Rev. B* **55**, 15552–15561 (1997). doi: 10.1103/PhysRevB.55.15552. URL http://link.aps.org/doi/10.1103/PhysRevB.55.15552.

22. C. Wang, N. R. Cooper, B. I. Halperin and A. Stern, Particle-hole symmetry in the fermion-Chern–Simons and Dirac descriptions of a half-filled Landau level, *Phys. Rev. X.* **7**, 031029 (2017). doi: 10.1103/PhysRevX.7.031029. URL https://link.aps.org/doi/10.1103/PhysRevX.7.031029.

23. J. H. Smet, K. von Klitzing, D. Weiss and W. Wegscheider, dc transport of composite fermions in weak periodic potentials, *Phys. Rev. Lett.* **80**, 4538–4541 (1998). doi: 10.1103/PhysRevLett.80.4538. URL http://link.aps.org/doi/10.1103/PhysRevLett.80.4538.

24. R. L. Willett, K. W. West and L. N. Pfeiffer, Geometric resonance of composite fermion cyclotron orbits with a fictitious magnetic field modulation, *Phys. Rev. Lett.* **83**, 2624–2627 (1999). doi: 10.1103/PhysRevLett.83.2624. URL http://link.aps.org/doi/10.1103/PhysRevLett.83.2624.

25. S. D. M. Zwerschke and R. R. Gerhardts, Positive magnetoresistance of composite fermion systems with a weak one-dimensional density modulation, *Phys. Rev. Lett.* **83**, 2616–2619 (1999). doi: 10.1103/PhysRevLett.83.2616. URL http://link.aps.org/doi/10.1103/PhysRevLett.83.2616.

26. J. H. Smet, S. Jobst, K. von Klitzing, D. Weiss, W. Wegscheider and V. Umansky, Commensurate composite fermions in weak periodic electrostatic potentials: Direct evidence of a periodic effective magnetic field, *Phys. Rev. Lett.* **83**, 2620–2623 (1999). doi: 10.1103/PhysRevLett.83.2620. URL http://link.aps.org/doi/10.1103/PhysRevLett.83.2620.

27. D. Kamburov, Y. Liu, M. A. Mueed, M. Shayegan, L. N. Pfeiffer, K. W. West and K. W. Baldwin, What determines the Fermi wave vector of composite fermions? *Phys. Rev. Lett.* **113**, 196801 (2014). doi: 10.1103/PhysRevLett.113.196801. URL http://link.aps.org/doi/10.1103/PhysRevLett.113.196801.

28. A. K. C. Cheung, S. Raghu and M. Mulligan, Weiss oscillations and particle-hole symmetry at the half-filled Landau level, *ArXiv e-prints* (2016).

29. A. Mitra and M. Mulligan, Fluctuations and magnetoresistance oscillations near the half-filled Landau level, *arXiv:1901.08070* (2019).

30. S. H. Simon and B. I. Halperin, Finite-wave-vector electromagnetic response of fractional quantized Hall states, *Phys. Rev. B* **48**, 17368–17387 (1993). doi: 10.1103/PhysRevB.48.17368.

31. D. X. Nguyen and D. T. Son, Algebraic approach to fractional quantum Hall effect, *Phys. Rev. B* **98**, 241110 (2018). doi: 10.1103/PhysRevB.98.241110. URL https://link.aps.org/doi/10.1103/PhysRevB.98.241110.

32. A. C. Balram, C. Tőke and J. K. Jain, Luttinger theorem for the strongly correlated Fermi liquid of composite fermions, *Phys. Rev. Lett.* **115**, 186805 (2015). doi: 10.1103/PhysRevLett.115.186805. URL https://link.aps.org/doi/10.1103/PhysRevLett.115.186805.

33. D. X. Nguyen, S. Golkar, M. M. Roberts and D. T. Son, Particle-hole symmetry and composite fermions in fractional quantum Hall states, *Phys. Rev. B* **97**, 195314 (2018). doi: 10.1103/PhysRevB.97.195314. URL https://link.aps.org/doi/10.1103/PhysRevB.97.195314.

34. M. Levin and D. Thanh Son, Particle-Hole Symmetry and Electromagnetic Response of a Half-Filled Landau Level, *ArXiv e-prints* (2016).

35. S. D. Geraedts, M. P. Zaletel, R. S. K. Mong, M. A. Metlitski, A. Vishwanath and O. I. Motrunich, The half-filled Landau level: The case for Dirac composite fermions, *Science.* **352**, 197–201 (2016). doi: 10.1126/science.aad4302.

36. N. Read, Theory of the half-filled Landau level, *Semiconductor Science and Technology.* **9**(11S), 1859 (1994). URL http://stacks.iop.org/0268-1242/9/i=11S/a=002.

37. N. Read, Recent progress in the theory of composite fermions near even-denominator filling factors, *Surface Science.* **361–362**, 7–12 (1996). ISSN 0039-6028. doi: https://doi.org/10.1016/0039-6028(96)00318-4. URL http://www.sciencedirect.com/science/article/pii/0039602896003184.

38. R. Shankar and G. Murthy, Towards a field theory of fractional quantum Hall states, *Phys. Rev. Lett.* **79**, 4437–4440 (1997). doi: 10.1103/PhysRevLett.79.4437. URL http://link.aps.org/doi/10.1103/PhysRevLett.79.4437.

39. N. Read, Lowest-Landau-level theory of the quantum Hall effect: The Fermi-liquid-

like state of bosons at filling factor one, *Phys. Rev. B* **58**, 16262–16290 (1998). doi: 10.1103/PhysRevB.58.16262. URL `http://link.aps.org/doi/10.1103/PhysRevB.58.16262`.

40. V. Pasquier and F. Haldane, A dipole interpretation of the $\nu = 1/2$ state, *Nuclear Physics B* **516**(3), 719–726 (1998). ISSN 0550-3213. doi: http://dx.doi.org/10.1016/S0550-3213(98)00069-8. URL `http://www.sciencedirect.com/science/article/pii/S0550321398000698`.

41. D.-H. Lee, Neutral fermions at filling factor $\nu = 1/2$, *Phys. Rev. Lett.* **80**, 4745–4748 (1998). doi: 10.1103/PhysRevLett.80.4745. URL `http://link.aps.org/doi/10.1103/PhysRevLett.80.4745`.

42. G. Murthy and R. Shankar, $\nu = \frac{1}{2}$ Landau level: Half-empty versus half-full, *Phys. Rev. B* **93**, 085405 (2016). doi: 10.1103/PhysRevB.93.085405. URL `http://link.aps.org/doi/10.1103/PhysRevB.93.085405`.

43. D. Bohm and D. Pines, A collective description of electron interactions: Iii. Coulomb interactions in a degenerate electron gas, *Phys. Rev.* **92**, 609–625 (1953). doi: 10.1103/PhysRev.92.609. URL `https://link.aps.org/doi/10.1103/PhysRev.92.609`.

44. B. I. Halperin and A. Stern, Comment on "Towards a field theory of fractional quantum Hall states", *Phys. Rev. Lett.* **80**, 5457–5457 (1998). doi: 10.1103/PhysRevLett.80.5457. URL `https://link.aps.org/doi/10.1103/PhysRevLett.80.5457`.

45. A. Stern, B. I. Halperin, F. von Oppen and S. H. Simon, Half-filled Landau level as a Fermi liquid of dipolar quasiparticles, *Phys. Rev. B* **59**, 12547–12567 (1999). doi: 10.1103/PhysRevB.59.12547. URL `http://link.aps.org/doi/10.1103/PhysRevB.59.12547`.

46. B. I. Halperin, Statistics of quasiparticles and the hierarchy of fractional quantized Hall states, *Phys. Rev. Lett.* **52**, 1583–1586 (1984). doi: 10.1103/PhysRevLett.52.1583. URL `https://link.aps.org/doi/10.1103/PhysRevLett.52.1583`.

47. S. D. Geraedts, J. Wang, E. H. Rezayi and F. D. M. Haldane, Berry phase and model wave function in the half-filled Landau level, *Phys. Rev. Lett.* **121**, 147202 (2018). doi: 10.1103/PhysRevLett.121.147202. URL `https://link.aps.org/doi/10.1103/PhysRevLett.121.147202`.

48. J. Wang, S. D. Geraedts, E. H. Rezayi and F. D. M. Haldane, Lattice Monte Carlo for quantum Hall states on a torus, *Phys. Rev. B* **99**, 125123 (2019). doi: 10.1103/PhysRevB.99.125123. URL `https://link.aps.org/doi/10.1103/PhysRevB.99.125123`.

49. S. Pu, M. Fremling and J. K. Jain, Berry phase of the composite-fermion Fermi sea: Effect of Landau-level mixing, *Phys. Rev. B* **98**, 075304 (2018). doi: 10.1103/PhysRevB.98.075304. URL `https://link.aps.org/doi/10.1103/PhysRevB.98.075304`.

50. C. Wang and T. Senthil, Composite Fermi liquids in the lowest Landau level, *Phys. Rev. B* **94**(24), 245107 (2016). doi: 10.1103/PhysRevB.94.245107.

51. N. Seiberg, T. Senthil, C. Wang and E. Witten, A duality web in 2+1 dimensions and condensed matter physics, *Ann. Phys..* **374**, 395–433 (2016). ISSN 0003-4916. doi: https://doi.org/10.1016/j.aop.2016.08.007. URL `http://www.sciencedirect.com/science/article/pii/S0003491616301531`.

52. C. Dasgupta and B. I. Halperin, Phase transition in a lattice model of superconductivity, *Phys. Rev. Lett.* **47**, 1556–1560 (1981). doi: 10.1103/PhysRevLett.47.1556. URL `http://link.aps.org/doi/10.1103/PhysRevLett.47.1556`.

53. M. E. Peskin, Mandelstam-'t Hooft duality in abelian lattice models, *Ann. Phys.* **113**(1), 122–152 (1978). ISSN 0003-4916. doi: https://doi.org/10.1016/0003-4916(78)90252-X. URL `http://www.sciencedirect.com/science/article/pii/000349167890252X`.

54. C. Wang and T. Senthil, Half-filled Landau level, topological insulator surfaces, and three-dimensional quantum spin liquids, *Phys. Rev. B* **93**, 085110 (2016). doi: 10.1103/PhysRevB.93.085110. URL `http://link.aps.org/doi/10.1103/PhysRevB.93.085110`.

55. D. F. Mross, A. Essin and J. Alicea, Composite Dirac liquids: Parent states for symmetric surface topological order, *Phys. Rev. X.* **5**, 011011 (2015). doi: 10.1103/PhysRevX.5.011011. URL `http://link.aps.org/doi/10.1103/PhysRevX.5.011011`.

56. M. A. Metlitski and A. Vishwanath, Particle-vortex duality of 2D Dirac fermion from electric-magnetic duality of 3D topological insulators, *Phys. Rev. B* **93**, 245151 (2016).

57. C. Wang and T. Senthil, Dual Dirac liquid on the surface of the electron topological insulator, *Phys. Rev. X.* **5**, 041031 (2015). doi: 10.1103/PhysRevX.5.041031. URL `http://link.aps.org/doi/10.1103/PhysRevX.5.041031`.

58. M. A. Metlitski, S-duality of $u(1)$ gauge theory with $\theta = \pi$ on non-orientable manifolds: Applications to topological insulators and superconductors, *ArXiv e-prints* arXiv: 1509.00355 (2015).

59. A. Karch and D. Tong, Particle-vortex duality from 3D bosonization, *Phys. Rev. X.* **6**, 031043 (2016). doi: 10.1103/PhysRevX.6.031043. URL `https://link.aps.org/doi/10.1103/PhysRevX.6.031043`.

60. T. Senthil, D. T. Son, C. Wang and C. Xu, Duality between (2+1)D quantum critical points, *Phys. Rep.* **827**, 1–48 (2019). ISSN 0370-1573. doi: https://doi.org/10.1016/j.physrep.2019.09.001. URL `http://www.sciencedirect.com/science/article/pii/S0370157319302637`. Duality between (2+1)D quantum critical points.

61. H. Fukuyama, P. M. Platzman and P. W. Anderson, Two-dimensional electron gas in a strong magnetic field, *Phys. Rev. B* **19**, 5211–5217 (1979). doi: 10.1103/PhysRevB.19.5211. URL `https://link.aps.org/doi/10.1103/PhysRevB.19.5211`.

62. A. A. Koulakov, M. M. Fogler and B. I. Shklovskii, Charge density wave in two-dimensional electron liquid in weak magnetic field, *Phys. Rev. Lett.* **76**, 499–502 (1996). doi: 10.1103/PhysRevLett.76.499. URL `http://link.aps.org/doi/10.1103/PhysRevLett.76.499`.

63. M. M. Fogler and A. A. Koulakov, Laughlin liquid to charge-density-wave transition at high Landau levels, *Phys. Rev. B* **55**, 9326–9329 (1997). doi: 10.1103/PhysRevB.55.9326. URL `http://link.aps.org/doi/10.1103/PhysRevB.55.9326`.

64. M. P. Lilly, K. B. Cooper, J. P. Eisenstein, L. N. Pfeiffer and K. W. West, Evidence for an anisotropic state of two-dimensional electrons in high Landau levels, *Phys. Rev. Lett.* **82**, 394–397 (1999). doi: 10.1103/PhysRevLett.82.394. URL `http://link.aps.org/doi/10.1103/PhysRevLett.82.394`.

65. E. Fradkin and S. A. Kivelson, Liquid-crystal phases of quantum Hall systems, *Phys. Rev. B* **59**, 8065–8072 (1999). doi: 10.1103/PhysRevB.59.8065. URL `https://link.aps.org/doi/10.1103/PhysRevB.59.8065`.

66. E. Fradkin, S. A. Kivelson, E. Manousakis and K. Nho, Nematic phase of the two-dimensional electron gas in a magnetic field, *Phys. Rev. Lett.* **84**, 1982–1985 (2000). doi: 10.1103/PhysRevLett.84.1982. URL `https://link.aps.org/doi/10.1103/PhysRevLett.84.1982`.

67. F. von Oppen, B. I. Halperin and A. Stern, Conductivity tensor of striped quantum Hall phases, *Phys. Rev. Lett.* **84**, 2937–2940 (2000). doi: 10.1103/PhysRevLett.84.2937. URL `https://link.aps.org/doi/10.1103/PhysRevLett.84.2937`.

68. K. Cooper, M. Lilly, J. Eisenstein, T. Jungwirth, L. Pfeiffer and K. West,

An investigation of orientational symmetry-breaking mechanisms in high Landau levels, *Solid State Communications.* **119**(2), 89–94 (2001). ISSN 0038-1098. doi: https://doi.org/10.1016/S0038-1098(01)00212-5. URL http://www.sciencedirect.com/science/article/pii/S0038109801002125.

69. C. Wexler and A. T. Dorsey, Disclination unbinding transition in quantum Hall liquid crystals, *Phys. Rev. B* **64**, 115312 (2001). doi: 10.1103/PhysRevB.64.115312. URL https://link.aps.org/doi/10.1103/PhysRevB.64.115312.

70. L. Radzihovsky and A. T. Dorsey, Theory of quantum Hall nematics, *Phys. Rev. Lett.* **88**, 216802 (2002). doi: 10.1103/PhysRevLett.88.216802. URL https://link.aps.org/doi/10.1103/PhysRevLett.88.216802.

71. Q. Qian, J. Nakamura, S. Fallahi, G. C. Gardner and M. J. Manfra, Possible nematic to smectic phase transition in a two-dimensional electron gas at half-filling, *Nature Communications.* **8**, 1536 (2017). doi: 10.1038/s41467-017-01810-y. URL https://doi.org/10.1038/s41467-017-01810-y.

72. D. X. Nguyen, A. Gromov and D. T. Son, Fractional quantum Hall systems near nematicity: Bimetric theory, composite fermions, and Dirac brackets, *Phys. Rev. B* **97**, 195103 (2018). doi: 10.1103/PhysRevB.97.195103. URL https://link.aps.org/doi/10.1103/PhysRevB.97.195103.

73. Y. You, G. Y. Cho and E. Fradkin, Nematic quantum phase transition of composite Fermi liquids in half-filled Landau levels and their geometric response, *Phys. Rev. B* **93**, 205401 (2016). doi: 10.1103/PhysRevB.93.205401. URL https://link.aps.org/doi/10.1103/PhysRevB.93.205401.

74. J. Xia, V. Cvicek, J. P. Eisenstein, L. N. Pfeiffer and K. W. West, Tilt-induced anisotropic to isotropic phase transition at $\nu = 5/2$, *Phys. Rev. Lett.* **105**, 176807 (2010). doi: 10.1103/PhysRevLett.105.176807. URL https://link.aps.org/doi/10.1103/PhysRevLett.105.176807.

75. Y. You, G. Y. Cho and E. Fradkin, Theory of nematic fractional quantum Hall states, *Phys. Rev. X.* **4**, 041050 (2014). doi: 10.1103/PhysRevX.4.041050. URL https://link.aps.org/doi/10.1103/PhysRevX.4.041050.

76. X. Wan and K. Yang, Striped quantum Hall state in a half-filled Landau level, *Phys. Rev. B* **93**, 201303 (2016). doi: 10.1103/PhysRevB.93.201303. URL https://link.aps.org/doi/10.1103/PhysRevB.93.201303.

77. X.-G. Wen, Topological orders and edge excitations in fractional quantum Hall states, *Adv. Phys.* **44**, 405 (1995).

78. X. G. Wen and A. Zee, Shift and spin vector: New topological quantum numbers for the Hall fluids, *Phys. Rev. Lett.* **69**, 953–956 (1992). doi: 10.1103/PhysRevLett.69.953. URL https://link.aps.org/doi/10.1103/PhysRevLett.69.953.

79. N. Read, Non-abelian adiabatic statistics and Hall viscosity in quantum Hall states and $p_x + i p_y$ paired superfluids, *Phys. Rev. B* **79**, 045308 (2009). doi: 10.1103/PhysRevB.79.045308. URL https://link.aps.org/doi/10.1103/PhysRevB.79.045308.

80. K. Yang, Acoustic wave absorption as a probe of dynamical geometrical response of fractional quantum Hall liquids, *Phys. Rev. B* **93**, 161302 (2016). doi: 10.1103/PhysRevB.93.161302. URL https://link.aps.org/doi/10.1103/PhysRevB.93.161302.

81. M. Ippoliti, R. N. Bhatt and F. D. M. Haldane, Geometry of flux attachment in anisotropic fractional quantum Hall states, *Phys. Rev. B* **98**, 085101 (2018). doi: 10.1103/PhysRevB.98.085101. URL https://link.aps.org/doi/10.1103/PhysRevB.98.085101.

82. F. D. M. Haldane, The origin of holomorphic states in Landau levels from non-commutative geometry and a new formula for their overlaps on the torus, *Journal of Mathematical Physics.* **59**(8), 081901 (2018). doi: 10.1063/1.5046122. URL https://doi.org/10.1063/1.5046122.

83. S.-F. Liou, F. D. M. Haldane, K. Yang and E. H. Rezayi, Chiral gravitons in fractional quantum Hall liquids, *Phys. Rev. Lett.* **123**, 146801 (2019). doi: 10.1103/PhysRevLett.123.146801. URL https://link.aps.org/doi/10.1103/PhysRevLett.123.146801.

84. G. Y. Cho, Y. You and E. Fradkin, Geometry of fractional quantum Hall fluids, *Phys. Rev. B* **90**, 115139 (2014). doi: 10.1103/PhysRevB.90.115139. URL https://link.aps.org/doi/10.1103/PhysRevB.90.115139.

85. A. Gromov, G. Y. Cho, Y. You, A. G. Abanov and E. Fradkin, Framing anomaly in the effective theory of the fractional quantum Hall effect, *Phys. Rev. Lett.* **114**, 016805 (2015). doi: 10.1103/PhysRevLett.114.016805. URL https://link.aps.org/doi/10.1103/PhysRevLett.114.016805.

86. N. R. Cooper, B. I. Halperin and I. M. Ruzin, Thermoelectric response of an interacting two-dimensional electron gas in a quantizing magnetic field, *Phys. Rev. B* **55**, 2344–2359 (1997). doi: 10.1103/PhysRevB.55.2344. URL http://link.aps.org/doi/10.1103/PhysRevB.55.2344.

87. X. Liu, T. Li, P. Zhang, L. N. Pfeiffer, K. W. West, C. Zhang and R.-R. Du, Thermopower and Nernst measurements in a half-filled lowest Landau level, *Phys. Rev. B* **97**, 245425 (2018). doi: 10.1103/PhysRevB.97.245425. URL https://link.aps.org/doi/10.1103/PhysRevB.97.245425.

88. D. E. Khmel'nitskiĭ, Quantization of Hall conductivity, *JETP Lett.* **38**, 552–556 (1983).

89. A. M. M. Pruisken, Universal singularities in the integral quantum Hall effect, *Phys. Rev. Lett.* **61**, 1297–1300 (1988). doi: 10.1103/PhysRevLett.61.1297. URL https://link.aps.org/doi/10.1103/PhysRevLett.61.1297.

90. S. Kivelson, D.-H. Lee and S.-C. Zhang, Global phase diagram in the quantum Hall effect, *Phys. Rev. B* **46**, 2223–2238 (1992). doi: 10.1103/PhysRevB.46.2223. URL https://link.aps.org/doi/10.1103/PhysRevB.46.2223.

91. I. Burmistrov, S. Bera, F. Evers, I. Gornyi and A. Mirlin, Wave function multifractality and dephasing at metal-insulator and quantum Hall transitions, *Annals of Physics.* **326**(6), 1457–1478 (2011). ISSN 0003-4916. doi: https://doi.org/10.1016/j.aop.2011.01.005. URL http://www.sciencedirect.com/science/article/pii/S0003491611000273.

92. J. T. Chalker and P. D. Coddington, Percolation, quantum tunnelling and the integer Hall effect, *Journal of Physics C: Solid State Physics.* **21**(14) 2665 (1988). URL http://stacks.iop.org/0022-3719/21/i=14/a=008.

93. H. Obuse, I. A. Gruzberg and F. Evers, Finite-size effects and irrelevant corrections to scaling near the integer quantum Hall transition, *Phys. Rev. Lett.* **109**, 206804 (2012). doi: 10.1103/PhysRevLett.109.206804. URL https://link.aps.org/doi/10.1103/PhysRevLett.109.206804.

94. W. Nuding, A. Klümper and A. Sedrakyan, Localization length index and subleading corrections in a Chalker–Coddington model: A numerical study, *Phys. Rev. B* **91**, 115107 (2015). doi: 10.1103/PhysRevB.91.115107. URL https://link.aps.org/doi/10.1103/PhysRevB.91.115107.

95. I. C. Fulga, F. Hassler, A. R. Akhmerov and C. W. J. Beenakker, Topological quantum number and critical exponent from conductance fluctuations at the quantum Hall plateau transition, *Phys. Rev. B* **84**, 245447 (2011). doi: 10.1103/PhysRevB.84.245447. URL https://link.aps.org/doi/10.1103/PhysRevB.84.245447.

96. M. R. Zirnbauer, The integer quantum Hall plateau transition is a current algebra after all, *Nuclear Physics B* **941**, 458–506 (2019). ISSN 0550-3213. doi: https://doi.org/10.1016/j.nuclphysb.2019.02.017. URL http://www.sciencedirect.com/science/article/pii/S0550321319300458.

97. W. Li, C. L. Vicente, J. S. Xia, W. Pan, D. C. Tsui, L. N. Pfeiffer and K. W. West, Scaling in plateau-to-plateau transition: A direct connection of quantum Hall systems with the Anderson localization model, *Phys. Rev. Lett.* **102**, 216801 (2009). doi: 10.1103/PhysRevLett.102.216801. URL https://link.aps.org/doi/10.1103/PhysRevLett.102.216801.

98. I. A. Gruzberg, A. Klümper, W. Nuding and A. Sedrakyan, Geometrically disordered network models, quenched quantum gravity, and critical behavior at quantum Hall plateau transitions, *Phys. Rev. B* **95**, 125414 (2017). doi: 10.1103/PhysRevB.95.125414. URL https://link.aps.org/doi/10.1103/PhysRevB.95.125414.

99. A. Klümper, W. Nuding and A. Sedrakyan, Random network models with variable disorder of geometry, *Phys. Rev. B* **100**, 140201 (2019). doi: 10.1103/PhysRevB.100.140201. URL https://link.aps.org/doi/10.1103/PhysRevB.100.140201.

100. D. Shahar, D. C. Tsui, M. Shayegan, E. Shimshoni and S. L. Sondhi, A different view of the quantum Hall plateau-to-plateau transitions, *Phys. Rev. Lett.* **79**, 479–482 (1997). doi: 10.1103/PhysRevLett.79.479. URL https://link.aps.org/doi/10.1103/PhysRevLett.79.479.

101. P. Kumar, Y. B. Kim and S. Raghu, Self-duality of the integer quantum Hall to insulator transition: a composite fermion description, *preprint* (2019).

102. B. P. Dolan, Modular invariance, universality and crossover in the quantum Hall effect, *Nuclear Physics B* **554**(3), 487–513 (1999). ISSN 0550-3213. doi: https://doi.org/10.1016/S0550-3213(99)00326-0. URL http://www.sciencedirect.com/science/article/pii/S0550321399003260.

103. M. Barkeshli, M. Mulligan and M. P. A. Fisher, Particle-hole symmetry and the composite Fermi liquid, *Phys. Rev. B* **92**(16), 165125 (2015). doi: 10.1103/PhysRevB.92.165125.

104. G. Moore and N. Read, Nonabelions in the fractional quantum Hall effect, *Nuclear Physics B* **360**(2–3), 362–396 (1991). ISSN 0550-3213. doi: http://dx.doi.org/10.1016/0550-3213(91)90407-O. URL http://www.sciencedirect.com/science/article/pii/055032139190407O.

105. N. Read and D. Green, Paired states of fermions in two dimensions with breaking of parity and time-reversal symmetries and the fractional quantum Hall effect, *Phys. Rev. B* **61**, 10267–10297 (2000). doi: 10.1103/PhysRevB.61.10267. URL https://link.aps.org/doi/10.1103/PhysRevB.61.10267.

106. M. Levin, B. I. Halperin and B. Rosenow, Particle-hole symmetry and the Pfaffian state, *Phys. Rev. Lett.* **99**, 236806 (2007). doi: 10.1103/PhysRevLett.99.236806. URL http://link.aps.org/doi/10.1103/PhysRevLett.99.236806.

107. S.-S. Lee, S. Ryu, C. Nayak and M. P. A. Fisher, Particle-hole symmetry and the $\nu = \frac{5}{2}$ quantum Hall state, *Phys. Rev. Lett.* **99**, 236807 (2007). doi: 10.1103/PhysRevLett.99.236807. URL http://link.aps.org/doi/10.1103/PhysRevLett.99.236807.

108. R. H. Morf, Transition from quantum Hall to compressible states in the second Landau level: New light on the $\nu = 5/2$ enigma, *Phys. Rev. Lett.* **80**, 1505–1508 (1998). doi: 10.1103/PhysRevLett.80.1505. URL https://link.aps.org/doi/10.1103/PhysRevLett.80.1505.

109. A. E. Feiguin, E. Rezayi, K. Yang, C. Nayak and S. Das Sarma, Spin polarization of the $\nu = 5/2$ quantum Hall state, *Phys. Rev. B* **79**, 115322 (2009). doi: 10.1103/PhysRevB.79.115322. URL https://link.aps.org/doi/10.1103/PhysRevB.79.115322.

110. H. Wang, D. N. Sheng and F. D. M. Haldane, Particle-hole symmetry breaking and the $\nu=(5)/(2)$ fractional quantum Hall effect, *Phys. Rev. B* **80**(24), 241311 (2009). doi: 10.1103/PhysRevB.80.241311.

111. M. Storni, R. H. Morf and S. Das Sarma, Fractional quantum Hall state at $\nu = \frac{5}{2}$ and the Moore–Read Pfaffian, *Phys. Rev. Lett.* **104**, 076803 (2010). doi: 10.1103/PhysRevLett.104.076803. URL https://link.aps.org/doi/10.1103/PhysRevLett.104.076803.

112. E. H. Rezayi and S. H. Simon, Breaking of particle-hole symmetry by Landau level mixing in the $\nu = 5/2$ quantized Hall state, *Phys. Rev. Lett.* **106**, 116801 (2011). doi: 10.1103/PhysRevLett.106.116801. URL https://link.aps.org/doi/10.1103/PhysRevLett.106.116801.

113. Z. Papić, F. D. M. Haldane and E. H. Rezayi, Quantum phase transitions and the ν=5/2 fractional Hall state in wide quantum wells, *Phys. Rev. Lett.* **109**(26), 266806 (2012). doi: 10.1103/PhysRevLett.109.266806.

114. M. P. Zaletel, R. S. K. Mong, F. Pollmann and E. H. Rezayi, Infinite density matrix renormalization group for multicomponent quantum Hall systems, *Phys. Rev. B* **91**, 045115 (2015). doi: 10.1103/PhysRevB.91.045115. URL https://link.aps.org/doi/10.1103/PhysRevB.91.045115.

115. E. H. Rezayi, Landau level mixing and the ground state of the $\nu = 5/2$ quantum Hall effect, *Phys. Rev. Lett.* **119**(2), 026801 (2017). doi: 10.1103/PhysRevLett.119.026801.

116. S. H. Simon, M. Ippoliti, M. P. Zaletel and E. H. Rezayi, Energetics of Pfaffian-antipfaffian domains, *arXiv:1909.12844* (2019).

117. M. V. Milovanović, Paired states in half-filled Landau levels, *Phys. Rev. B* **95**(23), 235304 (2017). doi: 10.1103/PhysRevB.95.235304.

118. P. T. Zucker and D. E. Feldman, Stabilization of the particle-hole Pfaffian order by Landau-level mixing and impurities that break particle-hole symmetry, *Phys. Rev. Lett.* **117**, 096802 (2016). doi: 10.1103/PhysRevLett.117.096802. URL http://link.aps.org/doi/10.1103/PhysRevLett.117.096802.

119. Y. Kim, A. C. Balram, T. Taniguchi, K. Watanabe, J. K. Jain and J. H. Smet, Even denominator fractional quantum Hall states in higher Landau levels of graphene, *Nature Physics.* **15**, 154–158 (2019). doi: 10.1038/s41567-018-0355-x. URL https://doi.org/10.1038/s41567-018-0355-x.

120. J. K. Jain, Incompressible quantum Hall states, *Phys. Rev. B* **40**, 8079–8082 (1989). doi: 10.1103/PhysRevB.40.8079. URL https://link.aps.org/doi/10.1103/PhysRevB.40.8079.

121. R. L. Willett, M. J. Manfra, L. N. Pfeiffer and K. W. West, Confinement of fractional quantum Hall states in narrow conducting channels, *Appl. Phys. Lett.* **91**(5), 052105 (2007). doi: 10.1063/1.2762299. URL https://doi.org/10.1063/1.2762299.

122. R. L. Willett, K. Shtengel, C. Nayak, L. N. Pfeiffer, Y. J. Chung, M. L. Peabody, K. W. Baldwin and K. W. West, Interference measurements of non-abelian e/4 and abelian e/2 quasiparticle braiding, *arXiv:1905.10248* (2019).

123. B. Rosenow, B. I. Halperin, S. H. Simon and A. Stern, Bulk-edge coupling in the non-abelian $\nu = 5/2$ quantum Hall interferometer, *Phys. Rev. Lett.* **100**, 226803 (2008). doi: 10.1103/PhysRevLett.100.226803. URL https://link.aps.org/doi/10.1103/PhysRevLett.100.226803.

124. Y. A. Bychkov, S. V. Iordanskii and G. M. Eliashberg, *JETP Lett.* **33**, 143 (1981).

125. C. Kallin and B. I. Halperin, Excitations from a filled Landau level in the two-dimensional electron gas, *Phys. Rev. B* **30**, 5655–5668 (1984). doi: 10.1103/PhysRevB.30.5655. URL https://link.aps.org/doi/10.1103/PhysRevB.30.5655.

126. S. L. Sondhi, A. Karlhede, S. A. Kivelson and E. H. Rezayi, Skyrmions and the crossover from the integer to fractional quantum Hall effect at small Zeeman energies, *Phys. Rev. B* **47**, 16419–16426 (1993). doi: 10.1103/PhysRevB.47.16419. URL https://link.aps.org/doi/10.1103/PhysRevB.47.16419.

127. H. A. Fertig, L. Brey, R. Côté and A. H. MacDonald, Charged spin-texture excitations and the Hartree–Fock approximation in the quantum Hall effect, *Phys. Rev. B* **50**, 11018–11021 (1994). doi: 10.1103/PhysRevB.50.11018. URL https://link.aps.org/doi/10.1103/PhysRevB.50.11018.

128. S. E. Barrett, G. Dabbagh, L. N. Pfeiffer, K. W. West and R. Tycko, Optically pumped NMR evidence for finite-size skyrmions in GaAs quantum wells near Landau level filling $\nu = 1$, *Phys. Rev. Lett.* **74**, 5112–5115 (1995). doi: 10.1103/PhysRevLett.74.5112. URL https://link.aps.org/doi/10.1103/PhysRevLett.74.5112.

129. M. Rasolt, F. Perrot and A. H. MacDonald, New gapless modes in the fractional quantum Hall effect of multicomponent fermions, *Phys. Rev. Lett.* **55**, 433–436 (1985). doi: 10.1103/PhysRevLett.55.433. URL https://link.aps.org/doi/10.1103/PhysRevLett.55.433.

130. M. Rasolt, B. I. Halperin and D. Vanderbilt, Dissipation due to a "valley wave" channel in the quantum Hall effect of a multivalley semiconductor, *Phys. Rev. Lett.* **57**, 126–129 (1986). doi: 10.1103/PhysRevLett.57.126. URL https://link.aps.org/doi/10.1103/PhysRevLett.57.126.

131. T. Chakraborty and P. Pietiläinen, Fractional quantum Hall effect at half-filled Landau level in a multiple-layer electron system, *Phys. Rev. Lett.* **59**, 2784–2787 (1987). doi: 10.1103/PhysRevLett.59.2784. URL https://link.aps.org/doi/10.1103/PhysRevLett.59.2784.

132. D. Yoshioka, A. H. MacDonald and S. M. Girvin, Fractional quantum Hall effect in two-layered systems, *Phys. Rev. B* **39**, 1932–1935 (1989). doi: 10.1103/PhysRevB.39.1932. URL https://link.aps.org/doi/10.1103/PhysRevB.39.1932.

133. H. A. Fertig, Energy spectrum of a layered system in a strong magnetic field, *Phys. Rev. B* **40**, 1087–1095 (1989). doi: 10.1103/PhysRevB.40.1087. URL https://link.aps.org/doi/10.1103/PhysRevB.40.1087.

134. S. Q. Murphy, J. P. Eisenstein, G. S. Boebinger, L. N. Pfeiffer and K. W. West, Many-body integer quantum Hall effect: Evidence for new phase transitions, *Phys. Rev. Lett.* **72**, 728–731 (1994). doi: 10.1103/PhysRevLett.72.728. URL https://link.aps.org/doi/10.1103/PhysRevLett.72.728.

135. J. P. Eisenstein and A. H. MacDonald, Bose-Einstein condensation of excitons in bilayer systems, *Nature.* **432**, 691 (2004). doi: 10.1038/nature03081.

136. J. Eisenstein, Exciton condensation in bilayer quantum Hall systems, *Ann. Rev. Condensed Matter Physics.* **5**(1), 159–181 (2014). doi: 10.1146/annurev-conmatphys-031113-133832. URL https://doi.org/10.1146/annurev-conmatphys-031113-133832.

137. I. Sodemann, I. Kimchi, C. Wang and T. Senthil, Composite fermion duality for half-filled multicomponent Landau levels, *Phys. Rev. B* **95**, 085135 (2017). doi: 10.1103/PhysRevB.95.085135. URL https://link.aps.org/doi/10.1103/PhysRevB.95.085135.

138. G. Möller, S. H. Simon and E. H. Rezayi, Paired composite fermion phase of quantum Hall bilayers at $\nu = \frac{1}{2} + \frac{1}{2}$, *Phys. Rev. Lett.* **101**, 176803 (2008). doi: 10.1103/PhysRevLett.101.176803. URL https://link.aps.org/doi/10.1103/PhysRevLett.101.176803.

139. G. Möller, S. H. Simon and E. H. Rezayi, Trial wave functions for $\nu = \frac{1}{2} + \frac{1}{2}$ quantum Hall bilayers, *Phys. Rev. B* **79**, 125106 (2009). doi: 10.1103/PhysRevB.79.125106. URL https://link.aps.org/doi/10.1103/PhysRevB.79.125106.

140. M. V. Milovanović, E. Dobardžić and Z. Papić, Meron deconfinement in the quantum Hall bilayer at intermediate distances, *Phys. Rev. B* **92**, 195311 (Nov,

2015). doi: 10.1103/PhysRevB.92.195311. URL https://link.aps.org/doi/10.1103/PhysRevB.92.195311.

141. I. B. Spielman, M. Kellogg, J. P. Eisenstein, L. N. Pfeiffer and K. W. West, Onset of interlayer phase coherence in a bilayer two-dimensional electron system: Effect of layer density imbalance, *Phys. Rev. B* **70**, 081303 (2004). doi: 10.1103/PhysRevB.70.081303. URL https://link.aps.org/doi/10.1103/PhysRevB.70.081303.

142. J. P. Eisenstein, L. N. Pfeiffer and K. W. West, Precursors to exciton condensation in quantum Hall bilayers, *Phys. Rev. Lett.* **123**, 066802 (2019). doi: 10.1103/PhysRevLett.123.066802. URL https://link.aps.org/doi/10.1103/PhysRevLett.123.066802.

143. X. Liu, J. I. A. Li, K. Watanabe, T. Taniguchi, J. Hone, B. I. Halperin, C. Dean and P. Kim, Crossover between strongly coupled and weakly coupled exciton superfluids, *in preparation* (2019).

Chapter 3

Probing Composite Fermions Near Half-Filled Landau Levels

Mansour Shayegan

*Department of Electrical Engineering, Princeton University,
Princeton, NJ 08544, USA*
shayegan@princeton.edu

Composite fermions (CFs), exotic particles formed by pairing an even number of flux quanta to each electron, provide a fascinating description of phenomena exhibited by interacting two-dimensional electrons at high perpendicular magnetic fields and low temperatures. At and near Landau level half-fillings, CFs occupy a Fermi sea. Here we probe this Fermi sea via geometric resonance measurements, manifesting minima in the magnetoresistance when the CFs' cyclotron orbit diameter becomes commensurate with the period of a periodic potential imposed on the plane. The measured positions of the geometric resonance minima exhibit an asymmetry with respect to the field at $\nu = 1/2$, and suggest that the Fermi sea area is determined by the density of the *minority* carriers in the lowest Landau level, namely electrons for $\nu < 1/2$ and holes for $\nu > 1/2$. In spite of the observed asymmetry, the positions of the geometric resonance minima do exhibit particle-hole symmetry to a high degree when properly analyzed. We also report measurements of CF Fermi sea shape, tuned by the application of either parallel magnetic field or uniaxial strain. The strain-induced results reveal that the Fermi sea anisotropy for CFs (α_{CF}) is less than the anisotropy of their low-field hole (fermion) counterparts (α_F), and closely follows the relation $\alpha_{CF} = \alpha_F^{1/2}$. We also see evidence for fully spin-polarized CFs near $\nu = 1/4$ in the lowest Landau level, as well as near $\nu = 5/2$ in the excited Landau level. The latter data are consistent with the 5/2 fractional quantum Hall effect being a topological p-wave paired state of CFs. Finally, we review measurements in bilayer systems in wide quantum wells and double quantum wells. In wide wells, even when the system hosts a fractional quantum Hall state at $\nu = 1/2$, we observe a CF Fermi sea that is consistent with the total carrier density, favoring a single-component state. In the double quantum well system, we use the CF geometric resonances observed in one layer to probe a Wigner crystal state in the other layer which has a much lower density and filling factor.

Contents

1. Introduction

Low-disorder two-dimensional electron systems (2DESs), such as those confined to selectively-doped semiconductor structures, provide nearly ideal systems for studying electron-electron interaction phenomena. At low temperatures and in high magnetic fields, when the electrons' Coulomb energy dominates over their kinetic energy, the ground state of such a "clean," interacting system is often no longer a simple system of randomly-moving electrons. Rather, various highly correlated electron states, such as the fractional quantum Hall liquid, Wigner crystal, and striped phases are possible.[1,2] Studies of these states have been at the forefront of condensed matter research for the past four decades.

The physics of an interacting 2DES at and near the Landau level (LL) filling factor $\nu = 1/2$ is particularly fascinating. The composite fermion (CF) formulation[2-4] provides a beautiful and powerful description of this physics and the observed phenomena. Exploiting the transmutability of the statistics in 2D, a gauge transformation that binds an even number of magnetic flux quanta to each electron, maps the 2DES at half-filled LLs to a system of CFs at a vanishing *effective* magnetic field. ($\Phi_0 = h/e$ is the flux quantum, where h is the Planck constant and e is charge of an electron.) Such transformation also elegantly maps the *fractional* quantum Hall states (FQHSs) observed in the 2D *electron* system to the *integer* quantum Hall states for the CF system. Moreover, since the effective magnetic field is zero at filling factor $\nu = 1/2$, the CFs possess certain Fermi-liquid-like properties. Most notably, they have a Fermi sea with a well-defined Fermi contour and Fermi wave vector (Fig. 1), and are able to support phenomena such as ballistic transport. This has been verified over the last 25 years in several experiments where the commensurability, i.e. the geometric resonance (GR), of the CF cyclotron orbit diameter with certain lateral features imposed on the potential seen by the CFs is measured (see Fig. 2).[5-26]

Here we present a brief review of our recent experiments[12-24] which directly probe, through GR measurements, the size and shape of the CF Fermi sea near

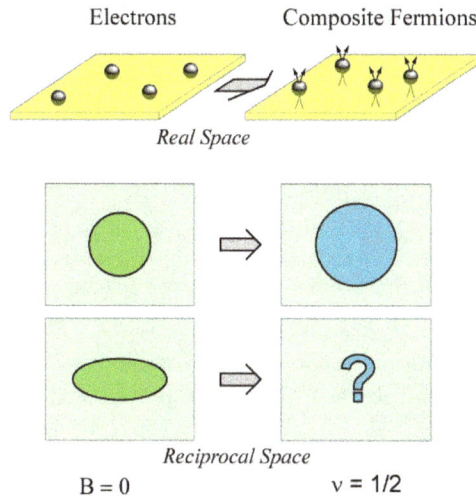

Fig. 1.   Top panel: Schematic diagrams of 2D electrons and CFs in real space. Lower panels: Schematic diagrams of isotropic and anisotropic electron (left) and corresponding CF (right) Fermi contours.

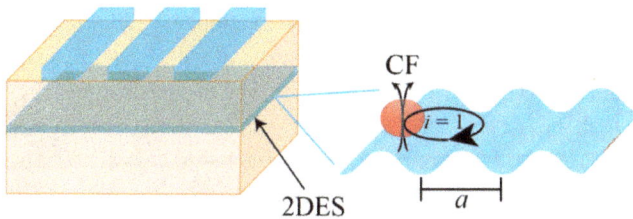

Fig. 2.   Schematic figure showing a sample whose surface is patterned with stripes of negative electron-beam resist with period a, in order to impose a strain-induced, periodic density modulation on the 2DES. Near a half-filled LL, and CFs execute cyclotron motion whose orbit diameter is determined by their Fermi wave vector. When this cyclotron orbit becomes commensurate with the period of the modulation, the magnetoresistance exhibits a minimum, signaling a geometric resonance (GR). We use the magnetic field position of the GR to determine the CF Fermi wave vector. [Note Ref. 23.]

filling factors $\nu = 1/2$, $3/2$, $5/2$, as well as near $\nu = 1/4$ where four-flux CFs are prevalent. In Sec. 2 we discuss our experimental technique for GR measurements. Our main focus for data on standard (single-layer) 2D systems in the *lowest* LL will be on the $\nu = 1/2$ CFs (Sec. 3) where we address two fundamental questions: (1) What determines the Fermi wave vector of CFs? (2) In a 2DES with an *anisotropic* Fermi sea, do the CFs inherit some anisotropy in their Fermi sea? We also examine, in Sec. 4, the CF Fermi sea near $\nu = 3/2$ and find that the CFs can be either fully or partially spin polarized, depending on the 2DES density and the width of the quantum well (QW) in which they are confined. In Sec. 5, we present GR data near $\nu = 1/4$ where four-flux CFs are present.

We also present, in Sec. 6, CF data in the *first excited* LL, near $\nu = 5/2$. Unlike in the lowest LL, where the electrons are in a *compressible* state at $\nu = 1/2$, $1/4$, or $3/2$, the 2D electrons at $\nu = 5/2$ condense into an *incompressible* FQHS. Nevertheless, our GR data show that, very near $\nu = 5/2$, there is indeed a CF Fermi sea whose measured Fermi wave vector is consistent with the CFs being fully spin polarized.[23] Such full spin polarization is a pre-requisite for the 5/2 FQHS being non-Abelian and therefore of use in topological quantum computing.

Finally, we discuss data in two other systems. First, in Sec. 7, we show that CF GRs can also be observed near $\nu = 1/2$ in *wide* GaAs QWs where the ground state at $\nu = 1/2$ in the lowest LL is an incompressible FQHS.[18,19] The data reveal that, even though the charge distribution in the wide QW is bilayer-like, the GR positions are consistent with a CF Fermi sea size that corresponds to the *total* electron density. This suggests that, at filling factors very close to $\nu = 1/2$ where the incompressible FQHS is formed, the CFs behave as a single-layer system.

Second, in Sec. 8, we present data in a bilayer system where the electrons are confined to *two* separate, but closely spaced, GaAs QWs.[25,26] The bilayer system has unequal layer densities: The top ("probe") layer is near its $\nu = 1/2$ and hosts a sea of CFs while the bottom layer has a very low density and is at very low fillings ($\nu \ll 1$) where a Wigner crystal (an ordered array of electrons) is expected to form. The magnetoreistance data in this system exhibit GRs of the CF layer, induced by the periodic potential of Wigner crystal electrons in the other layer, and provide a unique, direct glimpse at the symmetry of the Wigner crystal, its lattice constant, and melting. They also demonstrate a striking example of how one can probe an exotic many-body state of 2D electrons using equally exotic quasiparticles of another many-body state.[25–27]

2. Experimental Geometry

The samples used in all our studies contain 2D electron (or hole) systems confined to GaAs QWs grown by molecular beam epitaxy. The QW in each sample is flanked on its sides by undoped (spacer) and doped layers. The typical 2D density is $\sim 1.5 \times 10^{11}$ cm^{-2}, and the low-temperature mobility is $\sim 10^7$ cm^2/Vs (for 2D electron systems) or $\sim 10^6$ cm^2/Vs (for 2D hole systems). The magnetotransport measurements are mostly performed in ^{3}He cryostats with a base temperature of 0.3 K or dilution refrigerators with base temperatures of ~ 30 mK, and using a low-frequency lock-in technique. The high magnetic fields are provided by either superconducting magnets (up to 18 T), or resistive and/or resistive-superconducting (hybrid) magnets (up to 45 T), available at the National High Magnetic Field Laboratory in Tallahassee, Florida.

In order to experimentally measure the *shapes* of the Fermi seas for the zero-field particles or for CFs, we use GR (also known as commensurability, or Weiss) oscillations in samples patterned with a unidirectional periodic potential modula-

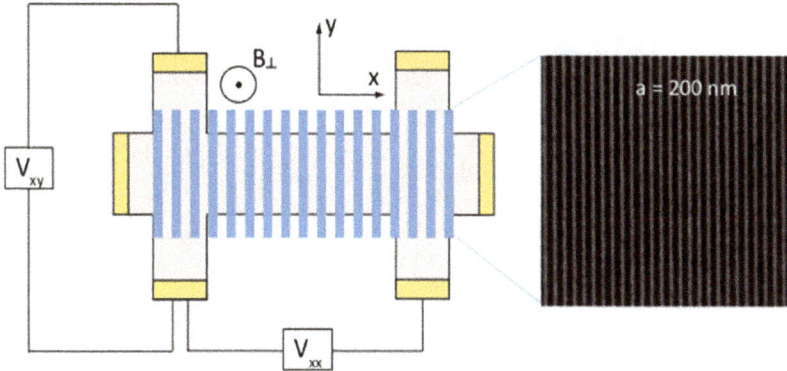

Fig. 3. Sample geometry used for GR measurements: A Hall bar with stripes of negative electron-beam resist placed on its surface in order to impose a strain-induced, periodic density modulation in the 2DES. On the right, the micrograph of a typical, patterned, sample surface (with a period of $a = 200$ nm).

tion.[28–32] The geometry is depicted schematically in Figs. 2 and 3. The sample is patterned into the shape of a Hall bar, and a grating which consists of stripes of negative electron-beam resist, with a period of $a \sim 200$ nm, is deposited on its surface. The negative electron-beam resist imparts a periodic, unidirectional strain to the sample surface which, thanks to the piezo-electric effect in GaAs, causes a periodic modulation of the 2D electron (or hole) density.[12–24,33–39] In a perpendicular magnetic field ($B_\perp$), the magnetoresistance of such a sample with a unidirectional density modulation exhibits minima at the commensurability or *geometric resonance* condition[8–24,28–39]:

$$2R_C/a = i \pm 1/4 \quad (i = 1, 2, 3, \dots), \tag{1}$$

where $2R_C = 2\hbar k_F/eB_\perp$ is the cyclotron diameter *along* the modulation direction and a is the period of the potential modulation; k_F is the Fermi wave vector *perpendicular* to the modulation direction.[40] The $-1/4$ phase factor corresponds to an *electrostatic* potential modulation, which is applicable to charged particles near zero magnetic field, while $+1/4$ applies to a *magnetic* modulation, which is relevant to CFs[8–24]; see Sec. 3.2 for a more detailed discussion of the GR condition for CFs. The $B_\perp$ positions of the magnetoresistance minima therefore provide a direct measure of k_F. Moreover, as we describe later, the anisotropy of k_F can also be directly determined from measurements along the two perpendicular arms of an L-shaped Hall bar.

In Fig. 4 we show an example of the GR features observed in one of our high-mobility 2DES samples whose surface is patterned with stripes of electron-beam resist. The trace shows a pronounced positive magnetoresistance at very low magnetic fields, characteristic of 2DESs with a periodic density modulation.[28–32,37] At higher fields, between $\simeq 0.04$ and 0.25 T, numerous minima are seen whose positions agree very well with the expected GR resistance minima according to the electro-

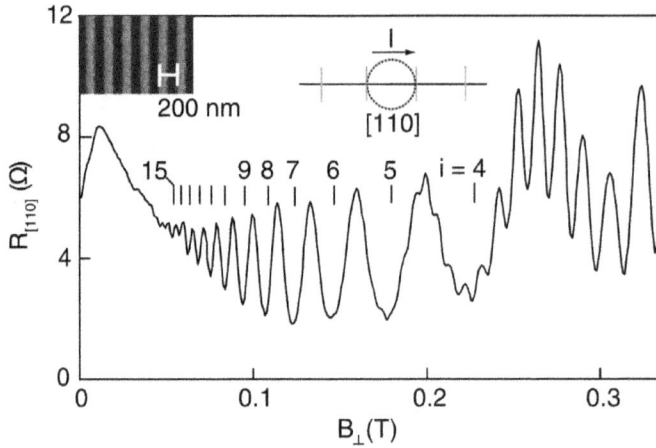

Fig. 4. Low-field magnetoresistance, measured at $T = 0.30$ K, in a very high-mobility 2DES along the [110] direction, showing pronounced GR oscillations. The 2D electron density is $n \simeq 2.84 \times 10^{11}$ cm^{-2}, and the mobility is $\mu = 18.4 \times 10^6$ cm^2/Vs. Vertical lines mark the positions of the expected GR resistance minima according to the electrostatic GR condition $2Rc/a = i - 1/4$; see Eq. (1). Shubnikov–de Haas oscillations are visible on top of the GR oscillations above 0.2 T. The large number of GR minima (up to $i \simeq 17$) attests to the very high quality of the sample and the periodic modulation. Inset: A scanning electron microscopic image of the 200-nm-period grating of negative electron-beam resist.[37]

static GR condition $2R_C/a = i - 1/4$; see Eq. (1). Above $\simeq$0.2 T, Shubnikov–de Haas oscillations are visible on top of the GR oscillations. The large number of GR minima (up to $i \simeq 17$) attests to the very high quality of the sample and the periodic modulation.

3. Geometric Resonance Data Near $\nu = 1/2$

3.1. Electron-flux composite fermion data: Overview

Figure 5 shows a magnetoresistance trace at $\theta = 0$ ($B_\parallel = 0$) along the [110] Hall bar of a 40-nm-wide QW which contains a high-mobility 2DES, and whose surface has stripes of electron-beam resist with an $a = 200$ nm period. It exhibits prominent GR features near $\nu = 1/2$, including a characteristic, V-shaped, resistance dip at $\nu = 1/2$, followed by several resistance minima on each side of $\nu = 1/2$ [marked by $i = 1, 2, 3$ in Fig. 5(b)] and flanked by regions of rapidly rising resistance. The positions of the resistance minima follow closely those expected from the *magnetic* commensurability condition [Eq. (1)] for fully *spin-polarized* CFs with *circular* Fermi sea, namely $2R_C^*/a = i + 1/4$, where $2R_C^* = 2k_F^*/eB_\perp^*$ is the CF cyclotron orbit diameter, $k_F^* = (4\pi n^*)^{1/2}$ is the CF Fermi wave vector, $n^* = n$ is the CF density, and $B_\perp^* = B_\perp - B_{\perp,\nu=1/2}$ is the *effective* magnetic field seen by the CFs near $\nu = 1/2$. (Throughout this article, we use "*" to denote CF parameters.) This

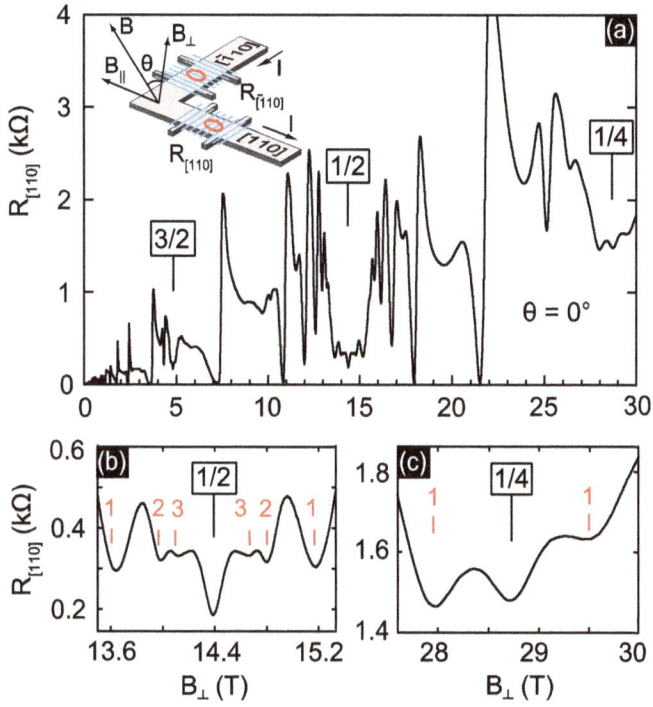

Fig. 5.   Inset: An L-shaped Hall bar with a periodic superlattice of negative electron-beam resist. (a) Magnetoresistance trace, measured at $T = 0.3$ K, from the [110] Hall bar of a 40-nm-wide, GaAs QW containing a 2DES at density $n = 1.74 \times 10^{11}$ cm^{-2} and a very small ($< 1\%$) periodic density modulation. (b) and (c) Prominent GR resistance minima are seen near $\nu = 1/2$ and $1/4$. The positions of the resistance minima expected for GR of fully spin-polarized CFs with a circular Fermi contour are marked with indexed vertical lines (see text). [Note Ref. 14.]

is consistent with previous reports of GR features for CFs near $\nu = 1/2$,[8–11] except that here we see the additional $i = 2$ and 3 minima, attesting to the very high quality of the sample and the periodic potential. Also, there are subtle deviations of the positions of these minima from the traditionally expected values, as we discuss in Sec. 3.2. Note that GR resistance minima are also seen near $\nu = 1/4$ at very high fields [Fig. 5(c)], providing evidence for the direct observation of ballistic transport and GR for *four-flux* CFs.[24] The positions of these resistance minima also match closely the expected values based on a *magnetic, fully-spin-polarized* GR condition equivalent to the one near $\nu = 1/2$.[24] We will discuss the GR data for four-flux CFs in more detail in Sec. 5.

3.2. *What determines the Fermi wave vector of composite fermions?*

In earlier experimental studies of the CF Fermi sea through GR measurements, it was assumed that the density of the CFs (n^*) near $\nu = 1/2$ is fixed and is equal to the 2D electron density n.[5–13] This assumption, combined with the assumption of the validity of Eq. (1), leads to *symmetric* field positions of the GR minima with respect to the position of $\nu = 1/2$; the positions of the red markers in Fig. 5(b) are based on these assumptions. Given the quality of the earlier experimental data and the accuracy of the positions of the GR minima, these assumptions sounded reasonable, as the agreement with the experimental data was acceptable. The data of Fig. 5(b), however, give a clear hint that there is a pronounced *asymmetry* in the magnetic field positions of the GR minima with respect to the field at $\nu = 1/2$: while the positions of the GR minima on the right-hand-side of $\nu = 1/2$ agree very well with assuming $n^* = n$, on the left-hand-side the observed minima are closer to the position of $\nu = 1/2$. A trivial possibility is that the 2D electron density changes slightly with increasing magnetic field. However, this is ruled out by the fact that the field positions of the FQHSs observed in the same sample are quite consistent with those expected based on the filling factors and the 2D electron density, as seen by the vertical dotted lines in Fig. 6(a).[15] This is true for the FQHSs observed on *both* sides of $\nu = 1/2$.

The asymmetry in the field positions of the GR minima [Figs. 5(b) and 6(b)], prompted Kamburov *et al.* to conduct a systematic study of its details.[15] A similar

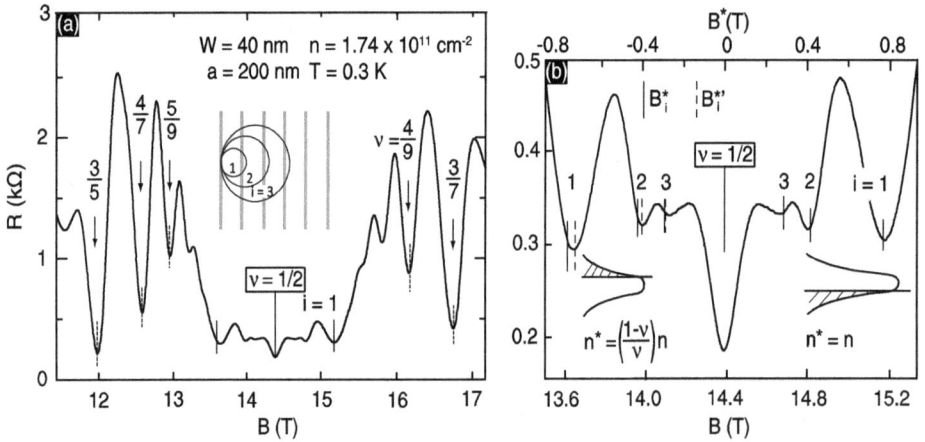

Fig. 6. (a) Magnetoresistance trace for a 2DES with density $n = 1.74 \times 10^{11}$ cm^{-2} and subjected to a periodic potential modulation, exhibiting strong CF GR minima near $\nu = 1/2$. The inset schematically shows the CR condition of the quasi-classical CF cyclotron orbits, marked as $i = 1$, 2, and 3, with a periodic potential modulation. Dotted vertical lines mark the expected positions of the FQHSs, based on the 2D electron density. (b) The CF GR minima are shown in greater detail. Vertical solid lines mark the expected positions of the GR resistance minima when the CF density (n^*) is assumed to be equal to the electron density; these positions are symmetric about $\nu = 1/2$. If n^* equals the density of *minority* carriers in the lowest LL, then the expected positions for the $B < B_{1/2}$ side are those shown with dashed vertical lines; the positions for $B > B_{1/2}$ are unchanged. The schematic insets indicate the basis of the CF minoirity density model which assumes that CFs are formed by the minority carriers in the lowest LL (hatched parts of the broadened level). It is clear that the dashed lines match the observed minima much more closely, supporting the "minority" density model. [Note Ref. 15.]

asymmetry was observed in a number of 2D electron *and* 2D hole samples, with different QW width, 2D carrier densities, and modulation periods.[15] Moreover, assuming that Eq. (1) is valid, it was found that this unexpected asymmetry is consistent with the CFs' Fermi wave vector being determined by the *minority* carriers in the lowest LL,[15] namely, if we assume that the density of CFs, n^*, is equal to the density of occupied states (*electrons*) for $\nu < 1/2$ while, for $\nu > 1/2$, n^* is equal to the density of unoccupied states (*holes*) (see Fig. 6(b) and also Ref. 15). This would suggest a subtle breaking of the particle-hole symmetry for CFs near $\nu = 1/2$. (Note that the observed GR minima are *also asymmetric* in their filling factor positions with respect to $\nu = 1/2$, i.e. the minima are *not* observed at ν and $(1 - \nu)$, when the magnetic field is varied with n and a held fixed; see Ref. 15 and the *Note Added* at the end of this chapter.)

The results described above stirred a bit of excitement, and stimulated theoretical work, trying to understand the relation between predictions of different CF theories and questions of symmetry about $\nu = 1/2$, and more generally assessing the validity of particle-hole symmetry for CFs.[41–59] One study[41] suggests that our observation implies the existence of an "anti-CF liquid," a new state of matter distinct from the "CF liquid". Numerical calculations by Balram *et al.*,[44,45] on the other hand, appear to indicate that, similar to the experimental findings, k_F^* is determined by the density of *minority* carriers in the lowest LL, namely $k_F^* = (4\pi n)^{1/2}$ for $\nu < 1/2$, while $k_F^* = (4\pi p)^{1/2}$ for $\nu > 1/2$ where p is the density of holes in the lowest LL.

More generally, partly inspired by questions of particle-hole symmetry, Son[43] introduced an alternate to the standard CF theory, in which CFs have the character of Dirac fermions and their density is determined by the magnetic flux density rather than the density of electrons. Combined with Eq. (1), this leads to an asymmetry in the positions of the magnetoresistance minima, when one varies the magnetic field about the $\nu = 1/2$ value, as is done in our experiments.

A full prediction for the magnetoresistance traces, based on Son's Dirac theory was carried out by Cheung *et al.*[52] Their results, and comparison with the experimental data, are shown in Fig. 7. The calculations exhibit an asymmetry which is qualitatively similar to the one observed in the experimental data: the field positions of resistance minima for $\nu < 1/2$ are farther away from the $\nu = 1/2$ field than the $\nu > 1/2$ minima are. This is because in the calculations the CF wave vector for the CFs is not fixed as a function of magnetic field but varies as the inverse of the magnetic length: $k_F^* \sim l_B^{-1} \sim B_\perp^{1/2}$.[52]

Although the predictions of the Dirac theory are different from what might have been expected based on a simple application of the usual CF theory based on Ref. 4, it was shown in Refs. 51 and 52 that the two theories actually make the same predictions for the positions of the magnetoresistance GR minima, when they are properly evaluated. The reason is that the electric field at finite wave vector q produced by an imposed grating is not completely screened by the electrons, and

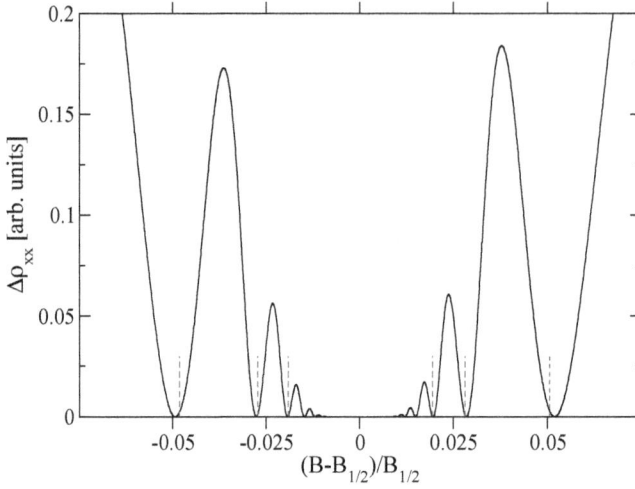

Fig. 7. Calculated oscillations in resistivity ρ_{xx} for electrons near half filling as predicted by the Dirac CF theory.[52] Similar to the experiments, the electron density n is fixed and the external magnetic field is varied. The vertical dashed bars correspond to the positions of the minima found in the experiments.[15] [Note Ref. 52.]

CFs, in the description of Ref. 4, feel the residual electric field as well as the effective magnetic field produced by the density modulation. Although effects of the electric field are smaller, by a factor of order q/k_F, than effects of the magnetic field, they are sufficient to give a small correction to the phase shift of $+1/4$ in Eq. (1), with opposite sign for filling factors larger or smaller than $1/2$. This correction brings the predictions for the positions of the magnetoresistance minima of the standard CF theory into agreement with the predictions of the Son's Dirac theory. In Ref. 52, the authors calculate the positions of the GR minima using both theories and indeed confirm that the positions agree to a very high degree of accuracy (0.002%).

However, note that the calculated minima positions in Fig. 7, while qualitatively consistent with the asymmetry seen in experimental data, disagree with the experimental data on both sides of $\nu = 1/2$. Most notably, in all the experimental traces, the position of the principal ($i = 1$) GR minimum on the $\nu < 1/2$ side agrees quite well with $k_F^* = (4\pi n)^{1/2}$ (see Fig. 6, and also Fig. 4 of Ref. 15), while according to the calculations this minimum would always be at slightly larger $B_\perp^*$ [because k_F^* grows slightly larger than $(4\pi n)^{1/2}$]. In a more recent theoretical study, Mitra and Mulligan[59] report that the inclusion of gauge-field fluctuations in the calculations leads to a better agreement with the experimental data.

It should be noted that in the above discussions, the asymmetries about $\nu = 1/2$ are defined in terms of magnetic field positions under the experimentally relevant conditions of fixed electron density and varying the magnetic field. By contrast, discussions of particle-hole symmetry in the theoretical literature most frequently assume varying electron density and constant magnetic field. The magnetic field asymmetries measured in our experiments are interpreted as a *qualitative* verification of particle-hole symmetry under the theorists' conditions.

In closing this section, it is worth emphasizing and reiterating our remarkable experimental observation: The asymmetric positions of the GR minima can be explained based on very simple assumptions, namely that Eq. (1) is valid, and that the density of CFs is fixed and equal to the density of *minority* carriers in the lowest LL, i.e. $k_F^* = (4\pi n)^{1/2}$ for $\nu < 1/2$, while $k_F^* = (4\pi p)^{1/2}$ for $\nu > 1/2$ where p is the density of holes in the lowest LL. This observation is true for all the samples we have studied, including 2D holes whose large effective mass implies severe LL mixing.[15] (The LL mixing parameter, κ, defined as the ratio of the Coulomb to cyclotron energies, is $\simeq 4$ for the 2D hole samples in Ref. 15.) Recently, we extended our studies to GaAs 2DESs with very low densities (down to $n \simeq 2.5 \times 10^{10}$ cm^{-2}) where LL mixing is also significant ($\kappa \simeq 2$) and the above observation still holds true, namely, the positions of the GR minima agree well (to within $\simeq 3\%$ in B^*) with the values expected from the above simple assumptions. Our experimental conclusion that the density of CFs appears to be fixed and equal to the density of *minority* carriers in the lowest LL is consistent with the results of numerical calculations of Ref. 44. However, according to the other theoretical works discussed in previous paragraphs, the observed positions of the GR minima can only be explained quantitatively if one uses sophisticated theories that take into account subtle phenomena and corrections such as the effect of the electrostatic contribution of the modulation, gauge-field fluctuations, etc. This brings up an interesting question. If at the end these theories indeed agree quantitatively with the experimental observations, is it simply a fortuitous coincidence that the results of such sophisticated theories lead to a simple conclusion, namely that the GR minima positions are consistent with a picture where the CF Fermi wave vector is determined by the electron density for $\nu < 1/2$, and hole density for $\nu > 1/2$? Or can there be a simple reason for this?

3.3. *Can composite fermions have an anisotropic Fermi sea?*

A natural question that arises when considering CFs is depicted in Fig. 1: in a 2D system with an *anisotropic* electron effective mass and Fermi sea (at $B_\perp = 0$), do the CFs retain such anisotropy? The answer to this question is not obvious.[60–62] One might argue that, because the CFs are a manifestation of the electron-electron interaction, their physical properties should depend only on the magnitude of the magnetic field that quantifies this interaction, and not on the electrons' zero-field properties. On the other hand, if the effective mass of electrons is anisotropic, the interaction could be anisotropic and thus lead to an anisotropic CF Fermi contour. This question was first addressed experimentally[62] in 2010 in a study of 2DESs in AlAs QWs where the zero-field electrons occupy conduction-band valleys with anisotropic effective mass. This anisotropy leads to anisotropic transport (mobility anisotropy).[63] It was found that, qualitatively similar to their $B_\perp = 0$ electron counterparts, CFs also exhibit a *transport* anisotropy, namely, the resistance at $\nu = 1/2$ is larger along the low-mobility (larger effective mass) axis of the $B_\perp = 0$ electron Fermi sea compared to the resistance along the high-mobility axis.[62] While this observation suggests that the CFs might possess an anisotropic Fermi sea, it does not provide conclusive or quantitative evidence for such anisotropy; e.g. an anisotropy in CF scattering time would also lead to anisotropic transport.

Since 2010, there has been a surge of interest in studies of anisotropy in interacting 2DESs and in particular the FQHSs and CFs.[13,14,17,20,22,24,64−81] This has also been partly motivated by a recent revelation by Haldane[67] that the FQHSs can possess a geometric degree of freedom intimately linked to the underlying anisotropy of the 2D system. (FQHSs, associated with Laughlin's wave function, had been historically considered to be isotropic and rotationally invariant.) Of fundamental interest is how such anisotropy affects properties of the CFs and the FQHSs.

Here we review our measurements of CF Fermi sea anisotropy in different GaAs 2DESs and 2D hole systems under varying experimental conditions. In Sec. 3.4 we describe the anisotropy induced by a parallel magnetic field. Section 3.5 deals with how the *"warping"* of the Fermi sea in a 2D hole system confined to a wide GaAs QW is transferred to CFs. In Sec. 3.6, we review our tuning and probing of the Fermi sea for 2D holes confined to a GaAs QW which is subjected to uniaxial strain. In this case, the results allow us to make a quantitative comparison: we find that the Fermi sea anisotropy for CFs (α_{CF}) is less than the anisotropy of their low-field hole (fermion) counterparts (α_F), and closely follows a remarkably simple relation: $\alpha_{CF} = \alpha_F^{1/2}$. (We define α as the ratio of Fermi wave vectors along the two principal directions of the anisotropic Fermi sea.)

3.4. *Composite fermion Fermi sea anisotropy induced by a parallel magnetic field*

In a 2D carrier system with finite layer thickness, a parallel magnetic field ($B_\parallel$) couples to the carriers' out-of-plane motion and leads to a deformation of the Fermi sea and the cyclotron orbit.[36−38,82] As illustrated schematically in Figs. 8(a)–8(d), the $B_\parallel$ component of the field elongates the cyclotron orbit along the direction of $B_\parallel$. Since the Fermi contour and the cyclotron orbit have the same shape but are rotated by 90°, the Fermi sea is then elongated in the direction perpendicular to $B_\parallel$.[83] In our experiments, we measure the CF GR resistance minima near filling factor $\nu = 1/2$ along two perpendicular arms of an L-shaped Hall bar, shown in Fig. 8(e), at different tilt angles (θ) where θ is the angle between the sample normal and the magnetic field.

The data for a 2DES confined to a 40-nm-wide GaAs QWs are shown in Figs. 9(a) and 9(b) for the two Hall bar directions.[14] When the magnetic field is purely perpendicular ($\theta = 0$), as shown in the bottom traces of Fig. 9 panels, the observed positions of the GR minima agree with the positions anticipated for a circular CF Fermi sea (dashed green lines in Fig. 9). As the sample is tilted to increase $B_\parallel$, the

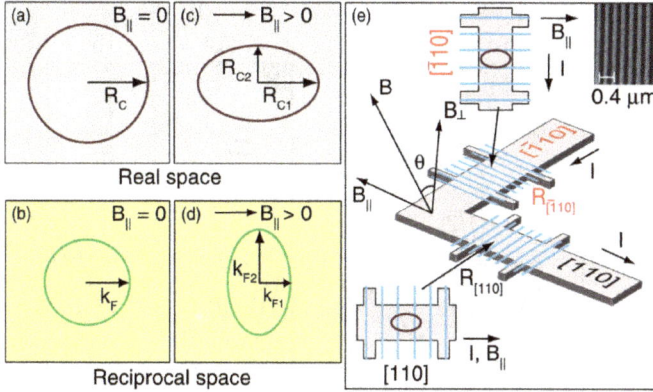

Fig. 8. (a), (b) The cyclotron orbit and the Fermi contour are shown, respectively, for an isotropic 2D system when $B_\parallel = 0$. (c), (d) If the 2D system has a finite (nonzero) thickness, applying $B_\parallel > 0$ distorts the cyclotron orbit and the Fermi contour. (e) The sample used for the Fermi sea anisotropy measurements has two Hall bars along two perpendicular directions, and $B_\parallel$ is introduced along the [110] direction by tilting the sample with respect to the magnetic field direction. The electron-beam resist grating covering the top surface of each Hall bar is shown as blue stripes. The orientation of the Hall bars and the resist gratings are chosen to probe the Fermi wave vectors in the [$\bar{1}$10] and [110] directions. The cyclotron orbits, given with brown lines, are shown for the case when the orbit diameter fits the grating period a in the [$\bar{1}$10] direction but is larger than a in the [110] direction. Inset: scanning electron microscope image of the electron-beam resist grating with a period $a = 200$ nm. [Note Ref. 13.]

minima move away [Fig. 9(a)] or *toward* [Fig. 9(b)] the magnetic field at $\nu = 1/2$ position, depending on the orientation of the Hall bar. These shifts are a direct measure of the changes in the size of the CF Fermi contour wave vectors along and perpendicular to $B_\parallel$.[14]

We use the positions of the GR resistance minima along the [$\bar{1}$10] and [110] directions to directly extract the magnitude of the CF Fermi wave vectors along [110] and [$\bar{1}$10], respectively. (Note that the GR condition along a given modulation direction gives the size of k_F^* in the direction *perpendicular* to the modulation direction.[40]) From the GR condition near $\nu = 1/2$ with $i = 1$ [see Eq. (1)], $k_F^* = (5/8)(eaB_\perp^*/\hbar)$. Here $B_\perp^*$ indicates the effective CF magnetic field at which the $i = 1$ resistance minimum is observed. Using this relation we convert the $B_\perp^*$ positions of the resistivity minima seen in Fig. 9 to the size of the CF k_F^* along the [$\bar{1}$10] and [110] directions. The results, normalized to k_{F0}^*, the value of k_F^* at $B_\parallel = 0$, are summarized in Fig. 10(a) for $B_\perp^* > 0$. Clearly, with increasing $B_\parallel$, the CF Fermi wave vector along [$\bar{1}$10] increases while along [110] it decreases. The data indicate a severe $B_\parallel$-induced anisotropy of the CF Fermi sea, up to a factor of $\simeq 2$ at $B_\parallel \simeq 25$ T for this 40-nm-wide GaAs QW.

To probe the role of the layer thickness on the anisotropy of the CF Fermi sea, similar measurements were performed on 30- and 50-nm-wide GaAs QWs.[14] The results, also summarized in Fig. 10(a), reveal that the wider the QW, the larger

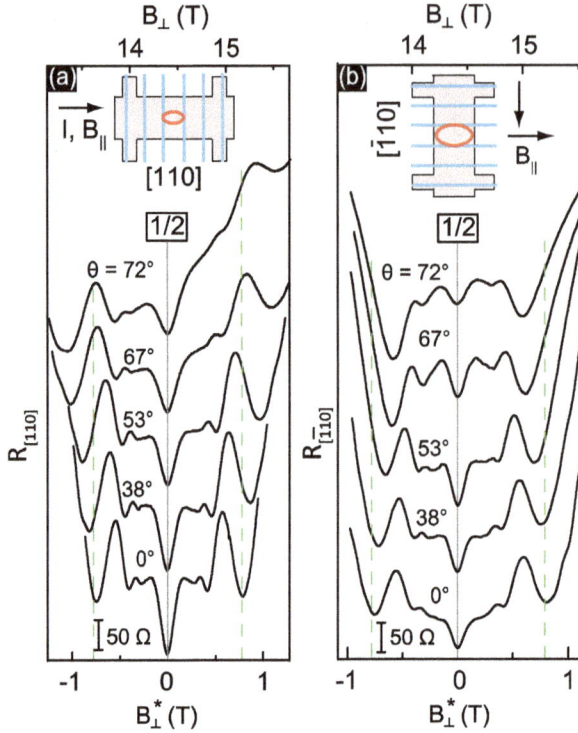

Fig. 9. Evolution of the magnetoresistance near $\nu = 1/2$ for 2D electrons confined to a 40-nm-wide QW sample, measured at $T = 0.3$ K along the [110] and [$\bar{1}$10] Hall bars. The tilt angle θ is given for each trace, and the traces are shifted vertically for clarity. The vertical green dashed lines mark the expected positions of the primary GR resistance minima if the CF cyclotron orbit were circular. In both panels, the scale for the applied *external* field $B_\perp$ is shown on top while the bottom scale is the *effective* magnetic field $B_\perp^*$ felt by the CFs. As the sample is tilted at an angle θ to introduce $B_\parallel$ along [110], the resistance minima for the [110] Hall bar in (a) move *away* from $\nu = 1/2$ while those in the [$\bar{1}$10] Hall bar in (b) move *toward* $\nu = 1/2$, revealing the distortion of the CF Fermi sea. [Note Ref. 14.]

the anisotropy. (In the 50-nm-wide QW sample, at large $B_\parallel$ (>15 T), the sample resistance dramatically rises near $\nu = 1/2$ and precludes us from measuring the CF GR minima.) The correlation between the larger anisotropy and QW width is best seen in Fig. 10(b) which shows the ratio of the measured k_F^* along [$\bar{1}$10] and [110] for each QW. This ratio is getting close to 1.6 at $B_\parallel = 15$ T for the 50-nm-wide QW, indicating a severe distortion as a result of $B_\parallel$. In contrast, the ratio at $B_\parallel = 15$ T is $\simeq 1.3$ for the 40-nm-wide QW, and only $\simeq 1.1$ for the 30-nm-QW.

Next we discuss the shape of the CF Fermi sea. Since in our experiments we measure k_F^* only along two specific, perpendicular directions, we cannot rule out a complicated shape. However, our data are consistent with nearly elliptical CF Fermi contours. As seen in Fig. 10(b) inset, the geometric mean of the two k_F^*'s we measure along [$\bar{1}$10] and [110], divided by k_{F0}^*, the wave vector expected for a

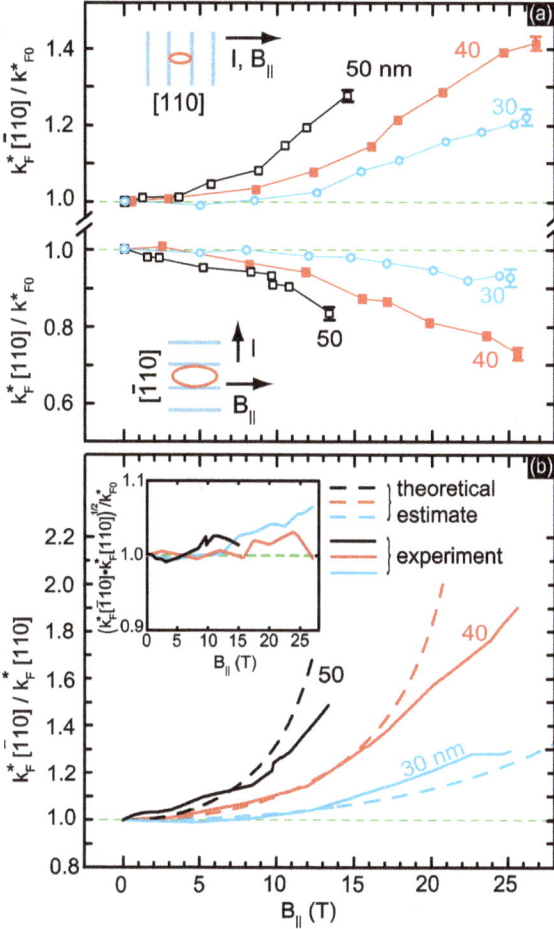

Fig. 10.   (a) Deduced values of the CF Fermi wave vectors k_F^* from the $B_\perp^* > 0$ positions of the $i = 1$ GR minima along the $[\bar{1}10]$ and $[110]$ directions, normalized to k_{F0}^* (see text). Data shown with open circles (blue), filled squares (red), and open squares (black) are from the 30-, 40-, and 50-nm-wide QWs, respectively. Representative error bars are included for each data set. (b) Solid lines: Relative anisotropy of the CF Fermi contours in each QW deduced from dividing the (interpolated) measured values of k_F^* along $[\bar{1}10]$ by those along $[110]$. Dashed lines: Theoretical estimate of the anisotropy using a perturbative model (see text). Inset: Geometric mean of the measured values of k_F^* along the two directions normalized to k_{F0}^* for each QW. [Note Ref. 14.]

circular CF Fermi contour whose enclosed area is equal to the density of CFs, is close to unity (to better than 7%) in the entire range of $B_\parallel$. This implies that the Fermi sea areas enclosed by elliptical contours with major and minor k_F^*'s equal to those we measure indeed enclose the area needed to account for the CFs.

The data presented in Figs. 9 and 10 provide direct and quantitative evidence that the CF Fermi sea becomes anisotropic with the application of $B_\parallel$. The strong dependence of the distortion on $B_\parallel$ and on the QW width implies that the origin of

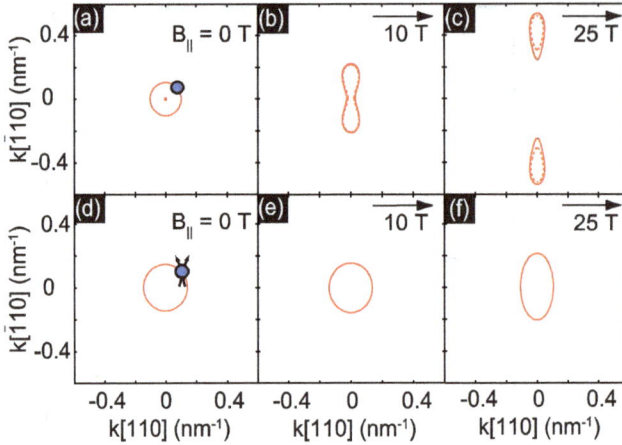

Fig. 11. (a)–(c) Calculated *electron* Fermi contours for a 40-nm-wide GaAs QW with density $n = 1.7 \times 10^{11}$ cm^{-2} at $B_\parallel = 0$, 10, 25 T. The solid and dotted contours correspond to the majority- and minority-spin electrons, respectively. (d)–(f) Evolution of the distortion of the CF Fermi contour with $B_\parallel$ *measured* in the 40-nm-wide QW. [Note Ref. 14.]

this anisotropy is the coupling between $B_\parallel$ and the out-of-plane motion of the CFs. Such coupling is known to severely distort the Fermi sea of *low-field* carriers.[36–38] Indeed, we have experimentally measured the Fermi sea for low-field electrons as a function of $B_\parallel$ and the data, which are overall in good agreement with the calculations, show severe Fermi sea distortions.[37,38] Figure 11 provides comparisons between the Fermi seas for the electrons [Figs. 11(a)–11(c)] in a 40-nm-wide GaAs QW sample at $B_\parallel = 0$, 10, and 25 T, calculated self-consistently based on an 8×8 Kane Hamiltonian,[37,38,84] and the CF Fermi seas at the corresponding $B_\parallel$, deduced from our measurements [Figs. 11(d)–11(f)]. For both electrons and CFs, the Fermi seas become significantly distorted with increasing $B_\parallel$, but the distortion is much more severe for the electrons. At sufficiently large $B_\parallel$ (>12 T), the electrons in fact split into a bilayer-like system with a *disconnected* Fermi sea; see, e.g. Fig. 11(c).[37,38] *Remarkably, however, the CF Fermi contour remains connected up to the highest* $B_\parallel = 25$ T.

To date, there are no theoretical calculations that treat the anisotropy of CF Fermi sea in the presence of $B_\parallel$. This is partly because it is theoretically challenging to treat rigorously and quantitatively the coupling of $B_\parallel$ to the quasi-2D electron system with a finite layer thickness.[72] Furthermore, besides its thickness, other parameters of the quasi-2D carrier system, such as the details of the band structure and effective mass, as well as the character of the LL where the CFs are formed, also play important roles in determining the anisotropy of the CF Fermi sea in a strong $B_\parallel$. For example, *hole-flux* CFs in a 17.5-nm-wide GaAs QW[13] exhibit a Fermi sea anisotropy of $\simeq 1.2$ at $B_\parallel = 15$ T. This is comparable to the anisotropy we observe for *electron-flux* CFs in the much wider 30- and 40-nm-wide QW samples [Fig. 10(b)], indicating that layer thickness is not the only parameter that determines the CF Fermi sea anisotropy.

There is, however, a simple model that can explain qualitatively the striking difference between the distortions of the Fermi seas for CFs and zero-field particles seen in Fig. 11. The model is based on a perturbative treatment of the role of $B_\parallel$ in modifying the Fermi sea of a 2DES confined to QW with finite (non-zero) width.[14,82] The model predicts that the Fermi distortion is determined by the ratio of the carriers' effective mass in the confinement direction (m_z) and parallel to the 2D plane ($m_\parallel$): a small $m_z/m_\parallel$ ratio leads to a small distortion and vice versa. Now it is well known that, thanks to interaction, $m_\parallel$ for CFs is significantly enhanced relative to $m_\parallel$ for the zero-field particles,[3,4,85,86] while m_z is not. Indeed, taking $m_\parallel^* \simeq 1$ and $m_z^* = m_z = 0.067$ for CFs, where both masses are in units of the free-electron mass, we find the distortions indicated by dashed lines in Fig. 10(b), in reasonable agreement with the experimental data (see Ref. 14 for details). Note that for the *zero-field electrons* in a GaAs QW, $m_\parallel = m_z = 0.067$ so that $m_z/m_\parallel = 1$; in this case the perturbative scheme does not apply and a much more severe distortion would be expected, consistent with the self-consistent calculations and the experimental observations.

The above simple model can also explain why the distortion of the Fermi sea is more severe for the CFs in 2D *hole* systems. As detailed in Ref. 13, a 2D hole system confined to a GaAs QW which is only 17.5 nm thick exhibits a Fermi sea anisotropy of $\simeq 1.2$ at $B_\parallel = 15$ T. This is comparable to the anisotropy observed for electron-flux CFs in the much wider 30- and 40-nm-wide QW samples (Fig. 10), indicating that layer thickness is not the only parameter that determines the CF Fermi sea anisotropy. It turns out that the GaAs 2D holes have a larger m_z ($\simeq 0.6$ instead of 0.067 for electrons),[84] so that the ratio $m_z^*/m_\parallel^*$ for hole-flux CFs is relatively large (compared to electron-flux CFs). This means that the simple model would predict a more sever Fermi sea distortion for hole CFs, qualitatively consistent with the experimental observations.[13,14]

We emphasize that there is no rigorous microscopic justification for the simple model described above. A quantitative explanation of the experimental data based on rigorous, theoretical calculations of the CF Fermi sea in the presence of $B_\parallel$ is therefore certainly desirable and would add to our understanding of CFs and their dynamics in $B_\parallel$.

In Secs. 3.5 to 3.7, we present results for 2D *hole* systems whose Fermi sea can be rendered anisotropic without the application of $B_\parallel$. As we will show, such data are particularly valuable as they can be compared to the predictions of rigorous theoretical calculations.

3.5. *Hole-flux composite fermion data: Overview*

The 2D hole system (2DHS) confined to a GaAs QW provides an interesting platform to study the CFs. The energy-band structure for 2DHSs is more complex.[84]

The 2D holes typically possess a much larger effective mass, and the bands can be anisotropic (warped), especially in wide QWs and at large 2D hole densities. Moreover, the energy bands are very sensitive to in-plane strain and can be made significantly anisotropic with the application of a minute amount of uniaxial strain (of the order of 10^{-4}). The low-temperature mobility of 2D holes, on the other hand, is typically about 10 times smaller than the 2DES at similar densities. Nevertheless, the 2DHS samples have sufficiently high quality to exhibit CF ballistic phenomena such as GRs.[12,13,17,22] Indeed, the CFs in a 2DHS were the first CF system for which an anisotropic Fermi sea was reported in parallel magnetic fields.[13] As we show below, the 2DHS allows us to learn novel properties of CFs.

Figure 12 illustrates the first observation of GR for hole-flux CFs as reported in Ref. 12. The 2DHS is confined to a GaAs QW of width 17.5 nm, and has a low-temperature mobility of $\simeq 1.2 \times 10^6$ cm^2/Vs. Similar to high-quality 2DESs, the (black) trace taken for the patterned section of the Hall bar exhibits pronounced CF GR features near $\nu = 1/2$ whose positions are consistent with Eq. (1), and assuming a fully spin-polarized, circular CF Fermi sea. Note in Fig. 12(d) that, here too, the position of the observed resistance minimum for $\nu < 1/2$ agrees very well with the position expected based on the density of CFs being equal to the density of holes, while for $\nu > 1/2$ the minimum is slightly closer to the $\nu = 1/2$ magnetic field and is more consistent with the CF density being equal to the density of minority carriers in the lowest LL. There are also hints of GR features near $\nu = 3/2$ [Fig. 12(c)], as well as clear indications of GRs (up to $i = 5$) for 2D holes near zero-magnetic field [Fig. 12(b)].

3.6. *Fermi sea anisotropy induced by uniaxial strain: Transference of anisotropy to composite fermions*

Figure 13(a) provides an example of the tunability, as well as complexity, of the 2D hole bands.[17,22,36,39,84] Here we show the results of self-consistent calculations for the Fermi contours of 2D holes confined to a (001) GaAs square QW of width 17.5 nm, and subjected to uniaxial, in-plane strain (ε). The splitting between the two $p+$ and $p-$ contours as well as the anisotropy of these contours is clear in Fig. 13(a) plots; the splitting comes from the strong spin-orbit interaction in 2DHSs. As is evident in Fig. 13(a), the Fermi contours and the corresponding cyclotron orbit trajectories of 2D holes become significantly anisotropic for minute values of ε, of the order of 10^{-4} [see Fig. 13(a)]. In our experiments we tune the Fermi sea anisotropy and measure it for both 2D holes[39] and hole-flux CFs.[22] Following the technique described in Ref. 87, we glue the sample to one side of a stacked piezo actuator whose shape can be tuned via applying voltage (V_P) to its leads, thus exerting the strain to the sample [Fig. 13(b)]. To monitor the Fermi sea shape, similar to the 2D electron case, we made L-shaped Hall bar samples with surface gratings that lead to a small, unidirectional, periodic potential (density) modulation for the 2DHS [Fig. 13(c)]. The samples reveal GR minima for both 2D holes and CFs, and we

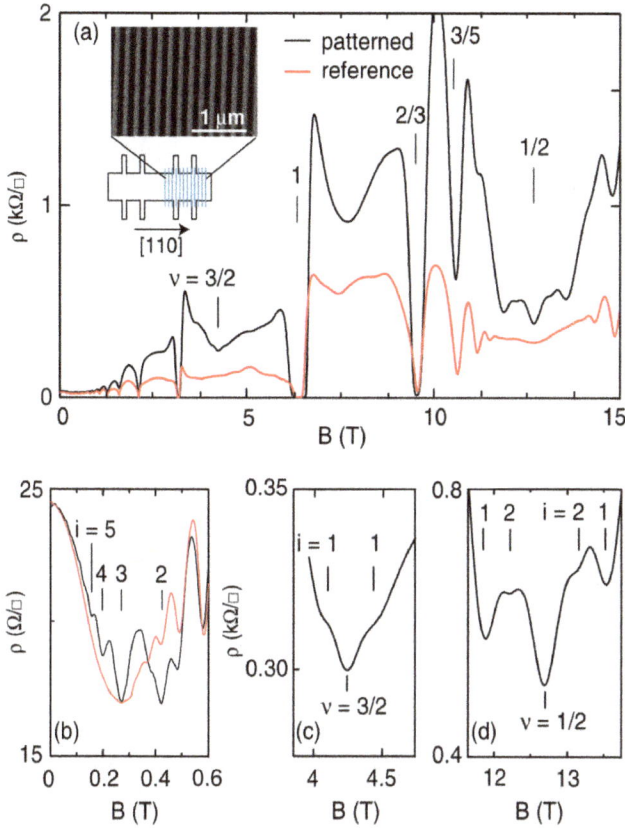

Fig. 12. (a) Magnetotransport data at $T = 0.3$ K from the patterned (upper, black trace) and reference (lower, red trace) regions of a GaAs 2D *hole* sample with a density of $p = 1.53 \times 10^{11}$ cm^{-2}. The modulation period for this sample is $a = 175$ nm. The inset shows a scanning electron micrograph of the patterned region of a different hole sample with $a = 200$ nm. (b) Low-field magnetoresistance of the patterned and unpatterned (reference) regions. The (black) trace for the patterned region shows GRs for holes. The expected field positions of the GR minima are marked with indexed vertical lines ($i = 2, 3, \ldots$), based on Eq. (1) and assuming electrostatic modulation. These minima are absent in the (red) trace for the reference region of the sample. Both traces show Shubnikov–de Haas oscillations starting around 0.4 T. (c), (d) GR features for CFs near $\nu = 3/2$ and $1/2$ and their expected positions (vertical marks, $i = 1, 2$) according to Eq. (1) and assuming magnetic modulation, and assuming a CF density equal to the density of the 2D holes at zero magnetic field. [Note Ref. 12.]

analyze the positions of the minima to determine the shape and anisotropy of the ballistic cyclotron orbits and Fermi contours. We find the data for 2D holes to be in quantitative agreement with the results of calculations shown in Fig. 13(a), although we cannot resolve the spin-orbit-split $p+$ and $p-$ spin subbands.[39]

In Fig. 14 we present magnetoresistance traces taken along the two perpendicular arms of the L-shaped Hall bar shown in Fig. 13(c) for a strain of $\varepsilon = -1.8 \times 10^{-4}$. For the two arms, the strain shifts the positions of the GR resistance minima observed on

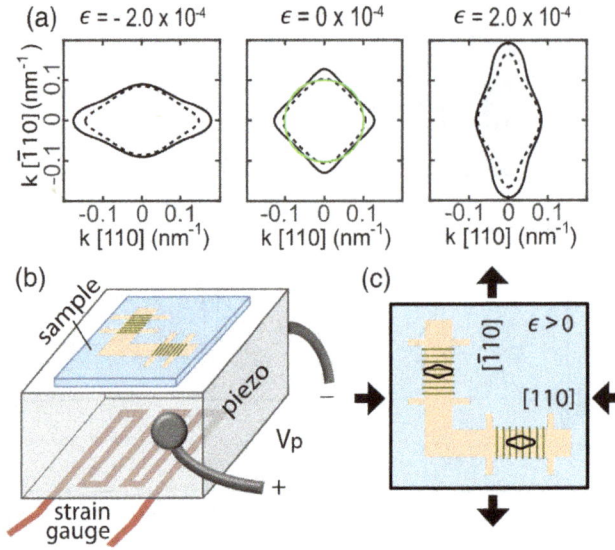

Fig. 13. (a) Calculated Fermi contours of GAAs holes at density $p = 1.8 \times 10^{11}$ cm^{-2} as a function of strain ε along the [$\bar{1}$10] direction. The solid and dashed contours represent two spin subbands, split by the spin-orbit interaction; the green circle with radius $k_0 = (2\pi p)^{1/2}$ shows a spin-degenerate, circular Fermi contour at the same density. (b) Schematic of the experimental setup showing a thinned GaAs wafer glued on a piezo actuator. A strain gauge mounted underneath measures the strain along [$\bar{1}$10]. (c) Sample fabricated to an L-shaped Hall bar has regions with electron-beam resist gratings on the surface. Thick arrows indicate the deformation of the crystal when a positive voltage V_P is applied to the piezo. The resulting deformed cyclotron orbits are shown in black; note that these are rotated by 90° with respect to the Fermi contours in reciprocal space.[83] The shapes of the orbits and therefore the Fermi contours are determined via GR measurements. [Note Ref. 22.]

the flanks of $\nu = 1/2$ in opposite directions, reflecting the CF Fermi sea anisotropy. This shift is more clearly seen in Fig. 14(c). From the positions of the minima, we deduce the CF Fermi wave vectors along the two perpendicular directions, as shown in Fig. 14(e). The Fermi wave vectors for the 2D holes, deduced from the low-field data [Fig. 14(b)], are summarized in Fig. 14(d), and show good agreement with the results of band calculations. Note that if we assume an elliptical Fermi contour for the CF Fermi contour as depicted in Fig. 14(e), its enclosed area indeed accounts for the density of CFs in the system (assuming a fully-spin-polarized system).

Figure 15 illustrates the highlight of the study in Ref. 22: comparison of the strain-induced Fermi sea anisotropy for CFs and holes. The measured anisotropy for CFs (α_{CF}), defined as the ratio of the Fermi wave vectors along the principle axes of the Fermi sea is shown by black circles, and the *square root* of the equivalent anisotropy for holes (α_F), by an orange curve. Remarkably, the measured α_{CF} for CFs essentially coincides with $\alpha_F^{1/2}$ over the entire range of strains applied in the experiments. This is particularly impressive because there are no fitting or adjustable parameters.

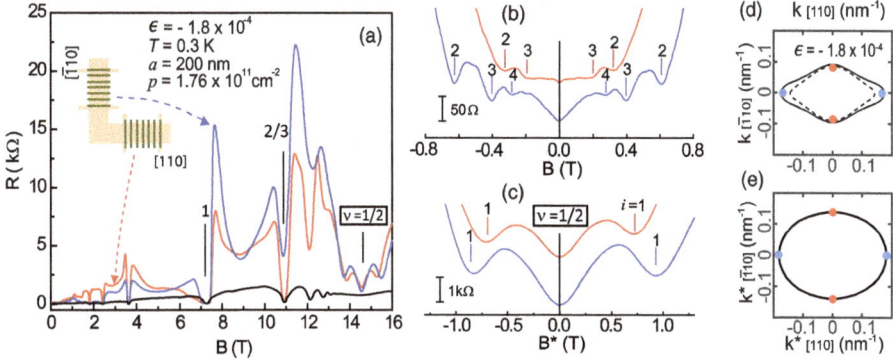

Fig. 14.   (a) Magnetoresistance traces taken from different regions of the Hall bar when strain $\varepsilon = -1.8 \times 10^{-4}$ is applied along [$\bar{1}$10]. The blue and red traces are for the patterned regions along the [$\bar{1}$10] and [110] arms, while the black trace is for an unpatterned region. (b), (c) The blue and red traces in (a) are shown enlarged, exhibiting GR features for: (b) holes at low fields, and (c) CFs at high fields near LL filling factor $\nu = 1/2$. The effective field for CFs in (c) is shown as B^*. The vertical lines in (b) and (c) indicate the positions of minima satisfying the GR conditions for holes and CFs (see text). (d) Calculated Fermi contours of spin-split holes at $p = 1.8 \times 10^{11}$ cm^{-2} and $\varepsilon = -1.8 \times 10^{-4}$. Red and blue dots represent the measured Fermi wave vectors along the [$\bar{1}$10] and [110] directions, using the traces in (b). (e) The elliptical Fermi contour for CFs, deduced from the measured Fermi wave vectors using the traces in (c). [Note Ref. 22.]

The data presented in Fig. 15 indicate that the Fermi sea anisotropy for CFs is less than the anisotropy of their low-field hole (fermion) counterparts, and closely follows a simple relation $\alpha_{CF} = \alpha_F^{1/2}$. This contradicts some of the theories which predict that α_F and α_{CF} should be the same.[61,74]   On the other hand, the results of recent density-matrix renormalization group calculations indicate that, for Coulomb interaction, the $\alpha_{CF} = \alpha_F^{1/2}$ relation is indeed obeyed.[76] Inspired by these findings, the question of Fermi sea anisotropy renormalization in interacting 2D Fermi systems was addressed more generally in a very recent study.[81] Using various theoretical techniques, Ref. 81 concludes that only for Dirac fermions with long-range Coulomb interaction there is a universal square-root decrease of the CF Fermi sea anisotropy, as observed in Ref. 22 experiments.

We would like to emphasize here that, in Refs. 62 and 22, some simple, heuristic and intuitive arguments were provided for a qualitative justification of the $\alpha_{CF} = \alpha_F^{1/2}$ relationship. These arguments rely on two simple and plausible assumptions: (i) a 2D system (at zero magnetic field) with an anisotropic Fermi sea can be mapped to an isotropic system with anisotropic Coulomb interaction, and (ii) in such a system the CF effective mass, which in general is expected to scale with the Coulomb interaction,[3,4] has the same anisotropy as the Coulomb interaction.[22,62]

An interesting related question is the impact of anisotropy on the strength of FQHSs. This was addressed in Ref. 22, focusing on the energy gap for the $\nu = 2/3$ FQHS. The sample used for the measurements has a slightly lower density $p = 1.3 \times 10^{11}$ cm^{-2} and, using a different cooldown procedure, larger strain values (ε up

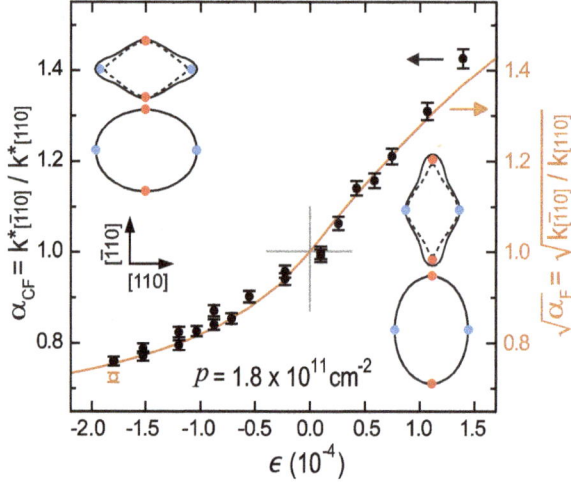

Fig. 15.   Strain-dependent Fermi sea anisotropy of holes (α_F) and hole-flux CFs (α_{CF}). Black circles are the measured α_{CF}. The orange curve represents the *square root* of the calculated α_F. The open orange circle shows the measured α_F for holes as described in Fig. 14. The left and right insets show the hole and CF Fermi contour shapes for $\varepsilon = -1.5$ and 1.4×10^{-4}. [Note Ref. 22.]

to 5.5×10^{-4}) and anisotropy (α_F as large as 3.3) were achieved, as shown in Fig. 16. The measured energy gap Δ, determined from the expression $R(T) \sim e^{-\Delta/2T}$, is 2.1 K for $\varepsilon = 0$, and it decreases only to 2.0 K even for a large anisotropy $\alpha_F = 3.3$. The small decrease of Δ is consistent with recent theoretical predictions,[74] suggesting that the FQHSs in the lowest LL are quite robust against anisotropy.

3.7.   *Composite fermions with a warped Fermi sea*

In this section we review data revealing that when the 2D carrier system at zero magnetic field possesses a severe "warping" in its Fermi sea, the warping is partially transferred to the CFs.[17] The 2D holes confined to a relatively wide GaAs QW provide a nearly ideal system for the measurements. Figures 17(b) and 17(c) show the calculated energy band dispersions and the Fermi contours of 2D holes in a wide, symmetric, GaAs QW, based on an 8×8 Kane Hamiltonian[84] which combines the Dresselhaus spin-orbit coupling and the non-parabolicity of the 2D hole bands. As seen in Fig. 17(c), the Fermi contour is 4-fold symmetric but significantly warped as a result of severe mixing between the heavy-hole and light-hole states. Spin-orbit coupling also causes the Fermi contours of two different spin species to split. As a result of warping, the Fermi wave vectors for both majority $(p+)$ and minority $(p-)$ spin contours along $[\bar{1}10]$ and $[110]$ are larger than the Fermi wave vector of a circular Fermi contour (shown as a red circle in Fig. 17(c)) which contains the same density of (spin-unpolarized) 2D holes.

   Figure 17(a) illustrates the schematic sample geometry used to probe the warped Fermi sea. Measurements on 2D holes at low magnetic field in this wide QW sample

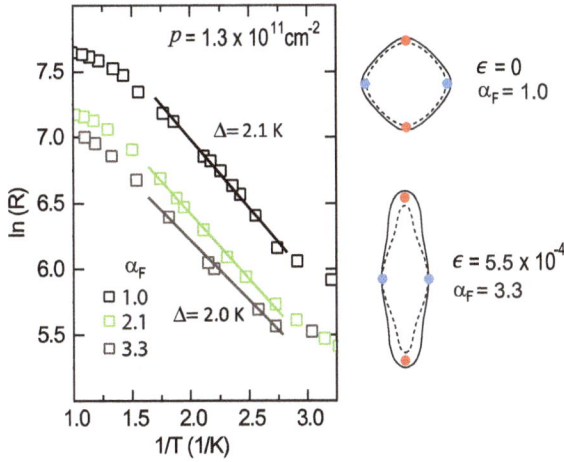

Fig. 16. (a) Longitudinal resistance at $\nu = 2/3$ is recorded for an unpatterned region of the 2D hole sample at different temperatures for energy gap (Δ) measurements, in cases of $\alpha_F = 1.0$, 2.1, and 3.3. The y-axis scale is $ln(R)$, where R is measured in Ohms. The slopes of the straight lines yield the energy gaps. The calculated 2DHS Fermi contour shapes are shown on the right, where the red dots indicatge the measured k along [$\bar{1}$10] and blue dots are determined based on calculations for $p = 1.3 \times 10^{11}$ cm^{-2}, respectively. [Note Ref. 22.]

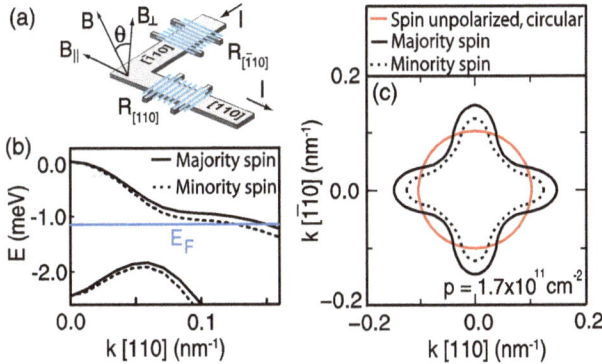

Fig. 17. (a) Schematics of the experimental geometry used to measure warped Fermi contours of 2D holes and hole-flux CFs. The electron-beam resist grating covering the top surface of each Hall bar arm is shown as blue stripes. (b) Dispersions for the two lowest, majority ($p+$) and minority ($p-$) spin subbands along [110] for a symmetric, 25-nm-wide, GaAs QW, calculated self-consistently at $B = 0$. (c) Calculated Fermi contours, exhibiting significant warping. For comparison, a circular Fermi contour containing the same density of (spin-unpolarized) 2D holes is also shown. [Note Ref. 17.]

show GR minima whose field positions are *not* consistent with a circular Fermi sea, but rather suggest a warped sea similar to that shown in Fig. 17(c). The magnetoresistance data at high magnetic fields, presented in Fig. 18 also clearly indicate a significant warping of the CF Fermi sea.[17] In this figure, the expected

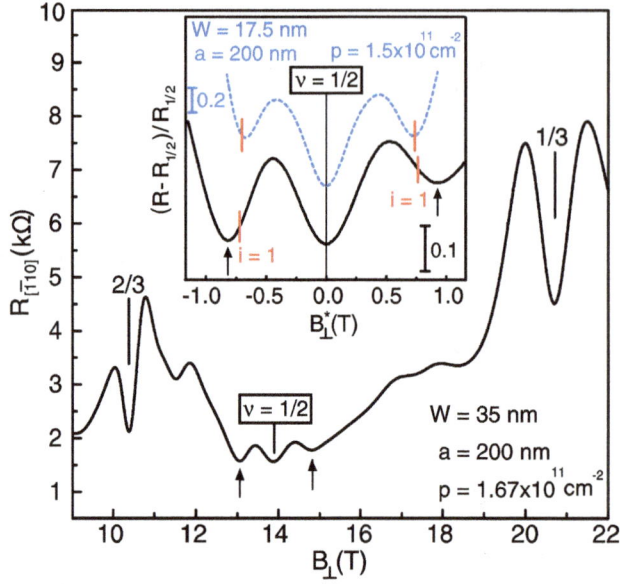

Fig. 18. (a) Magnetoresistance trace, taken at $T = 0.3$ K, from the [$\bar{1}$10] Hall bar for a 35-nm-wide GaAs QW containing a 2DHS. The two prominent minima near $\nu = 1/2$ are signatures of GR of CF cyclotron orbit diameter with the period ($a = 200$ nm) of the density modulation. Inset: Enlarged trace near $\nu = 1/2$ shows that the positions of the minima are measurably farther from $B_\perp^* = 0$ than expected for a circular Fermi contour of fully-spin-polarized CFs (marked by vertical red tick marks). For comparison, a trace (dashed blue) from a narrower 2DHS, which is confined to 17.5-nm-wide GaAs QW, is also included. In this case, similar to their 2D hole counterparts near zero magnetic field, CFs also show essentially no warping in their Fermi contour; this is evinced by the GR minima near $\nu = 1/2$, whose positions agree well with the red tick marks that are based on a circular Fermi contour. [Note Ref. 17.]

positions of the $i = 1$ GR minimum for CFs with a *circular* Fermi is marked by red marks (see Fig. 18 inset). For the narrow (17.5-nm-wide) QW sample, whose trace is shown by a dashed blue curve in the inset, the marked positions agree well with the observed minima. [More extensive data are presented for the narrow QW sample in Secs. 3.5 and 3.6; also, as shown in Fig. 13(a), the 2D holes in this sample do not possess a significant warping.] In contrast, the observed GR minima for the 35-nm-QW data (black trace) are clearly farther from the $\nu = 1/2$ position than expected for a circular Fermi sea, providing evidence for a warped CF Fermi sea.

It is worth noting that based on Fig. 18 data, and also on similar data taken on three other samples, it was concluded in Ref. 17 that the Fermi sea warping inherited by the CFs is smaller than the warping of their zero-field counterparts. Here "warping" is defined as k_F^* along [110] divided by k_F^* for a circular Fermi sea; see [Fig. 17(c)]. This conclusion is corroborated by the results of a recent theoretical study that also finds that the CFs in a warped band with 4-fold rotational symmetry inherit a finite, but small warping.[78]

4. Geometric Resonance Data Near $\nu = 3/2$

Compared to the $\nu = 1/2$ CFs, there are very few reports of GR for CFs near $\nu = 3/2$.[88] At $\nu = 3/2$, the lower-energy spin state of the lowest $(N = 0)$ LL, the $|0\uparrow\rangle$ state, is fully occupied, while its higher-energy spin state, $|0\downarrow\rangle$, is half filled. In a simple picture, one might expect the $\nu = 3/2$ CFs to be fully spin-polarized with the same spin $(|\downarrow\rangle)$ as the level they are formed in $(|0\downarrow\rangle)$. However, the CF spin polarization depends on the interplay between the Coulomb energy $(E_C = e^2/4\pi\varepsilon l_B)$ and the Zeeman energy $(E_Z = g\mu_B B)$, where $l_B = (\hbar/eB_\perp)^{1/2}$ is the magnetic length, ε is the dielectric constant, g is the Lande g factor, e is the electron charge, and μ_B is the Bohr magneton.[89–97] When E_C fully dominates over E_Z, the CFs can be partially spin polarized. This is indeed evinced by transitions observed between FQHSs with different spin polarizations as E_Z/E_C is varied.[89–97] The CF spin polarization near $\nu = 3/2$ can also be deduced directly from measurements of the size of the CF Fermi wave vector. However, such experiments had been elusive.[88] Here we summarize the results of our recent GR measurements which directly probe the size and shape of the $\nu = 3/2$ CF Fermi sea.[16] The results also have implications for the spin polarization of CFs near $\nu = 3/2$.

The measurement technique is again similar to the one used in the previous sections of this article. For this study, GR data for several GaAs 2DES samples with varying 2D electron density and QW width were measured.[16] In Fig. 19, we show the magnetoresistance traces at zero tilt angle $(B_\| = 0)$ along the two Hall bar arms of a 40-nm-wide QW sample with $n = 2.71 \times 10^{11}$ cm^{-2}. They exhibit prominent GR features near $\nu = 3/2$, including a characteristic, V-shaped, resistance dip, followed by resistance minima on the sides of $\nu = 3/2$ and flanked by regions of rapidly rising resistance. This behavior is qualitatively similar to the GR features near $\nu = 1/2$ shown throughout this article. GR features are absent in the trace from the unpatterned region, shown with a dashed line in Fig. 19.

In Fig. 19, the positions of the expected $i = 1$ GR minima are marked for fully-spin-polarized $(f = 1)$ CFs, assuming either magnetic $(+)$ or electrostatic $(-)$ periodic modulation [see Eq. (1)]. Note that the effective magnetic field seen by the CFs near $\nu = 3/2$ is $B^* = 3(B - B_{3/2})$, and the positions of the expected minima are calculated based on the simple assumption that the CF density is $n^* = n/3$.[16] It is clear that the observed minima agree well with the magnetic modulation, similar to the case of $\nu = 1/2$ CFs.

Figure 20 summarizes similar data taken for samples with different QW width (W) and density (n), as indicated below the self-consistently calculated charge distributions (in the absence of magnetic field) shown on the right ride of the figure. The density of the top trace, shown in Fig. 20 in blue, is significantly higher compared to the other traces. In the vicinity of $\nu = 3/2$, each trace exhibits GR features; their positions are marked with vertical black arrows. The GR features in the $W = 30$ nm and the lower-density $W = 40$ nm samples (bottom two traces in Fig. 20) are rather weak. As the QW width and/or the electron density increase,

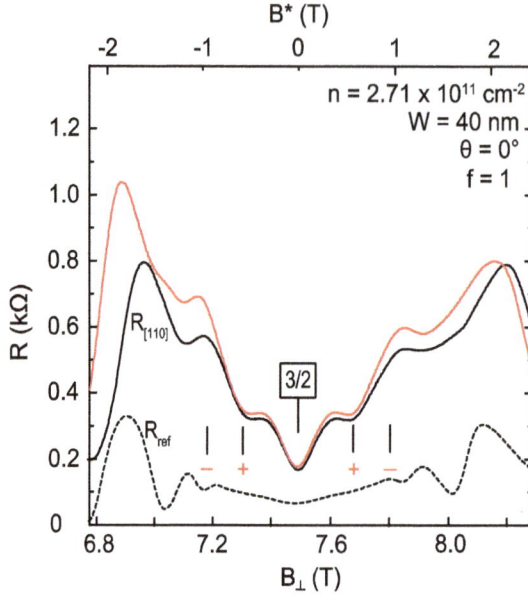

Fig. 19. Magnetoresistance traces (solid lines), taken at $T = 0.3$ K on a 2DES confined to a 40-nm-wide GaAs QW and patterned with periodic negative electron-beam resist stripes of period $a = 200$ nm, showing strong GR in the vicinity of $\nu = 3/2$. The black trace was taken along [110] and the red trace along [$\bar{1}$10]. Such features are absent in the reference (unpatterned) region data (dashed trace). Two sets of vertical lines mark the expected resistance minima for magnetic (+) and electrgostatic (−) GR based on Eq. (1) with $n^* = n/3$ and $B_\perp^* = 3(B - B_{3/2})$, where $B_{3/2}$ is the field at $\nu = 3/2$, and assuming fully-spin-polarized CFs ($f = 1$). The observed minima agree with the *magnetic* GR condition. [Note Ref. 16.]

the features develop into pronounced resistivity minima and move outward from $\nu = 3/2$. The positions of the minima are slightly asymmetric with respect to $\nu = 3/2$, with those on the $B^* < 0$ side being closer to $\nu = 3/2$.

A detailed analysis of the positions of the GR features seen in Fig. 20 is provided in Ref. 16. The main conclusions are two-fold. First, the slight asymmetries observed in the positions of GR features with respect to $B_{3/2}$ suggest that, similar to the $\nu = 1/2$ CFs, the density of the CFs near $\nu = 3/2$ is determined by the density of minority carriers.[16] Second, and more importantly, at a fixed electron density of $\simeq 1.8 \times 10^{11}$ cm^{-2}, as the QW width increases from 30 to 60 nm, the CFs show increasing spin polarization. This can be seen in Fig. 20: the positions for the GR features for the 60-nm-wide QW agree with the expected positions for fully-spin-polarized CFs ($f = 1$, solid vertical green and red lines), while for the narrower wells the minima are observed at smaller values of $|B^*|$, as expected for spin-unpolarized CFs ($f = 0.5$, dashed vertical green and red lines). This dependence of CF spin polarization on QW width is attributed to the enhancement of the Zeeman energy relative to the Coulomb energy in wider wells where the latter is softened because of the larger electron layer thickness. The CF data of Fig. 20

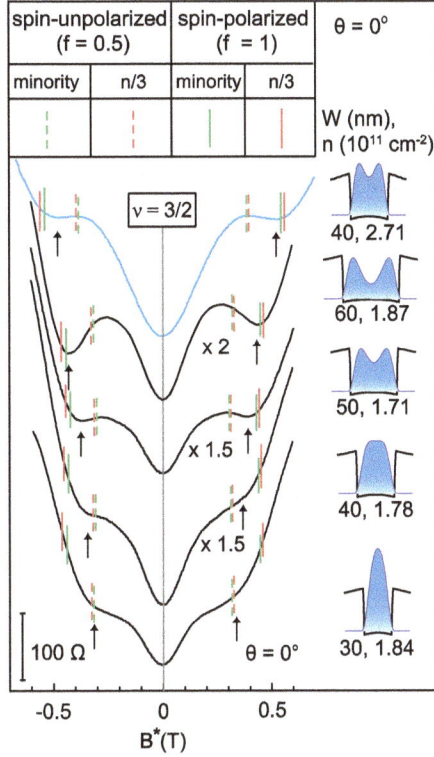

Fig. 20. (a) Magnetoresistance traces taken at $T = 0.3$ K on 2DESs confined to GaAs QWs of width W and density n, as indicated to the right of each trace. The top trace, shown in blue, has a much higher density than the other traces. Each sample is patterned with periodic negative electron-beam resist stripes of period $a = 200$ nm. In the vicinity of $\nu = 3/2$ the data show strong GR features whose positions are marked by black vertical arrows. Also shown for each sample are different sets of vertical marks, indicating the expected positions of the $i = 1$ GR resistance minimum, based on different assumptions. The red marks assume that the CF density is $n^* = n/3$, while the green lines are based on the assumption that n^* equals the density of minority carriers in the half-filled LL. Solid lines are for full spin polarization ($f = 1$) and dashed ones are for the spin-unpolarized case ($f = 0.5$). The charge distribution and the potential profile, calculated self-consistently at zero magnetic field, for each sample are shown to the right of the corresponding trace. [Note Ref. 16.]

are indeed quantitatively consistent with the results of independent measurements of the spin polarization of FQHSs near $\nu = 3/2$.[16,97]

5. Geometric Resonance Data Near $\nu = 1/4$

Qualitatively similar to the case of $\nu = 1/2$, at $\nu = 1/4$ electrons merge with an even number of (in this case *four*) flux quanta and form a *four-flux* CF (^{4}CF) Fermi sea. Unlike $\nu = 1/2$, there is no particle-hole symmetry at $\nu = 1/4$, although Ref. 57 proposes theories for Dirac CFs and an emergent reflection symmetry about

$\nu = 1/4$. This provides motivation for studies of ^{4}CFs whose physics could be distinct from two-flux CFs (^{2}CFs) that we have been discussing so far. However, measurements of ^{4}CFs are very scarce,[98–100] partly because they require very high magnetic fields, and also because of the proximity of $\nu = 1/4$ to the Wigner crystal formation near $\nu = 1/5$.[101–104] Therefore, many fundamental questions have remained unanswered: Do ^{4}CFs have properties similar to the ^{2}CFs? Do ^{4}CFs show an asymmetry in the field positions of the GR minima similar to ^{2}CFs discussed Sec. 3.2? What happens to the ^{4}CF Fermi sea when the Fermi sea for zero-field electrons is highly anisotropic? Our GR measurements reported here provide answers to these fundamental questions, and reveal surprises for ^{4}CFs.[24]

Our experimental platform is a 2DES, with density $n = 1.78 \times 10^{11}$ cm^{-2} and low-temperature mobility 1.4×10^7 cm^2/Vs, confined to a modulation-doped, 40-nm-wide, GaAs QW.[24] The sample surface has a grating of electron-beam resist [Fig. 21(a)], similar to the other samples discussed in this article. In Fig. 21(b) we show a representative magnetoresistance trace, exhibiting well-developed GR features flanking symmetrically a deep V-shaped minimum at $\nu = 1/4$. Figure 21(c) zooms in around $\nu = 1/4$. From the period of the modulation, $a = 240$ nm, we determine the expected positions for the primary $i = 1$ GR resistance minima according to $B^*_{i=1} = 2\hbar k^*_F / ea(1 + 1/4)$ where $B^*_{i=1} = B_{i=1} - B_{\nu=1/4}$. We assume a fully spin-polarized CF sea and mark the expected positions for $B_{i=1}$ considering two possibilities: (i) black solid lines for $k^*_F = (4\pi n)^{1/2}$, and (ii) orange dashed lines for k^*_F changing according to the magnetic length,[4,43,51,52,59] i.e. $k^*_F = (4\pi n)^{1/2} \times (B/B_{\nu=1/4})^{1/2}$; see Sec. 3.2. The difference between the expected $B^*_{i=1}$ for the two assumptions is very small and cannot be resolved in our experiments. From Fig. 21(c), it is clear that the observed GR minima positions are in excellent agreement with the expected $B^*_{i=1}$, confirming that the ^{4}CFs near $\nu = 1/4$ are fully spin polarized. More importantly, unlike the ^{2}CF GRs flanking $\nu = 1/2$ (Fig. 6), the GR features for ^{4}CFs are quite symmetric around $\nu = 1/4$. This is reasonable, considering that the *minority* carrier density, which was found experimentally in Ref. 15 to determine k^*_F for ^{2}CFs, is the same on the two sides of $\nu = 1/4$ and is equal to n.

In our work, we also addressed the question of how an anisotropy in the Fermi sea of the electrons at zero field affects the ^{4}CF Fermi sea. To address this question, we used the method described in Sec. 3.4, namely, we apply an in-plane magnetic field ($B_\parallel$) which, through its coupling to the out-of-plane motion of the electrons in a quasi-2D system, severely distorts the Fermi sea of the low-field electrons. We then determine the subsequent anisotropy of the ^{4}CF cyclotron orbit via measuring the positions of the CF GR minima along the two perpendicular arms of the L-shaped Hall bar [inset of Fig. 22(b)]. In our experiments, we tilt the sample so that $B_\parallel$ is always along [110], with θ denoting the angle between the field direction and the normal to the 2D plane [Fig. 22(b), inset]. As seen in Fig. 22, the application of $B_\parallel$ affects the positions of the ^{4}CF GR minima. Traces for the two arms of the

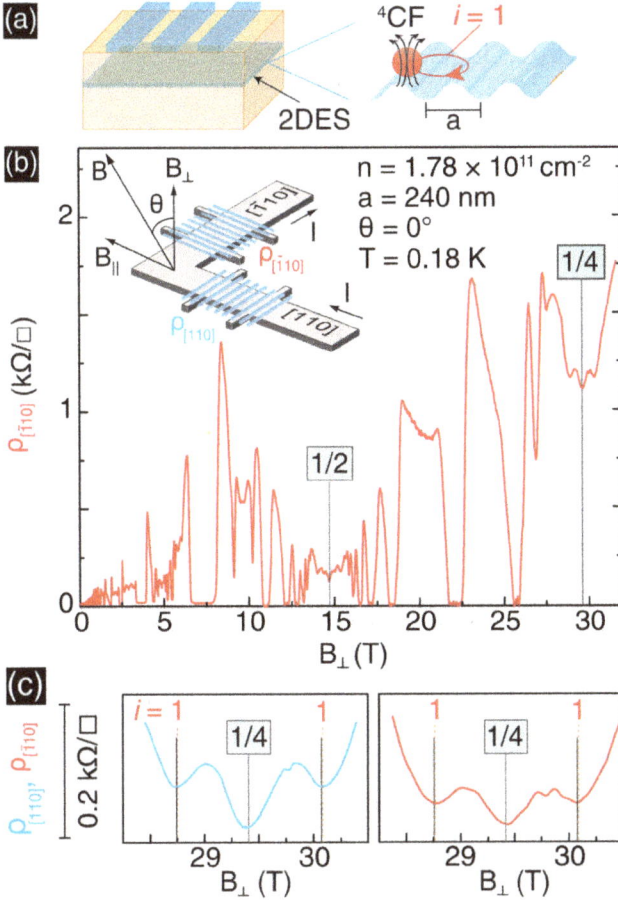

Fig. 21. GR features for four-flux composite fermions (^{4}CFs) near $\nu = 1/4$. (a) Lateral surface superlattice of period a, inducing a periodic density perturbation in the 2DES. When the ^{4}CFs' cyclotron orbit becomes commensurate with the period of the perturbation, the $i = 1$ GR occurs. (b) Magnetoresistance trace revealing GR features near $\nu = 1/4$ and $\nu = 1/2$. Inset: The L-shaped Hall bar along [$\bar{1}$10] and [110] directions used for the measurements. (c) Magnetoresistance near $\nu = 1/4$ demonstrating the $i = 1$ ^{4}CF GR features, resistance minima flanking $\nu = 1/4$. Black solid and orange dashed lines mark the *expected* positions for the $i = 1$ GR for fully spin-polarized ^{4}CFs with circular Fermi contour assuming $k_F^* = (4\pi n)^{1/2}$ and $k_F^* = (4\pi n)^{1/2} \times (B/B_{\nu=1/4})^{1/2}$, respectively. The extra minimum near $B_\perp = 29.75$ T stems from the $i = 2$ GR. [Note Ref. 24.]

Hall bar along [110] and [$\bar{1}$10] are shown in Figs. 22(a) and 22(b). In both panels, the vertical dotted lines mark the expected positions of the $i = 1$ GR minima of fully-spin-polarized ^{4}CF with a circular Fermi sea, i.e. $B_{i=1}^* = 2\hbar k_F^*/ea(1 + 1/4)$. These lines match the observed positions of the resistance minima for the bottom traces of Fig. 22, which were taken at $\theta = 0$. When we increase θ and thereby $B_\parallel$, for the [110] arm [Fig. 22(a)], the positions of the two GR minima shift away from $\nu = 1/4$ to larger values of $|B_\perp^*|$. In contrast, the GR minima for the [$\bar{1}$10]

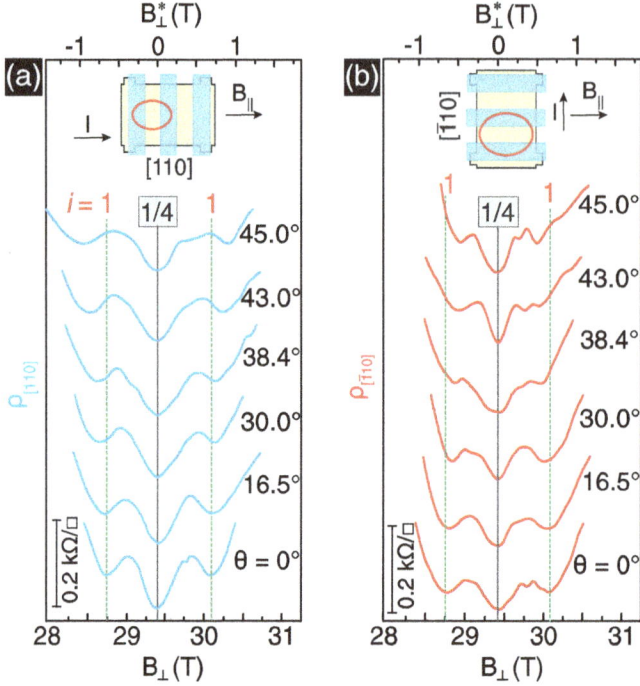

Fig. 22.   Tilt evolution of the ^{4}CF GR features near $\nu = 1/4$ along (a) [110] and (b) [$\bar{1}$10] directions. The insets show the orientation of the Hall bars, and the ^{4}CF cyclotron orbit for the $i = 1$ GR. Magnetoresistance traces are vertically offset for clarity; the tilt angle θ is given for each trace. The *expected* positions for the $i = 1$ ^{4}CF GRs are marked with vertical dotted lines assuming that $k_F^* = (4\pi n)^{1/2}$. In both panels, the scale for the applied external field $B_\perp$ is shown on the bottom axis while the top scale is the effective magnetic field $B_\perp^*$ experienced by the ^{4}CFs. [Note Ref. 24.]

arm [Fig. 22(b)] move toward smaller $|B_\perp^*|$. Using the field positions of the GR minima along the [110] and [$\bar{1}$10] directions, we directly extract the magnitude of the Fermi wave vector k_F^* along [$\bar{1}$10] and [110], respectively; we use the expression $k_F^* = B_{i=1}^* ea(1 + 1/4)/2\hbar$.

It is clear from the shifts of the GR minima with increasing tilt angle (Fig. 22) that the ^{4}CF Fermi sea is becoming progressively more anisotropic. The deduced anisotropy is shown in the lower panels (e)–(h) of Fig. 23. In the upper panels (a)–(d), the corresponding anisotropy is shown for zero-field fermions (electrons). Qualitatively similar to the case of ^{2}CF (Fig. 11), the anisotropy of the of ^{4}CF Fermi sea is significantly less severe than that of electrons. More importantly, similar to the ^{2}CF case, the ^{4}CF Fermi sea remains connected even at the largest applied values of $B_\parallel$, while for the 2D electrons the Fermi contour splits into two tear-shaped sections. The connectivity of the ^{4}CF Fermi sea at large $B_\parallel$ can be attributed to the much larger in-plane effective mass of CFs compared to their out-of-plane mass, similar to the case of ^{2}CF;[14] see Sec. 3.4.

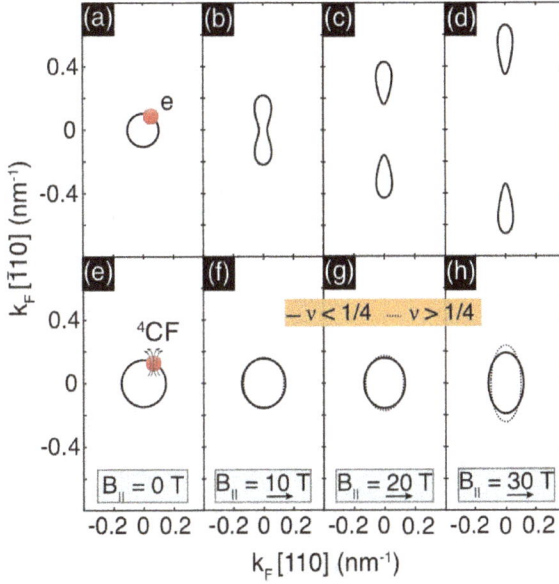

Fig. 23. Comparison between the evolution with $B_{||}$ of the calculated Fermi contour of electrons [(a)–(d)] and measured Fermi contour of ^{4}CFs near $\nu = 1/4$ [(e)–(h)]. For simplicity, in (a)–(d) only the majority-spin contour is shown. In (e)–(h), solid and dotted contours denote the ^{4}CF Fermi contours for $\nu < 1/4$ and $\nu > 1/4$ respectively. Even though the electron Fermi sea completely splits at large $B_{||}$, the ^{4}CF Fermi sea near $\nu = 1/4$ remains intact. [Note Ref. 24.]

A surprising finding of our study is highlighted in Figs. 23(e)–23(h): The measured $B_{||}$-induced ^{4}CF Fermi sea anisotropy is significantly smaller for $\nu < 1/4$ compared to $\nu > 1/4$. This can be seen directly from the traces in Fig. 22 where, when the sample is tilted in the magnetic field, the positions of the ^{4}CF GR minima are highly *asymmetric* with respect to the position of $\nu = 1/4$. This differs from the ^{2}CF Fermi sea at $\nu = 1/2$ where both sides show similar anisotropy with increasing $B_{||}$ (see Fig. 9 and Ref. 14). The different behavior for $\nu = 1/4$ and $\nu = 1/2$ is particularly puzzling considering that ^{4}CFs and ^{2}CFs form in the same LL. As discussed in detail in Ref. 24, the origin of this difference might be that ^{4}CF have different in-plane effective masses ($m_{||}$) on the two sides of $\nu = 1/4$, namely that $m_{||}$ is larger for $\nu < 1/4$. A similar conclusion was reached in Ref. 100 from the measured FQHS energy gap data, and was attributed to the proximity of the $\nu < 1/4$ filling range to where a magnetic-field-induced Wigner crystal is typically observed.[101–104]

6. Geometric Resonance Data Near the $\nu = 5/2$ FQHS

Compared to the lowest ($N = 0$) LL discussed so far, the ground state at and near the even-denominator filling in the first-excited ($N = 1$) LL, e.g. at $\nu = 5/2$, is very enigmatic. Instead of a compressible Fermi sea of CFs, there is an incompressible FQHS whose origin has been controversial since its discovery in 1987.[105] The initial explanation[105] for the 5/2 FQHS was based on a two-component, spin-unpolarized, Ψ_{331} wave function.[106] Following this argument, Haldane and Rezayi[107] constructed a spin-unpolarized, CF wave function based on the hollow-core model. On the other hand, Moore and Read[108] presented a spin-polarized, p-wave-paired, CF ground state, known as the Moore–Read Pfaffian. Spectacularly, this state should harbor quasiparticles obeying non-Abelian statistics whose interchange takes the system from one of its many ground states to another, whereas the interchange of ordinary Abelian quasiparticles only adds a phase to their wave functions. Numerical investigations[109] support such a non-Abelian ground state at $\nu = 5/2$. As a consequence, the 5/2 FQHS has attracted much attention as a promising platform for topological quantum computation.[110]

Despite the enormous interest in the 5/2 FQHS, little conclusive experimental evidence is available for its origin,[111] although some of the recent quasiparticle tunneling and interference studies lend support to its being a non-Abelian state.[111–116] Early surface acoustic wave experiments[111,117] showed hints of Fermi sea effects at $\nu = 5/2$, but there had been no compelling evidence for the existence of CFs near $\nu = 5/2$ until the recent measurements[23] that we describe below were reported. Also, a crucial prerequisite for the 5/2 FQHS to be non-Abelian is that it is single-component (fully spin polarized). However, the spin polarization of the 5/2 FQHS had remained an open question[111] even 30 years after its discovery. Initial tilted-field measurements[118] indicated the $\nu = 5/2$ FQHS to be spin unpolarized, a conclusion which is also favored by the photoluminescence[119] experiments. On the other hand, density-dependence,[120] and later tilt-dependence[121] activation gap measurements, as well as inelastic light scattering data[122] suggested possible full spin polarization. These observations were supported in recent nuclear magnetic resonance[123,124] and tunneling[125] experiments.

Here, we provide a brief summary of very recent GR experiments[23] addressing two fundamental questions regarding the origin of the 5/2 FQHS: Are CFs present near $\nu = 5/2$, and are they fully spin polarized? These GR measurements, which directly probe the Fermi sea and wave vector of CFs, provide unambiguous and conclusive positive answers to these questions. It is worth emphasizing that the technique is simple, and yet most direct and quantitative; also, it does not rely on any fitting schemes or parameters. Moreover, our measurements do not involve exposure of the sample to illumination or radiation, or to in-plane magnetic fields.

The most important finding of Ref. 23 is highlighted in Fig. 24. The figure shows a magnetotransport trace in the first-excited LL for a very clean 2DES, with density 1.46×10^{11} cm^{-2} and low-temperature mobility 1.3×10^7 cm^2/Vs, confined to a modulation-doped, 30-nm-wide, GaAs QW. Similar to the other samples discussed in this article, the specimen has a one-dimensional, periodic array of negative electron-beam resist, inducing a very small periodic density modulation, the estimated magnitude of which is $\ll 1\%$.[126] The magnetotransport trace in Fig. 24 exhibits two well-developed minima (marked with two red arrows) flanking

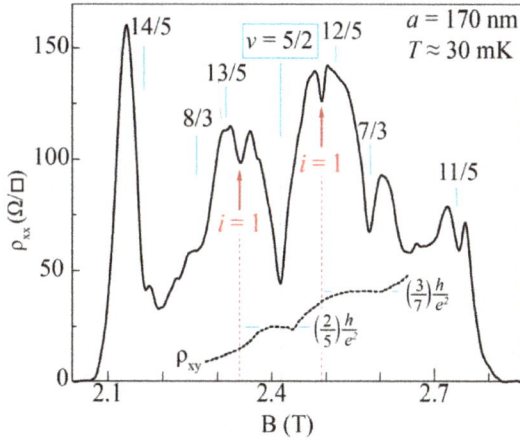

Fig. 24. Magnetoresistance trace taken at $T = 30$ mK on a 2DES confined to a 30-nm-wide GaAs QW and patterned with periodic negative electron-beam resist stripes of period $a = 170$ nm, showing strong GR minima in the vicinity of $\nu = 5/2$. The expected positions of the primary $i = 1$ GR minima for fully-spin-polarized CFs near $\nu = 5/2$ are marked with red arrows flanking $\nu = 5/2$. These minima do not coincide with any of the expected FQHS minima, labeled by blue markers. Blue horizontal lines mark the expected quantized values of the Hall plateau for $\nu = 5/2$ and $7/3$ in the ρ_{xy} trace. [Note Ref. 23.]

the deep V-shaped minimum at $\nu = 5/2$. As discussed in Ref. 23, the positions of these minima are quantitatively consistent with the GR of CFs occupying a fully-spin-polarized Fermi sea. Their observation therefore confirms the existence of a well-defined CF Fermi wave vector, thus, providing direct evidence for a CF Fermi sea near 5/2.

Several features of Fig. 24 data are noteworthy. First, it is clear from Fig. 24 that the positions of the GR features do not coincide with any of the observed or expected odd-denominator FQHS minima (marked with blue lines). Second, the FQHS at $\nu = 5/2$ is well developed as is evident from the strong minimum in ρ_{xx} and the corresponding plateau at $2/5(h/e^2)$ in the Hall (ρ_{xy}) trace. The coexistence of the FQHS at $\nu = 5/2$ and fully-spin-polarized CFs on its sides argues in favor of a fully-spin-polarized $\nu = 5/2$ FQHS. As mentioned above, this provides crucial support to the model of the 5/2 FQHS as a topological p-wave paired state of CFs. It is also remarkable that the CFs near half-filling in the $N = 1$ LL are fully spin-polarized at a modest magnetic field of about 2.4 T, while CFs in the $N = 0$ LL are not polarized even at significantly higher fields (see, e.g. Fig. 20; the field for $\nu = 3/2$ in the lowest trace of Fig. 20 is about 5 T, and yet the CFs are not fully polarized. Third, as discussed in detail in Ref. 23, the CFs near $\nu = 5/2$ appear to be much more fragile than CFs near $\nu = 1/2$. Near $\nu = 5/2$, the CF GR minima are observed only when the amplitude of the potential modulation is extremely small. Moreover, they disappear very quickly as the temperature is increased, and are absent at $T = 0.3$ K, the temperature at which a robust GR of CFs $\nu = 1/2$ is seen (see, e.g. Fig. 6).

7. Geometric Resonance Data Near the $\nu = 1/2$ FQHS in the Lowest Landau Level in Wide GaAs Quantum Wells

The GR measurements can be extended to probe CFs in wide QW systems with a bilayer-like charge distribution near filling factor $\nu = 1/2$. In these systems, for appropriate values of density or tilt angle, there is a FQHS at $\nu = 1/2$ in the lowest LL.[127–132] The origin of this FQHS has been enigmatic. This is partly because the energies of the different FQH ground states at $\nu = 1/2$ in these wide QWs are very close to each other. Early theories suggested that, because of the finite interlayer tunneling, the $\nu = 1/2$ FQHS in a wide QW is a one-component (Pfaffian) FQHS,[133,134] implying that it could support non-Abelian quasi-particles. Later measurements and calculations, on the other hand, favored a two-component (Ψ_{331}), Abelian FQHS which should be the ground state in the limit of very small or no tunneling.[135,136]

In order to shed light on the origin of the $\nu = 1/2$ FQHS, we performed GR measurements on both 2DESs and 2DHSs confined to wide GaAs QWs.[18,19] The sample parameters, namely the carrier density and QW width, were carefully chosen so that a FQHS is observed at $\nu = 1/2$; see the phase diagram in Fig. 25(a) for the 2DES. The magnetoresistance traces for the reference (*unpatterned*) and *patterned* sections of the sample are shown in Figs. 25(b) and 25(c). The patterned section has

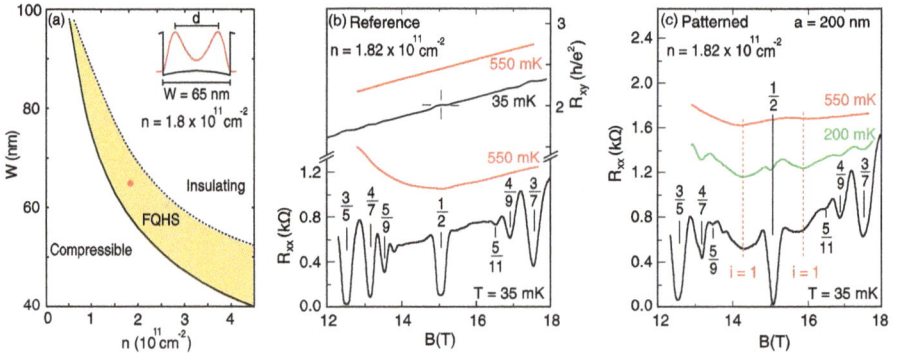

Fig. 25. (a) The well width (W) versus density (n) phase diagram for the different ground states of a 2DES at $\nu = 1/2$ in symmetric, wide, GaAs QWs.[129] The yellow region is where the FQHS is observed. The ground state changes to "compressible" and "insulating" to the left and right of the FQHS region, respectively. The red circle marks the position of the 2DES sample used in the GR measurements. Inset: Results of self-consistent calculations of the charge distribution and Hartree potential for the 2DES sample at $B = 0$. (b) Longitudinal (R_{xx}) and Hall (R_{xy}) magnetoresistance traces from the reference (*unpatterned*) region of the sample. A strong FQHS with a well-developed Hall plateau is observed at $\nu = 1/2$ at $T = 35$ mK. (c) R_{xx} traces from the *patterned* region of the same sample. Unlike the trace for the unpatterned sample, here resistance minima are observed on the flanks of $\nu = 1/2$, and they persist up to a high temperature of 500 mK. The positions of the minima coincide with the expected positions for the GR of CFs with a density equal to the *total* density of the 2DES (see the vertical, dotted, red lines marked $i = 1$). In both (b) and (c), traces taken at different temperatures are shifted vertically for clarity. [Note Ref. 18.]

(a)

W (nm)

50

40

30

d

W = 35 nm

$p = 1.7 \times 10^{11}$ cm^{-2}

Insulating

FQHS

Compressible

0   1   2   3

p (10^{11} cm^{-2})

(b) Reference   35 mK   2.5

R_{xx} (kΩ)

3

2

1

0

$\frac{2}{3}$ $\frac{3}{5}$ $\frac{4}{7}$

$\frac{1}{2}$

$\frac{4}{7}$ $\frac{7}{9}$

$\frac{3}{5}$

$\frac{5}{11}$

$\frac{5}{9}$

$p = 1.68 \times 10^{11}$ cm^{-2} T = 35 mK

R_{xy} (h/e^2)

2.5

2.0

1.5

1.0

10   12   14   16   18

B(T)

(c) Patterned   a = 200 nm

R_{xx} (kΩ)

8

7

6

5

4

3

2

1

0

$\frac{2}{3}$ $\frac{3}{5}$

$\frac{2}{5}$

300 mK

$\frac{1}{2}$

T = 35 mK

$i = 1$ $i = 1$

$p = 1.68 \times 10^{11}$ cm^{-2}

10   12   14   16   18

B(T)

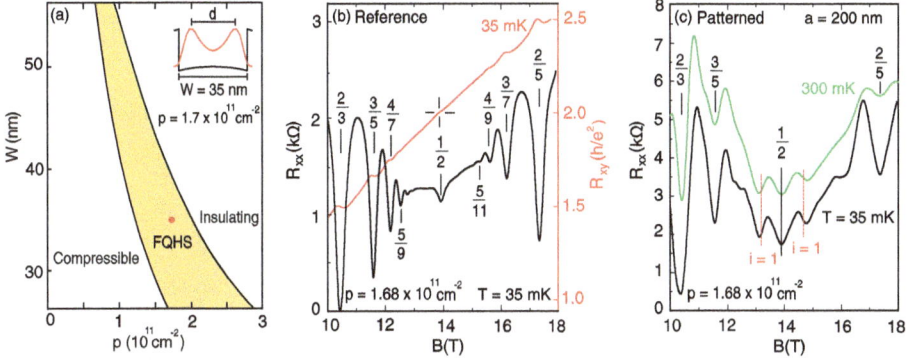

Fig. 26. (a) Phase diagram for the different ground states at $\nu = 1/2$ of a 2DHS confined to symmetric, wide GaAs QWs.[130] The red circle marks the position of the studied 2DHS sample in the phase diagram. Inset: Self-consistently calculated charge distribution and Hartree potential for the 2DHS sample at $B = 0$. (b) R_{xx} and R_{xy} traces from the reference (unpatterned) region of the sample. (c) R_{xx} traces from the patterned region, showing GR of CFs. As in Fig. 25, the positions of the observed resistance minima are consistent with the *total* 2DHS density (considering the expected warped Fermi contour in this wide QW; see Fig. 17 and Sec. 3.7). Traces at different temperatures are shifted vertically for clarity. [Note Ref. 18.]

a periodic density modulation with a period of $a = 200$ nm.[18] As seen in Fig. 25(c), the trace for the patterned section exhibits minima consistent with the GR of CFs' cyclotron orbit with the period of the modulation. In Fig. 26, data are shown for a 2DHS confined to a wide GaAs QW. Similar to the 2DES data, the patterned section shows GR of CFs.

The significance of the results presented in Figs. 25 and 26 is that the CFs on the flanks of the $\nu = 1/2$ FQHS appear single-layer-like (single component) as the positions of the GR minima correspond to the *total* density of the 2DES (or 2DHS) confined to the wide QW. In contrast, similar measurements[19] performed near the QHS at $\nu = 1$ reveal that the CFs' GR minima are consistent with *half* the electron density in the QW, implying that CFs prefer to stay in separate layers and exhibit a two-component behavior. We conclude that these GR experiments, which directly probe the Fermi sea of CFs near the $\nu = 1/2$ FQHS in wide QWs, support the notion that the system is one-component near $\nu = 1/2$. This is consistent with the results of recent, accurate, numerical calculations which strongly favor a one-component, non-Abelian FQHS in wide QWs with proper parameters.[137]

8. Composite Fermions Dance to the Tune of a Wigner Crystal!

When the kinetic energy of a 2DES is quenched at very high perpendicular magnetic fields so that the Coulomb repulsion dominates, the electrons are expected to condense into an ordered array, forming a quantum Wigner crystal (WC). Although this exotic state has long been suspected[138–140] in high-mobility 2DESs at very low LL fillings ($\nu \ll 1$), its direct observation has been elusive. The WC, being pinned by the ubiquitous residual disorder, manifests as an insulating phase in dc electronic transport[102-104] and exhibits resonances in its ac (microwave) transport which strongly suggest collective motions of the electrons.[101,104,142,143] Various other measurements, such as thermopower,[144] nuclear magnetic resonance,[145] tunneling resonance,[146] and screening efficiency,[147] also support a WC ground state at very low fillings.

Here we briefly describe a novel technique that provides a direct probe of the WC order and its lattice constant using CFs in an adjacent layer.[25,26] The technique consists of measuring magnetotransport traces in a bilayer electron system with unequal layer densities. The bottom layer has a very low density and is in the WC regime ($\nu \ll 1$), while the top ("probe") layer is near its $\nu = 1/2$ and hosts a sea of CFs [Fig. 27(a)]. The idea is that the CFs in the top layer would feel a periodic potential modulation from the bottom WC layer's charge density. As the magnetic field is swept near the half-filling of the top layer, the CFs in this layer should exhibit GR phenomena whenever their cyclotron orbit encircles a certain integer number of the WC lattice points [Fig. 27(a)]. Note that for such a GR, one would expect the magentoresistance to exhibit a *maximum*, similar to what is observed in a system of 2D electrons[148-151] or CFs[152] with a periodic array of "anti-dots".

The magentoresistance data for such a bilayer system are shown in Figs. 27(c) and 27(d). When the bottom-layer is completely depleted by applying a sufficiently large negative back-gate bias, the trace shows the normal behavior near the half-filling of the top layer [top trace in Fig. 27(c)]. Once 2D electrons with a very small density are induced in the bottom layer, however, small features become discernible in the magnetoresistance [middle trace Fig. 27(c)]. These features are seen clearly in Fig. 27(d) where the traces are shown expanded near the top layer half-filling. In Fig. 27(d) the blue triangles mark the positions of magnetoresistance maxima *expected* for the $i = 1$, 3, and 7 GR of CFs in the top layer, due to the periodic potential imposed by a triangular WC potential in the bottom layer, when the size of the CF orbit is such as to enclose $i = 1$, 3, or 7 electrons in the WC, as indicated in Fig. 27(b). Despite some clear discrepancies, in Fig. 27(d) there is overall good agreement between the positions of the observed and expected resistance maxima. It is worth emphasizing that the positions of the triangles are based on the measured top- and bottom-layer densities and do not rely on any fitting parameters. Note also the disappearance of the oscillatory features above $T \approx 200$ mK, which can be associated with the melting of the WC.

The data in Fig. 27 provide evidence that the CFs in the top layer are indeed dancing to the tune of a WC formed at low temperatures in the bottom layer. They also present a unique, direct glimpse at the symmetry of the WC, its lattice constant, and melting. More generally, they demonstrate a striking example of how one can probe an exotic many-body state of 2D electrons using equally exotic quasi-particles of another many-body state. More recent measurements[27] reveal that the CF layer, in turn, affects the WC layer by shifting the frequency of its pinning-mode resonances.

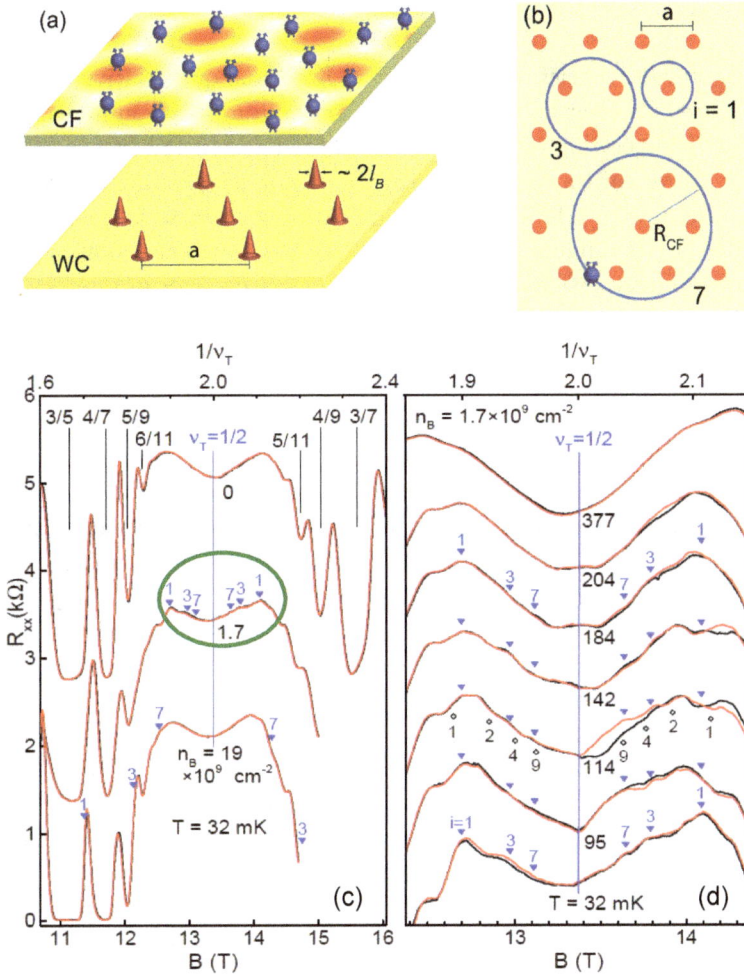

Fig. 27.   Overview of the bilayer experimental setup to probe a Wigner crystal (WC) using composite fermions (CFs). (a) The bilayer system has a high-density top layer which hosts a CF Fermi sea at high magnetic fields near its filling factor $\nu_T = 1/2$. The bottom layer has a much lower density so that it is at a very small filling factor ($\nu_B \ll 1$), thus allowing a WC to form. The top layer feels a periodic potential modulation from the bottom WC layer's charge density; l_B is the magnetic length. (b) As the magnetic field is swept near the top layer's $1/2$ filling, the CFs in the top layer execute cyclotron motion, leading to GR maxima in the magnetoresistance of the top layer when the CF cyclotron orbit encircles 1, 3, 7 lattice points. (c) Magnetoresistance traces for the bilayer sample, measured as the bottom-layer density n_B is reduced from 19 to 1.7, and then to zero (in units of 10^9 cm^{-2}) via applying negative back-gate voltage. The top-layer density n_T is fixed at $\simeq 1.6 \times 10^{11}$ cm^{-2}. The middle trace ($n_B = 1.7 \times 10^9$ cm^{-2}) exhibits small additional features near $\nu_T = 1/2$ which are better seen in the expanded figure shown in (d). (d) Temperature dependence of the magnetoresistance traces at $n_B = 1.7 \times 10^9$ cm^{-2}. The triangles mark the *expected* positions for resistance maxima when the $i = 1, 3, 7$ CF cyclotron orbit becomes commensurate with a triangular WC potential [see (b)]. For the $T = 114$ mK trace, we also show the expected ($i = 1, 2, 4, 9$) GR resistance maxima position if the WC had a square symmetry. In (c) and (d), two traces are shown for each condition: The black is for the up sweep of B and the red is for the down sweep. The origin of the slight hysteresis seen in some of the traces is unknown. Traces for different conditions are shifted vertically for clarity. [Note Ref. 25.]

9. Summary

This article provides a brief review of recent, direct and quantitative measurements of the CF Fermi sea near half-filled LLs in 2D electron or hole systems confined to GaAs QWs. The very high quality of these carrier systems allows us to probe the CF Fermi wave vector through the ballistic transport of CFs in GR measurements. The data near $\nu = 1/2$ filling factor reveal subtleties such as an asymmetry in the positions of GR minima relative to the half filling, raising questions regarding particle-hole symmetry and what determines the wave vector of the CF Fermi sea. The experimental data strongly suggest that the magnitude of the Fermi wave vector is determined by the density of minority carriers in the lowest LL, and that particle-hole symmetry is preserved near $\nu = 1/2$ (see also *Note added* below).

The bulk of the article addresses the question of how an anisotropy in the Fermi sea of zero-field particles is transferred to the CFs. The main experimental finding, through measurements of strain-induced anisotropy, is that the Fermi sea anisotropy for CFs (α_{CF}) is less than the anisotropy of their low-field (fermion) counterparts (α_F), and closely follows the simple relation $\alpha_{CF} = \alpha_F^{1/2}$. This relation has found theoretical justification. Fermi sea anisotropies induced by applying a parallel magnetic field are more complex, but can be qualitatively explained through a simple, heuristic model based on the heavy, in-plane CF effective mass. A rigorous theoretical description, however, is lacking.

We also observe GR of CFs near $\nu = 3/2$. The data provide information regarding the spin-polarization of CFs. We find that, if the 2D electrons have low densities and are confined to relatively narrow GaAs QWs, then the CF Fermi sea is not fully spin polarized. At high densities and in wide QWs, the GR minima are consistent with fully-spin-polarized CFs.

Using GR measurements we can also probe the ballistic transport and Fermi sea of *four-flux* CFs near $\nu = 1/4$. As expected, these are fully spin polarized. A surprise, however, is that the magnetic field positions of GR minima on the two sides of $\nu = 1/4$ become asymmetric (with respect to the $\nu = 1/4$ position) in large parallel fields, suggesting an asymmetry in the CF effective mass.

The data near $\nu = 5/2$, in the first excited LL, are particularly interesting as they show the presence of fully-spin-polarized CFs, even though the ground state at 5/2 is a FQHS. This observation is consistent with 5/2 FQHS being a topological *p*-wave paired state of CFs, and therefore non-Abelian and of potential use in quantum computing.

The GR measurements can also probe CFs in systems with a bilayer-like charge distribution near the FQHS at filling factor $\nu = 1/2$. The experimental data provide evidence for a Fermi sea which is single-component, namely its measured Fermi wave vector is consistent with the *total* electron density in the wide QW.

Finally, the results for bilayer electron or hole systems in double-QW structures with very different densities reveal how one can probe a WC state in one layer using CFs in the other layer and vice versa.

Note added:

Very recently, Hossain *et al.*[153] revisited the problems of what determines the CF Fermi wave vector and the particle-hole symmetry near $\nu = 1/2$ by extending the experiments of Ref. 15 to very low density samples (from $n \simeq 10.2 \times 10^{10}$ cm^{-2} down to $\simeq 2.5 \times 10^{10}$ cm^{-2}). Based on their data, they arrive at the following noteworthy conclusions:

(i) Similar to the data of Ref. 15, the observed positions of the GR minima are asymmetric in magnetic field with respect to $\nu = 1/2$. The minima positions are in excellent agreement with the simple assumption that the CF Fermi wave vector is determined by the density of *minority carriers* in the lowest LL, i.e. $k_F^* = (4\pi n_{min})^{1/2}$, where n_{min} is equal to the electron density for $\nu < 1/2$, and hole density for $\nu > 1/2$.

(ii) The minima positions observed in the experiments disagree with the predictions of the Dirac theory,[52] and the disagreement grows at the lowest densities.

(iii) Similar to the data of Ref. 15, the observed GR minima are *also asymmetric* in their filling factor positions with respect to $\nu = 1/2$, i.e. the minima are *not* observed at ν and $(1-\nu)$, when the magnetic field is varied with n and a held fixed. Naively, this may suggest a breaking of particle-hole symmetry. However, if the data are plotted appropriately, they are indeed consistent with particle-hole symmetry. The key point is that the GR features for $\nu > 1/2$ and $\nu < 1/2$ are observed at two different values of B. As a result, for a sample with a fixed n and a, the values of a/l_B are different at ν and $(1-\nu)$. This can be accounted for by plotting the data as a function of a/l_B; Hossain *et al.*[153] showed that their data plotted in this fashion indeed exhibit symmetry about $\nu = 1/2$.

(iv) The low-density data of Hossain *et al.*[153] are also important from another vantage point. They allow testing a fundamental property of CFs, namely whether or not their Fermi sea satisfies the Luttinger theorem,[154] which postulates that the Fermi sea area should be independent of the interaction between the fermions so long as the interaction does not cause a phase transition. In the case of CFs, even though the non-perturbative part of the electron-electron interaction is already used in making the CFs, the residual interaction between the CFs increases substantially in the low-density regime thanks to the increased mixing between the LLs.[2] This effect can be quantified by the LL mixing parameter κ defined as the ratio of the Coulomb to cyclotron energies. In Ref. 153 experiments, κ reaches values as large as ~ 2 at low densities, implying significant interaction between the CFs. Data of Ref. 153, however, show that the CF Fermi wave vector, which determines the Fermi sea area, follows the same expression $k_F^* = (4\pi n_{min})^{1/2}$ over the whole electron density range and LL mixing, making a convincing case that the Luttinger theorem is obeyed in a strongly interacting CF system. Moreover, the symmetries about $\nu = 1/2$ observed down to the lowest densities in plots of the GR minima vs. a/l_B indicate that particle-hole symmetry is also robust even for interacting CFs.

(v) As the electron density is lowered below $n = 4.21 \times 10^{10}$ cm^{-2}, Hossain *et al.*[155] found that the CFs near $\nu = 1/2$ lose their full magnetization and become partially spin polarized. This is expected[2] and can be explained by the relative enhancement of the Coulomb energy over the Zeeman energy. Remarkably, however, as the density is further reduced, below 3.51×10^{10} cm^{-2}, CFs make a sudden transition and become fully magnetized again. This spontaneous magnetization of CFs closely resembles the Bloch ferromagnetism,[156] theoretically predicted over 90 years ago for a sufficiently dilute, interacting electron system (at zero magnetic field) when the exchange energy gained in aligning the spins exceeds the enhancement in the kinetic energy. The theoretical calculations presented in Ref. 155 provide a semi-quantitative understanding of the full magnetization of the CFs at very low densities in terms of an enhanced interaction between CFs due to LL mixing, which is significant at small n. It is also remarkable in the data of Ref. 155 that, in the narrow density range where the CFs are partially spin polarized, the sum of the Fermi sea areas for the two spin species equals (within the experimental accuracy) the Fermi sea area expected for fully-spin-polarized CFs.

Acknowledgments

I would like to thank all my students, post-docs, and colleagues whose work this article is based on: Doby Kamburov, M. A. Mueed, Md. Shafayat Hossain, Insun Jo, Hao Deng, Roland Winkler, Yang Liu, Meng K. Ma, Kevin A. Villegas Rosales, and Medini Padmanabhan. I am also thankful to Loren Pfeiffer, Ken West, Kirk Baldwin, and Yoon Jang (Edwin) Chung for growing, via molecular beam epitaxy, the fantastic GaAs/AlGaAs wafers that were used for the measurements. Many thanks to Lloyd Engel, Matthew Fisher, Michael Mulligan (who kindly provided Fig. 7), Matteo Ippoliti, and Ravin Bhatt for illuminating discussions. I am particularly grateful to Jainendra Jain and Bert Halperin for careful readings of this manuscript and making detailed suggestions, especially for Sec. 3.2.

I also thank S. Hannahs, T. Murphy, A. Suslov J. Park, G. Jones, and H. Baek, at the National High Magnetic Field Laboratory in Tallahassee, FL, for valuable technical support during the measurements, and A. Endo and Y. Iye for useful discussions and advice regarding the use of negative electron-beam resist for fabricating samples with modulated density.

The work at Princeton University has been supported primarily by the National Science Foundation (Grant Nos. DMR 1709076, DMR 1305691, DMR 0904117, ECCS 1906253, and ECCS 1508925), the Department of Energy through the Basic Energy Sciences (Grant No. DE-FG02-00-ER45841), the Gordon and Betty Moore Foundation (Grant No. GBMF4420), and the Keck Foundation. The research was also supported in part by QuantEmX travel grants from the Institute for Complex Adaptive Matter (ICAM) and the Gordon and Betty Moore Foundation through Grant No. GBMF5305. The National High Magnetic Field Laboratory is supported by the National Science Foundation Cooperative Agreement No. DMR 1644779, by the State of Florida, and the U.S. Department of Energy.

Finally, I would like to dedicate this article to the memory of Millie Dresselhaus for a lifetime of mentorship and kindness.

References

1. M. Shayegan, Flatland electrons in high magnetic fields, in *High Magnetic Fields: Science and Technology*, Vol. 3, edited by F. Herlach and Noboru Miura (World Scientific, Singapore, 2006), pp. 31–60. [cond-mat/0505520]
2. J. K. Jain, *Composite Fermions* (Cambridge University Press, Cambridge, 2007).
3. J. K. Jain, Composite fermion approach for the fractional quantum Hall effect, *Phys. Rev. Lett.* **63**, 199 (1989).
4. B. I. Halperin, P. A. Lee and N. Read, Theory of the half filled Landau level, *Phys. Rev. B* **47**, 7312 (1993).
5. R. L. Willett, R. R. Ruel, K. W. West and L. N. Pfeiffer, Experimental demonstration of a Fermi surface at one-half filling of the lowest Landau level, *Phys. Rev. Lett.* **71**, 3846 (1993).
6. W. Kang, H. L. Stormer, L. N. Pfeiffer, K. W. Baldwin and K. W. West, How real are composite fermions? *Phys. Rev. Lett.* **71**, 3850 (1993).
7. V. J. Goldman, B. Su and J. K. Jain, Detection of composite fermions by magnetic focusing, *Phys. Rev. Lett.* **72**, 2065 (1994).
8. J. H. Smet, D. Weiss, R. H. Blick, G. Lutjering, K. von Klitzing, R. Fleischmann, R. Ketzmerick, T. Geisel and G. Weimann, Magnetic focusing of composite fermions through arrays of cavities, *Phys. Rev. Lett.* **77**, 2272 (1996).
9. S. D. M. Zwerschke and R. R. Gerhardts, Positive magnetoresistance of composite fermion systems with a weak one-dimensional density modulation, *Phys. Rev. Lett.* **83**, 2616 (1999).
10. J. H. Smet, S. Jobst, K. von Klitzing, D. Weiss, W. Wegscheider and V. Umansky, Commensurate composite fermions in weak periodic electrostatic Potentials: Direct evidence of a periodic effective magnetic field, *Phys. Rev. Lett.* **83**, 2620 (1999).
11. R. L. Willett, K. W. West and L. N. Pfeiffer, Geometric resonance of composite fermion cyclotron orbits with a fictitious magnetic field modulation, *Phys. Rev. Lett.* **83**, 2624 (1999).
12. D. Kamburov, M. Shayegan, L. N. Pfeiffer, K. W. West and K. W. Baldwin, Commensurability oscillations of hole-flux composite fermions, *Phys. Rev. Lett.* **109**, 236401 (2012).
13. D. Kamburov, Y. Liu, M. Shayegan, L. N. Pfeiffer, K. W. West and K. W. Baldwin, Composite fermions with tunable Fermi contour anisotropy, *Phys. Rev. Lett.* **110**, 206801 (2013).
14. D. Kamburov, M. A. Mueed, M. Shayegan, L. N. Pfeiffer, K. W. West, K. W. Baldwin, J. J. D. Lee and R. Winkler, Fermi contour anisotropy of GaAs electron-Flux composite fermions in parallel magnetic fields, *Phys. Rev. B* **89**, 085304 (2014).
15. D. Kamburov, Y. Liu, M. A. Mueed, M. Shayegan, L. N. Pfeiffer, K. W. West and K. W. Baldwin, What determines the Fermi wave vector of composite fermions? *Phys. Rev. Lett.* **113**, 196801 (2014).
16. D. Kamburov, M. A. Mueed, I. Jo, Y. Liu, M. Shayegan, L. N. Pfeiffer, K. W. West, K. W. Baldwin, J. J. D. Lee and R. Winkler, Determination of Fermi contour and spin polarization of $\nu = 3/2$ composite fermions via ballistic commensurability measurements, *Phys. Rev. B* **90**, 235108 (2014).

17. M. A. Mueed, D. Kamburov, Y. Liu, M. Shayegan, L. N. Pfeiffer, K. W. West, K. W. Baldwin and R. Winkler, Composite fermions with a warped Fermi contour, *Phys. Rev. Lett.* **114**, 176805 (2015).
18. M. A. Mueed, D. Kamburov, S. Hasdemir, M. Shayegan, L. N. Pfeiffer, K. W. West and K. W. Baldwin, Geometric resonance of composite fermions near the $\nu = 1/2$ fractional quantum Hall state, *Phys. Rev. Lett.* **114**, 236406 (2015).
19. M. A. Mueed, D. Kamburov, L. N. Pfeiffer, K. W. West, K. W. Baldwin and M. Shayegan, Geometric resonance of composite fermions near bilayer quantum Hall states, *Phys. Rev. Lett.* **117**, 246801 (2016).
20. M. A. Mueed, D. Kamburov, S. Hasdemir, L. N. Pfeiffer, K. W. West, K. W. Baldwin and M. Shayegan, Anisotropic composite fermions and fractional quantum Hall effect, *Phys. Rev. B* **93**, 195436 (2016).
21. M. A. Mueed, D. Kamburov, Md. Shafayat Hossain, L. N. Pfeiffer, K. W. West, K. W. Baldwin and M. Shayegan, Search for composite fermions at filling factor 5/2: Role of Landau level and subband index, *Phys. Rev. B* **95**, 165438 (2017).
22. I. Jo, K. A. Villegas Rosales, M. A. Mueed, L. N. Pfeiffer, K. W. West, K. W. Baldwin, R. Winkler, M. Padmanabhan and M. Shayegan, Transference of Fermi contour anisotropy to composite fermions, *Phys. Rev. Lett.* **119**, 016402 (2017).
23. Md. Shafayat Hossain, M. K. Ma, M. A. Mueed, L. N. Pfeiffer, K. W. West, K. W. Baldwin and M. Shayegan, Direct observation of composite fermions and their fully spin polarized Fermi sea near $\nu = 5/2$, *Phys. Rev. Lett.* **120**, 256601 (2018). Erratum: *Phys. Rev. Lett.* **121**, 209901 (2018).
24. Md. Shafayat Hossain, Meng K. Ma, M. A. Mueed, D. Kamburov, L. N. Pfeiffer, K. W. West, K. W. Baldwin, R. Winkler and M. Shayegan, Geometric resonance of four-flux composite fermions, *Phys. Rev. B* (Rapid Communications) **100**, 041112(R) (2019).
25. H. Deng, Y. Liu, I. Jo, L. N. Pfeiffer, K. W. West, K. W. Baldwin and M. Shayegan, Commensurability oscillations of composite fermions induced by the periodic potential of a Wigner crystal, *Phys. Rev. Lett.* **117**, 096601 (2016).
26. Insun Jo, Hao Deng, Y. Liu, L. N. Pfeiffer, K. W. West, K. W. Baldwin and M. Shayegan, Evidence for cyclotron orbits of composite fermions in the fractional quantum Hall regime, *Phys. Rev. Lett.* **120**, 016802 (2018).
27. A. T. Hatke, H. Deng, Y. Liu, L. W. Engel, L. N. Pfeiffer, K. W. West, K. W. Baldwin and M. Shayegan, Wigner solid pinning modes tuned by fractional quantum Hall states of a nearby layer, *Science Advances* **5**, eaao2848 (2019).
28. D. Weiss, K. von Klitzing, K. Ploog and G. Weimann, Magnetoresistance oscillations in a two-dimensional electron gas induced by a submicrometer periodic potential, *Europhys. Lett.* **8**, 179 (1989).
29. R. W. Winkler, J. P. Kotthaus and K. Ploog, Landau band conductivity in a two-dimensional electron system modulated by an artificial one-dimensional superlattice potential, *Phys. Rev. Lett.* **62**, 1177 (1989).
30. R. R. Gerhardts, D. Weiss and K. von Klitzing, Novel Magnetoresistance oscillations in a periodically modulated two-dimensional electron gas, *Phys. Rev. Lett.* **62**, 1173 (1989).
31. C. W. J. Beenakker, Guiding-center-drift resonance in a periodically modulated two-dimensional electron gas, *Phys. Rev. Lett.* **62**, 2020 (1989).
32. P. H. Beton, E. S. Alves, P. C. Main, L. Eaves, M. W. Dellow, M. Henini, O. H. Hughes, S. P. Beaumont and C. D. W. Wilkinson, Magnetoresistance of a two-dimensional electron gas in a strong periodic potential, *Phys. Rev. B* **42**, 9229 (1990).
33. E. Skuras, A. R. Long, I. A. Larkin, J. H. Davies and M. C. Holland, Anisotropic piezoelectric effect in lateral surface superlattices, *Appl. Phys. Lett.* **70**, 871 (1997).

34. A. Endo, S. Katsumoto and Y. Iye, Envelope of Commensurability magnetoresistance oscillation in unidirectional lateral superlattices, *Phys. Rev. B* **62**, 16761 (2000).
35. D. Kamburov, H. Shapourian, M. Shayegan, L. N. Pfeiffer, K. W. West, K. W. Baldwin and R. Winkler Ballistic transport of (001) GaAs 2D holes through a strain-induced lateral superlattice, *Phys. Rev. B* **85**, 121305(R) (Rapid Communications) (2012).
36. D. Kamburov, M. Shayegan, R. Winkler, L. N. Pfeiffer, K. W. West and K. W. Baldwin, Anisotropic Fermi contour of (001) GaAs holes in parallel magnetic fields, *Phys. Rev. B* **86**, 241302(R) (Rapid Communications) (2012).
37. D. Kamburov, M. A. Mueed, M. Shayegan, L. N. Pfeiffer, K. W. West, K. W. Baldwin, J. J. D. Lee and R. Winkler, Anisotropic Fermi contour of (001) GaAs electrons in parallel magnetic fields, *Phys. Rev. B* **88**, 125435 (2013).
38. M. A. Mueed, D. Kamburov, M. Shayegan, L. N. Pfeiffer, K. W. West, K. W. Baldwin and R. Winkler, Splitting of the Fermi contour of quasi-2D electrons in parallel magnetic fields, *Phys. Rev. Lett.* **114**, 236404 (2015).
39. I. Jo, M. A. Mueed, L. N. Pfeiffer, K. W. West, K. W. Baldwin, R. Winkler, M. Padmanabhan and M. Shayegan, Tuning of Fermi contour anisotropy in GaAs (001) 2D holes via strain, *Appl. Phys. Lett.* **110**, 252103 (2017).
40. O. Gunawan, Y. P. Shkolnikov, E. P. De Poortere, E. Tutuc and M. Shayegan, Ballistic transport in AlAs two-dimensional electrons, *Phys. Rev. Lett.* **93**, 246603 (2004).
41. M. Barkeshli, M. Mulligan and M. P. A. Fisher, Particle-hole symmetry and the composite Fermi liquid, *Phys. Rev. B* **92**, 165125 (2015).
42. S. Kachru, M. Mulligan, G. Torroba and H. Wang, Mirror symmetry and the half-filled Landau level, *Phys. Rev. B* **92**, 235105 (2015).
43. D. T. Son, Is the composite fermion a Dirac particle? *Phys. Rev. X* **5**, 031027 (2015).
44. A. C. Balram, C. Tőke and J. K. Jain, Luttinger theorem for the strongly correlated Fermi liquid of composite fermions, *Phys. Rev. Lett.* **115**, 186805 (2015).
45. A. C. Balram and J. K. Jain, Nature of composite fermions and the role of particle-hole symmetry: A microscopic account, *Phys. Rev. B* **93**, 235152 (2016).
46. C. Wang and T. Senthil, Half-filled Landau level, topological insulator surfaces, and three-dimensional quantum spin liquids, *Phys. Rev. B* **93**, 085110 (2016).
47. Z. Wang and S. Chakravarty, Pairing of particle-hole symmetric composite fermions in half-filled Landau level, *Phys. Rev. B* **94**, 165138 (2016).
48. M. Mulligan, S. Raghu and M. P. A. Fisher, Emergent particle-hole symmetry in the half-filled Landau level, *Phys. Rev. B* **94**, 075101 (2016).
49. S. D. Geraedts, M .P. Zaletel, Roger S. K. Mong, M. A. Metlitski, A. Vishwanath and O .I. Motrunich, The half-filled Landau level: The case for Dirac composite fermions, *Science* **352**, 197 (2016).
50. P. Zucker and D. E. Feldman, Stabilization of the particle-hole Pfaffian order by Landau-level mixing and impurities that break particle-hole symmetry, *Phys. Rev. Lett.* **117**, 096802 (2016).
51. C. Wang, N. R. Cooper, B. I. Halperin and A. Stern, Particle-hole symmetry in the Fermion-Chern-Simons and Dirac descriptions of a half-Filled Landau level, *Phys. Rev. X* **7**, 031029 (2017).
52. A. K. C. Cheung, S. Raghu and M. Mulligan, Weiss oscillations and particle-hole symmetry at the half-filled Landau level, *Phys. Rev. B* **95**, 235424 (2017).
53. Y. Yang, X. Luo and Y. Yu, Nonperturbative emergence of the Dirac fermion in a strongly correlated composite Fermi liquid, *Phys. Rev. B* **95**, 035123 (2017).
54. A. C. Balram and J. K. Jain, Fermi wave vector for the partially spin-polarized composite-fermion Fermi sea, *Phys. Rev. B* **96**, 235102 (2017).

55. G. J. Sreejith, Y. Zhang and J. K. Jain, Surprising robustness of particle-hole symmetry for composite-fermion liquids, *Phys. Rev. B* **96**, 125149 (2017).

56. D. X. Nguyen, S. Golkar, M. M. Roberts and D. T. Son, Particle-hole symmetry and composite fermions in fractional quantum Hall states, *Phys. Rev. B* **97**, 195314 (2018).

57. H. Goldman and E. Fradkin, Dirac composite fermions and emergent reflection symmetry about even-denominator filling fractions, *Phys. Rev. B* **98**, 165137 (2018).

58. D. T. Son, The Dirac composite fermion of the fractional quantum Hall effect, *Ann. Rev. Conden. Mat. Phys.* **9**, 397 (2018).

59. A. Mitra and M. Mulligan, Fluctuations and magnetoresistance oscillations near the half-filled Landau level, *Phys. Rev. B* **100**, 165122 (2019).

60. J. D. Nickila, Fractional quantum Hall effect in an anisotropic two-dimensional system, *Phys. Stat. Sol. (b)* **187**, K57 (1995).

61. D. B. Balagurov and Y. E. Lozovik, Fermi surface of composite fermions and one-particle excitations at $\nu = 1/2$: Effect of band-mass anisotropy, *Phys. Rev. B* **62**, 1481 (2000).

62. T. Gokmen, M. Padmanabhan and M. Shayegan, Transference of transport anisotropy to composite fermions, *Nature Phys.* **6**, 621 (2010).

63. M. Shayegan, E. P. De Poortere, O. Gunawan, Y. P. Shkolnikov, E. Tutuc and K. Vakili, Two-dimensional electrons occupying multiple valleys in AlAs, *Phys. Stat. Sol. (b)* **243**, 3629–3642 (2006).

64. M. Mulligan, C. Nayak and S. Kachru, Isotropic to anisotropic transition in a fractional quantum Hall state, *Phys. Rev. B* **82**, 085102 (2010).

65. M. Mulligan, C. Nayak and S. Kachru, Effective field theory of fractional quantized Hall nematics, *Phys. Rev. B* **84**, 195124 (2011).

66. J. Xia, J. P. Eisenstein, L. N. Pfeiffer and K. W. West, Evidence for a fractionally quantized Hall state with anisotropic longitudinal transport, *Nature Physics* **7**, 845 (2011).

67. F. D. M. Haldane, Geometrical description of the fractional quantum Hall effect, *Phys. Rev. Lett.* **107**, 116801 (2011).

68. R. Z. Qiu, F. D. M. Haldane, X. Wan, K. Yang and S. Yi, Model anisotropic quantum Hall states, *Phys. Rev. B* **85**, 115308 (2012).

69. B. Yang, Z. Papic, E. H. Rezayi, R. N. Bhatt and F. D. M. Haldane, Band mass anisotropy and the intrinsic metric of fractional quantum Hall systems, *Phys. Rev. B* **85**, 165318 (2012).

70. H. Wang, R. Narayanan, X. Wan and F. Zhang, Fractional quantum Hall states in two-dimensional electron systems with anisotropic interactions, *Phys. Rev. B* **86**, 035122 (2012).

71. Y. Liu, S. Hasdemir, M. Shayegan, L. N. Pfeiffer, K. W. West and K. W. Baldwin, Evidence for a $\nu = 5/2$ fractional quantum Hall nematic state in parallel magnetic fields, *Phys. Rev. B* **88**, 035307 (2013).

72. Z. Papic, Fractional quantum Hall effect in a tilted magnetic field, *Phys. Rev. B* **87**, 245315 (2013).

73. K. Yang, Geometry of compressible and incompressible quantum Hall states: Application to anisotropic composite-fermion liquids, *Phys. Rev. B* **88**, 241105(R) (2013).

74. A. C. Balram and J. K. Jain, Exact results for model wave functions of anisotropic composite fermions in the fractional quantum Hall effect, *Phys. Rev. B* **93**, 075121 (2016).

75. Z. Zhu, I. Sodemann, D. N. Sheng and L. Fu, Anisotropy-driven transition from the Moore-Read state to quantum Hall stripes, *Phys. Rev. B* **95**, 201116(R) (2017).

76. M. Ippoliti, S. D. Geraedts and R. N. Bhatt, Numerical study of anisotropy in a composite Fermi liquid, *Phys. Rev. B* **95**, 201104(R) (2017).
77. M. Ippoliti, S. D. Geraedts and R. N. Bhatt, Connection between Fermi contours of zero-field electrons and $\nu = 1/2$ composite fermions in two-dimensional systems, *Phys. Rev. B* **96**, 045145 (2017).
78. M. Ippoliti, S. D. Geraedts and R. N. Bhatt, Composite fermions in bands with N-fold rotational symmetry, *Phys. Rev. B* **96**, 115151 (2017).
79. M. Ippoliti, R. N. Bhatt and F. D. M. Haldane, Geometry of flux attachment in anisotropic fractional quantum Hall states, *Phys. Rev. B* **98**, 085101 (2018).
80. Md. Shafayat Hossain, M. K. Ma, Y. J. Chung, L. N. Pfeiffer, K. W. West, K. W. Baldwin and M. Shayegan, Unconventional anisotropic even-denominator fractional quantum Hall state in a system with mass anisotropy, *Phys. Rev. Lett.* **121**, 256601 (2018).
81. J. N. Leaw, H.-K. Tang, M. Trushin, F. F. Assaad and S. Adam, Universal Fermi-surface anisotropy renormalization for interacting Dirac fermions with long-range interactions, *Proc. Natl. Acad. Sci. (USA)* **116**, 26431 (2019).
82. F. Stern, Transverse Hall effect in the electric quantum limit, *Phys. Rev. Lett.* **21**, 1687 (1968).
83. N. W. Ashcroft and N. D. Mermin, *Solid State Physics* (Holt, Rinehart and Winston, Philadelphia, 1976), Chapter 12.
84. R. Winkler, *Spin-Orbit Coupling Effects in Two-Dimensional Electron and Hole Systems* (Springer, Berlin, 2003).
85. R. R. Du, H. L. Stormer, D. C. Tsui, L. N. Pfeiffer and K. W. West, Experimental evidence for new particles in the fractional quantum Hall effect, *Phys. Rev. Lett.* **70**, 2944 (1993).
86. H. C. Manoharan, M. Shayegan and S. J. Klepper, Signatures of a novel Fermi liquid in a two-dimensional composite particle metal, *Phys. Rev. Lett.* **73**, 3270 (1994).
87. M. Shayegan, K. Karrai, Y. Shkolnikov, K. Vakili, E. P. De Poortere and S. Manus, Low-temperature, *in-situ*-tunable, uniaxial stress measurements in semiconductors using a piezoelectric actuator, *Appl. Phys. Lett.* **83**, 5235 (2003).
88. A. Endo, M. Kawamura, S. Katsumoto and Y. Iye, Magnetotransport of $\nu = 3/2$ composite fermions under periodic effective magnetic-field modulation, *Phys. Rev. B* **63**, 113310 (2001).
89. R. G. Clark, S. R. Haynes, A. M. Suckling, J. R. Mallett, P. A. Wright, J. J. Harris and C. T. Foxon, Spin configurations and quasiparticle fractional charge of fractional quantum Hall effect ground states in the $N = 0$ Landau level, *Phys. Rev. Lett.* **62**, 1536 (1989).
90. J. P. Eisenstein, H. L. Stormer, L. Pfeiffer and K. W. West, Evidence for a phase transition in the fractional quantum Hall effect, *Phys. Rev. Lett.* **62**, 1540 (1989).
91. L. W. Engel, S. W. Hwang, T. Sajoto, D. C. Tsui and M. Shayegan, Fractional quantum Hall effect at $\nu = 2/3$ and 3/5 in tilted magnetic fields, *Phys. Rev. B* **45**, 3418 (1992).
92. R. R. Du, A. S. Yeh, H. L. Stormer, D. C. Tsui, L. N. Pfeiffer and K. W. West, Fractional quantum Hall effect around $\nu = 32$: Composite fermions with a spin, *Phys. Rev. Lett.* **75**, 3926 (1995).
93. K. Park and J. K. Jain, *Phys. Rev. Lett.* **80**, 4237 (1998).
94. I. V. Kukushkin, K. v. Klitzing and K. Eberl, Spin polarization of composite fermions: Measurements of the Fermi energy, *Phys. Rev. Lett.* **82**, 3665 (1999).
95. S. C. Davenport and S. H. Simon, Spinful composite fermions in a negative effective field, *Phys. Rev. B* **85**, 245303 (2012).

96. V. V. Vanovsky, V. S. Khrapai, A. A. Shashkin, V. Pellegrini, L. Sorba and G. Biasiol, Spin transition in the fractional quantum Hall regime: Effect of the extent of the wave function, *Phys. Rev. B* **87**, 081306(R) (2013).

97. Y. Liu, S. Hasdemir, A. Wojs, J. K. Jain, L. N. Pfeiffer, K. W. West, K. W. Baldwin and M. Shayegan, Spin polarization of composite fermions and particle-hole symmetry breaking, *Phys. Rev. B* **90**, 085301 (2014).

98. D. R. Leadley, R. J. Nicholas, C. T. Foxon and J. J. Harris, Measurements of the effective mass and scattering times of composite fermions from magnetotransport analysis, *Phys. Rev. Lett.* **72**, 1906 (1994).

99. A. S. Yeh, H. L. Stormer, D. C. Tsui, L. N. Pfeiffer, K. W. Baldwin and K. W. West, Effective mass and g-factor of four-flux-quanta composite fermions, *Phys. Rev. Lett.* **82**, 592 (1999).

100. W. Pan, H. L. Stormer, D. C. Tsui, L. N. Pfeiffer, K. W. Baldwin and K. W. West, Effective mass of the four-flux composite fermion at $\nu = 1/4$, *Phys. Rev. B* **61**, R5101(R) (2000).

101. E. Y. Andrei, G. Deville, D. C. Glattli, F. I. B. Williams, E. Paris and B. Etienne, Observation of a magnetically induced Wigner solid, *Phys. Rev. Lett.* **60**, 2765 (1988).

102. H. W. Jiang, R. L. Willett, H. L. Stormer, D. C. Tsui, L. N. Pfeiffer and K. W. West, Quantum liquid versus electron solid around $\nu = 1/5$ Landau-level filling, *Phys. Rev. Lett.* **65**, 633 (1990).

103. V. J. Goldman, M. Santos, M. Shayegan and J. E. Cunningham, Evidence for two-dimensional quantum Wigner crystal, *Phys. Rev. Lett.* **65**, 2189 (1990).

104. For an early review, see M. Shayegan, Case for the magnetic-field-induced two-dimensional Wigner crystal, in *Perspectives in Quantum Hall Effects*, edited by S. D. Sarma and A. Pinczuk (Wiley, New York, 1997), pp. 343–383.

105. R. L. Willett, J. P. Eisenstein, H. L. Stormer, D. C. Tsui, A. C. Gossard and J. H. English, Observation of an even-denominator quantum number in the fractional quantum Hall effect, *Phys. Rev. Lett.* **59**, 1776 (1987).

106. B. I. Halperin, Theory of the quantized Hall conductance, *Helv. Phys. Acta* **56**, 75 (1983).

107. F. D. M. Haldane and E. H. Rezayi, Spin-singlet wave function for the half-integral quantum Hall effect, *Phys. Rev. Lett.* **60**, 956 (1988).

108. G. Moore and N. Read, Nonabelions in the fractional quantum Hall effect, *Nucl. Phys. B* **360**, 362 (1991).

109. R. Morf, Transition from quantum Hall to compressible states in the second Landau level: New light on the $\nu = 5/2$ enigma, *Phys. Rev. Lett.* **80**, 1505 (1998).

110. C. Nayak, S. H. Simon, A. Stern, M. Freedman and S. D. Sarma, Non-abelian anyons and topological quantum computation, *Rev. Mod. Phys.* **80**, 1083 (2008).

111. R. L. Willett, The quantum Hall effect at 5/2 filling factor, *Rep. Prog. Phys.* **76**, 076501 (2013).

112. I. P. Radu, J. B. Miller, C. M. Marcus, M. A. Kastner, L. N. Pfeiffer and K. W. West, Quasi-particle properties from tunneling in the $\nu = 5/2$ fractional quantum Hall state, *Science* **320**, 899 (2008).

113. X. Lin, C. Dillard, M. A. Kastner, L. N. Pfeiffer and K. W. West, Measurements of quasiparticle tunneling in the $\nu = 5/2$ fractional quantum Hall state, *Phys. Rev. B* **85**, 165321 (2012).

114. R. L. Willett, L. N. Pfeiffer and K. W. West, Measurement of filling factor 5/2 quasiparticle interference with observation of charge $e/4$ and $e/2$ period oscillations, *Proc. Natl. Acad. Sci. (USA)* **106**, 8853 (2009).

115. R. L. Willett, C. Nayak, K. Shtengel, L. N. Pfeiffer and K. W. West, Magnetic-field-

tuned Aharonov-Bohm oscillations and evidence for non-Abelian anyons at $\nu = 5/2$, *Phys. Rev. Lett.* **111**, 186401 (2013).

116. R. L. Willett, K. Shtengel, C. Nayak, L. N. Pfeiffer, Y. J. Chung, M. L. Peabody, K. W. Baldwin and K. W. West, Interference measurements of non-Abelian $e/4$ & Abelian $e/2$ quasiparticle braiding, cond-mat arXiv:1905.10248.

117. R. L. Willett, K. W. West and L. N. Pfeiffer, Experimental demonstration of Fermi surface effects at filling factor 5/2, *Phys. Rev. Lett.* **88**, 066801 (2002).

118. J. P. Eisenstein, R. Willett, H. L. Stormer, D. C. Tsui, A. C. Gossard and J. H. English, Collapse of the even-denominator fractional quantum Hall effect in tilted fields, *Phys. Rev. Lett.* **61**, 997 (1988).

119. M. Stern, P. Plochocka, V. Umansky, D. K. Maude, M. Potemski and I. Bar-Joseph, Optical probing of the spin polarization of the $\nu = 5/2$ quantum Hall state, *Phys. Rev. Lett.* **105**, 096801 (2010).

120. W. Pan, H. L. Stormer, D. C. Tsui, L. N. Pfeiffer, K. W. Baldwin and K. W. West, Experimental evidence for a spin-polarized ground state in the $\nu = 5/2$ fractional quantum Hall effect, *Solid State Commun.* **119**, 641 (2001).

121. C. Zhang, T. Knuuttila, Y. Dai, R. R. Du, L. N. Pfeiffer and K. W. West, $\nu = 5/2$ Fractional quantum Hall effect at 10 T: Implications for the Pfaffian state, *Phys. Rev. Lett.* **104**, 166801 (2010).

122. U. Wurstbauer, K. W. West, L. N. Pfeiffer and A. Pinczuk, Resonant inelastic light scattering investigation of low-lying gapped excitations in the quantum fluid at $\nu = 5/2$, *Phys. Rev. Lett.* **110**, 026801 (2013).

123. Tiemann, G. Gamez, N. Kumada and K. Muraki, Unraveling the spin polarization of the $\nu = 5/2$ fractional quantum Hall state, *Science* **335**, 828 (2012).

124. M. Stern, B. A. Piot, Y. Vardi, V. Umansky, P. Plochocka, D. K. Maude and I. Bar-Joseph, NMR probing of the spin polarization of the $\nu = 5/2$ quantum Hall state, *Phys. Rev. Lett.* **108**, 066810 (2012).

125. J. P. Eisenstein, L. N. Pfeiffer and K. W. West, Quantum Hall spin diode, *Phys. Rev. Lett.* **118**, 186801 (2017).

126. M. A. Mueed, Md. Shafayat Hossain, L. N. Pfeiffer, K. W. West, K. W. Baldwin and M. Shayegan, Reorientation of the stripe phase of 2D electrons by a minute density modulation, *Phys. Rev. Lett.* **117**, 076803 (2016).

127. Y. W. Suen, L. W. Engel, M. B. Santos, M. Shayegan and D. C. Tsui, Observation of a $\nu = 1/2$ fractional quantum Hall state in a double layer electron system, *Phys. Rev. Lett.* **68**, 1379 (1992).

128. Y. W. Suen, H. C. Manoharan, X. Ying, M. B. Santos and M. Shayegan, Origin of the 1/2 fractional quantum Hall state in wide single quantum wells, *Phys. Rev. Lett.* **72**, 3405 (1994).

129. J. Shabani, Y. Liu, M. Shayegan, L. N. Pfeiffer, K. W. West and K. W. Baldwin, Phase diagrams for the stability of the $\nu = 1/2$ fractional quantum Hall effect in electron systems confined to symmetric, wide GaAs quantum wells, *Phys. Rev. B* **88**, 245413 (2013).

130. Y. Liu, A. L. Graninger, S. Hasdemir, M. Shayegan, L. N. Pfeiffer, K. W. West, K. W. Baldwin and R. Winkler, Fractional quantum Hall effect at $\nu = 1/2$ in hole systems confined to GaAs quantum wells, *Phys. Rev. Lett.* **112**, 046804 (2014).

131. Y. Liu, S. Hasdemir, A. L. Graninger, D. Kamburov, M. Shayegan, L. N. Pfeiffer, K. W. West, K. W. Baldwin and R. Winkler, Even-denominator fractional quantum Hall effect at a Landau level crossing, *Phys. Rev. B* **89**, 165313 (2014).

132. S. Hasdemir, Y. Liu, H. Deng, M. Shayegan, L. N. Pfeiffer, K. W. West and K. W. Baldwin, $\nu = 1/2$ fractional quantum Hall effect in tilted magnetic fields, *Phys. Rev. B* **91**, 045113 (2015).

133. M. Greiter, X. G. Wen and F. Wilczek, Paired Hall state at half filling, *Phys. Rev. Lett.* **66**, 3205 (1991).

134. M. Greiter, X. G. Wen and F. Wilczek, Paired Hall states in double-layer electron systems, *Phys. Rev. B* **46**, 9586 (1992).

135. S. He, S. Das Sarma and X. C. Xie, Quantized Hall effect and quantum phase transitions in coupled two-layer electron systems, *Phys. Rev. B* **47**, 4394 (1993).

136. M. R. Peterson and S. Das Sarma, Quantum Hall phase diagram of half-filled bilayers in the lowest and the second orbital Landau levels: Abelian versus non-Abelian incompressible fractional quantum Hall states, *Phys. Rev. B* **81**, 165304 (2010).

137. W. Zhu, Z. Liu, F. D. M. Haldane and D. N. Sheng, Fractional quantum Hall bilayers at half filling: Tunneling-driven non-Abelian phase, *Phys. Rev. B* **94**, 245147 (2016).

138. Y. E. Lozovik and V. I. Yudson, Crystallization of a two-dimensional electron gas in a magnetic field, *JETP Lett.* **22**, 11 (1975).

139. P. K. Lam and S. M. Girvin, Liquid-solid transition and the fractional quantum- Hall effect, *Phys. Rev. B* **30**, 473 (1984).

140. D. Levesque, J. J. Weis and A. H. MacDonald, Crystallization of the incompressible quantum-fluid state of a two-dimensional electron gas in a strong magnetic field, *Phys. Rev. B* **30**, 1056 (1984).

141. Y. P. Li, T. Sajoto, L. W. Engel, D. C. Tsui and Shayegan, Low frequency noise in the reentrant insulating phase around the 1/5 fractional quantum Hall liquid, *Phys. Rev. Lett.* **67**, 1630 (1991).

142. C.-C. Li, L. W. Engel, D. Shahar, D. C. Tsui and M. Shayegan, Microwave conductivity resonance of two-dimensional hole systems, *Phys. Rev. Lett.* **79**, 1353 (1997).

143. Y. P. Chen, G. Sambandamurthy, Z. H. Wang, R. M. Lewis, L. W. Engel, D. C. Tsui, P. D. Ye, L. N. Pfeiffer and K. W. West, Melting of a 2D quantum electron solid in high magnetic field, *Nature Physics* **2**, 452 (2006).

144. V. Bayot, X. Ying, M.B. Santos and M. Shayegan, Thermopower measurements in the reentrant insulating phase of a two-dimensional hole system, *Europhys. Lett.* **25**, 613 (1994).

145. L. Tiemann, T. D. Rhone, N. Shibata and K. Muraki, NMR profiling of quantum electron solids in high magnetic fields, *Nature Physics* **10**, 648 (2014).

146. J. Jang, B. M. Hunt, L. N. Pfeiffer, K. W. West and R. C. Ashoori, Sharp tunnelling resonance from the vibrations of an electronic Wigner crystal, *Nature Physics* **13**, 340 (2017).

147. H. Deng, L. N. Pfeiffer, K. W. West, K. W. Baldwin and M. Shayegan, Probing the melting of a two-dimensional quantum Wigner crystal via its screening efficiency, *Phys. Rev. Lett.* **122**, 116601 (2019).

148. D. Weiss, M. L. Roukes, A. Menschig, P. Grambow, K. von Klitzing and G. Weimann, Electron pinball and commensurate orbits in a periodic array of scatterers, *Phys. Rev. Lett.* **66**, 2790 (1991).

149. T. Yamashiro, J. Takahara, Y. Takagaki, K. Gamo, S. Namba, S. Takaoka and K. Murase, Commensurate classical orbits on triangular lattices of anti-dots, *Solid State Commun.* **79**, 885 (1991).

150. D. Weiss, K. Richter, E. Vasiliadou and G. Lutjering, Magnetotransport in anti-dot arrays, *Surf. Sci.* **305**, 408 (1994).

151. S. Meckler, T. Heinzel, A. Cavanna, G. Faini, U. Gennser and D. Mailly, Commensurability effects in hexagonal antidot lattices with large antidot diameters, *Phys. Rev. B* **72**, 035319 (2005).

152. W. Kang, H. L. Stormer, L. N. Pfeiffer, K. W. Baldwin and K. W. West, How real are composite fermions? *Phys. Rev. Lett.* **71**, 3850 (1993).

153. Md. Shafayat Hossain, M. A. Mueed, M. K. Ma, K. A. Villegas Rosales, Y. J. Chung, L. N. Pfeiffer, K. W. West, K. W. Baldwin and M. Shayegan, A precise experimental test of the Luttinger theorem and particle-hole symmetry for a strongly correlated fermionic system, unpublished.
154. J. M. Luttinger, Fermi surface and some simple equilibrium properties of a system of interacting fermions, *Phys. Rev.* **119**, 1153 (1960).
155. Md. Shafayat Hossain, Tongzhou Zhao, Songyang Pu, M. A. Mueed, M. K. Ma, K. A. Villegas Rosales, Y. J. Chung, L. N. Pfeiffer, K. W. West, K. W. Baldwin, J. K. Jain and M. Shayegan, Bloch ferromagnetism of composite fermions, unpublished.
156. F. Bloch, Bemerkung zur Elektronentheorie des Ferromagnetismus und der elektrischen Leitfaehigkeit, *Z. Phys.* **57**, 545 (1929).

Chapter 4

Edge Probes of Topological Order

Moty Heiblum

Department of Condensed Matter Physics, Weizmann Institute of Science,
Rehovot, Israel

D. E. Feldman

Department of Physics, Brown University, Providence, RI 02912, USA

According to the bulk-edge correspondence principle, the physics of the gapless edge in the quantum Hall effect determines the topological order in the gapped bulk. As the bulk is less accessible, the last two decades saw the emergence of several experimental techniques that invoke the study of the compressible edge. We review the properties of the edge, and describe several experimental techniques that include shot noise and thermal noise measurements, interferometry, and energy (thermal) transport at the edge. We pay special attention to the filling factor 5/2 in the first excited Landau level (in two-dimensional electron gas in GaAs), where experimental evidence of a non-Abelian topological order was found. A brief discussion is devoted to recent interferometry experiments that uncovered unexpected physics in the integer quantum Hall effect. The chapter also addresses the theory of edge states, for systems with Abelian and non-Abelian topological orders.

Contents

1. Introduction

The fractional quantum Hall effect (FQHE) famously gives rise to quasiparticles whose charges are lower than an electron charge. Their exchange statistics is neither Fermi nor Bose. Roughly speaking, the knowledge of the topological order in FQHE systems consists in the understanding of the charges and the statistics of those fractionally charged anyons. Since the anyons exist in the gapped bulk of 2D electron liquid, experiments probing the bulk may *a priori* seem the best way to test the order of the state. Yet, such probes, being difficult to realize, are often not useful in respect to the latter. Hence, the focus of this chapter is on edge physics. The electron liquid is gapless at the sample's edges, where the filling factor changes from the bulk value to zero, and the bulk topological order, protected by the bulk gap, is absent. Naively, this makes the edge a wrong place to look for signatures of the topological order. Nevertheless, most of our knowledge of the topological order in the FQHE regime comes from probing the edge.

The most basic question involves the charges of the anyons. Edge physics can shed light on this question, e.g., through measuring shot noise. Weak tunneling between opposite edges of the sample, brought to nearby proximity by means of a narrow constriction, leads to a universal noise that is proportional to the carriers' charge. Additional information comes from strong tunneling experiments. This physics is reviewed in Sec. 3.

Since the exchange statistics of anyons is defined in terms of particles moving around each other, anyonic interferometry is a natural tool to understand the states' topological orders. In turn, interferometry is studied by splitting the incident chiral current emanating from the source into two (or more) different paths, and merging them again before the drain. We address interferometry in Sec. 5, focusing on the integer QHE (IQHE). Understanding interferometry in the IQHE regime is mandatory before one can rely on its interpretation in the FQHE. Recent developments, such as the discovery of the pairing effect and the lobe structure, underscore how challenging interferometry is even in the IQHE.

Another approach has recently emerged as a powerful probe of topological orders: Experiments involving energy transport. We first address the physics of neutral modes in Sec. 4. The simplest quantum Hall edges carry only downstream charged channels (the chirality is dictated by the magnetic field). However, more typically, charged channels may coexist with one or more (often upstream, namely, with opposite chirality) neutral modes. Since the neutral modes (not carrying net charge) weakly interact with external probes, accessing them is a difficult task. A breakthrough in that direction came in 2010, when the existence of upstream neutral modes was proven experimentally via shot noise measurements. An even more important development consisted in the measurements of the thermal conductance of quantum Hall edge modes (Sec. 7). In particular, a thermal conductance experiment gave strong support to non-Abelian statistics at filling factor $\nu = 5/2$.

Section 6 focuses on that enigmatic $\nu = 5/2$ filling factor. Its very existence was a surprise. The most intuitive way to think of FQHE states is based on composite fermions (CFs). Odd-denominator fractions can be interpreted as IQHE gapped states of non-interacting CFs. On the other hand, a picture of weakly interacting CFs suggests gapless states at even-denominator filling factors. This agrees with the experiments in the first two spin-split Landau levels (LLs), but not in higher LLs. The apparent solution of this paradox lies in a picture of Cooper-pairing of CFs. Yet, the solution gives rise to another conundrum: Numerous ways exist to build Cooper pairs and instead of 'no obvious candidates for the $\nu = 5/2$ topological liquid' we now have way too many. Until new experimental probes of neutral modes were developed, the scant amount of available experimental data shifted the focus to numerical calculations. Different orders were leading at different times, and most recently numerics gave support to the Pfaffian and anti-Pfaffian topological orders, exhibiting the beautiful property of non-Abelian statistics. This feature is of current great interest for quantum computing.

A very recent thermal conductance experiment does support a non-Abelian liquid at $\nu = 5/2$; however the topological order appears to be neither of the two leading numerical candidates. The best fit to the existing data comes from the PH-Pfaffian topological order. Presumably, disorder effects, neglected in numerics, may be responsible for this order. More work is needed until the order of the $\nu = 5/2$ liquid and its universality are settled.

In addition to key Secs. 3–7 and a summary, this chapter reviews some background information: Edge models in Sec. 2 and in Appendix A, bulk-edge correspondence at $\nu = 5/2$ in Appendix B, and Coulomb effects in interferometry in Appendix C.

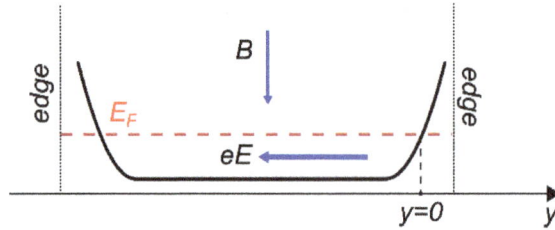

Fig. 1. Electrons near the edge at $y = 0$ experience a perpendicular magnetic field B in the z-direction, and an electrostatic force eE due to the confining potential in the y-direction. The thick line shows a simplified confining potential.

2. Edge Channels

Halperin was the first to suggest that edge-channels transport is responsible for the conduction mechanism in the QHE.[1] According to this successful model, current flows along the edges of the sample, with electrons performing classical chiral 'skipping orbits'. In the quantum mechanical language, current flows at the crossing of the Landau levels and the Fermi energy near the physical edges of the sample. In the IQHE regime, electron correlations are weak, and currents flow in non-interacting 1D-like channels. Alternatively, in the FQHE regime, electron correlations are strong, leading the edge channels to possess chiral Luttinger liquid-like properties.[2] Due to the chirality, backscattering in wide samples is minimized and the edge channels tend to propagate 'ballistically' for a long distance.[3] Since the interpretation of the experiments discussed here depends on the inner structure of the edge channels, we start with a brief introduction to edge channel models (excellent discussions can be found in the reviews by Kane and Fisher[4] and by Wen[3]).

We begin with the simplest spin-polarized non-interacting state with filling factor $\nu = 1$ (the lowest spin-split Landau level). Our starting point is a two-dimensional electron gas (2DEG) in an external potential $V(y)$, with an edge at $y = 0$ and a perpendicular magnetic field in the z-direction (Fig. 1).

The system is infinite in the x-direction. In the Landau gauge, $A_x = By$, $A_y = A_z = 0$, the infinitely degenerate wave-functions of the lowest Landau level are,

$$\psi_k(x, y) = \exp\left(ikx - \frac{[y - y_0]^2}{2l_B^2}\right),\tag{1}$$

where $y_0 = c\hbar k/|e|B$, and the magnetic length $l_B = \sqrt{\hbar c/|e|B}$ ($\sqrt{\hbar/|e|B}$ in the MKS units). These wave-functions describe charge density in narrow strips running along the x-direction and with width $\sim l_B$ along the y-direction. When the potential $V(y)$ changes on a much larger scale than l_B, it does not affect the wave-functions but lifts their degeneracy. Only the states below the Fermi energy $E_F = \frac{\hbar|e|B}{2mc} + V(y = 0)$ are filled, that is, the states with $y_0 < 0$ (corresponding to a positive edge velocity). Hence, our model describes an edge channel at $y = 0$ with a small width, where the charge density changes from zero at the edge to the bulk value inside the liquid.

We address that limit by linearizing the potential near $y = 0$ and expressing the Hamiltonian as,

$$H = \sum_k v k \psi_k^+ \psi_k \,, \tag{2}$$

where the fermion operators create and annihilate electrons in the states $\psi_k(x, y)$. The states are parametrized by the momentum k in the x-direction, and the problem reduces effectively to 1D with all the excitations propagating downstream with the same drift velocity v. The physical origin of the chirality is apparent in Fig. 1, with charges drifting at the edge in perpendicular to electric and magnetic fields. This is also evident from the semi-classical picture of skipping orbits.[5] The Hamiltonian H can be seen as a chiral part of a Hamiltonian of a non-chiral quantum wire. A similar Hamiltonian describes the opposite edge of the Hall bar, where excitations move in the opposite direction (at the same chirality). The Landauer conductance e^2/h of a quantum wire agrees with what one expects in the IQHE.

Even though electrons are fermions, it is often convenient to rewrite the edge model in terms of Bose operators. This approach does not make much difference in a non-interacting problem, but has advantages if Coulomb interactions at the edge are taken into account. While in terms of electron edge transport, the FQHE regime is a hopelessly complicated problem of strongly interacting fermions, many of the difficulties magically disappear in the Bose language, providing an easy way to describe the low-energy excitations, which are bosonic plasmons.[3]

We focus on the 1D charge density $\rho(x)$ along the edge, obtained by integrating the charge distribution along the y-direction. Defining the Bose field $\varphi(x)$ according to $\rho(x) = e\partial_x\varphi(x)/2\pi$, we get an edge-action:

$$\frac{S}{\hbar} = -\frac{1}{4\pi} \int dx\, dt [\partial_t \varphi \partial_x \varphi + v(\partial_x \varphi)^2]\,, \tag{3}$$

which we further briefly discuss in Appendix A. This action is a minimal model of the edge in the IQHE regime, as additional terms are unimportant at low energies. When long-range Coulomb interactions are important, another term, $\int dx_1 dx_2 \rho(x_1)\rho(x_2)/\epsilon|x_1 - x_2|$, must be included. The interaction is responsible particularly for a high edge velocity on an ungated edge and for a 'striped' structure of an edge, which may reconstruct in realistic samples due to Coulomb interaction. As was predicted in Refs. 6 and 7, and observed in Ref. 8, the edge may consist of compressible and incompressible stripes that form multiple parallel conducting channels, resulting from edge-reconstruction,[9] thus necessitating the addition to the action of the contributions from pairs of counter-propagating identical channels. Such pairs of channels are subject to Anderson localization via impurities on relatively short distances.[10] An extension of the action in Eq. (3) to integer states with filling $\nu > 1$ is easy: Each filled Landau level leads to one edge channel.

2.1. *Particle-states in the FQHE regime*

We now turn to the FQHE and discuss the basics of the chiral Luttinger liquid theory, introduced by Wen.[3] For Laughlin states at $\nu = 1/(2p + 1)$, there are only minute changes in the expressions from those in the IQHE. This can be understood from Jain's Composite Fermion (CF) picture.[11] With a CF formed after inserting an even number $2p$ of flux quanta for each electron ($2p\Phi_0 = 2ph/e$), each CF 'feels' a reduced effective magnetic field (in the mean-field approximation). Alternatively, it is legitimate to think of each electron as a combination of $2p$ flux quanta with a single CF.[11,12] Consequently, at $\nu = 1/2$, the CF is immersed in a zero effective magnetic field. At $\nu = 1/3$, its effective filling factor is 1, and a single edge channel is expected in the above action. This maintains the general structure of Eq. (3), with the only difference being the prefactor due to a different electrical conductance, $\nu e^2/h$ (see Appendix A):

$$\frac{S}{\hbar} = -\frac{1}{4\pi\nu} \int dx\, dt [\partial_t \varphi \partial_x \varphi + v(\partial_x \varphi)^2] \tag{4}$$

The physics is richer at other fractional filling factors. Consider $\nu = 2/5$ as an example. The CF description with $p = 1$ leads to two filled Landau levels of CFs, and thus two downstream edge channels. The simplest model is thus a sum of the two actions of the type described in Eqs. (3) and (4), but with added terms representing the Coulomb interaction between the two channels and inter-channel charge tunneling. We postpone a discussion of charge tunneling to Sec. 4.

2.2. *Hole-conjugate states in the FQHE regime*

The $\nu = 2/3$ state and other states in the range $1/2 < \nu < 1$ (the 'hole-conjugate' states) behave quite differently than the particle-states. The $\nu = 2/3$ liquid can be thought as an FQHE state of $\nu = 1/3$ 'holes' on top of an IQHE electron liquid at $\nu = 1$. Imagine that the charge density looks roughly as depicted in Fig. 2.

As the picture suggests, there are two counter-propagating edge channels. Since the electrostatic potential follows the electron density, the direction of the confining electric field is opposite at the two interfaces: One separating $\nu = 1$ from $\nu = 0$ (at the sample's edge), and the other separating $\nu = 1$ from $\nu = 2/3$ (equivalently, $\nu = 0$ of holes from $\nu = 1/3$ of holes). Two oppositely propagating channels emerge: A downstream channel identical to that of $\nu = 1$ and an upstream channel of $\nu = 1/3$.[13] On a short scale, inter-channel equilibration does not take place with the theoretically expected (but never observed) two-terminal conductance of $4e^2/3h$ [Fig. 2(c)].[14] At longer edges, after charge equilibration, the two-terminal conductance drops to $2e^2/3h$ (as is always experimentally observed; see also later). Edge-reconstruction may lead to additional pairs of counter-propagating modes on the edges of realistic samples. In particular, a pair of modes of conductance $e^2/3h$ was found,[15,16] replacing the modes' picture of MacDonald.[13] A general way to

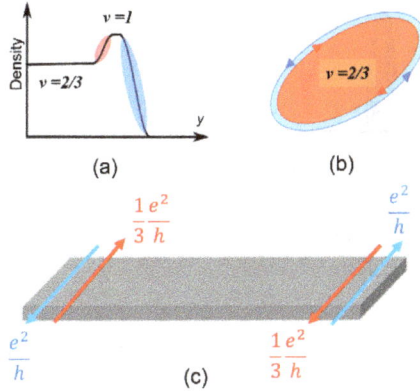

Fig. 2.   MacDonald's clean edge configuration, with the edge modes of the $\nu = 2/3$ state.
(a) Potential (and density) distribution near the edge.   (b) Chirality is counter-clockwise, the
$\nu = 1$ channel (blue) moves downstream, while the $\nu = 1/3$ channel (red) moves upstream.
(c) Two-terminal conductance $G = (1 + 1/3)\, e^2/h$. With inter-mode scattering, the downstream
charge channel has conductance $G = 2e^2/3h$ being accompanied by an upstream neutral mode
(not shown).

construct edge actions is known as the K-matrix formalism.[3] We briefly summarize
it in Appendix A.

3. Shot Noise and Charge in the FQHE

3.1. *Theoretical considerations*

One of the most striking features of the FQHE is its excitations that carry a fraction of an electron charge. Such excitations are an inevitable consequence of the
coexistence of the quantized Hall conductance $\sigma = \nu e^2/h$ with an energy gap in
the bulk.[17] Following Laughlin, consider an FQHE system pierced by a solenoid
in the middle of a plaquette in a lattice on which electrons move (electrons cannot
enter the solenoid and are not affected by it as long as the magnetic field through
the solenoid is zero). Next, the magnetic flux slowly increases within the solenoid
by one flux quantum Φ_0. The time-dependent flux gives rise to a circular electric
field E around the solenoid and its attendant electric current in the bulk away from
the solenoid. Charge moves away from the pierced plaquette to the edge of the
sample. The total transferred charge is $\Delta q = \sigma \int dt dl E = \nu e$, where the integration
extends over a circle inside the 2D electron gas, with the center at the solenoid. The
depleted charge is νe — being the charge of an excitation. Indeed, at the end of
the process, we are left with one flux quantum through the piercing solenoid. Such
flux is invisible to electrons in the 2D gas and can be eliminated by a large gauge
transformation. Hence, the final charge distribution corresponds to an eigenstate of
the initial Hamiltonian without the added flux, and the depleted charge is indeed
an excitation charge. Thus, the observed fractional quantized conductance leads,
though not in a direct way, to a fractional excitation charge.

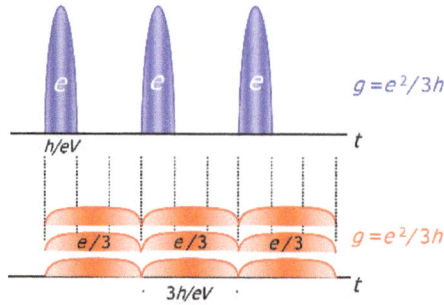

Fig. 3. Trains of electrons (top) and fractional charges (bottom), with both leading to the same conductance. After partitioning of charges, e.g., with a partly pinched QPC, the shot noise is proportional to the particle charge.

Note that it is not necessarily the minimal possible charge. The reason is that the system is made of electrons, and hence, there are always excitations of the charge e. For $\nu = m/n$, with coprimes m and n, one can always build an excitation of charge e/n from several charges e and νe. For an even n, the minimal charge was argued to be even lower, $e/2n$.[18]

Edge physics can be used for direct evidence of charge fractionalization by measuring shot noise. The idea is a century old and was employed by Schottky in the early 20th century to measure the charge of an electron in vacuum tubes.[19-21]

Imagine charge transport through a barrier by rare tunneling events. The current $I = ep$, where p is the tunneling rate, is transporting the total charge $Q = ept$ during the time t. Its mean-square-fluctuation is $\langle \Delta Q^2 \rangle = e^2 pt$ and the spectral density, defined as $S_i = 2\langle \Delta Q^2 \rangle / t$, in units of A^2/Hz is equal to:

$$S_i = \int dt \langle I(t)I(0) + I(0)I(t) \rangle = 2eI , \qquad (5)$$

where we substituted $Q = \int I(t)dt$. Thus, measurements of the noise and the current yield the carrier's charge. Nothing above implies that the carriers must be electrons. If the current comes from rare tunneling events of charge e^* quasiparticles, then $S_i = 2e^*I$ (Fig. 3).

Experimentally, a quantum point contact (QPC) constriction brings two opposite edge channels to nearby proximity, allowing charge to tunnel between the channels. The minimal tunneling charge is set by the bulk, being the charge of a quasiparticle.[22,23] We thus turn to shot noise experiments in FQHE.

3.2. *Measurements of shot noise*

Shot noise in mesoscopic conductors has proved to be a powerful experimental probe.[19,22-24] The noise provides information that is not necessarily available from the time-average flux of the particles (see Fig. 3), such as the particles' charge, their correlations, and even their temperature.

At zero temperature, an unpartitioned edge channel is noiseless — a direct consequence of Fermi statistics.[24-26] At a finite temperature, the channel carries also thermal noise, which is a property of any conductor, independent of its microscopic details and the carriers' charge. The spectral density of the added noise takes the form $S_T = 4k_BTg$, with k_B the Boltzmann's constant, T the temperature, and g the conductance of the edge channel. The so-called 'quantum shot noise' (the excess noise above the thermal noise) differs from the classical shot noise by reflecting the noise-free property of the emitting reservoir and the finite tunneling strength.[24-26] This was demonstrated first in the simplest mesoscopic system — a QPC.[27,28] At zero temperature, the contribution of the partitioned p-th channel (on a multi-channel edge), and a weakly back-scattering QPC ($t_p \sim 1$) is,

$$S_i(0) = 2e^*Vg_pt_p(1 - t_p), \qquad (6)$$

where $S_i(0)$ is the 'zero frequency' spectral density ($f \ll e^*V/h$), V the applied source voltage, g_p the conductance of the fully transmitted p-th channel, e^* the quasiparticle charge, and t_p the effective transmission coefficient of the p-th channel (with $I_p = Vg_pt_p$) in the QPC.[24,29] Note, that Eq. (6) can also be used when back-scattering is strong ($t_p \ll 1$), but with e^* replaced by e.[22,23] The first limit in Eq. (6) corresponds to a constriction between two channels with the FQHE bulk filling, while the second limit emerges when the two channels are separated by a $\nu \sim 0$ insulating region (a nearly depleted constriction). Clearly, only electrons are allowed to tunnel in the second case.

To make the process of charge determination clearer, the fully reflected inner channels and fully transmitted outer ones do not contribute to the shot noise; only the partitioned channel does (Fig. 4). For example, at $\nu = 2/5$ with $p = 2$, consider the excess noise generated by partitioning the inner CF channel, which separates $\nu = 2/5$ from $\nu = 1/3$. The effective transmission of the inner channel is $t_{inner} = (g - g_{1/3})/(g_{2/5} - g_{1/3})$, with the transmitted current it carries being $I_t = V(g_{2/5} - g_{1/3})t_{inner}$, where g is the actual conductance of the partly pinched QPC, while $g_{2/5}$ and $g_{1/3}$ are the corresponding quantum conductances of the fractional states.

At a finite temperature, the expression for the spectral density of non-interacting fermions of charge e^* takes the form,[24]

$$S = 2e^*V g_pt_p(1 - t_p) \left[\coth\left(\frac{e^*V}{2k_BT} \right) - \frac{2k_BT}{e^*V} \right] + 4k_BTg \qquad (7)$$

Since the tunneling particles are interacting fermions, the choice of Eq. (7) to fit the FQHE data may seem unjustified. We justify it at the end of this section. At any rate, when $V \gg VT \sim 2k_BT/e^*$, the noise approaches a linear dependence on V (and I), as predicted by the zero temperature expression [Eq. (6)]. It should be noted that different approaches to determine the quasiparticles charge were employed. They include, for example, tunneling with a single electron transistor to a localized puddle of quasiparticles,[30] or resonant tunneling of quasiparticles into an isolated island (or an impurity).[31-33]

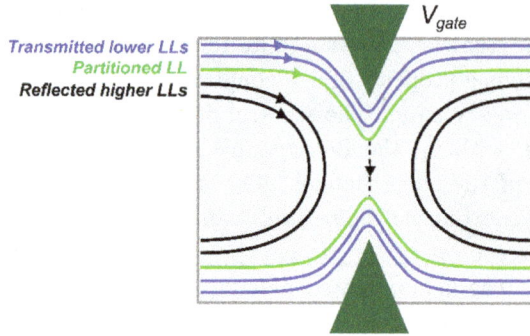

Fig. 4.   Partitioning of edge modes by a partly pinched QPC.

3.3. *Experimental techniques*

Heterostructures, hosting 2DEG with typical carrier density $n_s = (1-2) \cdot 10^{11}$ cm^{-2} and mobility $\mu = (2-30) \cdot 10^6$ cm^2/V-s at 0.3 K, were employed for a variety of measurements conducted in a dilution refrigerator with electrons' temperature of 10–50 mK. Achieving the electron temperature close to the lattice temperature was aided by 'cold grounding' of most of the ohmic contacts (by connecting the contacts with short wires to the 'cold finger'). The noise was filtered by an *LC* circuit with the center frequency ∼1 MHz and bandwidth 30–100 kHz (depending on the sample's resistance). Reflected and transmitted currents from the QPC were collected by different terminals. Voltage fluctuations were amplified by a low-noise homemade preamplifier, cooled to 4.2 K. Its typical voltage noise is less than 2.5×10^{-19}V^2/Hz and the current noise ∼10^{-28}A^2/Hz. The signal was amplified by a room temperature amplifier and measured by a spectrum analyzer. Since the accurate magnitude of the noise signal is important, careful calibration of the total gain is routinely performed by measuring a known charge or temperature.[34,35]

3.4. *Weak versus strong backscattering*

We start with noise measurements of a single composite fermion channel at bulk filling $\nu = 1/3$ and $g_{1/3} = e^2/3h$. When we use a QPC constriction, even a relatively weak backscattering potential may lead, at the lowest temperature (10–20 mK), to a highly nonlinear transmission coefficient with a non-universal dependence on the source voltage[36–39] (not agreeing with the chiral Luttinger liquid model[40–43]). Such nonlinearity leads to a non-universal excess noise with a larger Fano factor, $F = e^*/e > 1/3$. Only at extremely weak backscattering, the transmission coefficient is nearly independent of the source voltage, with the Fano factor $F = e^*/e = 1/3$[36] (Fig. 5).

What about the excess noise at $\nu = 2/5$? With nearly unity effective transmission, $t_{\text{inner}} = \frac{g-g_{1/3}}{g_{2/5}-g_{1/3}}$, the Fano factor increases smoothly with lowering the

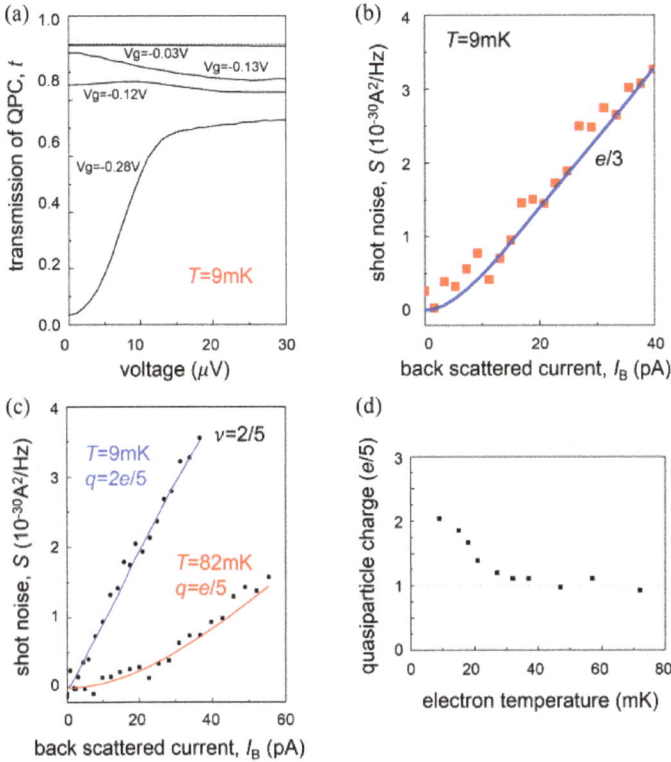

Fig. 5. The nonlinear transmission and the excess shot noise in a QPC for fractional fillings $\nu = 1/3$ and $\nu = 2/5$. (a) $\nu = 1/3$. Dependence of the transmission on the applied voltage. Drastic disagreement in the weak back-scattering regime with the chiral Luttinger liquid prediction. A better agreement with CLL at strong backscattering. (b) $\nu = 1/3$. Excess shot noise measured at extremely weak backscattering, ~ 0.01 (see in (a) a constant transmission. Note the extremely small excess noise). (c, d) $\nu = 2/5$. Noise measurements at very weak backscattering at different temperatures. The measured charge increases to $e^* = 2e/5$ at the lowest temperature (from Ref. 36).

temperature; with $F = 2/5$ at $T \sim 9$ mK and $F = 1/5$ at $T \sim 82$ mK [Figs. 5(a) and 5(b)]. A similar trend of the Fano factor is found also at $\nu = 3/7$.[36]

In the strong backscattering limit, the quasiparticle charge is expected to approach the electron charge. An example of the measured shot noise at $\nu = 1/3$ as a function of the transmission is summarized in Fig. 6, where measurements taken with a few samples collapse onto one curve.[44]

3.5. *The puzzle of the noise formula*

One piece of the shot noise story has been left out: What is the justification for the fitting formulae in Eqs. (6) and (7)? Empirically, the justification is that they work. A microscopic approach to the noise builds on the chiral Luttinger liquid model with a point impurity,[22,23,40–43] which allows an exact solution for the

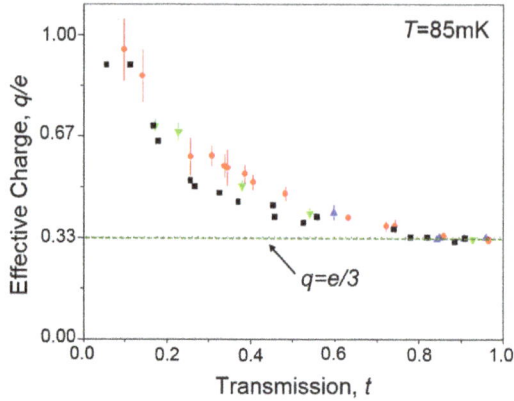

Fig. 6.   Evolution of quasiparticle charge as a function of the backscattering strength at $\nu = 1/3$. At weak backscattering strength the measured charge is $e^* = e/3$, increasing as the QPC pinches, approaching the charge of an electron. Different colors represent measurements in different devices (from Ref. 44).

I–V characteristic and the shot noise. However, the solution does not agree with the data obtained with a partitioning QPC. While this is not a surprise, as the simplest integrable model misses much of the edge physics (Sec. 2), the surprise is that simplistic Eqs. (6) and (7) work.

The key to this success is related to the Johnson–Nyquist (J–N) noise $4k_B T g$ in Eq. (7) — a consequence of the general fluctuation–dissipation theorem (FDT). While the J–N formula applies only in thermal equilibrium, its non-equilibrium generalization was discovered from the exact expression for the noise in the integrable model,[22,23,40–43] as well as in non-integrable models.[45–47]

The equilibrium FDT is based on two ingredients: The linear response theory and the Gibbs distribution. Away from equilibrium, only the first ingredient remains; hence, the FDT crumbles. However, in chiral systems the linear response theory is based on the causality principle in the temporal and space domains; namely, any events have consequences only downstream. This stronger causality opens a way to derive rigorously numerous model-independent FDT-like theorems for quantum Hall edge channels.[45–47] At weak backscattering, the FDTs agrees quite well with Eq. (7), as long as the transmission is weakly nonlinear; yet, Eq. (7) has been also successful for relatively strong backscattering (with e replacing e^* for QPC near pinch-off). This is harder to address since no model-independent equation could be derived in that regime, and only a crude, approximate expression exists.[48] Its analytical structure is rather different from Eq. (7), but amazingly, it is almost numerically indistinguishable from Eq. (7), thus explaining its success. Yet, it also shows that the determined charge e^* can only be interpreted as the charge of the tunneling carriers (electrons or quasiparticles) for transmissions near zero or near one. In other cases, e^* might be seen as some average of multiple charges, involved in transport.

As a simplistic illustration of the latter, imagine, for example, that both quasi-particle and electron tunneling is possible. In a model with rare independent tunneling events of the two carrier types, the current $I = e^*p^* + ep$, where p and p^* are the tunneling rates. The noise $S = 2(e^*)^2p^* + 2e^2p = 2qI$, where the effective charge $q = \frac{(e^*)^2p^* + e^2p}{e^*p^* + ep}$ is between the charges of an electron and a quasiparticle.

4. Neutral Modes

4.1. *Theory*

The edge models of the states at fillings $\nu = 1$, $1/3$, $2/5$, etc. are consistent with the quantization of the bulk conductance (Sec. 2); however, the simplest model of the $\nu = 2/3$ state is not. The most basic edge model includes two non-interacting, counter-propagating channels of the downstream conductance e^2/h and upstream conductance $e^2/3h$. As discussed below, such a model leads to an incorrect conductance of $4e^2/3h$ (Fig. 2). The correct (measured) two-terminal conductance of $2e^2/3h$ is restored by edge equilibration processes.

The simplest model misses density–density interaction of the charges in the two counter-propagating channels. Yet, such interaction alone cannot change the incorrect value of the conductance. Indeed, in the thermodynamic limit, the total charge of each of the four edge channels on both sides of the Hall bar must be conserved separately [Fig. 2(c)]. Thus, one can assign separate electrochemical potentials to each of the channels. For the two channels, emanating from the grounded terminal, the electrochemical potential is zero, and for the other two channels, emanating from the source, it is eV. Charge conservation dictates that the current must be conserved in all the points along the edge, independently of the interaction strength along the edge. This yields incorrect conductance, and hence, something beyond Coulomb effects is needed to solve our problem.

The solution lies in inter-channel tunneling,[49,50] which equilibrates the electrochemical potentials of the counter-propagating channels on each edge. We briefly review the edge theory in the presence of the inter-channel tunneling in Appendix A. The discussion reveals a subtlety: Momentum mismatch between the two edge channels makes tunneling ineffective in equilibrating the channels in a clean sample. Thus, the quantized electrical conductance crucially depends on the presence of disorder and the resulting random tunneling.

The final outcome of the Kane–Fisher–Polchinski theory[49,50] is a picture of decoupled charge and neutral modes with the action,

$$\frac{S}{\hbar} = -\int dx\, dt \left\{ \frac{1}{4\pi}[-\partial_t\varphi_n\partial_x\varphi_n + v_n(\partial_x\varphi_n)^2] + \frac{3}{8\pi}[\partial_t\varphi_c\partial_x\varphi_c + v_c(\partial_x\varphi_c)^2] \right\}, \quad (8)$$

where the charge mode φ_c carries charge and energy downstream, and the neutral mode φ_n carriers energy without net charge in the opposite, upstream, direction. The coefficient $\frac{3}{8\pi}$ in the action of the charge mode encodes the correct electrical

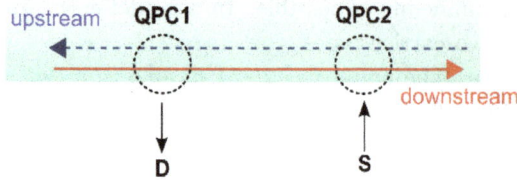

Fig. 7.   A theoretical proposal to excite and measure an upstream neutral mode. The excitation of the mode takes place in **S** downstream QPC2. The energy propagates upstream (broken line) towards QPC1, where it fragments into (or heats up) the downstream charge mode (full line), with measured shot noise like fluctuations in **D**.

conductance $2e^2/3h$. All the allowed perturbations that can be added to this action, such as random interaction of the charge and neutral modes, are irrelevant in the renormalization group sense.

Kane and Fisher extended the same picture to other states with upstream modes (hole-conjugate states); namely, states at the filling factors $\frac{p+1}{2p+1}$.[49] In particular, they discovered that these states can be described in terms of a single downstream charged channel and p decoupled upstream neutral modes.

What are the experimental consequences of the upstream neutral modes? If the edge is longer than the electrochemical equilibration length between the counter-propagating channels, electrical current can only flow downstream , that is, in the direction of the charge mode. Energy, though, can go in both directions. Unfortunately, probing the energy flow is more difficult than probing charge currents. It was first proposed theoretically,[51] and subsequently observed in experiments, that energy currents can be probed from excess charge noise with a zero net current.[52] We provide first a brief 'theoretical realization' of the two-QPC configuration[51] (Fig. 7), which allows an analytic calculation of the excess charge noise that results from upstream neutral modes.

Imagine that charge can tunnel into and from the edge through QPC1 and QPC2 as shown in Fig. 7. Contact QPC1 is upstream relative to QPC2, with the charge mode running from QPC1 to QPC2 along the QHE edge. In the absence of upstream neutral modes (dotted line), all edge channels run in the downstream direction, while any upstream mode runs from QPC2 to QPC1. Here, QPC2 is biased while QPC1 is connected to unbiased drain D. If all edge modes move downstream, no energy and no charge will reach QPC1. For an edge supporting upstream neutral modes, the energy injected into the QHE edge at QPC2 travels in both directions along the edge, and thus arrives at QPC1. The energy arriving at QPC1 from QPC2 gives rise to excess noise in drain D. One can think of that noise as resulting from fragmentation of neutral particles to electron-hole pairs,[52] or as an additional J–N noise due to the heating of QPC1 by the energy current.[53] Such excess noise is a smoking-gun evidence of an upstream neutral mode. While the exact setup of Fig. 6 allows analytic calculation of the excess noise,[51] somewhat different setups

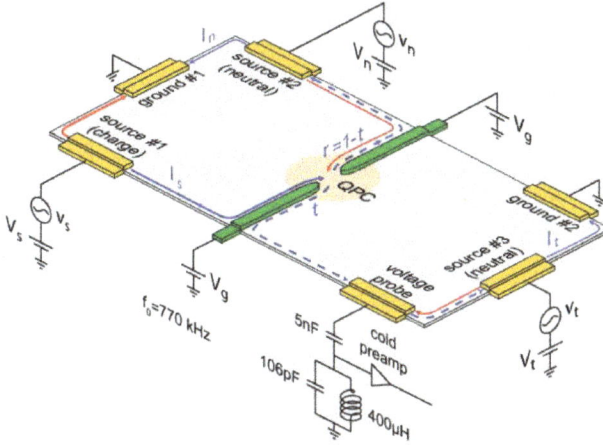

Fig. 8.   An actual realization of the first experiment designed to measure the upstream neutral modes. Source #2, charged with V_n, injects charge current downstream (blue line, I_n), and a neutral mode, excited by the 'hot spot', upstream (red line). It impinges on the QPC, and generates charge fluctuations moving downstream (broken blue line). Noise is measured at the amplifier. Other contacts allow different measurements (from Ref. 52).

have been employed to actually detect the upstream neutral modes, as discussed below.[52–57]

4.2. *Experiments*

Proposals of how to detect the neutral modes involve measuring tunneling exponents in constrictions[49]; measuring thermal transport[58]; looking for resonances in a long constriction[59]; or looking for heating effects on shot noise.[51,60]

The first established experimental method involved a QPC constriction located upstream from an energized ohmic contact. An excited neutral mode, emanating from the 'hot spot' at the source,[61] propagates upstream along the edge and impinges on a partly pinched QPC. This leads to observed current fluctuations (Fig. 8).[52]

Similar data was found in all hole-conjugate states and also in the $\nu = 5/2$ state.[55] The noise had a $t(1 - t)$ dependence, with t the transmission of the QPC (Fig. 9). This dependence was attributed to the fragmentation of the dipole-like neutral carriers that carried the upstream heat. One cannot rule out thermal noise that resulted from heating up the constriction.[53]

With a downstream charge mode and an upstream neutral mode impinging simultaneously on the QPC, the shot noise exhibited a higher particle temperature and a smaller Fano factor.[52]

While a direct impingement of an upstream neutral mode on a large drain contact was not expected to lead to charge fluctuations, they were found in a subsequent experiment.[10] Simple heating cannot account for this noise, as the contact is macroscopic and, thus, hardly should heat up. Yet, this phenomenon can be understood

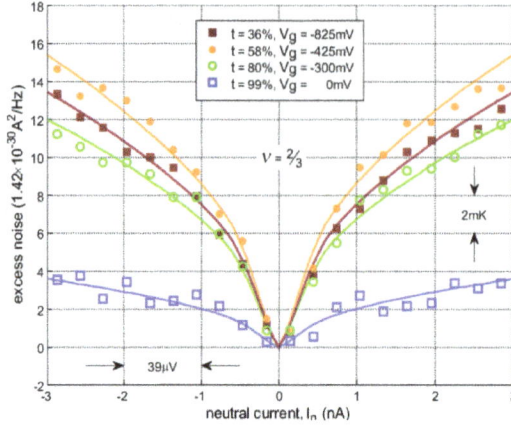

Fig. 9.   Detection of an upstream neutral mode at filling $\nu = 2/3$. Shown is the excess noise, measured by the amplifier as a function of the current I_n and the transmission t of the QPC (see Fig. 8). The noise is proportional to I_n and approximately to $t(1-t)$ (from Ref. 52).

if one takes into account the cold downstream channel that counter-propagates with respect to the upstream neutral mode. Inter-mode particle-hole excitations are created along the path of the neutral mode, and close to the hot-spot of the source contact, the charge excitations will be reabsorbed by the source. However, close enough to the drain, the fluctuations will be absorbed by the drain.[62]

Since neutral modes carry heat, the most direct way to detect them is to measure the temperature increase at the measurement point upstream from the heat source. This can be done, for example, with quantum dot (QD) thermometry, that is, by allowing a neutral mode impinge on a QD and measuring the 'Coulomb blockaded' conductance peaks.[54,56] Moreover, employing QDs, one can convert the temperature difference between the input and output QPCs into charge current.[56]

It is worth emphasizing the fact that neutral modes were found along short edges in many particle-like states (such as $\nu = 1/3, 2/5, 1$), which are not expected to support topologically protected upstream modes.[10,63] Such upstream modes are not topological, and emerge due to edge-reconstruction, which, in turn, depends on the softness of the edge potential.[9,15] Efforts to suppress these modes by sharpening the edge potential failed so far.[63]

5. Interference of Edge Modes

The most direct way to ascertain the statistics of a fractional state is to braid its quasiparticles via interference experiments. Exploiting the chiral motion of the electrons in edge channels, an electronic analogue of the ubiquitous optical Mach–Zehnder interferometer (MZI)[64,65] and an analogue of the Fabry–Perot interferometer (FPI)[66−69] were constructed (Fig. 10). While the MZI is a two-path interferometer, the FPI interferes multiple circulating paths. The two geometries

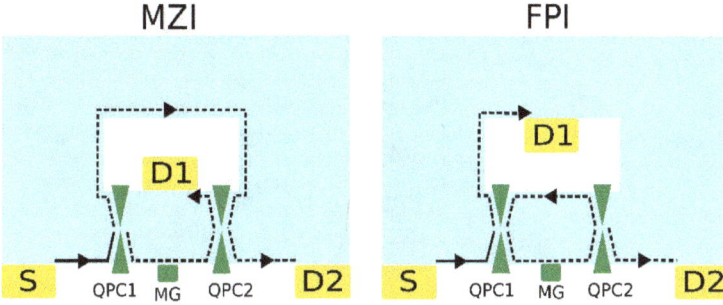

Fig. 10. Schematic realizations of a Mach–Zehnder Interferometer (MZI) and a Fabry–Perot Interferometer (FPI). Blue regions are the 2DEG, green regions are the QPCs, and white regions are etched (vacuum). Note different topology due to the location of drain D1.

provide complimentary experimental approaches. An FPI is simpler, but an MZI is more robust to effects of the bulk between the edges of the interferometer, as discussed in Secs. 5 and 6.

Interferometry in FQHE appears to be challenging. Some of the results, observed mostly in the FPI, pose difficulties in their interpretation. In fact, even the interpretation of interference in the IQHE proved challenging. In this chapter, we focus on the IQHE. Two puzzling regimes will be highlighted later.

5.1. *Mach–Zehnder interferometry*

Schematics of the electronic MZI is depicted in Fig. 10.[64] Two QPC constrictions split and recombine the impinging edge channel, and two ohmic contacts, **D1** and **D2,** serve as drains. In an actual realization, with its schematics shown in Fig. 11, the inner contact, **D2** is connected to the outside circuit via an 'air bridge'. A phase difference ϕ between the two paths is introduced via the Aharonov–Bohm (AB) effect, $\phi_{AB} = 2\pi AB/\Phi_0$, where A is the area enclosed by the two paths. A modulation gate, **MG**, is used to tune the area A.

We review briefly the operation of the interferometer. The conductance, from source to drain, is determined by the corresponding transmission probability T_{SD}, which, in turn, depends on the interference between the two paths. When the system is tuned to filling $v = 1$, a single edge channel carries the current. Neglecting decoherence processes, with transmission (reflection) amplitude t_i (r_i) of the ith QPC fulfilling $|r_i|^2 + |t_i|^2 = 1$, the collected currents at **D1** and **D2** are, respectively,

$$I_1 \propto T_{SD1} = t_1^2 t_2^2 + r_1^2 r_2^2 + 2|t_1 t_2 r_1 r_2| \cos(\phi_{AB} + \gamma), \tag{9}$$

with I_2 the complementary current, $T_{SD1} + T_{SD2} = 1$, and the phase γ independent of the magnetic field. The visibility is defined as: Visibility $= (I_{max} - I_{min})/(I_{max} + I_{min})$, and for $|t_2|^2 = 0.5$, visibility $= 2|t_1|(1 - |t_1|^2)^{1/2}$. We address below a few highlights, which reveal unexpected behavior in the integer regime.

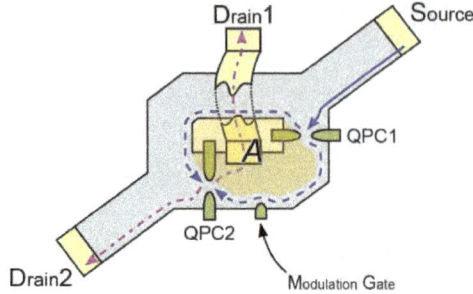

Fig. 11.   A typical structure of an MZI. Note the inner contact (D1, in the orange region of the 2DEG), connected with an air bridge to an outside ground. A modulation gate is used to change the area enclosed by the two interfering trajectories (dotted lines).

5.1.1. Lobe structure

The highest visibility obtained in an MZI was at bulk filling $1 < \nu_B < 2$. In the nonlinear regime, with a DC voltage added to a small AC signal, the visibility evolves in an unexpected fashion. Figure 12 shows the visibility and the phase of the AB oscillations at $\nu_B = 2$ as a function of the applied DC voltage. The visibility oscillates between a maximum value and zero — in a lobe fashion.[64] Three striking features are observed: (i) The lobes scale in the source DC voltage is $V_S \sim 15\mu V$ (diminishing with lowering the magnetic field). The scale of the overall decaying envelope is $V_S \sim 30$–$50\mu V$. These energy scales are far smaller than any other energy in the IQHE regime. Moreover, diluting the impinging current on the MZI (with an external QPC) quenches the lobes pattern; (ii) The phase of the AB oscillations is rigid in each lobe, being independent of the energy (DC voltage). Crossing the zero visibility, at a certain value of V_S, the phase flips by π. (iii) The visibility shows exponential decay with increasing temperature and length of the MZI arms.[65,70–72] These results, which cannot be explained by a single-particle model, are worth a further consideration.

Naively, when treating the interfering integer edge channel as a non-interacting Fermi gas, one would expect a rather weak dependence of the visibility on the source voltage. However, inter-channel interaction, and charge fluctuations (due to partitioning the incoming current with the QPCs) may complicate the simple non-interacting picture.[73–77]

Two leading theoretical models catch some of the features of the lobes structure. One considers short-range inter-channel interaction at filling factor two, as well as long-range intra-edge interactions. Inter-channel interaction gives birth to two modes, slow and fast. Roughly speaking,[78] the scale of the lobes is comparable to $eV = \hbar v/L$, with v the velocity of the slow mode. The overall envelope is related to the fluctuations in the number of electrons in each arm of the MZI.[79,80] However, the model fails to explain the phase rigidity away from $L_1 = L_2$, and especially at filling one.

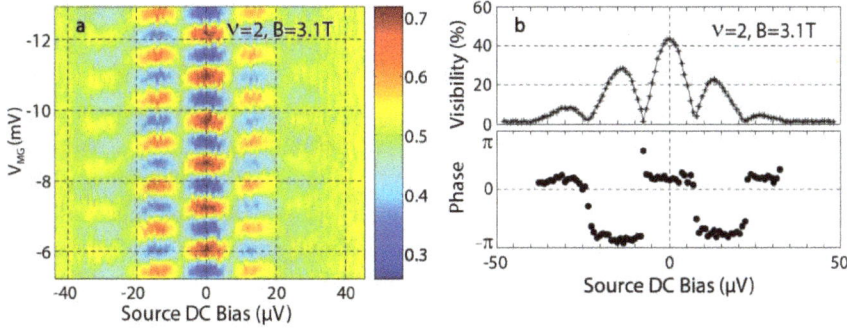

Fig. 12. Aharonov–Bohm (AB) interference in an MZI. (a) Oscillations of the conductance as a function of the area (tuned by the modulation gate, V_{MG}) and an added DC source bias (arbitrary scale: red — high visibility, blue — low visibility). (b) The visibility of the AB oscillations as a function of the DC bias. Note the lobe structure, with a constant phase of the oscillations throughout each lobe and with an abrupt reversal at the nodes (from Ref. 65).

Another approach assumes ad-hoc bunching of two electrons with interference of all possible trajectories.[81] While this assumption seems unjustified, such bunching was found in experiments with FPI (see Sec. 5.2.3).

5.1.2. *Phase averaging versus decoherence*

Interference can be quenched in two different ways: First, decoherence, e.g. spin flips in one of the two paths. Second, phase averaging: Particles interfere but each with a different phase (due to fluctuations of the phase in time or phase dependence on the energy). These two mechanisms can be distinguished by measuring the shot noise of the interfering signal. Assume a phenomenological parameter that accounts for decoherence, k, with $t_1^2 = 0.5$; then, $I_1 \propto T_{SD1} = 0.5 + k[t_2^2(1 - t_2^2)]^{1/2}\langle\cos\varphi_{AB}\rangle$, where the angle brackets denote phase averaging. Obviously, either when $k = 0$ (total decoherence) or when φ_{AB} spans a 2π range, the interference term (second term) is expected to vanish. However, the shot noise $S_1 \propto T_{SD1}(1 - T_{SD1}) = 1/4 - k^2 t_2^2(1 - t_2^2)\langle\cos^2\varphi_{AB}\rangle$, will have the following dependence on t_2: (i) For $k = 0$, S_1 is independent of t_2; (ii) For $k > 0$, the $\langle\cos^2\varphi_{AB}\rangle$ term will not vanish upon averaging, leading to $S_1 = 1/4 - k^2 t_2^2(1 - t_2^2)/2$. These two extreme behaviors may help in distinguishing between the dephasing mechanisms. Thus far, only $k > 0$ was found (e.g., Ref. 64).

5.1.3. *Melting of the interference*

An observation of interference of anyons would be a step towards understanding their anyonic statistics. However, the complex edge structures of FQHE states, hosting counter-propagating charge and neutral channels, have been suspected to prevent the observation of the much sought after interference of anyons. Specifically, the interference of the outer-most edge channel in the MZI was found to diminish gradually as the filling was lowered towards $\nu_B \sim 1$, followed by an apparent increase of the excitation of a 'non-topological' upstream neutral mode.[63,81] The simultaneous emergence of a neutral mode and a $\nu_{QPC} = 1/3$ conductance plateau in a partly pinched QPC suggested a reconstruction of the edge mode into two modes, $\nu = 2/3$ and $\nu = 1/3$, which leads to the formation of an upstream neutral mode. The $\nu_{QPC} = 1/3$ conductance plateau was found to persist for the range of bulk fillings $1/2 < \nu_B < 1.5$. Uncharacteristically, shot noise was measured on the conductance plateau, resulting from the partitioned upstream neutral mode.[16,63] Moreover, and even more surprisingly, particle-like states in the range of $\nu_B < 1/2$ were also accompanied by upstream neutral modes — independently of efforts to sharpen the edge-potential.[10,63]

5.2. *Fabry–Perot interferometry*

An FPI is a large quantum dot (QD). Like an MZI, it is formed by two QPCs but without an internal drain contact. The transmitted and reflected currents are absorbed by two remote detectors. Interference takes place among multiple trajectories that circulate around the inner perimeter of the FPI (Figs. 10 and 13).

In the lowest order, when the backscattering in both QPCs is weak, the interference takes place only between two dominant partial waves, with I_1 the transmitted current,

$$I_1 \sim T_{SD1} \approx t_1^2 t_2^2 + 2 t_1 t_2 r_1 r_2 \cos(\varphi_{AB} + \gamma), \tag{10}$$

where the parameters are defined above (I_2 is the complimentary reflected current).

Like in the MZI, the relative phase between consecutive rounds is controlled by varying the enclosed AB flux (with the magnetic field or a modulation gate). When Coulomb interaction is dominant, the enclosed charge tends to be constant, maintaining charge neutrality. This regime is called the Coulomb Dominated (CD) regime.[67–69,82,83]

5.2.1. *Aharonov–Bohm and Coulomb-dominated interference*

In the purely AB regime Coulomb interactions are negligible, and the 'flux containing area' remains nearly constant with changing magnetic field. Since the degeneracy of the LLs increases, charge is added to the liquid and the periodicity is expected to be that of a flux quantum. Here, since screening is highly effective, we assume that no quasiparticles are added to the bulk. When interactions dominate, the behavior of the FPI in the CD regime is quite different. For a general filling factor, there are fully reflected, partially transmitted, and fully transmitted edge channels in the QPCs that form the FPI. The number of fully reflected channels is the same as the number of circulating channels within the interior of the FPI. Numbering the fully reflected channels by f_R and the fully transmitted ones by f_T, the total number of participating channels is $f_{edge} = f_T + 1 + f_R$ — being deter-

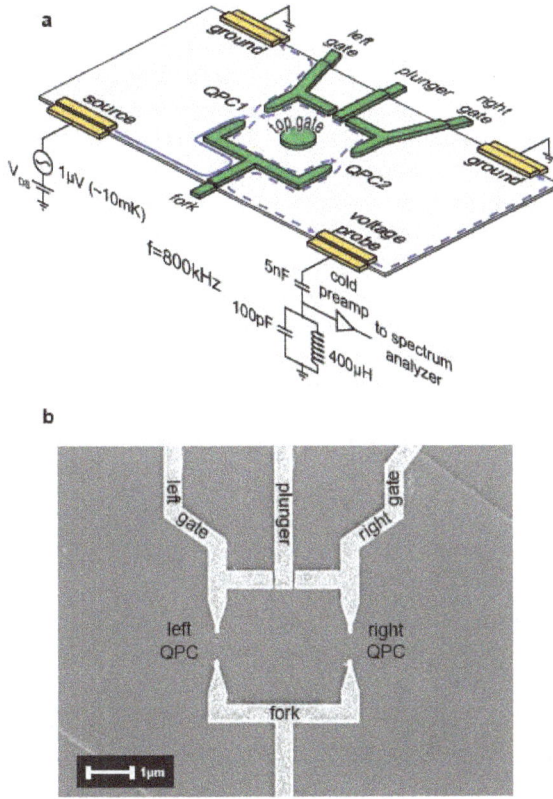

Fig. 13. A typical FPI. (a) A two-QPC schematic configuration of the FPI, with a single source and two drains (one at the amplifier and one at the ground). The broken lines represent the partitioned channel. Multiple trajectories circulate around the interior of the FPI (not shown). A plunger gate is used to change the enclosed area of the trajectories. A top gate is sometimes added for screening and testing. (b) An SEM micrograph of a typical FPI. The light gray areas are the gates forming the QPCs and the plunger gate (from Ref. 69).

mined by the bulk filling factor ν (for IQHE $f_{\text{edge}} = \nu$). Each configuration can be denoted as (ν, f_T). The experiments[69] revealed the following:

1. Each of the edge channels could be made to interfere independently of the others;
2. The periodicities in the magnetic field ΔB and the modulation-gate voltage ΔV_{MG} do not directly depend on the magnetic field (or the filling factor) and the transmission amplitudes t_1 and t_2, but only on A and f_T:

$$\Delta B(B, V_{\text{MG}}, f_T, t_1, t_2) = \Delta B(f_T) = -\frac{\phi_0}{A \cdot f_T}, \tag{11}$$

as was experimentally found to hold for a wide range of QPC's transmission. Similarly, the periodicity in the modulation gate voltage depends only on f_T:

$$\Delta V_{\text{MG}}(B, V_{\text{MG}}, f_T, t_1, t_2) = \Delta V_{\text{MG}}(f_T), \tag{12}$$

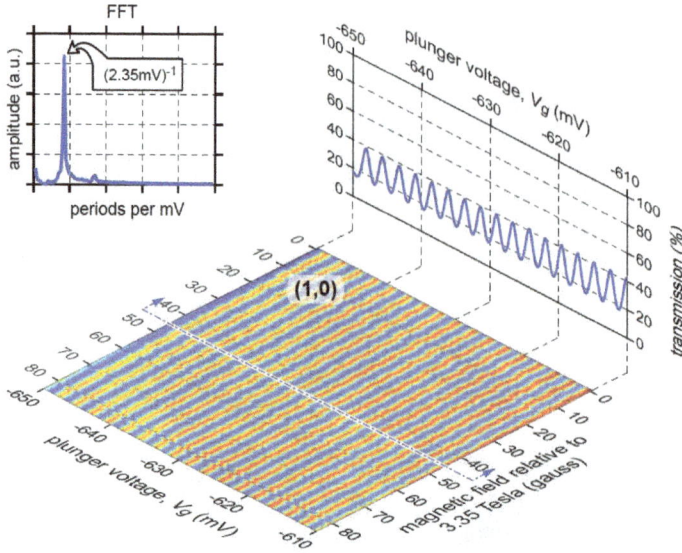

Fig. 14.   Interference of the edge channel at filling $\nu = 1$ in the FPI shown in Fig. 13. The interference is dominated by Coulomb interactions. While changing the area results in nice oscillations
(with a periodicity corresponding to adding/removing an electron), no magnetic field dependence
is observed (from Ref. 69).

with ΔV_{MG} monotonically increasing with increasing f_{T} (see an exception in
Ref. 69);

3. For $f_{\mathrm{T}} = 0$, the enclosed flux is independent of the magnetic field, indicating
   that the area of the FPI is proportional to $1/B$ (Fig. 14);
4. For $f_{\mathrm{T}} > 0$, the enclosed flux decreases with increased magnetic field, indicating
   that the area of the FPI shrinks faster than $1/B$.

In Appendix C a more detailed elaboration of this behavior is presented.

5.2.2.   The AB – CD mixed regime

Several ways to overcome the CD behavior were suggested: (1) Covering the sample
with a top gate in order to increase the total capacitance [Fig. 15(a)],[68,69,85] or
adding an internal screening layer.[86] (2) Placing an ohmic contact in the bulk to
break its isolation and allow carrier drainage [Fig. 15(b)].[84,87] This method was
found to be more effective than placing a top gate. (3) An intermediate coupling
regime was achieved by placing a large ohmic contact outside the perimeter of
the interferometer.[88] A checker-board pattern of the conductance (as a function
of B and V_{MG}), with periodicities that correspond to both AB and CD regimes,
is typically observed (Fig. 16).[88] This configuration is promising, since it allows
for the construction of small interferometers; yet, it provides information on the

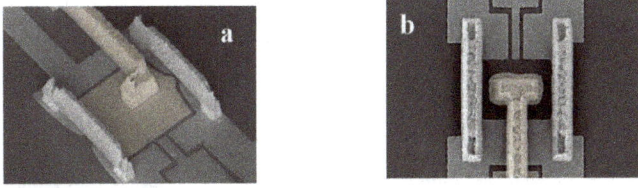

Fig. 15.   Two realizations of screened FPIs. (a) A grounded top gate screens the interior of the
FPI, allowing adding or subtracting charge with little energy cost. (b) A grounded ohmic contact
in the center of the FPI eliminates the restrictions forced by Coulomb interactions. Charge can
be added or drained from the bulk with no energy cost. While the configuration in (b) always
exhibits AB interference, configuration (a) exhibits AB interference only above a critical area (from
Ref. 84).

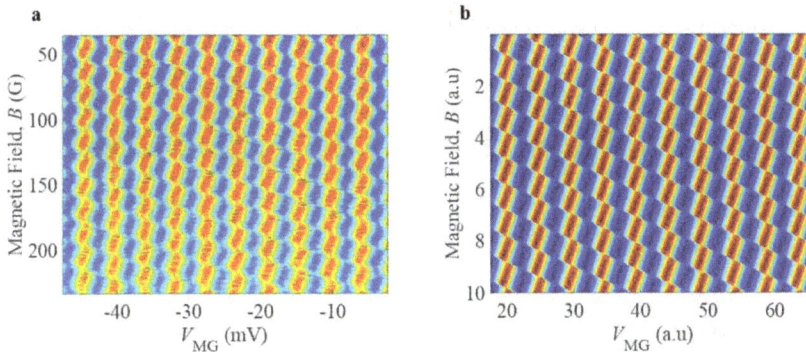

Fig. 16.   The mixed AB-CD regime. (a) Measured checker-board pattern of the interference.
(b) Calculated checkered-board (from Ref. 88).

coherent component of the interference pattern. Note, the physics behind the partial
suppression of the Coulomb interaction is not yet understood.

5.2.3. *The pairing effect*

Electron 'pairing' is a rare phenomenon appearing only in a few unique physical
systems, e.g. superconductors and Kondo-correlated quantum dots.[89,90] Yet, in an
FPI in the AB regime (with suppressed Coulomb interactions), an unexpected but
robust electron 'pairing' was found in the IQHE regime. The pairing took place
within the outermost interfering edge channel in a wide range of bulk filling factors, $2 < \nu < 5$ (Fig. 17).[84,86,87] The main observations were: (i) High visibility
AB conductance oscillations with the magnetic flux periodicity of $\Phi_0/2 = h/2e$
(instead of the ubiquitous h/e periodicity); (ii) The interfering quasiparticle charge
was $e^* \sim 2e$, as determined by shot noise measurements; (iii) The $h/2e$ periodicity
was fully dephased when the first inner (adjacent to the outermost) edge channel was dephased (while keeping the interfering edge channel intact). This was a
clear observation of inter-channel entanglement; and (iv) The AB flux periodicity

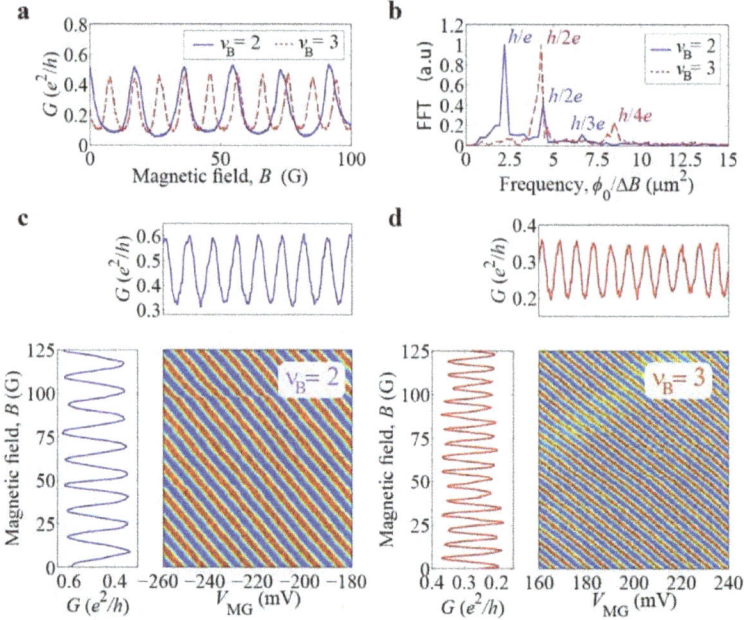

Fig. 17. The pairing phenomenon detected in the screened FPI shown in Fig. 15(a). (a) Interference oscillation of the outer-most edge channel at $\nu = 2$ and $\nu = 3$. (b) FFT of the oscillations shows an exact factor of two in the periodicities. (c) and (d) show the 2D plot of the conductance as a function of B and V_{MG}, with constant phase lines proving pure AB interference (from Ref. 87).

was exclusively determined by the enclosed area of the first inner (adjacent to the outermost) channel and not by the outermost interfering channel.

An extensive set of measurements suggests the formation of a neutral edge mode due to strong interaction between the two outer edge channels. The neutral mode plays a crucial role in the pairing phenomenon; however, the exact mechanism for this phenomenon is not clear. The pairing may not be a curious isolated phenomenon, but one of many manifestations of unexpected edge physics in the quantum Hall regime (see the above lobe-structure).

6. Non-Abelian Order of the $\nu = 5/2$ State

6.1. Fractional statistics

While the fractional Hall conductance goes hand-in-hand with the fractional filling, the fractional charge goes hand-in-hand with fractional statistics, which is neither Bose nor Fermi. We illustrate this point with an example of the Laughlin state at $\nu = 1/3$ in an MZI. The elementary excitations carry the charge $e/3$ and the edge consists of a single chiral mode. Imagine that the FQHE liquid fills the blue region of an MZI (Fig. 18 and Sec. 5). A gedanken experiment with the MZI can demonstrate that the $e/3$ quasiparticles cannot be fermions or bosons.

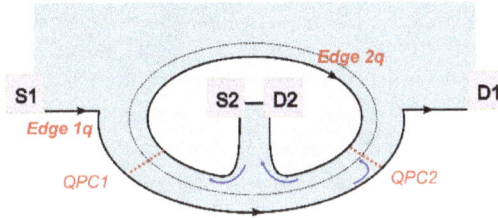

Fig. 18.   A simplified schematic picture of a Mach–Zehnder interferometer. Quasiparticles tunnel between the inner and outer edges at QPC1 and QPC2. Drains D1 and D2 absorb the interfering quasiparticles.

Recall Eq. (9), with the phase-dependent term in the transmission, $t_{SD1} = \eta \cos(\phi_{AB} + \gamma)$, with $\phi_{AB} = 2\pi\phi/\varphi_{1/3}$ and $\varphi_{1/3} = 3\Phi_0$. The transmission probability, and hence, apparently, the current is periodic in the magnetic flux ϕ with the period of three flux quanta. This contradicts the rigorous Byers–Yang theorem,[91] which states that the period cannot exceed Φ_0. The theorem expresses the fact that an FQHE liquid is made of electrons, and any integer number of flux quanta is invisible to electrons that make a circle around the enclosed flux.

In order to solve the paradox one must recall the notion of fractional statistics. Specifically, the quasiparticle wave function accumulates a non-trivial phase when encircling another quasiparticle. Then the statistical phase γ in Eq. (9) in combination with φ_{AB} leads to a total phase that is proportional to the number n of $e/3$ quasiparticles inside the dashed circle (Fig. 18). Each tunneling event changes n, since the tunneling quasiparticles are absorbed by drain D2 inside the dashed loop. Hence, the oscillating transmission depends on $t_{SD2} = \eta \cos(\varphi_{AB} + \frac{2\pi n}{3})$ returning to its initial value after three consecutive tunneling events of quasiparticles. The three events transfer to D2 the combined charge of one electron. This is crucial for the "single-electron" periodicity, demanded by the Byers–Yang theorem. The drain current can be found as the ratio of that charge and the average time it takes for three tunneling events, $I_{SD2} = \frac{e}{\frac{1}{p(1)} + \frac{1}{p(2)} + \frac{1}{p(3)}}$, where $p(i)$ is the transmission probability after i consecutive tunneling events (modulo $i = 3$). This result[92] is consistent with the Byers–Yang theorem.[91]

6.2. *Non-Abelian statistics and the 5/2 liquid*

In the previous section we considered an example of Abelian statistics: Particles accumulate non-trivial phases when they go around each other, but combining (or fusing in the usual terminology) any two anyons results in a unique particle type, as two fermions always fuse into a boson. However, non-Abelian anyons have multiple fusion channels.[93,94] Two identical anyons can form multiple bound states, which can be understood as composite particles — and different composite particles can exhibit at least two different statistics. Such property is seen as valuable for quantum computing.[94,95] The existence of non-Abelian anyons was first conjectured at the FQHE filling of $\nu = 5/2$, consequently, attracting intense interest from experimentalists in recent years.

In studying this fractional state, geometric resonance measurements observed cyclotron motion of CFs,[96,97] suggesting that CFs are present at $\nu = 5/2$. This leads to a theoretical challenge. Indeed, in a mean-field description, CFs of even-denominator states move in an effective zero magnetic field. This seems to conflict with the observed energy gaps at $\nu = 5/2$ and $\nu = 7/2$, since the liquids of weakly interacting fermions in a zero field are gapless, as is indeed the case at $\nu = 1/2$ and $\nu = 3/2$. To explain the observed gapped states, one has to assume that CFs form Cooper pairs.[98]

The language of Cooper pairs allows a prediction of the quasiparticle's charge. In a rather simplistic picture, the Cooper pairs are interpreted as independent bosonic particles of charge $2e$, while the two filled integer Landau levels are ignored. Taking $\nu = 1/2$ for electrons translates into the filling factor $\nu' = 1/8$ for the charge-$2e$ bosons. Just as the quasiparticle charge is νe in the Laughlin state of the filling factor ν for electrons, now the quasiparticles are of charge $2e\nu' = e/4$. In other words, the paired state allows insertion of a half-vortex with the corresponding charge $e/4$. This quasiparticle charge is consistent with several experiments.[99–102]

What is the exchange statistics of the $e/4$ anyons? Following Laughlin's argument (Sec. 3), the insertion of one flux quantum generates an excitation of charge $e/2$ at filling $\nu = 1/2$. Thus, one might naively think that one just needs to insert two flux quanta to get an electron with charge e. However, the argument for the exchange phase here is different. The exchange of electrons generates an exchange phase of π; achieved also by moving one electron along half a circle with the center at the other electron. Yet, such half-encircling of the two inserted flux quanta in the center leads to an accumulated Aharonov–Bohm phase of 2π — suggesting that flux insertion generates a charge-e boson (instead of an electron). Hence, a combination of a hole of charge $-e$ with such boson yields a neutral fermion. In the composite fermion language, this is just an unpaired composite fermion. Such fermions can be combined with charge-$e/2$ excitations, which thus come in two species. In other words, we can build two combinations of two $e/4$ anyons. Two logical possibilities are now open: (1) The system has more than one sort of $e/4$ particles; (2) only one sort of $e/4$ anyons exists. In the first case, two types of $e/2$ anyons could be built from different combinations of elementary anyons. In the second case, there must be two ways to fuse two identical particles. This is a signature of non-Abelian statistics.

To refine the list of possibilities, we turn to the edge physics. Naively, the edge action is very similar to the actions at $\nu = 1$ and $\nu = 1/3$.

$$\frac{S}{\hbar} = -\frac{2}{4\pi} \int dx\, dt [\partial_t \varphi \partial_x \varphi + v(\partial_x \varphi)^2], \tag{13}$$

with the coefficient of 2 reflecting the electrical conductance $e^2/2h$ of the half-filled Landau level. To identify the electron operator at the edge, an operator that destroys charge e must be constructed. This follows the logic of Appendix A and yields the operator $\exp(2i\varphi(x))$. Alas, this answer cannot possibly be correct, as it satisfies bosonic commutation relations instead of the expected fermionic ones. Hence, we have to assume that the edge has an additional neutral fermionic mode (or modes, ψ_k),

$$\frac{S}{\hbar} = -\frac{2}{4\pi}\int dx\,dt[\partial_t\varphi\partial_x\varphi + v(\partial_x\varphi)^2] + \int dx\,dt\sum i\psi_k(\partial_t + v_k\partial_x)\psi_k, \quad (14)$$

with electron operators constructed as $\psi_k \exp(2i\varphi(x))$. The operators ψ_k do not have to be complex Dirac fermions. They can also be real Majorana fermions, which are real or imaginary parts of Dirac fermions. Indeed, ψ_k account for the fermionic parity of the system, while the total number of fermions is counted by the Bose field $\varphi(x)$ Moreover, since a Dirac fermion is just a combination of two Majorana fermions, then, without loss of generality, ψ_k can represent Majorana operators.

Opposite signs of v_k correspond to opposite chiralities of edge modes. With v_k having opposite signs, pairs of counter-propagating Majorana modes would be localized (gapped) by disorder. Thus, we can assume the same sign for all v_k. Different allowed edge structures are distinguished by the Chern number ν_C, that is, by the net number of the Majorana edge modes.[103] ν_C is negative if the propagation direction of the Majorana modes is upstream, while the central charge, proportional to the fully equilibrated thermal conductance, is $c = 1 + \nu_C/2$.

According to Kitaev's classification,[95] all topological orders can be distinguished by the Chern number. In particular, an even ν_C corresponds to Abelian statistics while an odd ν_C to a non-Abelian topological order. To see the origin of this rule, the principle of bulk-edge correspondence is discussed, though heuristically, in Appendix B. A rigorous approach is addressed in detail in Refs. 95 and 103.

6.3. *Proposed 5/2 states*

Not all choices of the Chern number are equally interesting or likely to occur in nature. All the realistic possibilities in the literature have a small ν_C. Moreover, according to Kitaev's 16-fold way,[95] fractional statistics depends only on ν_C mod 16. Table 1 lists the states that received most attention.

The non-Abelian PH-Pfaffian order is particularly interesting due to its particle-hole symmetry.[106] The filling factor of $1/2$ can be interpreted in two ways: As a half-filled Landau level of electrons or a half-filled Landau level of holes. The

Table 1. Proposed states with low Chern numbers.

ν_C	-3	-2	-1	0	1	2	3
order	anti-Pfaffian	113	PH-Pfaffian	$K = 8$	Pfaffian	331	$SU(2)_2$
Refs.	[58, 104]	[105]	[104, 106–109]	[110]	[93]	[111]	[112, 113]

particle-hole (PH) transformation changes electrons into holes and vice versa. In general, the PH-transformation changes the topological order since the topological orders for the electrons and holes do not have to be the same. We proceed in the spirit of Fig. 2, which illustrates the effect of the PH-transformation on the edge modes of the 2/3-liquid of electrons, which can be also thought of as Laughlin's $\nu = 1/3$ state of holes. The figure shows how the $\nu = 2/3$ edge mode hosts a downstream integer $\nu = 1$ channel and an upstream $\nu = 1/3$ channel. The same trick applies at $\nu = 5/2$, where the downstream integer channels are ignored. For example, start with the Pfaffian edge, which contains a downstream charged bosonic mode (with conductance $e^2/2h$) and one downstream Majorana mode. The PH-transformation yields a downstream integer channel (with conductance e^2/h), an upstream charged bosonic mode (with conductance $e^2/2h$), and an upstream Majorana mode. The same arguments as presented in Appendix A show that the two bosonic modes (a downstream integer and an upstream with conductance $e^2/2h$) reorganize into a downstream charge mode with conductance $e^2/2h$, and an upstream neutral mode (preserving the central charge). The latter can be fermionized and represented as two upstream Majorana modes. Accounting for one more upstream Majorana, we get $\nu_C = -3$, that is, the anti-Pfaffian state. The PH-Pfaffian state is unique in not changing its structure under the PH transformation. The edge contains a downstream charged mode of conductance $e^2/2h$ and an upstream Majorana fermion. The PH transformation results in a downstream mode of conductance e^2/h, upstream mode of conductance $e^2/2h$, and a downstream neutral Majorana mode. The two charge modes reorganize into a single downstream charged mode and an upstream neutral boson. The boson can be fermionized and represented as a combination of two upstream Majorana modes. Interaction with the downstream Majorana gaps one of them out. We are left with the initial edge structure of the PH-Pfaffian liquid.

Past experiments did not provide a way to determine the topological order of the $\nu = 5/2$ state, and the focus had been on numerical calculations. Numerical evidence was produced for several orders,[114–116] and eventually, the numerical debate concentrated on the Pfaffian ($\nu_C = 1$) and anti-Pfaffian ($\nu_C = -3$) states.[116–118] Yet, as shown below, experimental evidence has increasingly pointed towards the PH-Pfaffian order ($\nu_C = -1$).[107]

6.4. Experiments at $\nu = 5/2$

As discussed above (Sec. 6.2), multiple groups have obtained results compatible with the quasiparticle charge of $e/4$. This does not shed light on the topological order since the same charge is predicted for all orders of the 16-fold way. Other existing probes of bulk properties unfortunately shed little light on topological order. More useful information comes from edge probes.[103] The observation of upstream heat transport at $\nu = 5/2$ provides evidence of a negative Chern number and is inconsistent with numerically supported Pfaffian order. As discussed in Ref. 107, tunneling experiments appear inconsistent with the numerically supported anti-Pfaffian state.[119-122] On the other hand, those experiments seem compatible with the PH-Pfaffian order. A natural way to distinguish the PH-Pfaffian state from numerically supported candidates is based on thermal conductance measurements, discussed in the next section. Another approach is interferometry.

The most obvious and deterministic way to measure the braiding statistics of quasiparticles involves interferometry. For non-Abelian states, the FPI is expected to exhibit the even-odd effect.[123,124] In other words, the contribution of the interference to the current, which follows $\sim\cos(\varphi + \gamma)$, can only be observed if the total number of $e/4$ quasiparticles inside the interferometer is even. This can be understood in the following way: Let the number of trapped anyons be odd, say one. The trapped anyon has two fusion channels with the anyon at the edge. The probabilities of the two fusion outcomes turn out to be the same, while the statistical phases differ by π. This leads to total destructive interference of the two fusion channels and no magnetic flux dependence of the current. The argument does not apply for systems with an even number of trapped anyons, where the interference should be restored because then there is only one fusion possibility. As an example, the existence of a single fusion channel is most obvious when the number of trapped anyons is zero since then the edge anyon can fuse with nothing in only one way. This beautiful effect is a striking signature of a non-Abelian topological order; yet, it cannot distinguish among different non-Abelian orders. Surprisingly, under certain conditions it might not be able to distinguish non-Abelian orders from Abelian ones, since the existence of two types of Abelian $e/4$ particles can mimic two fusion channels of non-Abelian anyons.[125]

Interference experiments performed with a small FPI with the $\nu = 5/2$ state are consistent with non-Abelian statistics[126]; yet, their theoretical interpretation has proved difficult.[127] The MZI[64] is free from some of the challenges of the FPI. Hence, it should offer unique signatures for various topological orders.[92,128-133] The theory of the MZI as an anyonic interferometer is considerably more involved than that in the FPI case.[92] Here, we only mention a particularly striking behavior, predicted for the PH-Pfaffian state[107]: The transmitted current through the MZI is not expected to depend on the magnetic field, while the noise of that current diverges at some values of the field. The very different behavior of the current and noise in the MZI reflects a long-term memory effect (Sec. 6.1), absent in the FPI.

7. Thermal Transport

7.1. *Theoretical arguments*

It has long been known that in the QHE both electrical and thermal conductance are quantized.[134-136] Yet, the first observation of quantized heat current in the QHE regime came more than three decades after the discovery of quantized electric transport reflecting the difficulty of probing neutral modes.

The fundamental limit of quantum heat flow was discovered around the same time as FQHE.[134–136] The simplest example of the quantization is a 1D quantum wire. Let $v(k) = \frac{d\varepsilon(k)}{\hbar dk}$ be the velocity of non-interacting fermions in a 1D conductor. Assume that they enter a 1D ballistic quantum wire from the left at a temperature T_L and from the right, where the fermions are at a temperature T_R. The heat flow through the wire is,

$$J_T = \int \frac{dk}{2\pi} \varepsilon(k) v(k) [n(T_L) - n(T_R)] = \frac{\pi^2 k_B^2 (T_L^2 - T_R^2)}{6h}, \qquad (15)$$

with $n(T)$ being the Fermi distribution at the temperature T. This yields the quantized thermal conductance $\kappa_0 = \frac{dJ_T}{d\Delta T} = \frac{\pi^2 k_B^2 T}{3h}$, where $\Delta T = T_L - T_R \to 0$ (linear regime). In contrast to the quantization of the electrical conductance in quantum wires, the result in Eq. (15) does not hold in the presence of inter-mode interaction among counter-propagating channels. Evidently, inter-channel equilibration will have backscatter heat via the upstream channels. At the same time, quantization of the heat conductance in the QHE regime is still expected as long as the propagation length is substantially longer than the thermal equilibration length among all channels. Moreover, no heat exchange between the two spatially separated edges of the sample should be allowed. Note that the same quantization of heat transport as in IQHE channels holds for Abelian FQHE channels.[135,137] The reason is simple: The action given in Eq. (4) for an FQHE channel can be reduced to the action of IQHE in Eq. (3) by a rescaling of the Bose field φ. The same argument would not work for charge transport since the rescaling would affect the definition of the charge density. No such problem is encountered for the energy density since it is nothing else but the Hamiltonian density.

For the case of several non-interacting co-propagating chiral downstream edge channels, their thermal conductances simply add up:

$$\kappa = n\kappa_0, \qquad (16)$$

where n is the total number of the channels. Essentially, the same argument as for charge transport in Sec. 4 shows that interaction between co-propagating modes cannot change Eq. (16). In the presence of counter-propagating channels for a long distance, n becomes the difference between the numbers of the downstream and upstream channels (that is, the net downstream number of channels).

While the heat conductance does not distinguish states in the IQHE from Abelian states in the FQHE, it distinguishes non-Abelian orders from Abelian ones. Indeed, for non-Abelian orders, n in Eq. (16) is no longer an integer.[136] The example of Majorana fermions is most important in this context. A Majorana fermion is a real or imaginary part of a usual complex fermion, and hence, can be thought of as a half of a Dirac fermion. Thus, the thermal conductance of a Majorana channel is one-half of the thermal conductance of a Dirac channel, thus, contributing $\kappa_0/2$ to the thermal conductance. This suggests a tempting way to probe the topological order of the $\nu = 5/2$ state: All needed to know is the Chern number, and the thermal conductance is proportional to it.

7.2. *Experimental results*

A pioneering measurement of the thermal conductance in the IQHE regime was performed by Jezouin *et al.*[138] This work was extended to the FQHE regime by Banerjee *et al.*,[139] proving the universality of the thermal conductance in 'particle-like' fractional states. Moreover, measurements were extended to the more intriguing 'hole-conjugate' states (i.e. $\nu = 2/3$, $3/5$, and $4/7$), which support counter-propagating charge and neutral modes. Having this baseline, the developed methods allowed exploring the states in the second Landau level, and, in particular, the intricate $\nu = 5/2$ state. The latter was found to support a fractional thermal conductance.[140]

More specifically, the $\nu = 1/3$ Laughlin state, with conductance $e^2/3h$ and quasi-particle charge $e^* = e/3$, was found to carry one unit of thermal conductance $1\kappa_0 T$, without a factor $1/3$.[139] The 'hole-like' states, $\nu = 3/5$ and $\nu = 4/7$, supporting a larger number of upstream chiral neutral modes (see Sec. 2.2), had a negative thermal conductance determined by the **net** chirality of all their edge modes (the negative sign of the heat flow was not measured).[135] The most explored $\nu = 2/3$ state is special. It has a downstream charge mode with conductance $2e^2/3h$ and an upstream neutral mode, so that the net number of the downstream modes is $n = 0$. Since full equilibration takes place only at infinite propagation length (thermal transport is diffusive), the lowest thermal conductance was ~$0.25\kappa_0 T$ at some 40 mK.[139]

Similar measurements were performed in the second Landau level, testing the three major fractions, $\nu = 7/3$, $5/2$, and $8/3$. While the odd denominator states were found to conduct heat as expected, the thermal conductance of the $\nu = 5/2$ state was found to be **2.5κ_0**; suggesting the PH-Pfaffian order.[140] This order is not expected from numerical calculations (see discussion below).

7.3. *Experimental method*

The experimental setups adopted heating a small floating metallic contact, which in turn emitted edge modes towards a colder (grounded) contact.[138–140] Measuring the temperature of the floating contact was sufficient for the determination of the total thermal conductance.

A DC input, provided by a source current, heated the floating contact. The outgoing current was split into N similar wide arms (mesas), with n chiral channels propagating along the two edges of each arm. The dissipated power in the floating contact raised the contact's temperature in equilibrium to $T_{\rm m}$, with the dissipated power being equal to the emitted power. The latter was carried away from the floating contact by phonons (to the bulk) and by the edge modes, $P_{\rm diss} = \Delta P_{\rm ph} + \Delta P_{\rm e}$, respectively. Consequently, $P_{\rm diss} = 0.5 \cdot N \cdot n \cdot \kappa_o \cdot (T_m^2 - T_0^2) + \beta\,(T_m^5 - T_0^5)$, where T_0 is the electron temperature in the 'cold' contacts, and β is the prefactor of the phonon emission (depends on the floating contact size).[141] The phonon contribution

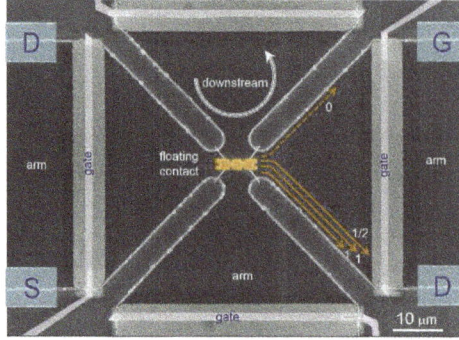

Fig. 19. The heart of the device used to measure thermal transport at filling $\nu = 5/2$. The small ohmic contact (area 12 μm^2) serves as the heated floating reservoir and injects currents into four arms. The effective propagation length (to a cold contact, not shown) in each arm is $\sim$150 μm. Arms can be pinched-off by negatively charging surface gates. As an example, the energy carrying edge modes that correspond to the PH-Pfaffian order are shown ('cold modes' are not shown). The solid orange arrows represent downstream charge modes; each carries $\kappa_0 T$ heat flux. The dashed orange arrow represents an upstream Majorana mode, carrying $0.5\kappa_0 T$ heat flux (from Ref. 140).

became negligible in comparison with the electronic contribution for $T_m < 30$ mK and $T_0 = 10$ mK (depending on β, Ref. 139). At higher temperatures, the phonon contribution can be subtracted out by changing the participating number of the edge modes emanating from the floating contact. The contact temperature T_m is deduced by measuring the downstream thermal noise (Fig. 19).

7.4. Experimental considerations

Measuring the thermal conductance is more demanding and less accurate than measuring the electrical conductance. Here are a few points that highlight a few of the difficulties: (i) Charge should fully equilibrate in the heated small floating contact; (ii) The floating contact must be 'close to ideal'; namely, with a negligible reflection coefficient (to minimize producing excess shot noise); (iii) The outgoing current must split equally between all the open arms; (iv) Bulk heat conductance must be negligible (in particular for the $\nu = 5/2$ samples, where the doping scheme can leave parallel heat conduction through the doping regions[140,142,143]); (v) The temperature must be stable during prolonged measurements (in particular at the lowest temperatures); (vi) The gain of the amplification chain must be carefully calibrated, as it is crucial in the determination of the temperature; (vii) Energy loss to the environment (e.g. due to Coulomb interaction with metallic gates) must be minimized.

7.5. Inter-channel equilibration

Deviations from the expected thermal conductance were found at the filling factors $\nu = 2/3$,[139] and $\nu = 8/3$.[140] At $\nu = 5/2$, the observed thermal conductance was $\sim$2.5$\kappa_0 T$ at a range of temperatures, but increased to $\sim$2.75$\kappa_0 T$ at a lower temperature. It is likely that the reason in all three cases is incomplete inter-mode thermal equilibration.

We start with a discussion of the physics at $\nu = 2/3$. Here, one upstream mode coexists with one downstream mode. They emanate from different reservoirs and can have different temperatures T_u and T_d. In the absence of interactions, the heat current through each channel is,

$$J(T) = \frac{\pi^2 k_B^2 T^2}{6h}, \tag{17}$$

where T_u and T_d should be substituted in place of T for the upstream and down-stream channels. Electron interactions result in energy exchange between the channels and make their temperatures coordinate-dependent. In the simplest model, the inter-channel heat current per unit length can be expressed in the spirit of Newton's law of cooling,

$$\dot{j}_{ud} = \frac{\pi^2 k_B^2 (T_d^2 - T_u^2)}{6h\xi}, \tag{18}$$

where ξ is the temperature-equilibration length. The energy balance equation becomes

$$\frac{dJ(T_u[x])}{dx} = \frac{dJ(T_d[x])}{dx} = -\dot{j}_{ud}[x]. \tag{19}$$

Its solution reveals that the observed thermal conductance is diffusive, and thus far from the naive zero value at full equilibrium (at the sample length $L \gg \xi$). It takes the form

$$\kappa \sim \frac{1}{1 + L/\xi}. \tag{20}$$

Detailed analysis[140,144] shows that the equilibration length increases at lower temperatures, and hence, the thermal conductance grows as the temperature decreases — as observed.

The above discussion crucially depends on the equal numbers of the upstream and downstream channels at $\nu = 2/3$. In the states with an unequal number of up- and down-stream channels, finite size corrections to the equilibrium thermal conductance are much smaller than in Eq. (20) and exhibit an exponential dependence on L/ξ.[39,140] At first sight, this makes puzzling the observed deviations from the expected equilibrated thermal conductance at $\nu = 8/3$. The observed behavior can be explained by decoupling of some of the channels from the rest of the system.[144] Strong Coulomb interaction between integer and fractional edge channels makes it meaningless to assign different temperatures to integer and fractional channels. The right language to use is 'weakly versus strongly interacting modes'. In this case, the appropriate weakly interacting modes are the overall downstream charged mode of the conductance $\frac{8e^2}{3h}$, the spin mode, the upstream neutral mode, and one more neutral downstream mode.[144] Since the downstream charged mode is much faster than the neutral modes, the thermal excitations of that mode rapidly travel through the system before they can exchange their energy with the neutral modes. This example gives a hint of the importance of equilibration among modes in thermal measurements.

The observed increase of the thermal conductance at low temperatures at $\nu = 5/2$ cannot be easily explained in the same way. With the assumption that the state is a PH-Pfaffian liquid, it is crucial that the PH-Pfaffian equilibration length diverges faster at low temperatures than it does in the other possible states.[140] A lack of equilibration was also used to explain the observed thermal conductance at $\nu = 5/2$ assuming the anti-Pfaffian order, however, these explanations face difficulties.[144–147]

8. Conclusion

In the first twenty years of the quantum Hall effect the focus of the experimental research was on bulk physics. In the late 1990s, the focus began shifting towards probing the edges. The first major success in the fractional regime was the detection of fractional charges via shot noise.[34,35] followed by the Nobel Prize to Stormer, Tsui, and Laughlin in 1998. Experiments concentrated in the following years on the properties of exotic states that harbor multiple and counter-propagating edge modes. Hand in hand, material quality improved substantially, mainly in GaAs-AlGaAs heterojunctions.[143,148,149] Follow up work has concentrated on deducing bulk topological orders from edge behavior.

One major research direction has been QHE interferometry, which may serve as a beacon to the anyonic statistics, be it Abelian or non-Abelian. Fabry–Perot (FPI)[66–69] and Mach–Zehnder (MZI)[64,65] interferometers were employed. The FPI can be constructed in smaller dimensions, yet, it tends to suffer from strong Coulomb interaction, which hides the Aharonov–Bohm (AB) interference — making the interpretation of the data more difficult. The MZI, on the other hand, being larger (harboring an ohmic contact in its interior), shows always a non-interacting, two-path, AB interference. Interference in the integer regime is relatively easy; yet, new and unexpected effects have been discovered (electron pairing[84] and lobe-structure of the visibility[65]) — likely due to interactions in the integer regime. These unexpected effects prove the absence of full understanding even in the integer regime. On the other hand, experimental evidence of AB interference in the fractional regime is rather scarce, preventing thus far an assured demonstration of the anyonic nature of the quasiparticles.[86,126,150] Neutral modes (either topological or due to spontaneous edge-reconstruction) may be responsible for the dephasing in the fractional regime.[63]

Upstream neutral edge modes, though theoretically predicted earlier,[50] were observed only more recently.[52–57] Indeed, measuring energy transport is more difficult than measuring the transport of charge. The results gave a strong confirmation of the existing theory of Abelian topological orders. Even more interesting results were obtained for the $\nu = 5/2$ state, where a non-Abelian topological order has long been suspected.

Numerical works have produced preponderance of evidence in favor of the non-Abelian Pfaffian and anti-Pfaffian orders.[116–118] At the same time, experimental evidence seems to point towards a related, but distinct, PH-Pfaffian order.[107] A recent thermal conductance experiment,[140] performed in an extremely high-mobility 2DEG buried in GaAs-AlGaAs heterostructures, showed results consistent with the PH-Pfaffian order. If one accepts the PH-Pfaffian order, then what is wrong with the numerical prediction?

Numerical works treat Landau level mixing (LLM) in an approximate manner, and it was suggested that LLM may be responsible for a different topological order in realistic samples.[151] More attention has focused on disorder-based explanations.[107,152,153] Indeed, until a recent preprint,[154] no attempts were made to include disorder in simulations of the 5/2 physics. Under appropriate conditions, LLM and disorder may result in the macroscopic PH-Pfaffian order. It was proposed that a system with long correlations in its disorder can split into microscopic domains of Pfaffian and anti-Pfaffian liquids with a PH-Pfaffian emerging as a coherent average.[152,153] A different approach was taken by interpreting the data in terms of the anti-Pfaffian state, assuming partial equilibration.[144] Yet, this interpretation must satisfy a few stringent constraints.[146]

One thing is clear: Whatever future developments await us, edge probes of topological orders will play a crucial role.

Acknowledgments

MH acknowledges W. Yang and V. Umansky for their extremely valuable help with the manuscript; the partial support of the Israeli Science Foundation (ISF), the Minerva foundation, and the European Research Council under the European Community's Seventh Framework Program (FP7/2007– 2013)/ERC Grant agreement 339070. DEF acknowledges partial support of the NSF under grant No. DMR-1607451.

Appendix A. Edge actions

A.1. $\nu = 1$

We briefly address the origin of the action in Eq. (3):

$$\frac{S}{\hbar} = -\frac{1}{4\pi} \int dx \, dt [\partial_t \varphi \partial_x \varphi + v(\partial_x \varphi)^2] \tag{A.1}$$

The second term in the action, quadratic in $\partial_x \varphi$, is the energy cost of changing the charge density away from the minimal-energy configuration. To justify the first term and the coefficient v in front of the second term, we make two observations: (i) The equation of motion for the action,

$$\partial_t \partial_x \varphi + v \partial_x^2 \varphi = 0 \,, \tag{A.2}$$

reveals chiral transport with the speed v, exactly as in the fermionic formulation; (ii) The action gives the correct conductance. One can check it by adding the contribution of the chemical potential $\int dx\, dt \rho V$, minimizing the action with respect to the charge density, and finally computing the current $j = ev\partial_x \varphi / 2\pi$ from the equation of motion. The structure of the action determines the commutation relations for the field φ. Indeed, the action combines the Hamiltonian with a piece, containing the time derivative of φ. This piece tells us what variable is canonically conjugated to φ. The corresponding commutation relation is $[\varphi(x), \partial_y \varphi(y)] = -2\pi i \delta(x - y)$, or equivalently, $[\varphi(x), \varphi(y)] = \pi i\, \mathrm{sign}(x - y)$. This allows identifying the electron operator ψ, which is a Fermi operator that changes the total charge $Q = \int dx \rho$ by an electron charge e. The above commutation relations show that any operator of the form $\exp(i[2n + 1]\varphi)$ satisfies the Fermi anti-commutation relation. Finally, the effect of electron annihilation on the charge of the edge implies the commutation relation $[\psi, Q] = e\psi$, being consistent with $\psi(x) = \exp(i\varphi(x))$.

A.2. A general Abelian edge

A systematic way to construct edge actions is known as the K-matrix formalism.[3] The topological properties of the bulk are encoded by an integer symmetric matrix K and an integer column t. Each bulk excitation corresponds to an integer column l_k. The filling factor equals $t^T K^{-1} t$, the quasiparticle charge is $e l_k^T K^{-1} t$ and the statistical phase accumulated by an anyon of type l_k making a full circle around an anyon of type l_m is $2\pi l_k^T K^{-1} l_m$. Unimodular (determinant $= 1$) integer matrices W define equivalent descriptions of the same phase with $K \to WKW^T$, $t \to Wt$. According to the 'bulk-edge correspondence' principle, topological properties of the edge are determined by those of the bulk. In particular, the simplest edge theory is given by the action,

$$\frac{S}{\hbar} = -\frac{1}{4\pi} \int dx\, dt \sum_{km} [K_{km} \partial_t \varphi_k \partial_x \varphi_m + V_{km} \partial_x \varphi_k \partial_x \varphi_m] \,, \tag{A.3}$$

where the first term contains the bulk K-matrix and the second term V describes interactions. To identify edge channels one needs to diagonalize simultaneously the symmetric matrices K and V. The signs of the eigenvalues of K determine the propagation directions of the modes. One can deduce the K matrices at $\nu = 1, 1/3$, and $2/3$ from the discussion above and in Sec. 2.

A.3. *Disorder on the edge at $\nu = 2/3$*

The simplest edge model of the hole-conjugate $\nu = 2/3$ state has the action,

$$\frac{S}{\hbar} = -\frac{1}{4\pi} \int dx\, dt [\partial_t \varphi_1 \partial_x \varphi_1 + v_1(\partial_x \varphi_1)^2]$$

$$+ \frac{3}{4\pi} \int dx\, dt [\partial_t \varphi_{1/3} \partial_x \varphi_{1/3} - v_{1/3}(\partial_x \varphi_{1/3})^2], \tag{A.4}$$

where the fields φ_1 and $\varphi_{1/3}$ define the charge densities within the two channels. As discussed in Sec. 4, such an action results in an incorrect electrical conductance. The goal of this section is to incorporate the missing inter-channel interaction and tunneling into the model. This allows fixing the conductance issue.

The new model represents the total charge density at the edge as $e\partial_x \varphi_c / 2\pi = e\partial_x (\varphi_1 + \varphi_{1/3})/2\pi$, accompanied by a neutral mode $\varphi_n = (\varphi_1 + 3\varphi_{1/3})/\sqrt{2}$. The form of φ_n is motivated by: (i) This choice simplifies the expression for the tunneling operator that transfers an electron between the two edge channels: $T = \exp(\pm\sqrt{2}i\varphi_n)$; and (ii) The K-matrix contribution to the action in Eq. (A.3) remains diagonal in the variables φ_n and φ_c. The action then becomes,

$$\frac{S}{\hbar} = -\int dx\, dt \left\{ \frac{1}{4\pi}[-\partial_t \varphi_n \partial_x \varphi_n + v_n(\partial_x \varphi_n)^2] + \frac{3}{8\pi}[\partial_t \varphi_c \partial_x \varphi_c + v_c(\partial_x \varphi_c)^2] \right.$$

$$\left. + \frac{w_{nc}}{4\pi} \partial_x \varphi_n \partial_x \varphi_c + [W(x)\exp(\sqrt{2}i\varphi_n) + \text{H.c.}] \right\}, \tag{A.5}$$

where w_{nc} describes the interaction between the neutral and charged modes, and $W(x)$ is the position-dependent tunneling amplitude.

The mode velocities v_c and v_n are determined by the Coulomb interaction, similarly to w_{nc}, and it is expected that $v_c > v_n, w_{nc}$. On an etched edge, where the Coloumb interaction is not screened by defining gates, v_c is dominated by the long-range part of the Coloumb force, and the velocity of the low-energy excitations of the charge mode diverges logarithmically with the wavelength.

$W(x)$ depends on the microscopic details. In a clean sample, $W(x) \sim \exp(ix\Delta k)$ carries information about the momentum mismatch Δk of the edge modes φ_1 and $\varphi_{1/3}$. Indeed, Fig. 2 shows that these modes are spatially separated, while Eq. (1) shows that spatial separation translates into momentum difference in a strong perpendicular magnetic field. This leads us to a new difficulty: In the low-energy limit, no tunneling processes satisfy both energy and momentum conservation. Thus, charge tunneling is ineffective and we still get a wrong conductance. The crucial observation is the role of disorder,[50] since in any realistic sample, $W(x)$ is a random function of the coordinate.

The role of the random tunneling term can be understood from its behavior under renormalization group (RG) transformations.[49,50] In the lowest order, the RG procedure consists of two steps: (i) Using the quadratic part of the action [Eq. (A.5)] to integrate out the large wave-vector part of $\varphi_n(x)$ from the tunneling operator $\exp(\sqrt{2}i\varphi_n)$; and (ii) Rescaling x and t. For a coordinate-independent

$W(x)$, the rescaling of x and t by a factor of s would multiply W by s^2. Since W is spatially random, the rescaling of x generates a factor of $\sqrt{s}$, so that the total rescaling factor is $s^{3/2}$. Assuming $v_c \gg w_{nc}$, one finds that the renormalized tunneling amplitude grows as $W \sim l^{1/2}$, where the length l is the running RG ultraviolet cut-off scale. Thus, even arbitrarily small W results in strong inter-channel tunneling at low energies on a sufficiently long edge.

Since RG flows into the strong coupling regime, it is not obvious, what exactly happens at low energies. Kane, Fisher, and Polchiski[50] found an answer by fermionizing the action [Eq. (A.5)], leading to decoupled charge and neutral modes:

$$\frac{S}{\hbar} = -\int dx\, dt \left\{ \frac{1}{4\pi}[-\partial_t \varphi_n \partial_x \varphi_n + v_n(\partial_x \varphi_n)^2] + \frac{3}{8\pi}[\partial_t \varphi_c \partial_x \varphi_c + v_c(\partial_x \varphi_c^2)] \right\} \quad \text{(A.6)}$$

This action describes the RG fixed point. In general, irrelevant perturbations are also present.

Appendix B. Bulk-edge correspondence at $\nu = 5/2$

The goal of this section is to explain why an odd number of edge Majorana modes corresponds to non-Abelian statistics in the bulk, and an even number of edge Majorana modes corresponds to Abelian statistics (Sec. 6.2). Our approach is heuristic. A more rigorous discussion is beyond the scope of this chapter. Our discussion builds on bulk-edge correspondence.

We interpret the bulk anyons as small holes in a 2D electron gas, with each hole having a perimeter and hence carrying edge modes with an action of the type of Eq. (14). Such picture does not reflect the microscopic details of realistic samples, since propagation of these holes would involve dramatic destruction and reconstruction of chemical bonds in the material. Still, it is sensible to expect that this simplistic model captures the universal topological properties. Besides, the 2D gas can be locally depleted by a scanning tip. The resulting "hole" in the gas can bind anyons.

Let us start with the Chern number $\nu_C = 2$. The two neutral edge modes around the hole can be combined into a single Dirac fermion. For a large hole, the edge spectrum is almost continuous. Yet, in the limit of a small hole, most edge states have a high energy and drop out from low-energy physics. In the simplest model, exactly one fermionic state survives in the low-energy limit. The state can be either filled or empty. This corresponds to an almost trivial Hamiltonian $H = \varepsilon\Psi^+\Psi$, where Ψ is a complex fermion operator, represented as a combination of two real fermions a and b: $\Psi = a + ib$. Since a^2 and b^2 are c-numbers, the Hamiltonian reduces to,

$$H = 2i\varepsilon ab + \text{const.} \quad \text{(B.1)}$$

Effectively, two neutral Majorana fermions live in the hole. With two holes in close proximity in a $\nu_C = 1$ system and each hole carrying a single Majorana mode, there are two of them altogether. Consequently, it is sensible to expect that the low-energy physics is similar to that of a single hole at $\nu_C = 2$, and that the Hamiltonian is still the one in Eq. (B.1). When one moves the two holes far from each other, the general structure of the Hamiltonian remains; however, the distant neutral excitations cannot interact anymore (thus, $\varepsilon \to 0$). The system hence possesses two degenerate states of different total fermionic parity $(a - ib)(a + ib)$. In other words, the two holes can form states of two different statistics. The holes are anyons in our model, and so we have discovered two fusion possibilities.

To establish non-Abelian statistics, each hole should represent the same anyon type irrespectively of the fusion channel. If this were not the case, some local observable would be different in the two fusion channels. Local observables are Hermitian Bose-operators built from a only or from b only. Any such operator is an even power of a Majorana fermion, and hence, a trivial constant. In other words, no local measurements can distinguish the states of the holes in the two fusion channels, and both fusion outcomes emerge from a combination of the same anyons.

Appendix C. Coulomb dominated regime in the FQHE regime

For $f_T = 0$ in the integer regime, increasing B leads to area shrinking, keeping the threading flux constant, and to imbalance between electrons and ionized donors. Eventually, the imbalance relaxes (when interactions reach the charging energy), by adding a quasiparticle within the interfering Landau level. The area returns to its initial state with the addition of a flux quantum — being invisible in the interference pattern. For $f_T > 0$, with increasing B the interfering channel loses electrons to the f_T lower Landau levels (whose density increases with field), and the threading flux reduces (opposite to the behavior in the pure AB regime).

We discuss now the periodicity in the gate voltage ΔV_{MG}, as it provides an insight to the charge of the interfering quasiparticle. We assume that the capacitance C between the modulation gate and the interfering channel depends only on f_T, namely, $C = C(f_T)$. The interfering channel flows at the interface between two areas with different filling factors, ν_{out} and ν_{in}, each with quasiparticle charge $q_i = e\nu_i$. As long as the sole function of biasing the plunger gate by δV_{MG} is to move the interface between the two filling factors by area δA, and thus expel a δq charge from within the interferometer, we may write:

$$\delta q = C\delta V_{\mathrm{MG}} = \frac{B \cdot \delta A}{\phi_0} \cdot e(\nu_{\mathrm{in}} - \nu_{\mathrm{out}}) \tag{C.1}$$

Assuming that the change in the area does not change the number of quasiparticles enclosed by the interfering loop (large energy is required to induce quasiparticles or quasiholes), the change of the area leads to a change in the AB phase,

$$\delta\varphi = 2\pi \frac{e^*}{e} \frac{B \cdot \delta A}{\phi_0} \tag{C.2}$$

Combining Eqs. (C.1) and (C.2) we get a relation between the interfering quasiparticle charge e^* and the periodicity ΔV_{MG}:

$$\frac{e^*}{e} = \frac{\nu_{\mathrm{in}} - \nu_{\mathrm{out}}}{\Delta V_{\mathrm{MG}}/\Delta V_e}, \tag{C.3}$$

where ΔV_e is the gate voltage needed to expel one electron.

Indeed, in the cases (1/3, 0) and (2/5, 0), the periodicities in the gate voltage
are the same as for $f_T = 0$ in the integer cases; namely, the expelled charge per
period must be in both cases e. Since the interfering edge channel in both cases
belongs to the 1/3 fractional state, $\nu_{\mathrm{in}} - \nu_{\mathrm{out}} = 1/3$, the interfering quasiparticle
charge must be $e^* = e/3$ [Eq. (C.3)].[69]

In the case (2/5, 1/3), where the interfering channel belongs to the 2/5 fractional
state, the observed[69] periodicity is nearly 1/3 of the period in the integer cases, and
thus the expelled charge must be $e/3$. As $\nu_{\mathrm{in}} - \nu_{\mathrm{out}} = 1/15$ the interfering charge
is $e^* = e/5$. This is a striking example of an expelled charge $e/3$ per period of gate
voltage, while the interfering quasiparticles carried charge $e/5$.[69]

References

1. B. I. Halperin, Quantized Hall conductance, current-carrying edge states, and the
   existence of extended states in a two-dimensional disordered potential, *Phys. Rev. B*
   25(4), 2185–2190 (1982).
2. X. G. Wen, Chiral Luttinger liquid and the edge excitations in the fractional quantum
   Hall states, *Phys. Rev. B* **41**(18), 12838–12844 (1990).
3. X.-G. Wen, Quantum field theory of many-body systems: From the origin of sound
   to an origin of light and electrons (Oxford University Press on Demand 2004).
4. C. L. Kane and M. P. A. Fisher, Edge-State Transport, in: S. Das Sarma, A.
   Pinczuk (Eds.) *Perspectives in Quantum Hall Effects: Novel Quantum Liquids in
   Low-Dimensional Semiconductor Structures* (John Wiley, New York, 1996).
5. M. Buttiker, The quantum Hall-effect in open conductors, in: M. Reed (ed.) *Semiconductors and Semimetals, Vol 35: Nanostructured Systems* (Elsevier Academic
   Press Inc, San Diego, 1992).
6. D. B. Chklovskii, B. I. Shklovskii and L. I. Glazman, Electrostatics of edge channels,
   Phys. Rev. B **46**(7), 4026–4034 (1992).
7. B. Y. Gelfand and B. I. Halperin, Edge electrostatics of a mesa-etched sample and
   edge-state-to-bulk scattering rate in the fractional quantum Hall regime, *Phys. Rev.
   B* **49**(3), 1862–1866 (1994).
8. N. Pascher, C. Rossler, T. Ihn, K. Ensslin, C. Reichl and W. Wegscheider, Imaging
   the conductance of integer and fractional quantum Hall edge states, *Phys. Rev. X*
   4(1), 011014 (2014).
9. C. Chamon and X. G. Wen, Sharp and smooth boundaries of quantum Hall liquids,
   Phys. Rev. B **49**(12), 8227–8241 (1994).
10. H. Inoue, A. Grivnin, Y. Ronen, M. Heiblum, V. Umansky and D. Mahalu, Pro-
    liferation of neutral modes in fractional quantum Hall states, *Nat. Comm.* **5**, 4067
    (2014).
11. J. K. Jain, Composite-fermion approach for the fractional quantum Hall effect, *Phys.
    Rev. Lett.* **63**(2), 199–202 (1989).

12. J. K. Jain, S. A. Kivelson and D. J. Thouless, Proposed measurement of an effective flux quantum in the fractional quantum Hall effect, *Phys. Rev. Lett.* **71**(18), 3003–3006 (1993).

13. A. H. MacDonald, Edge states in the fractional-quantum-Hall-effect regime, *Phys. Rev. Lett.* **64**(2), 220–223 (1990).

14. Y. Cohen, Y. Ronen, W. Yang, D. Banitt, J. Park, M. Heiblum, A. D. Mirlin, Y. Gefen and V. Umansky, Synthesizing a $\nu = 2/3$ fractional quantum Hall effect edge state from counter-propagating $\nu = 1$ and $\nu = 1/3$ states, *Nat. Comm.* **10**, 1920 (2019).

15. Y. Meir, Composite edge states in the $\nu = 2/3$ fractional quantum Hall regime, *Phys. Rev. Lett.* **72**(16), 2624–2627 (1994).

16. R. Sabo, I. Gurman, A. Rosenblatt, F. Lafont, D. Banitt, J. Park, M. Heiblum, Y. Gefen, V. Umansky and D. Mahalu, Edge reconstruction in fractional quantum Hall states, *Nat. Phys.* **13**(5), 491–496 (2017).

17. R. B. Laughlin, Anomalous quantum Hall effect — An incompressible quantum fluid with fractionally charged excitations, *Phys. Rev. Lett.* **50**(18), 1395–1398 (1983).

18. M. Levin and A. Stern, Fractional topological insulators, *Phys. Rev. Lett.* **103**(19), 196803 (2009).

19. Y. M. Blanter and M. Buttiker, Shot noise in mesoscopic conductors, *Phys. Rep.* **336**(1-2), 1–166 (2000).

20. W. Schottky, Über spontane stromschwankungen in verschiedenen elektrizitätsleitern, *Ann. Phys.* **362**(23), 541–567 (1918).

21. W. Schottky, On spontaneous current fluctuations in various electrical conductors, *J. Micro. Nanolith. MEM.* **17**(4), 1–11 (2018).

22. C. Chamon, D. E. Freed and X. G. Wen, Tunneling and quantum noise in one-dimensional Luttinger liquids, *Phys. Rev. B* **51**(4), 2363–2379 (1995).

23. C. L. Kane and M. P. Fisher, Nonequilibrium noise and fractional charge in the quantum Hall effect, *Phys. Rev. Lett.* **72**(5), 724–727 (1994).

24. T. Martin and R. Landauer, Wave-packet approach to noise in multichannel mesoscopic systems, *Phys. Rev. B* **45**(4), 1742–1755 (1992).

25. G. B. Lesovik, Excess quantum noise in 2D ballistic point contacts, *J. Exp. Theor. Phys.* **49**(9), 592–594 (1989).

26. V. A. Khlus, Current and voltage fluctuations in microjunctions between normal metals and superconductors, *J. Exp. Theor. Phys.* **66**(6), 2179 (1987).

27. M. Reznikov, M. Heiblum, H. Shtrikman and D. Mahalu, Temporal correlation of electrons: Suppression of shot noise in a ballistic quantum point contact, *Phys. Rev. Lett.* **75**(18), 3340–3343 (1995).

28. A. Kumar, L. Saminadayar, D. C. Glattli, Y. Jin and B. Etienne, Experimental test of the quantum shot noise reduction theory, *Phys. Rev. Lett.* **76**(15), 2778–2781 (1996).

29. M. Reznikov, E. De Picciotto, M. Heiblum, D. C. Glattli, A. Kumar and L. Saminadayar, Quantum shot noise, *Superlattice Microst.* **23**(3-4), 901–915 (1998).

30. S. Ilani, J. Martin, E. Teitelbaum, J. H. Smet, D. Mahalu, V. Umansky and A. Yacoby, The microscopic nature of localization in the quantum Hall effect, *Nature* **427**(6972), 328–332 (2004).

31. J. A. Simmons, S. W. Hwang, D. C. Tsui, H. P. Wei, L. W. Engel and M. Shayegan, Resistance fluctuations in the integral-quantum-Hall-effect and fractional-quantum-Hall-effect regimes, *Phys. Rev. B* **44**(23), 12933–12944 (1991).

32. V. J. Goldman and B. Su, Resonant tunneling in the quantum Hall regime: Measurement of fractional charge, *Science* **267**(5200), 1010–1012 (1995).

33. J. D. F. Franklin, I. Zailer, C. J. B. Ford, P. J. Simpson, J. E. F. Frost, D. A. Ritchie, M. Y. Simmons and M. Pepper, The Aharonov–Bohm effect in the fractional quantum Hall regime, *Surf. Sci.* **361**(1-3), 17–21 (1996).

34. R. dePicciotto, M. Reznikov, M. Heiblum, V. Umansky, G. Bunin and D. Mahalu, Direct observation of a fractional charge, *Nature* **389**(6647), 162–164 (1997).

35. L. Saminadayar, D. C. Glattli, Y. Jin and B. Etienne, Observation of the $e/3$ fractionally charged Laughlin quasiparticle, *Phys. Rev. Lett.* **79**(13), 2526–2529 (1997).

36. Y. C. Chung, M. Heiblum and V. Umansky, Scattering of bunched fractionally charged quasiparticles, *Phys. Rev. Lett.* **91**(21), 216804 (2003).

37. Y. C. Chung, M. Heiblum, Y. Oreg, V. Umansky and D. Mahalu, Anomalous chiral Luttinger liquid behavior of diluted fractionally charged quasiparticles, *Phys. Rev. B* **67**(20), 201104 (2003).

38. E. Comforti, Y. C. Chung, M. Heiblum and A. V. Umansky, Multiple scattering of fractionally charged quasiparticles, *Phys. Rev. Lett.* **89**(6), 066803 (2002).

39. E. Comforti, Y. C. Chung, M. Heiblum, V. Umansky and D. Mahalu, Bunching of fractionally charged quasiparticles tunnelling through high-potential barriers, *Nature* **416**(6880), 515–518 (2002).

40. P. Fendley, A. W. Ludwig and H. Saleur, Exact nonequilibrium dc shot noise in Luttinger liquids and fractional quantum Hall devices, *Phys. Rev. Lett.* **75**(11), 2196–2199 (1995).

41. P. Fendley, A. W. Ludwig and H. Saleur, Exact conductance through point contacts in the $\nu = 1/3$ fractional quantum Hall effect, *Phys. Rev. Lett.* **74**(15), 3005–3008 (1995).

42. P. Fendley, A. W. W. Ludwig and H. Saleur, Exact nonequilibrium transport through point contacts in quantum wires and fractional quantum Hall devices, *Phys. Rev. B* **52**(12), 8934–8950 (1995).

43. P. Fendley and H. Saleur, Nonequilibrium de noise in a Luttinger liquid with an impurity, *Phys. Rev. B* **54**(15), 10845–10854 (1996).

44. T. G. Griffiths, E. Comforti, M. Heiblum, A. Stern and V. V. Umansky, Evolution of quasiparticle charge in the fractional quantum Hall regime, *Phys. Rev. Lett.* **85**(18), 3918–3921 (2000).

45. C. J. Wang and D. E. Feldman, Fluctuation-dissipation theorem for chiral systems in nonequilibrium steady states, *Phys. Rev. B* **84**(23), 235315 (2011).

46. C. Wang and D. E. Feldman, Chirality, causality, and fluctuation-dissipation theorems in nonequilibrium steady states, *Phys. Rev. Lett.* **110**(3), 030602 (2013).

47. C. J. Wang and D. E. Feldman, Fluctuation theorems without time-reversal symmetry, *Int. J. Mod. Phys. B* **28**(7), 1430003 (2014).

48. D. E. Feldman and M. Heiblum, Why a noninteracting model works for shot noise in fractional charge experiments, *Phys. Rev. B* **95**(11), 115308 (2017).

49. C. L. Kane and M. P. Fisher, Impurity scattering and transport of fractional quantum Hall edge states, *Phys. Rev. B* **51**(19), 13449–13466 (1995).

50. C. L. Kane, M. P. Fisher and J. Polchinski, Randomness at the edge: Theory of quantum Hall transport at filling $\nu = 2/3$, *Phys. Rev. Lett.* **72**(26), 4129–4132 (1994).

51. D. E. Feldman and F. F. Li, Charge-statistics separation and probing non-Abelian states, *Phys. Rev. B* **78**(16), 161304 (2008).

52. A. Bid, N. Ofek, H. Inoue, M. Heiblum, C. L. Kane, V. Umansky and D. Mahalu, Observation of neutral modes in the fractional quantum Hall regime, *Nature* **466**(7306), 585–590 (2010).

53. Y. Gross, M. Dolev, M. Heiblum, V. Umansky and D. Mahalu, Upstream neutral

modes in the fractional quantum Hall effect regime: Heat waves or coherent dipoles, *Phys. Rev. Lett.* **108**(22), 226801 (2012).

54. V. Venkatachalam, S. Hart, L. Pfeiffer, K. West and A. Yacoby, Local thermometry of neutral modes on the quantum Hall edge, *Nat. Phys.* **8**(9), 676–681 (2012).

55. M. Dolev, Y. Gross, R. Sabo, I. Gurman, M. Heiblum, V. Umansky and D. Mahalu, Characterizing neutral modes of fractional states in the second Landau level, *Phys. Rev. Lett.* **107**(3), 036805 (2011).

56. I. Gurman, R. Sabo, M. Heiblum, V. Umansky and D. Mahalu, Extracting net current from an upstream neutral mode in the fractional quantum Hall regime, *Nat. Comm.* **3**, 1289 (2012).

57. A. Rosenblatt, F. Lafont, I. Levkivskyi, R. Sabo, I. Gurman, D. Banitt, M. Heiblum and V. Umansky, Transmission of heat modes across a potential barrier, *Nat. Comm.* **8**(1), 2251 (2017).

58. M. Levin, B. I. Halperin and B. Rosenow, Particle-hole symmetry and the Pfaffian state, *Phys. Rev. Lett.* **99**(23), 236806 (2007).

59. B. J. Overbosch and C. Chamon, Long tunneling contact as a probe of fractional quantum Hall neutral edge modes, *Phys. Rev. B* **80**(3), 035319 (2009).

60. E. Grosfeld and S. Das, Probing the neutral edge modes in transport across a point contact via thermal effects in the Read-Rezayi non-Abelian quantum Hall states, *Phys. Rev. Lett.* **102**(10), 106403 (2009).

61. U. Klass, W. Dietsche, K. Vonklitzing and K. Ploog, Imaging of the dissipation in quantum-Hall effect experiments, *Z. Phys. B Con. Mat.* **82**(3), 351–354 (1991).

62. C. Spånslätt, J. Park, Y. Gefen and A. Mirlin, *Topological Classification of Shot Noise on Fractional Quantum Hall Edges*, arXiv:1906.05623, (2019).

63. R. Bhattacharyya, M. Banerjee, M. Heiblum, D. Mahalu and V. Umansky, Melting of interference in the fractional quantum Hall effect: Appearance of neutral modes, *Phys. Rev. Lett.* **122**(24), 246801 (2019).

64. Y. Ji, Y. Chung, D. Sprinzak, M. Heiblum, D. Mahalu and H. Shtrikman, An electronic Mach–Zehnder interferometer, *Nature* **422**(6930), 415–418 (2003).

65. I. Neder, M. Heiblum, Y. Levinson, D. Mahalu and V. Umansky, Unexpected behavior in a two-path electron interferometer, *Phys. Rev. Lett.* **96**(1), 016804 (2006).

66. C. D. C. Chamon, D. E. Freed, S. A. Kivelson, S. L. Sondhi and X. G. Wen, Two point-contact interferometer for quantum Hall systems, *Phys. Rev. B* **55**(4), 2331–2343 (1997).

67. J. A. Folk, C. M. Marcus and J. S. Harris, Jr., Decoherence in nearly isolated quantum dots, *Phys. Rev. Lett.* **87**(20), 206802 (2001).

68. A. Kou, C. M. Marcus, L. N. Pfeiffer and K. W. West, Coulomb oscillations in antidots in the integer and fractional quantum Hall regimes, *Phys. Rev. Lett.* **108**(25), 256803 (2012).

69. N. Ofek, A. Bid, M. Heiblum, A. Stern, V. Umansky and D. Mahalu, Role of interactions in an electronic Fabry–Perot interferometer operating in the quantum Hall effect regime, *Proc. Natl. Acad. Sci.* **107**(12), 5276–5281 (2010).

70. P. Roulleau, F. Portier, D. C. Glattli, P. Roche, A. Cavanna, G. Faini, U. Gennser and D. Mailly, Finite bias visibility of the electronic Mach–Zehnder interferometer, *Phys. Rev. B* **76**(16), 161309 (2007).

71. L. V. Litvin, A. Helzel, H. P. Tranitz, W. Wegscheider and C. Strunk, Edge-channel interference controlled by Landau level filling, *Phys. Rev. B* **78**(7), 075303 (2008).

72. E. Bieri, M. Weiss, O. Goktas, M. Hauser, C. Schonenberger and S. Oberholzer, Finite-bias visibility dependence in an electronic Mach–Zehnder interferometer, *Phys. Rev. B* **79**(24), 245324 (2009).

73. I. Neder, M. Heiblum, D. Mahalu and V. Umansky, Entanglement, dephasing, and phase recovery via cross-correlation measurements of electrons, *Phys. Rev. Lett.* **98**(3), 036803 (2007).

74. I. Neder, F. Marquardt, M. Heiblum, D. Mahalu and V. Umansky, Controlled dephasing of electrons by non-Gaussian shot noise, *Nat. Phys.* **3**(8), 534-537 (2007).

75. P. Roulleau, F. Portier, P. Roche, A. Cavanna, G. Faini, U. Gennser and D. Mailly, Tuning decoherence with a voltage probe, *Phys. Rev. Lett.* **102**(23), 236802 (2009).

76. P. Roulleau, F. Portier, P. Roche, A. Cavanna, G. Faini, U. Gennser and D. Mailly, Noise dephasing in edge states of the integer quantum Hall regime, *Phys. Rev. Lett.* **101**(18), 186803 (2008).

77. H. Inoue, A. Grivnin, N. Ofek, I. Neder, M. Heiblum, V. Umansky and D. Mahalu, Charge fractionalization in the integer quantum Hall effect, *Phys. Rev. Lett.* **112**(16), 166801 (2014).

78. A. Helzel, L. V. Litvin, I. P. Levkivskyi, E. V. Sukhorukov, W. Wegscheider and C. Strunk, Counting statistics and dephasing transition in an electronic Mach–Zehnder interferometer, *Phys. Rev. B* **91**(24), 245419 (2015).

79. I. P. Levkivskyi and E. V. Sukhorukov, Dephasing in the electronic Mach–Zehnder interferometer at filling factor $\nu = 2$, *Phys. Rev. B* **78**(4), 045322 (2008).

80. M. J. Rufino, D. L. Kovrizhin and J. T. Chalker, Solution of a model for the two-channel electronic Mach–Zehnder interferometer, *Phys. Rev. B* **87**(4), 045120 (2013).

81. D. L. Kovrizhin and J. T. Chalker, Multiparticle interference in electronic Mach–Zehnder interferometers, *Phys. Rev. B* **81**(15), 155318 (2010).

82. B. Rosenow and B. I. Halperin, Influence of interactions on flux and back-gate period of quantum Hall interferometers, *Phys. Rev. Lett.* **98**(10), 106801 (2007).

83. B. I. Halperin, A. Stern, I. Neder and B. Rosenow, Theory of the Fabry-Perot quantum Hall interferometer, *Phys. Rev. B* **83**(15), 155440 (2011).

84. H. K. Choi, I. Sivan, A. Rosenblatt, M. Heiblum, V. Umansky and D. Mahalu, Robust electron pairing in the integer quantum Hall effect regime, *Nat. Comm.* **6**, 7435 (2015).

85. Y. M. Zhang, D. T. McClure, E. M. Levenson-Falk, C. M. Marcus, L. N. Pfeiffer and K. W. West, Distinct signatures for Coulomb blockade and Aharonov–Bohm interference in electronic Fabry-Perot interferometers, *Phys. Rev. B* **79**(24), 241304 (2009).

86. J. Nakamura, S. Fallahi, H. Sahasrabudhe, R. Rahman, S. Liang, G. C. Gardner and M. J. Manfra, Aharonov–Bohm interference of fractional quantum Hall edge modes, *Nat. Phys.* **15**(6), 563–569 (2019).

87. I. Sivan, R. Bhattacharyya, H. K. Choi, M. Heiblum, D. E. Feldman, D. Mahalu and V. Umansky, Interaction-induced interference in the integer quantum Hall effect, *Phys. Rev. B* **97**(12), 125405 (2018).

88. I. Sivan, H. K. Choi, J. Park, A. Rosenblatt, Y. Gefen, D. Mahalu and V. Umansky, Observation of interaction-induced modulations of a quantum Hall liquid's area, *Nat. Comm.* **7**, 12184 (2016).

89. E. Sela, Y. Oreg, F. von Oppen and J. Koch, Fractional shot noise in the Kondo regime, *Phys. Rev. Lett.* **97**(8), 086601 (2006).

90. O. Zarchin, M. Zaffalon, M. Heiblum, D. Mahalu and V. Umansky, Two-electron bunching in transport through a quantum dot induced by Kondo correlations, *Phys. Rev. B* **77**(24), 241303 (2008).

91. N. Byers and C. N. Yang, Theoretical considerations concerning quantized magnetic flux in superconducting cylinders, *Phys. Rev. Lett.* **7**(2), 46 (1961).

92. K. T. Law, D. E. Feldman and Y. Gefen, Electronic Mach–Zehnder interferometer as a tool to probe fractional statistics, *Phys. Rev. B* **74**(4), 045319 (2006).

93. G. Moore and N. Read, Nonabelions in the fractional quantum Hall effect, *Nucl. Phys. B* **360**(2-3), 362–396 (1991).

94. C. Nayak, S. H. Simon, A. Stern, M. Freedman and S. Das Sarma, Non-Abelian anyons and topological quantum computation, *Rev. Mod. Phys.* **80**(3), 1083–1159 (2008).

95. A. Kitaev, Anyons in an exactly solved model and beyond, *Ann. Phys-New York* **321**(1), 2–111 (2006).

96. R. L. Willett, K. W. West and L. N. Pfeiffer, Experimental demonstration of Fermi surface effects at filling factor 5/2, *Phys. Rev. Lett.* **88**(6), 066801 (2002).

97. M. S. Hossain, M. K. Ma, M. A. Mueed, L. N. Pfeiffer, K. W. West, K. W. Baldwin and M. Shayegan, Direct observation of composite fermions and their fully-spin-polarized Fermi sea near $\nu = 5/2$, *Phys. Rev. Lett.* **120**(25), 256601 (2018).

98. N. Read and D. Green, Paired states of fermions in two dimensions with breaking of parity and time-reversal symmetries and the fractional quantum Hall effect, *Phys. Rev. B* **61**(15), 10267–10297 (2000).

99. M. Dolev, M. Heiblum, V. Umansky, A. Stern and D. Mahalu, Observation of a quarter of an electron charge at the $\nu = 5/2$ quantum Hall state, **452**(7189), 829–834 (2008).

100. J. B. Miller, I. P. Radu, D. M. Zumbuhl, E. M. Levenson-Falk, M. A. Kastner, C. M. Marcus, L. N. Pfeiffer and K. W. West, Fractional quantum Hall effect in a quantum point contact at filling fraction 5/2, *Nat. Phys.* **3**(8), 561–565 (2007).

101. M. Dolev, Y. Gross, Y. C. Chung, M. Heiblum, V. Umansky and D. Mahalu, Dependence of the tunneling quasiparticle charge determined via shot noise measurements on the tunneling barrier and energetics, *Phys. Rev. B* **81**(16), 161303 (2010).

102. R. L. Willett, L. N. Pfeiffer and K. W. West, Measurement of filling factor 5/2 quasiparticle interference with observation of charge $e/4$ and $e/2$ period oscillations, *Proc. Natl. Acad. Sci.* **106**(22), 8853–8858 (2009).

103. K. K. W. Ma and D. E. Feldman, The sixteenfold way and the quantum Hall effect at half-integer filling factors, *Phys. Rev. B* **100**(3), 035302 (2019).

104. S. S. Lee, S. Ryu, C. Nayak and M. P. Fisher, Particle-hole symmetry and the $\nu = 5/2$ quantum Hall state, *Phys. Rev. Lett.* **99**(23), 236807 (2007).

105. G. Yang and D. E. Feldman, Experimental constraints and a possible quantum Hall state at $\nu = 5/2$, *Phys. Rev. B* **90**(16), 161306 (2014).

106. D. T. Son, Is the composite fermion a Dirac particle? *Phys. Rev. X* **5**(3), 031027 (2015).

107. P. T. Zucker and D. E. Feldman, Stabilization of the particle-hole pfaffian order by Landau-Level mixing and impurities that break particle-hole symmetry, *Phys. Rev. Lett.* **117**(9), 096802 (2016).

108. L. Fidkowski, X. Chen and A. Vishwanath, Non-Abelian topological order on the surface of a 3D topological superconductor from an exactly solved model, *Phys. Rev. X* **3**(4), 041016 (2013).

109. P. Bonderson, C. Nayak and X.-L. Qi, A time-reversal invariant topological phase at the surface of a 3D topological insulator, *J. Stat. Mech-Theory E.* **2013**(09), P09016 (2013).

110. B. J. Overbosch and X. G. Wen, *Phase Transitions on the Edge of the $\nu = 5/2$ Pfaffian and Anti-Pfaffian Quantum Hall State*, arXiv:0804.2087, (2008).

111. B.I. Halperin, Theory of the quantized Hall conductance, *Helv. Phys. Acta* **56**(1-3), 75–102 (1983).

112. X. G. Wen, Non-Abelian statistics in the fractional quantum Hall states, *Phys. Rev. Lett.* **66**(6), 802–805 (1991).

113. J. K. Jain, Incompressible quantum Hall states, *Phys. Rev. B* **40**(11), 8079–8082 (1989).

114. F. D. Haldane and E. H. Rezayi, Spin-singlet wave function for the half-integral quantum Hall effect, *Phys. Rev. Lett.* **60**(10), 956–959 (1988).

115. A. H. MacDonald, D. Yoshioka and S. M. Girvin, Comparison of models for the even-denominator fractional quantum Hall effect, *Phys. Rev. B* **39**(11), 8044–8047 (1989).

116. R. H. Morf, Transition from quantum Hall to compressible states in the second Landau Level: New light on the $\nu = 5/2$ enigma, *Phys. Rev. Lett.* **80**(7), 1505–1508 (1998).

117. K. Pakrouski, M. R. Peterson, T. Jolicoeur, V. W. Scarola, C. Nayak and M. Troyer, Phase diagram of the $\nu = 5/2$ fractional quantum Hall effect: Effects of Landau-Level mixing and nonzero width, *Phys. Rev. X* **5**(2), 021004 (2015).

118. E. H. Rezayi, Landau Level mixing and the ground state of the $\nu = 5/2$ quantum Hall effect, *Phys. Rev. Lett.* **119**(2), 026801 (2017).

119. S. Baer, C. Rossler, T. Ihn, K. Ensslin, C. Reichl and W. Wegscheider, Experimental probe of topological orders and edge excitations in the second Landau level, *Phys. Rev. B* **90**(7), 075403 (2014).

120. I. P. Radu, J. B. Miller, C. M. Marcus, M. A. Kastner, L. N. Pfeiffer and K. W. West, Quasi-particle properties from tunneling in the $\nu = 5/2$ fractional quantum Hall state, *Science* **320**(5878), 899–902 (2008).

121. X. Lin, C. Dillard, M. A. Kastner, L. N. Pfeiffer and K. W. West, Measurements of quasiparticle tunneling in the $v = 5/2$ fractional quantum Hall state, *Phys. Rev. B* **85**(16), 165321 (2012).

122. H. Fu, P. Wang, P. Shan, L. Xiong, L. N. Pfeiffer, K. West, M. A. Kastner and X. Lin, Competing $\nu = 5/2$ fractional quantum Hall states in confined geometry, *Proc. Natl. Acad. Sci.* **113**(44), 12386–12390 (2016).

123. A. Stern and B. I. Halperin, Proposed experiments to probe the non-Abelian $\nu = 5/2$ quantum Hall state, *Phys. Rev. Lett.* **96**(1), 016802 (2006).

124. P. Bonderson, A. Kitaev and K. Shtengel, Detecting non-Abelian statistics in the $\nu = 5/2$ fractional quantum Hall state, *Phys. Rev. Lett.* **96**(1), 016803 (2006).

125. A. Stern, B. Rosenow, R. Ilan and B. I. Halperin, Interference, Coulomb blockade, and the identification of non-Abelian quantum Hall states, *Phys. Rev. B* **82**(8), 085321 (2010).

126. R. L. Willett, K. Shtengel, C. Nayak, L. N. Pfeiffer, Y. J. Chung, M. L. Peabody, K. W. Baldwin and K. W. West, *Interference Measurements of Non-Abelian e/4 & Abelian e/2 Quasiparticle Braiding*, arXiv:1905.10248, (2019).

127. N. Jiang and X. Wan, Recent progress on the non-Abelian $\nu = 5/2$ quantum Hall state, *AAPPS Bulletin* **29**(1), 58–64 (2019).

128. D. E. Feldman, Y. Gefen, A. Kitaev, K. T. Law and A. Stern, Shot noise in an anyonic Mach–Zehnder interferometer, *Phys. Rev. B* **76**(8), 085333 (2007).

129. D. E. Feldman and A. Kitaev, Detecting non-Abelian statistics with an electronic Mach–Zehnder interferometer, *Phys. Rev. Lett.* **97**(18), 186803 (2006).

130. V. V. Ponomarenko and D. V. Averin, Mach–Zehnder interferometer in the fractional quantum Hall regime, *Phys. Rev. Lett.* **99**(6), 066803 (2007).

131. C. J. Wang and D. E. Feldman, Identification of 331 quantum Hall states with Mach–Zehnder interferometry, *Phys. Rev. B* **82**(16), 165314 (2010).

132. G. Campagnano, O. Zilberberg, I. V. Gornyi, D. E. Feldman, A. C. Potter and

Y. Gefen, Hanbury Brown-Twiss interference of anyons, *Phys. Rev. Lett.* **109**(10), 106802 (2012).

133. G. Yang, Probing the $\nu = 5/2$ quantum Hall state with electronic Mach–Zehnder interferometry, *Phys. Rev. B* **91**(11), 115109 (2015).
134. J. B. Pendry, Quantum limits to the flow of Information and entropy, *J. Phys. A-Math. Gen.* **16**(10), 2161–2171 (1983).
135. C. L. Kane and M. P. A. Fisher, Quantized thermal transport in the fractional quantum Hall effect, *Phys. Rev. B* **55**(23), 15832–15837 (1997).
136. A. Cappelli, M. Huerta and G. R. Zemba, Thermal transport in chiral conformal theories and hierarchical quantum Hall states, *Nucl. Phys. B* **636**(3), 568–582 (2002).
137. L. G. C. Rego and G. Kirczenow, Fractional exclusion statistics and the universal quantum of thermal conductance: A unifying approach, *Phys. Rev. B* **59**(20), 13080–13086 (1999).
138. S. Jezouin, F. D. Parmentier, A. Anthore, U. Gennser, A. Cavanna, Y. Jin and F. Pierre, Quantum limit of heat flow across a single electronic channel, *Science* **342**(6158), 601–604 (2013).
139. M. Banerjee, M. Heiblum, A. Rosenblatt, Y. Oreg, D. E. Feldman, A. Stern and V. Umansky, Observed quantization of anyonic heat flow, *Nature* **545**(7652), 75–79 (2017).
140. M. Banerjee, M. Heiblum, V. Umansky, D. E. Feldman, Y. Oreg and A. Stern, Observation of half-integer thermal Hall conductance, *Nature* **559**(7713), 205–210 (2018).
141. F. C. Wellstood, C. Urbina and J. Clarke, Hot-electron effects in metals, *Phys. Rev. B* **49**(9), 5942–5955 (1994).
142. V. Umansky and M. Heiblum, MBE growth of high-mobility 2DEG, *Molecular Beam Epitaxy* (Elsevier Science BV, Netherlands, 2013).
143. V. Umansky, M. Heiblum, Y. Levinson, J. Smet, J. Nubler and M. Dolev, MBE growth of ultra-low disorder 2DEG with mobility exceeding 35×10^6 cm^2/V s, *J. Cryst. Growth*, **311**(7), 1658–1661 (2009).
144. K. K. W. Ma and D. E. Feldman, Partial equilibration of integer and fractional edge channels in the thermal quantum Hall effect, *Phys. Rev. B* **99**(8), 085309 (2019).
145. S. H. Simon, Interpretation of thermal conductance of the $\nu = 5/2$ edge, *Phys. Rev. B* **97**(12), 121406 (2018).
146. D. E. Feldman, Comment on "Interpretation of thermal conductance of the $\nu = 5/2$ edge", *Phys. Rev. B* **98**(16), 167401 (2018).
147. S. H. Simon and B. Rosenow, *Partial Equilibration of the Anti-Pfaffian Edge Due to Majorana Disorder*, arXiv:1906.05294, (2019).
148. L. Pfeiffer and K. W. West, The role of MBE in recent quantum Hall effect physics discoveries, *Physica E* **20**(1-2), 57–64 (2003).
149. M. Samani, A. V. Rossokhaty, E. Sajadi, S. Luscher, J. A. Folk, J. D. Watson, G. C. Gardner and M. J. Manfra, Low-temperature illumination and annealing of ultrahigh quality quantum wells, *Phys. Rev. B* **90**(12), 121405 (2014).
150. P. V. Lin, F. E. Camino and V. J. Goldman, Electron interferometry in the quantum Hall regime: Aharonov–Bohm effect of interacting electrons, *Phys. Rev. B* **80**(12) (2009).
151. L. Antonić, J. Vučičević and M. V. Milovanović, Paired states at 5/2: Particle-hole Pfaffian and particle-hole symmetry breaking, *Phys. Rev. B* **98**(11), 115107 (2018).
152. C. Wang, A. Vishwanath and B. I. Halperin, Topological order from disorder and the quantized Hall thermal metal: Possible applications to the $\nu = 5/2$ state, *Phys. Rev. B* **98**(4), 045112 (2018).

153. D. F. Mross, Y. Oreg, A. Stern, G. Margalit and M. Heiblum, Theory of disorder-induced half-integer thermal Hall conductance, *Phys. Rev. Lett.* **121**(2), 026801 (2018).
154. W. Zhu and D. N. Sheng, *Disorder-Driven Transition and Intermediate Phase for* $\nu = 5/2$ *Fractional Quantum Hall Effect*, arXiv:1809.04776, (2018).

https://doi.org/10.1142/9789811217494_0005

Chapter 5

Exploring Quantum Hall Physics at Ultra-Low Temperatures and at High Pressures

Gábor A. Csáthy

Department of Physics and Astronomy, Purdue University,
525 Northwestern Avenue, West Lafayette, IN 47907, USA
gcsathy@purdue.edu

The use of ultra-low temperature cooling and high hydrostatic pressure techniques has significantly expanded our understanding of two-dimensional electron gas confined to GaAs/AlGaAs structures. This chapter reviews a selected set of experiments employing these specialized techniques in the study of the fractional quantum Hall states and the charged ordered phases, such as the re-entrant integer quantum Hall states and the quantum Hall nematic. Topics discussed include a successful cooling technique used, novel odd denominator fractional quantum Hall states, new transport results on even denominator fractional quantum Hall states and on re-entrant integer quantum Hall states, and phase transitions observed in half-filled Landau levels.

Contents

1. Introduction

The two-dimensional electron gas (2DEG) is one of the richest model systems in condensed matter physics. Indeed, an impressive number of new phenomena were discovered in this system and several novel theoretical concepts were introduced to explain these phenomena. The integer[1] and fractional quantum Hall effect[2] are among the most important discoveries in the 2DEG and work on these Hall effects precipitated ideas on emergent quasiparticles[3–7] and on topological concepts in condensed matter.[8,9]

A large number of fractional quantum Hall states (FQHSs), especially the ones forming in the lowest Landau level, are well understood.[10,11] Their properties are accounted for by Laughlin's wavefunction[3] and Jain's theory of composite fermions.[4,5] There are, however, a handful of fractional quantum Hall states which are thought to harbor more intricate topological order. These states, sometimes referred to as exotic fractional quantum Hall states,[12] remain a focus of current interest.

Historically, the two most studied 2DEGs were confined to MOSFETs and GaAs/AlGaAs heterostructures.[13] In addition to these two examples, 2DEGs are supported by various material hosts. Examples are AlAs/GaAlAs,[14] CdTe/CdMgTe,[15] Si/SiGe,[16] Ge/SiGe,[17,18] ZnO/MgZnO,[19] hydrogen passivated Si surface,[20] and electrons on the surface of superfluid Helium.[21] The study of the 2DEG enjoyed a resurgence of interest with pioneering work on graphene[22] and other layered materials, such as transition metal dichalcogenides[23] and black phosphorus.[24] Work on the totality of 2DEGs highlighted some of the universal, host-independent physics. Examples are the formation of Landau levels in the integer quantum Hall regime and composite fermions in the fractional quantum Hall regime. In addition, each of these hosts enriched the physics of the 2DEG. New physics resulted from 2DEGs with novel internal degrees of freedom, such as the valley and the pseudospin quantum number, and also from 2DEGs possessing an inherent anisotropy. Furthermore, the generation of Moiré lattices in layered van der Waals structures[25] led to Hofstadter physics[26] and, most recently, to magic angle superconductivity.[27]

Even though 2DEGs confined to GaAs/AlGaAs heterostructures have been studied for more than three decades, this system continues to play a privileged role. A large number of phases were first seen in GaAs/AlGaAs. This is because numerous innovations in the Molecular Beam Epitaxy (MBE) growth technique culminated in 2DEGs of record high mobilities approaching 4×10^7 cm^2/Vs or record long mean free paths in excess of 0.3 mm, when measured at temperatures below 1 Kelvin.[13] Such high carrier mobilities are possible because of the exceedingly low defect levels and also because of innovative sample structures. Growth efforts to further improve this material system continue.[29–35] Historical milestones in the evolution of the GaAs MBE technology can be seen in Fig. 1.

One area in which 2DEGs in GaAs/AlGaAs excel when compared to those in other hosts is the support of both topological and traditional Landau phases. In

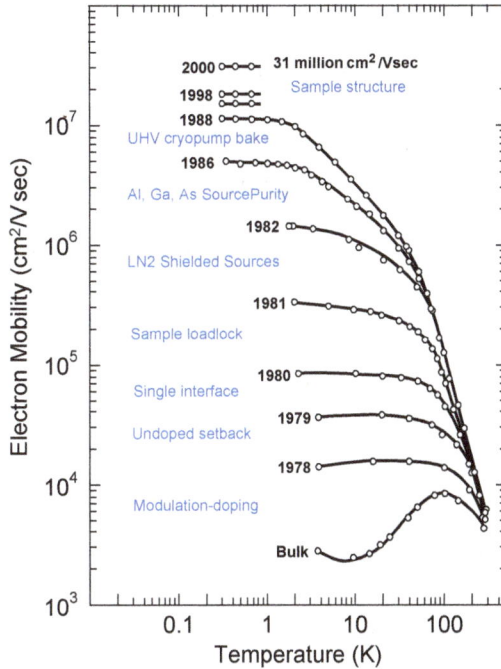

Fig. 1.   Milestones in the evolution of the quality of 2DEGs in the GaAs/AlGaAs system, as measured by the electron mobility. Adapted from Ref. 13.

contrast to topological phases, traditional Landau phases may be characterized by an order parameter. Charge ordered phases of the 2DEG are examples of such traditional Landau phases. The most well-known example of charge ordered phase is the Wigner crystal.[36] However, high quality 2DEGs allow for a more intricate charge ordering.[37–39] The phase at half-filled Landau levels with a strong resistance anisotropy is commonly associated with the electronic nematic, whereas the so-called re-entrant integer quantum Hall states (RIQHSs) are thought to be identical to the electronic bubble phases.[37–41] Recently an increasing amount of attention is lavished on the study of these charge ordered phases.

The 2DEG was probed with various ingenious techniques. However, electric transport played a special role among these techniques as it historically was used to reveal new electronic phases. Over the last decade or so, transport experiments performed at ultra-low temperature have been especially fertile in providing new insight into the physics of the 2DEG. In the following, we present a personal view on some of these experiments, focusing mainly on the second Landau level of the 2DEG confined to GaAs/AlGaAs hosts. In Sec. 2 the reader will find experimental details of the ultra-low temperature cooling technique based on the He-3 immersion cell. Section 3 discusses recently discovered FQHSs, all of which are at odd denominators, followed by an examination of possible origins of these states. In Sec. 4, recent results of transport experiments on the even denominator FQHSs are discussed. Topics include phase transitions at the Landau level filling factor $\nu = 5/2$, a discussion of the energy gap of the FQHS at this filling factor, and its behavior in the presence of short-range alloy disorder. Section 5 contains results on RIQHSs in the second Landau level, such as magnetoresistive fingerprints of these states in high mobility samples, a discussion of the precursors of these states, and a summary of new, developing RIQHSs. Finally, the pressure-induced phase transition at $\nu = 5/2$ from a FQHS to the quantum Hall nematic is discussed in Sec. 6.

2. Cooling Electrons Below 10 mK and Sample State Preparation

Lowering the electronic temperature typically allows for the resolution of states of reduced energy scales. In other words, unless the energy spectrum is already disorder dominated, a lower temperature may reveal increasingly more fragile electronic ground states. Reducing the electron temperature, however, is not a trivial task. While modern dilution refrigerators routinely generate mixing chamber temperatures below 10 mK, similarly low electronic temperatures in semiconductor nanostructures are often difficult to achieve. This is due to the combination of reduced electron-phonon coupling at milliKelvin temperatures and of minute amounts of uncontrolled radiofrequency power traveling on the measurement wires; under such conditions, the electron temperature in a transport setup is often higher than that of the phonons and also of the coldest spot of the refrigerator.

In this section, we describe a successful electron thermalization setup based on a He-3 immersion cell. Such an immersion cell was first built by Jian-Sheng Xia and coauthors at the MicroKelvin Facility of the National High Magnetic Field Laboratory in Gainesville.[47] In this setup each ohmic contact of an electrically conductive sample, such as a 2DEG, is soldered onto individual wire heatsinks that consist of a silver wire surrounded by silver sinter. Constrained by the geometry of the superconductive magnet bore, one may achieve a surface area of the order of a m^2 for the sinter of each wire heatsink. In order to take advantage of such a large surface area for thermalization purposes, the 2DEG and the wire heatsinks are immersed into liquid He-3 that ensures both thermalization and electrical isolation. This thermalization setup was used to examine the quantization at $\nu = 5/2$,[44] to discover the $\nu = 12/5 = 2+2/5$ FQHS,[46] to study metallic behavior in a hole gas,[48] and to understand the plateau-to-plateau transition in the integer quantum Hall regime.[49] A schematic of a He-3 immersion cell is shown in Fig. 2. Xia *et al.* have also built a He-3 immersion cell with a hydraulically driven rotator stage for tilted field measurements.[50] Since cooling electronic systems to ultra-low temperatures is increasingly important for the study of fragile electronic order, besides immersion cell technology there are also efforts to develop alternative cooling techniques,[51-54] some of which are designed on cryogen free platforms.

For our He-3 immersion cell we adopted the design from Ref. 47. In contrast to the setup in Gainesville, instead of using a dilution refrigerator equipped with a

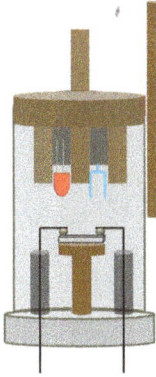

Fig. 2. Schematic of a He-3 immersion cell for ultra-low temperature transport measurements of 2DEGs. Adapted from Ref. 55.

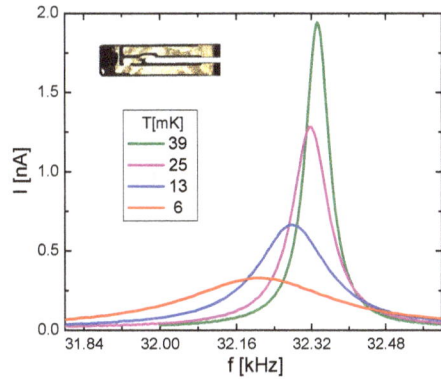

Fig. 3. A quartz tuning fork and the temperature dependence of its response when immersed into a He-3 bath. Adapted from Ref. 55.

nuclear demagnetization stage, we have attached our immersion cell to a modified dilution refrigerator with a 5 mK base temperature. Construction details can be found in Ref. 55. Microwave filters installed on measurement wires are a critical part of the setup. We used a three-stage filter on each wire. First, a set of capacitive filters mounted on the top of the refrigerator, as part of a D-sub connector, is used at room temperature. Second, on their way to the sample, the measurement wires are well heatsunk at each stage of the refrigerator and passed through silver epoxy embedded along about the one foot length of tail connecting the immersion cell to the mixing chamber. Finally, through the skin effect, silver sinters of each wire heatsink mounted within the immersion cell will also efficiently dissipate microwaves.[55] Additional low-pass RC filters with a cut-off frequency of 50 kHz and mounted on the still did not make a difference in electron thermalization, therefore they were later removed.

Parallel with the development of the cooling technology, there was also a flurry of activities in thermometry. Examples are thermometers based on resistive elements,[56] on Johnson noise measurements,[57–59] and on the measurement of the tunneling conductance.[59–63] The extreme environment of temperatures below 10 mK and strong magnetic fields limit the choice of thermometers. The widely used RuO resistive sensors become unreliable for thermometry below about 20 mK, especially in strong fields. Paramagnetic susceptibility thermometers are not suitable for operation in strong magnetic fields. While He-3 melting curve thermometers could be used, we were deterred by the additional effort needed for handling the He-3 at high pressures. However, we already had liquid He-3 in the immersion cell for thermalization purposes, albeit the He-3 was not at the high pressures needed for operation at the liquid-solid phase boundary required in a melting curve thermometer. Under such conditions the viscosity of the He-3 liquid provides a convenient way for temperature monitoring from about 100 mK down to the superfluid onset temperature (not within the reach of our instrument). Since viscosity is independent of the magnetic field, it is ideally suited for the demanding low temperature and strong magnetic field environment of measurements in the quantum Hall regime. We opted for a quartz tuning fork based viscometer.[55] An example of a quartz tuning fork and the temperature dependence of the response curve of the tuning fork immersed into liquid He-3 are shown Fig. 3.

Finally, the electronic state of 2DEGs confined to GaAs/AlGaAs is often prepared by a brief low temperature illumination using either a red light emitting diode facing the 2DEG[64,65] or light guided towards the sample using fiber optics.[66] One effect of such illumination is the increase of the electron density that is desirable for robust ground states. In addition, illumination may also improve the homogeneity of the electron gas.

3. Recently Discovered Fractional Quantum Hall States

In this section we discuss new FQHSs discovered in the 2DEG in the GaAs/AlGaAs system. Two of these, the FQHSs at $\nu = 2+6/13$ and $\nu = 3+1/3$, were seen in the region of the second Landau level, whereas the FQHSs at $\nu = 4/11$ and $\nu = 5/13$ develop in the lowest Landau level. These FQHSs are fragile, hence the use of the immersion cell technology played a central role in their discovery. We show that an analysis of the energy gaps reveals valuable insight into the nature of these FQHSs.

Ground states with a second Landau level character are the strongest in 2DEGs with the optimal electron density. Indeed, as a rule of thumb, the energy gap of a given FQHS increases with the value of the magnetic field at which it forms. Therefore in order to maximize the energy gap of a FQHS, one must use a 2DEG with the largest possible density. However, because the finite thickness of the electronic wavefunction in the direction perpendicular to the plane of the 2DEG, past a critical density the Fermi level moves to the lowest Landau level of the second electric sub-band. As shown by Shayegan and coauthors[67,68] and also supported by theory,[69] such a population of the second electrical sub-band will have a strong influence on the ground states: the orbital part of the single particle wavefunction changes from a second Landau level character to a lowest Landau level character. The most dramatic consequence of populating the second electrical sub-band is the rapid collapse of the FQHS at $\nu = 5/2$.[67,68] The optimal density is therefore the largest density at which the Fermi level falls into the second Landau level and at which ground states retain a second Landau level character. For a quantum well of a 30 nm width, the optimal density is about 3.0×10^{11} cm^{-2}.

Figure 4 shows magnetoresistance traces with a particularly rich structure over the full width of the second Landau level, i.e. both the lower and upper spin branches, in another 2DEG near its optimal density.[70,71] FQHSs are shaded in blue, while RIQHSs in yellow. At first blush, magnetoresistance in the lower spin branch of the second Landau level shown in Fig. 4 is similar to that in Ref. 46.

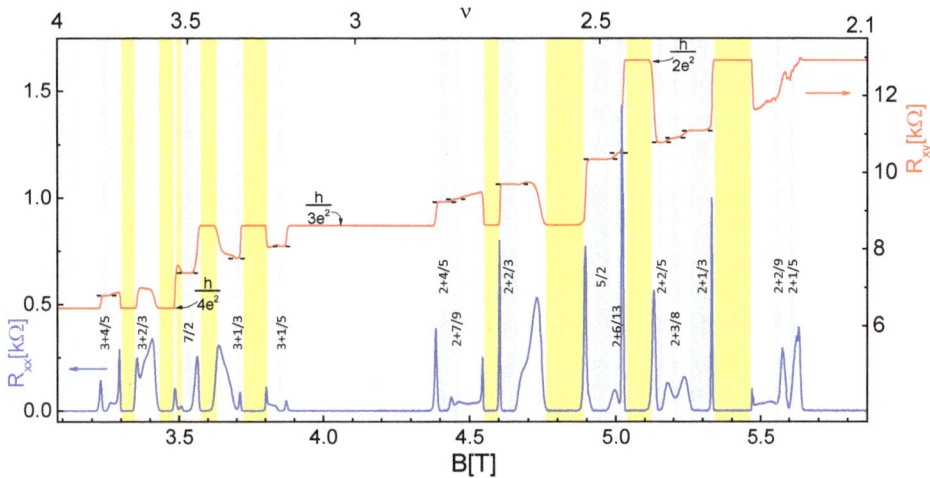

Fig. 4. A particularly rich magnetotransport trace measured at 6.9 mK in the second Landau level. Both spin branches are shown. Blue shades mark fractional quantum Hall states (FQHSs) at the labeled filling factors, whereas yellow shades show various re-entrant integer quantum Hall states (RIQHSs). Adapted from Ref. 71.

However, a more careful examination reveals a notable difference in the sharpness of several magnetoresistance peaks. For example, the width at half height of the peak in R_{xx} shown in Fig. 4 near $B = 5.02$ T is 6.6 mT, while that near $B = 5.33$ T is merely 4.2 mT. In contrast, the corresponding peaks in Ref. 46 are considerably wider. The differences in the two traces are surprising, given that samples and measurement conditions were comparable: the electron densities, mobilities, and the fridge temperatures were $n = 3.1 \times 10^{11}$ cm^{-2}, $\mu = 31 \times 10^6$ cm^2/Vs, $T = 9$ mK in Ref. 46, and $n = 3.0 \times 10^{11}$ cm^{-2}, $\mu = 32 \times 10^6$ cm^2/Vs, $T = 6.9$ mK in Refs. 70 and 71. The difference in the sharpness of the resistance peaks is likely attributed to a difference in sample homogeneity. Indeed, in the presence of a density variation, the magnetoresistance will be a convolution of magnetoresistances corresponding to the range of electron densities and therefore a density inhomogeneity will lead to a broadening of sharp resistive features. Density inhomogeneity in high quality GaAs/AlGaAs samples may be estimated from the widths of sharp magnetoresistance peaks,[72] from an analysis of quantum lifetime measurements,[73] and it is also accessible with the micro-photoluminescence technique on length-scales larger than the laser spotsize.[74,75] We believe that the improved sample homogeneity results from a combination of improved sample growth and sample illumination techniques.

3.1. $\nu = 2 + 6/13$ and $\nu = 2 + 2/5$ fractional quantum Hall states

A particularly interesting region within the second Landau level is that of $2+1/3 < \nu < 2+2/3$, which was conjectured to host FQHSs with unusual topological order.[76] Several FQHSs in this region may have exotic fractional correlations.[6,7,77–83] The

FQHSs at the two endpoints of this range may be either Laughlin states or states radically different from it.[84–92] Until 2004, FQHSs in this region were detected, using a He-3 immersion cell, at $\nu = 5/2$, $\nu = 2 + 2/5$, and $\nu = 2 + 3/8$.[46] An account of FQHSs in GaAs/AlGaAs, including the ones in the second Landau level, was published in 2008.[93]

Another immersion cell experiment confirmed a fully quantized $\nu = 2 + 2/5$ FQHS, with an energy gap of 80 mK[70] in the $2 + 1/3 < \nu < 2 + 2/3$ range. It also yielded a new FQHS in this range at $\nu = 2 + 6/13$.[70] As seen in Fig. 4, this FQHS is located between the $\nu = 5/2$ FQHS and a RIQHS. Since its first observation, the $\nu = 2 + 6/13$ FQHS has been seen in an increasing number of experiments.[32,94–96]

The observation of a FQHS at $\nu = 2 + 6/13$ was a surprise. Indeed, Jain's model of non-interacting composite fermions[4] offers a concise and elegant framework for the understanding of FQHSs developing at filling factors of the form $p/(2p\pm1)$, with $p = 1, 2, 3, ...$. One may notice that the subset of $\nu = 2 + 1/3$, $2 + 2/5$, and $2 + 6/13$ filling factors, at which there are FQHSs in the second Landau level, also belongs to the Jain sequence of the form $2 + p/(2p + 1)$, with $p = 1, 2$, and 6, respectively. However, other FQHSs at intermediate values of $p = 3, 4$, and 5 are not present in Fig. 4. As shown in Fig. 5, the missing FQHSs are absent not only at the lowest temperatures, but for a range of temperatures.[97] Because of these absent FQHSs with $p = 3, 4$, and 5, it was unexpected to see a FQHS of an unusually high order $p = 6$ at $\nu = 2 + 6/13$. We note that even though in other experiments local minima in R_{xx} were seen at $\nu = 2 + 4/9$ in Refs. 31 and 98 and at $\nu = 2 + 3/7$ in Ref. 98, in the the absence of a Hall plateau the identification of a FQHS at these filling factors is inconclusive.[31,98]

At this time it is not known why the $\nu = 2 + 3/7, 2 + 4/9$, and $2 + 5/11$ FQHSs

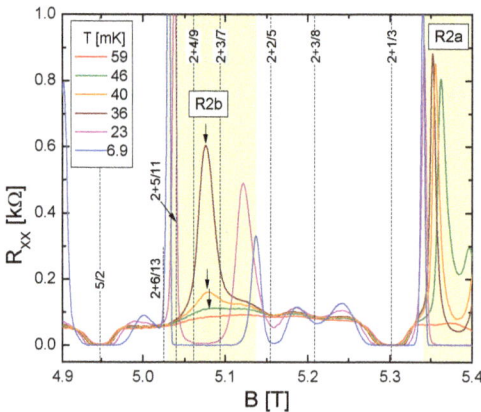

Fig. 5. Details of the temperature dependence of the magnetoresistance near the RIQHS R2b. Vertical arrows mark the precursor of the RIQHS R2b. Adapted from Ref. 97.

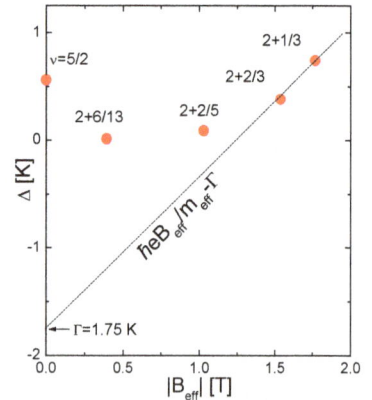

Fig. 6. Energy gaps of FQHSs in the $2 + 1/3 < \nu < 2 + 2/3$ range versus $|B_{\text{eff}}|$. Adapted from Ref. 70.

are not present in Fig. 5. One explanation for the missing FQHSs at $\nu = 2 + 3/7, 2 + 4/9$, and $2 + 5/11$ may be due to a competition with the RIQHS labeled *R2b*. If the energies of the competing RIQHS and FQHSs are comparable, one may expect incipient FQHSs near the onset temperature of the RIQHS. However, magnetotransport did not exhibit such features near the onset of the RIQHS *R2b*.[97] Indeed, gentle curvatures of the magnetoresistance seen in Fig. 5 near $\nu = 2 + 3/7$ and $2 + 4/9$ in traces measured at $T = 59$, 46, and 40 mK could not be assigned to incipient FQHSs; instead these features have been associated with magnetoresistive fingerprints of the precursor of the RIQHS.[97] This suggests that the RIQHS labeled *R2b* is considerably more stable than the FQHSs and therefore it is strongly favored from an energetic point of view. According to a second argument presented in the following, if the FQHSs at $\nu = 2 + 2/5$ and $2 + 6/13$ have topological order different from that of the model of non-interaction composite fermions, then FQHSs at the intermediate filling factors $\nu = 2 + 3/7, 2 + 4/9$, and $2 + 5/11$ do not have to necessarily exist.

The development of FQHSs at filling factors belonging to the Jain sequence $2 + p/(2p + 1)$ does not necessarily mean that these FQHSs can be described by the model of non-interacting composite fermions. Indeed, there are ideas proposed according to which in the presence of strong residual interactions between the composite fermions, the nature of a FQHS may be different from that predicted by the model of non-interacting composite fermions. An examination of the energy gaps of these FQHSs provided early insight in this regard.[70]

In the following we discuss energy gaps of the odd denominator FQHSs of the second Landau level. We will use a widely used phenomenological model.[99,100] The purpose of this model is to bridge the difference between energy gaps obtained from numerical experiments and those from measurements. The former are calculated in simulations that do not include any disorder, while the latter are extracted from measured data in real 2DEGs that necessarily have some disorder present. According to this simple model, the measured energy gap Δ is reduced from the expected theoretical value in the limit of no disorder, also called the intrinsic gap Δ^{int}, by an amount due to disorder broadening Γ:

$$\Delta = \Delta^{int} - \Gamma. \tag{1}$$

The intrinsic gap Δ^{int} contains effects of the finite width of the wavefunction in the direction perpendicular to the plane of the 2DEG and of Landau level mixing. Since disorder effects are not included into Δ^{int}, it may therefore be directly compared to gaps from numerical experiments, such as the ones from exact diagonalization. Equation (1) therefore offers a simple way to deal with the effects of the disorder on the energy gap, a task that remains challenging for the theory. The concept of disorder broadening was used to understand early gap measurements of FQHSs[99,100] and also of the series of FQHSs in the lowest Landau level.[101] In the latter work, the dependence of gaps on the effective magnetic field was found linear. The slope of this linear dependence yields the cyclotron mass of the composite fermions.[101]

Furthermore, it was found that energy gaps extrapolate to a negative offset at zero effective magnetic field.[101] This negative offset was identified with a filling factor independent Γ, with values between 1 and 2 K.

In the second Landau level the number of FQHSs is considerably less than that in the lowest Landau level. Nonetheless, as shown in Fig. 6, the simplest possible analysis involving a reflection of the FQHS at $\nu = 2 + 2/3$ to positive effective magnetic fields shows that a linear dependence of the gaps does no longer hold in the second Landau level.[70] The complex dependence of the energy gaps on the effective magnetic field indicates that residual interactions of the composite fermions are significant in the second Landau level and, as a result, at least some of the FQHSs may be beyond the model of non-interacting composite fermions.[70] If in addition the effective mass of composite fermions in the second Landau level is assumed to be the same as of those forming in the lowest Landau level, one finds that the energy gaps of the $\nu = 2 + 1/3$ and $2 + 2/3$ FQHSs are consistent with the model of free composite fermions when the disorder broadening is $\Gamma = 1.75$ K.[70] The expected energy gaps of FQHSs of free composite fermions under these assumptions are shown by a dashed line in Fig. 6. The proximity of the measured gaps at $\nu = 2 + 1/3$ and $2 + 2/3$ to the dashed line shown in Fig. 6 suggests that FQHSs at $\nu = 2 + 1/3$ and $2 + 2/3$ may be of the Laughlin type. Similar conclusions on the nature of the $\nu = 2 + 1/3$ and $2 + 2/3$ FQHSs were also drawn in studies of the neutral modes at $\nu = 2+1/3$ and $2+2/3$,[103] edge-to-edge tunneling experiments at $\nu = 2+2/3$,[104] and magnetoroton measurements at $\nu = 2 + 1/3$.[105] Nonetheless, overlap calculations of the numerically obtained wave function with the Laughlin state and other theory work suggest that the FQHS at $\nu = 2+1/3$ is not yet satisfactorily understood.[84–92] The analysis presented above is the simplest possible one; an alternative model for the energy gaps at odd denominators may be found in Ref. 102.

Under the assumptions described above, the measured energy gaps of both the $\nu = 2 + 6/13$ and $\nu = 2 + 2/5$ FQHSs are significantly larger than the expected values of a model of non-interacting composite fermions.[70] This discrepancy of the energy gaps, also shown in Fig. 6, constituted an early experimental evidence that the FQHSs at $\nu = 2 + 6/13$ and $2 + 2/5$ are not similar to their lowest Landau level counterparts forming at $\nu = 6/13$ and $2/5$, but they are likely to be of exotic, perhaps non-Abelian nature. The argument according to which the interactions between the composite fermions at $\nu = 2 + 6/13$ are significant enough to change the topological order of the FQHSs at this filling factor is not unreasonable, since similar interactions at the nearby filling factor $\nu = 5/2$ are known to drastically change the ground state from a Fermi sea of composite fermions to a FQHS. In fact the FQHSs at $\nu = 2 + 6/13$ and $\nu = 5/2$ may be closely related. Indeed, according to a recent proposal, FQHSs at $\nu = 2 + 6/13$ and $\nu = 5/2$ as well as other FQHSs of the second Landau level may be accounted for within the same description based on parton wavefunctions.[106–108]

The above analysis of the gaps has implications not only for the FQHS at $\nu =$

$2 + 6/13$, but also for the FQHS at $\nu = 2 + 2/5$. According to this analysis, the $\nu = 2+2/5$ FQHS is also an exotic FQHS.[70] Proposals for the description of the $\nu = 2+2/5$ FQHS include the Read–Rezayi parafermion construction,[7] the particle-hole conjugate of this state,[78] the Bonderson–Slingerland state,[79] the Gaffnian,[77] the Levin–Halperin Abelian construction,[81] multipartite composite fermion states,[82,83] and a parton state.[108] Numerical work favors the Read–Rezayi description,[7,109–112] which however is in close competition with the Bondenson–Slingerland state.[113] It appears that no single theory among the ones predicting exotic behavior at $\nu = 2 + 2/5$ and $2+6/13$ can account for FQHSs at these two filling factors in a natural way and therefore one may surmise that these two FQHSs have fundamentally different origins.

While at $\nu = 2 + 6/13$ there is a visible FQHS, a FQHS at the particle-hole symmetry related filling factor of $\nu = 2+7/13$ is conspicuously missing.[70] A similar lack of particle-hole symmetry is also observed at partial filling $2/5$. Indeed, as already discussed, the ground state at $\nu = 2+2/5$ is a FQHS, whereas at $\nu = 2+3/5$ a FQHS does not form. The lack of particle-hole symmetry at $\nu = 2 + 7/13$ and $\nu = 2 + 3/5$ affects FQHSs differently than the RIQHSs.[97] It is interesting to note that a related particle-hole asymmetry was also observed in a high quality bilayer graphene.[114] In this experiment, a weak FQHS is seen at $\nu = 2 + 7/13$, but not at $\nu = 2 + 6/13$. However, in Ref. 114 a competition with a RIQHS cannot be invoked. This result indicates that a particle-hole symmetry breaking effect, such as Landau level mixing, is likely at play and it favors one set of FQHSs, while suppresses fractional correlations at the particle-hole symmetric filling factor.

Finally, we note that data discussed above are all in the single layer limit, i.e. when the second electrical sub-band is not occupied. Recent work on the $\nu = 2+2/5$ and $2+3/5$ FQHSs in a wide quantum well in a GaAs/AlGaAs system in which the second sub-band is occupied suggests that these two FQHSs belong to the model of non-interacting composite fermions.[67] Once the orbital wavefunctions acquire a lowest Landau level character in samples with the second electrical sub-band occupied, particle-hole symmetry of the FQHSs appears to be restored.[67]

3.2. $\nu = 3 + 1/3$ *fractional quantum Hall state*

The upper spin branch of the second Landau level, i.e. the $3 < \nu < 4$ region, is at a lower magnetic field and therefore hosts fewer FQHSs than the lower spin branch. Early work on this region established the FQHSs at $\nu = 7/2$, $\nu = 3 + 1/5$, and $\nu = 3 + 4/5$ as well as four RIQHSs.[45] The FQHS discovered at $\nu = 3 + 1/3$ is the most recently seen ground state in this region.[71] A FQHS as at this filling factor is identified by a vanishing magnetoresistance and a Hall resistance quantized to $h/(3 + 1/3)e^2$.

Even though in Fig. 4 a distinct minimum in the magnetoresistance is also observed at $\nu = 3 + 2/3$, a FQHS at this filling factor is conspicuously missing.[71] Indeed, an examination of the temperature dependence of the magnetoresistance revealed that the local minimum at $\nu = 3+2/3$ does not follow the usual decreasing trend with a decreasing temperature. Therefore the opening of an energy gap, a defining property of FQHSs, could not be established at $\nu = 3+2/3$. Furthermore, the Hall resistance at $\nu = 3+2/3$ was not quantized, in fact its value was not close to the expected value of $h/(3+2/3)e^2$ for a FQHS. Taken together, the existence of a fractional quantum Hall ground state at $\nu = 3+2/3$ so far could not be established.[71]

In order to gain insight into FQHSs at partial filling 1/3, one may plot the energy gaps of these FQHSs against the Coulomb energy E_c. As shown in Fig. 7, the energy gaps of the FQHSs at $\nu = 2+1/3$, $2+2/3$, and $3+1/3$ plotted this way follow a linear trend. This suggests that these three FQHSs are similar in nature and, in the sample studied, are not close to a possible spin transition.[115]

The relationship between gaps and the Coulomb energy shown in Fig. 7 may be relevant to the absence of FQHSs in the third Landau level. As shown in Fig. 7, the energy gaps at the Coulomb energies calculated at $\nu = 3+2/3$, $4+1/3$, and $4+2/3$ extrapolate to negative values. Within the phenomenological model embodied in Eq. (1), a disorder broadening exceeding the intrinsic gap results in a negative measured gap. Under such circumstances the formation of a FQHS is forbidden. Based on such an argument, in the sample under investigation disorder effects are too strong for the observation of the opening of an energy gap at $\nu = 3+2/3$, $4+1/3$, and $4+2/3$. According to the Hartree–Fock theory,[37,38] exact diagonalization,[116] and density matrix renormalization group calculations,[117] the formation of the RIQHSs in the third Landau level is usually attributed to their favorable cohesion energy as compared to that of FQHSs at partial filling 1/3. However, it is also possible that FQHS of the third Landau level are suppressed solely because of the presence of disorder.

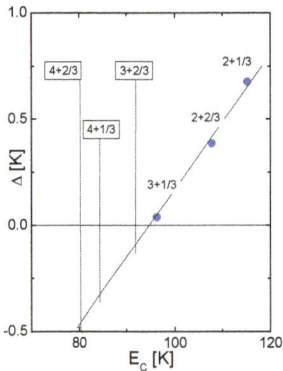

Fig. 7. Energy gaps in the second Landau level at partial filling 1/3 versus the Coulomb energy. Data taken from Refs. 70 and 71.

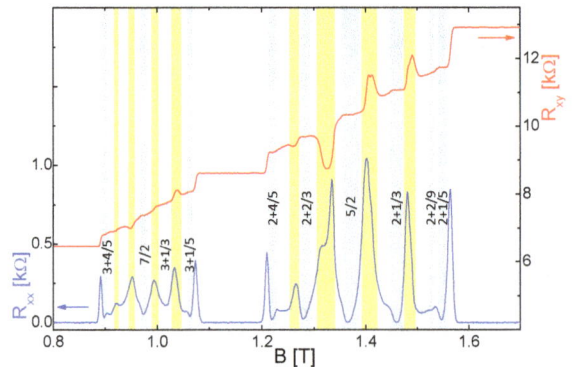

Fig. 8. Magnetotransport in the second Landau level in a sample of low density. Blue shades mark FQHSs at the labeled filling factors, whereas yellow shades show precursors of eight RIQHSs. The lower spin branch data between $B = 1.1$ and 1.7 T is adapted from Ref. 176.

The most surprising feature of magnetotransport in the upper spin branch is the relative robustness of the FQHS at $\nu = 3 + 1/5$ when compared to that of the FQHS at $\nu = 3 + 1/3$.[71] This result is consistent with earlier work in which a FQHS was seen at $\nu = 3 + 1/5$, but not at $\nu = 3 + 1/3$.[45] The energy gaps were found to obey the $\Delta_{3+1/3} < \Delta_{3+1/5}$ relation.[71] This relationship is very unusual since an opposite inequality $\Delta_{2+1/3} > \Delta_{2+1/5}$ holds in the lower spin branch of the second Landau level[44,70,98,118,119] and $\Delta_{1/3} > \Delta_{1/5}$ in the lowest Landau level.[42,120] The latter two relationships are often ascribed to the robustness of flux-two composite fermions when compared to the higher order flux-four objects.[11] To summarize, the expected relationship between the energy gaps of the FQHSs at partial filling factors $1/3$ and $1/5$ is reversed in the upper spin branch of the second Landau level.[71] This anomalous gap reversal indicates an unanticipated difference between the prominent odd denominator FQHSs forming in the second Landau level.[71]

Since the upper spin branch forms at higher values of the Landau level mixing parameter than the lower spin branch, the mixing effect is likely of importance. However, details of the influence of Landau level mixing are not understood. To illustrate this, magnetoresistance is shown in Fig. 8 for a sample of a low electron density $n = 8.3 \times 10^{10}$ cm^{-2}. There are prominent minima developed in the magnetoresistance and a strong Hall quantization at $\nu = 3 + 1/5$ and $3 + 4/5$, but only a weak minimum and a lack of Hall quantization at $\nu = 3 + 1/3$. These results suggest that FQHSs at partial fillings $1/5$ stay strong in the upper spin branch of the second Landau level even at significantly large values of the Landau level mixing parameter.

The above described anomalous energy gaps of the prominent odd denominator FQHSs in the upper spin branch of the second Landau level highlight the lack of understanding of these FQHSs and even elicit the provocative possibility that some of these FQHSs may be of exotic origin.[71]

3.3. $\nu = 4/11$ and $5/13$ fractional quantum Hall states

One of the regions of interest for the observation of novel FQHSs in the lowest Landau level is that between $1/3 < \nu < 2/5$. The observation of a local minimum in the magnetoresistance in this range at $\nu = 4/11$ and of less pronounced features at filling factors such as $\nu = 5/13$, $6/17$, and $3/8$ was an early indicator of possible fractional quantum Hall ground states.[121] Initially the FQHS at $\nu = 4/11$ was interpreted as a fractional quantum Hall effect of composite fermions.[121] However, theoretical work suggested that, owing to the residual interactions between the composite fermions, at $\nu = 4/11$ the ground state is a FQHS with an unusual topological order.[122–124] This idea received strong support from a composite fermion diagonalization study over an extended Hilbert space.[125]

While the theory results favoring a FQHS with a novel topological order at $\nu = 4/11$ are compelling,[123–125] the lack of observation of an energy gap in the excitation spectrum, i.e. of incompressibility, at this and other nearby filling factors raised the question whether fractional quantum Hall ground states can be experimentally realized. A necessary condition for the opening of a gap is a longitudinal magnetoresistance R_{xx} that decreases with temperature.

The most recent experimental work in the $1/3 < \nu < 2/5$ range focused on searching for an activated behavior. There are two reports that confirmed the development of Hall quantization and the opening of an energy gap at $\nu = 4/11$, of magnitudes 7 mK[127] and 15 mK,[128] respectively. Magnetotransport data in the $1/3 < \nu < 2/5$ range from Ref. 128 are shown in Fig. 9. Temperature dependence of magnetotransport at $\nu = 5/13$ was also found to be consistent with the development of an incipient gap.[128] These measurements established fractional quantum Hall ground states at $\nu = 4/11$ and $5/13$ and opened up the possibility for novel topological order in the $1/3 < \nu < 2/5$ region.

Despite considerable progress at $\nu = 4/11$ and $5/13$, transport data at other filling factors of interest, such as $\nu = 3/8$ and $6/17$, did not conclusively establish the opening of an energy gap.[127,128] Magnetoresistance at $\nu = 3/8$ in both experiments was found to increase with a lowering temperature, precluding therefore activation. Furthermore, the filling factor $\nu = 6/17$ is extremely close to the quantized plateau associated with the $\nu = 1/3$ FQHS. A recent numerical study of the FQHS at $\nu = 6/17$ found it is described by the model of non-interacting composite fermions.[126] For now, the behavior of the magnetoresistance at $\nu = 3/8$ and $6/17$ cannot be conclusively associated with the opening of an energy gap despite the existence of a depression in R_{xx} at these filling factors.

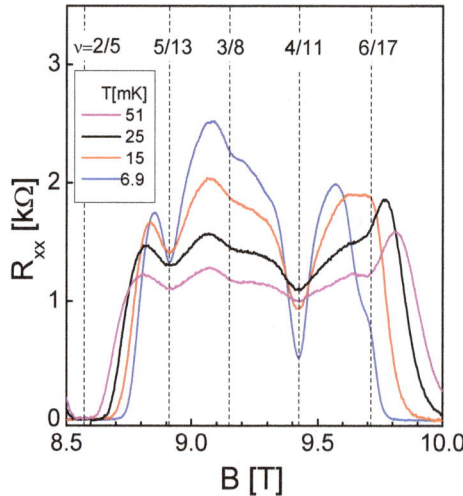

Fig. 9.   Magnetotransport in the range of filling factors $1/3 < \nu < 2/5$ of the lowest Landau level. Developing FQHSs are seen at $\nu = 4/11$ and $5/13$. Adapted from Ref. 128.

4. Even Denominator Fractional Quantum Hall States

The FQHS at $\nu = 5/2$ is the most notable FQHS beyond the model of non-interacting composite fermions. This FQHS was discovered[42] and its full quantization was reported in Refs. 43 and 44. Related FQHSs develop at other filling factors of even denominator in the GaAs/AlGaAs system at $\nu = 7/2$[45] and $\nu = 2 + 3/8$.[46] Even denominator FQHSs have also been seen in ZnO/MgZnO[129] and in bilayer[114,130–132] and, most recently, in monolayer[133,134] graphene. However, it is not yet established whether or not these latter FQHSs and the $\nu = 5/2$ FQHS in GaAs/AlGaAs share the same origin.

The instability of the Fermi sea of composite fermions and the existence of an energy gap at this filling factor is naturally explained by a Cooper-like pairing of the composite fermions.[6,135–140] Because of constraints on the spin degree of freedom, such a pairing must necessarily be p-wave pairing.[136] The Pfaffian is the earliest description of the FQHS at $\nu = 5/2$ which is consistent with such a pairing.[6] The Pfaffian is of considerable interest since at least some of its quasiparticles are predicted to obey exotic non-Abelian braiding statistics. However, at this filling factor there are also other topologically distinct candidates, such as the anti-Pfaffian,[141,142] the (3,3,1) Abelian state,[143] a variational wave function based on an anti-symmetrized bilayer state,[144] the particle-hole symmetric Pfaffian,[145,146] a stripe-like alternation of the Pfaffian and anti-Pfaffian,[147] and other exotic states.[148,149]

An intense effort is focused on unraveling the properties of the even denominator FQHSs. Some aspects of this effort were reviewed in Refs. 150 and 151. Edge-to-edge tunneling,[104,152–154] quasiparticle interferometry,[155] upstream neutral modes,[103] edge heat conduction measurements[156] probed the structure of the edge states at $\nu = 5/2$. However, results from these measurements do not yet offer a consensus on the origin of the $\nu = 5/2$ FQHS. Since the physics of the edge may be considerably more complicated than that of the bulk, bulk probes such as transport and heat capacity measurements[157] remain important in the study of the FQHS at $\nu = 5/2$. In this section we discuss results on the $\nu = 5/2$ FQHS that were learnt from recent transport data. Several experiments suggest that away from the optimal density phase transitions are allowed in the FQHS at $\nu = 5/2$. We will also discuss the energy gap and the disorder broadening of the FQHS at this filling factor in the highest quality samples. The last subsection contains results of a systematic study of the behavior of the $\nu = 5/2$ FQHS in the presence of short-range alloy disorder.

4.1. $\nu = 5/2$ *fractional quantum Hall state at low electron densities*

The regime of low densities for the $\nu = 5/2$ FQHS emerged as an area of interest because of the possibility of phase transitions occurring here. In the following we discuss a spin transition[158] and a topological phase transition at $\nu = 5/2$.[159] In addition, a transition from the $\nu = 5/2$ FQHS towards the quantum Hall nematic will be discussed in Sec. 6.[160–162]

A possible phase transition in the $\nu = 5/2$ FQHS is a spin transition, from a fully spin polarized to a partially polarized state. According to experimental work performed at large magnetic fields and hence large densities, the $\nu = 5/2$ FQHS is fully spin polarized.[163,164] However, these experiments do not rule out a partially spin polarized FQHS at low electron densities. Recent experimental work on a series of samples, including samples of density as low as $n = 4.1 \times 10^{10}$ cm^{-2}, found a region of densities at which the energy gap nearly closes.[158] These results have been interpreted as being consistent with a spin transition in the $\nu = 5/2$ FQHS.[158] While these results[158] are suggestive of a phase transition, a spin transition is only one of the possibilities. Indeed, while changing the density other parameters of the system may also change. Examples of such parameters are the Landau level mixing parameter, defined as the ratio of Coulomb and cyclotron energies, and the width of the quantum well. Therefore orbitally driven phase transitions may also occur when the density is changed.

As discussed earlier, theory finds that for $\nu = 5/2$ FQHS several topologically different candidate states are allowed.[6,141–149] If more than one of these states can be stabilized, intriguing topological phase transitions may occur at $\nu = 5/2$ between pairs of such distinct FQHSs. Such phase transitions may be driven by a parameter of the 2DEG, such as the Landau level mixing parameter κ. As an example, it was argued that a direct topological phase transition between the Pfaffian and the anti-Pfaffian may occur.[141,142] According to numerical work, the Pfaffian and anti-Pfaffian ground states may compete as the Landau level mixing is tuned.[87,165–173] However, due to difficulties stemming from the non-perturbative nature of the calculations and due to limited computational resources, details of a possible transition between the Pfaffian and the anti-Pfaffian at large Landau level mixing could not be firmly established.[172] Nonetheless, the regime of low densities or large Landau level mixing has emerged as a region of interest for a possible topological phase transition in the $\nu = 5/2$ FQHS.

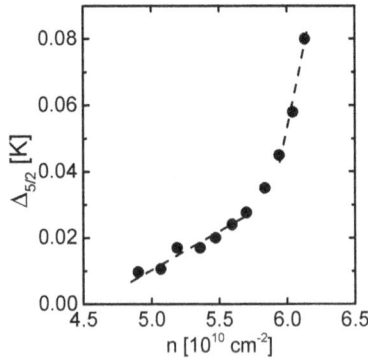

Fig. 10.   Energy gap of the $\nu = 5/2$ FQHS versus density in a high quality low density sample. The anomalous trend of the energy gap at $\nu = 5/2$ versus density may be interpreted as being due to a topological phase transition. Adapted from Ref. 159.

In a recent experiment on a density-tuned 2DEG,[159] an anomalously sharp change in the density dependence of the energy gap of the $\nu = 5/2$ FQHS was reported in the vicinity of $\kappa = 2.6$ or $n = 5.8 \times 10^{10}$ cm^{-2}. This behavior is shown in Fig. 10. The origins of the observed anomalous dependence of the energy gap are not known; one possibility is a topological phase transition in the $\nu = 5/2$ FQHS.[159] We note that the energy gap at the apparent transition point in this experiment did not close. While the closure of the gap is generally believed to be a necessary condition for a topological phase transition, one may also envision topological transitions in which the gap does not fully close, for example in a first order phase transition. A topological phase transition in the $\nu = 5/2$ FQHS may also be induced in confined geometries by changing the confinement potential.[154,174]

Another interesting possibility for a phase transition at $\nu = 5/2$ from a FQHS towards the quantum Hall nematic was recently revealed by measurements at high hydrostatic pressures.[160,161] Further details of this phase transition can be found in Sec. 6.

4.2. *Energy gap at $\nu = 5/2$ in pristine samples*

Unraveling the effects of the disorder is an important endeavor in contemporary condensed matter physics. Disorder is well understood in the single particle regime for example in connection with Anderson localization and the universal plateau-to-plateau transition in the integer quantum Hall effect.[49,175] In contrast, understanding disorder in correlated electron systems, such as the 2DEG in the fractional quantum Hall regime, continues to pose serious challenges.

In the following, we will refer to samples of the highest possible mobility and, therefore, the least amount of disorder as pristine samples. The highest mobility under given growth conditions is a function of the electron density.[13]

Efforts in understanding the energy gap of the $\nu = 5/2$ FQHS examined its relationship to electron mobility μ and quantum lifetime τ_q. In the most general case, the energy gap at $\nu = 5/2$ in pristine samples did not correlate with the mobility.[29,93,118,119,176–179] Similarly, the energy gap at $\nu = 5/2$ did not scale with the quantum lifetime.[73,176] In fact, in gated samples it was found that the quantum lifetime is approximately constant over the density range at which the energy gap at $\nu = 5/2$ decreased from its largest value to zero.[73,119] These results are perhaps not surprising since, in contrast to the energy gap at $\nu = 5/2$, both the mobility and the quantum lifetime are parameters measured near zero magnetic field, in a regime that may be understood within a single electron description.

The phenomenological model described earlier in Sec. 3.1 offers a framework for analyzing the gap of the $\nu = 5/2$ FQHS. The use of this model for gaps measured in the second Landau level, however, poses challenges as these gaps were suspected to be smaller than the disorder broadening. In contrast, Γ at $\nu = 1/3$ is negligible as compared to Δ^{int} and the intrinsic gap may be estimated from the measured value of the gap, $\Delta^{\mathrm{int}} \simeq \Delta$ in samples of high density.[99,100] Therefore in order to estimate

Δ^{int} at $\nu = 5/2$, one needs to measure independently two quantities: both Δ and Γ. Hence a quantitative knowledge of the disorder broadening plays a significant role in understanding the energy gaps of the exotic FQHSs. However, Γ is not directly accessible from gap measurements at $\nu = 5/2$.

To resolve this impasse, Morf and d'Ambrumenil[180] proposed an analysis based on the measurement of two independent quantities: the energy gaps of FQHSs at both $\nu = 5/2$ and $\nu = 7/2$ in a sample of given electron density. In addition, it was assumed that these two FQHSs have the same dimensionless intrinsic gap $\delta^{int} = \Delta^{int}/E_c$, where E_c is the Coulomb energy. The dimensionless intrinsic gap δ^{int} and disorder broadening Γ may be obtained by plotting the measured gaps at $\nu = 5/2$ and $7/2$ against the Coulomb energy and by extracting the slope and the intercept with the vertical scale of the line passing through the two points. Following this recipe,[180] an analysis of the gaps in a sample of electron density $n = 3.0 \times 10^{11}$ cm^{-2} from Ref. 45 yielded values $\delta^{int} = 0.014$ and $\Gamma = 1.24$ K. A similar fit, shown in Fig. 11(a), on the sample from Ref. 70 yielded $\delta^{int} = 0.019$ and $\Gamma = 1.5$ K.[176]

The analysis of the gaps proposed by Morf and d'Ambrumenil[180] was later extended for samples of low densities,[176] down to $n = 8.3 \times 10^{10}$ cm^{-2}. The intrinsic gaps as plotted against the density, shown in Fig. 11(b), were in reasonable agreement with a numerical simulation that accounted for Landau level mixing within the random phase approximation.[119] However, such an agreement can only be considered crude at best since the assumption of equal δ^{int} for $\nu = 5/2$ and $\nu = 7/2$

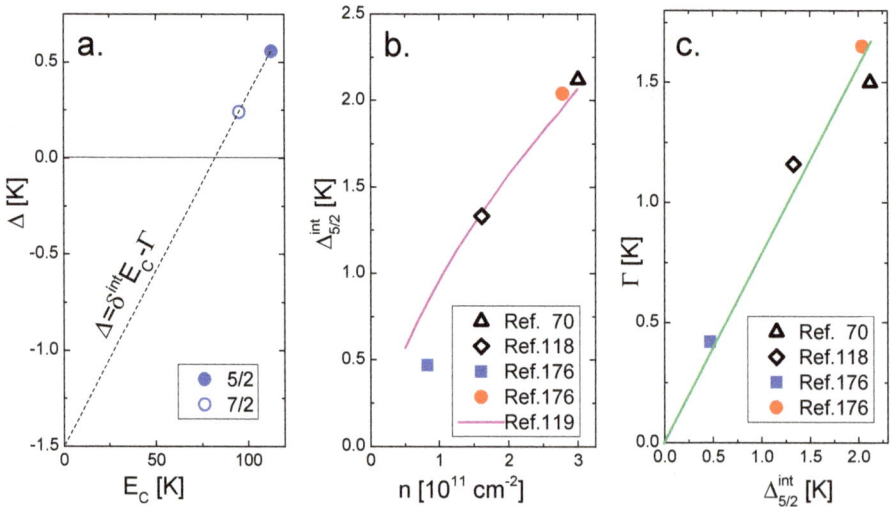

Fig. 11.   (a) A fit to the measured gaps at even denominators used to extract the dimensionless intrinsic gap δ^{int} and the disorder broadening Γ. Data is from Ref. 176. (b) A comparison of intrinsic gaps at $\nu = 5/2$ obtained from measurements (data points) and numerics (line). Data taken from Ref. 176. (c) The dependence of Γ on $\Delta^{int}_{5/2}$.

FQHSs is only approximate in measurements due to the slightly different Landau level mixing parameters at these filling factors.

It is instructive to plot the calculated δ^{int} against the Landau level mixing parameter κ. The decreasing trend obtained is shown in Fig. 12. It was found that a linear fit to the data extrapolates to $\delta^{int} = 0.032$ in the limit of $\kappa = 0$. This extrapolated value is in good agreement with the numerically obtained values at zero Landau level mixing: $0.02 - 0.04$ from an exact diagonalization on finite size systems,[181] 0.030 from a DMRG calculation[182] and 0.031 from an exact diagonalization study adjusted for the finite width of the quantum well.[119] The latter two figures are from extrapolations to the thermodynamic limit. Since simulations with no Landau level mixing are significantly easier than the ones that include it,[172] the agreement of δ^{int} from these simulations and δ^{int} extrapolated to $\kappa = 0$ is an important confidence test of the analysis above.

Another extrapolation of interest is to large κ. The functional form for this extrapolation is not known; in Ref. 176 a linear extrapolation was used. The extrapolated intrinsic gap at $\nu = 5/2$ was found to close near $\kappa \simeq 3$.[176] This result is consistent with the totality of observations of electron gases at low densities in the GaAs/AlGaAs system.[158,176] This result is also consistent with observations in p-type carriers, i.e. two-dimensional hole gases in GaAs/AlGaAs in which values of κ are significantly larger than 3 and in which an even denominator FQHS was not observed. As an example, in Fig. 13 we show the $2 < \nu < 3$ range for a hole gas with $\kappa = 14$, in which an odd denominator FQHS develops, but an even denominator FQHS is not present.[183] We note that it is widely appreciated that, besides Landau level mixing, energy gaps also depend on the width of the quantum well in its dimensionless form w/l_B. For the samples on which Fig. 12 is based, these dimensionless widths were fortuitously nearly constant.[176]

Progress in MBE technology has resulted in a dramatic improvement of the

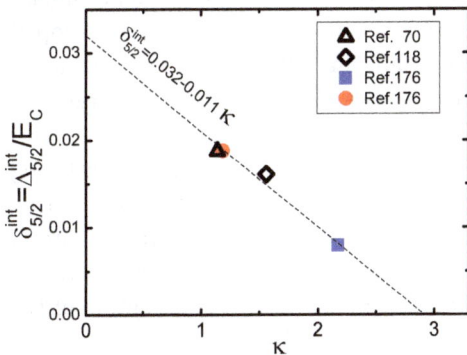

Fig. 12. Dimensionless intrinsic gaps at $\nu = 5/2$ as function of the Landau level mixing parameter κ. Dotted line is a fit through the data. Adapted from Ref. 176.

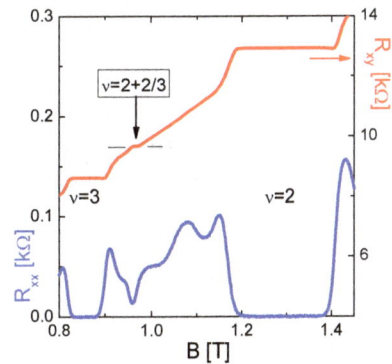

Fig. 13. Magnetotransport in the $2 < \nu < 3$ range of a p-type sample. A FQHS is seen at $\nu = 2 + 2/3$. Adapted from Ref. 183.

quality of 2DEG confined not only to GaAs/AlGaAs,[13] but also to ZnO/MgZnO structures.[19] In such 2DEGs of increased mobility there is a recent report[129] of even denominator FQHSs. Measured parameters of the 2DEG in this experiment show that these even denominator FQHSs occur near $\kappa \simeq 15$.[129] According to the extrapolation to large κ shown in Fig. 12, at such large values of κ a FQHS at $\nu = 5/2$ cannot be stabilized in GaAs/AlGaAs samples. This result thus raises the interesting possibility that the even denominator FQHSs in GaAs/AlGaAs and in ZnO/MnZnO may have a different origin.

Finally, it is interesting to note that, as seen in Fig. 11(c), the disorder broadening Γ in these samples,[176] extracted by the technique of Morf and d'Ambrumenil, is found to be linear with the intrinsic gap Δ^{int}. The slope of a linear fit to data in Fig. 11(c) passing through the origin is 0.78.

4.3. Energy gap at $\nu = 5/2$ in samples with alloy disorder

Progress in MBE growth of GaAs/AlGaAs structures has produced sufficiently high quality 2DEGs to observe quantum Hall phenomena even when there was disorder intentionally added to the 2DEG. Adding disorder during the MBE growth process allowed for a high degree of its control. So far only alloy disorder was systematically studied.[184,185] The particular type of alloy disorder investigated consisted of minute amounts of neutral Al atoms added to the GaAs region supporting the 2DEG. Experiments on such alloy samples had a strong impact on the understanding of the plateau-to-plateau transition in the integer quantum Hall regime,[49,175] mapping the electronic wavefunction along a direction perpendicular to the plane of the 2DEG,[186] and contributed to the understanding of pinning of Wigner crystals.[187]

Another successful experiment was performed on alloy samples that enabled the study of the FQHS at $\nu = 5/2$ in the presence of alloy disorder.[185,188] These samples were based on modern sample structures assuring the high density, near $n \simeq 3 \times 10^{11}$ cm^{-2}, and high quality necessary for robust second Landau level fractional quantum Hall states. Specifically, samples were modulation-doped $Al_{0.24}Ga_{0.76}As/Al_xGa_{1-x}As/Al_{0.24}Ga_{0.76}As$ quantum wells, where the molar Al fraction x in the quantum well was significantly less than that in the barriers.[185] The dependence of the scattering rate on the amount of Al introduced to the quantum well can be seen in Fig. 14. The linear dependence and slope of the scattering rate on the amount of Al characterizes the alloy potential and was found to be consistent with earlier results.[184] Moreover, because of the small scattering rate in the pristine sample with no added impurities to the quantum well, alloy scattering exceeded the residual scattering rate in all these samples, with the exception of the one with the lowest non-zero x.[185]

As shown in Fig. 15, an increasing amount of alloy disorder decreases the mobility and it also suppresses the energy gap of the $\nu = 5/2$ FQHS.[188] A similar behavior was found in numerical work on the $\nu = 1/3$ FQHS, in which the strength of a Gaussian correlated white noise was increased.[189–193] A peculiarly interesting

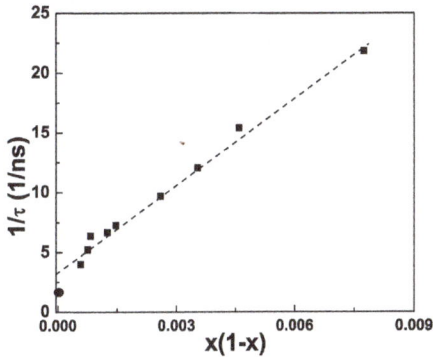

Fig. 14. The dependence of scattering rate $1/\tau$ on alloy disorder in a series of 30 nm quantum well based alloy samples. Here x is the Al molar concentration. Adapted from Ref. 185.

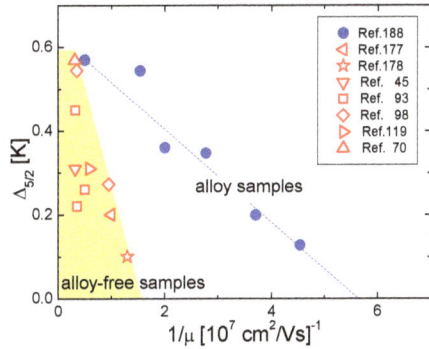

Fig. 15. The dependence of the energy gap of the $\nu = 5/2$ FQHS on inverse mobility in a set of pristine and a series of alloy samples near the density 3×10^{11} cm^{-2}. Adapted from Ref. 188.

feature of the data is revealed when the alloy samples are compared to high quality pristine, i.e. alloy-free samples. Because the mobility is not independent of the sample density, in Fig. 15 only pristine samples from the literature are shown that have their density in the $2.65 \times 10^{11} \le n \le 3.2 \times 10^{11}$ cm^{-2} range, i.e. close to that of the alloy samples. Energy gaps for the pristine samples are clustered in the area shaded in yellow in Fig. 15, but show no correlation with $1/\mu$. In contrast, the energy gap in the series of alloy samples shows a linear functional dependence on the inverse mobility $1/\mu$. This suggests that when in a series of similar samples one particular type of disorder dominates, such as in the series of alloy samples, the energy gap and and the mobility appear to be correlated.[188] However, in samples which have different types of dominating disorder, a correlation between the energy gap and and the mobility is not expected.[188]

An interesting feature of the data shown in Fig. 15 is that a strong $\nu = 5/2$ FQHS with $\Delta_{5/2} = 127$ mK develops in the alloy sample with $\mu = 2.2 \times 10^6$ cm^2/Vs. This is surprising, since at such a low mobility a $\nu = 5/2$ FQHS has never been observed in pristine, alloy-free samples. Indeed, for pristine, alloy-free samples with density near $n = 3 \times 10^{11}$ cm^{-2}, the threshold mobility value for the observation of a $\nu = 5/2$ FQHS is $\mu_c \simeq 7 \times 10^6$ cm^2/Vs. The mobility threshold for a fully quantized $\nu = 5/2$ FQHS is therefore significantly lowered in the presence of alloy disorder and, therefore, the $\nu = 5/2$ FQHS is robust to the presence of alloy disorder.[188] Alloy disorder does not appear to be as detrimental to the development of the $\nu = 5/2$ FQHS as the residual disorder is unintentionally added during sample growth.[188] The gap $\Delta_{5/2}$ in the series of alloy samples studied closes at an extrapolated threshold of $\mu_c^{\text{alloy}} \simeq 1.8 \times 10^6$ cm^2/Vs.

Al is a neutral impurity and it perturbs the GaAs crystal on a subnanometer length scale. Alloy disorder is thus a type of disorder that generates a short-range scattering potential. Other types of disorder, due to either short- or long-range scattering potentials, have not yet been systematically studied.

5. Studies of Re-Entrant Integer Quantum Hall States

RIQHSs are collectively localized ground states associated with electronic bubble phases.[37-41] They were discovered in the third and higher Landau levels[40,41] and later also seen in the second Landau level.[45] There are also two reports of RIQHSs in the lowest Landau level, however the nature of these RIQHSs may be different from those forming in higher Landau levels.[194,195] In this section we discuss recent results on RIQHSs in the second Landau level. Topics include the magnetoresistive fingerprints of the RIQHSs in high mobility samples, a discussion of the precursors of the RIQHSs, and a summary of new RIQHSs.

Transport signatures of RIQHSs are $R_{xx} = 0$ and $R_{xy} = h/ie^2$, where $i =$ integer.[40,41] However, in contrast to integer quantum Hall states, RIQHSs are centered at a filling factor other than an integer. A characteristic of the RIQHSs is that they are separated from the nearby integer quantum Hall plateaus by distinct resistive signatures.[40,41] Regions shaded in yellow in Figs. 4 and 16 mark several RIQHSs forming in the second Landau level. The above transport signatures of RIQHSs indicate a pinned insulator and the re-entrant property was argued to distinguish RIQHSs from an Anderson insulator.[45,196]

Following the discovery of RIQHSs, efforts on these states were focused on establishing their fundamental signatures in magnetotransport, on their filling factors of formation,[45,95,97,197] on gaining an understanding of their energy scale,[95,197] on their microwave pinning modes,[198-200] on thermopower,[201] on their electrical breakdown,[196,202-206] and most recently, on their attenuation of surface acoustic waves.[207,208]

RIQHSs develop at temperatures higher than 100 mK in the third Landau level and at tens of mK in the second Landau level. Therefore the use of a He-3 immersion cell to generate ultra-low temperatures may seem unnecessary in their study. Nonetheless, immersion cells have contributed to the RIQHSs in an unexpected way: the large heat capacity of the He-3 liquid provided enhanced temperature stability. Since the magnetoresistance of RIQHS near their onset changes extremely rapidly with temperature, the high temperature stability afforded by the immersion cell setup allowed for a careful mapping of these temperature sensitive features.[95,97,197]

5.1. *Resistive fingerprints of re-entrant integer quantum Hall states*

The vanishing longitudinal magnetoresistance and Hall quantization to an integer value were transport features of RIQHSs clearly identified at their discovery.[40,41,45] It later became apparent that samples with improved quality exhibit additional transport features associated with RIQHSs. One such feature is the flanking of the longitudinal magnetoresistance by sharp peaks at both ends of the magnetic field.

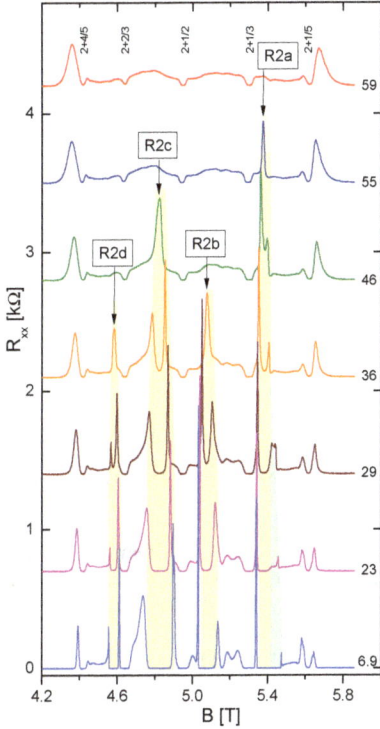

Fig. 16. Waterfall plot of magnetoresistance in the lower spin branch of the second Landau level, as measured at different temperatures. Traces are labeled by temperatures in mK. Adapted from Ref. 97.

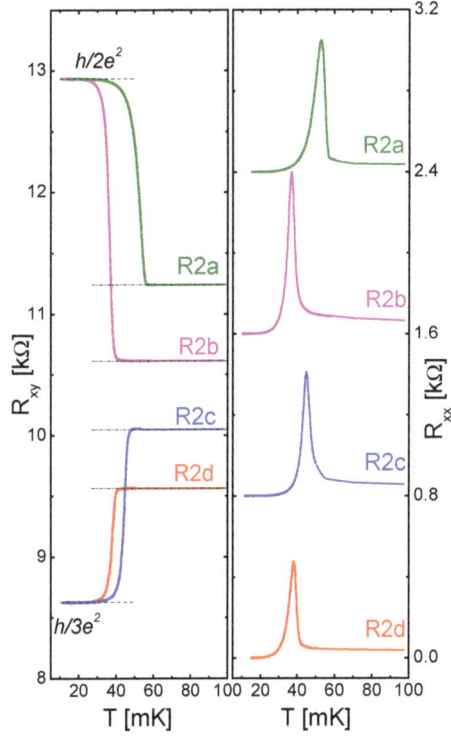

Fig. 17. Temperature dependence of R_{xx} and R_{xy} measured at the central filling factor of the RIQHSs labeled $R2a$, $R2b$, $R2c$, and $R2d$. Sharp peaks in R_{xx} mark the onset temperatures of the RIQHSs. Adapted from Ref. 95.

In other words, the region of vanishing magnetoresistance of a RIQHS is delimited by two sharp peaks. As seen in Fig. 4, these two sharp resistance peaks in the flanks of RIQHSs are present in both spin branches of the second Landau level.[95] Similar sharp peaks are also delimiting the RIQHSs in the third Landau level (not shown).[197]

As the temperature is raised, the two sharp peaks in R_{xx} in the flanks of RIQHSs persist.[95] However, the two sharp peaks in R_{xx} move closer to each other and, as a result, the width of the vanishing R_{xx} plateau shrinks. Such a trend may be seen in Fig. 16 for several RIQHSs and may be examined in detail in the middle row of panels of Fig. 18 for one of the RIQHSs labeled $R2b$. At the temperature of $T = 32.6$ mK the magnetoresistance shown in Fig. 18(b) no longer vanishes, but it consists of two peaks with a non-zero local minimum in between them. The location in filling factor of this minimum is T-independent to a good degree and defines the central filling factor ν_c of the RIQHS. At $T = 35.7$ mK the two spikes of $R_{xx}(\nu)$

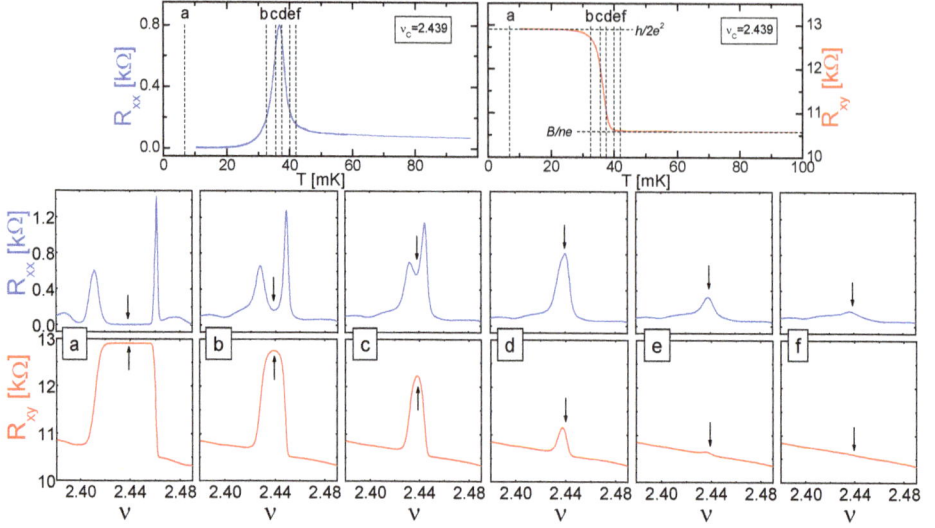

Fig. 18. Correlation of cuts along different axes in the $R_{xx}(\nu, T)$ and $R_{xy}(\nu, T)$ manifolds in the region of the RIQHS labeled as R2b. The two top panels show the temperature dependence at $\nu = \nu_c$, whereas the two lower rows of panels show ν dependence at fixed temperatures. Data in panels a, b, c, d, e, and f are collected at the temperatures $T = 6.9, 32.6, 35.7, 37.7, 40.0$, and 41.7 mK. Arrows mark the central filling factor $\nu = \nu_c$. Precursor of the RIQHS are seen in panels d, e, and f. Data taken from Ref. 95.

move closer to each other; between them there is still a local minimum, albeit with a large resistance.

A second transport feature associated with the RIQHSs becomes apparent by measuring the temperature dependence $R_{xx}(T)|_{\nu}$ at a fixed ν or B-field.[95] One may visualize such a $R_{xx}(T)|_{\nu}$ curve as a cut in the $R_{xx}(\nu, T)$ manifold at a given filling factor of choice ν. Such $R_{xx}(T)|_{\nu}$ and $R_{xy}(T)|_{\nu}$ curves measured at $\nu = \nu_c$ are shown in Fig. 17 for the RIQHSs of the lower spin branch of the second Landau level. It was found that, as the temperature is lowered, the Hall resistance $R_{xy}(T)|_{\nu=\nu_c}$ changes from its classical value to its quantized value, either $h/2e^2$ or $h/3e^2$. At the same time the curve $R_{xx}(T)|_{\nu=\nu_c}$ exhibits a sharp peak. The inflection point in $R_{xy}(T)|_{\nu=\nu_c}$ and the sharp peak in $R_{xx}(T)|_{\nu=\nu_c}$ coincide and may be interpreted as the onset temperature of the RIQHS. An analysis of the onset temperatures of the RIQHSs in the second Landau level obtained this way found that they scale with the Coulomb energy, demonstrating the collective nature of these RIQHSs.[95] Figure 18 may be used to correlate data collected at constant $\nu = \nu_c$ with data at constant temperature for both R_{xx} and R_{xy}.

5.2. The precursor of re-entrant integer quantum Hall states

When comparing the $T = 35.7$ and 37.7 mK traces shown in Figs. 18(c) and 18(d), respectively, one notices that a modest increase in T of only 2 mK results in a

qualitative change in the magnetoresistance of a RIQHS from a double peak structure to a single peak in $R_{xx}(\nu)$. As the temperature is further raised, this single
peak rapidly decreases until it merges into a low resistance background. Since the
single peaks in R_{xx} marked by arrows in Figs. 18(d), 18(e), and 18(f) may also be
associated with RIQHSs, even though the magnetoresistance does not vanish and
the Hall resistance is far from quantization. These single peaks are thus signatures
of RIQHSs at the highest temperatures and they may be associated with precursors of the RIQHSs. Simultaneously with the described changes in R_{xx}, R_{xy} evolves
from the quantized value $h/2e^2$ at the lowest temperatures toward its classical value
B/ne.

Precursors of RIQHSs, as defined above, can be seen for each RIQHS both in the
second Landau level[95] (see Fig. 16) and also the third Landau level.[197] The concept
of precursor of a RIQHS also provides a natural explanation for the observation
of single peaks in R_{xx} in several experiments in the regions of the RIQHSs.[98,118]
Furthermore, eight precursors of RIQHSs also develop in the second Landau level
of a sample[176] of very low density $n = 8.3 \times 10^{10}$ cm^{-2}; these precursors are marked
by yellow shading in Fig. 8.

Recently RIQHSs have also been observed in high quality graphene.[209] These
results highlight the universal, host-independent physics at play in the formation
of the RIQHSs. It is interesting to note that in Ref. 209 only the RIQHS labeled
$R6a$ is fully developed, i.e. for which $R_{xx} = 0$ and Hall resistance quantized to an
integer. At other symmetry related filling factors in the regions labeled $R6b$, $R7a$,
and $R7b$, a peak in R_{xx} was observed which, following the terminology discussed,
is the signature of precursors of RIQHSs.

5.3. *Proliferation of re-entrant integer quantum Hall states in the second Landau level*

At the time of the discovery of the RIQHSs in the second Landau level, eight such
RIQHSs were observed.[45] These RIQHSs are some of the most fragile, as their
onset temperatures do not exceed 55 mK.[95,197] Besides these eight RIQHSs, in the
second Landau level there are three other magnetoresistance features that can be
associated with additional RIQHSs.

One such additional RIQHSs was discovered as a split-off phase of the RIQHS
in the second Landau level developing at the largest magnetic field.[46] This split-
off phase, labeled $R2\tilde{a}$ in Refs. 95 and 97, is fully developed, since at the lowest
temperatures accessed $R_{xx} = 0$ and $R_{xy} = h/2e^2$. This RIQHS is also shown in
Fig. 19(a).

Another developing RIQHS at $2 + 1/5 < \nu < 2 + 2/9$ was identified in Ref. 197
and it is labeled as $R2r$ in Fig. 19(a). Using the terminology introduced earlier, the
magnetoresistive behavior of $R2r$ is consistent with that associated with a precursor
of a RIQHS. Indeed, the peak in R_{xx} increases and the value of R_{xy} moves closer
to $h/2e^2$ as the temperature is lowered. The nature of $R2r$ is uncertain, but it

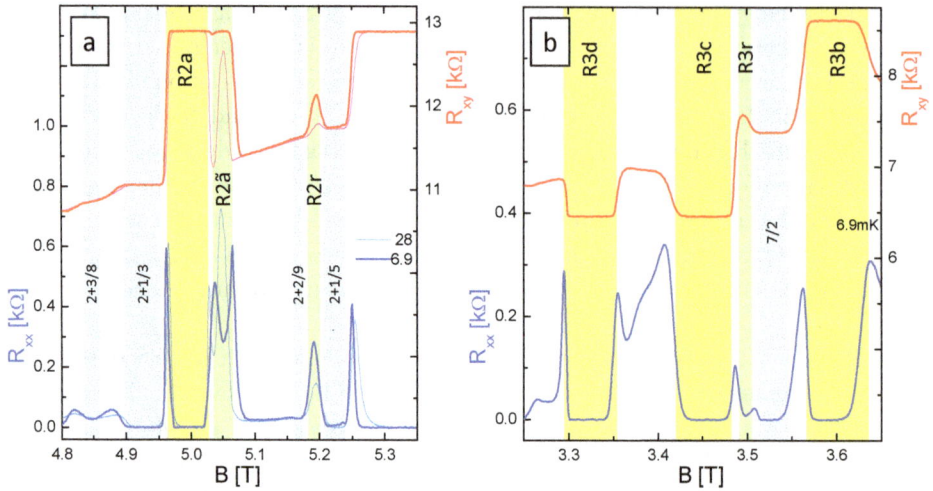

Fig. 19.   Magnetotransport in the second Landau level exhibiting an increased number of RIQHSs. RIQHSs discovered in Ref. 45 are shaded in yellow, any additional RIQHSs are shaded in green, and FQHSs are shaded in blue. The $T = 6.9$ mK traces in panel a are from Ref. 197. Traces in panel b are from Ref. 71.

may be related to the re-entrant insulating phase of the lowest Landau level that forms in the range between $1/5 < \nu < 2/9$ range and which was associated with the re-entrant Wigner crystal.[210] $R2r$ could also be related to the Wigner crystal forming in the flanks of integer plateaus.

In addition, an interesting incipient RIQHS was also detected in the upper spin branch of the second Landau level, near $\nu = 7/2$.[71] This incipient RIQHS is labeled with $R3r$ in Fig. 19(b). The magnetoresistance in this region has a distinct local minimum. An intriguing feature of this ground state is that instead of a Hall resistance moving toward $h/4e^2$, the Hall resistance appears to move towards $h/3e^2$. This latter property suggests that $R3r$ is different from other RIQHSs nearby, such as the one labeled $R3c$ in Fig. 19(b).

The development of an increasing number of RIQHs in the second Landau level indicates that the competition between bubble phases and other ground states in this region is more intricate than previously thought.

6. High Pressure Studies of the Second Landau Level

Hydrostatic pressure is widely used in probing condensed matter systems, such as superconductors and correlated electron systems. Pressure decreases the lattice constant of a crystal. This has profound effects on crystal properties, since the Bloch wavefunctions supported by the crystal are changed. As a consequence, both the electronic and the phononic degrees of freedom are affected. Of particular importance is the impact of high pressure on the band parameters of an electronic system.

Examples of quantities tuned by pressure are the dielectric constant, effective mass, Landé g-factor, band energies, and donor energy levels.

The effect of hydrostatic pressure on 2DEGs in GaAs/AlGaAs structures is well-documented. Information of the effective mass, band energies, and pressure dependence of the electron density is found in the early literature.[211] In addition, high pressure experiments manipulating the g-factor yielded significant knowledge on the spin polarization of FQHSs forming in the lowest Landau level.[212]

More recent use of hydrostatic pressures has contributed to the physics of half-filled Landau level. This section contains a brief account of these experiments. It is well known that, depending on the number of filled Landau levels, half-filled single layer 2DEGs hosted in GaAs/AlGaAs have three distinct ground states. In the lowest Landau level, at $\nu = 1/2$ and $3/2$, there is a featureless Fermi sea of composite fermions.[101] In other Landau levels ordered ground states develop. Indeed, in the second Landau level, at $\nu = 5/2$ and $7/2$, fractional quantum Hall states were found.[42,45] Finally, in high Landau levels, at $\nu = 9/2, 11/2, 13/2, \ldots$, the ground state is the quantum Hall nematic.[40,41] The quantum Hall nematic[39] is closely related to the stripe phase predicted by the Hartree–Fock theory.[37,38]

A peculiar feature of the ordered ground states forming at half filling is that, until recently, in clean enough samples and at low enough temperatures only one type of order seemed to develop. Thus a phase transition between the two different ordered phases, the FQHS and the quantum Hall nematic, could not be realized in purely perpendicular magnetic fields. This was surprising for two reasons. First, early theoretical work indicated that a phase transition between a FQHS and the stripe phase is allowed.[213] Second, experimental work has clearly indicated that the $\nu = 5/2$ FQHS is close to a nematic phase. Indeed, experiments in tilted magnetic fields have shown that the isotropic FQHS at $\nu = 5/2$ is superseded by an anisotropic nematic phase at relatively modest tilt angles.[214,215] Nonetheless, during the three-decade long history of experimental studies at $\nu = 5/2$, a nematic phase at this filling factor has never been seen in magnetic fields applied perpendicularly to the 2DEG. As a result, in a large body of numerical work focused at $\nu = 5/2$, the ground state obtained was typically compared to the Pfaffian; the quantum Hall nematic, with a few exceptions,[169,213] was not considered a viable ground state at $\nu = 5/2$.

A first indication that in the second Landau level order other than the topological order of a FQHS may be present came from an experiment on a low density sample[158] in which an incipient anisotropy was reported at $\nu = 7/2$. The observed resistance anisotropy was 2. Subsequent work on 2DEG at high hydrostatic pressures revealed resistance anisotropy at both $\nu = 5/2$[160,161] and $7/2$.[162] Furthermore, similarly to the levels of anisotropy measured at $\nu = 9/2$, the anisotry observed in these latter experiments[160–162] reached extreme values exceeding 1000.

In Fig. 20 we summarize salient features of the longitudinal resistance as measured at three different pressures. Since the GaAs/AlGaAs sample was mounted in

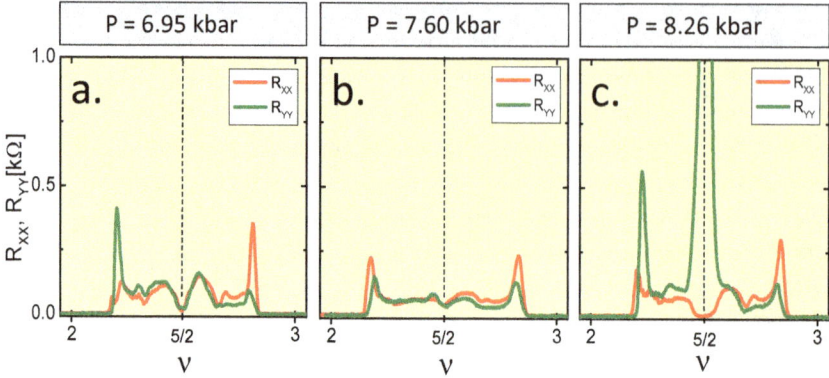

Fig. 20. Longitudinal magnetoresistance at $T \approx 12$ mK as measured in the second Landau level along two mutually perpendicular crystallographic axes of the GaAs. Green traces show R_{xx} measured along the [1$\bar{1}$0] crystal direction, whereas red traces show R_{yy} as measured along the [110] crystal direction. As the pressure is increased, at $\nu = 5/2$ the following sequence of ground states is observed: an isotropic FQHS (a), a nearly isotropic Fermi liquid (b), and the quantum Hall nematic (c). Adapted from Ref. 160.

a commercial pressure clamp cell, electron thermalization to the base temperature of the refrigerator was no longer possible. The electronic temperature in these experiments was estimated to be $T \approx 12$ mK.[160] Figures 20(a)–(c) show R_{xx} and R_{yy}, magnetoresistance data collected along two mutually perpendicular crystal axes of GaAs at the pressures of 6.95, 7.60, and 8.26 kbar, respectively. Data shown are consistent with the following sequence of ground states of the 2DEG at $\nu = 5/2$: a rotationally invariant FQHS at $P = 6.95$ kbar, a ground state close to an isotropic Fermi fluid at $P = 7.60$ kbar, and a quantum Hall nematic at $P = 8.26$ kbar. Because at $T \approx 12$ mK the isotropic liquid is observed in an extremely narrow range of pressures, it was argued that data from Fig. 20 are suggestive of a direct quantum phase transition from the FQHS to the quantum Hall nematic in the limit of zero temperatures.[160] This quantum phase transition occurs in the sample studied at the critical pressure of $P_c \simeq 7.8$ kbar. Since the electron density is pressure dependent, the density at the critical pressure was $n_c = 1.1 \times 10^{11}$ cm^{-2} in this experiment.[160] Later work mapping out the temperature dependence of the phases strengthened the argument of a direct phase transition in the limit of zero temperatures.[161]

There are several reasons to believe that the nematic seen at $\nu = 5/2$ at high pressures, shown in Fig. 20(c), is likely similar in nature to the quantum Hall nematic developing at $\nu = 9/2$ and other high value filling factors in samples in the ambient.[160] First, they are both centered at half-integer filling factors. Second, the temperature dependence of the magnetoresistance is similar in both cases: it is isotropic above the nematic onset temperature and it has a very abrupt, almost exponential onset. Third, both the nematic near both $\nu = 5/2$ and $\nu = 9/2$ develop over a relatively narrow range of filling factors. Indeed, the range of fillings for the nematic in both cases is about $\Delta\nu \simeq 0.15$. This value is in sharp contrast to the significantly larger $\Delta\nu \simeq 0.6$ range of filling factors of nematicity developing near $\nu = 5/2$ in tilted magnetic fields. This is significant, because an ordered phase induced by an external intensive parameter is not necessarily identical to a spontaneously ordered phase. For example, the magnetized phase induced in a paramagnet when placed in an external magnetic field is not identical to the spontaneous ferromagnetic phase, as the two do not share the same correlation functions. In tilted field experiments on 2DEGs the symmetry breaking field favoring nematicity is the in-plane component of the magnetic field. While the nematic phase at $\nu = 5/2$ forming in tilted magnetic fields is likely related to the nematic forming at $\nu = 5/2$ at high pressure but in the absence of any symmetry breaking fields, the exact nature of this relationship is yet to be determined.

The importance of data shown in Fig. 20 is two-fold. First, it was established that the quantum Hall nematic may be stabilized at $\nu = 5/2$ in the absence of an in-plane magnetic field. Second, it was pointed out that the phase transition from a FQHS to the quantum Hall nematic is of a special type.[160] Indeed, this phase transition involves a FQHS, which is a topological phase, and the quantum Hall nematic, which is a transitional Landau phase with a broken symmetry. Phase transitions between two Landau phases are well known. Furthermore, recent intense investigations of phase transitions between two topologically distinct phases have significantly contributed to their understanding. However, phase transitions between the two distinct classes of phases, such as the transition from a fractional quantum Hall state to the nematic, remain rare and present an opportunity for further theoretical development.[216–221]

One may gain further insight into the transition by investigating it in the upper spin branch of the second Landau level. Such studies confirmed a qualitatively similar phase transition at filling factor $\nu = 7/2$. It was shown that in a purely perpendicular magnetic field only even denominator FQHSs may be involved in the FQHS-to-nematic phase transition, highlighting therefore an interesting competition between paired states of composite fermions and nematicity.[162] Furthermore, in this study it was also shown that the role of the pressure is to tune the electron-electron interaction; data suggest that a FQHS-to-nematic phase transition may also be induced by other means of tuning the electron-electron iteration.[162]

7. Conclusions

In this chapter we addressed a few topics on the electron gas confined to GaAs/AlGaAs hosts. The study of this system presents an opportunity to learn about the effects of the Coulomb interaction on the various topological phases and their competition with charge ordered phases in a clean environment. We discussed recently discovered FQHSs, the effect of the disorder on the even denominator FQHSs, novel transport features of the RIQHSs, and recently discovered phase transitions at even denominator filling factors. Many of these results were enabled by the use of ultra-low temperature and of high pressure techniques. However, the topics covered barely scratched the surface of the ongoing research on this and related systems. As the purity of different materials improves, the effects of interactions in those materials will come to the fore. It is then expected that the interplay of interactions and topology will lead to new physics beyond that of single-electron band structure in the growing family of topological materials. Other results, such as the competition of pairing and nematicity in half-filled Landau levels, revealed a strong connection between the 2DEG and other strongly correlated materials. Research on the 2DEG in the fractional quantum Hall regime will certainly impact efforts on these materials and will continue to present future opportunities for discovery.

Acknowledgments

I would like to thank Dan Tsui for introducing me to fractional quantum Hall physics and Jian-Sheng Xia for sharing information about the He-3 immersion cell. Measurements would have not been possible without the samples of exceptional quality grown by Loren Pfeiffer and Kenneth West at Princeton and by my colleague at Purdue, Michael Manfra. I am grateful to Ashwani Kumar, Nodar Samkharadze, Nianpei Deng, Ethan Kleinbaum, Katherine Schreiber, and Vidhi Shingla for their tireless work in the lab and for their original contributions to the topics discussed. Last but not least, I have benefited from numerous discussions with Nicholas d'Ambrumenil, Rudro Biswas, Rui-Rui Du, Eduardo Fradkin, James Eisenstein, Lloyd Engel, Jainendra Jain, Koji Muraki, Wei Pan, Zlatko Papić, Steve Simon, Jurgen Smet, and Michael Zudov. This work was supported by the NSF-DMR 1904497 and the DOE BES award DE-SC0006671.

References

1. K. von Klitzing, G. Dorda, and M. Pepper, New method for high-accuracy determination of the fine-structure constant based on quantized Hall resistance, *Phys. Rev. Lett.* **45**, 494–497 (1980).
2. D.C. Tsui, H.L. Stormer, and A.C. Gossard, Two-dimensional magnetotransport in the extreme quantum limit, *Phys. Rev. Lett.* **48**, 1559–1562 (1982).
3. R.B. Laughlin, Anomalous quantum Hall effect: An incompressible quantum fluid with fractionally charged excitations, *Phys. Rev. Lett.* **50**, 1395–1398 (1983).
4. J.K. Jain, Composite-fermion approach for the fractional quantum Hall effect, *Phys. Rev. Lett.* **63**, 199–202 (1989).
5. B.I. Halperin, P.A. Lee, and N. Read, Theory of the half-filled Landau level, *Phys. Rev. B* **47**, 7312–7343 (1993).
6. G. Moore and N. Read, Non-Abelions in the fractional quantum Hall effect, *Nucl. Phys. B* **360**, 362–396 (1991).
7. N. Read and E. Rezayi, Beyond paired quantum Hall states: Parafermions and incompressible states in the first excited Landau level, *Phys. Rev. B* **59**, 8084–8092 (1999).

8. X.G. Wen, *Quantum Field Theory of Many-Body Systems* (Oxford University Press, Oxford, UK, 2004).
9. T.H. Hansson, M. Hermanns, S.H. Simon, and S.F. Viefers, Quantum Hall physics: Hierarchies and conformal field theory techniques, *Rev. Mod. Phys.* **89**, 025005 (2017).
10. H.L. Stormer, D.C. Tsui, and A.C. Gossard, The fractional quantum Hall effect, *Rev. Mod. Phys.* **71**, S298–S305 (1999).
11. J.K. Jain, *Composite Fermions* (Cambridge University Press, Cambridge, England, 2007).
12. J.K. Jain, Composite fermion theory of exotic fractional quantum Hall effect, *Ann. Rev. Cond. Matter Phys.* **6**, 39–62 (2015).
13. L. Pfeiffer and K.W. West, The role of MBE in recent quantum Hall effect physics discoveries, *Physica E* **20**, 57–64 (2003).
14. T.S. Lay, J.J. Heremans, Y.W. Suen, M.B. Santos, K. Hirakawa, M. Shayegan, and A. Zrenner, High-quality two-dimensional electron system confined in an AlAs quantum well, *Appl. Phys. Lett.* **62**, 3120–3122 (1993).
15. B. A. Piot, J. Kunc, M. Potemski, D. K. Maude, C. Betthausen, A. Vogl, D. Weiss, G. Karczewski, and T. Wojtowicz, Fractional quantum Hall effect in CdTe, *Phys. Rev. B* **82**, 081307 (2010).
16. K. Lai, W. Pan, D.C. Tsui, S. Lyon, M. Mühlberger, and F. Schäffler, Two-flux composite fermion series of the fractional quantum Hall states in strained Si, *Phys. Rev. Lett.* **93**, 156805 (2004).
17. Q. Shi, M.A. Zudov, C. Morrison, and M. Myronov, Spinless composite fermions in an ultrahigh-quality strained Ge quantum well, *Phys. Rev. B* **91**, 241303 (2015).
18. O.A. Mironov, N. d'Ambrumenil, A. Dobbie, D.R. Leadley, A.V. Suslov, and E. Green, Fractional quantum Hall states in a Ge quantum well, *Phys. Rev. Lett.* **116**, 176802 (2016).
19. J. Falson and M. Kawasaki, A review of the quantum Hall effects in MgZnO/ZnO heterostructures, *Rep. Prog. Phys.* **81**, 056501 (2018).
20. T.M. Kott, B. Hu, S.H. Brown, and B.E. Kane, Valley-degenerate two-dimensional electrons in the lowest Landau level, *Phys. Rev. B* **89**, 041107 (2014).
21. Y. Monrakha and K. Kono, *Two-Dimensional Coulomb Liquids and Solids* (Springer-Verlag, Berlin Heidelberg, 2004).
22. A.K. Geim and K.S. Novoselov, The rise of graphene, *Nat. Mater.* **6**, 183–191 (2007).
23. S. Manzeli, D. Ovchinnikov, D. Pasquier, O.V. Yazyev, and Andras Kis, 2D transition metal dichalcogenides, *Mat. Rev. Mater.* **2**, 1–15 (2017).
24. F. Yang, Z. Zhang, N.Z. Wang, G.J. Ye, W. Lou, X. Zhou, K. Watanabe, T. Taniguchi, K. Chang, X.H. Chen, and Y. Zhang, Quantum Hall effect in electron-doped black phosphorus field-effect transistors, *Nano Lett.* **18**, 6611–6616 (2018).
25. M. Yankowitz, J. Xue, D. Cormode, J.D. Sanchez-Yamagishi, K. Watanabe, T. Taniguchi, P. Jarillo-Herrero, P. Jacquod, and B.J. LeRoy, Emergence of superlattice Dirac points in graphene on hexagonal boron nitride, *Nature Physics* **8**, 382–386 (2012).
26. C.R. Dean, L. Wang, P. Maher, C. Forsythe, F. Ghahari, Y. Gao, J. Katoch, M. Ishigami, P. Moon, M. Koshino, T. Taniguchi, K. Watanabe, K.L. Shepard, J. Hone, and P. Kim, Hofstadter's butterfly and the fractal quantum Hall effect in Moiré superlattices, *Nature* **497**, 598–602 (2013).
27. Y. Cao, V. Fatemi, S. Fang, K. Watanabe, T. Taniguchi, E. Kaxiras, and P. Jarillo-Herrero, Unconventional superconductivity in magic-angle graphene superlattices, *Nature* **556**, 43–50 (2018).

28. D.G. Schlom and L.N. Pfeiffer, Upward mobility rocks!, *Nat. Mater.* **9**, 881–883 (2010).

29. V. Umansky, M. Heiblum, Y. Levinson, J. Smet, J. Nübler, and M. Dolev, MBE growth of ultra-low disorder 2DEG with mobility exceeding 35×10^6 cm^2/Vs, *J. Cryst. Growth* **311**, 1658–1661 (2009).

30. M.J. Manfra, Molecular beam epitaxy of ultra-high-quality AlGaAs/GaAs heterostructures: Enabling physics in low-dimensional electronic systems, *Annu. Rev. Condens. Matter Phys.* **5**, 347–373 (2014).

31. C. Reichl, J. Chen, S. Baer, C. Rössler, T. Ihn, K. Ensslin, W. Dietsche, and W. Wegscheider, Increasing the $\nu = 5/2$ gap energy: An analysis of MBE growth parameters, *New J. Phys.* **16**, 023014 (2014).

32. J.D. Watson, G.A. Csáthy, and M.J. Manfra, Impact of heterostructure design on transport properties in the second Landau level of in situ back-gated two-dimensional electron gases, *Phys. Rev. Applied* **3**, 064004 (2015).

33. D. Kamburov, K.W. Baldwin, K.W. West, M. Shayegan, and L.N. Pfeiffer, Interplay between quantum well width and interface roughness for electron transport mobility in GaAs quantum wells, *Appl. Phys. Lett.* **109**, 232105 (2016).

34. G.C. Gardner, S. Fallahi, J.D. Watson, and M.J. Manfra, Modified MBE hardware and techniques and role of Gallium purity for attainment of two-dimensional electron gas mobility $> 35 \times 10^6$ cm^2/Vs in AlGaAs/GaAs quantum wells grown by MBE, *J. Cryst. Growth* **441**, 71–77 (2016).

35. Y.J. Chung, K.W. Baldwin, K.W. West, M. Shayegan, and L.N. Pfeiffer, Surface segregation and the Al problem in GaAs quantum wells, *Phys. Rev. Mater.* **2**, 034006 (2018).

36. M. Shayegan, Case for the magnetic-field-induced two-dimensional Wigner crystal, in *Perspectives in Quantum Hall Effects* 343–384 (Wiley, 2004).

37. A.A. Koulakov, M.M. Fogler, and B.I. Shklovskii, Charge density wave in two-dimensional electron liquid in weak magnetic field, *Phys. Rev. Lett.* **76**, 499–502 (1996).

38. R. Moessner and J.T. Chalker, Exact results for interacting electrons in high Landau levels, *Phys. Rev. B* **54**, 5006–5015 (1996).

39. E. Fradkin and S.A. Kivelson, Liquid-crystal phases of quantum Hall systems, *Phys. Rev. B* **59**, 8065–8072 (1999).

40. M.P. Lilly, K.B. Cooper, J.P. Eisenstein, L.N. Pfeiffer, and K.W. West, Evidence for an anisotropic state of two-dimensional electrons in high Landau levels, *Phys. Rev. Lett.* **82**, 394–397 (1999).

41. R.R. Du, D.C. Tsui, H.L. Stormer, L.N. Pfeiffer, K.W. Baldwin, and K.W. West, Strongly anisotropic transport in higher two-dimensional Landau levels, *Solid State Commun.* **109**, 389–894 (1999).

42. R. Willett, J.P. Eisenstein, H.L. Störmer, D.C. Tsui, A.C. Gossard, and J.H. English, Observation of an even-denominator quantum number in the fractional quantum Hall effect, *Phy. Rev. Lett.* **59**, 1776–1779 (1987).

43. J.P. Eisenstein, R.L. Willett, H.L. Stormer, L.N. Pfeiffer, and K.W. West, Activation energies for the even-denominator fractional quantum Hall effect, *Surf. Sci.* **229**, 31–33 (1990).

44. W. Pan, J.-S. Xia, V. Shvarts, D.E. Adams, H.L. Stormer, D.C. Tsui, L.N. Pfeiffer, K.W. Baldwin, and K.W. West, Exact quantization of the even-denominator fractional quantum Hall state at $\nu = 5/2$ Landau level filling factor, *Phys. Rev. Lett.* **83**, 3530–3533 (1999).

45. J.P. Eisenstein, K.B. Cooper, L.N. Pfeiffer, and K.W. West, Insulating and fractional quantum Hall states in the first excited Landau level, *Phys. Rev. Lett.* **88**, 076801 (2002).
46. J.S. Xia, W. Pan, C.L. Vicente, E.D. Adams, N.S. Sullivan, H.L. Stormer, D.C. Tsui, L.N. Pfeiffer, K.W. Baldwin, and K.W. West, Electron correlation in the second Landau level: A competition between many nearly degenerate quantum phases, *Phys. Rev. Lett.* **93**, 176809 (2004).
47. J.S. Xia, E.D. Adams, V. Shvarts, W. Pan, H.L. Stormer, and D.C. Tsui, Ultra-low-temperature cooling of two-dimensional electron gas, *Physica B* **280**, 491–492 (2000).
48. J. Huang, J.S. Xia, D.C. Tsui, L.N. Pfeiffer, and K.W. West, Disappearance of metal-like behavior in GaAs two-dimensional holes below 30 mK, *Phys. Rev. Lett.* **98**, 226801 (2007).
49. W. Li, C.L. Vicente, J.S. Xia, W. Pan, D.C. Tsui, L.N. Pfeiffer, and K.W. West, Scaling in plateau-to-plateau transition: A direct connection of quantum Hall systems with the Anderson localization model, *Phys. Rev. Lett.* **102**, 216801 (2009).
50. J.S. Xia, E.D. Adams, N.S. Sullivan, W. Pan, H.L. Stormer, and D.C. Tsui, Sample cooling and rotation at ultra-low temperatures and high magnetic fields, *Int. J. Mod. Phys. B* **16**, 2986–2989 (2002).
51. A.C. Clark, K.K. Schwarzwälder, T. Bandi, D. Maradan, and D.M. Zumbühl, Method for cooling nanostructures to microkelvin temperatures, *Rev. Sci. Instrum.* **81**, 103904 (2010).
52. M. Palma, D. Maradan, L. Casparis, T.-M. Liu, N.M. Froning, and D.M. Zumbühl, Magnetic cooling for microkelvin nanoelectronics on a cryofree platform, *Rev. Sci. Instrum.* **88**, 043902 (2017).
53. D.I. Bradley, A.M. Guénault, D. Gunnarsson, R.P. Haley, S. Holt, A.T. Jones, Yu.A. Pashkin, J. Penttilä, J.R. Prance, M. Prunnila, and L. Roschier, On-chip magnetic cooling of a nanoelectronic device, *Scient. Rep.* **7**, 45566 (2017).
54. N. Yurttagül, M. Sarsby, and A. Geresdi, Indium as a high cooling power nuclear refrigerant for quantum nanoelectronics, *Phys. Rev. Appl.* **12**, 011005 (2019).
55. N. Samkharadze, A. Kumar, M.J. Manfra, L.N. Pfeiffer, K.W. West and G.A. Csáthy, Integrated electronic transport and thermometry at milliKelvin temperatures and in strong magnetic fields, *Rev. Sci. Instrum.* **82** 053902 (2011).
56. N. Samkharadze, A. Kumar, and G.A. Csáthy, A new type of carbon resistance thermometer with excellent thermal contact at millikelvin temperatures, *J. Low Temp. Phys.* **160**, 246–253 (2010).
57. A. Shibahara, O. Hahtela, J. Engert, H. van der Vliet, L.V. Levitin, A. Casey, C.P. Lusher, J. Saunders, D. Drung, and Th. Schurig, Primary current-sensing noise thermometry in the millikelvin regime, *Phil. Trans. Royal Soc. A* **374**, 20150054 (2016).
58. E. Kleinbaum, V. Shingla, and G.A. Csáthy, SQUID-based current sensing noise thermometry for quantum resistors at dilution refrigerator temperatures, *Rev. Sci. Instrum.* **88**, 034902 (2017).
59. Z. Iftikhar, A. Anthore, S. Jezouin, F.D. Parmentier, Y. Jin, A. Cavanna, A. Ouerghi, U. Gennse, and F. Pierre, Primary thermometry triad at 6 mK in mesoscopic circuits, *Nature Commun.* **7**, 12908 (2016).
60. D. Maradan, L. Casparis, T.-M. Liu, D.E.F. Biesinger, C.P. Scheller, D.M. Zumbühl, J.D. Zimmerman, and A.C. Gossard, GaAs quantum dot thermometry using direct transport and charge sensing, *J. Low Temp. Phys.* **175**, 784–798 (2014).

61. A.V. Feshchenko, L. Casparis, I.M. Khaymovich, D. Maradan, O.-P. Saira, M. Palma, M. Meschke, J.P. Pekola, and D.M. Zumbühl, Tunnel-junction thermometry down to millikelvin temperatures, *Phys. Rev. Appl.* **4**, 034001 (2015).

62. D.I. Bradley, R.E. George, D. Gunnarsson, R.P. Haley, H. Heikkinen, Yu.A. Pashkin, J. Penttilä, J.R. Prance, M. Prunnila, L. Roschier, and M. Sarsby, Nanoelectronic primary thermometry below 4 mK, *Nature Commun.* **7**, 10455 (2016).

63. M. Palma, C.P. Scheller, D. Maradan, A.V. Feshchenko, M. Meschke, and D.M. Zumbühl, On-and-off chip cooling of a Coulomb blockade thermometer down to 2.8 mK, *Appl. Phys. Lett.* **111**, 253105 (2017).

64. H.L. Störmer, R. Dingle, A.C. Gossard, W. Wiegmann, and M.D. Sturge, Two-dimensional electron gas at a semiconductor-semiconductor interface, *Solid State Commun.* **29**, 707–709 (1979).

65. H.L. Störmer, A.C. Gossard, W. Wiegmann, and K. Baldwin, Dependence of electron mobility in modulation-doped GaAs-(AlGa)As heterojunction interfaces on electron density and Al concentration, *Appl. Phys. Lett.* **39**, 912–915 (1981).

66. M. Samani, A.V. Rossokhaty, E. Sajadi, S. Lüscher, J.A. Folk, J.D. Watson, G.C. Gardner, and M.J. Manfra, Low-temperature illumination and annealing of ultrahigh quality quantum wells, *Phys. Rev. B* **90**, 121405 (2014).

67. J. Shabani, Y. Liu, and M. Shayegan, Fractional quantum Hall effect at high fillings in a two-subband electron system, *Phys. Rev. Lett.* **105**, 246805 (2010).

68. Y. Liu, D. Kamburov, M. Shayegan, L.N. Pfeiffer, K.W. West, and K.W. Baldwin, Anomalous robustness of the $\nu = 5/2$ fractional quantum Hall state near a sharp phase boundary, *Phys. Rev. Lett.* **107**, 176805 (2011).

69. Z. Papić, F.D.M. Haldane, and E.H. Rezayi, Quantum phase transitions and the $\nu = 5/2$ fractional Hall state in wide quantum wells, *Phys. Rev. Lett.* **109**, 266806 (2012).

70. A. Kumar, G.A. Csáthy, M.J. Manfra, L.N. Pfeiffer, and K.W. West, Nonconventional odd-denominator fractional quantum Hall states in the second Landau level, *Phys. Rev. Lett.* **105**, 246808 (2010).

71. E. Kleinbaum, A. Kumar, L.N. Pfeiffer, K.W. West, and G.A. Csáthy, Gap reversal at filling factors $3+1/3$ and $3+1/5$: Towards novel topological order in the fractional quantum Hall regime, *Phys. Rev. Lett.* **114**, 076801 (2015).

72. W. Pan, J.S. Xia, H.L. Stormer, D.C. Tsui, C.L. Vicente, E.D. Adams, N.S. Sullivan, L.N. Pfeiffer, K.W. Baldwin, and K.W. West, Quantization of the diagonal resistance: Density gradients and the empirical resistance rule in a 2D system, *Phys. Rev. Lett.* **95**, 066808 (2005).

73. Q. Qian, J. Nakamura, S. Fallahi, G.C. Gardner, J.D. Watson, S. Lüscher, J.A. Folk, G.A. Csáthy, and M.J. Manfra, Quantum lifetime in ultrahigh quality GaAs quantum wells: Relationship to $\Delta_{5/2}$ and impact of density fluctuations, *Phys. Rev. B* **96**, 035309 (2017).

74. D. Kamburov, K.W. Baldwin, K.W. West, S. Lyon, L.N. Pfeiffer, and A. Pinczuk, Use of micro-photoluminescence as a contactless measure of the 2D electron density in a GaAs quantum well, *Appl. Phys. Lett.* **110**, 262104 (2017).

75. Y.J. Chung, K.W. Baldwin, K.W. West, N. Haug, J. van de Wetering, M. Shayegan, and L.N. Pfeiffer, Spatial mapping of local density variations in two-dimensional electron systems using scanning photoluminescence, *Nano Lett.* **19**, 1908–1913 (2019).

76. A. Wójs, Transition from Abelian to non-Abelian quantum liquids in the second Landau level, *Phys. Rev. B* **80**, 041104 (2009).

77. S.H. Simon, E.H. Rezayi, N.R. Cooper, and I. Berdnikov, Construction of a paired wave function for spinless electrons at filling fraction $\nu = 2/5$, *Phys. Rev. B* **75**, 075317 (2007).

78. W. Bishara, G.A. Fiete, and C. Nayak, Quantum Hall states at $\nu = 2/(k+2)$: Analysis of the particle-hole conjugates of the general level-k Read–Rezayi states, *Phys. Rev. B* **77**, 241306 (2008).

79. P. Bonderson and J.K. Slingerland, Fractional quantum Hall hierarchy and the second Landau level, *Phys. Rev. B* **78**, 125323 (2008).

80. B.A. Bernevig and F.D.M. Haldane, Properties of non-Abelian fractional quantum Hall states at filling $\nu = k/r$, *Phys. Rev. Lett.* **101**, 246806 (2008).

81. M. Levin and B.I. Halperin, Collective states of non-Abelian quasiparticles in a magnetic field, *Phys. Rev. B* **79**, 205301 (2009).

82. G.J. Sreejith, C. Tőke, A. Wójs, and J.K. Jain, Bipartite composite fermion states, *Phys. Rev. Lett.* **107**, 086806 (2011).

83. G.J. Sreejith, Y.-H. Wu, A. Wójs, and J.K. Jain, Tripartite composite fermion states, *Phys. Rev. B* **87**, 245125 (2013).

84. N. d'Ambrumenil and A.M. Reynolds, Fractional quantum Hall states in higher Landau levels, *J. Phys. C* **21**, 119–132 (1988).

85. V.W. Scarola and J.K. Jain, Phase diagram of bilayer composite fermion states, *Phys. Rev. B* **64**, 085313 (2001).

86. C. Tőke, M.R. Peterson, G.S. Jeon, and J.K. Jain, Fractional quantum Hall effect in the second Landau level: The importance of inter-composite-fermion interaction, *Phys. Rev. B* **72**, 125315 (2005).

87. Z. Papić, N. Regnault, and S. Das Sarma, Interaction-tuned compressible-to-incompressible phase transitions in quantum Hall systems, *Phys. Rev. B* **80**, 201303 (2009).

88. A.C. Balram, Y.-H. Wu, G.J. Sreejith, A. Wójs, and J.K. Jain, Role of exciton screening in the 7/3 fractional quantum Hall effect, *Phys. Rev. Lett.* **110**, 186801 (2013).

89. S. Johri, Z. Papić, R.N. Bhatt, and P. Schmitteckert, Quasiholes of 1/3 and 7/3 quantum Hall states: Size estimates via exact diagonalization and density-matrix renormalization group, *Phys. Rev. B* **89**, 115124 (2014).

90. M.R. Peterson, Y.-L. Wu, M. Cheng, M. Barkeshli, Z. Wang, and S. Das Sarma, Abelian and non-Abelian states in $\nu = 2/3$ bilayer fractional quantum Hall systems, *Phys. Rev. B* **92**, 035103 (2015).

91. T. Jolicoeur, Shape of the magnetoroton at $\nu = 1/3$ and $\nu = 7/3$ in real samples, *Phys. Rev. B* **95**, 075201 (2017).

92. J.-S. Jeong, H. Lu, K.H. Lee, K. Hashimoto, S.B. Chung, and K. Park, Competing states for the fractional quantum Hall effect in the 1/3-filled second Landau level, *Phys. Rev. B* **96**, 125148 (2017).

93. W. Pan, J.S. Xia, H.L. Stormer, D.C. Tsui, C. Vicente, E.D. Adams, N.S. Sullivan, L.N. Pfeiffer, K.W. Baldwin, and K.W. West, Experimental studies of the fractional quantum Hall effect in the first excited Landau level, *Phys. Rev. B* **77**, 075307 (2008).

94. C. Zhang, C. Huan, J.S. Xia, N.S. Sullivan, W. Pan, K.W. Baldwin, K.W. West, L.N. Pfeiffer, and D.C. Tsui, Spin polarization of the $\nu = 12/5$ fractional quantum Hall state, *Phys. Rev. B* **85**, 241302 (2012).

95. N. Deng, J.D. Watson, L.P. Rokhinson, M.J. Manfra, and G.A. Csáthy, Contrasting energy scales of re-entrant integer quantum Hall states, *Phys. Rev. B* **86**, 201301 (2012).

96. Q. Qian, J. Nakamura, S. Fallahi, G.C. Gardner, and M.J. Manfra, Possible nematic to smectic phase transition in a two-dimensional electron gas at half-filling, *Nat. Commun.* **8**, 1536 (2017).

97. V. Shingla, E. Kleinbaum, A. Kumar, L.N. Pfeiffer, K.W. West, and G.A. Csáthy, Finite-temperature behavior in the second Landau level of the two-dimensional electron gas, *Phys. Rev. B* **97**, 241105 (2018).

98. H.C. Choi, W. Kang, S. Das Sarma, L.N. Pfeiffer, and K.W. West, Activation gaps of fractional quantum Hall effect in the second Landau level, *Phys. Rev. B* **77**, 081301 (2008).

99. G.S. Boebinger, H.L. Stormer, D.C. Tsui, A.M. Chang, J.C.M. Hwang, A.Y. Cho, C.W. Tu, and G. Weimann, Activation energies and localization in the fractional quantum Hall effect, *Phys. Rev. B* **36**, 7919–29 (1987).

100. R.L. Willett, H.L. Stormer, D.C. Tsui, A.C. Gossard, and J.H. English, Quantitative experimental test for the theoretical gap energies in the fractional quantum Hall effect, *Phys. Rev. B* **37**, 8476–8479 (1988).

101. R.R. Du, H.L. Stormer, D.C. Tsui, L.N. Pfeiffer, and K.W. West, Experimental evidence for new particles in the fractional quantum Hall effect, *Phys. Rev. Lett.* **70**, 2944–2947 (1993).

102. H.C. Manoharan, M. Shayegan, and S.J. Klepper, Signatures of a novel Fermi liquid in a two-dimensional composite particle metal, *Phys. Rev. Lett.* **73**, 3270–3773 (1994).

103. M. Dolev, Y. Gross, R. Sabo, I. Gurman, M. Heiblum, V. Umansky, and D. Mahalu, Characterizing neutral modes of fractional states in the second Landau level, *Phys. Rev. Lett.* **107**, 036805 (2011).

104. S. Baer, C. Rössler, T. Ihn, K. Ensslin, C. Reichl, and W. Wegscheider, Experimental probe of topological orders and edge excitations in the second Landau level, *Phys. Rev. B* **90**, 075403 (2014).

105. U. Wurstbauer, A.L. Levy, A. Pinczuk, K.W. West, L.N. Pfeiffer, M.J. Manfra, G.C. Gardner, and J.D. Watson, Gapped excitations of unconventional fractional quantum Hall effect states in the second Landau level, *Phys. Rev. B* **92**, 241407 (2015).

106. A.C. Balram, S. Mukherjee, K. Park, M. Barkeshli, M.S. Rudner, and J.K. Jain, Fractional quantum Hall effect at $\nu = 2 + 6/13$: The Parton paradigm for the second Landau level, *Phys. Rev. Lett.* **121**, 186601 (2018).

107. A.C. Balram, M. Barkeshli, and M.S. Rudner, Parton construction of a wave function in the anti-Pfaffian phase, *Phys. Rev. B* **98**, 035127 (2018).

108. A.C. Balram, M. Barkeshli, and M.S. Rudner, Parton construction of particle-hole-conjugate Read–Rezayi parafermion fractional quantum Hall states and beyond, *Phys. Rev. B* **99**, 241108 (2019).

109. E.H. Rezayi and N. Read, Non-Abelian quantized Hall states of electrons at filling factors 12/5 and 13/5 in the first excited Landau level, *Phys. Rev. B* **79**, 075306 (2009).

110. W. Zhu, S.S. Gong, F.D.M. Haldane, and D.N. Sheng, Fractional quantum Hall states at $\nu = 13/5$ and 12/5 and their non-Abelian nature, *Phys. Rev. Lett.* **115**, 126805 (2015).

111. K. Pakrouski, M. Troyer, Y.-L. Wu, S. Das Sarma, and M.R. Peterson, Enigmatic 12/5 fractional quantum Hall effect, *Phys. Rev. B* **94**, 075108 (2016).

112. R.S.K. Mong, M.P. Zaletel, F. Pollmann, and Z. Papić, Fibonacci anyons and charge density order in the 12/5 and 13/5 quantum Hall plateaus, *Phys. Rev. B* **95**, 115136 (2017).

113. P. Bonderson, A.E. Feiguin, G. Möller, and J.K. Slingerland, Competing topological orders in the $\nu = 12/5$ quantum Hall state, *Phys. Rev. Lett.* **108**, 036806 (2012).

114. A.A. Zibrov, C. Kometter, H. Zhou, E.M. Spanton, T. Taniguchi, K. Watanabe, M.P. Zaletel, and A.F. Young, Tunable interacting composite fermion phases in a half-filled bilayer-graphene Landau level, *Nature* **549** 360–364, (2017).

115. W. Pan, K.W. Baldwin, K.W. West, L.N. Pfeiffer, and D.C. Tsui, Spin transition in the $\nu = 8/3$ fractional quantum Hall effect, *Phys. Rev. Lett.* **108**, 216804 (2012).

116. D.M. Haldane, E.H. Rezayi, and K. Yang, Spontaneous breakdown of translational symmetry in quantum Hall systems: Crystalline order in high Landau levels, *Phys. Rev. Lett.* **85**, 5396–5399 (2000).

117. N. Shibata and D. Yoshioka, Ground-State phase diagram of 2D electrons in a high Landau level: A density-matrix renormalization group study, *Phys. Rev. Lett.* **86**, 5755–5758 (2001).

118. C.R. Dean, B.A. Piot, P. Hayden, S. Das Sarma, G. Gervais, L.N. Pfeiffer, and K.W. West, Intrinsic gap of the $\nu = 5/2$ fractional quantum Hall state, *Phys. Rev. Lett.* **100**, 146803 (2008).

119. J. Nuebler, V. Umansky, R. Morf, M. Heiblum, K. von Klitzing, and J. Smet, Density dependence of the $\nu = 5/2$ energy gap: Experiment and theory, *Phys. Rev. B* **81**, 035316 (2010).

120. J.R. Mallett, R.G. Clark, R.J. Nicholas, R. Willett, J.J. Harris, and C.T. Foxon, Experimental studies of the $\nu = 1/5$ hierarchy in the fractional quantum Hall effect, *Phys. Rev. B* **38**, 2200–2203 (1988).

121. W. Pan, H.L. Stormer, D.C. Tsui, L.N. Pfeiffer, K.W. Baldwin, and K.W. West, Fractional quantum Hall effect of composite fermions, *Phys. Rev. Lett.* **90**, 016801 (2003).

122. P. Sitko, K.-S. Yi, and J.J. Quinn, Composite fermion hierarchy: Condensed states of composite fermion excitations, *Phys. Rev. B* **56**, 12417 (1997).

123. A. Wójs, K.-S. Yi, and J.J. Quinn, Fractional quantum Hall states of clustered composite fermions, *Phys. Rev. B* **69**, 205322 (2004).

124. A. Wójs, G. Simion, and J.J. Quinn, Spin phase diagram of the $\nu_e = 4/11$ composite fermion liquid, *Phys. Rev. B* **75**, 155318 (2007).

125. S. Mukherjee, S.S. Mandal, Y.-H. Wu, A. Wójs, and J.K. Jain, Enigmatic 4/11 state: A prototype for unconventional fractional quantum Hall effect, *Phys. Rev. Lett.* **112**, 016801 (2014).

126. A.C. Balram, Interacting composite fermions: Nature of the $4/5, 5/7, 6/7$, and $6/17$ fractional quantum Hall states, *Phys. Rev. B* **94**, 165303 (2016).

127. W. Pan, K.W. Baldwin, K.W. West, L.N. Pfeiffer, and D.C. Tsui, Fractional quantum Hall effect at Landau level filling $\nu = 4/11$, *Phys. Rev. B* **91**, 041301 (2015).

128. N. Samkharadze, I. Arnold, L.N. Pfeiffer, K.W. West, and G.A. Csáthy, Observation of incompressibility at $\nu = 4/11$ and $\nu = 5/13$, *Phys. Rev. B* **91**, 081109 (2015).

129. J. Falson, D. Maryenko, B. Friess, D. Zhang, Y. Kozuka, A. Tsukazaki, J.H. Smet, and M. Kawasaki, Even-denominator fractional quantum Hall physics in ZnO, *Nat. Phys.* **11**, 347–351 (2015).

130. D.-K. Ki, V.I. Falko, D.A. Abanin, and A.F. Morpurgo, Observation of even denominator fractional quantum Hall effect in suspended bilayer graphene, *Nano Lett.* **14**, 2135–2139 (2014).

131. Y. Kim, D.S. Lee, S. Jung, V. Skákalová, T. Taniguchi, K. Watanabe, J.S. Kim, and J.H. Smet, Fractional quantum Hall states in bilayer graphene probed by transconductance fluctuations, *Nano Lett.* **15**, 7445–7451 (2015).

132. J.I.A. Li, C. Tan, S. Chen, Y. Zeng, T. Taniguchi, K. Watanabe, J. Hone, and C.R. Dean, Even-denominator fractional quantum Hall states in bilayer graphene, *Science* **358**, 648–652 (2017).

133. A.A. Zibrov, E.M. Spanton, H. Zhou, C. Kometter, T. Taniguchi, K. Watanabe, and A.F. Young, Even-denominator fractional quantum Hall states at an isospin transition in monolayer graphene, *Nat. Phys.* **14**, 930–935 (2018).

134. Y. Kim, A.C. Balram, T. Taniguchi, K. Watanabe, J.K. Jain, and J.H. Smet, Even denominator fractional quantum Hall states in higher Landau levels of graphene, *Nature Phys.* **15**, 154–158 (2019).

135. M. Greiter, X.-G. Wen, and F. Wilczek, Paired Hall state at half filling, *Phys. Rev. Lett.* **66**, 3205–3208 (1991).

136. N. Read and D. Green, Paired states of fermions in two dimensions with breaking of parity and time-reversal symmetries and the fractional quantum Hall effect, *Phys. Rev. B* **61**, 10267-10297 (2000).

137. V.W. Scarola, K. Park, and J.K. Jain, Cooper instability of composite fermions, *Nature* **406**, 863–865 (2000).

138. E.H. Rezayi and F.D.M. Haldane, Incompressible paired Hall state, stripe order, and the composite fermion liquid phase in half-filled Landau levels, *Phys. Rev. Lett.* **84**, 4685–4688 (2000).

139. H. Lu, S. Das Sarma, and K. Park, Superconducting order parameter for the even-denominator fractional quantum Hall effect, *Phys. Rev. B* **82**, 201303 (2010).

140. S.A. Parameswaran, S.A. Kivelson, S.L. Sondhi, and B.Z. Spivak, Weakly coupled Pfaffian as a Type I quantum Hall liquid, *Phys. Rev. Lett.* **106**, 236801 (2011).

141. M. Levin, B.I. Halperin, and B. Rosenow, Particle-hole symmetry and the Pfaffian state, *Phys. Rev. Lett.* **99**, 236806 (2007).

142. S.-S. Lee, S. Ryu, C. Nayak, and M.P.A. Fisher, Particle-hole symmetry and the $\nu = 5/2$ quantum Hall state, *Phys. Rev. Lett.* **99**, 236807 (2007).

143. B.I. Halperin, Theory of the quantized Hall conductance, *Helv. Phys. Acta* **56**, 75–102 (1983).

144. J.-S. Jeong and K. Park, Bilayer mapping of the paired quantum Hall state: Instability toward anisotropic pairing, *Phys. Rev. B* **91**, 195119 (2015).

145. D.T. Son, Is the Composite fermion a Dirac particle?, *Phys. Rev. X* **5**, 031027 (2015).

146. P.T. Zucker and D.E. Feldman, Stabilization of the particle-hole Pfaffian order by Landau-level mixing and impurities that break particle-hole symmetry, *Phys. Rev. Lett.* **117**, 096802 (2016).

147. X. Wan and Kun Yang, Striped quantum Hall state in a half-filled Landau level, *Phys. Rev. B* **93**, 201303 (2016).

148. X.G. Wen and Q. Niu, Ground-state degeneracy of the fractional quantum Hall states in the presence of a random potential and on high-genus Riemann surfaces, *Phys. Rev. B* **41**, 9377–9396 (1990).

149. X.G. Wen, Non-Abelian statistics in the fractional quantum Hall states, *Phys. Rev. Lett.* **66**, 802–805 (1991).

150. R.L. Willett, The quantum Hall effect at 5/2 filling factor, *Rep. Prog. Phys.* **76**, 076501 (2013).

151. X. Lin, R.R. Du, and X. Xie, Recent experimental progress of fractional quantum Hall effect: 5/2 filling state and graphene, *Nat. Sci. Rev.* **1**, 564–579 (2014).

152. I.P. Radu, J.B. Miller, C.M. Marcus, M.A. Kastner, L.N. Pfeiffer, and K.W. West, Quasi-particle properties from tunneling in the $\nu = 5/2$ fractional quantum Hall state, *Science* **320**, 899–902 (2008).

153. X. Lin, C. Dillard, M.A. Kastner, L.N. Pfeiffer, and K.W. West, Measurements of quasiparticle tunneling in the $\nu = 5/2$ fractional quantum Hall state, *Phys. Rev. B* **85**, 165321 (2012).

154. H. Fu, P. Wang, P. Shan, L.N. Pfeiffer, K.W. West, M.A. Kastner, and X. Lin, Competing $\nu = 5/2$ fractional quantum Hall states in confined geometry, *Proc. Nat. Acad. Sci.* **113**, 12386–12390 (2016).

155. R.L. Willett, C. Nayak, K. Shtengel, L.N. Pfeiffer, and K.W. West, Magnetic-field-tuned Aharonov–Bohm oscillations and evidence for non-Abelian anyons at $\nu = 5/2$ *Phys. Rev. Lett.* **111**, 186401 (2013).

156. M. Banerjee, M. Heiblum, V. Umansky, D.E. Feldman, Y. Oreg, and A. Stern, Observation of half-integer thermal Hall conductance, *Nature* **559**, 205–210 (2018).

157. B.A. Schmidt, K. Bennaceur, S. Gaucher, G. Gervais, L.N. Pfeiffer, and K.W. West, Specific heat and entropy of fractional quantum Hall states in the second Landau level, *Phys. Rev. B* **95**, 201306 (2017).

158. W. Pan, A. Serafin, J.S. Xia, L. Yin, N.S. Sullivan, K.W. Baldwin, K.W. West, L.N. Pfeiffer, and D.C. Tsui, Competing quantum Hall phases in the second Landau level in the low-density limit, *Phys. Rev. B* **89**, 241302 (2014).

159. N. Samkharadze, D. Ro, L.N. Pfeiffer, K.W. West, and G.A. Csáthy, Observation of an anomalous density-dependent energy gap of the $\nu = 5/2$ fractional quantum Hall state in the low-density regime, *Phys. Rev. B* **96**, 085105 (2017).

160. N. Samkharadze, K.A. Schreiber, G.C. Gardner, M.J. Manfra, E. Fradkin, and G.A. Csáthy, Observation of a transition from a topologically ordered to a spontaneously broken symmetry phase, *Nature Phys.* **12**, 191–195 (2016).

161. K.A. Schreiber, N. Samkharadze, G.C. Gardner, R.R. Biswas, M.J. Manfra, and G.A. Csáthy, Onset of quantum criticality in the topological-to-nematic transition in a two-dimensional electron gas at filling factor $\nu = 5/2$, *Phys. Rev. B* **96**, 041107 (2017).

162. K.A. Schreiber, N. Samkharadze, G.C. Gardner, Y. Lyanda-Geller, M.J. Manfra, L.N. Pfeiffer, K.W. West, and G.A. Csáthy, Electron-electron interactions and the paired-to-nematic quantum phase transition in the second Landau level, *Nature Commun.* **9**, 2400 (2018).

163. L. Tiemann, G. Gamez, N. Kumad, and K. Muraki, Unraveling the spin polarization of the $\nu = 5/2$ fractional quantum Hall state, *Science* **335**, 828–831 (2012).

164. U. Wurstbauer, K.W. West, L.N. Pfeiffer, and A. Pinczuk, Resonant inelastic light scattering investigation of low-lying gapped excitations in the quantum fluid at $\nu = 5/2$, *Phys. Rev. Lett.* **110**, 026801 (2013).

165. A. Wójs and J.J. Quinn, Landau level mixing in the $\nu = 5/2$ fractional quantum Hall state, *Phys. Rev. B* **74**, 235319 (2006).

166. M.R. Peterson, Th. Jolicoeur, and S. Das Sarma, Finite-layer thickness stabilizes the Pfaffian state for the 5/2 fractional quantum Hall effect: Wave function overlap and topological degeneracy, *Phys. Rev. Lett.* **101**, 016807 (2008).

167. A. Wójs, C. Tőke, and J.K. Jain, Landau-level mixing and the emergence of Pfaffian excitations for the 5/2 fractional quantum Hall effect, *Phys. Rev. Lett.* **105**, 096802 (2010).

168. E.H. Rezayi and S.H. Simon, Breaking of particle-hole symmetry by Landau level mixing in the $\nu = 5/2$ quantized Hall state, *Phys. Rev. Lett.* **106**, 116801 (2011).

169. A. Tylan-Tyler and Y. Lyanda-Geller, Phase diagram and edge states of the $\nu = 5/2$ fractional quantum Hall state with Landau level mixing and finite well thickness, *Phys. Rev. B* **91**, 205404 (2015).

170. M.P. Zaletel, R.S.K. Mong, F. Pollmann, E.H. Rezayi, Infinite density matrix renormalization group for multicomponent quantum Hall systems, *Phys. Rev. B.* **91**, 045115 (2015).

171. K. Pakrouski, M.R. Peterson, Th. Jolicoeur, V.W. Scarola, C. Nayak, and M. Troyer, Phase diagram of the $\nu = 5/2$ fractional quantum Hall effect: Effects of Landau-level mixing and nonzero width, *Phys. Rev. X* **5**, 021004 (2015).

172. E.H. Rezayi, Landau level mixing and the ground state of the $\nu = 5/2$ quantum Hall effect, *Phys. Rev. Lett.* **119**, 026801 (2017).

173. W. Luo and T. Chakraborty, Pfaffian state in an electron gas with small Landau level gaps, *Phys. Rev. B* **96**, 081108 (2017).

174. Bo Yang, Three-body interactions in generic fractional quantum Hall systems and impact of Galilean invariance breaking, *Phys. Rev. B* **98**, 201101 (2018).

175. W. Li, G.A. Csáthy, D.C. Tsui, L.N. Pfeiffer, and K.W. West, Scaling and universality of integer quantum Hall plateau-to-plateau transitions, *Phys. Rev. Lett.* **94**, 206807 (2005).

176. N. Samkharadze, J.D. Watson, G. Gardner, M.J. Manfra, L.N. Pfeiffer, K.W. West, and G.A. Csáthy, Quantitative analysis of the disorder broadening and the intrinsic gap for the fractional quantum Hall state, *Phys. Rev. B* **84**, 121305 (2011).

177. W. Pan, N. Masuhara, N.S. Sullivan, K.W. Baldwin, K.W. West, L.N. Pfeiffer, and D.C. Tsui, Impact of disorder on the 5/2 fractional quantum Hall state, *Phys. Rev. Lett.* **106**, 206806 (2011).

178. G. Gamez and K. Muraki, $\nu = 5/2$ fractional quantum Hall state in low-mobility electron systems: Different roles of disorder, *Phys. Rev. B* **88**, 075308 (2013).

179. Q. Qian, J. Nakamura, S. Fallahi, G.C. Gardner, J.D. Watson, and M.J. Manfra, High-temperature resistivity measured at $\nu = 5/2$ as a predictor of the two-dimensional electron gas quality in the $N = 1$ Landau level, *Phys. Rev. B* **95**, 241304 (2017).

180. R.H. Morf and N. d'Ambrumenil, Disorder in fractional quantum Hall states and the gap at $\nu = 5/2$, *Phys. Rev. B.* **68**, 113309 (2003).

181. R.H. Morf, Transition from quantum Hall to compressible states in the second Landau level: New light on the $\nu = 5/2$ enigma, *Phys. Rev. Lett.* **80**, 1505–1508 (1998).

182. A.E. Feiguin, E. Rezayi, C. Nayak, and S. Das Sarma, Density matrix renormalization group study of incompressible fractional quantum Hall states, *Phys. Rev. Lett.* **100**, 166803 (2008).

183. A. Kumar, N. Samkharadze, G.A. Csáthy, M.J. Manfra, L.N. Pfeiffer, and K.W. West, Particle-hole asymmetry of fractional quantum Hall states in the second Landau level of a two-dimensional hole system, *Phys. Rev. B* **83**, 201305 (2011).

184. W. Li, G.A. Csáthy, D.C. Tsui, L.N. Pfeiffer, and K.W. West, Direct observation of alloy scattering of two-dimensional electrons in $Al_xGa_{1-x}As$, *Appl. Phys. Lett.* **83**, 2823–2834 (2003).

185. G.C. Gardner, J.D. Watson, S. Mondal, N. Deng, G.A. Csáthy, and M.J. Manfra, Growth and electrical characterization of $Al_{0.24}Ga_{0.76}As/Al_xGa_{1-x}As/Al_{0.24}Ga_{0.76}As$ modulation-doped quantum wells with extremely low x, *Appl. Phys. Lett.* **102**, 252103 (2013).

186. C. Reichl, W. Dietsche, T. Tschirky, T. Hyart, and W. Wegscheider, Mapping an electron wave function by a local electron scattering probe, *New J. Phys.* **17**, 113048 (2015).

187. B.-H. Moon, L.W. Engel, D.C. Tsui, L.N. Pfeiffer, and K.W. West, Microwave pinning modes near Landau filling $\nu = 1$ in two-dimensional electron systems with alloy disorder, *Phys. Rev. B* **92**, 035121 (2015).

188. N. Deng, G.C. Gardner, S. Mondal, E. Kleinbaum, M.J. Manfra, and G.A. Csáthy, $\nu = 5/2$ fractional quantum Hall state in the presence of alloy disorder, *Phys. Rev. Lett.* **112**, 116804 (2014).

189. D.N. Sheng, X. Wan, E.H. Rezayi, K. Yang, R.N. Bhatt, and F.D.M. Haldane, Disorder-driven collapse of the mobility gap and transition to an insulator in the fractional quantum Hall effect, *Phys. Rev. Lett.* **90**, 256802 (2003).

190. X. Wan, D.N. Sheng, E.H. Rezayi, K. Yang, R.N. Bhatt, and F.D.M. Haldane, Mobility gap in fractional quantum Hall liquids: Effects of disorder and layer thickness, *Phys. Rev. B* **72**, 075325 (2005).

191. Z. Liu and R.N. Bhatt, Evolution of quantum entanglement with disorder in fractional quantum Hall liquids, *Phys. Rev. B* **96**, 115111 (2017).

192. Z. Liu and R.N. Bhatt, Quantum entanglement as a diagnostic of phase transitions in disordered fractional quantum Hall liquids, *Phys. Rev. Lett.* **117**, 206801 (2016).

193. B.A. Friedman, Short and long ranged impurities in fractional quantum Hall systems, *Int. J. Mod. Phys. B* **32**, 1850338 (2018).

194. W. Li, D.R. Luhman, D.C. Tsui, L.N. Pfeiffer, and K.W. West, Observation of re-entrant phases induced by short-range disorder in the lowest Landau level of $Al_xGa_{1x}As/Al_{0.32}Ga_{0.68}As$ heterostructures, *Phys. Rev. Lett.* **105**, 076803 (2010).

195. Y. Liu, C.G. Pappas, M. Shayegan, L.N. Pfeiffer, K.W. West, and K.W. Baldwin, Observation of re-entrant integer quantum Hall states in the lowest Landau level, *Phys. Rev. Lett.* **109**, 036801 (2012).

196. K.B. Cooper, M.P. Lilly, J.P. Eisenstein, L.N. Pfeiffer, and K.W. West, Insulating phases of two-dimensional electrons in high Landau levels: Observation of sharp thresholds to conduction, *Phys. Rev. B* **60**, 11285 (1999).

197. N. Deng, A. Kumar, M.J. Manfra, L.N. Pfeiffer, K.W. West, and G.A. Csáthy, Collective nature of the re-entrant integer quantum Hall states in the second Landau level, *Phys. Rev. Lett.* **108**, 086803 (2012).

198. R.M. Lewis, P.D. Ye, L.W. Engel, D.C. Tsui, L.N. Pfeiffer, and K.W. West, Microwave resonance of the bubble phases in 1/4 and 3/4 filled high Landau levels, *Phys. Rev. Lett.* **89**, 136804 (2002).

199. R.M. Lewis, Y. Chen, L.W. Engel, D.C. Tsui, P.D. Ye, L.N. Pfeiffer, and K.W. West, Evidence of a first-order phase transition between Wigner–Crystal and bubble phases of 2D electrons in higher Landau levels, *Phys. Rev. Lett.* **93**, 176808 (2004).

200. R.M. Lewis, Y.P. Chen, L.W. Engel, D.C. Tsui, L.N. Pfeiffer, and K.W. West, Microwave resonance of the re-entrant insulating quantum Hall phases in the first excited Landau level, *Phys. Rev. B* **71**, 081301 (2005).

201. W.E. Chickering, J.P. Eisenstein, L.N. Pfeiffer, and K.W. West, Thermoelectric response of fractional quantized Hall and re-entrant insulating states in the $N = 1$ Landau level, *Phys. Rev. B* **87**, 075302 (2013).

202. X. Wang, H. Fu, L. Du, X. Liu, P. Wang, L.N. Pfeiffer, K.W. West, R.R. Du, and X. Lin, Depinning transition of bubble phases in a high Landau level, *Phys. Rev. B* **91**, 115301 (2015).

203. S. Baer, C. Rössler, S. Hennel, H.C. Overweg, T. Ihn, K. Ensslin, C. Reichl, and W. Wegscheider, Nonequilibrium transport in density-modulated phases of the second Landau level, *Phys. Rev. B* **91**, 195414 (2015).

204. A.V. Rossokhaty, Y. Baum, J.A. Folk, J.D. Watson, G.C. Gardner, and M.J. Manfra, Electron-hole asymmetric chiral breakdown of re-entrant quantum Hall states, *Phys. Rev. Lett.* **117**, 166805 (2016).

205. K. Bennaceur, C. Lupien, B. Reulet, G. Gervais, L.N. Pfeiffer, and K.W. West, Competing charge density waves probed by nonlinear transport and noise in the second and third Landau levels, *Phys. Rev. Lett.* **120**, 136801 (2018).

206. B. Friess, V. Umansky, K. von Klitzing, and J.H. Smet, Current flow in the bubble and stripe phases, *Phys. Rev. Lett.* **120**, 137603 (2018).

207. M.E. Msall and W. Dietsche, Acoustic measurements of the stripe and the bubble quantum Hall phase, *New J. Phys.* **17**, 043042 (2015).

208. B. Friess, Y. Peng, B. Rosenow, F. von Oppen, V. Umansky, K. von Klitzing, and J.H. Smet, Negative permittivity in bubble and stripe phases, *Nature Phys.* **13**, 1124–1129 (2017).

209. S. Chen, R. Ribeiro-Palau, K. Yang, K. Watanabe, T. Taniguchi, J. Hone, M.O. Goerbig, and C.R. Dean, Competing fractional quantum Hall and electron solid phases in graphene, *Phys. Rev. Lett.* **122**, 026802 (2019).

210. H.W. Jiang, R.L. Willett, H.L. Stormer, D.C. Tsui, L.N. Pfeiffer, and K.W. West, Quantum liquid versus electron solid around $\nu = 1/5$ Landau-level filling, *Phys. Rev. Lett.* **65**, 633–636 (1990).

211. D.K. Maude and J.C. Portal, Parallel transport in low-dimensional semiconductor structures, in *Semiconductors and Semimetals* **55**, 1–43 (1998).

212. H. Cho, J.B. Young, W. Kang, K.L. Campman, A.C. Gossard, M. Bichler, and W. Wegscheider, Hysteresis and spin transitions in the fractional quantum Hall effect, *Phys. Rev. Lett.* **81**, 2522–2525 (1998).

213. E.H. Rezayi and F.D.M. Haldane, Incompressible paired Hall state, stripe order, and the composite fermion liquid phase in the half-filled Landau levels, *Phys. Rev. Lett.* **84**, 4685–4688 (2000).

214. W. Pan, R.R. Du, H.L. Stormer, D.C. Tsui, L.N. Pfeiffer, K.W. Baldwin, and K.W. West, Strongly anisotropic electronic transport at Landau level filling factor $\nu = 9/2$ and $\nu = 5/2$ under a tilted magnetic field, *Phys. Rev. Lett.* **83**, 820–823 (1999).

215. M.P. Lilly, K.B. Cooper, J.P. Eisenstein, L.N. Pfeiffer, and K.W. West, Anisotropic states of two-dimensional electron systems in high Landau levels: Effect of an in-plane magnetic field, *Phys. Rev. Lett.* **83**, 824–827 (1999).

216. Y. You, G.Y. Cho, and E. Fradkin, Nematic quantum phase transition of composite Fermi liquids in half-filled Landau levels and their geometric response. *Phys. Rev. B* **93**, 205401 (2016).

217. A. Zhu, I. Sodemann, D.N. Sheng, and L. Fu, Anisotropy-driven transition from the Moore–Read state to quantum Hall stripes, *Phys. Rev. B* **95**, 201116 (2017).

218. A. Mesaros, M.J. Lawler, and E.-A. Kim, Nematic fluctuations balancing the zoo of phases in half-filled quantum Hall systems, *Phys. Rev. B* **95**, 125127 (2017).

219. K. Lee, J. Shao, E.-A. Kim, F.D.M. Haldane, and E.H. Rezayi, Pomeranchuk instability of composite Fermi liquid, *Phys. Rev. Lett.* **121**, 147601 (2018).

220. D.X. Nguyen, A. Gromov, and D.T. Son, Fractional quantum Hall systems near nematicity: Bimetric theory, composite fermions, and Dirac brackets, *Phys. Rev. B* **97**, 195103 (2018).

221. L.H. Santos, Y. Wang, and E. Fradkin, Pair-density-wave order and paired fractional quantum Hall fluids, *Phys. Rev. X* **9**, 021047 (2019).

https://doi.org/10.1142/9789811217494_0006

Chapter 6

Correlated Phases in ZnO-Based Heterostructures

Joseph Falson* and Jurgen H. Smet

*Max Planck Institute for Solid State Research,
Heisenbergstrasse 1, 70569 Stuttgart, Germany*

Over the past decade, zinc oxide based heterostructures have emerged as a high mobility platform. At low temperature, they are host to a wide array of quantum Hall features in which the role of a tunable spin susceptibility is prominent. In this chapter, we describe the background of these heterostructures, introduce the parameter space they occupy, and the exotic correlated electronic phases they unveil.

Contents

1. Introduction

The past decade has seen a number of unanticipated advances in the field of quantum Hall physics. Among these, the impact of realizing true two-dimensional materials, which may be isolated in atomically thin layers and combined into van der Waals

---

*Current address: Department of Applied Physics and Materials Science, California Institute of Technology, 1200 E. California Blvd, Pasadena, California, 91125, USA.

heterostructures with remarkable flexibility, cannot be overstated. However, great leaps have also been made in the field of epitaxy, which is the more traditional means for realizing two-dimensional systems through electrostatic confinement. One example is zinc oxide (ZnO), the focus of this chapter, which has risen from a defect plagued oxide to a versatile heterostructure for studying quantum transport of correlated electrons.

ZnO has been a focal point of oxide materials research outside the field of the quantum Hall effect for a long period of time. Owing to its wide bang-gap (3.4 eV), large exciton binding energy (60 meV), room-temperature electrical conductivity, as well as strong spontaneous and piezoelectric polarization attributes, it has been widely studied as a transparent conducting oxide and optoelectronic material. It also finds applications in cosmetics and other industrial branches. Its study has spawned a dense library of literature focused on both fundamental physical phenomena and industrial applications. From a general point of view, the strong ionicity of oxides typically renders their chemistry and crystallography much more complex than their covalently bonded semiconductor counterparts. This typically limits their crystalline quality, which may be of limited consequence in the aforementioned fields of research. In the realm of quantum Hall science, however, it is detrimental, as only samples of highest crystallinity are known to display the delicate correlated phases of interest. ZnO may therefore be considered an unlikely candidate to emerge among a new generation of ultra-high mobility low-dimensional electron systems. A number of fortuitous circumstances have facilitated its rise. Firstly, its Würzite crystal structure is at a deep thermodynamic minimum, resulting in no other stable crystal structures forming at ambient conditions. Furthermore, the relevant building blocks (Zn^{2+}, Mg^{2+} and O^{2-}) do not take multiple ionic numbers, simplifying the chemistry of donor/acceptor balancing which is a significant issue in other complex oxides.

The seeds for quantum Hall research in ZnO were sown with the advent of single crystal thin films with high purity.[1] A prime motivation for initializing such research was to achieve light emitting diodes operating in the ultraviolet region of the electromagnetic spectrum.[2] These devices, initially crafted using pulsed laser deposition (PLD) were eventually superseded by molecular beam epitaxy grown films,[3] which importantly utilized single crystal ZnO substrates, as opposed to sapphire or $ScAlMgO_4$ which imposed a finite lattice mismatch on the grown ZnO films. More generally, molecular beam epitaxy has clearly emerged as a powerful tool for synthesizing high quality oxide films.[4] In retrospect, the availability of high quality ZnO substrates was the single most important factor in enabling the dramatic improvements in film quality which occurred in the years after 2010.

In this chapter, we will provide an overview of the rapid development of this system. After introducing the state-of-the-art growth techniques, we will review general transport phenomena, including an analysis of the relevant material parameters such as the effective mass and spin susceptibility. Finally, we will delve into the fractional quantum Hall effect, which has proven to be quite rich in this system. The large spin susceptibility of the charge carriers is particularly instrumental to gain control over the spin degree of freedom. It offers a convenient paradigm to tune between correlated ground states, including fractional quantum Hall, charge density wave and compressible phases.

2. ZnO-Based Heterostructures

When interfacing two materials with quantitatively different electric polarizations in thin film epitaxy, a discontinuous polarization field (**P**) appears at their heterojunction. The system will attempt to screen this discontinuity through a number of means, including the formation of defects and alloy gradient profiles. If these mechanisms are however efficiently suppressed in the sample preparation stage, another option is to capture itinerant carriers, which form a 2DES. This is for instance the case in AlGaN/GaN heterostructures,[6] and also the ZnO based heterostructures described here. The crystal structure of ZnO is shown in Fig. 1(a). It has an inversion asymmetric Würtzite structure and hosts a spontaneous $\mathbf{P} \approx 5 \ \mu C/cm^2$

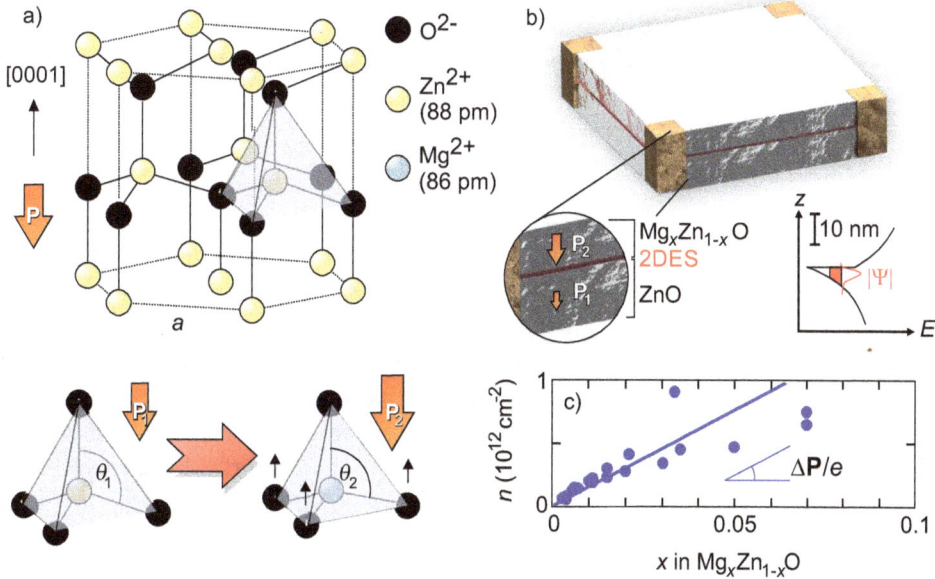

Fig. 1. (a) The Würzite crystal structure of ZnO with a local tetrahedron highlighted. The substitution of Mg^{2+} into this tetrahedron results in a modified bond angle and hence amplified non-centrosymmetricity of the crystal. This culminates in an altered polarization of the $Mg_xZn_{1-x}O$ alloy. (b) Schematic overview of a $Mg_xZn_{1-x}O/ZnO$ sample in van der Pauw geometry. The 2DES is induced at the interface of ZnO and $Mg_xZn_{1-x}O$ and exhibits a wavefunction[5] (Ψ) with a spread in the growth direction of approximately 5 nm for $x = 0.014$. (c) Charge density n as a function of x in $Mg_xZn_{1-x}O$.

in the c-axis direction.[7,8] The origin of this polarization is subtle,[9] and is comprised of both electronic and ionic contributions. For the sake of understanding our heterostructure, we can isolate the problem into Zn^{2+} (or Mg^{2+}) and O^{2-} sublattices. Upon inspection of local tetrahedra, the ionic contribution can be understood by the displacement of Zn^{2+} ions towards the basal plane. By substituting Zn^{2+} with Mg^{2+}, the latter does not act as an electron donor as it is isovalent. However, the bond angles are modified and the non-centrosymmetricity is amplified, due to a preference of MgO to form octahedral co-ordinated bonds. This process modifies $\mathbf{P}$ in the $Mg_xZn_{1-x}O$ phase and when a $Mg_xZn_{1-x}O$ slab is epitaxially grown on top of ZnO a polarization mismatch $\Delta\mathbf{P}$ with a roughly linear dependence on x is generated at the heterointerface. The heterostructure is schematically shown in Fig. 1(b). The $Mg_xZn_{1-x}O$ layer is not directly deposited on top of the ZnO substrate, but rather on an epitaxially grown ZnO buffer layer covering the substrate as the grown film has a lower impurity content than the substrates received from the manufacturer. The 2DES wavefunction resides primarily within the ZnO layer with a tail in $Mg_xZn_{1-x}O$. It has a width of approximately 5 nm in the growth direction for $x = 0.014$, as depicted in Fig. 1 from self-consistent calculations.[5] Decreasing (increasing) x has the effect of widening (narrowing) the electronic wavefunction. The second electronic subband of the interfacial potential well is observed[10] to be occupied at charge carrier densities of $n \approx 4 \times 10^{12}$ cm^{-2}, which corresponds to $x \approx 0.27$. The charge density of the accumulated 2DES is correlated with the magnitude of $\Delta\mathbf{P}$ divided by the elementary charge e and hence also rises approximately linearly with x as illustrated in Fig. 1(c).

2.1. *Molecular beam epitaxy*

Here we briefly describe the molecular beam epitaxy growth of devices that will later be addressed in more detail. The procedure stems from the University of Tokyo where the growth took place.[11,12] The growth proceeds at a base pressure of 5×10^{-9} Pa when the chamber is cooled with liquid nitrogen. It is performed on single crystal Zn-polar [0001] ZnO substrates. These are highly conducting ($\rho \leq 4$ Ωcm) which is reflective of a low residual lithium impurity content that would otherwise compensate native defects present in the crystal. The substrates are prepared for epitaxy by cleaning in an acid solution as described in Ref. 13. The growth chamber hosts two Zn (7N5) and two Mg (6N) effusion cells for metallic flux. Importantly, state-of-the-art samples employ pure ozone (O_3) as an oxidizing agent. This is a very low energy and highly pure process as the ozone is generated and distilled in its liquid phase external to the MBE apparatus. The high purity O_3 gas is injected into the chamber through a feedback controlled piezoelectric leak valve designed to maintain constant O_3 pressure flow. The O_3 is transported to the vicinity of the substrate using a temperature controlled gas cell held at 80°C. The substrate is rotated during the growth at a speed of 7 revolutions per minute to enhance the uniformity of the deposited film.[14] In the typical heterostructure, shown in Fig. 1(b), the homoepitaxial ZnO layer and the alloyed $Mg_xZn_{1-x}O$ layer have both a thickness of about 500 nm. When $x \geq 0.02$, it is possible to determine the Mg content and the total thickness of the $Mg_xZn_{1-x}O/ZnO$ MBE layer using X-ray diffraction techniques. However, the highest mobility samples have $x \leq 0.02$, rendering this technique ineffective. Therefore, the thickness of the total grown film is most easily evaluated by a needle surface profiler relative to the edge of the substrate that is shielded during the growth by the substrate holder. For $x \leq 0.02$, photoluminescence may be employed to measure the excitonic energies originating at the $Mg_xZn_{1-x}O/ZnO$ interface to determine x.[15]

2.2. Transport characteristics

At high temperature, the $Mg_xZn_{1-x}O/ZnO$ samples display bulk conductivity due to electrons donated by native defects within the grown films and substrate.[20] However, upon cooling to $T \approx 100$ K, these mobile carriers are localized and the heterointerface becomes the dominating contributor of electrical conductivity. We therefore relegate the remainder of this chapter to temperatures $T \leq 2$ K where the bulk is strongly insulating and the effects of phonon scattering are largely mitigated. In this regime, the low frequency transport of carriers can be described classically by the Drude theory of dc conductivity given by $\sigma = en\mu$. Here, the electron mobility μ emerges and is determined by elastic scattering events. A number of 2DES now display fantastic electron mobilities exceeding 3×10^7 cm^2/Vs in GaAs/AlGaAs quantum wells,[21–23] 2×10^6 cm^2/Vs in AlAs/AlGaAs quantum wells,[24] 2×10^6 cm^2/Vs in Si/SiGe quantum wells,[25] and 1×10^6 cm^2/Vs in InAs quantum wells surrounded by either InGaAs[26] or AlGaSb.[27] For these systems with extremely clean interfaces, it is generally the background impurity content that is unintentionally incorporated during the growth which limits the final low temperature mobility.

The chronological advances of μ for the $Mg_xZn_{1-x}O/ZnO$ system are displayed in Fig. 2. The 2007 films benefited from optimized growth conditions using pulsed laser deposition and were the first samples to display the integer quantum Hall effect.[16] Samples reported since 2009 are fabricated using MBE.[17] The data for 2010 were obtained for sample growth with an oxygen radical cell as the oxidizing agent. These films displayed fractional quantum Hall features, lifting the material system into the realm of correlated electron physics.[18] Subsequently, purified ozone has been used exclusively for growth. This change immediately resulted in an increase in peak mobility to 700,000 cm^2/Vs.[11] It is suspected that this increase is partially due to the mitigation of Si impurities. The latter were detectable in secondary ion mass spectroscopy (SIMS) for films grown using the oxygen radical source. The cell crucibles are made of quartz and were therefore likely the origin. This issue was however entirely eliminated by switching to ozone. The data for 2016 represents the state-of-the-art at the time of writing this chapter. A number of improvements covering the choice of substrate holder material, heterostructure design, and growth conditions

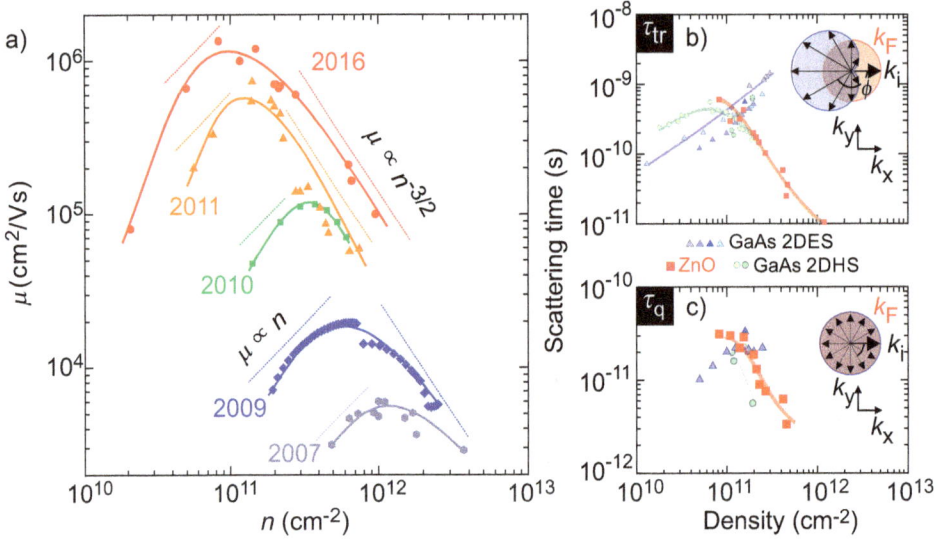

Fig. 2. Historical development of the electron mobility μ as a function of charge density n. Data are taken from references: 2007,[16] 2009,[17] 2010,[18] 2011[11] and 2016.[12] Comparison of (b) transport (τ_{tr}) and (c) quantum (τ_{q}) scattering times for ZnO and GaAs-based 2DES and 2DHS as a function of density. Adapted from Refs. 12 and 19. The insets schematically represent the Fermi circle (shaded pink) with the weighting factors (blue shading) which distinguish backscattering transport (b) and isotropic quantum scattering rates (c). Lines are guides for the eye.

were responsible for this advance, as detailed in Ref. 12. The record mobility is currently 1.3×10^6 cm^2/Vs. Dotted guidelines in Fig. 2 illustrate mobility limiting mechanisms for each growth generation. On the low density side the mobility is mainly governed by charged impurity scattering setting the boundary to $\mu \propto n$ due to enhanced screening of disorder as the kinetic energy is increased. Towards high densities, interface roughness and/or alloy scattering prevails and $\mu \propto n^{-3/2}$ due to scattering off the disordered Mg$_x$Zn$_{1-x}$O cap as x is made larger. The interface roughness scattering limitation has proven difficult to overcome, as it results from the intrinsic alloying of Mg required to generate a 2DES. It inevitably induces disorder at the interface. Much more remarkable has been the increase in μ for the dilute n regime where we may now realize metallic samples with $n \approx 0.2 \times 10^{11}$ cm^{-2}. This is no doubt a manifestation of the significant suppression of impurities with each successive sample generation.

Comparing systems based on mobility alone is however insufficient for multiple reasons. Firstly, the raw mobility incorporates material specific parameters such as the effective mass m^* through the relationship $\tau_{\mathrm{tr}} = \mu m^*/e$, where τ_{tr} is the time between scattering events in the Drude picture. Secondly, the mobility is primarily reduced by backscattering and therefore it is skewed towards large angle scattering involving a large change in in-plane momentum, as per Eq. (1),

$$\frac{1}{\tau_{tr}} = \frac{m^*}{\hbar^3} \int_0^{2\pi} \frac{d\phi}{2\pi} W(q)(1 - \cos\phi). \tag{1}$$

Here, $q = 2k_F \sin(\phi/2)$ is the in-plane momentum transferred by scattering, $\hbar$ is the reduced Planck constant, $k_F = (2\pi n)^{1/2}$ the Fermi wave number, $W(q)$ the Fourier component of the correlation function of the screened disorder potential, and ϕ the scattering angle. The $(1 - \cos\phi)$ factor reflects that small-angle scattering with $\cos\phi \ll 1$ is less relevant for the mobility as no backscattering takes place.

An alternative figure of merit is the single particle lifetime or quantum scattering time (τ_q) which essentially describes the time spent in an eigenstate before scattering.[28] It is expressed as Eq. (1) but with the factor $(1 - \cos\phi)$ replaced by unity. This change implies that all scattering events are weighted equally, regardless of the angle. Experimentally, τ_q may be quantified through an analysis of the low field Shubnikov–de Haas (SdH) oscillations, since the broadening of Landau levels (Γ) from disorder is related to τ_q according to $\Gamma = \hbar/2\tau_q$. A comparison between different material systems is presented in Fig. 2. Figure 2(b) plots the transport scattering time τ_{tr} and Fig. 2(c) the quantum scattering time τ_q (see Ref. 19 for a full analysis). In GaAs, the ratio of these two time scales is large indicating predominant small angle scattering. In ZnO the opposite is true. As a result, transport scattering times in ZnO are substantially lower than in GaAs, however the quantum scattering times that can be achieved are almost identical and this, as will be shown later, is reflected in excellent fractional quantum Hall data on ZnO based heterostructures. For the sake of completeness we note that while τ_q may be a far better metric to quantify quality, open issues remain as it does not necessarily for instance correlate to the strength of the incompressible ground state at filling factor $\nu = 5/2$ in GaAs.[22,29,30]

3. Interaction Effects in the Fermi-Liquid

Interacting metals may be accurately described by the Fermi-liquid framework introduced by Landau.[31–34] In contrast to a non-interacting system where an occupation number of either 1 or 0 as $T \to 0$ above or below the Fermi momentum k_F defines eigenstates of the system, a distribution of fractional occupation numbers describes an ensemble of a Fermi liquid's long lived excited quasiparticles. Residual interaction between these quasiparticles results in a renormalization of the systems parameters, including the effective mass, spin susceptibility and heat capacity. A full review of this topic is beyond the goal of this chapter. Here we present an experimental analysis of the effective mass and spin susceptibility in order to gain insight into the strength of correlation effects in the system.

In weakly interacting systems, the Landau parameters defining the strength of interaction between quasiparticles remain approximately constant and therefore the renormalization of band parameters is a weak effect often to be neglected. This is not the case in ZnO, which owing to its relatively small dielectric constant ($\epsilon = 8.5\epsilon_0$) and heavy effective mass ($m^* = 0.3m_0$), has a Coulomb interaction strength much larger than the kinetic energy of carriers. The strength of interaction in parabolically dispersing two-dimensional systems may be understood through the dimensionless parameter r_s given by,

$$r_s = \frac{1}{(\pi n)^{1/2} a_B}. \tag{2}$$

Here, $a_B = 4\pi\epsilon\hbar^2/m^* e^2$ is the effective Bohr radius of the material. In the case of a single-valley ($g_v = 1$), spinful ($g_s = 2$) system, this is equivalent to the ratio of the Coulomb to kinetic energy. As per Eq. (2), the strength of interactions is amplified when the charge density is reduced. In monolayer graphene, $r_s = e^2/\epsilon\hbar v_F$, where v_F is the Fermi velocity of the Dirac cone, resulting in $r_s \leq 2.2$.[35] However, as shown in Fig. 6(a), r_s readily exceeds 5 in ZnO and approaches 30 in the dilute limit. Accordingly, strong renormalization of the system's band parameters is encountered. This renormalization can be experimentally captured by measuring the effective mass (m^*) and spin susceptibility ($g^* m^*/m_0$) of carriers in transport.

Table 1. Band parameters of prominent parabolic two-dimensional systems, the value of r_s is calculated for a charge density of 1×10^{11} cm^{-2} under the assumptions listed in the table under Notes.

System	m^*/m_0	ϵ/ϵ_0	g_v	r_s	Notes
GaAs electron	0.069	13	1	1.8	
ZnO electron	0.03	8.5	1	11.2	
AlAs electron	0.46	10	1,2	15.6	
GaAs hole	0.2–0.39	13	1	5.2	Assuming $m^* = 0.2m_0$
SiGe/Si electron	0.19	12.6	2	10	
Bilayer graphene	0.03	2.5	2	8.5	On SiO$_2$

	v_F	ϵ/ϵ_0	g_v	r_s	
Monolayer graphene	1×10^6 m/s	2.5	2	0.8	On SiO$_2$

Owing to the dependence of system parameters on the charge density, we will construct this chapter by calling upon a range of samples. These are summarized in Table 2.

3.1. Effective mass

The effective mass may be obtained through temperature dependent measurements of the Shubnikov–de Haas oscillations,[19] and also by microwave resonance techniques where the incident radiation energy equals that required for an excitation between Landau levels split by the cyclotron energy. These excitations emerge in multiple experiments which either yield the bare band mass or the interaction renormalized effective mass. The most widely employed detection method

Table 2. Summary of the charge density (n), spin susceptibility (g^*m^*/m_0), and electron mobility (μ) of MgZnO/ZnO samples presented in this chapter.

Sample	n (10^{11}cm^{-2})	g^*m^*/m_0	μ (cm^2/Vs)	Fig.	Ref.
S1	2.05	2	300,000	4	36
S2	1.8	2.1	300,000	5	
S3	1.6	2.4	1,200,000	7	19
S4	1.9	2.1	600,000	8	37
S5	2.9	1.65	450,000	9	
S6	5.8	1.35	190,000	10, 15, 19	38
S7	2.3	1.9	530,000	11, 12, 13, 14	39
S8	4.5	1.48	350,000	15, 16, 17, 18, 19	38

relies on bolometric heating of the 2DES due to resonant absorption of radiation near the cyclotron resonance (CR) condition where the angular frequencies coincide $\omega_{\text{cyc}} = \omega_{\text{mw}}$. The dissipation at the resonance generates a resistance peak in the longitudinal magnetoresistance. Alternatively, the drop in the transmitted radiation through the sample at the resonance may be detected on the backside. Both approaches deliver the bare band mass, because they address the cyclotron resonance in the limit $q \to 0$ in a uniform 2DES for homogeneous incident radiation for which Kohn's theorem is applicable.[41] To access the renormalization effects, it is required to probe the system at large momentum $q \to \infty$. In addition to traditional transport methods, this condition is satisfied in the so-called "microwave induced resistance oscillations" (MIRO) setting. These oscillations occur primarily in high mobility[42] and also in high density GaAs samples,[43] and more recently

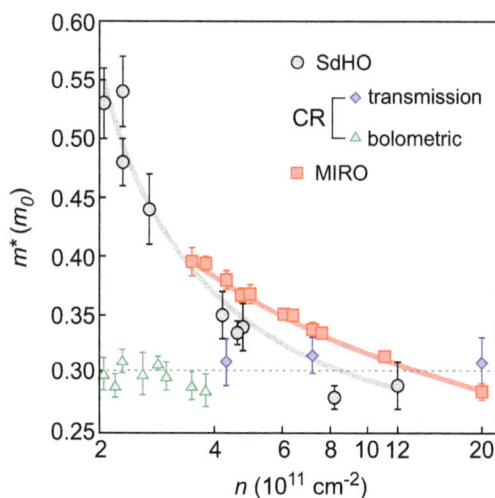

Fig. 3. The effective mass (m^*) as a function of n extracted from four distinct experiments in MgZnO/ZnO-heterostructures: (1) temperature dependence of the Shubnikov–de Haas oscillations,[19] (2) microwave transmission near the CR, (3) bolometric response in the longitundinal resistance near the CR, (4) microwave induced resistance oscillations.[40]

Ge/SiGe heterostructures.[44] They involve individual scattering events of quasiparticles which are subject to renormalization effects.[40,45–47] A detailed analysis has
been performed in Ref. 40, which we summarize in Fig. 3. The CR response in both
transmission and resistively detected bolometric studies offer a convenient reference
level of the bare mass close to the ZnO band mass $m^* = 0.3m_0$, while the MIRO
and SdHO reveal a steady increase in m^* as n is reduced and interaction effects get
amplified.

3.2. The coincidence method

A powerful experimental technique, frequently employed in the study of ground
states and spin properties of two-dimensional systems, is the selective enhancement
of the Zeeman energy (E_Z) over the cyclotron energy (E_{cyc}).[48] The experimental
configuration is depicted in Fig. 4(c). The method exploits the reduced dimen-
sionality of the conducting layer. When tilting the sample away from the plane
perpendicular to the total magnetic field vector with amplitude B_t, the orbital degree of freedom of the charge carriers only responds to the magnetic field component
perpendicular to the two-dimensional system, B_p. The spin degree of freedom, however, is affected by the total field B_t. Here, $B_p = B_t \cdot \cos\theta$. This results in different
functional dependencies of the Zeeman energy and cyclotron energy, so that level
crossings can be induced. The discretized energy levels are given by

$$E_N = \left(N + \frac{1}{2}\right) E_{cyc} \pm \frac{1}{2} E_Z = \left(N + \frac{1}{2}\right) \frac{\hbar e B_p}{m^*} + g\mu_B B_t \sigma, \tag{3}$$

with g^* the electron g-factor, μ_B the Bohr magneton and σ the spin of carriers
$(+\frac{1}{2}, -\frac{1}{2})$, denoted as $\downarrow, \uparrow$ throughout this text. N defines the Landau level index,
whose role will be extensively explored in Sec. 5. The evolution of this ladder of
spin split Landau levels as a function of tilt angle θ (or more precisely $1/\cos\theta$) is
plotted in Fig. 4(b) for a fixed value of B_p. The levels are identified by their orbital
quantum number (N) and spin orientation (up or down). As the angle θ is adjusted
in experiment, the Zeeman energy gets selectively boosted while maintaining a
constant cyclotron energy. Eventually levels of opposite spin and different orbital
index cross. This occurs at the condition

$$\frac{E_Z}{E_{cyc}} = \frac{g^* m^*}{2m_0 \cos\theta} = j, \tag{4}$$

where j is the level crossing index. It is equal to the difference between the orbital
indices of the coinciding levels.

The ability to induce level crossings by simply rotating the sample is a powerful
addition to our toolbox for studying 2D systems. As evident from the above expression, it immediately enables to quantify the spin susceptibility $g^* m^*/m_0$ when
the levels coincide. However, an important prerequisite is that $g^* m^*/m_0$ is large
enough, i.e. the ratio of the Zeeman to cyclotron energy should at least be of order
0.1 as otherwise the coincidence is hard to reach with typically available dc magnetic

fields.[49] This condition is satisfied in the case of ZnO as well as a number of other 2D charge carrier systems including but not limited to accumulation layers in Si,[50] AlAs quantum wells,[51] $Si/Si_{1-x}Ge_x$ heterostructures[52] and p-GaAs quantum wells, although the latter is complicated due to the anisotropic g-factor of holes.[53] We recall that the Landau level filling $\nu = hn/eB_p$ also depends only on the perpendicular component of the applied magnetic field. Therefore, in the experiment depicted in Fig. 4(b) the filling factor remains constant. If the uppermost occupied Landau level is only partially filled, the chemical potential (μ_ν) will trace this uppermost level as it takes on different orbital indices and spin orientation past each crossing. This is very instrumental in the fractional quantum Hall regime to explore the role of the orbital wave function, as we will discuss in Sec. 5.

The data in Fig. 4(a) depicts such a coincidence experiment performed at low magnetic field on sample S1 with a density of $n = 2.05\times10^{11}$ cm^{-2}.[36,54] The color rendering represents the longitudinal resistance (R_{xx}) in the B_p vs $1/\cos\theta$ plane. The temperature was set at $T = 450$ mK to suppress correlated ground states which will be discussed in later sections. Firstly, a non-monotonic longitudinal resistance at integer fillings of $\nu = 6$ and 5, as shown in Figs. 4(d) and 4(e), is observed as a function of $1/\cos\theta$. This is associated with a sudden increase in the density of

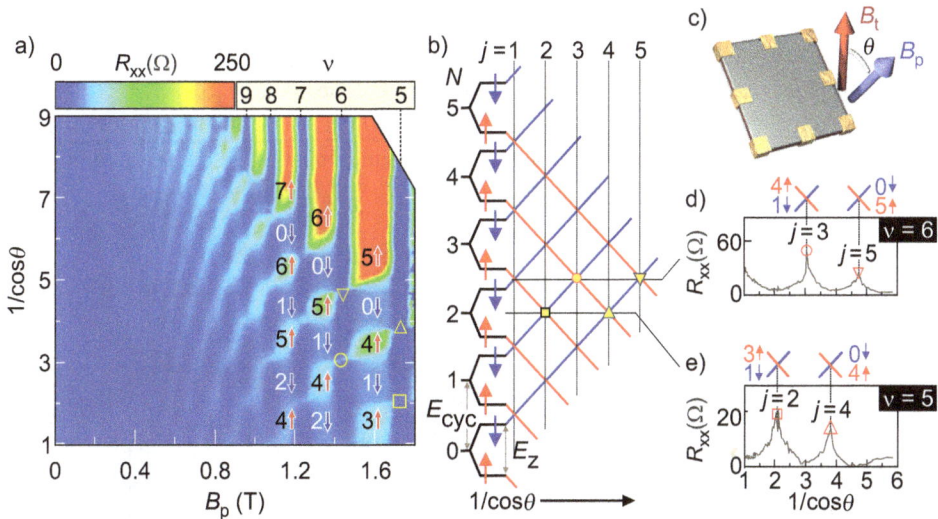

Fig. 4.   Overview of the coincidence method, demonstrated using sample S1 ($n = 2.05\times10^{11}$ cm^{-2}). (a) Map of the recorded longitudinal resistance as a function of B_p and $1/\cos\theta$. (b) Evolution of the fan of spin split Landau levels for a fixed value of B_p but with increasing $1/\cos\theta$. At specific angles, opposing spin branches of levels with different orbital indices will intersect. The difference in the value of the orbital indices is marked at the top ($j = 1, 2, \ldots$). (c) MgZnO/ZnO heterostructure tilted by an angle θ relative to the orientation of the total magnetic field vector with amplitude B_t. (d and e) R_{xx} as a function of $1/\cos\theta$ at filling factor $\nu = 6$ and 5 displays a sharp spike in the resistance and a breakdown of the integer quantum Hall state at the coincidence of Landau levels with opposite spins and different orbital indices. Adapted from Ref. 36.

states at the chemical potential μ_ν when levels coincide, as shown schematically using the Landau fan. More spectacular is the clear "chequer-board" pattern that appears.[36] It is associated with the spin orientation ($\uparrow, \downarrow$) of the carriers occupying the uppermost partially filled level at μ_ν.[54] By comparing the data with the Landau level fan presented in Fig. 4(b), it becomes apparent that whenever μ_ν is pinned in a level with majority spin orientation, the resistance is high relative to the case where the adjacent minority spin level is partially filled at the same angle. This rule allows us to assign to each tile of the chequerboard the quantum numbers of the level that is currently partially filled. These quantum numbers have been superimposed on top of the color map presented in Fig. 4(a). Some coincidences for $j = 1, 2$ and 3 have been marked by symbols in Figs. 4(a) and 4(b). For the filling factor range covered in the experiment of Fig. 4(b) all coincidences at a given j occur approximately at the same tilt angle. At lower filling however a distortion of the chequerboard pattern may occur due to the filling factor dependent spin population difference. The large swing in the degree of electron spin polarization at low filling causes a filling factor dependent exchange enhancement.

Given sufficiently weak spin-orbit coupling (as is the case in ZnO, see Ref. 55), the process of level crossings is expected to be an Ising-like spin-flop in the Hartree–Fock framework with no intermediate coherence between the opposing spin levels. Disorder related density fluctuations will then inevitably give rise to a pattern of domains with two distinct polarization degrees. This ferromagnetic behavior leaves clear signatures in magnetotransport as displayed in Fig. 5(a) in the vicinity of $\nu = 2$ at the $j = 1$ coincidence position recorded on sample S2 ($n = 1.8 \times 10^{11}$ cm^{-2}). A resistance peak appears and exhibits hysteresis as the magnetic field direction is reversed. This behavior has also been reported in other 2D material

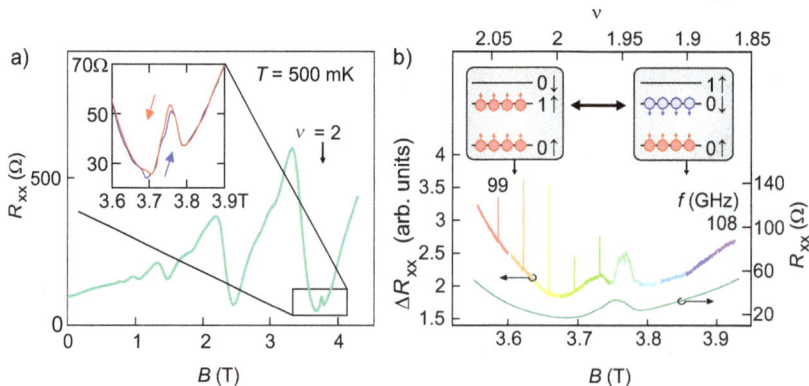

Fig. 5.   Ferromagnetic domains and electron spin resonance across a Landau level crossing demonstrated using sample S2 ($n = 1.8 \times 10^{11}$ cm^{-2}). (a) Hysteretic magnetotransport features in the vicinity of $\nu = 2$ at $j = 1$, with the up- and down-sweeps highlighted in the inset. (b) Magnetotransport under microwave illumination, showing the difference between light and dark traces (ΔR_{xx}) at frequencies from 99~108 GHz in 1 GHz steps. The schematics display the energy levels for the high and low spin polarization configurations across the level crossing.

systems.[56,57] The modified spin configuration of the electronic system beyond the level crossing can also be detected in the vicinity of the resistance spike feature in an electron spin resonance experiment as illustrated in Fig. 5(b). At magnetic fields below the resistance peak, the electron system is fully spin polarized and microwave radiation with a photon energy matching the bare Zeeman energy ($\hbar\omega_{MW} = g\mu_B B_t$) can excite electrons to the unpopulated Landau levels with opposite spin while preserving the orbital index of the initial electronic state. This causes sharp peaks in the resistance. Note that the g-factor in the above expression is the bare spin precession factor not modified by renormalization effects as the resonance is probed in the limit $q \to 0$. The discrete shift in the electron spin polarization beyond the coincidence towards an unpolarized configuration is reflected in the absence of any electron spin resonance features, since Pauli blocking and the need to conserve the orbital index suppresses any spin flip transitions.

3.3. *Spin susceptibility*

Contrary to the ESR features, the coincidence peak in transport is governed by the renormalized quantities g^* and m^* and permits a quantification of the spin susceptibility of the charge carriers. We plot g^*m^*/m_0 as a function of n in Fig. 6. We interpret a large portion of the renormalization of g^*m^*/m_0 to originate from the increase in m^*, discussed previously in Sec. 3.1. We note that some quantitative corrections to g^*m^*/m_0 are necessary when the total spin polarization P of the filled level increases. More accurately, Ref. 36 found a correction of $g^*m^*/m_0 = g^*m^*/m_0|_{P=0} + 0.38P$, for a sample S1 with $r_s = 8$ ($n = 2\times10^{11}$ cm^{-2}), where $g^*m^*/m_0|_{P=0}$ is the extrapolated spin susceptibility when $P = 0$. In Fig. 6 we typically neglect this correction as we usually perform our analysis in circumstances where P remains small.

We also note that the application of a field parallel ($B_{||}$) to the 2DES plane may be used to determine g^*m^*/m_0. Under the condition $B_{||} = 2E_F/g^*\mu_B$, the Zeeman energy is enhanced to a level where it equals twice the zero field kinetic energy. Hence, it becomes energetically preferable for the system to polarize its spin at the cost of increasing E_F. The reduction of the density of states causes an increase in the resistivity and the experimental signature for full spin polarization is a saturating R_{xx} with $B_{||}$, as no further modification to the density of states occurs in the absence of cyclotron motion.[58] Such an analysis in our heterostructures produces values of g^*m^*/m_0 in agreement with the results of the coincidence method.

The n-dependence of g^*m^*/m_0 has a profound impact in the remaining sections of this chapter. By referring to Eq. (4) and Fig. 6, it is clear that not only the application of an in-plane magnetic field can produce coincidences between spin split Landau levels. Also tuning of the charge carrier density n can modify the energy landscape. This is presented schematically using the panels on the right-hand side of Fig. 6. Each time g^*m^*/m_0 takes on an even-integer value, j too passes through

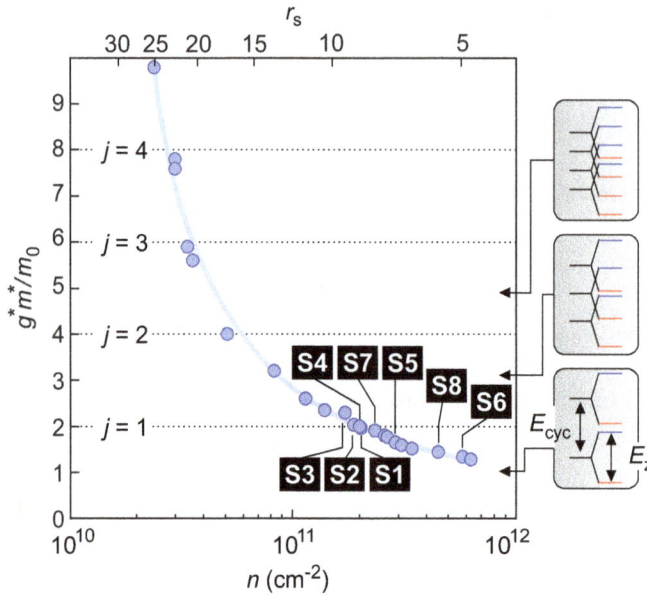

Fig. 6.   Spin susceptibility extracted with the coincidence method described in Sec. 3.2 as a function of n and r_s calculated using the band parameters $m^* = 0.3\ m_0$ and $\epsilon = 8.5\epsilon_0$. The panels on the right-hand side schematically present spin-split Landau levels illustrating the magnitude of cyclotron to Zeeman energy for portions of the parameter space with $g^* m^*/m_0$ between the horizontal dotted lines. Samples S1–S8 utilized in this chapter are indicated.

an integer value, signifying a crossing of the opposing spin branches of neighboring Landau levels, as per Eq. (4). The data in Fig. 6 also point to a divergent spin susceptibility and effective mass in the dilute limit. This is an indication that the Landau parameters are drifting toward a critical value for the appearance of a Stoner-type instability,[59] which remains an exciting research frontier.

4. The Quantum Hall Effects in ZnO

The integer quantum Hall effect in MgZnO heterostructures was first observed in 2007 in films grown by pulsed laser deposition, as was briefly discussed in Sec. 2.1.[16] Apart from the orbital degree of freedom, charge carriers only possess real spin as an additional quantum number and therefore integer quantum Hall behavior resembles that of GaAs based 2D electron systems except for an altered strength of Shubnikov–de Haas and quantum Hall minima in the longitudinal resistance at odd and even filling numbers of the spin split Landau levels. For an electron density of approximately 2×10^{11} cm^{-2}, the gap between Landau levels with the same orbital index but opposite spin is larger than the gap between adjacent Landau levels with different orbital index. This is schematically shown in the bottom inset to Fig. 6, and follows from Eq. (4) using the values of $g^* m^*/m_0$ presented in Fig. 6. Minima at even filling factor will therefore disappear first as temperature is increased

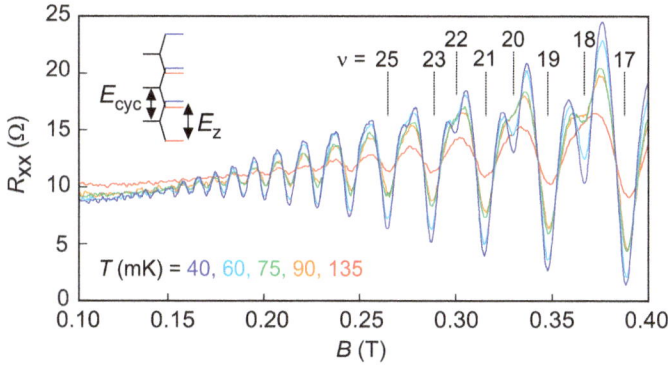

Fig. 7. Temperature dependent low field transport of sample S3 ($n = 1.6\times10^{11}$ cm^{-2}), displaying the superior robustness of integer quantum Hall states and SdH oscillations at odd integer filling factors compared to even fillings. The inset displays the schematic Landau levels split by the Zeeman term which slightly exceeds the cyclotron gap.

or the magnetic field lowered. An example is illustrated in Fig. 7 for sample S3 ($n = 1.6\times10^{11}$ cm^{-2}), which results in an energy diagram where E_Z slightly exceeds E_{cyc}, but still produces a constellation where the gap between Landau levels with differing orbital index and spin is much smaller than that of the same orbital index with opposing spin, as shown in the inset. Having established the dependency of g^*m^*/m_0 on n, we will utilize multiple samples to explore the parameter space of this effect from this point on. These samples were summarized in Table 2. Since no other material specific peculiarities exist, the remainder of this section will focus on the fractional quantum Hall effect.

Figure 8 plots the longitudinal resistance for sample S4 up to a perpendicular magnetic field of 32.5 T. This trace provides a good overview of the observable fractional quantum Hall states in state-of-the-art samples. The composite fermion model has been a particularly instrumental description of the fractional quantum Hall effect.[60] In a nutshell, composite fermions are topological bound states of electrons and an even number $2m$ of quantized vortices, whose formation takes care of Coulomb interaction so that only weak residual interactions remain among these quasi-particles. In a mean field approximation they experience a drastically reduced magnetic field. For example at half filling of the lowest Landau level, i.e. $\nu = 1/2$, composite fermions with two bound vortices ($m = 1$) experience no field and form a compressible Fermi liquid in the lowest Landau level ($N = 0$).[61] Upon deviation from exact half filling, they execute cyclotron motion in an effective field $\Delta B = B - B_{\nu=1/2}$ as proven in the past in a series of experiments highlighting the commensurability between the cyclotron radius and some externally imposed length scale such as for instance one-dimensional and two-dimensional periodic modulations or orifices separated at the micrometer scale.[62–70]

The quantization of the composite fermion cyclotron motion gives rise to a fan of composite fermion Landau levels. Complete filling of such levels generates

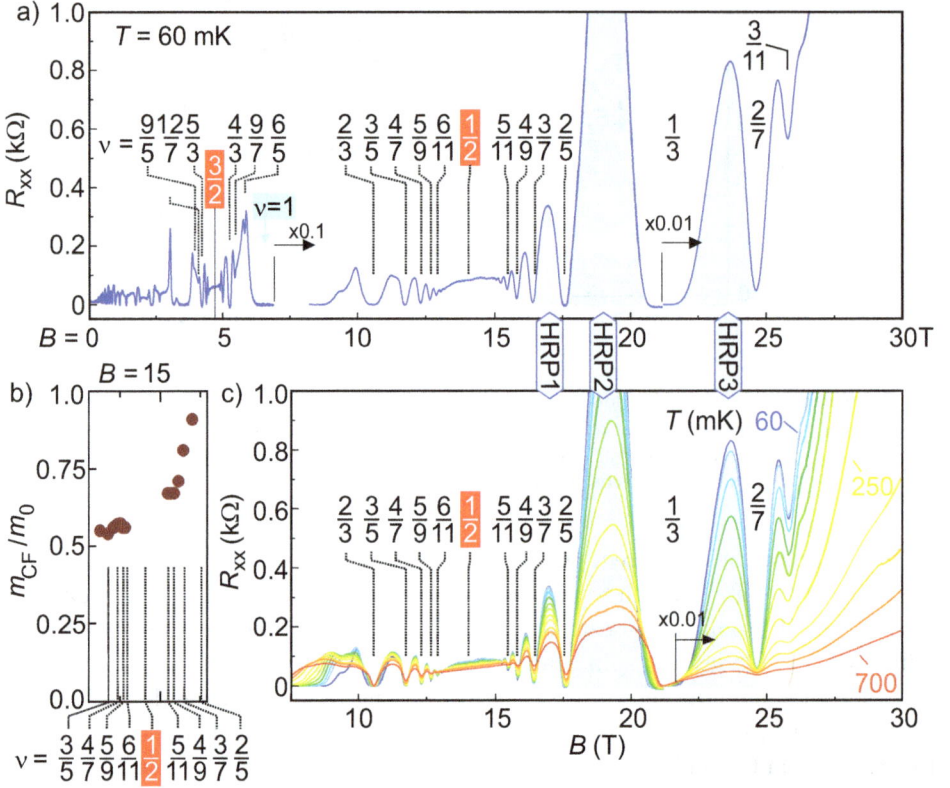

Fig. 8. Overview of the transport properties of a high mobility MgZnO/ZnO heterostructure (sample S4, $n = 1.9 \times 10^{11}$ cm^{-2}) in high magnetic fields. (a) Base temperature magnetotransport as a function of perpendicular field. Notable fractional quantum Hall features are indicated. (b) Composite fermion mass as a function of ν. (c) Temperature dependence of transport highlighting high resistance phases 1, 2 and 3. We refer the reader to Fig. 9 for a close-up view of transport for $\nu > 1$. Adapted from Ref. 37.

Shubnikov–de Haas oscillations and integer quantum Hall minima in the longitudinal resistance that coincide with the fractional quantum Hall effect where the liquid condenses into a gapped ground state at rational filling factors

$$\nu = \frac{p}{2mp \pm 1}. \tag{5}$$

Here, p is the integer number of completely filled composite fermion levels. The minus sign in the denominator corresponds to when the effective field is anti-parallel to the real field. Focusing on magnetotransport around $\nu = 1/2$ in Fig. 8 using sample S4, the sequence of fractional quantum Hall features brings no surprises and is seen to be well described by Eq. (5), with $\nu = 2/3$ and $\nu = 2/5$ corresponding to $p = 2$ but in opposing directions of the effective magnetic field. A similar exercise can be performed to identify higher order fractional quantum hall states, which can be clearly observed around the half filled lowest Landau level at fillings up to

$\nu = 6/13$, corresponding to $p = 6$. An incipient 7/15-state ($p = 7$) may also be discerned. A similar sequence of fractional features is seen when electrons capture an increased number of vortices (i.e. $m = 2$). These correspond to "four-flux" composite fermions. The quantization of their orbital motion is responsible for the states at $\nu = 2/7$ ($p = 2$) and 3/11 ($p = 3$).

A similar scenario of compressibility occurs at $\nu = 3/2$ where holes in the upper spin branch of the lowest Landau level form the basis for dual vortex attachment. They experience a reduced field of $\Delta B = 3(B - B_{\nu=3/2})$ with the factor of 3 reflecting that only one third of the carriers participate in transport at the chemical potential μ_ν. While some fractional quantum Hall states around $\nu = 3/2$ can be seen in Fig. 8 using sample S4, we point the reader towards a data set recorded on the state-of-the-art sample S5 in Fig. 9. It presents the region of interest and reveals fractional quantum Hall behavior up to filling $\nu = 14/7$ and 13/9 corresponding to 5 filled composite fermion levels ($p = 5$). The fractional quantum Hall effect and the applicability of the composite fermion model in higher Landau levels is more subtle and will be covered later in Sec. 4.2.

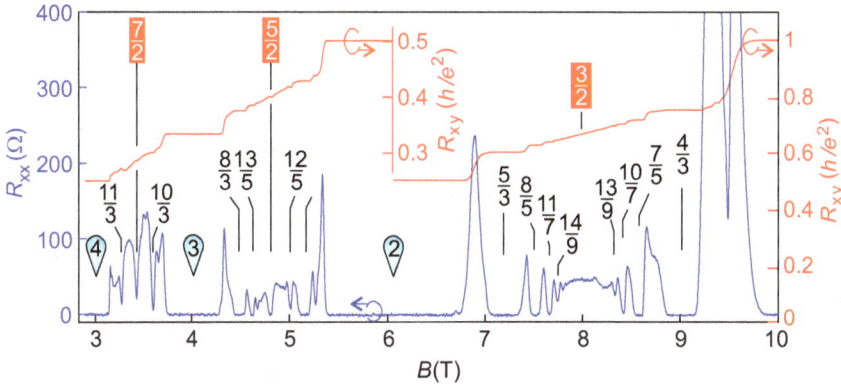

Fig. 9. A portrait of the fractional quantum Hall effect in ZnO for $\nu > 1$ recorded on sample S5 ($n = 2.9 \times 10^{11}$ cm^{-2}). Taken at $T \approx 10$ mK, the partial fillings of $\nu = 7/2$ and 5/2 are seen to condense into gapped ground states with quantized Hall resistances, while $\nu = 3/2$ remains compressible.

By interpreting the oscillations around $\nu = 1/2$ as the Shubnikov–de Haas effect of composite fermions, an effective mass (m_{CF}) may be extracted from their temperature dependent behavior. Such an analysis has been carried out in Ref. 37. The effective mass is similar in magnitude to that of electrons, although this is of course incidental as it bears no relation to the dispersion of the original single particle electronic band structure. The effective mass m_{CF} is reflective of the coulomb interaction strength instead. It should scale in proportion to $E_c = e^2/\epsilon l_B$, where $l_B = \sqrt{\hbar/eB_p}$ is the magnetic length. While the mass indeed steadily increases as the field and, hence, the Coulomb interaction gains in strength, the studies on ZnO[37,71] fail to confirm a $\sqrt{B_p}$ dependence. They rather point to a linear field dependence. This is currently not well understood, although it may be a result of strong mixing among the Landau levels, as is briefly discussed in Sec. 4.3.

4.1. Insulating phases in the quantum limit

Apart from incompressible fractional quantum Hall states associated with two and four flux composite fermions, the lowest Landau level also hosts insulating phases at higher magnetic fields. These were anticipated to be the ground state of an interacting low disorder 2DES for $\nu < 1$ prior to the discovery of the fractional quantum Hall effect.[72,73] The insulating behavior originates from correlation driven crystallization of the electron liquid into a Wigner lattice in conjunction with pinning.[37,74–76] An example can be seen in Fig. 8.[37] Within a high resistance background, it is possible to identify fractional quantum Hall states at $\nu = 2/5, 1/3, 2/7$ and $3/11$. Between each of these states, a high resistance phase (HRP) is resolved that grows in prominence as temperature is reduced. The activated temperature behavior points to an insulating ground state. Interestingly, this occurs at higher ν in ZnO compared with GaAs two-dimensional electron systems, which may be due to a stronger effect of Landau level mixing,[77] as discussed in Sec. 4.3.

4.2. Higher Landau levels

Experimental studies on GaAs based 2DES indicate that fractional quantum Hall states tend to become weaker as we move to the $N = 1$ Landau level, where they also share territory with charge density wave states that are distinct from the Wigner crystal in the lowest Landau level.[78] From the $N = 2$ Landau level onwards charge density wave physics even subdues the fractional quantum Hall effect all together. This dramatic reorganization of the ground states for $N \geq 1$ Landau levels is also anticipated in MgZnO-heterostructures. However, the ability to alter the order of the Landau levels in these heterostructures by applying an in-plane magnetic field enables to confirm in a more rigorous manner that not the filling factor but the orbital index of the partially filled Landau level is relevant for the strength of the fractional quantum Hall ground states. Before addressing such experiments pertaining to MgZnO/ZnO-based 2DES, it is instructive to provide an intuitive assessment for why fractional quantum Hall states become energetically unfavorable whereas charge density wave physics proliferates as the orbital index increases as observed in GaAs 2DES.

This reorganization can be understood at a hand waving level by regarding the binding of vortices to form composite fermions as equivalent to screening the Coulomb interaction.[79] In the lowest Landau level this approach is extraordinarily effective and the residual interactions among composite fermions are indeed weak. For higher Landau levels the wave function expands and develops nodes. This alters the Coulomb interaction as evident from the changes in the Haldane pseudopotentials.[80] The interaction is effectively softened and the identical operation of forming composite fermions in the $N = 1$ Landau level will therefore overscreen the Coulomb interaction resulting in a net attractive interaction. In higher Landau levels, the overscreening is even more severe and the viability of the composite fermion model needs to be questioned all together. This accounts for the successive weakening of the fractional quantum Hall effect. At the same time charge density wave states become energetically more favorable. This is intuitive when choosing the symmetric gauge for which the electron wave functions have a ring-like shape with a diameter increasing in size with the orbital index of the underlying level. For larger rings the overlap between rings barely changes even when their centers come closer. The additional Hatree energy penalty as electrons approach is small and is offset by a reduction in exchange energy. This culminates in charge clustering and the formation of "stripe" or "bubble"-regions with a local higher electron or hole density corresponding to the nearest integer filling.[78] First experimental signatures for these charge density wave phases were reported later on in transport experiments on GaAs 2DES.[81,82] The bubble phase manifests itself in incompressible and re-entrant integer quantum Hall behavior, while the stripe phase shows up as a strong resistance anisotropy depending on the direction of the current flow through the sample. The spatial periodicity of the stripe phase induced at $\nu = 9/2$ was found to be 3.6 times the cyclotron radius in a surface acoustic wave study[83] and 1.5 ± 0.4 times the cyclotron radius in a nuclear magnetic resonance study of the tilt-induced anisotropic phase at $\nu = 5/2$ in GaAs.[84]

The physics in the $N = 1$ Landau level has traditionally been the richest as this level accommodates the full potpourri of ground states, not only charge density wave states and conventional odd denominator fractional quantum Hall states but also an incompressible ground state at half filling appeared. Since its discovery[85] in GaAs at $\nu = 5/2$ and later $7/2$,[86] a number of candidate wavefunctions have been conceived to describe this incompressible behavior. Even though it has been experimentally challenging to supply conclusive evidence, it is now generally considered a manifestation of pairing of composite fermions mediated by the residual attraction from the overscreening of the Coulomb interaction. Historically, bosonization into a paired state in a Bardeen–Cooper–Schrieffer-like manner, either with spin-singlet pairs of even-parity[87] or spin triplet pairs with odd-parity[88] was considered first. The pairing of composite fermions is theoretically described by a Pfaffian[88] wave function or its hole partner the anti-Pfaffian.[89,90] In the context of the Dirac composite fermion description of the composite fermion Fermi sea also particle hole symmetric pairing described by the so-called particle-hole symmetric Pfaffian has been put forward.[91] For the sake of completeness, it should be noted that in the presence of one or more additional degrees of freedom incompressible behavior at even denominator filling can be facilitated by the formation of a multi-component many-body Laughlin-like state such as the 331-state put forward by Halperin.[61] The phases observed in wide quantum wells and bilayer heterostructures are believed to belong to this classification.[92,93]

4.3. *Level mixing*

The discussion so far has relied on Landau-levels that are well defined with minimal inter-level virtual excitations. This condition is only marginally true in experimentally realized systems and it is worthwhile to compare the importance of Landau level mixing in existing 2DES. A gauge for the relevance of Landau level mixing in a given 2DES is the ratio between cyclotron and Coulomb energy terms or $\kappa = E_{cyc}/E_c$. Equation (6) compares κ for a number of prominent two-dimensional systems.

$$\kappa = \begin{cases} \dfrac{e^2}{\hbar\omega_c\epsilon l_B} \sim & 2.6/\sqrt{B} \quad \text{GaAs electron} \\ & 16.6/\sqrt{B} \quad \text{ZnO electron} \\ & 22.5/\sqrt{B} \quad \text{AlAs electron} \\ & 14.6/\sqrt{B} \quad \text{GaAs hole} \\ \\ e^2/\epsilon v_F \hbar \sim & 2.2 \qquad\qquad \text{suspended graphene} \\ & 0.9 \qquad\qquad \text{graphene on } SiO_2 \\ & 0.5 - 0.8 \quad \text{graphene on boron nitride} \end{cases} \tag{6}$$

In parabolically dispersing systems κ is determined by the cyclotron frequency $\omega_c = eB_p/m^*$, ϵ the material dielectric constant and $l_B = \sqrt{\hbar/eB_p}$ the magnetic length. It therefore primarily depends on the system's effective mass and dielectric constant. In graphene the Fermi velocity v_F takes the role of the effective mass and the field dependence vanishes, $\kappa = e^2/\epsilon v_F\hbar$. For monolayer graphene, the dielectric constant depends on the substrate chosen,[94] and will vary accordingly between a value of $\kappa = 0.5$–2.2. The situation in bilayer graphene is much more complicated due to an orbital degeneracy in the $N = 0$ Landau level. Mixing effects are different between the lowest Landau level and higher levels.[95,96] Following this metric, MgZnO-heterostructures are in the upper range and Landau level mixing is bound to be a key player. From a theoretical point of view, the mixing between levels is, apart from disorder, considered a prominent mechanism in suppressing the activation gaps of fractional quantum Hall states.[80,94,97] There is also interest around how Landau level mixing may influence the topology of ground states, including the $\nu = 5/2$ fractional quantum Hall state in the $N = 1$ Landau level.[89,90] In-situ tuning of the strength of the mixing parameter is possible in parabolically dispersing 2D systems in gate-tunable samples, although this remains challenging for MgZnO/ZnO-heterostructures. Some tunability has been achieved. A field-effect transistor was produced in Refs. 18 and 76 by fabricating a top gate on an atomic layer deposition produced aluminum oxide as the dielectric layer. Good tunability of n was achieved $(2\times10^{11}$ cm^{-2}/V$)$ from depletion to full accumulation of the intrinsic charge density given by x in $Mg_x Zn_{1-x}O$ (see Sec. 1) while maintaining gate voltages less than approximately 3 V. However it was seen that the quality of transport was reduced in these devices compared with pristine chips of the same wafer. We have tentatively attributed this to the formation of charged traps at the interface between the deposited dielectric and MgZnO layers. The need for a dielectric layer was circumvented in Ref. 98 by producing an "air-gap" transistor. This is a somewhat cumbersome but a rather effective means of producing highly tunable devices, with gating efficiencies of about 1×10^9 cm^{-2}/V and the ability to fully deplete samples by applying approximately 200 V to the gate electrode. The wafer under study is placed up-side-down upon a passivated silicon wafer. The two-dimensional electron system is capacitively coupled to the doped silicon through the capping MgZnO layer (approximately 500 nm thick), the free space in between the two chips (approximately 5 μm), and the silicon oxide layer (300 nm). An alternative method is to simply use an electrode which is evaporated on the back-surface of the substrate. The typical gating efficiency through a 350 μm thickness substrate layer is however reduced to 2×10^8 cm^{-2}/V. We have only deployed this configuration on devices that are intrinsically dilute ($n \leq 0.3 \times 10^{11}$ cm^{-2}), where we can deplete the device enough to induce an insulator with the application of approximately -50 V to the gate.

5. Orbital Index and Spin Transitions

The discussion and collected background in the previous section on the ground states that develop in traditional AlGaAs/GaAs heterostructures depending on the orbital quantum number of the partially filled Landau level turns out particularly instrumental for the understanding of the magnetotransport behavior in MgZnO/ZnO heterostructures when a rearrangement of the Landau levels is enforced. This is either accomplished by tuning the density or by selectively enhancing the Zeeman energy upon application of an in-plane magnetic field by rotating the sample as discussed previously in Sec. 3.2 using the coincidence method. Such a rearrangement modifies the overall degree of spin polarization as well as the spin orientation and the orbital character of the charge carriers at the chemical potential as discussed in that section and illustrated in Fig. 4 for sample S1 ($n = 2 \times 10^{11}$ cm^{-2} and $g^*m^*/m_0 \approx 2$).

Figure 10(a) serves as another example of a color rendition of the longitudinal resistance in the ($\nu, 1/\cos\theta$)-plane recorded at 450 mK on sample S6, which has a higher density of $n = 5.8 \times 10^{11}$ cm^{-2} and hence smaller $g^*m^*/m_0 \approx 1.35$. We have chosen to present data taken on this sample as the smaller g^*m^*/m_0 also implies a smaller E_Z/E_{cyc} ratio in the absence of an in-plane field, as per Eq. (4). As a consequence, the $j = 1$ coincidence is pushed to a higher value of θ where we may resolve transport both well before and well after the first level crossing which is not the case for the data recorded on sample S1 shown in Fig. 4. The Landau level crossings on this sample are representative of all devices with $n \geq 2 \times 10^{11}$ cm^{-2}, which corresponds to $g^*m^*/m_0 \leq 2$. Strong similarities between the data presented

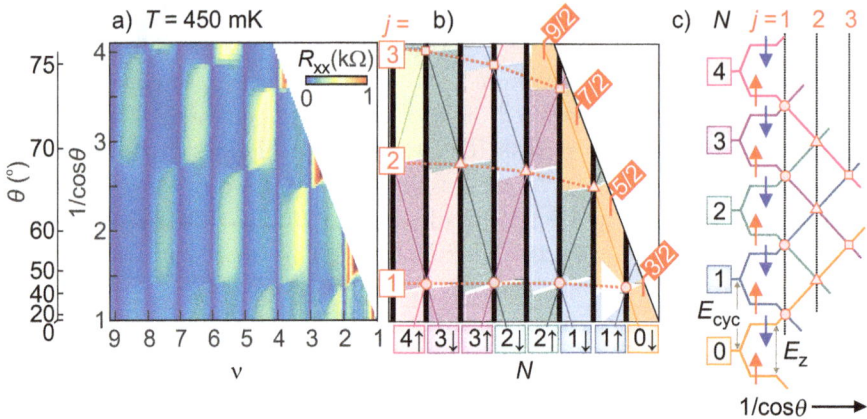

Fig. 10. (a) Level crossings observed in the longitudinal resistance of sample S6 (with $n = 5.8 \times 10^{11}$ cm^{-2} and $g^* m^* / m_0 = 1.35$) as the quantum limit is approached. (b) Color coded map and legend of the orbital and spin quantum number of the partially occupied Landau at the chemical potential matching the experimental data in (a). Black regions correspond to incompressible integer quantum Hall behavior when the chemical potential lies in a gap. Dashed lines mark the level coincidences beyond which spins reorient and the orbital index changes. The change in the orbital index $j = 1, 2$ and 3 during the level crossings is indicated. (c) Color coded Landau level fan schematically presenting the level crossings encountered in experiment. Adapted from Ref. 38.

in Fig. 4 are observable, namely it is possible to resolve a chequerboard pattern. As previously explained, this immediately enables to identify the spin orientation of the carriers at the chemical potential as well as transitions where both the spin gets reversed and the orbital index changed. Whereas in Fig. 4 it was absent, sample S6 ensures that also the first coincidence ($j = 1$) is visible and postponed to higher tilt angles. We may assign each portion of the parameter space the (N, σ)-pair of quantum numbers of the partially filled level, an exercise that has been carried out in Fig. 4(b), using the Landau fan presented in Fig. 4(c). The coincidences are marked by labeled and dashed lines. In the following sections, we will focus on the development and transitions among compressible and incompressible ground states surrounding the even denominator fillings in the order 7/2, 3/2 and finally 5/2, where the physics is particularly rich, as spin split Landau levels are brought into coincidence and cross.

5.1. *Ground state transitions near $\nu = 7/2$*

Provided that $g^* m^* / m_0 < 2$ at $\nu = 7/2$, the spin split Landau levels are ordered in standard fashion in the absence of an in-plane magnetic field and μ_ν lies in the $(1, \downarrow)$ level. Incompressible behavior due to composite fermion pairing is then anticipated at $\nu = 7/2$. According to Fig. 6, this condition should be met for samples with densities exceeding $n = 2 \times 10^{11}$ cm^{-2}, as shown in the right-hand schematic of that figure. Figure 9 displays magnetotransport at zero tilt angle for

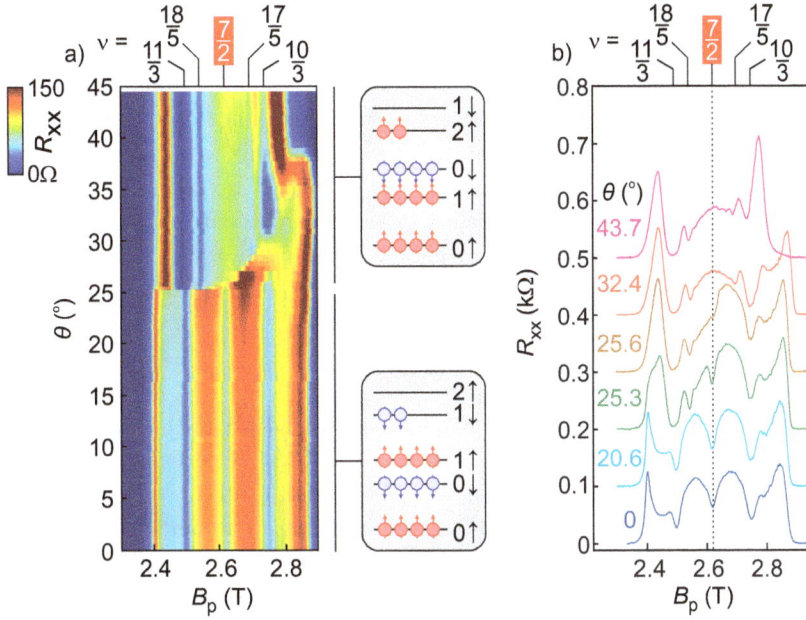

Fig. 11. Incompressibile behavior and transition to a compressible state after the $j = 1$ coincidence at $\nu = 7/2$. Data is taken using sample S7 with $n = 2.3 \times 10^{11}$ cm^{-2} and $g^*m^*/m_0 \approx 1.9$. (a) Color rendition of the longitudinal resistance R_{xx} as a function of B_p and θ. Schematics on the right indicate the two-level configurations encountered through the experimental parameter space. (b) Line cuts at discrete tilt angles starting from $\theta = 0°$ up to $43.7°$. Adapted from Ref. 39.

sample S5 with a density $n = 2.9 \times 10^{11}$ cm^{-2} and indeed incompressible behavior is observed. In Fig. 11 we present more extensive data taken on sample S7 that has $n = 2.3 \times 10^{11}$ cm^{-2} and $g^*m^*/m_0 \approx 1.9$.[39] This results in a qualitatively similar energy diagram to that presented in Fig. 10, however with quantitative corrections to where the jth coincidence occurs due to the modified g^*m^*/m_0. Here in sample S7, the $j = 1$ coincidence position occurs at approximately $\theta = 25°$. The device presented in Fig. 11 was the first example demonstrating incompressibility at $\nu = 7/2$ outside the realm of ultra-high mobility GaAs-based heterostructures. Two additional fractional quantum Hall states are observed at $\nu = 11/3$ and $10/3$, which remain weak in GaAs-heterostructures where the "bubble" charge density wave states compete and dominate in transport.[99] The development of the 7/2-state as well as these conventional two-flux composite fermion FQH states are shown in a color map of the longitudinal resistance across part of the (B_p, θ)-plane in (a) as well as a series of traces at discrete angles in (b). Near $\nu = 7/2$, a sharp transition from incompressible to compressible behavior becomes apparent at a rotation angle of approximately $\theta = 25°$ where the $(1, \downarrow)$ level is replaced by the $(2, \uparrow)$ level at the chemical potential. The level crossing is confirmed in high temperature data (not shown, see Ref. 39 for details) and by utilizing Eq. (4). This confirms experiments on GaAs where the paired composite fermion state exclusively emerges in an $N = 1$

Landau level. However, contrary to expectation, the two-flux composite fermion states at filling 11/3 and 10/3 persist and even gain in strength. In addition, their 18/5 and 17/5 companions emerge as well. This contradicts common wisdom on the dependence of the strength of FQH states on the orbital index. It may be considered a unique experimental signature for significant level mixing among the $(2,\uparrow)$ and the nearest occupied $(0,\downarrow)$ level, which would indeed favor the proliferation of these FQH states. At the same time such a level mixing is still fully in line with the absence of both a paired composite fermion state as well as the suppression of charge density wave physics.

5.2. *Ground state transitions at* $\nu = 3/2$

A similar study has been carried out on the same sample (S7) in the filling factor interval $1 < \nu < 2$ and is summarized in Fig. 12.[39] At zero tilt angle μ_ν is pinned in the $(0,\downarrow)$ level promoting a compressible ground state at $\nu = 3/2$ as well as a presumably lengthy series of odd-denominator two flux composite fermion FQH states. This is confirmed in the experimental trace of (a) showing FQH states up to 16/11 and 17/11. According to the level diagram of Fig. 10, valid also for this sample, electrons are transferred from the $(0,\downarrow)$-level to the $(1,\uparrow)$-level upon tilting the sample offering the prospect of observing an even denominator FQH state equivalent to the 5/2-state but at filling 3/2 as the chemical potential gets pinned in the $(1,\uparrow)$. A line trace beyond the transition at an angle above $40°$ is plotted in (b) and indeed incompressible behavior appears in the longitudinal resistance together with a properly quantized plateau in the Hall resistance at filling 3/2.[39] Concomitantly, the odd-denominator fractional quantum Hall states have become less numerous.

The color rendition in (c) maps this transition in greater detail. The 3/2 state appears here as the blue vertical dash in the region $\theta = 38 \sim 42°$. Figure 13 compares the behavior at $\nu = 3/2$ and 7/2 in the course of the $j = 1$ coincidence. While the $\nu = 7/2$ state (a) is rapidly suppressed as the transition takes place (black triangle), the $\nu = 3/2$ state (b) only emerges beyond it (green triangle). The previous assertion that the odd-denominator fractional quantum Hall states weaken in the vicinity of $\nu = 3/2$ as the level crossing is traversed and μ_ν enters the $(1,\uparrow)$ Landau level is confirmed in (c) which plots the activation energy (Δ_μ) of these states beyond approximately $\theta = 32°$. A rapid decline is clearly visible.

Ultimately, at higher θ, the system becomes fully spin polarized with the $(1,\uparrow)$ and $(0,\uparrow)$-levels occupied. As seen in Fig. 4 as well as in Fig. 10, which were taken on different samples but reflect analogous physics, full spin polarization is always accompanied by a dramatic increase in the sample resistance. This is reminiscent of the effect of a fully-spin polarized electron gas in the presence of a strong in-plane magnetic field.[58] There, the loss of spin degeneracy is found to increase the scattering rate of carriers due to an impaired screening of disorder. We conjecture that a similar mechanism is at work at $\nu = 3/2$ at high θ as this condition is

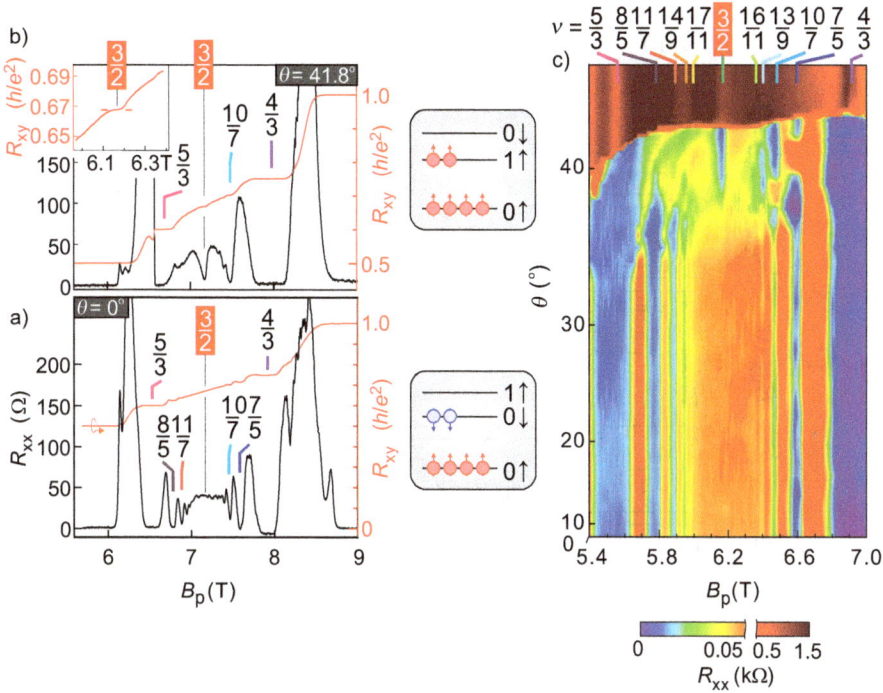

Fig. 12. Tuning of compressibility at $\nu = 3/2$ in sample S7 with $n = 2.3 \times 10^{11}$ cm^{-2}. (a) and (b) comparison of magnetotransport at $\theta = 0°$ and $41.8°$. (c) Mapping of magnetotransport as a function of B_p and θ in the vicinity of $\nu = 3/2$. $T \approx 15$ mK for all data. Adapted from Ref. 39.

analogous to a fully polarized electron gas. The rise in the overall resistance level goes along with a drastic degradation of the fractional quantum Hall states, although minima are still resolved at rational filling factors, as shown in Fig. 14. Here, sweeps at three temperatures are displayed at a tilt angle of $45°$. Intriguingly, as T is reduced the Hall resistance now displays clear signs of re-entrant integer quantum Hall behavior normally associated with CDW bubble states, although the re-entrant behavior at filling $\nu > 3/2$ is anomalous as the Hall resistance is anticipated to drop to 0.5 h/e^2 while it actually retraces to h/e^2. This is at present not understood. All in all, the overall qualitative behavior seems to match the anticipated orbital index dictated development and evolution of ground states.

Closer inspection of the color map in Fig. 12(c) reveals a complex series of gap closing events for the odd denominator FQH-states in the tilt angle range of approximately $35° < \theta < 42°$. A detailed analysis of these transitions has been carried out in the supplementary information of Ref. 39 and assigns the gap closures to the crossings between two composite fermion levels, each stemming from a different fan originating from the $(0, \downarrow)$ and $(1, \uparrow)$ electron Landau level respectively. The transitions are accompanied in each case by a change in the spin polarization. This analysis enables to identify $\theta \approx 40°$ as the point at which the single-particle electron

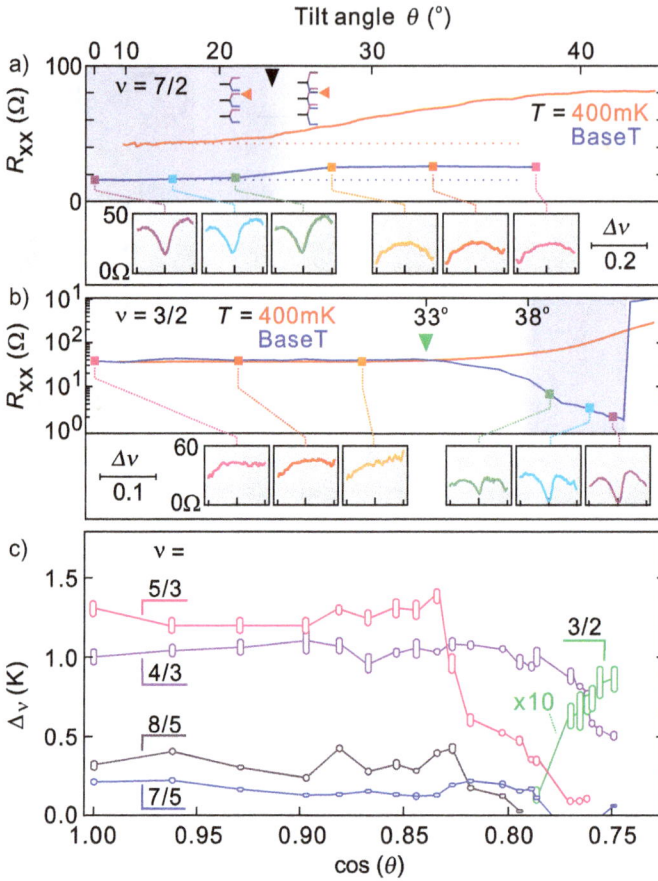

Fig. 13.   Comparison between $\nu = 7/2$ and $3/2$ in sample S7 with $n = 2.3 \times 10^{11}$ cm^{-2}. Longitudinal resistance at $T = 400$ (red) and base temperature (≈ 15 mK, blue) at $\nu = 7/2$ (a) and $\nu = 3/2$ (b) as a function of θ. The black and the green arrows indicate the approximate location of the single-particle level crossings at adjacent integer filling. The insets display the magnetoresistance around half-filling at discrete tilt angles. Purple shading represents portions of the parameter space where significant $N = 1$ character is anticipated. (c) Activation energy (Δ_μ) of prominent fractional quantum Hall phases around $\nu = 3/2$ as a function of θ. Adapted from Ref. 39.

Landau levels are degenerate at $\nu = 3/2$. This puts the gapped even-denominator in the vicinity of the level crossing and both levels would be partially populated. Two distinct possibilities may then account for the gapped ground state. Firstly, pairing in the $N = 1$ Landau level may occur for sufficient spin polarization, as such even-denominator states have been shown to be fairly resilient to a depolarization process.[96,100,101] It physically involves pairing in one component while the other forms a Fermi surface. Recent theoretical works propose such a state is energetically favorable under intermediate polarizations and corresponds to a "Z_2 exciton metal".[100,101] The second choice is pairing in a multi-component state.[102] This would be distinct from previous observations of two-component states as both spin

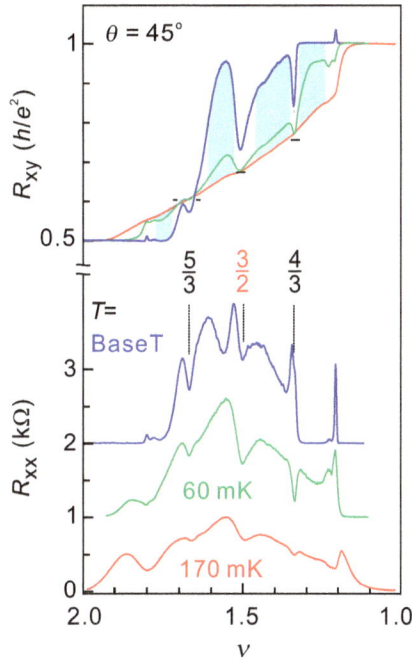

Fig. 14. Data recorded around $\nu = 3/2$ using sample S7 ($n = 2.3 \times 10^{11}$ cm^{-2}) after the system is fully spin polarized at 45° at three temperatures. While minima at rational fractional fillings are still resolved, strong re-entrant integer quantum Hall behavior is observed across a vast portion of the partial filling factor. Adapted from Ref. 39.

and orbital quantum numbers differ both for the coinciding levels for the case at hand. Reference 103 however found no such coherence in numerical studies which evaluated the level crossing at $\nu = 3/2$ in ZnO. There, a sharp-first order transition was always encountered and hence the present state of affairs favors a $N = 1$ paired composite fermion state equivalent to the 5/2-state in GaAs.

5.3. *Ground state transitions at* $\nu = 5/2$

In Fig. 9, covering magnetotransport data between filling factor 1 through 4 on sample S5 ($n = 2.9 \times 10^{11}$ cm^{-2}, $g^*m^*/m_0 = 1.65$), an accurately quantized $\nu = 5/2$ state with vanishing longitudinal resistance is observed. This at first fits our expectation for charge carriers occupying the $(1,\uparrow)$-level at the chemical potential and the incompressible ground state would be analogous to the commonly observed $\nu = 5/2$ state in GaAs-based samples.[85,104] The discussion above would suggest that this state is suppressed readily by selectively increasing E_Z until μ_ν gets pinned into the $(0,\downarrow)$-level. The experiments reported in Ref. 38 however reveal that the physics surrounding this partial filling is much more complex and intimately depends on the proximity of the $(1,\uparrow)$ and $(0,\downarrow)$-levels. At the neighboring integer fillings a first order spin-flop transition is encountered as seen in Fig. 5 on sample S2 and 18

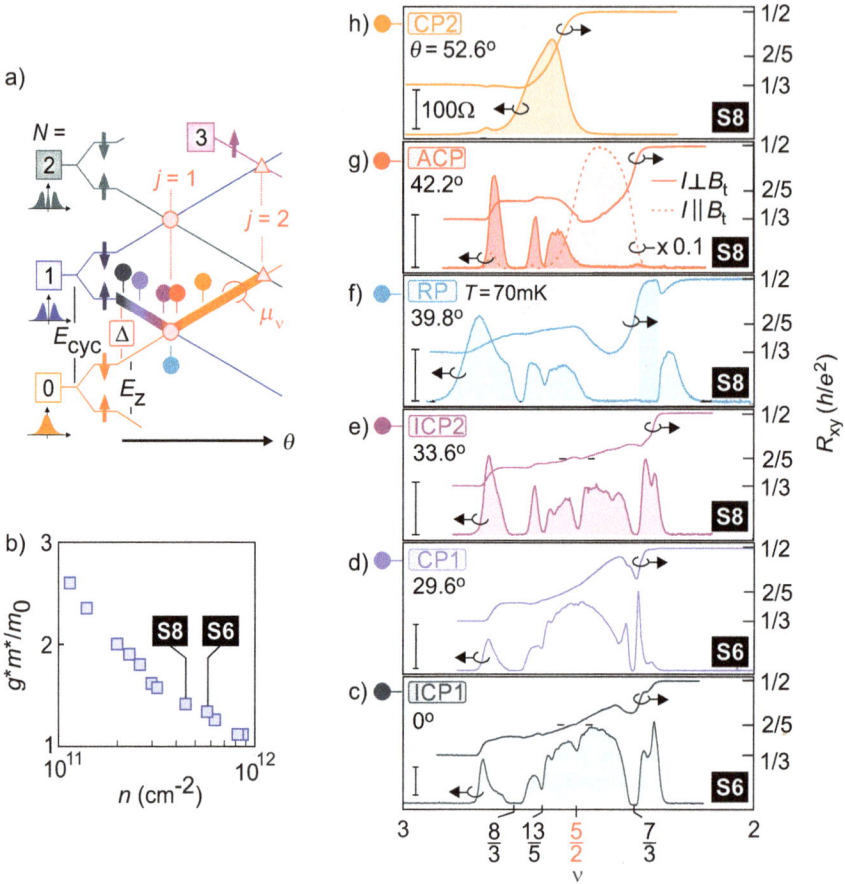

Fig. 15.   Overview of the phases that develop at or near $\nu = 5/2$ as the tilt angle is varied. (a) Schematic single-particle energy level ladders given $g^*m^*/m_0 < 2$. Thicker line sections trace the chemical potential (μ_ν) across the $j = 1$ transition at fixed filling $\nu = 5/2$. This level diagram applies to all samples for which the $(0, \downarrow)$-level is still below the $(1, \uparrow)$ at zero tilt angle. For these samples the value of θ at which $j = 1$ coincidence occurs will just vary systematically according to the g^*m^*/m_0 value. (b) g^*m^*/m_0 as a function of n, highlighting samples S6 and S8 used in the analysis. (c–h) Representative trace for each distinct phase identified across the level crossing. The acronyms CP, RP, ACP and ICP refer to compressible, re-entrant, incompressible and anisotropic compressible phase, respectively. Here, $T \approx 20$ mK, except (f) where $T = 70$ mK. Adapted from Ref. 38.

on sample S8. Near $\nu = 5/2$ at least six distinct electronic phases appear as charge carriers are transferred between the levels. Figure 15 summarizes these phases which develop in the course of the level crossing as a function of the tilt angle θ. The incompressible state at $\nu = 5/2$ illustrated in Fig. 9 is merely one of these and will be properly identified among the six phases that appear at the end of this section.

Figure 15(a) plots the energy level diagram relevant for all those samples in which the standard ordering of the spin resolved Landau levels applies at zero tilt. In the absence of Landau level mixing, the separation Δ between the adjacent $(1, \uparrow)$- and $(0, \downarrow)$-levels equals

$$\Delta = E_{(1,\uparrow)} - E_{(0,\downarrow)} = \hbar\omega_{\rm c} + \frac{e^2}{\epsilon l_{\rm B}}\epsilon_{\rm e}(1) - E_{\rm Z}. \tag{7}$$

Here, $\epsilon_{\rm e}(1) \approx -0.63$ (see Refs. 34 and 38). This splitting (Δ) is composed of two parts: a single particle term and a term purely related to exchange interaction. The latter term has not been included in the diagram.

The development of μ_ν with tilt angle for filling $\nu = 5/2$ has been highlighted. The six different phases that emerge across the transition at $\nu = 5/2$ are illustrated in (c) through (h). Experimental details unfortunately require the use of more than one sample in order to obtain completely developed examples for each of the ground states that can appear. For these line traces two samples have been used. Sample S6, previously deployed to compile Fig. 10, has a density of $n = 5.8{\times}10^{11}$ cm^{-2} and a small $g^*m^*/m_0 = 1.35$. This pushes the $j = 1$ coincidence to a relatively high θ, so that transport in the $N = 1$ Landau level at $\nu = 5/2$ can be monitored across a broader range of θ. The use of a high n however places an upper bound on the highest achievable θ at $\nu = 5/2$ due to the maximum field obtainable in the laboratory. It also suppresses the mobility μ due to the higher Mg content. We therefore also rely on sample S8 in addition, whose $n = 4.4{\times}10^{11}$ cm^{-2} and $g^*m^*/m_0 = 1.48$ allows us to explore more of the parameter space after the levels have crossed owing to the smaller required B-fields. It has a higher μ, and therefore reveals cleaner transport details of these phases. We will conclude this section by presenting data taken on a set of five devices to illustrate the reproducibility of the observed phases.

The line traces with different color in these panels have been recorded at the angles marked with the same color dot in the level diagram of (a) in order to appreciate the approximate location of these phases with respect to the $j = 1$ crossing. In accordance with the expectation of a gapped state when μ_ν lies in $N = 1$ Landau level, a developing $\nu = 5/2$ state is observed in (c) which we denote ICP1 (sample S6). Upon sample rotation, this state is suppressed and the ground state at $\nu = 5/2$ turns compressible (CP1). In sample S8, the larger g^*m^*/m_0 results in a smaller Δ and renders the CP1 phase already stable at $\theta = 0°$ (not shown in this figure, but discussed later). Rotating this sample as to tune the levels closer to coincidence culminates in a second gap-opening event at $\nu = 5/2$ and the resulting incompressible state is referred to as the ICP2-phase. It is illustrated in (e). We attribute the high quality transport primarily to the higher mobility of this device compared to sample S6, however the reduced density prevents the clear observation of the ICP1 phase.

At slightly higher tilt angle, fractional quantum Hall physics seen in the ICP2 state is replaced by two distinct flavors of charge density wave physics. These phases are illustrated in Figs. 15(f) and 15(g). Noticeably, a highly anisotropic compressible phase (ACP) reminiscent of the stripe phase due to charge clustering is observed (g). Filling factor 5/2 is however only located at the periphery of the filling factor region where this anisotropic phase develops. Its center with the largest anisotropy between the longitudinal resistance recorded in two orthogonal directions is located at lower filling factor. Additional density wave physics proliferates in this range of tilt angles at lower filling factor as a re-entrant $\nu = 2$ integer quantum Hall state, normally associated with a pinned bubble phase. It will be referred to as the re-entrant phase RP. It is illustrated in (f). At the lowest temperature the Hall resistance plateau corresponding to filling 2 merely appears extended, but the re-entrant nature becomes apparent in data recorded at 70 mK shown in Fig. 15(f). As we continue to tilt the sample, the RP and ACP phases vanish. The system ends up compressible, once carriers are fully polarized in the $N = 0 \downarrow$ level (CP2). Filling factor 5/2 is located at the outskirts of this CP2 phase. At lower filling factor $\nu = 2$ integer quantum Hall behavior persists. At higher tilt angle to plateau even extends beyond filling factor 5/2. It is unclear whether this incompressible behavior is re-entrant in nature, since we have been unable to demonstrate that this broad plateau in the Hall resistance and minimum in the longitudinal resistance contains two integer quantum Hall phases punctuated by a region of finite resistance. Instead, the plateau and minimum just gets broader as temperature is lowered.

Figure 16 illustrates the evolution of all of these phases, with the exception of the ICP1 state, in the (ν,θ)-plane for sample S8. The red triangle at about $\theta = 41°$ indicates the approximate location of the level crossing at $\nu = 2$. (a) and (b) map the longitudinal resistance in two orthogonal crystal directions, with the aim of highlighting anisotropic features. The larger $g^* m^* / m_0$ of sample S8 renders

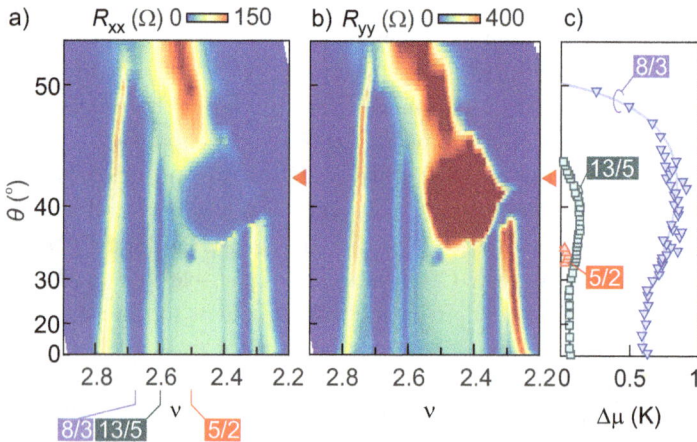

Fig. 16.   Magnetotransport data recorded on sample S8 ($n = 4.5 \times 10^{11}$ cm^{-2}) around $\nu = 5/2$ in the van der Pauw geometry. (a) and (b) Color maps of the longitudinal resistance as a function of ν and θ in two orthogonal crystal directions. The red triangle marks the location of the level crossing at $\nu = 2$. (c) Activation energy (Δ_μ) as a function of θ for identified prominent fractional quantum Hall features. Adapted from Ref. 38.

the neighboring $N = 1 \uparrow$ and $0 \downarrow$ levels in close proximity already without tilt. As a result, the ICP1 phase is absent and the CP1 phase is already stable at $\theta = 0°$. The ICP2 phase is observed in a narrow but finite range of tilt angles and is isotropic. The ACP phase is clearly resolved when comparing (a) and (b). The temperature dependence of the ACP phase is presented in Fig. 17 on both log and linear scales. The resistance at $\theta = 41.2°$ and $\nu = 2.4$ is measured simultaneously in orthogonal crystal directions while cooling. The anisotropy only emerges at low temperatures and transport turns back isotropic for $T > 200$ mK. This phase exhibits easy-axis transport features when the current is sent in the same direction as the total magnetic field. Evidently, the in-plane field component acts as a symmetry breaking field, since the anisotropic features will reorient by 90° if the in-plane field is rotated by this angle. Similar symmetry breaking physics has been observed in GaAs,[84,105–107] where also a resistance anisotropy develops covering a larger filling factor range centered around 5/2 compared to the stripe phases which are seen in higher Landau levels.[81,82] The proliferation of the RP phase below and above the crossing (red arrow) is visible in Fig. 16 as the blue low resistance region flanking the red anisotropic phase towards lower filling factors. At larger tilt angles, this blue region expands significantly and subdues all fractional quantum Hall features. Experimentally however it has not been possible to find clear evidence to discern between scenarios of this also being re-entrant behavior or rather just an anomalous extension of the $\nu = 2$ integer quantum Hall state.

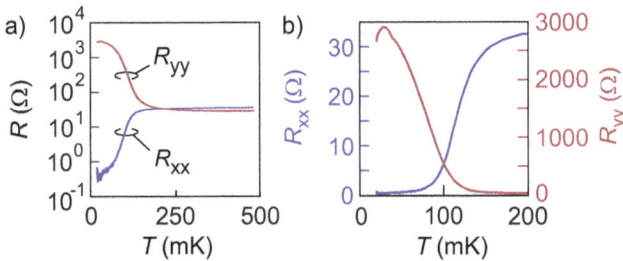

Fig. 17. Temperature dependence of the anisotropic phase (ACP) at $\theta = 41.2°$ and $\nu = 2.4$ on sample S8 ($n = 4.5 \times 10^{11}$ cm^{-2}), plotted on (a) log-scale and (b) linear scale. Note the different scales for R_{xx} and R_{yy} in (b). Adapted from Ref. 38.

The data plotted in panels (a) and (b) of Fig. 18 address the re-entrant incompressible phase (RP). Panel (a) displays the Hall resistance R_{xy} in the (ν,θ)-plane for sample S8. Next to the color map two line traces are shown that have been recorded below and above the $j = 1$ coincidence angle $\theta \approx 42°$ (red triangle). The Hall resistance clearly returns to its quantized value for the $\nu = 2$ integer quantum Hall state. This occurs in the red region resembling the shape of half a butterfly. In GaAs this re-entrant integer quantum Hall feature associated with "bubble" charge-density wave physics is prevalent in the $N = 1$ Landau level as well as in higher Landau levels.[78,81,82,104] Its robust observation near a level crossing

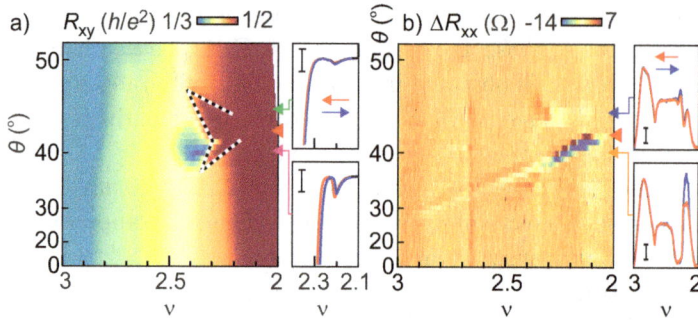

Fig. 18. Re-entrant integer quantum Hall behavior and hysteresis close to the $j = 1$ coincidence in sample S8 ($n = 4.5 \times 10^{11}$ cm^{-2}). (a) Hall resistance (R_{xy}) measured at $T = 90$ mK plotted as a function of ν and θ. The black-and-white dotted region indicates the range where re-entrant integer quantum Hall physics occurs. The panels on the right show single line traces of the Hall resistance at two selected angles above and below $\theta = 40°$ and $43°$. (b) Color map of the difference in the longitudinal resistance recorded during up and down sweeps of the magnetic field at $T = 200$ mK. This presentation highlights where magnetotransport is hysteretic. Panels on the right show individual line traces close to the $j = 1$ coincidence position (red triangle) and illustrate the existence of hysteresis. Adapted from Ref. 38.

is unanticipated and points to a localization of minority carriers at $\nu = 2 + \delta$. This re-entrant phase emerges in the parameter plane (ν, θ) where the longitudinal resistance exhibits weak hysteresis for up and down magnetic field sweeps as highlighted in Fig. 18(b). The color rendition shows the difference between data sets recorded when the B-field is swept from high-to-low and vice-versa. Single line traces for the up and down field sweep recorded at two different tilt angles marked by arrows have been included on the right. Similar to features introduced in Fig. 5, this hysteresis is indicative of a first-order spin-flop-like transition when the intersecting Landau levels are in close proximity. Its visibility is well resolved for filling factors close to $\nu = 2$, but disappears away from this filling.

Figure 19(a) summarizes the area in parameter space spanned by the tilt angle and the filling factor that is occupied by each of the above discussed competing phases. The tilt angle ordinate has however been replaced by $E_Z/E_{\mathrm{cyc}}|_{\nu=2}$, since the phases appear at different tilt angles for each sample due to the density dependent $g^* m^*/m_0$. This renormalization of the ordinate enables to draw a single phase diagram applicable for samples of different density. When E_Z/E_{cyc} equals 1, the $(1, \uparrow)$ and $(0, \downarrow)$ levels cross and this serves as the pivotal point in this diagram. Panel (a) is a composite representation utilizing the data taken from both samples S6 and S8. The portions of the diagram ascribed to a partially filled $(1, \uparrow)$ level and a partially filled $(0, \downarrow)$-level are color-coded. Also color coded are regions assigned to the ICP1, ICP2, ACP, RP and the conventional odd denominator (OD) fractional quantum Hall states. The dotted lines represent the parameter space where hysteretic transport features, indicative of a first-order phase transition, are observed at either $T = 90$ or 200 mK. Figure 19(b) plots the phase transitions

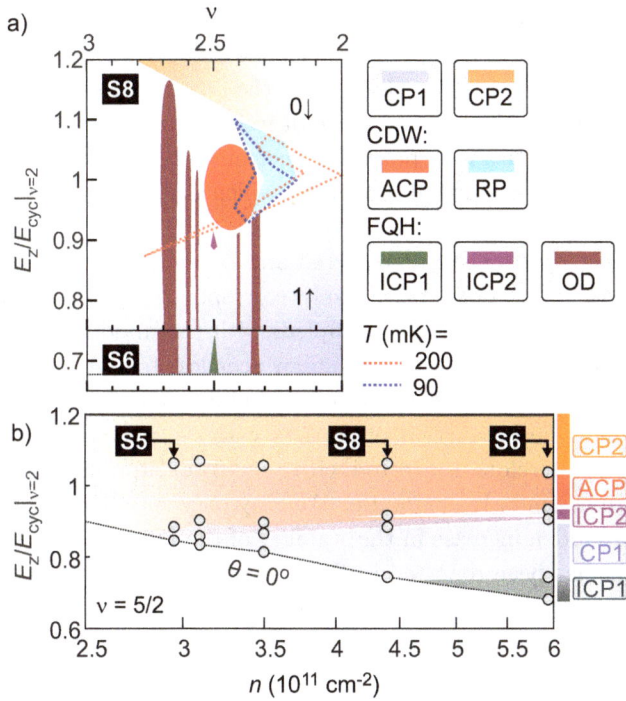

Fig. 19. Summary of phase transitions near $\nu = 5/2$. (a) A composite representation of phases observed in samples S6 and S8 in the $(\nu, E_Z/E_{cyc}|_{\nu=2})$-plane. The parameter space associated with level polarized compressible liquids (CP1 and CP2), charge density waves (CDW; ACP and RP) and fractional quantum Hall (FQH; ICP1, ICP2 and OD) ground states are color coded according to the boxes to the right of the panel. Regions with hysteretic transport features at either $T = 200$ or 90 mK are enveloped by dotted lines. (b) Phases at $\nu = 5/2$ as a function of n and $E_Z/E_{cyc}|_{\nu=2}$ for multiple samples studied. Adapted from Ref. 38.

at half-filling as a function of n. Owing to a suppression of g^*m^*/m_0 when n is increased, the portion of parameter space for which the $(1,\uparrow)$ Landau level is partially filled is larger, and more of these competing phases can be observed in one and the same sample. Unfortunately, with increasing density also the mobility drops (see Fig. 2) and hence a trade-off between the availability of all phases and the extent to which they are developed is inevitable. Open circles in Fig. 19 mark the boundary between the different phases. These data points have been compiled from a set of five samples. Clearly this sequence of phase transitions is highly reproducible and the ground state that emerges is intimately linked to the ratio $E_Z/E_{cyc}|_{\nu=2}$, i.e. the proximity of the coinciding levels.

We reiterate which of these states can be assigned with reasonable confidence. For instance, we associate the ICP1 phase with a paired state as it occurs when the chemical potential resides in the $(1,\uparrow)$ Landau level and far away from the coincidence of the $(0,\downarrow)$ and $(1,\uparrow)$ levels. Hence, its origin is similar to the $\nu = 7/2$ state discussed above and analogous to the $\nu = 5/2$ incompressible ground state in

GaAs or bilayer graphene.[85,88,96,104,108] Also the CP2 phase at high θ, on the other end of the phase diagram, is expected and understood with similar confidence. It appears when μ_ν lies in the $(0,\downarrow)$ level after the crossing has been traversed. The poor quality of transport, i.e. the absence of fractional quantum Hall features, is a result of the conversion to a fully polarized electron system as this is accompanied by increased Pauli blocking of virtual transitions which otherwise enhance the screening of static disorder.

The remaining phases were unexpected and therefore more difficult to conclusively account for. The CP1 phase likely corresponds to a condition where the coinciding levels are both partially occupied, with the majority of carriers hosted by the $(1,\uparrow)$ level. This would be a similar scenario to the regime immediately preempting the appearance of the $\nu = 3/2$ incompressible state discussed above as θ is increased and the proportion of carriers with $N = 1$ orbital character rises. Its demise may be therefore associated with the "Z_2 exciton metal" scenario.[100,101] The ICP2 phase is similarly enigmatic. We may only speculate on a few possibilities such as the sudden increase in the stability of the Moore–Read phase close to a level crossing, as has been reported in Ref. 109 or it may be that the ICP1 phase's transport features are washed out with increasing θ, and it suddenly reappears immediately prior to the level crossing. Another possibility is a two-component phase in analogy to the (3,3,1)-state of Halperin.[102] Yet we have no quantitative understanding for the viability of this state as it involves both different spin and orbital quantum numbers of the coinciding levels.

Finally, it is plausible that the anisotropic ACP state and the RP phase exhibiting re-entrant behavior correspond to symmetry breaking charge clustering into stripe and bubble phases. The stabilization of the stripe phase near the crossing of the $(1,\uparrow)$ and $(0,\downarrow)$ level, when it is likely that there is significant population of both levels, introduces the possibility of a novel state with inter-level coherence and built in nematic characteristics, similar to what has been discussed in graphene at integer levels.[110] We are not aware however of any studies addressing this feasibility in fractionally filled Landau levels.

We conclude this section by returning to Fig. 9 which displays transport of sample S5 with $n = 2.9 \times 10^{11}$ cm^{-2} at $\theta = 0°$. Locating such a device on the phase diagram presented in Fig. 19 puts the ICP2 phase as stable in the untilted measurement configuration. Hence, we attribute the strong minimum at $\nu = 5/2$ in Fig. 9 to the ICP2 phase rather than the ICP1 phase assigned to composite fermion pairing.

6. Summary and Outlook

Recent years have brought spectacular advances in the variety and quality of accessible two-dimensional electron systems. Within this frontier, ZnO finds a niche where the outcome of Fermi-liquid renormalization physics has impact within the fractional quantum Hall regime. Owing to the intrinsically large magnitude and experimental tunability of the spin susceptibility of carriers combined with excellent material quality, it is possible to induce a range of spin transitions in the fractional quantum Hall regime which have revealed new facets of known-phenomena as well as unexpected correlated phases. We anticipate a number of features identified to have emerged due to strong Landau level mixing in this system, which unfortunately remains to a large extent intangible and difficult to study both experimentally and theoretically. One enigmatic observation reported in Ref. 39 is an incipient minimum at $\nu = 9/2$ that is quickly suppressed by increasing the angle θ of the sample relative to the direction of the magnetic field. Subsequent experience now leads us to believe that this physics may be isolated to the vicinity of the level crossing, for example as discussed above near filling $\nu = 5/2$. We look forward to future studies where we may directly measure the spin polarization of carriers in an attempt to draw conclusions about their degree of orbital and spin polarization. One opportunity for such experiments is electron spin resonance, as briefly introduced in Fig. 5, although extended to the fractional quantum Hall regime at low temperatures. This is a challenging endeavor, as the technical requirements for the application of high frequency radiation at very low temperatures are quite rigorous.

We also anticipate ZnO to be a promising platform for studying the eventual breakdown of the Fermi-liquid paradigm in the dilute charge carrier density regime. Well-known scaling arguments exclude the viability of metallicity in two-dimensions whenever disorder is present.[111] Despite this, modern high mobility systems unveil metallic behavior down to millikelvin temperatures. In light of the possibility of a spontaneous phase transition occurring in this dilute limit, it will be interesting to observe the developments in high quality ZnO heterostructures as r_s approaches values of 30 and the metal-insulator transition is approached.[112–114]

Acknowledgments

We acknowledge extensive collaborations with T. Arima, S. Dorozhkin, B. Friess, D. Kärcher, M. Kawasaki, K. von Klitzing, Y. Kozuka, I. Kukushkin, T. Makino, D. Maryenko, I. Sodemann, D. Tabrea, T. Tambo, K. Tanaka, A. Tsukazaki, M. Uchida, and D. Zhang throughout the course of this research.

References

1. Y. Kozuka, A. Tsukazaki, and M. Kawasaki, Challenges and opportunities of ZnO-related single crystalline heterostructures, *Appl. Phys. Rev.* **1**(1), 011303 (2014). doi: 10.1063/1.4853535. URL https://doi.org/10.1063/1.4853535.
2. A. Tsukazaki, A. Ohtomo, T. Onuma, M. Ohtani, T. Makino, M. Sumiya, K. Ohtani, S. F. Chichibu, S. Fuke, Y. Segawa, H. Ohno, H. Koinuma, and M. Kawasaki, Repeated temperature modulation epitaxy for p-type doping and light-emitting diode based on ZnO, *Nat. Mater.* **4**(1), 42 (2005).

3. K. Nakahara, S. Akasaka, H. Yuji, K. Tamura, T. Fujii, Y. Nishimoto, D. Takamizu, A. Sasaki, T. Tanabe, H. Takasu, H. Amaike, T. Onuma, S. F. Chichibu, A. Tsukazaki, A. Ohtomo, and M. Kawasaki, Nitrogen doped $Mg_xZn_{1-x}O/ZnO$ single heterostructure ultraviolet light-emitting diodes on ZnO substrates, *Appl. Phys. Lett.* **97**(1), 013501 (2010). doi: 10.1063/1.3459139. URL https://doi.org/10.1063/1.3459139.

4. D. G. Schlom, Perspective: Oxide molecular-beam epitaxy rocks!, *APL Mater.* **3**(6), 062403 (2015). doi: 10.1063/1.4919763. URL https://doi.org/10.1063/1.4919763.

5. V. V. Solovyev, A. B. Van'kov, I. V. Kukushkin, J. Falson, D. Zhang, D. Maryenko, Y. Kozuka, A. Tsukazaki, J. H. Smet, and M. Kawasaki, Optical probing of MgZnO/ZnO heterointerface confinement potential energy levels, *Appl. Phys. Lett.* **106**(8), 082102 (2015). doi: 10.1063/1.4913313. URL https://doi.org/10.1063/1.4913313.

6. O. Ambacher, J. Smart, J. R. Shealy, N. G. Weimann, K. Chu, M. Murphy, W. J. Schaff, L. F. Eastman, R. Dimitrov, L. Wittmer, M. Stutzmann, W. Rieger, and J. Hilsenbeck, Two-dimensional electron gases induced by spontaneous and piezoelectric polarization charges in N- and Ga-face AlGaN/GaN heterostructures, *J. Appl. Phys.* **85**(6), 3222–3233 (1999). doi: 10.1063/1.369664. URL https://doi.org/10.1063/1.369664.

7. A. Dal Corso, M. Posternak, R. Resta, and A. Baldereschi, Ab initio study of piezoelectricity and spontaneous polarization in ZnO, *Phys. Rev. B.* **50**, 10715–10721 (1994). doi: 10.1103/PhysRevB.50.10715. URL https://link.aps.org/doi/10.1103/PhysRevB.50.10715.

8. A. Malashevich and D. Vanderbilt, First-principles study of polarization in $Zn_{1-x}Mg_xO$, *Phys. Rev. B.* **75**, 045106 (2007). doi: 10.1103/PhysRevB.75.045106. URL https://link.aps.org/doi/10.1103/PhysRevB.75.045106.

9. R. Resta, Macroscopic polarization in crystalline dielectrics: The geometric phase approach, *Rev. Mod. Phys.* **66**, 899–915 (1994). doi: 10.1103/RevModPhys.66.899. URL https://link.aps.org/doi/10.1103/RevModPhys.66.899.

10. V. V. Solovyev and I. V. Kukushkin, Two-subband occupation by 2D electrons in MgZnO/ZnO heterostructures, *Appl. Phys. Exp.* **12**(2), 021001 (2019). doi: 10.7567/1882-0786/aaf786. URL https://doi.org/10.7567%2F1882-0786%2Faaf786.

11. J. Falson, D. Maryenko, Y. Kozuka, A. Tsukazaki, and M. Kawasaki, Magnesium doping controlled density and mobility of two-dimensional electron gas in $Mg_x\ Zn_{1-x}O/ZnO$ Heterostructures, *Appl. Phys. Exp.* **4**(9), 091101 (2011). URL http://stacks.iop.org/1882-0786/4/i=9/a=091101.

12. J. Falson, Y. Kozuka, M. Uchida, J. H. Smet, T.-H. Arima, A. Tsukazaki, and M. Kawasaki, MgZnO/ZnO heterostructures with electron mobility exceeding 1×10^6 cm^2/Vs, *Scient. Rep.* **6**, 26598 (2016).

13. S. Akasaka, K. Nakahara, H. Yuji, A. Tsukazaki, A. Ohtomo, and M. Kawasaki, Preparation of an epitaxy-ready surface of a ZnO(0001) substrate, *Appl. Phys. Exp.* **4**(3), 035701 (2011). URL http://stacks.iop.org/1882-0786/4/i=3/a=035701.

14. M. Uchida, J. Falson, Y. Segawa, Y. Kozuka, A. Tsukazaki, and M. Kawasaki, Calibration and control of in-plane Mg doping distribution in $Mg_xZn_{1-x}O/ZnO$ heterostructures grown by molecular beam epitaxy, *Japanese J. Appl. Phys.* **54**(2), 028004 (2015). URL http://stacks.iop.org/1347-4065/54/i=2/a=028004.

15. Y. Kozuka, J. Falson, Y. Segawa, T. Makino, A. Tsukazaki, and M. Kawasaki, Precise calibration of Mg concentration in $Mg_xZn_{1-x}O$ thin films grown on ZnO substrates, *J. Appl. Phys.* **112**(4), 043515 (2012). doi: 10.1063/1.4748306. URL https://doi.org/10.1063/1.4748306.

16. A. Tsukazaki, A. Ohtomo, T. Kita, Y. Ohno, H. Ohno, and M. Kawasaki, Quantum Hall effect in polar oxide heterostructures, *Science.* **315**(5817), 1388–1391 (2007). ISSN 0036-8075. doi: 10.1126/science.1137430. URL http://science.sciencemag.org/content/315/5817/1388.

17. M. Nakano, A. Tsukazaki, A. Ohtomo, K. Ueno, S. Akasaka, H. Yuji, K. Nakahara, T. Fukumura, and M. Kawasaki, Electronic-field control of two-dimensional electrons in polymer-gated-oxide semiconductor heterostructures, *Adv. Mater.* **22**(8), 876–879 (2009). doi: 10.1002/adma.200902162. URL https://onlinelibrary.wiley.com/doi/abs/10.1002/adma.200902162.

18. A. Tsukazaki, S. Akasaka, K. Nakahara, Y. Ohno, H. Ohno, D. Maryenko, A. Ohtomo, and M. Kawasaki, Observation of the fractional quantum Hall effect in an oxide, *Nat. Mat.* **9**(11), 889 (2010).

19. J. Falson, Y. Kozuka, J. H. Smet, T. Arima, A. Tsukazaki, and M. Kawasaki, Electron scattering times in ZnO based polar heterostructures, *Appl. Phys. Lett.* **107**(8), 082102 (2015). doi: http://dx.doi.org/10.1063/1.4929381.

20. A. Janotti and C. G. Van de Walle, Native point defects in ZnO, *Phys. Rev. B.* **76**, 165202 (2007). doi: 10.1103/PhysRevB.76.165202. URL https://link.aps.org/doi/10.1103/PhysRevB.76.165202.

21. L. Pfeiffer and K. West, The role of MBE in recent quantum Hall effect physics discoveries, *Physica E: Low-dimensional Systems and Nanostructures*, **20**(1), 57–64 (2003). ISSN 1386-9477. doi: https://doi.org/10.1016/j.physe.2003.09.035. URL http://www.sciencedirect.com/science/article/pii/S1386947703005174.

22. V. Umansky, M. Heiblum, Y. Levinson, J. Smet, J. Nübler, and M. Dolev, MBE growth of ultra-low disorder 2DEG with mobility exceeding 35×10^6 cm^{2}V^{-1}s^{-1}, *J. Crystal Growth* **311**(7), 1658–1661 (2009). ISSN 0022-0248. doi: https://doi.org/10.1016/j.jcrysgro.2008.09.151. URL http://www.sciencedirect.com/science/article/pii/S0022024808009901. International Conference on Molecular Beam Epitaxy (MBE-XV).

23. G. C. Gardner, S. Fallahi, J. D. Watson, and M. J. Manfra, Modified MBE hardware and techniques and role of gallium purity for attainment of two-dimensional electron gas mobility $> 35 \times 10^6$ cm^{2}V^{-1}s^{-1} in AlGaAs/GaAs quantum wells grown by MBE, *J. Crystal Growth* **441**, 71–77 (2016). ISSN 0022-0248. doi: https://doi.org/10.1016/j.jcrysgro.2016.02.010. URL http://www.sciencedirect.com/science/article/pii/S0022024816300367.

24. Y. J. Chung, K. A. Villegas Rosales, H. Deng, K. W. Baldwin, K. W. West, M. Shayegan, and L. N. Pfeiffer, Multivalley two-dimensional electron system in an AlAs quantum well with mobility exceeding 2×10^6 cm^{2}V^{-1}s^{-1}, *Phys. Rev. Mater.* **2**, 071001 (2018). doi: 10.1103/PhysRevMaterials.2.071001. URL https://link.aps.org/doi/10.1103/PhysRevMaterials.2.071001.

25. M. Y. Melnikov, A. A. Shashkin, V. T. Dolgopolov, S.-H. Huang, C. W. Liu, and S. V. Kravchenko, Ultra-high mobility two-dimensional electron gas in a SiGe/Si/SiGe quantum well, *Appl. Phys. Lett.* **106**(9), 092102 (2015). doi: 10.1063/1.4914007. URL https://doi.org/10.1063/1.4914007.

26. A. T. Hatke, T. Wang, C. Thomas, G. C. Gardner, and M. J. Manfra, Mobility in excess of 10^6 cm^2/Vs in InAs quantum wells grown on lattice mismatched InP substrates, *Appl. Phys. Lett.* **111**(14), 142106 (2017). doi: 10.1063/1.4993784. URL https://doi.org/10.1063/1.4993784.

27. M. K. Ma, M. S. Hossain, K. A. Villegas Rosales, H. Deng, T. Tschirky, W. Wegscheider, and M. Shayegan, Observation of fractional quantum Hall effect in an InAs quantum well, *Phys. Rev. B.* **96**, 241301 (2017). doi: 10.1103/PhysRevB.96.241301. URL https://link.aps.org/doi/10.1103/PhysRevB.96.241301.

28. S. Das Sarma and F. Stern, Single-particle relaxation time versus scattering time in an impure electron gas, *Phys. Rev. B.* **32**, 8442–8444 (1985). doi: 10.1103/PhysRevB.32.8442. URL https://link.aps.org/doi/10.1103/PhysRevB.32.8442.

29. J. Nuebler, V. Umansky, R. Morf, M. Heiblum, K. von Klitzing, and J. Smet, Density dependence of the $\nu = \frac{5}{2}$ energy gap: Experiment and theory, *Phys. Rev. B.* **81**, 035316 (2010). doi: 10.1103/PhysRevB.81.035316. URL https://link.aps.org/doi/10.1103/PhysRevB.81.035316.

30. Q. Qian, J. Nakamura, S. Fallahi, G. C. Gardner, J. D. Watson, S. Lüscher, J. A. Folk, G. A. Csáthy, and M. J. Manfra, Quantum lifetime in ultrahigh quality GaAs quantum wells: Relationship to $\Delta_{5/2}$ and impact of density fluctuations, *Phys. Rev. B.* **96**, 035309 (2017). doi: 10.1103/PhysRevB.96.035309. URL https://link.aps.org/doi/10.1103/PhysRevB.96.035309.

31. L. D. Landau, The theory of a Fermi liquid, *Soviet Physics JETP.* **3**, 920 (1957).

32. L. D. Landau, Oscillations in a Fermi liquid, *Soviet Physics JETP.* **5**, 101 (1957).

33. L. D. Landau, On the theory of the Fermi liquid, *Soviet Physics JETP.* **8**, 70 (1959).

34. G. Giuliani and G. Vignale, *Quantum Theory of the Electron Liquid.* (Cambridge University Press, 2005). doi: 10.1017/CBO9780511619915.

35. S. Das Sarma, S. Adam, E. H. Hwang, and E. Rossi, Electronic transport in two-dimensional graphene, *Rev. Mod. Phys.* **83**, 407–470 (2011). doi: 10.1103/RevModPhys.83.407. URL https://link.aps.org/doi/10.1103/RevModPhys.83.407.

36. D. Maryenko, J. Falson, Y. Kozuka, A. Tsukazaki, and M. Kawasaki, Polarization-dependent Landau level crossing in a two-dimensional electron system in a MgZnO/ZnO heterostructure, *Phys. Rev. B.* **90**, 245303 (2014). doi: 10.1103/PhysRevB.90.245303. URL https://link.aps.org/doi/10.1103/PhysRevB.90.245303.

37. D. Maryenko, A. McCollam, J. Falson, Y. Kozuka, J. Bruin, U. Zeitler, and M. Kawasaki, Composite fermion liquid to Wigner solid transition in the lowest Landau level of zinc oxide, *Nat. Commun.* **9**(1), 4356 (2018).

38. J. Falson, D. Tabrea, D. Zhang, I. Sodemann, Y. Kozuka, A. Tsukazaki, M. Kawasaki, K. von Klitzing, and J. H. Smet, A cascade of phase transitions in an orbitally mixed half-filled Landau level, *Sci. Adv.* **4**, eaat8742 (2018).

39. J. Falson, D. Maryenko, B. Friess, D. Zhang, Y. Kozuka, A. Tsukazaki, J. H. Smet, and M. Kawasaki, Even denominator fractional quantum Hall physics in ZnO, *Nat. Phys.* **11**, 347–351 (2015).

40. D. Tabrea, I. A. Dmitriev, S. I. Dorozhkin, B. P. Gorshunov, A. Boris, Y. Kozuka, A. Tsukazaki, M. Kawasaki, K. von Klitzing, and J. Falson, Microwave response of interacting two-dimensional electrons in oxide heterostructures. Submitted (2019).

41. W. Kohn, Cyclotron resonance and de Haas–van Alphen oscillations of an interacting electron gas, *Phys. Rev.* **123**, 1242–1244 (1961). doi: 10.1103/PhysRev.123.1242. URL https://link.aps.org/doi/10.1103/PhysRev.123.1242.

42. M. A. Zudov, R. R. Du, J. A. Simmons, and J. L. Reno, Shubnikov–de Haas-like oscillations in millimeterwave photoconductivity in a high-mobility two-dimensional electron gas, *Phys. Rev. B.* **64**, 201311 (2001). doi: 10.1103/PhysRevB.64.201311. URL https://link.aps.org/doi/10.1103/PhysRevB.64.201311.

43. A. A. Bykov, A. K. Bakarov, D. R. Islamov, and A. I. Toropov, Giant magnetoresistance oscillations induced by microwave radiation and a zero-resistance state in a 2D electron system with a moderate mobility, *JETP Lett.* **84**(7), 391–394 (2006). ISSN 1090-6487. doi: 10.1134/S0021364006190076. URL https://doi.org/10.1134/S0021364006190076.

44. M. A. Zudov, O. A. Mironov, Q. A. Ebner, P. D. Martin, Q. Shi, and D. R. Leadley, Observation of microwave-induced resistance oscillations in a high-mobility two-dimensional hole gas in a strained Ge/SiGe quantum well, *Phys. Rev. B.* **89**, 125401 (2014). doi: 10.1103/PhysRevB.89.125401. URL https://link.aps.org/doi/10.1103/PhysRevB.89.125401.

45. A. T. Hatke, M. A. Zudov, J. D. Watson, M. J. Manfra, L. N. Pfeiffer, and K. W. West, Evidence for effective mass reduction in GaAs/AlGaAs quantum wells, *Phys. Rev. B.* **87**, 161307 (2013). doi: 10.1103/PhysRevB.87.161307. URL https://link.aps.org/doi/10.1103/PhysRevB.87.161307.

46. D. F. Kärcher, A. V. Shchepetilnikov, Y. A. Nefyodov, J. Falson, I. A. Dmitriev, Y. Kozuka, D. Maryenko, A. Tsukazaki, S. I. Dorozhkin, I. V. Kukushkin, M. Kawasaki, and J. H. Smet, Observation of microwave induced resistance and photovoltage oscillations in MgZnO/ZnO heterostructures, *Phys. Rev. B.* **93**, 041410 (2016). doi: 10.1103/PhysRevB.93.041410. URL https://link.aps.org/doi/10.1103/PhysRevB.93.041410.

47. A. V. Shchepetilnikov, D. D. Frolov, Y. A. Nefyodov, I. V. Kukushkin, and S. Schmult, Renormalization of the effective mass deduced from the period of microwave-induced resistance oscillations in GaAs/AlGaAs heterostructures, *Phys. Rev. B.* **95**, 161305 (2017). doi: 10.1103/PhysRevB.95.161305. URL https://link.aps.org/doi/10.1103/PhysRevB.95.161305.

48. F. F. Fang and P. J. Stiles, Effects of a tilted magnetic field on a two-dimensional electron gas, *Phys. Rev.* **174**, 823–828 (1968). doi: 10.1103/PhysRev.174.823. URL https://link.aps.org/doi/10.1103/PhysRev.174.823.

49. R. J. Nicholas, R. J. Haug, K. v. Klitzing, and G. Weimann, Exchange enhancement of the spin splitting in a GaAs-Ga$_x$Al$_{1-x}$As heterojunction, *Phys. Rev. B.* **37**, 1294–1302 (1988). doi: 10.1103/PhysRevB.37.1294. URL https://link.aps.org/doi/10.1103/PhysRevB.37.1294.

50. T. Neugebauer, K. von Klitzing, G. Landwehr, and G. Dorda, Surface quantum oscillations in (110) and (111) n-type silicon inversion layers, *Solid State Commun.* **17**(3), 295–300 (1975). ISSN 0038-1098. doi: https://doi.org/10.1016/0038-1098(75)90297-5. URL http://www.sciencedirect.com/science/article/pii/0038109875902975.

51. S. J. Papadakis, E. P. De Poortere, and M. Shayegan, Anomalous spin splitting of two-dimensional electrons in an AlAs quantum well, *Phys. Rev. B.* **59**, R12743–R12746 (1999). doi: 10.1103/PhysRevB.59.R12743. URL https://link.aps.org/doi/10.1103/PhysRevB.59.R12743.

52. P. Weitz, R. Haug, K. von Klitzing, and F. Schäffler, Tilted magnetic field studies of spin- and valley-splittings in Si/Si$_{1-x}$Ge$_x$ heterostructures, *Surface Science* **361-362**, 542–546 (1996). ISSN 0039-6028. doi: https://doi.org/10.1016/0039-6028(96)00465-7. URL http://www.sciencedirect.com/science/article/pii/0039602896004657.

53. R. Winkler, S. J. Papadakis, E. P. De Poortere, and M. Shayegan, Highly anisotropic *g*-factor of two-dimensional hole systems, *Phys. Rev. Lett.* **85**, 4574–4577 (2000). doi: 10.1103/PhysRevLett.85.4574. URL https://link.aps.org/doi/10.1103/PhysRevLett.85.4574.

54. D. Maryenko, J. Falson, M. S. Bahramy, I. A. Dmitriev, Y. Kozuka, A. Tsukazaki, and M. Kawasaki, Spin-selective electron quantum transport in nonmagnetic MgZnO/ZnO heterostructures, *Phys. Rev. Lett.* **115**, 197601 (2015). doi: 10.1103/PhysRevLett.115.197601. URL https://link.aps.org/doi/10.1103/PhysRevLett.115.197601.

55. Y. Kozuka, S. Teraoka, J. Falson, A. Oiwa, A. Tsukazaki, S. Tarucha, and M. Kawasaki, Rashba spin-orbit interaction in a $Mg_xZn_{1-x}O/ZnO$ two-dimensional electron gas studied by electrically detected electron spin resonance, *Phys. Rev. B.* **87**, 205411 (2013). doi: 10.1103/PhysRevB.87.205411. URL https://link.aps.org/doi/10.1103/PhysRevB.87.205411.

56. E. P. De Poortere, E. Tutuc, S. J. Papadakis, and M. Shayegan, Resistance spikes at transitions between quantum Hall ferromagnets, *Science* **290**(5496), 1546–1549 (2000). ISSN 0036-8075. doi: 10.1126/science.290.5496.1546. URL http://science.sciencemag.org/content/290/5496/1546.

57. T. Jungwirth and A. H. MacDonald, Resistance spikes and domain wall loops in Ising quantum Hall ferromagnets, *Phys. Rev. Lett.* **87**, 216801 (2001). doi: 10.1103/PhysRevLett.87.216801. URL https://link.aps.org/doi/10.1103/PhysRevLett.87.216801.

58. V. T. Dolgopolov and A. Gold, Magnetoresistance of a two-dimensional electron gas in a parallel magnetic field, *J. Exp. Theo. Phys. Lett.* **71**(1), 27–30 (2000). ISSN 1090-6487. doi: 10.1134/1.568270. URL https://doi.org/10.1134/1.568270.

59. N. Ashcroft and N. Mermin, *Solid State Phys.* Science: Physics, (Saunders College, 1976). ISBN 9780030493461. URL https://books.google.de/books?id=FRZRAAAAMAAJ.

60. J. K. Jain, *Composite fermions.* (Cambridge University Press, 2007).

61. B. I. Halperin, P. A. Lee, and N. Read, Theory of the half-filled Landau level, *Phys. Rev. B.* **47**, 7312–7343 (1993). doi: 10.1103/PhysRevB.47.7312. URL https://link.aps.org/doi/10.1103/PhysRevB.47.7312.

62. R. L. Willett, R. R. Ruel, K. W. West, and L. N. Pfeiffer, Experimental demonstration of a Fermi surface at one-half filling of the lowest Landau level, *Phys. Rev. Lett.* **71**, 3846–3849 (1993). doi: 10.1103/PhysRevLett.71.3846. URL https://link.aps.org/doi/10.1103/PhysRevLett.71.3846.

63. W. Kang, H. L. Stormer, L. N. Pfeiffer, K. W. Baldwin, and K. W. West, How real are composite fermions?, *Phys. Rev. Lett.* **71**, 3850–3853 (1993). doi: 10.1103/PhysRevLett.71.3850. URL https://link.aps.org/doi/10.1103/PhysRevLett.71.3850.

64. V. J. Goldman, B. Su, and J. K. Jain, Detection of composite fermions by magnetic focusing, *Phys. Rev. Lett.* **72**, 2065–2068 (1994). doi: 10.1103/PhysRevLett.72.2065. URL https://link.aps.org/doi/10.1103/PhysRevLett.72.2065.

65. J. H. Smet, D. Weiss, R. H. Blick, G. Lütjering, K. von Klitzing, R. Fleischmann, R. Ketzmerick, T. Geisel, and G. Weimann, Magnetic focusing of composite fermions through arrays of cavities, *Phys. Rev. Lett.* **77**, 2272–2275 (1996). doi: 10.1103/PhysRevLett.77.2272. URL https://link.aps.org/doi/10.1103/PhysRevLett.77.2272.

66. J. H. Smet, K. von Klitzing, D. Weiss, and W. Wegscheider, dc Transport of composite fermions in weak periodic potentials, *Phys. Rev. Lett.* **80**, 4538–4541 (1998). doi: 10.1103/PhysRevLett.80.4538. URL https://link.aps.org/doi/10.1103/PhysRevLett.80.4538.

67. J. H. Smet, S. Jobst, K. von Klitzing, D. Weiss, W. Wegscheider, and V. Umansky, Commensurate composite fermions in weak periodic electrostatic potentials: Direct evidence of a periodic effective magnetic field, *Phys. Rev. Lett.* **83**, 2620–2623 (1999). doi: 10.1103/PhysRevLett.83.2620. URL https://link.aps.org/doi/10.1103/PhysRevLett.83.2620.

68. R. L. Willett, K. W. West, and L. N. Pfeiffer, Geometric resonance of composite fermion cyclotron orbits with a fictitious magnetic field modulation, *Phys. Rev. Lett.* **83**, 2624–2627 (1999). doi: 10.1103/PhysRevLett.83.2624. URL https://link.aps.org/doi/10.1103/PhysRevLett.83.2624.

69. I. Kukushkin, J. Smet, K. Von Klitzing, and W. Wegscheider, Cyclotron resonance of composite fermions, *Nature* **415**(6870), 409 (2002).

70. D. Kamburov, Y. Liu, M. A. Mueed, M. Shayegan, L. N. Pfeiffer, K. W. West, and K. W. Baldwin, What determines the Fermi wave vector of composite fermions?, *Phys. Rev. Lett.* **113**, 196801 (2014). doi: 10.1103/PhysRevLett.113.196801. URL https://link.aps.org/doi/10.1103/PhysRevLett.113.196801.

71. D. Maryenko, J. Falson, Y. Kozuka, A. Tsukazaki, M. Onoda, H. Aoki, and M. Kawasaki, Temperature-dependent magnetotransport around $\nu = 1/2$ in ZnO heterostructures, *Phys. Rev. Lett.* **108**, 186803 (2012). doi: 10.1103/PhysRevLett.108.186803. URL https://link.aps.org/doi/10.1103/PhysRevLett.108.186803.

72. D. C. Tsui, H. L. Stormer, and A. C. Gossard, Two-dimensional magnetotransport in the extreme quantum limit, *Phys. Rev. Lett.* **48**, 1559–1562 (1982). doi: 10.1103/PhysRevLett.48.1559. URL https://link.aps.org/doi/10.1103/PhysRevLett.48.1559.

73. R. B. Laughlin, Anomalous quantum Hall effect: An incompressible quantum fluid with fractionally charged excitations, *Phys. Rev. Lett.* **50**, 1395–1398 (1983). doi: 10.1103/PhysRevLett.50.1395. URL https://link.aps.org/doi/10.1103/PhysRevLett.50.1395.

74. H. W. Jiang, R. L. Willett, H. L. Stormer, D. C. Tsui, L. N. Pfeiffer, and K. W. West, Quantum liquid versus electron solid around $\nu = 1/5$ Landau-level filling, *Phys. Rev. Lett.* **65**, 633–636 (1990). doi: 10.1103/PhysRevLett.65.633. URL https://link.aps.org/doi/10.1103/PhysRevLett.65.633.

75. M. B. Santos, Y. W. Suen, M. Shayegan, Y. P. Li, L. W. Engel, and D. C. Tsui, Observation of a re-entrant insulating phase near the 1/3 fractional quantum Hall liquid in a two-dimensional hole system, *Phys. Rev. Lett.* **68**, 1188–1191 (1992). doi: 10.1103/PhysRevLett.68.1188. URL https://link.aps.org/doi/10.1103/PhysRevLett.68.1188.

76. Y. Kozuka, A. Tsukazaki, D. Maryenko, J. Falson, S. Akasaka, K. Nakahara, S. Nakamura, S. Awaji, K. Ueno, and M. Kawasaki, Insulating phase of a two-dimensional electron gas in $Mg_xZn_{1-x}O/ZnO$ heterostructures below $\nu = \frac{1}{3}$, *Phys. Rev. B.* **84**, 033304 (2011). doi: 10.1103/PhysRevB.84.033304. URL https://link.aps.org/doi/10.1103/PhysRevB.84.033304.

77. J. Zhao, Y. Zhang, and J. K. Jain, Crystallization in the fractional quantum Hall regime induced by Landau-level mixing, *Phys. Rev. Lett.* **121**, 116802 (2018). doi: 10.1103/PhysRevLett.121.116802. URL https://link.aps.org/doi/10.1103/PhysRevLett.121.116802.

78. A. A. Koulakov, M. M. Fogler, and B. I. Shklovskii, Charge density wave in two-dimensional electron liquid in weak magnetic field, *Phys. Rev. Lett.* **76**, 499–502 (1996). doi: 10.1103/PhysRevLett.76.499. URL https://link.aps.org/doi/10.1103/PhysRevLett.76.499.

79. V. W. Scarola, K. Park, and J. Jain, Cooper instability of composite fermions, *Nature* **406**(6798), 863 (2000).
80. I. Sodemann and A. H. MacDonald, Landau level mixing and the fractional quantum Hall effect, *Phys. Rev. B.* **87**, 245425 (2013). doi: 10.1103/PhysRevB.87.245425. URL https://link.aps.org/doi/10.1103/PhysRevB.87.245425.
81. M. P. Lilly, K. B. Cooper, J. P. Eisenstein, L. N. Pfeiffer, and K. W. West, Evidence for an anisotropic state of two-dimensional electrons in high Landau levels, *Phys. Rev. Lett.* **82**, 394–397 (1999). doi: 10.1103/PhysRevLett.82.394. URL https://link.aps.org/doi/10.1103/PhysRevLett.82.394.
82. R. Du, D. Tsui, H. Stormer, L. Pfeiffer, K. Baldwin, and K. West, Strongly anisotropic transport in higher two-dimensional Landau levels, *Solid State Commun.* **109**(6), 389–394 (1999). ISSN 0038-1098. doi: https://doi.org/10.1016/S0038-1098(98)00578-X. URL http://www.sciencedirect.com/science/article/pii/S003810989800578X.
83. I. V. Kukushkin, V. Umansky, K. von Klitzing, and J. H. Smet, Collective modes and the periodicity of quantum Hall stripes, *Phys. Rev. Lett.* **106**, 206804 (2011). doi: 10.1103/PhysRevLett.106.206804. URL https://link.aps.org/doi/10.1103/PhysRevLett.106.206804.
84. B. Friess, V. Umansky, L. Tiemann, K. von Klitzing, and J. H. Smet, Probing the microscopic structure of the stripe phase at filling factor 5/2, *Phys. Rev. Lett.* **113**, 076803 (2014). doi: 10.1103/PhysRevLett.113.076803. URL https://link.aps.org/doi/10.1103/PhysRevLett.113.076803.
85. R. Willett, J. P. Eisenstein, H. L. Störmer, D. C. Tsui, A. C. Gossard, and J. H. English, Observation of an even-denominator quantum number in the fractional quantum Hall effect, *Phys. Rev. Lett.* **59**, 1776–1779 (1987). doi: 10.1103/PhysRevLett.59.1776. URL https://link.aps.org/doi/10.1103/PhysRevLett.59.1776.
86. J. P. Eisenstein, K. B. Cooper, L. N. Pfeiffer, and K. W. West, Insulating and fractional quantum Hall states in the first excited Landau level, *Phys. Rev. Lett.* **88**, 076801 (2002). doi: 10.1103/PhysRevLett.88.076801. URL https://link.aps.org/doi/10.1103/PhysRevLett.88.076801.
87. F. D. M. Haldane and E. H. Rezayi, Spin-singlet wave function for the half-integral quantum Hall effect, *Phys. Rev. Lett.* **60**, 956–959 (1988). doi: 10.1103/PhysRevLett.60.956. URL https://link.aps.org/doi/10.1103/PhysRevLett.60.956.
88. G. Moore and N. Read, Nonabelions in the fractional quantum Hall effect, *Nucl. Phys. B.* **360**(2), 362–396 (1991). ISSN 0550-3213. doi: https://doi.org/10.1016/0550-3213(91)90407-O. URL http://www.sciencedirect.com/science/article/pii/0550321391904070.
89. M. Levin, B. I. Halperin, and B. Rosenow, Particle-hole symmetry and the Pfaffian state, *Phys. Rev. Lett.* **99**, 236806 (2007). doi: 10.1103/PhysRevLett.99.236806. URL https://link.aps.org/doi/10.1103/PhysRevLett.99.236806.
90. S.-S. Lee, S. Ryu, C. Nayak, and M. P. A. Fisher, Particle-hole symmetry and the $\nu = \frac{5}{2}$ quantum Hall state, *Phys. Rev. Lett.* **99**, 236807 (2007). doi: 10.1103/PhysRevLett.99.236807. URL https://link.aps.org/doi/10.1103/PhysRevLett.99.236807.
91. D. T. Son, Is the composite fermion a Dirac particle?, *Phys. Rev. X.* **5**, 031027 (2015). doi: 10.1103/PhysRevX.5.031027. URL https://link.aps.org/doi/10.1103/PhysRevX.5.031027.
92. Y. W. Suen, L. W. Engel, M. B. Santos, M. Shayegan, and D. C. Tsui, Observation of a $\nu = 1/2$ fractional quantum Hall state in a double-layer electron system, *Phys. Rev. Lett.* **68**, 1379–1382 (1992). doi: 10.1103/PhysRevLett.68.1379. URL https://link.aps.org/doi/10.1103/PhysRevLett.68.1379.

93. J. P. Eisenstein, G. S. Boebinger, L. N. Pfeiffer, K. W. West, and S. He, New fractional quantum Hall state in double-layer two-dimensional electron systems, *Phys. Rev. Lett.* **68**, 1383–1386 (1992). doi: 10.1103/PhysRevLett.68.1383. URL https://link.aps.org/doi/10.1103/PhysRevLett.68.1383.

94. M. R. Peterson and C. Nayak, Effects of Landau level mixing on the fractional quantum Hall effect in monolayer graphene, *Phys. Rev. Lett.* **113**, 086401 (2014). doi: 10.1103/PhysRevLett.113.086401. URL https://link.aps.org/doi/10.1103/PhysRevLett.113.086401.

95. Z. Papić and D. A. Abanin, Topological phases in the zeroth Landau level of bilayer graphene, *Phys. Rev. Lett.* **112**, 046602 (2014). doi: 10.1103/PhysRevLett.112.046602. URL https://link.aps.org/doi/10.1103/PhysRevLett.112.046602.

96. A. Zibrov, C. Kometter, H. Zhou, E. Spanton, T. Taniguchi, K. Watanabe, M. Zaletel, and A. Young, Tunable interacting composite fermion phases in a half-filled bilayer-graphene Landau level, *Nature* **549**(7672), 360 (2017).

97. S. H. Simon and E. H. Rezayi, Landau level mixing in the perturbative limit, *Phys. Rev. B.* **87**, 155426 (2013). doi: 10.1103/PhysRevB.87.155426. URL https://link.aps.org/doi/10.1103/PhysRevB.87.155426.

98. T. Tambo, J. Falson, D. Maryenko, Y. Kozuka, A. Tsukazaki, and M. Kawasaki, Air-gap gating of MgZnO/ZnO heterostructures, *J. Appl. Phys.* **116**(8), 084310 (2014). doi: 10.1063/1.4894155. URL https://doi.org/10.1063/1.4894155.

99. E. Kleinbaum, A. Kumar, L. N. Pfeiffer, K. W. West, and G. A. Csáthy, Gap reversal at filling factors $3 + 1/3$ and $3 + 1/5$: Towards novel topological order in the fractional quantum Hall regime, *Phys. Rev. Lett.* **114**, 076801 (2015). doi: 10.1103/PhysRevLett.114.076801. URL https://link.aps.org/doi/10.1103/PhysRevLett.114.076801.

100. M. Barkeshli, C. Nayak, Z. Papić, A. Young, and M. Zaletel, Topological exciton Fermi surfaces in two-component fractional quantized Hall insulators, *Phys. Rev. Lett.* **121**, 026603 (2018). doi: 10.1103/PhysRevLett.121.026603. URL https://link.aps.org/doi/10.1103/PhysRevLett.121.026603.

101. M. P. Zaletel, S. Geraedts, Z. Papić, and E. H. Rezayi, Evidence for a topological "exciton Fermi sea" in bilayer graphene, *Phys. Rev. B.* **98**, 045113 (2018). doi: 10.1103/PhysRevB.98.045113. URL https://link.aps.org/doi/10.1103/PhysRevB.98.045113.

102. B. I. Halperin, Theory of the quantized Hall conductance, *Helvetica Physica Acta.* **56**, 75–102 (1983).

103. W. Luo and T. Chakraborty, Tilt-induced phase transitions in even-denominator fractional quantum Hall states at the ZnO interface, *Phys. Rev. B.* **94**, 161101 (2016). doi: 10.1103/PhysRevB.94.161101. URL https://link.aps.org/doi/10.1103/PhysRevB.94.161101.

104. W. Pan, J.-S. Xia, V. Shvarts, D. E. Adams, H. L. Stormer, D. C. Tsui, L. N. Pfeiffer, K. W. Baldwin, and K. W. West, Exact quantization of the even-denominator fractional quantum Hall state at $\nu = 5/2$ Landau level filling factor, *Phys. Rev. Lett.* **83**, 3530–3533 (1999). doi: 10.1103/PhysRevLett.83.3530. URL https://link.aps.org/doi/10.1103/PhysRevLett.83.3530.

105. W. Pan, R. R. Du, H. L. Stormer, D. C. Tsui, L. N. Pfeiffer, K. W. Baldwin, and K. W. West, Strongly anisotropic electronic transport at Landau level filling factor $\nu = 9/2$ and $\nu = 5/2$ under a tilted magnetic field, *Phys. Rev. Lett.* **83**, 820–823 (1999). doi: 10.1103/PhysRevLett.83.820. URL https://link.aps.org/doi/10.1103/PhysRevLett.83.820.

106. M. P. Lilly, K. B. Cooper, J. P. Eisenstein, L. N. Pfeiffer, and K. W. West, Anisotropic states of two-dimensional electron systems in high Landau levels: Effect of an in-plane magnetic field, *Phys. Rev. Lett.* **83**, 824–827 (1999). doi: 10.1103/PhysRevLett.83. 824. URL https://link.aps.org/doi/10.1103/PhysRevLett.83.824.

107. J. Xia, V. Cvicek, J. P. Eisenstein, L. N. Pfeiffer, and K. W. West, Tilt-induced anisotropic to isotropic phase transition at $\nu = 5/2$, *Phys. Rev. Lett.* **105**, 176807 (2010). doi: 10.1103/PhysRevLett.105.176807. URL https://link.aps.org/doi/10.1103/PhysRevLett.105.176807.

108. J. I. A. Li, C. Tan, S. Chen, Y. Zeng, T. Taniguchi, K. Watanabe, J. Hone, and C. R. Dean, Even denominator fractional quantum Hall states in bilayer graphene, *Science* **358**(6363), 648–652 (2017). ISSN 0036-8075. doi: 10.1126/science.aao2521. URL http://science.sciencemag.org/content/early/2017/10/04/science.aao2521.

109. Y. Liu, D. Kamburov, M. Shayegan, L. N. Pfeiffer, K. W. West, and K. W. Baldwin, Anomalous robustness of the $\nu = 5/2$ fractional quantum Hall state near a sharp phase boundary, *Phys. Rev. Lett.* **107**, 176805 (2011). doi: 10.1103/PhysRevLett. 107.176805. URL https://link.aps.org/doi/10.1103/PhysRevLett.107.176805.

110. R. Côté, J. Lambert, Y. Barlas, and A. H. MacDonald, Orbital order in bilayer graphene at filling factor $\nu = -1$, *Phys. Rev. B.* **82**, 035445 (2010). doi: 10.1103/PhysRevB.82.035445. URL https://link.aps.org/doi/10.1103/PhysRevB.82.035445.

111. E. Abrahams, P. W. Anderson, D. C. Licciardello, and T. V. Ramakrishnan, Scaling theory of localization: Absence of quantum diffusion in two dimensions, *Phys. Rev. Lett.* **42**, 673–676 (1979). doi: 10.1103/PhysRevLett.42.673. URL https://link.aps.org/doi/10.1103/PhysRevLett.42.673.

112. S. V. Kravchenko, G. V. Kravchenko, J. E. Furneaux, V. M. Pudalov, and M. D'Iorio, Possible metal-insulator transition at B=0 in two dimensions, *Phys. Rev. B.* **50**, 8039–8042 (1994). doi: 10.1103/PhysRevB.50.8039. URL https://link.aps.org/doi/10.1103/PhysRevB.50.8039.

113. E. Abrahams, S. V. Kravchenko, and M. P. Sarachik, Metallic behavior and related phenomena in two dimensions, *Rev. Mod. Phys.* **73**, 251–266 (2001). doi: 10.1103/RevModPhys.73.251. URL https://link.aps.org/doi/10.1103/RevModPhys.73.251.

114. S. V. Kravchenko and M. P. Sarachik, Metal–insulator transition in two-dimensional electron systems, *Rep. Prog. Phys.* **67**(1), 1–44 (2003). doi: 10.1088/0034-4885/67/1/r01. URL https://doi.org/10.1088%2F0034-4885%2F67%2F1%2Fr01.

Chapter 7

Fractional Quantum Hall Effects in Graphene

Cory Dean[*], Philip Kim[†], J. I. A. Li[‡] and Andrea Young[§]

[*]*Department of Physics, Columbia University, New York, NY 10027, USA*
[†]*Department of Physics, Harvard University, Cambridge, MA 02138, USA*
[‡]*Department of Physics, Brown University, Providence, RI 02912, USA*
[§]*Department of Physics, University of California, Santa Barbara, CA 93106, USA*

We review the most recent understanding of fractional quantum Hall effects and related phenomena observed in graphene-based van der Waals heterostructures.

Contents

1. Introduction

Many of the distinguishing features of graphene relative other material systems that host the fractional quantum Hall effect arise as a consequence of its honeycomb lattice structure. This gives rise to a relativistic band dispersion where the Landau levels (LLs) are no longer evenly spaced. These levels are characterized by spinor wave functions with contrasting orbital wave function on different carbon sublattices; this spinor structure results in interaction pseudopotentials with no analog in systems with scalar wave functions. In addition, the graphene crystal structure leads to a unique set of spin, valley, and orbital degeneracies. Experimentally, the carrier density can be varied *in-situ* over a range up to $\pm 10^{13}$ cm^{-2} using the electric field effect. This electrostatic doping does not significantly degrade the carrier mobility, providing experimental capability to explore the phase space of Landau level filling factor and applied magnetic field. These advantages make graphene a uniquely versatile material platform for studying the rich interplay between symmetry breaking and fractional quantum Hall (FQH) physics.

The ability to isolate individual graphene layers offers several additional practical advantages compared to traditional systems such as III-V semiconductor or oxide based quantum wells. The atomically thin nature removes finite well width effects, while weaker dielectric screening stabilizes the fractional quantum Hall effect (FQHE) at temperatures that can be an order of magnitude higher than conventional heterostructures. Perhaps most significant is the ability to integrate graphene layers into complex new van der Waals (vdW) heterostructures, again without loss of mobility. This technique has provided access to an exciting range of experiments without analog in conventional heterostructures. Examples include moire superlattices that arise from interlayer lattice mismatch[1,2] and double layer systems, which provide exceptional control over multicomponent FQH states with tunable interlayer correlations.[3-5] Finally, graphene represents the first high quality two-dimensional electron system in which the two-dimensional (2D) electrons are not confined to a remote interface, but instead are directly accessible, enabling surface probes and *in situ* modification of the disorder potential.[6] These last features may allow truly new understandings of the microscopic physics of the FQH effect using scanning probes.

In this chapter, we review experimental studies of the FQHE in graphene-based vdW heterostructures. The unique properties of the FQHE observed in graphene are intimately related to the underlying band structure. We therefore first provide a brief overview of electronic structure of mono- and bilayer graphene, including the structure of the Landau levels. We then discuss the nature of symmetry breaking, in particular quantum Hall ferromagnetism (QHFM), which manifests already in the integer quantum Hall effect (IQHE). The high degree of symmetry between spins and valleys within a single graphene LL allow for a rich set of symmetry broken phases—and *in situ* tunable transitions between them—that in turn influence and are influenced by FQH correlations. We then separately discuss experimental observations of the FQHE in monolayer and bilayer graphene. In both cases, our focus is on emphasizing characteristics of the observed FQH states that are unique to the graphene platform and distinct from more conventional heterostructures. We next discuss double layer graphene, which conceptually extends the concept of bilayer graphene but with tunable interlayer separation and, therefore, tunable interlayer correlations. This system closely resembles the double quantum well heterostructures studied extensively in GaAs, but with much greater flexibility in the available device parameters. Finally, we end the review by discussing the effect of moire superlattice potentials on the FQHE. This represents perhaps a unique opportunity to date provided by graphene in the context of FQHE studies, yielding entirely new phenomena that have not been realized in any prior system, such as the observation of fractional Chern insulators.

2. Integer Quantum Hall Effect in Graphene

2.1. *Single particle structure*

2.1.1. *Monolayer graphene*

Monolayer graphene consists of a single planar honeycomb net of carbon atoms (see Fig. 1). The Bravais lattice has unit lattice vectors

$$\vec{a}_1 = a \left(\frac{\sqrt{3}}{2}, \frac{1}{2} \right) \qquad \vec{a}_2 = a \left(\frac{\sqrt{3}}{2}, -\frac{1}{2} \right) \qquad (1)$$

where $a = 0.246$ nm. Each unit cell consists of two identical carbon atoms, termed the A and B sublattices. Each carbon atom has three nearest neighbors in the plane and four valence electrons. In-plane σ bonds form from the sp^2 hybridized in-plane valence electron orbitals, p_x, p_y and s. The σ bond is extremely strong, and contributes to the structural stability of graphene; however, the electrons are so tightly bound that they do not contribute to electronic transport. The remaining valence electron, consisting of the p_z orbital, forms the delocalized, covalent bond with the neighboring atoms that make up the π band, which determines the low energy electronic structure.

The π electrons are still sufficiently tightly bound that the electronic structure is well approximated by a tight-binding model.[7] In the case of monolayer graphene, the low energy band structure relevant to fractional quantum Hall physics is captured through the inclusion of only the nearest neighbor hopping between adjacent A and B sublattice sites, $\gamma_0 \simeq 3.16$ eV. The A and B sublattices are connected via three hopping vectors

$$\vec{\delta}_1 = \frac{a}{2} \left(\frac{1}{\sqrt{3}}, 1 \right) \qquad \vec{\delta}_2 = \frac{a}{2} \left(\frac{1}{\sqrt{3}}, -1 \right) \qquad \vec{\delta}_3 = \frac{a}{2} \left(-\frac{2}{\sqrt{3}}, 0 \right) \qquad (2)$$

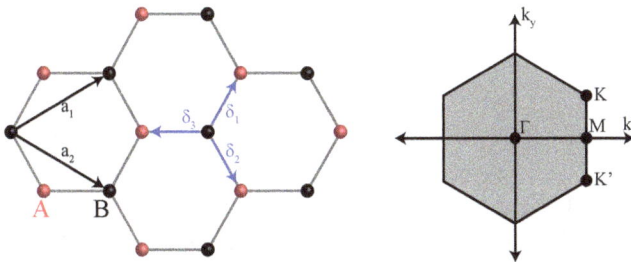

Fig. 1. (Left) Crystal structure of monolayer graphene with Bravais lattice vectors ($\vec{a}_1$ and $\vec{a}_2$) as well as nearest neighbor hopping vectors ($\vec{\delta}_1$, $\vec{\delta}_2$ and $\vec{\delta}_3$) indicated. (Right) Graphene first Brillouin zone, with several high symmetry points labeled.

and the tight binding Hamiltonian can be written in the basis of the wave function amplitudes on the two sublattices, φ_A and φ_B, as

$$\hat{H} = \gamma_0 \sum_{i=1}^{3} \begin{pmatrix} 0 & f(\vec{k}) \\ f^*(\vec{k}) & 0 \end{pmatrix} \tag{3}$$

where $f(\vec{k}) = e^{i\vec{k}\cdot\delta_i}$. The Hamiltonian Eq. (3) has eigenvalues

$$\varepsilon_\pm(\vec{k}) = \pm\gamma_0 \sqrt{3 + 4\cos\left(\frac{\sqrt{3}ak_x}{2}\right)\cos\left(\frac{ak_y}{2}\right) + 2\cos(ak_y)}. \tag{4}$$

The spectrum [Eq. (4)] is characterized by electron-hole symmetry of the bands and a degeneracy at the corners of the Brillouin zone, termed the K points, of which two examples are

$$\vec{K} = \frac{2\pi}{a}\left(\frac{1}{\sqrt{3}}, \frac{1}{3}\right) \qquad\qquad \vec{K}' = \left(\frac{1}{\sqrt{3}}, -\frac{1}{3}\right) \tag{5}$$

where $\varepsilon_\pm(\vec{K}) = \varepsilon_\pm(\vec{K}') = 0$. The electron-hole symmetry is in fact an artifact of the nearest neighbor approximation; however, the degeneracy is a fundamental property protected by the inversion symmetry of the honeycomb lattice, and survives to all orders as long as this remains a good symmetry. Each carbon atom contributes one electron to the π bands. As a result, in undoped graphene the bands are exactly half filled, with the Fermi surface consisting of two 'Fermi points' at $\vec{K}$ and $\vec{K}'$. After an appropriately chosen rotation of the relative wave function phase on the A and B sublattices, expanding the Hamiltonian about the K-points such that $q \equiv \vec{k} - \vec{K}$ yields the low energy effective Hamiltonian,

$$\hat{H} = \frac{\sqrt{3}a\gamma_0}{2}\begin{pmatrix} 0 & q_x + iq_y \\ q_x - iq_y & 0 \end{pmatrix} = \hbar v_F \vec{\sigma} \cdot \vec{q} \tag{6}$$

where $v_F \equiv \frac{\sqrt{3}\gamma_0 a}{2}$, $\vec{\sigma} = (\sigma_x, \sigma_y)$ is the vector of Pauli matrices. The Hamiltonian is identical in the two valleys with the basis corresponding to $\left(\varphi_{A,\vec{K}}, \varphi_{B,\vec{K}}\right)$ in the $\vec{K}$ and $\left(-\varphi_{B,\vec{K}'}, \varphi_{A,\vec{K}'}\right)$ in the $\vec{K}'$ valley. The spectrum is linear:

$$\varepsilon(\vec{q}) = \pm\hbar v_F |\vec{q}|. \tag{7}$$

The Hamiltonian (6) is formally identical to the theory of relativistic electrons first proposed by Dirac[8] in which the mass is exactly equal to zero, the physical speed of light has been replaced by the carrier Fermi velocity, and the physical electron spin—which leads to the spinor structure in theory of truly relativistic electrons—is replaced by a pseudospin, related to the graphene sublattices.

A magnetic field B normal to the sample plane is incorporated into the Dirac Hamiltonian through the minimal coupling of the vector potential $\vec{q} \rightarrow \vec{q} - \frac{e}{c}\vec{A}$ where $\vec{\nabla} \times \vec{A} = B$. With the aid of the standard definitions[9] for the creation and annihilation operators in the basis of the orbital Landau levels, we arrive at the Hamiltonian for monolayer graphene in a magnetic field,

$$\hat{H}^{ML} = \frac{\hbar v_F \sqrt{2}}{\ell_B} \begin{pmatrix} 0 & \hat{a}^\dagger \\ \hat{a} & 0 \end{pmatrix}, \tag{8}$$

where $\ell_B = \sqrt{\hbar/eB}$ is the magnetic length, the $\hat{a}$ and $\hat{a}^\dagger$ are operators in the space of scalar orbital Landau level wavefunctions (denoted $|N\rangle$) such that $\hat{a}^\dagger|N\rangle = \sqrt{N+1}|N+1\rangle$ and $\hat{a}|N\rangle = \sqrt{N}|N-1\rangle$, where N is a non-negative integer. As in conventional quantum Hall systems described by scalar Hamiltonians, the angular momentum (in the symmetric gauge) or guiding center coordinate (in the Landau gauge) is an integral of the motion, leading to the massive degeneracy of the Landau levels that ultimately gives rise to strongly correlated physics. The eigenenergies of Eq. (8) contain both negative and positive energy states, as well as a zero energy LL (ZLL) that does not disperse with magnetic field,

$$\varepsilon_{ML}(N) = \text{sgn}\,(N)\frac{\hbar v_F \sqrt{2|N|}}{\ell_B}. \tag{9}$$

Energy spectra for the two valleys are identical, making each LL twice again as degenerate, a fact tied again to the inversion symmetry of the honeycomb lattice. Combined with the spin degeneracy, there are four flavors of electron ($|K,\uparrow\rangle$, $|K,\downarrow\rangle$, $|K',\uparrow\rangle$, and $|K',\downarrow\rangle$, where $\uparrow$ and $\downarrow$ correspond to opposite real spin projections) allowing a quadruple occupancy of each cyclotron guiding center and a quadrupled LL degeneracy. The combination of the zero modes with the LL degeneracies leads to the anomalous observed[10,11] sequence of quantum Hall plateaus, with Hall conductance given by

$$\sigma_{xy}^{ML} = \pm 4\frac{e^2}{h}\left(n + \frac{1}{2}\right) \qquad\qquad n \in Z, n \geq 0. \tag{10}$$

In Eq. (10), the factor of $1/2$ reflects the presence of the ZLL, while the factor of four reflects the four-fold combined spin- and valley 'isospin' degeneracy. This degeneracy is not, in practice, perfect. The same magnetic field that creates the LLs cannot be prevented from exerting a Zeeman effect on the real spins, leading to a splitting of the otherwise degenerate LLs into two spin branches separated by $E_Z = g\mu_B B_T$. Here, $g = 2$ is the electron gyromagnetic ratio in graphene, and B_T is the total magnetic field, which need not be oriented in any particular way relative to the graphene plane. In monolayer graphene, an aligning field for the valley degree of freedom is represented (in the ZLL) by the Pauli operator σ_z in the $A-B$ sublattice space, equivalent to a potential modulation commensurate with the lattice. Such modulations can arise from the coupling of the graphene to a nearly commensurate substrate, most notably the hexagonal boron nitride ubiquitously used in producing high-quality heterostructures.[12,13] The high internal degeneracy of the monolayer graphene Landau levels is thus highly tunable, often *in situ*, making a variety of isospin phase transitions experimentally accessible in both the integer and fractional quantum Hall regime, as we discuss in subsequent sections.

Of great relevance to the fractional quantum Hall problem are the unusual form of the eigenvectors, which for the $\vec{K}$ valley can be written as

$$|N\rangle_{ML} = \begin{cases} \begin{pmatrix} |0\rangle \\ 0 \end{pmatrix} & N = 0 \\ \begin{pmatrix} ||N|\rangle \\ \mathrm{sgn}\,(N)\,||N|-1\rangle \end{pmatrix} & N \neq 0 \end{cases}. \tag{11}$$

Here we use the notation $|N\rangle$ to denote the Landau level wave functions of a free electron gas and we have suppressed the guiding center coordinate (or angular momentum, in the case of Landau level solutions in the symmetric gauge). The eigenvectors for $\vec{K}'$ valley can be obtained by considering symmetry. The eigenfunctions differ in two important respects from those in systems with scalar, massive Hamiltonians. First, for all levels except the ZLL, the wave functions contain orbital components of both the $|N|$ and $|N|-1$ conventional Landau levels. In the FQH problem, the orbital structure manifests in the form of the Haldane pseudopotentials, v_m, which describe the energy of two electrons with relative angular momentum m,

$$v_m = \int_0^\infty dq F_N(q) e^{-q^2} L_m(q^2), \tag{12}$$

where $F_N(q)$ is a Landau level dependent form factor, and L_m is the Laguerre polynomial. The Haldane pseudopotentials fully parameterize the Coulomb interaction after projecting the interaction to the lowest Landau level. In other words, interactions within an arbitrary Landau level can be simulated (ignoring the effects of Landau level mixing) by considering a set of Haldane pseudopotentials within the $N = 0$ landau level. The precise values of different Haldane pseudopotentials thus account for the differing behavior of the fractional quantum Hall effect between Landau levels with different orbital structures. As shown in Fig. 2(b), the pseudopotentials in monolayer graphene differ markedly from those in GaAs for $N \neq 0$. This has the effect of significantly changing the FQH sequences observed in the higher Landau levels as compared with GaAs, as discussed extensively in the theoretical and[14] and experimental literature.[15,16] Most notable is the relative strength of composite fermion states—and absence of even denominator states—in the $N = \pm 1$ LLs, and the report of a previously unobserved even denominator state in the $N = 3$ Landau level.[17]

The second relevant difference between graphene and GaAs is the sublattice structure of the wave functions. Unlike the physical spin, the valley degenerate wave functions have different microscopic orbital structure, encoded in the different probability amplitudes on the two sublattices in the different valleys for a given N. This effect is most pronounced in the ZLL. Whereas for the non-zero LLs, the electron is spread equally over the two sublattices (although the distribution can vary locally, $\langle x|N\rangle \neq \langle x|N-1\rangle$), in the ZLL the valley wavefunctions are completely localized on a single sublattice. This peculiarity is shown in Figs. 2(c) and 2(d). The LL wavefunctions have characteristic scale $\ell_B \gg a$; however, in the ZLL wavefunctions in (say) the K valley are spread over many unit cells but have

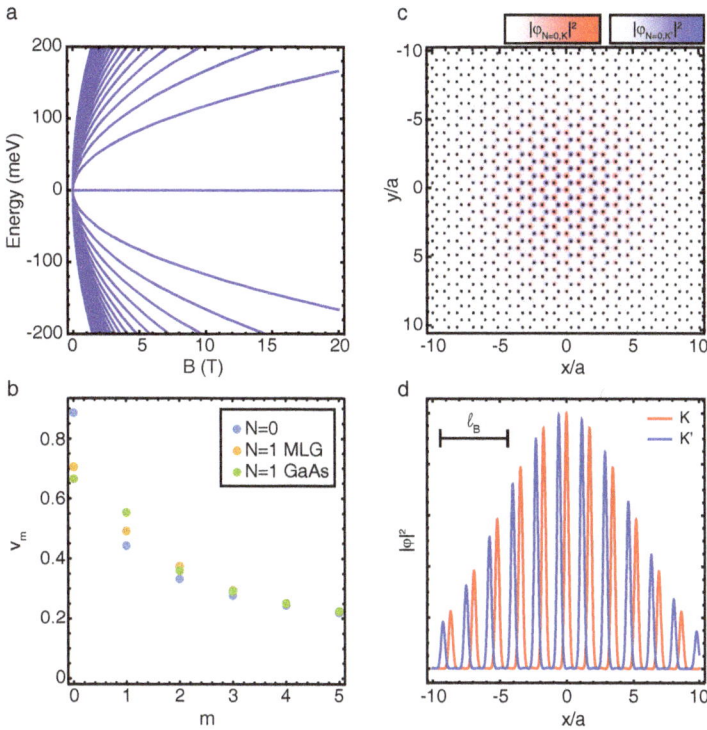

Fig. 2. **(a)** Landau level energies for monolayer graphene. **(b)** Comparison of Haldane pseudopotentials for monolayer graphene and GaAs. The $N = 0$ LL pseudopotentials are identical for
both systems, but the first excited level ($N = 1$) pseudopotentials differ significantly. **(c)** Two-
dimensional rendering of the probability distribution for an electron in the $N = 0$ Landau level
wave function in the $\vec{K}$ (red) and $\vec{K}'$ (blue) valleys. We take the symmetric gauge and plot wave
functions with angular momentum $m = 0$, and set $\ell_B = 5a$, an unrealistically small value, for
purposes of clarity. In reality, ℓ_B ranges from $15a$ (at $B = 45$ T) to $100a$ (at $B \approx 1$ T), the
magnetic field range over which the fractional quantum Hall effect is observed. **(d)** The $N = 0$,
$m = 0$ wave functions in the $\vec{K}$ and $\vec{K}'$ valleys plotted along the trajectory corresponding to $y = 0$
in (c).

zero probability of being found on the B sublattice. As discussed below, the valley-
contrasting anomalous sublattice structure of the ZLL has a significant influence on
the many-body physics pertaining to the valley- and spin symmetry breaking, with
experimentally observable consequences.

2.1.2. *Bilayer graphene*

The tight binding analysis for bilayer graphene is more complicated than that for
the monolayer, with significantly more parameters required to capture all features
of the spectrum over a comparable range of energies. Most experiments on FQH
to date are performed on Bernal stacked bilayer graphene. In Bernal—or '*A-B*'

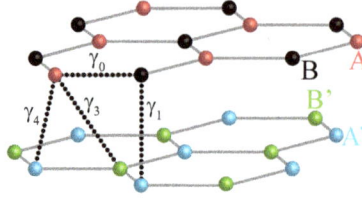

Fig. 3.   Crystal structure of Bernal stacked bilayer graphene with several hopping parameters indicated.

stacking—the monolayers are arranged so that the A sublattice of one lies directly on top of the B sublattice of the other (see Fig. 3). The unit cell in bilayer graphene consists of four carbon atoms, two in each layer, denoted A, B, A', and B'. The tight binding Hamiltonian incorporating γ_0, γ_1, γ_3, and γ_4, as well as an on-site energy Δ', is written[18,19] in the basis $(\varphi_A, \varphi_{B'}, \varphi_{A'}, \varphi_B)$ as

$$\hat{H} = \begin{pmatrix} \frac{u}{2} & \gamma_3 f(\vec{k}) & \gamma_4 f^*(\vec{k}) & \gamma_0 f^*(\vec{k}) \\ \gamma_3 f^*(\vec{k}) & -\frac{u}{2} & \gamma_0 f(\vec{k}) & \gamma_4 f(\vec{k}) \\ \gamma_4 f(\vec{k}) & \gamma_0 f^*(\vec{k}) & -\frac{u}{2} + \Delta' & \gamma_1 \\ \gamma_0 f(\vec{k}) & \gamma_4 f^*(\vec{k}) & \gamma_1 & \frac{u}{2} + \Delta' \end{pmatrix} \tag{13}$$

where the interlayer asymmetry parameter $u = v_1 - v_2$ has been introduced in anticipation of the existence of substrate and electric fields. While literature values differ slightly, a recent effort[20] to fix the parameters by comparison to density functional theory across a wide range of energies has yielded $\gamma_0 = 2.61$ eV, $\gamma_1 = 361$ meV, $\gamma_3 = -283$ meV, $\gamma_4 = -138$ meV, and $\Delta' = 15$ meV.

Expanding once again for low energies near the K and K' points and applying the substitutions of Eq. (13) applicable to high magnetic fields yields the low energy Hamiltonian in each valley,

$$\hat{H} = \begin{pmatrix} \xi\frac{u}{2} & \omega_3\hat{a} & \omega_4\hat{a}^\dagger & \omega_0\hat{a}^\dagger \\ \omega_3\hat{a}^\dagger & -\xi\frac{u}{2} & \omega_0\hat{a} & \omega_4\hat{a} \\ \omega_4\hat{a} & \omega_0\hat{a}^\dagger & -\xi\frac{u}{2} + \Delta' & \gamma_1 \\ \omega_0\hat{a} & \omega_4\hat{a}^\dagger & \gamma_1 & \xi\frac{u}{2} + \Delta' \end{pmatrix} \tag{14}$$

where for $\xi = \pm 1$ in valleys K and K'. The bases are $(\varphi_A, \varphi_{B'}, \varphi_{A'}, \varphi_B)$ for valley K and $(\varphi_{B'}, -\varphi_A, \varphi_B, -\varphi_{A'})$ for valley K', chosen to make the inversion symmetry between the valleys when $u = 0$ explicitly manifest. Here $\omega_0 = \hbar v_F \sqrt{2}/\ell_B$ is the cyclotron energy for monolayer graphene, $\omega_4 = \frac{\gamma_4}{\gamma_0}\omega_0$, and $\omega_3 = \frac{\gamma_3}{\gamma_0}\omega_0$.

The strongest modification to the linear bands of monolayer graphene arises from the strong coupling between the dimerized atoms in the two layers captured in the tight-binding model by γ_1. Qualitatively, γ_3 is responsible for trigonal warping, and is most important at low magnetic fields.[21,22] In what follows, we take $\gamma_3 = 0$, which significantly simplifies the eigenfunctions while introducing only minimal errors at the magnetic fields relevant for the FQHE. The primary role of γ_4 and Δ' are to break particle-hole symmetry in the $B = 0$ band structure. As we will see below, they also lift the accidental degeneracy of the $N = 0$ and $N = 1$ orbital Landau levels, with significant implications for the fractional quantum Hall effect. Along with γ_0, γ_1 is sufficient to produce the familiar features of the zero-magnetic field band structure of bilayer graphene, consisting of a parabolic-band-touching semimetal that becomes a semiconductor in the presence of an applied electric field. At low energies, the large scale of γ_1 as compared to the Fermi energy ensures that the wavefunction is primarily supported on the nondimerized A and B' sublattices, justifying an effective two-band model

$$\hat{H} = H_0 + H' \tag{15}$$

$$\hat{H}_0 = -\frac{\omega_0^2}{\gamma_1} \begin{pmatrix} u/2 & (\hat{a}^\dagger)^2 \\ (\hat{a})^2 & -u/2 \end{pmatrix} + \begin{pmatrix} u/2 & 0 \\ 0 & -u/2 \end{pmatrix} \tag{16}$$

$$\hat{H}' = \frac{\omega_0^2}{\gamma_1} \left(\frac{|\gamma_4|}{\gamma_0} + \frac{\Delta'}{\gamma_1} \right) \begin{pmatrix} \hat{a}^\dagger \hat{a} & 0 \\ 0 & \hat{a}\hat{a}^\dagger \end{pmatrix}, \tag{17}$$

where we have set $\gamma_3 = 0$. Consistent with inversion symmetry, the basis is $(\varphi_A, \varphi_{B'})$ in one valley and reversed (i.e. $(\varphi_{B'}, \varphi_A)$ in the other. $\hat{H}_0$ is straightforwardly diagonalized. As in the monolayer, bilayer graphene hosts zero energy Landau levels; unlike in the monolayer there are evidently two such levels, corresponding to the $N = 0$ and $N = 1$ orbital levels.

$$|0\rangle_{BLG} = \begin{pmatrix} |0\rangle \\ 0 \end{pmatrix} \qquad\qquad \varepsilon_{0,\xi} = \xi u/2 \tag{18}$$

$$|1\rangle_{BLG} = \begin{pmatrix} |1\rangle \\ 0 \end{pmatrix} \qquad\qquad \varepsilon_{1,\xi} = \xi u/2. \tag{19}$$

Absent an interlayer potential, then, there are eight degenerate levels at zero energy, with all other levels having the four-fold spin and valley degeneracy. The resulting Hall plateau sequence,

$$\sigma_{xy}^{BLG} = \pm \frac{4e^2}{h} n \qquad\qquad n \in Z, n \geq 1, \tag{20}$$

is indeed what was observed in the earliest reports.[23]

As is evident in Eqs. (18) and (19), the zero mode energies couple directly to the interlayer potential, u, which is proportional to the applied electric field and has contrasting sign in the different valleys. As discussed in later sections, electric field control of the valley degeneracy provides a versatile control knob for multicomponent fractional quantum Hall systems. For higher Landau levels, $|N| \geq 2$, the eigenfunctions have different Landau level orbitals on the different sublattices similar to high Landau levels in monolayer graphene. Although the wave functions are largely sublattice unpolarized overall, they are partially polarizable under applied

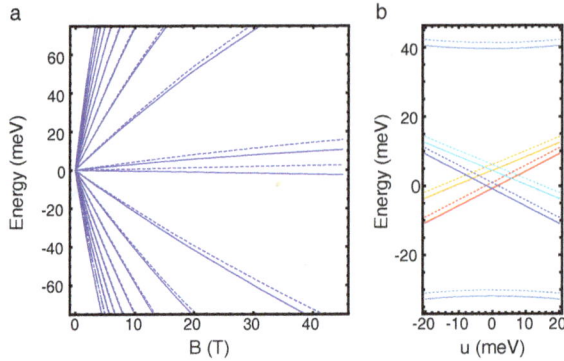

Fig. 4. (a) Landau level energies for bilayer graphene derived from the four-band model of Eq. (13) for $u = 0$. The spin splitting is included, with spin split levels indicated by dashed and solid lines, respectively. (b) Lowest several Landau level energies at $B = 15T$, as a function of the interlayer potential asymmetry u. Red and blue correspond to $N = 0$ orbital states, while orange and cyan correspond to $N = 1$ orbital states.

electric field:

$$|N\rangle_{BLG} = \frac{1}{\sqrt{2}} \begin{pmatrix} \left(1 + \mathrm{sgn}\,(N)\dfrac{u\gamma_1}{4\omega_0^2\sqrt{|N|(|N|-1)}}\right)|N\rangle \\ \mathrm{sgn}\,(N)\left(1 - \mathrm{sgn}\,(N)\dfrac{u\gamma_1}{4\omega_0^2\sqrt{|N|(|N|-1)}}\right)|N-2\rangle \end{pmatrix} \quad (21)$$

$$\varepsilon_{N,\xi} = \mathrm{sgn}\,(N)\frac{\omega_0^2}{\gamma_1}\sqrt{|N|(|N|-1) + \left(\frac{u\gamma_1}{2\omega_0^2}\right)^2}. \quad (22)$$

Note that energies are independent of the valley index ξ for $N \neq 0$ (see Fig. 4).

Equations (16)–(22) capture most of the essential physics of BLG relevant to the FQH effect. The system hosts an eight-fold degenerate zero energy Landau level (when the spin and valley degeneracy is accounted for), which includes an unusual orbital degeneracy between wave functions with differing Landau orbitals. As in monolayer graphene, the ZLL wave functions are strongly sublattice polarized, so that they resemble $N = 0$ and $N = 1$ orbitals of a scalar parabolic electron gas on a single sublattice with no support on the sub-dominant sublattice. Crucially, the sublattice structure of bilayer graphene, in which the dominant A and B' sublattices are positioned on separate layers, means that the energies of ZLL components in opposite valleys can be tuned relative to each other using an applied perpendicular electric field, which directly lifts the valley degeneracy. Finally, higher Landau levels consist of combinations of $|N\rangle$ and $|N - 2\rangle$ scalar Landau orbitals; for the FQH effect, these levels will realize pseudopotentials that are not realized in either scalar systems or in monolayer.

Not all of these features are robust. Generically, basic considerations of quantum mechanics suggest that levels do not remain degenerate unless that degeneracy is protected by a symmetry. In the case of BLG, the degeneracy of the $N = 0$ and

$N = 1$ levels is purely accidental and can be expected to lift as approximations are relaxed. In fact, the 0–1 degeneracy is already lifted in the two-band model by Eq. (17). This term leaves the energy of the $N = 0$ levels at exactly 0, but lifts the energy of the $N = 1$ level (for $u = 0$) to $\varepsilon_1 = \frac{\omega_0^2}{\gamma_1}\left(\frac{\gamma_4}{\gamma_0} + \frac{\Delta'}{\gamma_1}\right) \approx 0.12$ meV$\times B$ [T].

Other features of the BLG Landau levels cannot be captured without taking into consideration all four atoms in the unit cell, in a four-band model. Most important for the FQH effect is the precise orbital structure of the wave functions. In the case of $\gamma_3 = 0$, the four-band Landau level wave functions derived from Eq. (13) take on a simple form. The $N = 0$ wave function remains perfectly sublattice polarized (in the A or B' sublattice for the $\xi = \pm$ valley, respectively),

$$|N = 0\rangle = \begin{pmatrix} |0\rangle \\ 0 \\ 0 \\ 0 \end{pmatrix} \qquad \varepsilon_{0,\xi} = \xi u/2. \qquad (23)$$

However, the $N = 1$ orbital level differs significantly from that in the two-band approximation, acquiring finite weight on the dimerized sublattices

$$|N = 1\rangle = \begin{pmatrix} c_A |1\rangle \\ 0 \\ c_{A'} |0\rangle \\ c_B |0\rangle \end{pmatrix} \qquad \varepsilon_{1,\xi} \approx \frac{\omega_0^2}{\gamma_1}\left(\frac{\gamma_4}{\gamma_0} + \frac{\Delta'}{\gamma_1}\right) + \xi \alpha u/2 \qquad (24)$$

where $\alpha = |c_A|^2 + |c_B|^2 - |c_{A'}|^2$ reflects the now imperfect layer polarization of the $N = 1$ states. $c_B \sim O(\Delta'/\gamma_0, \gamma_4/\gamma_0)$ and is negligible in practice. However, $c_{A'} \approx \frac{\sqrt{3}}{2}\left(\frac{\gamma_0 a}{\gamma_1 \ell_B}\right)$ represents a significant correction: at $B = 30$ T, for instance, $|c_{A'}|/|c_A| \approx 0.3$. In other words, at reasonable magnetic fields the $N = 1$ orbital level in bilayer graphene contains a mixture of the lowest and first excited oscillator wavefunctions. This is reflected in the pseudopotentials governing electron-electron interactions within this level; even neglecting mixing with the nearby $N = 0$ level, at high magnetic fields the $N = 1$ Landau is not a perfect replica of the $|1\rangle$ oscillator states familiar from GaAs, but may contain FQH physics intermediate between that of the lowest and first excited levels.

2.2. *Quantum Hall ferromagnetism in graphene*

The Landau levels in both monolayer and bilayer graphene are characterized by an approximate degeneracy between levels with different spin and valley (and in the case of the bilayer graphene ZLL, orbital) quantum numbers. Naively, one might expect that the small terms that lift the degeneracies of the quartet or octet Landau levels would lead to an *all integer* quantum Hall effect, in contrast to the sequences of Eqs. (10) and (20). Indeed, early experiments showed integer quantum Hall plateaus at a high magnetic field outside of these sequences.[26,27] However, as first pointed out by Halperin[28] in the context of quasi spin-degenerate LLs in conventional GaAs

2D electron systems, the origin of these integer Hall plateaus is better understood in the frame of the quantum Hall ferromagnetism (QHFM). Here, QHFM refers to spontaneous symmetry breaking of multi-component quantum Hall system as an effect of Coulomb interactions, which tend to favor ferromagnetic polarization and consequently incompressible states at integer filling.

Indeed, graphene is exceptional among multicomponent quantum Hall systems[29,30] due to a near-perfect energetic hierarchy (see Fig. 5). Specifically, both cyclotron motion (parameterized by the cyclotron gaps E_N) and long-range interparticle Coulomb interactions (parameterized by $E_C = \frac{e^2}{\varepsilon \ell_B}$)—both spin- and valley-independent—dwarf explicit spin and valley symmetry breaking effects. These include the Zeeman effect as well as the anisotropy of the Coulomb interaction at short length scales, which, due to the peculiar lattice scale structure of the wavefunctions, is no longer valley symmetric. The long range Coulomb interaction is also considerably more important than the kinetic energy, parameterized by the Landau level broadening $\Gamma \approx 0.7$ meV for the cleanest graphene samples.[24,25] The combined four-flavor spin and valley degeneracy can, therefore, be thought of as a single SU(4) isospin.[31,32] The tendency towards ferromagnetism[32] is driven by the long range Coulomb, which favors antisymmetric orbital wavefunctions and is thus indifferent to the precise direction within the isospin space. However, regardless of the particular direction of polarization, ferromagnetism leads to the appearance of an energy gap for charged excitations and a robust quantum Hall effect for all integers, including those at partial filling of the quartet Landau levels. Historically, quantum Hall ferromagnetism in graphene was observed in greater detail whenever improvements in sample quality were made, or samples of similar quality were

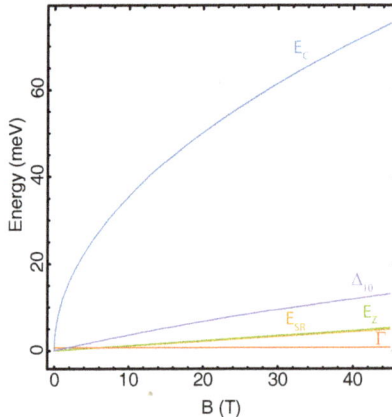

Fig. 5.   Energy scales in the graphene quantum Hall ferromagnet. The scale of the Coulomb energy (E_C), which does not distinguish between spin- and valley isospoin flavors, dwarfs anisotropic terms including the Zeeman effect (E_Z), lattice scale interactions (E_{SR}), and disorder ($\Gamma \lesssim 1$ meV[24,25] in the cleanest devices reported). For additional comparison, the $u = 0$ splitting between $N = 0$ and $N = 1$ orbitals in bilayer graphene, Δ_{01}, is also shown.

measured at ever higher magnetic fields.[15,26,33,34] This follows from a Stoner criterion[32] by which the kinetic energy cost of polarizing must be outweighed by the gain in exchange energy approximately $\Gamma/E_C \lesssim 1$. Figure 6(a) shows the evolution of the quantum Hall effect with magnetic field in monolayer graphene on a hBN substrate, marking one such sample improvement. Above fields of a few tesla, ρ_{xx} minima are visible at all integer filling within the experimental range. In addition, an insulating state develops at $\nu = 0$ [Fig. 6(b)] starting from 2–3 T.[27,35–37] These states each correspond to a different quantum Hall ferromagnet, with different polarization within the approximately SU(4) isospin space.

The *precise* SU(4) polarization for given experimental conditions depends on the interplay between anisotropies arising from the Zeeman effect (which can exist for both spins and valleys), lattice scale interactions, and disorder. The spin Zeeman effect is the most straightforwardly tunable by applying an in-plane magnetic field, and has provided the primary method for exploring the phase diagram of monolayer graphene QHFM. In particular, tilted magnetic field measurements of the activation gaps associated with magnetoresistance R_{xx} minima as well as the $\nu = 0$ insulator significantly constrain the order parameters. In the $N \neq 0$ LLs, a dominant Zeeman anisotropy leads to spin polarized ground states at half filling and valley textured excitations at quarter filling. In the ZLL, the situation is reversed: the $\nu = 0$ state is natively spin unpolarized, while $\nu = 1$ is spin and valley polarized and hosts real spin textured excitations.[38,39]

$\nu = 0$ presents the paradigmatic case for the richness of graphene quantum Hall ferromagnets. At $\nu = 0$ the ZLL is half filled with two electrons (or holes) per cyclotron guiding center. As a result, anisotropies must compete. Most simply, Pauli exclusion prevents simultaneous spin and valley polarization, and more generally restricts the space of two-particle isospin polarizations states available. In addition, the overlap between wavefunctions on the same guiding center makes the valley anisotropic short range electron interactions maximally important. Symmetry analysis[40] suggests four closely competing phases: a canted antiferromagnet, a sublattice polarized charge density wave, a bond-density wave with a Kekulé distortion, and a fully spin polarized state hosting symmetry protected edge states[41,42] [see Figs. 6(c) and 6(d)]. Guided by a large theoretical literature,[40,43–45] a series of experiments have laid bare the underlying dynamics in physically realized graphene devices. First, in-plane magnetic field has been shown to drive a transition from an insulator (ascribed to the canted antiferromagnetic phase) into a spin polarized state distinguished by helical edge transport.[46] Remarkably, the same effect can be achieved by selectively quenching the Coulomb interaction using strong dielectric screening,[47] which weakens the valley anisotropies relative to the (fixed) Zeeman energy. Finally, in devices aligned to hexagonal boron nitride,[12,13] an energy gap at neutrality opens up. In the quantum Hall regime, this amounts to a large valley Zeeman effect, and leads to a sublattice polarized quantum Hall ferromagnet. As discussed below, transitions between these different ferromagnetic orders play

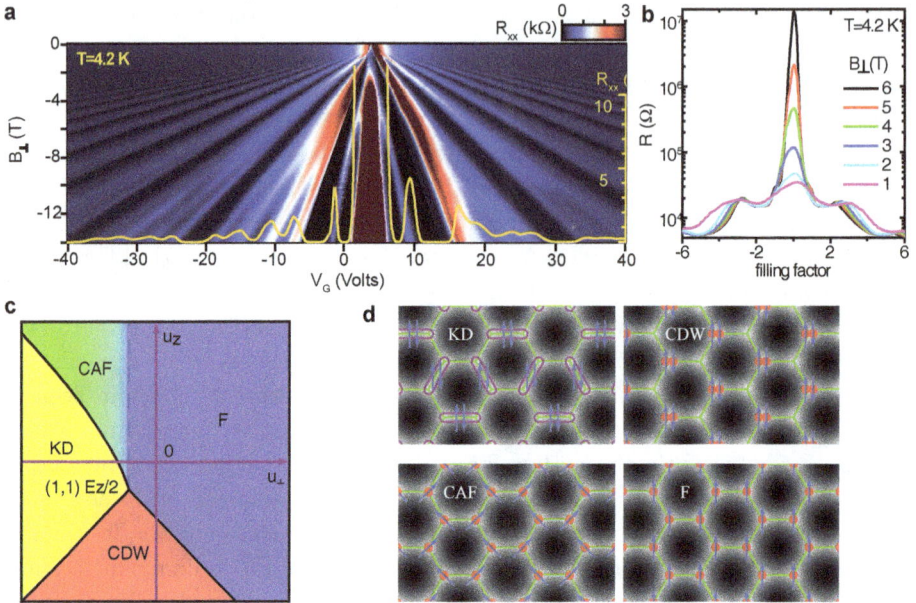

Fig. 6. All integer quantum Hall effect in graphene on h-BN.[38] (a) Landau fan from a monolayer graphene on h-BN device. Symmetry breaking of the Landau levels is visible from a few tesla, and at $B_\perp = 14T$ (superimposed) all integer filling factors feature minima in R_{xx}. The color scale for $-2 < \nu < 2$ has been expanded by a factor of 7. (b) Development of the $\nu = 0$ insulating state.[27,38] (c) Theoretical phase diagram of the $\nu = 0$ state, as a function of unknown coupling constants that parameterize the short range valley anisotropies. (d) Sublattice structure of the four mean-field states from the phase diagram in (c).

critical roles in the isospin polarizations of the FQH states near neutrality,[48,49] and similar dynamics are thought to be obtained in both higher LLs and in multilayer graphene systems.

The primary experimental feature of BLG is the ability to directly control the layer polarization—and by extension, the valley polarization—using perpendicular electric fields. In the ZLL, BLG thus offers direct *in situ* control of both spins and valleys, as follows from Eqs. (18) and (19). However, fully disentangling the nature of the all integer quantum Hall effect observed in clean BLG devices is complicated by the orbital degeneracy. First, even spin- and valley-control is insufficient to fully constrain symmetry breaking in an eight-fold degenerate space. Second, the nature of the orbital degeneracy differs from that of spins and valleys because the orbitals are distinct on the scale of ℓ_B. As a result, the long range Coulomb interaction is *not* symmetric with respect to the interchange of orbitals. This contrasts with the spin/valley isospin: whereas isospin symmetry breaking interactions are all weak compared to E_C, typically by a factor of a/ℓ_B, orbital degeneracy is directly lifted by the long range Coulomb interaction and the resulting energy splittings are naively of order E_C. Theoretical work[50,51] predicts that the long range Coulomb interaction favors filling pairs of states in sequence that differ in orbital quantum numbers but share the same spin and valley. In this case odd integer states are expected to have finite orbital polarization while even integer states are expected to be orbitally unpolarized (i.e. occupy both orbitals). However, a fully quantitative theoretical treatment of symmetry breaking in bilayer graphene has not been possible,[52–56] due to the fine energetic competition between states with different isospin and orbital polarizations, which can depend sensitively on poorly constrained experimental details. Experiments that probed the existence of gaps (and their response to average electric displacement field D and perpendicular magnetic field $B_\perp$) were consequently not able to definitively resolve the full phase diagram within the orbital, spin, and valley space.[57–65]

This problem was resolved by leveraging the finite interlayer separation of bilayer graphene, which leads to a directly detectable capacitive signature of layer polarization.[66] Combined with knowledge of the single particle wavefunction structure in the ZLL, as outlined in Figs. 7(a) and 7(b), this has permitted complete mapping of symmetry breaking in bilayer graphene QHFM.[67] The measurement takes advantage of the fact that in a dual-gate device geometry, the layer polarization p can be extracted by measuring the asymmetric capacitance C_A, defined as the difference between the capacitances of top and bottom gate electrodes of the graphene device. Figure 7(c) plots C_A in the 2D phase diagram of the ZLL. C_A falls into four discrete levels shown as blue, red, cyan, and orange in the color scale. These four values

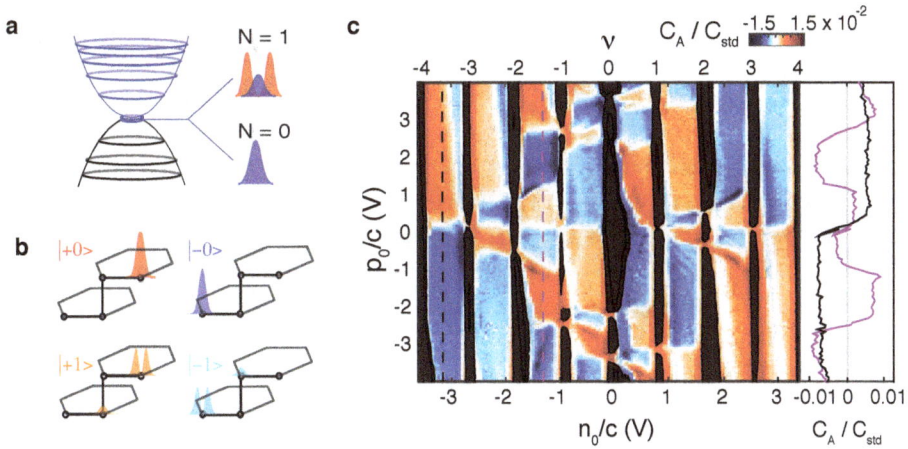

Fig. 7. (a) Energy diagram for $N = 0$ and $N = 1$ LLs in bilayer graphene. (b) Schematic depiction of the four single-particle wavefunctions $|\xi N\sigma\rangle$, showing their relative support on the four atomic sites A, B, A, and B of the bilayer graphene unit cell. While the $|\pm 0\sigma\rangle$ levels are fully polarized, wavefunction of $|\pm 1\sigma\rangle$ feature partial polarization. (c) Layer-antisymmetric capacitance C_A of bilayer graphene against a standard capacitor C_{std}, measured at $T = 300$ mK and $B = 31$ T as a function of polarization of the layer p_0 and carrier density n_0, normalized by average geometrical capacitance c. The color scheme highlights the four-tone contrast, interpreted as filling of $|\xi N\sigma\rangle = |+0\sigma\rangle$ (red), $|+1\sigma\rangle$ (orange), $|-0\sigma\rangle$ (blue), and $|-1\sigma\rangle$ (cyan) LLs.

are consistent with the expected layer polarization p of four distinct combinations of orbital and valley indices, which provide identification for the nature of the broken symmetry state. A cascade of phase transitions induced by displacement field is apparent in the phase diagram, demonstrating tunability between different orbital and valley indices.[67] Most importantly, the overall phase diagram, combined with the order of phase transitions, fully constrains symmetry breaking models, and can be quantitatively understood within a Hartree–Fock analysis that properly accounts for the single particle wavefunctions as well the interaction of electrons near the Fermi level with the sea of filled Landau levels. As discussed in subsequent sections, this map of symmetry breaking provides the basic understanding for the nature of the FQHE observed in cleaner devices.

3. FQHE in Monolayer Graphene

3.1. *Composite fermions in monolayer graphene*

We will first discuss single component FQHE in graphene, which can be achieved by aligning the crystal axis of graphene to that of the hBN substrate. The emergence of a moire superlattice lifts the valley degeneracy when $B \neq 0$, and gives rise to a transition in the underlying order of charge neutrality.[49] Breaking of the valley degeneracy results in a series of transitions in odd-denominator FQH state, while also stabilizing FQH states outside of the conventional single-component sequence, such as the robust incompressible states observed at even-denominator filling.[49] These even-denominator states can be described by the (331) wave-function, where two electron species are labeled by the valley index.[28]

In the single component composite fermion model, an even number of flux quanta are attached to each electron, transforming the strongly interacting electrons into a system of nearly independent composite fermion (CF) quasiparticles.[68] Owing to the flux attachment, each CF also moves in a reduced effective magnetic field, $B^* = B - an\varphi_0$, where n is the carrier density, $\varphi_o = h/e$ is the magnetic flux quantum, and a is the number of flux quanta attached to each electron. At filling fraction $\nu = 1/a$, the effective magnetic field is precisely zero, and the CFs behave as a metal with a well defined Fermi surface.[69] Away from these Fermi surfaces, the FQH states at fractional electron filling are reinterpreted instead as effective IQHE states of CFs, where the effective CF filling fraction, ν^*, is related to the real electron filling, ν, by the relation $\nu^* = \nu/(1 - a\nu)$. This remarkably simple construction makes it possible to interpret a wide range of complex behaviors associated with the correlated FQH states within the context of a non-interacting single particle picture.

Figure 8(b) shows bulk conductance versus filling fraction, measured in a monolayer graphene Corbino disk aligned with the hBN substrate at $B = 36$ T. Due to the Berry phase shift in graphene, the lowest LL for electrons (holes) begins at filling factor $\nu = -2(+2)$, in contrast to conventional zero-Berry phase materials, such as GaAs, in which the lowest LL begins at filling fraction $\nu = 0$. The IQH

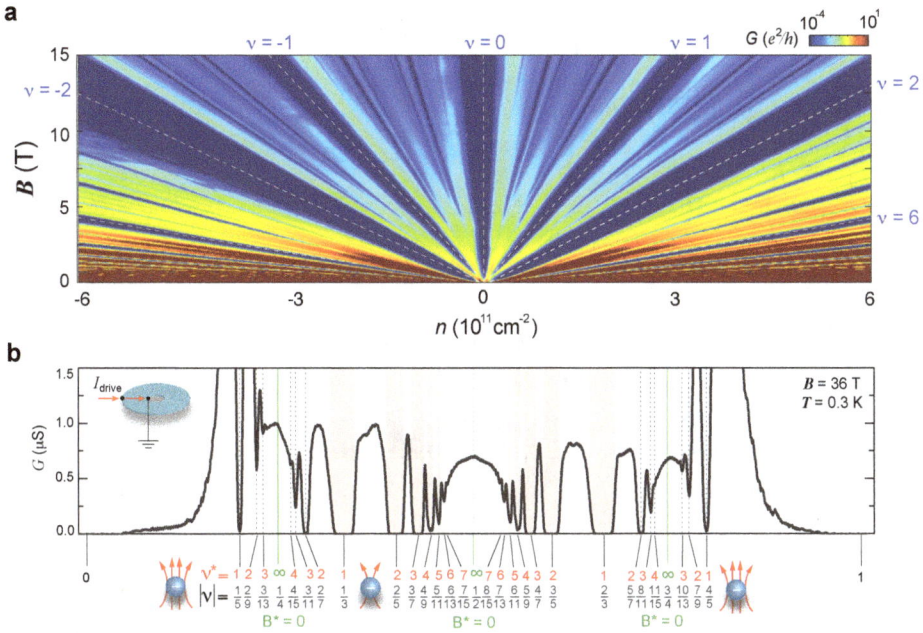

Fig. 8. (a) Bulk conductance G as a function of filling fraction ν and B field for monolayer graphene aligned with an hBN substrate in Corbino device geometry. (b) G versus ν for the lowest energy broken symmetry state measured at $B = 36$ T and $T = 0.3$ K. Various FQH states are denoted by black dashed line.

states at ± 2 are cyclotron gaps, whereas $\pm 1, 0$ states represent symmetry broken spin and valley IQHE states. The data range shown in this figure spans a single branch of the lowest LL between $\nu = 0$ to $\nu = 1$. A large number of FQH states are visible (fractional denominators as large as 15 are resolvable), confirming the excellent sample quality and transport resolution achievable in the Corbino device geometry. The top axis in Fig. 8(c) labels the filling fraction of the CF LLs (referred to as Λ-levels) calculated from the above relation for the 2-flux $(a = 2)$ and 4-flux $(a = 4)$ series. The sequence of states and their hierarchy are in excellent agreement with the non-interacting single-component CF model.[70]

3.2. *Multicomponent FQHE in monolayer graphene*

Figure 9(a) shows one of the earliest four-terminal Hall bar measurements of the FQHE reported for monolayer graphene, measured in high-mobility hBN-supported devices.[15] The data range spans the complete, four-fold degenerate, $N = 0$ LL. FQH states are observed at fractional fillings $\nu = 1/3, 2/3$ and $4/3$, identified by the usual transport signatures, namely quantization of R_{xy} to fractional values of $(1/\nu)h/e^2$, concomitant with minima in R_{xx}. Additionally, a weak minimum in R_{xx} at $\nu = 8/5$ is suggestive of an emerging FQH state at this filling. Notable is the absence of a

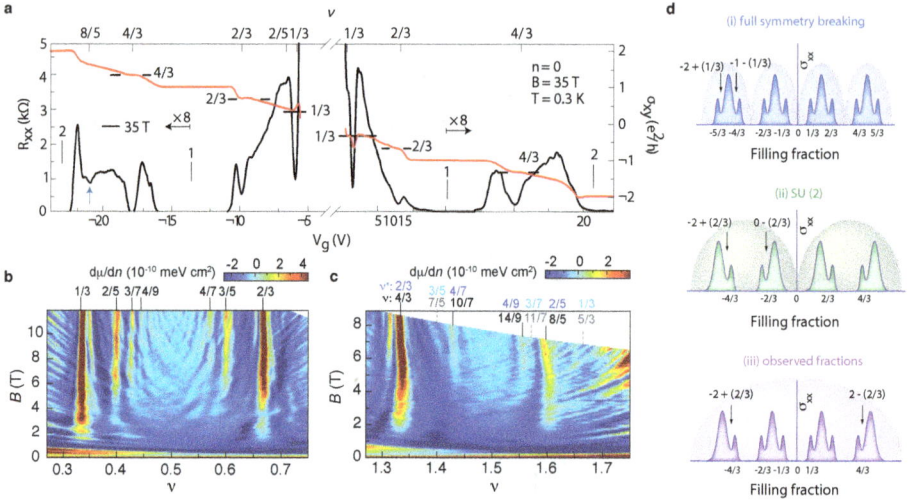

Fig. 9.   (a) Magnetoresistance (left axis) and Hall conductivity (right axis) in the $N = 0$ and $N = 1$ Landau levels at $B = 35$ T and $T \approx 0.3$ K. (b) Inverse compressibility $d\mu/dn$ measured as a function of filling factor and magnetic field in filling fraction range $0 < \nu < 1$. Incompressible states follow the standard composite fermion sequence. (c) $d\mu/dn$ between $\nu = 1$ and 2. Incompressible states occur only at filling fractions with even-numerators. (d) Schematic representation of the FQHE hierarchy observed in our sample. Shown is the expected electron-hole symmetry assuming (i) full lifting of all internal degeneracies due to, for example, coupling to external fields, (ii) full breaking of only one degeneracy, leaving a two-fold degenerate branch, (iii) schematic diagram of the experimental situation. Arrows label particle-hole conjugate pairs in each scenario.

FQH state appearing at $\nu = 5/3$. Due to the Berry phase shift in graphene, filling factor $\pm\frac{5}{3} = \pm2 \mp 1/3$ is expected to be the closest analog to the 1/3 state in conventional systems, making the non-observation of the 5/3 state conspicuous.

A distinguishing feature of graphene is the four-fold internal degeneracy, and in the presence of large magnetic fields, where the FQHE emerges, one may consider three simple scenarios: (i) all degeneracies are lifted by coupling to external fields, and FQH states cannot mix spin/valley branches; (ii) only one degeneracy, either spin or valley, is fully lifted, preserving an SU(2) symmetry in the remaining degenerate space and allowing, e.g. mixed pseudospin but not mixed spin FQH states; or (iii) both Zeeman and valley splitting terms are sufficiently small that Coulomb interactions mix all branches, allowing mixed spin/valley FQH states and excitations. The remaining degeneracy is expected to be reflected in the particle-hole symmetry of the FQH states, i.e. particle-hole conjugate FQH states within a single LL are expected to correspond to similar spin textures and gaps in the absence of any second order effects such as LL mixing.[71] The transport response does not simply reflect any of these scenarios with for example, the LL branches between filling $1 < |\nu| < 2$ appearing to manifest a different symmetry than the $0 < |\nu| < 1$.

Bulk compressibility measurement of suspended graphene devices, utilizing a scanning SET technique,[48,72] show a similar result to that observed in transport

(Fig. 9). In this case, substantially improved device response reveals a large number of fractions in each LL, with denominator reaching as high as 9. Capacitance measurement of the bulk compressibility provides a sensitive means to detect a bulk gap, but is unable to observe the Hall quantization plateau directly. Nonetheless, identification of each FQH state is possible from the trajectory of the compressibility minimum as a function of varying magnetic field and density where the FQH gaps follow a linear slope determined by the associated filling fraction. Replotting this response against the field and filling fractions, such as shown in Fig. 9(b), the FQH gaps appear as vertical features at fixed filling fractions. The high resolution compressibility map shown in Fig. 9(b) reveals that for filling fractions $0 < \nu < 1$, the fractional states follow the standard two-flux CF sequence,[68] i.e. states appear at filling fraction $\nu^* = p/(2p \pm 1)$, with the strength of the state monotonically decreasing as p increases. However, between $\nu = 1$ and 2, the response is markedly different, with only even numerator states observed, and no odd numerators FQHE occurring in this range.

The even numerator sequence evolves in steps of 2 [Fig. 9(c)], indicating that within this range of filling fraction the four-field spin and valley degeneracies of graphene are partially lifted and an approximate SU(2) symmetry is preserved. In the context of the composite Fermion theory, this response is analogous to the IQHE of real electrons observed in monolayer graphene, where the even integer states emerge from the otherwise four-fold degenerate LLs with stronger gaps than the odd-integer states.[33] One possibility for the origin of the even-numerator FQHE sequence is that the applied magnetic field couples to the spin-component via the Zeeman effect, lifting the spin degeneracy, while the valley symmetry remains intact. Within the CF framework, this can be viewed as resulting in two-fold degenerate CF Landau levels. This behavior is similar to previously studied multicomponent 2DEG's such as AlAs and strained silicon[71,73] where, due to competing spin and valley orders, weakened odd numerator FQH states are observed. However, there are notable differences between the systems. For example, in these semiconducting heterostructures, the Zeeman and Coulomb energies are comparable to the cyclotron gap, effectively freezing out the spin degree of freedom at large magnetic fields. By contrast, in graphene both Zeeman and Valley splitting terms are sufficiently small that Coulomb interactions likely mix branches, allowing mixed spin/valley FQH states and excitations. For example at magnetic fields as large as $B = 35$ T, the ratio between the Zeeman and Coulomb energies is still only $E_Z/E_C \sim 0.01\varepsilon$ (ε is the dielectric constant), and lattice scale interactions are only $a_0/l_B \sim 0.06$ (a_0 is the lattice constant).

The difference in the observed sequences within the different branches of the lowest energy LL (i.e. conventional CF hierarchy for fillings $|\nu| < 1$, but missing odd-numerator states for fillings $1 < |\nu| < 2$) suggests a complex interplay between electronic interactions and symmetry in graphene, the details of which remain not fully resolved. Whether an approximate SU(2) or SU(4) symmetry is assumed, a

fully spin and valley polarized ground state with low-lying valley Skyrmion excitations is theoretically expected at $\nu = 5/3$.[14,74-77] Quantitative calculations for
a corresponding valley Skyrmion gap $(0.03-0.043 \ e^2/\varepsilon l_B$[14,76]$)$ are comparable to
the $\sim 0.02-0.04 \ e^2/\varepsilon l_B$ LL broadening estimated for the device shown in Fig. 9(a)[15]
possibly explaining the non-observation of the 5/3. At even numerator filling in
the lowest LL, such as at 8/5 and 4/3, a valley-unpolarized ground state is theoretically predicted,[14,75,77,78] with larger-energy excitations than those occurring at
5/3, qualitatively consistent with experiment. The smallness of the Zeeman energy
in comparison with Coulomb theoretically allows for the possibility[79] of real spin
reversed excitations at $\nu = 1/3$, and may account for the relative strength of the
observed 1/3 in comparison with the absent 5/3. Alternatively, disorder may have
a different effect near the edge of the LL (5/3) than it does near the insulating
state at charge neutrality (1/3), possibly accounting for the appearance of different
electron-hole symmetries within the different LL branches. Further experimental
work that disambiguates the spin and valley ordering of each FQH state, such as
titled field measurements that isolate the Zeeman contribution, or valley-selective
spectroscopy could provide further insight into the details of symmetry breaking in
this system.

3.3. *FQHE gaps in monolayer graphene*

The quantitative value of the FQHE gaps measured in graphene are generally reported to be of the order of tens of Kelvin,[15,25,48,72,80,81] i.e. nearly a full order of
magnitude larger than typically observed in the highest mobility GaAs devices. This
large energy scale results in part due to the reduced dielectric screening in graphene
(taken to be 5 for graphene surrounded by vacuum[80,82]or 6.6 for fully graphene
fully encapsulated in hBN[25,67,81]) and near-zero width of the 2DES in comparison
to conventional semiconductor heterostructures, both of which increase the effective
strength of the electron interactions essential for the appearance of the FQHE. More
significant, however, is the ability to vary the carrier density in graphene by field
effect gating without substantial mobility degradation. Also, higher carrier densities can be achieved in graphene without population of higher sub-bands. These
advantages enable a wide range of FQH states to be realized at large magnetic fields
where the Coulomb energy, $E_c = e^2/\varepsilon l_B \propto \sqrt{B}$ can be maximized. The strongest
FQH gap in the lowest LL is generally the 1/3 state, with a measured intrinsic
gap value, after accounting for LL broadening found to be close to $0.01 \ e^2/\varepsilon l_B$ in
scaled Coulomb energy units in most devices.[15,25,48,72,80,81] This value is both in
excellent agreement with experimental measurement in GaAs heterostructures, and
the theoretical value calculated by exact diagonalization.[83,84]

   The capability to field-effect tune the density additionally allows measurement
of the FQHE gap evolution over a wide range of magnetic fields. Figure 10(d) shows
a plot of the activation energy gap, Δ, versus B, for the $\nu = 1/3$ state. A clear
kink in the trend is observed at $B \sim 8$ T below which the gap is best fit by a

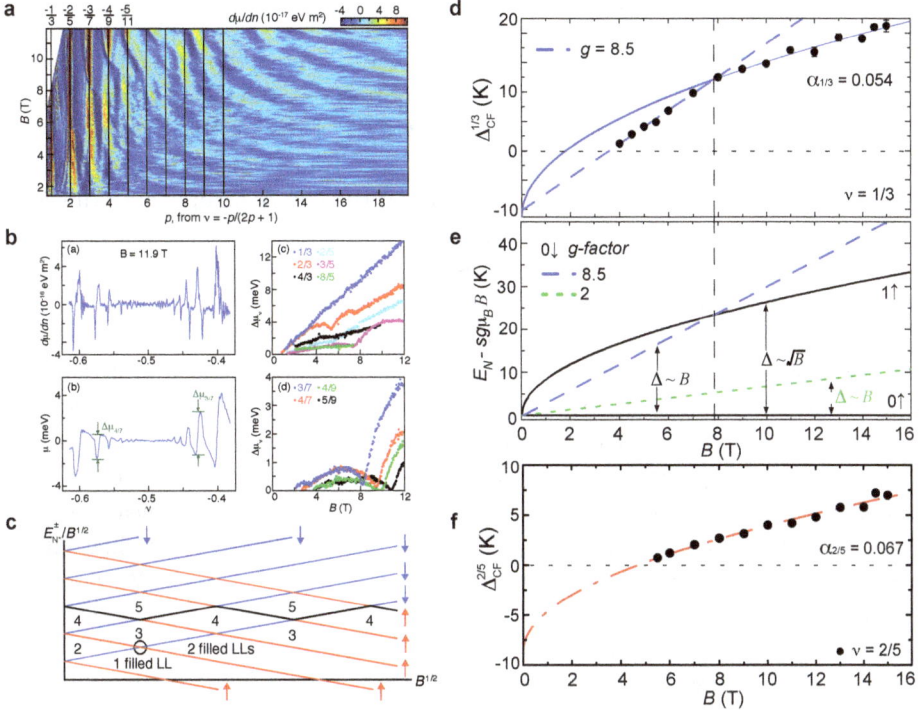

Fig. 10. (a) Inverse compressibility $d\mu = dn$ measured using a scanning single-electron transistor, as a function of filling factor ν and composite fermion Landau level inside P. (b) $d\mu = dn$ (upper left) and chemical potential μ relative to its value at $\nu = \frac{1}{2}$ (lower left) as a function of ν at $B = 11.9$ T. Steps in chemical potential $\Delta\mu$ (green labels in lower left panel) of FQH states as a function of B at measured multiples of $\nu = 1/3$ and $1/5$ (upper right), and $\nu = 1/7$ and $1/9$ (lower right). (c) Schematic of CF LL energies $E_p^\pm$ divided by and plotted against $B^{1/2}$. Crossings (black circle) between spin-up and -down CF LLs (colored arrows) correspond to phase transitions. (d)–(f) Activation energy gap measured using transport measurement in a Corbino-shaped graphene device. (d) Activation energy gap of the $\nu = 1/3$ FQH state. The blue solid curve is a $\sqrt{B}$ fit to data for $B > 8$ T. The blue dashed line is a linear fit to $B < 8$ T. (e) Energy level diagram. The levels are labeled for CF cyclotron orbits with different spin polarization and g-factor as a function of B field. The two lowest CF cyclotron levels are labeled by their CF orbital index, $N = 0$ and 1, and spin polarizations, $\uparrow$ and $\downarrow$. (f) Activation energy gap of the $\nu = 2/5$ FQH state. The red dash-dotted line is a $\sqrt{B}$ fit to the data.

linear B dependence (blue dashed line) and above which the gap transitions to a $\sqrt{B}$ dependence (blue solid curve). Notably, both the linear and square-root fits extrapolate to $\Delta \sim -10$ K at $B = 0$, similar to the value of disorder broadening estimated from the SdH behavior, providing a self-consistency validation of the fits.

The transition in the B dependence of the gap resembles similar behavior of the 1/3 FQH state in GaAs quantum wells, which was interpreted in the context of CF Landau levels with spin degrees of freedom.[85,86] In the CF picture, the size of the effective cyclotron gap that separates spin-degenerate CF LLs is given by,[86]

$\Delta_{CF}^{\text{cyclotron}} = \dfrac{\hbar e B^*}{m^*}$ where $B^* = B - B_{\nu=1/2}$ is the effective magnetic field for CFs and $m^* = \alpha m_e \sqrt{B}$ is the CF mass, m_e is the free electron mass and α depends on details of the quantum well. Allowing for spin degree of freedom, the CF LLs can split into spin branches, separated by the Zeeman energy $E_{CF}^{\text{Zeeman}} = \frac{1}{2}\mu_B g B$, where μ_B is the Bohr magneton, and g is the Lande g-factor. The transition in the gap-versus-B dependence results from a CF LL crossing when the CF Zeeman energy (linear in B), exceeds the CF cyclotron energy (square root in B), as illustrated in Fig. 10(e). This model well fits the data in the lowest LL. If we assume that the linear trend correlates to a real spin gap, the slope gives an estimate for the g-factor of 8.5. This is approximately four times larger than the bare electron g-factor ($g = 2$), and is indicative of strong exchange interaction and the potential existence of skyrmion spin textures.[38] In this picture, we imagine that the valley degree of freedom is frozen out[72] such that the square root region corresponds to the CF cyclotron gap. Fitting the above expression to this region gives a CF mass term of $\alpha = 0.054 \pm 0.004$. Including the projected disorder broadening of ~ 10 K, this gives a measure of the intrinsic gap to be $\Delta_{1/3} = (8.3 \pm 0.6)\sqrt{B[\text{T}]}$ [K], or $(0.084 \pm 0.004)\, e^2/\varepsilon l_B$ in Coulomb energy units, where we use $\varepsilon = 6.6$ for BN-encapsulated graphene.[67] We note that this result is in close agreement with the theoretical value of $0.1\, e^2/\varepsilon l_B$ calculated by exact diagonalization[83] without including any additional corrections.[87]

Figure 10(f) shows the B dependence of the $\nu = 2/5$ gap. In this case, the gap follows a $\sqrt{B}$-dependence over the entire accessible field range, projecting to a ~ -10 K at $B = 0$. The disorder broadening is consistent with the measurement of the $1/3$ gap and quantum oscillation analysis. The square root dependence is qualitatively consistent with the same CF picture as above in which the $\nu = 2/5$ represents a cyclotron gap of CF LLs. However, the lack of a transition is surprising (we would expect the CF cyclotron gap to show evidence of the same CF LL crossing that gives rise to the kink in the $1/3$ gap),[25] and may suggest that the exchange interaction for CFs is highly sensitive to composite fermion filling fraction.[67]

The single-particle valley degeneracy underlying the FQH is only approximate in real experiments, relating to the mean asymmetry between nominally identical graphene sublattices. Indeed, this degeneracy can be strongly lifted in hBN supported devices, with the substrate introducing a sublattice splitting as large as several tens of millielectron volts.[12,13,88]

3.4. *The role of disorder*

We conclude this section with a few remarks about the effects of sample quality. Progress in the study of the FQHE in graphene, like in other material platforms, has closely followed advancements in device fabrication. In general, reducing sample disorder enhances the ability to observe transport features of the QHE.[89] Indeed, the capability to fully resolve the IQHE and FQHE is routinely identified as an indicator of high sample quality[90,91] and has provided an important benchmark in the development of high-mobility 2D materials including graphene and related systems such as black phosphorous, InSe, and the semiconducting transition metal dichalcogenide (TMDs).[34,90–92] Since exfoliated graphene flakes are largely free of intrinsic impurities, improvements in device quality have largely focused on eliminating impurity scattering from the surrounding environment. The first experimental observation of the FQHE in graphene was reported in suspended devices where substrate impurity scattering was eliminated by removing the substrate altogether.[36,37] A more practical route to reducing extrinsic disorder however has been to replace the usual silicon oxide substrate and amorphous gate dielectrics with single crystal hBN,[34,93] an insulating analog of graphite. Fully encapsulating graphene in hBN, both top and bottom, results in device mobilities exceeding $> 10^6$ cm^2/Vs,[94] enabling observation and study of the FQHE in graphene in a versatile device geometry with wide gate tunability. A final improvement is obtained by replacing amorphous metals by graphite as the gate material,[95] resulting in exceptionally low charge inhomogeneity.

Despite these advancements, the ability to observe FQH states in a Hall bar geometry remains limited and is generally overshadowed by the resolution achievable in measurements that probe the bulk response.[48,95] A comparison of state-of-the-art measurements are illustrated in Fig. 11. Figure 11(b) shows the longitudinal (σ_{xx}) and Hall (σ_{xy}) conductivity at $B = 15$ T as a function of the top-gate voltage for an optimized Hall bar device consisting of graphite top and bottom gates,[95] in a geometry that allows the boundary of the Hall bar geometry to be electrostatically defined.[81] The result shows a high-quality transport response with a large number of FQH states observable in the first two LLs ($N = 0$ and $N = 1$). In addition, the onset of $1/5$ states can be observed in the longitudinal conductivity for the $N = 2$ LL. However, this level of detail in a Hall bar measurement is often limited to certain ranges of the magnetic field and charge carrier density, with integer and fractional QHE features randomly disappearing with varying B-field and density as shown in Refs. 25 and 96. Similar results are observed when measuring devices with etched defined Hall bars.

By comparison, Fig. 11(d) shows bulk conductance G_{xx} measured in a graphene Corbino device under similar conditions. In this case, detailed sequences of FQH states are resolvable in both the $N = 0$ LL and $N = 1$ LLs, and which continue to develop under increasing fields.[81] Whereas the Hall bar measurement primarily measures edge states, the Corbino measures the bulk response.[25,97–102] The Corbino response, therefore, suggests that transport measurements of the FQHE in graphene are not limited by bulk disorder, but rather the edge effects. This interpretation is further confirmed by magneto-capacitance measurement in monolayer graphene, where penetration field capacitance C_p is closely related to the compressibility of sample bulk and FQH states emerge as peaks in C_p.[2,49,95] FQH states are resolved with similar, if not better, quality compared to bulk conductance measurement, as shown in Fig. 11(f), C_p measurement.

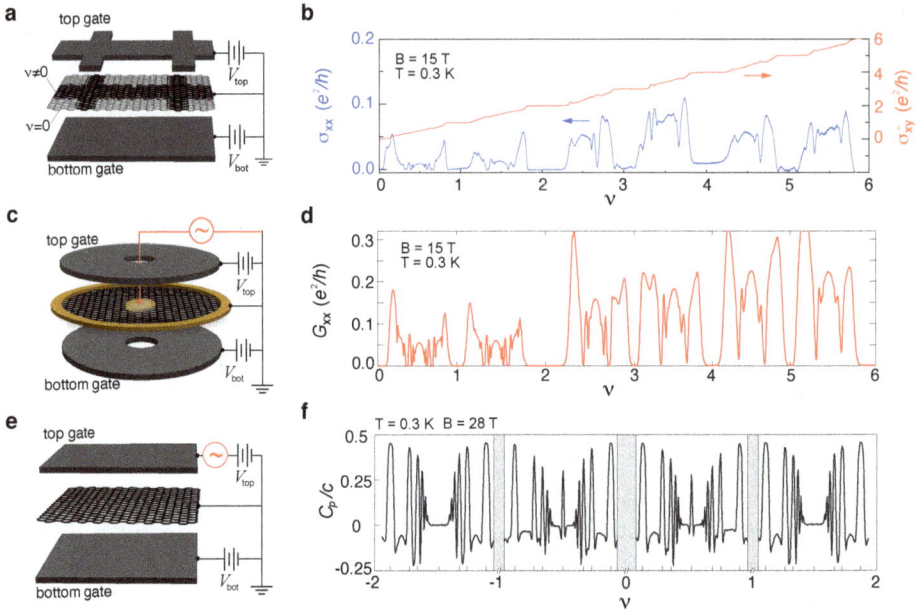

Fig. 11. (a) Cartoon schematic of the gate-defined device geometry. Dieletric BN layers, which separate each of the conducting layers, are omitted for clarity. (b) Longitudinal (left axis) and Hall conductivity (right axis) in the gate-defined regimen at $B = 15$ T. Only multiples of $\nu = 1/3$ states have been labeled for clarity. (c) Cartoon schematic of the graphene device with the Corbino device geometry. (d) Bulk conductance G_{xx} measured in the Corbino device as a function of filling fraction for the $N = 0$ and $N = 1$ LL, $6 < \nu < 6$. (e) Cartoon schematic of a graphene device used for penetration field capacitance measurement. A MLG flake is successively encapsulated in hexagonal boron nitride (hBN) dielectric and graphite gate layers. (f) Penetration field capacitance C_p taken at constant $B = 28.3$ T between filling factors $\nu = -2$ and 2. Incompressible features associated with the integer quantum Hall states are omitted to more clearly show the FQH features.

The capability to detect FQHE signatures with higher resolution using bulk probes, where bulk response dominates, compared to Hall bar geometries, where edge transport dominates, suggests that details of the sample edge play a significant role in the Hall bar response. This also suggests that the transport measurement in the conventional Hall geometry may be limited by difficulties related to probing the edge channels but not the bulk disorder. The origin of these difficulties may be two-fold. First, the Hall bar measurement requires good electrical contact,[103] since the leads should be well equilibrated to the edge modes in order to measure zero longitudinal resistance and accurate Hall plateau. This is a less stringent requirement in the Corbino geometry where QHE ground state appears as an insulating feature in bulk conductance, even for highly resistive contacts. Second, transport measurement of the edge state may be complicated by details of the potential profile near the graphene boundary,[101,104] edge disorder[105] and edge mode reconstruction.[106]

4. FQHE in Bilayer Graphene

4.1. *Odd and even denominator FQH states*

In BLG devices with dual graphite and hBN encapsulation, a host of FQH states are observed in the ZLL, both by transport measurements and penetration-field capacitance[62,63,95,96,107] (Fig. 12). At sufficiently large magnetic fields, the eight-fold degeneracy of the ZLL is fully lifted, such that each branch corresponds to a specific spin, valley and orbital index. On the basis of understanding of how this degeneracy is lifted (such as illustrated in Fig. 7), the observed FQH states can be associated with a specified ground-state order. While generally insensitive to which-valley or spin order, the observed FQHE sequence within each broken symmetry LL displays strong orbital dependence. For example, in LLs with orbital index

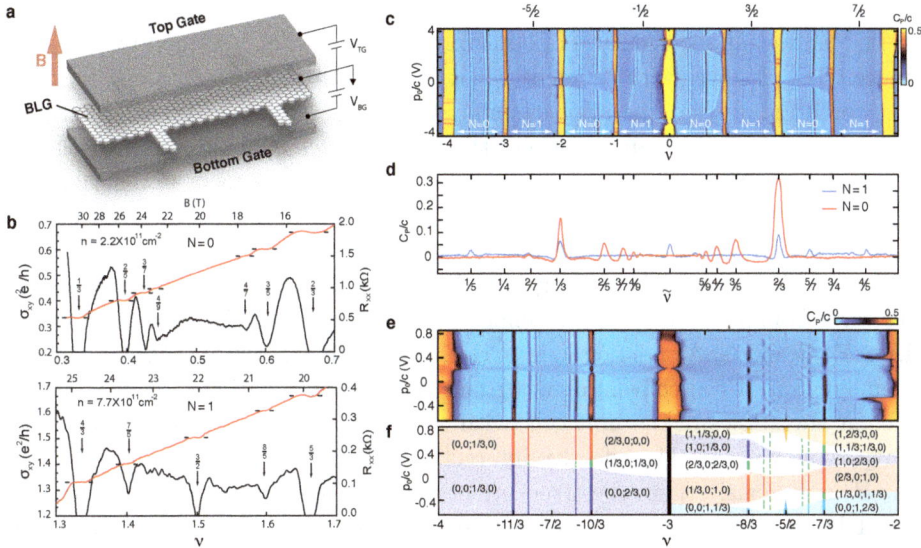

Fig. 12. (a) Cartoon schematic of a dual-gated device structure. (b) Magneto-transport acquired by sweeping B at $T = 0.3$ K and fixed carrier densities, $n = 2.2 \times 10^{11}$ cm^2 (top panel) and 7.7×10^{11} cm^{-2} (bottom panel). The range of filling fraction spans $0 < \nu < 1$ (top panel), and $1 < \nu < 2$ (bottom panel), corresponding to an $N = 0$ and $N = 1$ orbital branch of the LL, respectively. Bottom axis labels the filling fraction ν, with corresponding B values on the top axis. (c) Penetration-field capacitance C_P at $B = 12$ T measured as a function of filling fraction ν and layer polarization p_0 normalized by the total geometrical capacitance c. The plot spans the ZLL, showing incompressible quantum Hall states, manifesting as peaks in C_P, at all integer filling factors ν as well as at several rational ν. (d) FQH sequences in valence $N = 0$ and $N = 1$ regions as a function of the fractional part of ν, $\tilde{\nu} = \nu - [\nu]$ for $1 < \nu < 2$ (blue) and $2 < \nu < 3$ (red). The $N = 0$ levels are compressible at half-filling, whereas the $N = 1$ levels show incompressibility peaks. (e) Measured C_P for $-4 < \nu < -2$ near $D = 0$ at $B = 12$ T. (f) Annotated phase diagram for the range depicted in (c). Occupations of the four relevant orbitals $(\nu 0+, \nu 1+; \nu 0, \nu 1)$ are indicated for each fractional multiple of $1/3$. Shaded areas correspond to regions where the fractional filling lies entirely within one orbital.

$N = 0$, the FQHE appears only at odd-denominator filling and with a hierarchy that is consistent with the single-component CF model (Fig. 12). In contrast, in the $N = 1$ LLs, FQH states at odd-denominator filling are much weaker, with robust incompressible gaps observed only at $\nu = 1/3$ and $2/3$. Additionally, even denominator FQH states are observed at half-filling when the orbital index $N = 1$ (i.e. at $\nu = -5/2, -3/2, +1/2, +7/2$) but absent when the orbital index is $N = 0$ (Fig. 12). The differences between the FQHE in the $N = 0$ versus $N = 1$ orbitals closely resembles that observed in GaAs, suggesting a similar origin.

Near the displacement-field (D) value where the underlying order of LL changes abruptly, a cascade of phase transitions is observed in FQH states, as shown in Fig. 12(e). These transitions are associated with how charge carriers are transferred between two valleys with varying D-field, in the presence of an incompressible FQH state. Since valley polarization cannot be continuously changed without closing the energy gap, the process of depolarization occurs in abrupt steps so that each valley is always occupied by a robust incompressible state.

The multicomponent nature of electron wavefunction is described by the coexistence of FQH states with different valley indices, as demonstrated in Fig. 12(f). It is worth noting that the even denominator states display an unusual behavior at the onset of charge transfer, where the energy gap remains robust as polarization changes continuously.[95] It was suggested that the coexistence of polarizability and incompressibility indicates a gapless fractionalized insulator, such as an exciton phase with inter-valley pairing.[108]

Despite the superficial similarity, the $N=1$ orbital in BLG differs in two important ways from its counterpart in semiconductor quantum wells. First, the $N = 1$ LLs of BLG and GaAs are not strictly equivalent. The $N = 1$ LL of BLG includes a combination of the conventional $|0\rangle$ and $|1\rangle$ LL wavefunctions localized on the different sublattices of the unit cell, with the relative weight of the $|0\rangle$ wavefunction growing with B. The effective interaction depends on the orbital character so that B continuously tunes the structure of electron-electron interactions within an $N = 1$ level. At low B, the wavefunctions are purely $|1\rangle$-like, with comparatively soft interactions, while at high B, they are an equal admixture of $|0\rangle$ and $|1\rangle$ and interactions are consequently sharper. Numerical studies at $\tilde{\nu} = 1/2$ predict that a non-Abelian paired phase at lower B should give way to a gapless composite fermion liquid (CFL) at sufficiently high magnetic fields.[109,110] Indeed, experimental studies find that the $\tilde{\nu} = \frac{1}{2}$ gap changes non-monotonically with B [Figs. 13(b) and 13(c)], peaking around $B = 27$ T and then decaying with further increase of the magnetic field. The decrease of the $\tilde{\nu} = \frac{1}{2}$ gap despite an increase in the Coulomb scale $E_C \sim \sqrt{B}$ supports the scenario of a paired-to-CFL transition[111] at somewhat higher magnetic fields. Additionally, in the same range of field where the even denominator state weakens, the odd denominator states simultaneously transition to a more conventional hierarchy, typical of the $N = 0$ series, providing further evidence that the effective $N = 1$ interactions sharpen with magnetic field.

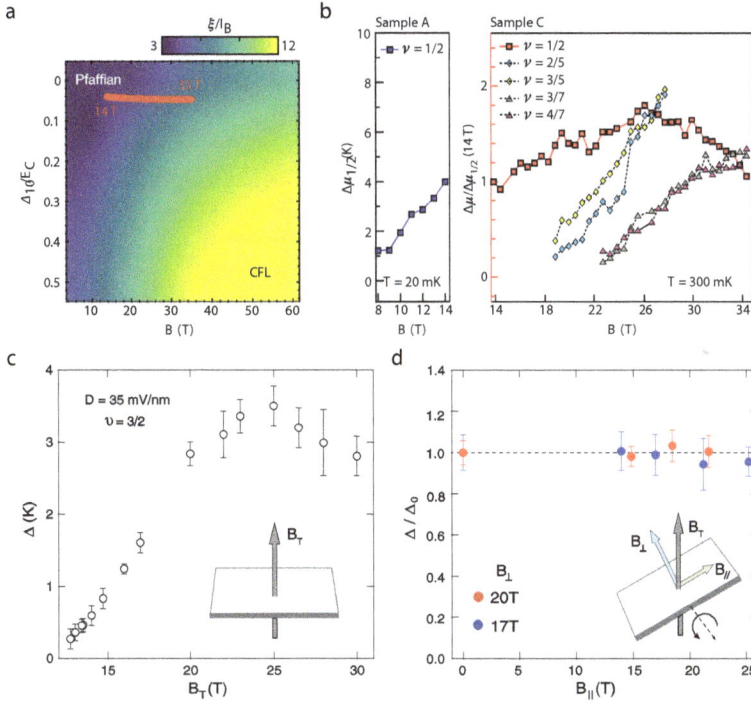

Fig. 13.   (a) DMRG calculation of the correlation length ξ at in the $N = 1$ level as a function of the energy splitting Δ_{10} between the $N = 0$ and $N = 1$ orbitals, in units of Coulomb energy EC, and the magnetic field B. In the lower right corner, the system transitions from the incompressible Pfaffian phase to the compressible CFL phase. The red line denotes an estimate 15 of the trajectory within the B10 plane corresponding to the data shown in (b). (b) The thermodynamic gap $\Delta\mu_{1/2}$ at different B in sample A (left) and sample C (right). Data in the right panel are scaled to the $\Delta\mu_{1/2}$ gap at $B = 14$ T. (c) Δ of the $\nu = \frac{3}{2}$ state as a function of magnetic field B, at $D = 35$ mV/nm and $T = 0.3$ K. (d) Energy gap of the $\nu = \frac{3}{2}$ state normalized to its $B_\parallel = 0$ value, versus parallel field for two fixed values of perpendicular field.

Second, particle-hole symmetry breaking differs in BLG as compared with GaAs. Within a single Landau level, the Pfaffian and anti-Pfaffian states, which can be understood as different pairing channels, are degenerate due to a particle-hole symmetry (effected by $\tilde{\nu} \leftrightarrow 1 - \tilde{\nu}$). Including scattering between LLs breaks this symmetry and determines the ground state. While the subject of long-standing debate, the recent numerical agreement between exact diagonalization and density matrix renormalization group (DMRG) methods suggests that the $\nu = \frac{5}{2}$ state of GaAs is in the anti-Pfaffian phase.[112–114] However, LL scattering is dramatically different in BLG: scattering between the ZLL and the $|N| \geq 2$ levels only breaks particle-hole symmetry weakly, while scattering within the ZLL breaks it strongly due to the small splitting $(\Delta_{10} \approx 0.1E_C)$ between $N = 0$ and $N = 1$ levels. In our experiment, particle-hole symmetry breaking manifests in the fractions observed in the $N = 1$ LL.

In addition to the even denominator states, the very high resolution achieved in capacitance measurements also revealed evidence of unusual odd denominator states in the $N = 1$ LLs, occurring at $\tilde{\nu} = \frac{7}{13}$ and $\frac{3}{5}$ [Fig. 12(d)]. These represent the particle-hole conjugates of the $\frac{6}{13}$ and $\frac{2}{5}$ states seen in GaAs.[115] DMRG calculations find that, in contrast to GaAs,[112–114] the even denominator state in BLG is more likely to be the Pfaffian over the anti-Pfaffian, in the experimentally accessible range.[95] Suggestively, $\frac{7}{13}$ (as well as $\frac{8}{17}$, where a weaker feature is also observed) is the predicted filling of the first "daughter" state of the Pfaffian phase.[116]

The spin-order of the even denominator ground state can be probed by applying a parallel magnetic field ($B_{\parallel}$) while keeping the perpendicular field ($B_{\perp}$) fixed. Figure 13(d) shows the energy gap of the 3/2 state versus the parallel field, measured for two different perpendicular field values. No parallel field dependence is seen up to the largest applied value of $B_{\parallel} \sim 25$ T. This indicates a spin polarized ground state, consistent with the Pfaffian/anti-Pfaffian, and rules out possible spin-singlet candidate wavefunctions that may give rise to an even denominator state, such as the Halperin (331).[28] This observation moreover supports the long held theoretical conjecture that the decrease with $B_{\parallel}$ of the gap of the 5/2 state in GaAs[117] is caused by transverse orbital effects owing to the finite quantum well width.[118,119] In BLG, by contrast, the crystal is less than 1 nm thick and, therefore, insensitive to these effects within accessible field ranges. Taken together these results provide compelling evidence that in the absence of some other, as yet unidentified, ground state, the even denominator FQH states observed here likely correspond to the Moore–Read Pfaffian, or its particle-hole conjugate.[107,114,120–123]

Among the four even denominator states, the size of the energy gap demonstrates an unexpected hierarchy [Fig. 14(a)], where Δ becomes smaller approaching the

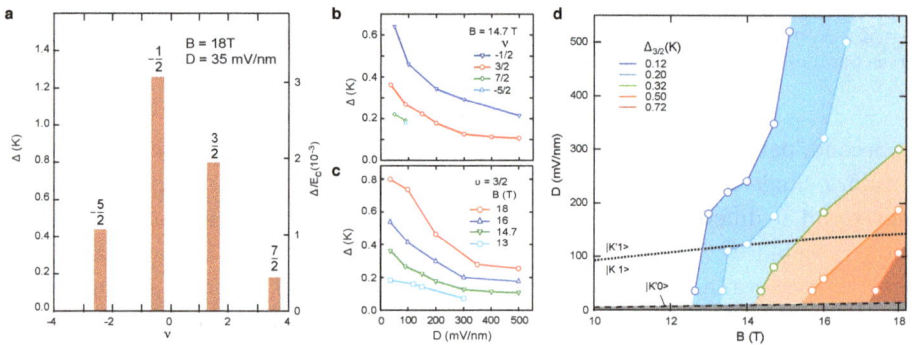

Fig. 14. (a) Δ for all even denominator states at $B = 18$ T and $D = 35$ mV/nm. (b) Δ as a function of D for $\nu = -\frac{1}{2}, \frac{3}{2}, -\frac{5}{2}$ and $\frac{7}{2}$ at $B = 14.7$ T. (c) Δ versus D for $\nu = \frac{3}{2}$ at various B. (d) Shown is contour line for $\Delta_{\frac{3}{2}}$ as a function of B and D. The valley polarization transition at non-zero D is shown as the black dotted line and the orbital polarization transition as the black dashed line. The even-denominator state disappears in the orbital 0 region (gray shaded area).

edge of the lowest energy Landau level (LLL) ($\Delta_{-\frac{1}{2}}$ is the largest and $\Delta_{\frac{7}{2}}$ is the smallest). Several effects may contribute to this observation, including variation in level mixing, electron interaction strength, and disorder, all of which may be filling fraction dependent within the LLL. Figure 14(b) shows the effect of the displacement field on the variation of Δ. For all of the even denominator states, the gap is observed to decrease continuously with increasing D, with no apparent correlation to the valley ordering phase diagram. The D-induced gap variation is significantly larger than theoretical predictions,[109] and the origin of this result is not known, but it was noted that also that the D dependence appears to vary with B [Fig. 14(c)]. This behavior is summarized in the experimentally measured B–D phase diagram for the 3/2 state gap, shown in Fig. 14(d). Also shown in this diagram are the D-field values for the valley (black dotted line) and orbital (black dashed line) transitions. Within the data resolution, there is no discernible dependence on the valley order. $\Delta_{\frac{3}{2}}$ is too small to measure in the white area and is absent altogether in the gray shaded area, where the ground state wavefunction has orbital index $N = 0$. We note that although a spin-singlet ground state origin for the even denominator state was ruled out, a valley singlet equivalent (331) state could be possible. However, this is also considered unlikely since the gap is observed to persist to large displacement fields, where the system is presumed to be fully layer polarized.[62,63,67]

Finally, we note that BLG has certain advantages over GaAs as a platform for interferometric detection of non-Abelian quasiparticles.[124] First, the large energy gap and small correlation length relative to GaAs may reduce bulk-edge coupling that is detrimental to interferometric probes,[125] while exponentially suppressing the density of thermally-activated quasiparticles. Second, hBN gate dielectrics can be made almost arbitrarily thin, allowing one to engineer edges and quantum point contacts using sharp electrostatic potentials. Recent experiments have demonstrated long coherence lengths in the quantum Hall regime along such gate-defined edges.[126] Finally, the putative Pfaffian state at $\nu = -\frac{1}{2}$ in BLG would have fewer edge modes than the anti-Pfaffian state at $\nu = \frac{5}{2}$ in GaAs, making the former a preferable candidate for interferometry. Even without phase coherent transport measurements, the thermodynamic measurements presented here, carried to lower temperatures, can be used to probe topological ground state degeneracy,[127] providing smoking-gun evidence for non-Abelian statistics in the near future.

5. Interlayer Correlated States in Double Layer Graphene

5.1. *Magneto exciton condensates in double layer graphene*

It has long been predicted that Coulomb correlation in a quantum Hall double-layer will bring about new physical structures that have no analog in a single-component system.[28,70] In the quantum Hall effect regime, the kinetic energy of electrons is quenched by the formation of LL. Therefore the energetic system is dominated by the competition between Coulomb interaction within each layer (intralayer) $E_{intra} = e^2/\varepsilon\ell_B$ and between different layers (interlayer) $E_{inter} = e^2/\varepsilon d$, where ℓ_B is magnetic length characterizing intralayer particle separation and d is the interlayer distance. Since interlayer Coulomb interaction can be sensitively tuned by varying the barrier thickness, double-layer structure provides a uniquely tunable platform to study the interplay between electron correlation, broken symmetry, and topological order in the quantum Hall effect regime.

Double-layer structure features a unique emergent phenomenon in the quantum Hall effect regime when electron density in each layer is tuned to fill half a LL. In this situation, Coulomb interaction between two layers binds empty electron states in one layer with electrons in the other layer, forming a system of excitons, which undergoes a phase transition to a Bose–Einstein condensate (BEC) phase at low temperature. This condensate phase is manifested in a variety of exotic phenomena under quantum transport measurement,[128] which has been observed in semiconductor double quantum wells and graphene double-layer structures.[3,129–132] In this section we will discuss the observation and experimental characterization of exciton condensate in double-layer graphene (DLG) structure.

In DLG, independent electrical contact on each graphene layer allows the dynamics of excitons to be probed using transport measurements of different geometries, such as Coulomb drag and counterflow, as shown in Fig. 15.[3,131,132] In Coulomb drag geometry, current flows through the drive layer while current is not allowed to flow in the passive drag layer which is held at ground potential only through just a single external contact. Interlayer correlation is characterized by the Hall and longitudinal resistance on both layers. In this geometry, exciton condensate is manifested in quantized Hall resistance, concomitant with zero longitudinal resistance on both drive and drag layers as shown in Fig. 15(a). In the counterflow geometry, current flows in opposite directions on different graphene layers, coupling directly

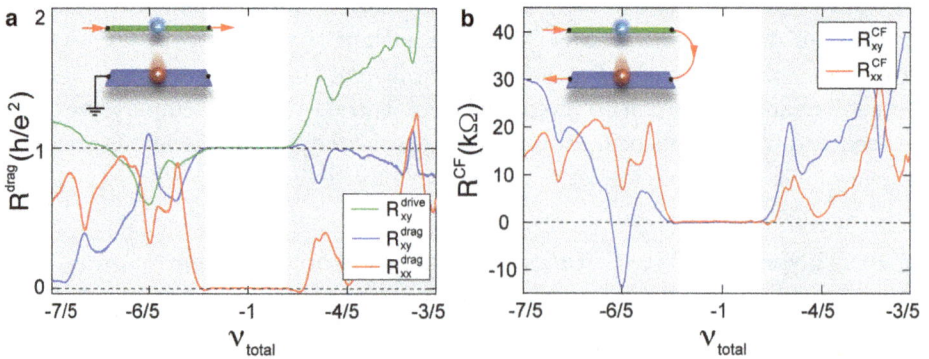

Fig. 15.   (a) Coulomb drag response of balanced double layers as a function of total filling fraction ν_{tot}. Exciton condensate occurs near $\nu = -1/2$, $\nu_{tot} = -1$. Inset, schematic of Coulomb drag geometry. (b) Counterflow measurement of the exciton condensate at $\nu_{tot} = -1$. Inset, schematic of the counterflow geometry.

to the dynamics of exciton transport owing to the unique charge distribution of electron-hole pairs. The condensate phase of excitons and the dissipationless nature of exciton flow are reflected by zero longitudinal and Hall resistance as shown in Fig. 15(b).

It has been demonstrated in double quantum well structures that exciton pairing remains robust in the presence of density mismatch across two quantum wells.[133] The same robustness is observed in double-layer structures consisting of monolayer and bilayer graphene, where exciton condensate is observed for large values of density mismatch, $\Delta \nu \sim 0.5$.[3,5,132,134] It has long been recognized that this robustness against density imbalance allows flux and local density to fluctuate coherently across two layers, giving rise to the spontaneously broken symmetry of pseudo-spin rotation inside the X–Y plane and a finite temperature phase transition of Kosterlitz–Thouless (KT) type.[128] A KT-type transition was recently observed by studying the current-voltage (IV) dependence of counterflow measurement in double-layer graphene structures. While the IV curve is linear above the transition temperature T_{KT}, it develops a nonlinear behavior $I \sim V^{\alpha}$ with the exponent $\alpha = 3$ at T_{KT}, when pairing between vortex and anti-vortex of superfluid phase is expected to occur. This finite temperature transition provides one of the unique characters of the exciton condensate, setting it apart from other emergent phenomena in the quantum Hall effect regime.

The topological nature of the condensate phase is also reflected by the temperature dependence of transport behavior above T_{KT}, where both longitudinal and Hall resistance display thermally activated behavior, with the activation energy gap shown to be much smaller than Coulomb energy.[3,132] This contrast in energy scale is indicative that thermal excitations of the condensate phase are unpaired vortex and anti-vortex, instead of free electrons and holes. Interestingly, energy gaps measured from Coulomb drag and counterflow geometry increase in the presence of density mismatch between graphene layers,[3] suggesting a more robust condensate phase in the presence of density imbalance. While such robustness appears universal in a range of double-layer structures, a theoretical explanation remained elusive to date.

The DLG structure offers a variety of experimental knobs to control electron correlation and the resulting emergent quantum phases. Most remarkably, pairing strength in such structure can be continuously tuned over a large range by varying the effective layer separation d/ℓ_B, establishing exciton condensate as a paradigm system to study the effect of fermionic pairing under extreme limits. In the weak coupling regime where $d/\ell_B \ll 1$, interlayer Coulomb interaction is small compared to the Fermi energy. In this regime, Fermi liquid behavior dominates, and the low temperature condensate is induced by a Cooper instability, where pairing occurs in momentum space between a small fraction of electrons near the Fermi surface.[135] The weak coupling regime features electron Cooper pairs with size much larger than inter-particle separation.[136] In the opposite limit of strong pairing interaction, fermions form tightly bound pairs, and the size of the pair is comparable to the average inter-particle separation. In this limit, the system behaves like bosonic gas of excitons, and the low temperature ground state is characterized by a BEC. The strength of pairing interaction is reflected by the existence, or lack thereof, of fermionic pairs above the condensate temperature.[136,137] In the strong pairing limit, a large population of excitons persists to temperature much higher than the condensate phase transition T_c, consistent with the BEC picture.[138] Whereas in the weak coupling limit, a temperature dependence similar to a BCS superconductor is observed, where exciton pairing vanishes simultaneously with the condensate phase at T_c.[136,137] The contrast in temperature dependence has provided indirect evidence of a crossover between strong and weak coupling regimes, which has been recently observed for exciton condensate in DLG structure.

5.2. *Fractional quantized states in double layers*

Apart from the exciton condensate phase, the layer degree of freedom in double-layer structure induces complex internal structures in quantum ground states at fractional filling, resulting in a series of fractional quantum Hall effect states described by the competition between intra- and interlayer Coulomb correlations.

The starting point for describing the complex order in a two-component structure is an expanded composite fermion (CF) model. As in the single-component system, the problem is transformed by attaching to the electrons flux quanta of an artificial Chern–Simons gauge field, to form quasiparticles (CFs), which one hopes will be only weakly interacting.[70] In the presence of two electron species, two types of flux quanta can be attached to each electron, one characterizing interaction within the same electron species and the other type seen by electrons of the opposite species. (We assume that only one spin and valley state need be considered in each layer.) This construction renormalizes both the intra- and interlayer interactions, giving rise to a system of CFs with reduced interactions. The effective magnetic field experienced by CFs is determined by the electron density in both graphene layers, $B_i^\dagger = B - (an_i + bn_j)\varphi_0$, where the subscript i denotes layer index, and a and b are the number of attached intralayer and interlayer flux quanta, respectively. In the case of matched layer density, the effective Λ-level filling fraction given by this model is[70]

$$\nu_i^\dagger = \nu_j^\dagger = \frac{\nu}{1 - (a+b)\nu}, \tag{25}$$

where $\nu \equiv \nu_1 = \nu_2$, and we use superscript "†" to distinguish from the effective filling fraction in a single component CF model.

We note several important features of this transformation: (i) The CFs retain their layer index, but the layers become nearly independent of one another; (ii) The CFs can experience different effective magnetic fields when the layer densities are not matched; (iii) While the intralayer flux attachment must be an even number, the interlayer flux can be any integer value (but no larger than a). For simplicity, a CF with a intra-layer and b inter-layer flux attachment will be referred to as an $(a + b)$-flux CF or $_b^a$CF.

As in the case of single-component systems, many of the prominent quantized Hall states can be obtained, with appropriate choices for the flux attachment parameters a and b, by choosing positive or negative integer values for the Λ-level filling fractions $\nu_i^\dagger$, as one would expect if one could completely neglect the residual interactions between CFs. (Negative values indicate that the effective magnetic field has opposite sign to the applied field.) As we shall see, however, there are also prominent quantized Hall states corresponding to half-integer values of $\nu_i^\dagger$. In these cases, the residual interaction between CFs is crucial, as the ground state description requires the formation of a condensate of CF excitons, pairs formed by a CF particle in one layer and a CF hole in the other.[4,5]

The set of possible ground states determined by the CF model, with $a = 2$ and $b = 0, 1, 2$ is illustrated in Figs. 16(a)–16(c).[68,70] Integer values of $\nu_i^\dagger$ are indicated

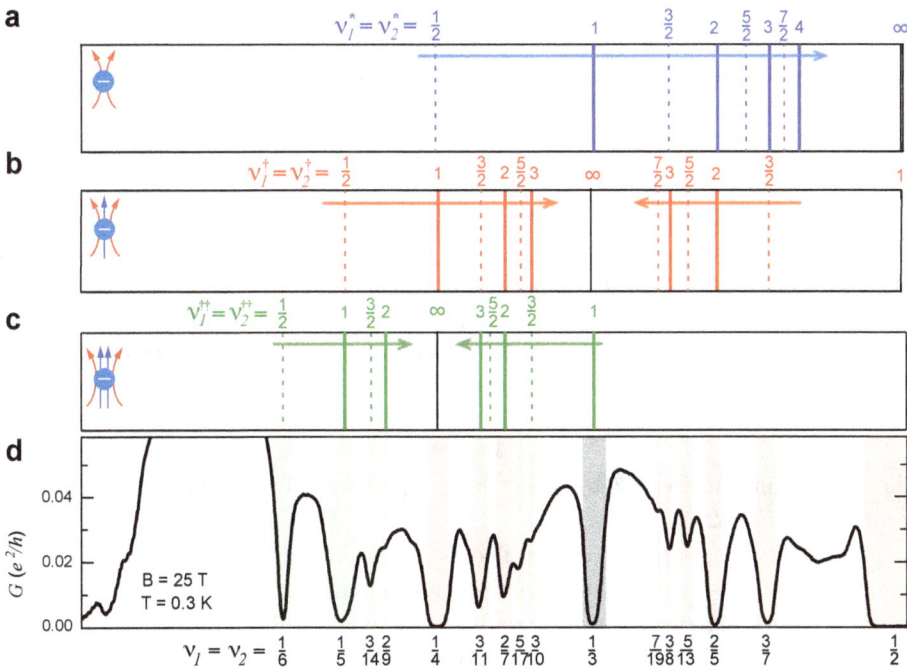

Fig. 16. Vertical lines in (a)–(c) mark expected sequences for $_0^2$CF, $_1^2$CF, $_2^2$CF. Schematic flux attachment for each CF construction is shown on the left of each panel, with red arrows denoting intra-species and blue arrows inter-species flux attachment. CFs form a Fermi surface at vertical black lines in (a)–(c), where effective magnetic field is zero according to Eq. (25). (d) Bulk conductance measured from graphene double-layer device with Corbino geometry as a function of filling fraction ν, along the equal density line at $B = 25$ T, $T = 0.3$ K. In $1/4 < \nu_1 = \nu_2 < 1/2$, both the filling fraction and hierarchy of observed insulating features match the expected behavior of $_1^2$CFs, whereas the collective behavior of incompressible states match $_2^2$CFs in $0 < \nu_1 = \nu_2 < 1/4$.

by solid vertical lines, while half-integer values are indicated by dashed lines. Results for the conductance G of a balanced double-layer sample in the Corbino geometry are shown, for comparison, in Fig. 16(d), in the region $0 < \nu_{\text{total}} < 1$. Quantized Hall states should be marked there (ideally) by a vanishing of G. It is seen that in the region $1/4 \leq \nu_1 = \nu_2 < 1/2$, the observed quantized states are (mostly) in excellent agreement with predictions of the model with $a = 2, b = 1$, ($\frac{2}{1}$CF states), while for $\nu_1 = \nu_2 < 1/4$, there is better agreement for $a = b = 2$, ($\frac{2}{2}$CF states). In several cases, however, the filling fraction seen in a measurement of G could be explained with several choices of b. As we shall see, below, definitive identifications can be obtained from measurements in the Hall bar drag geometry, where one can measure separately the Hall and longitudinal components of the drag resistance and drive resistance.

The emergence of $\frac{2}{2}$CF states indicates stronger inter-layer coupling or weaker intra-layer Coulomb screening at low charge carrier density near the edge of LL, since the number of attached flux reflects the strength of Coulomb interaction. We note that the coexistence of $\frac{2}{1}$CF and $\frac{2}{2}$CF in the same sample and constant magnetic field has a few important implications: (i) d/ℓ_B alone cannot fully describe the nature of effective Coulomb interaction, the effect of total charge carrier density should be considered as well; (ii) the observation of different CF constructions suggests that a wide range of effective Coulomb interaction can be achieved in DLG, making it possible to explore the tunability of electron correlation.

Since the dual-gated geometry allows density to be independently tuned on each layer in a double-layer graphene structure, FQH states emerge as patterns in a 2D phase diagram defined by ν_1 and ν_2. The evolution of FQH states away from the layer-balanced condition can be compared with the two-component CF model, offering more definitive evidence for the CF construction. In the presence of density imbalance, Eq. (25) can be generalized to

$$\nu_i^\dagger = \frac{\nu_i}{1 - a\nu_i - b\nu_j}. \tag{26}$$

The dashed lines in Fig. 17(b) reflect lines of constant individual layer CF filling, according to Eq. (26).

Figure 17(a) plots bulk conductance as a function of total electron filling fraction, $\nu_{\text{total}} = \nu_1 + \nu_2$, and filling fractions mismatch, $\Delta \nu = \nu_1 - \nu_2$. In the chosen color scale dark blue indicates low conductance, while white indicates high conductance. Figure 17(b) summarizes the main features of Fig. 17(a) where the solid lines identify prominent trajectories of low conductance. The region with $\nu_{\text{total}} > 1$ is related to $\nu_{\text{total}} < 1$ by a particle-hole transformation. We observe that states appearing at integer $\nu_i^\dagger$ values under density balance evolve along trajectories (solid blue lines) that match the expected trajectories for integer filling of $\frac{2}{1}$CF states in one layer (dashed blue lines). The intersection point between dashed blue lines, highlighted by blue circles in Fig. 17(b), corresponds to regions where both layers are expected to have integer-valued $\nu_i^\dagger$ according to Eq. (26), and therefore both layers should be insulating. We shall identify states at the intersection points as fully quantized, while states on the lines away from the intersection points will be denoted as *semi-quantized*. The distinctive transport properties of these states in a Hall drag geometry will be discussed below. It is worth noting that ground states corresponding to open circles in Fig. 17(b) are consistent with the layer-balanced ${}^2_1\text{CF}$ states that were identified in Fig. 17(a).

Figures 17(c) and 17(d) demonstrate robust insulating features at each of these crossing points, and show a well-defined hierarchy based on the depth of the con-ductance minimum: namely, the most robust ground states are observed at the four corners of the diamond shaped area, where Λ-level filling is the lowest, and the effective magnetic field is the strongest. The observed trajectories of the inte-ger $\nu_i^\dagger$ states with layer imbalance, and their hierarchical behavior, provide further evidence that these states are well described by the ${}^2_1\text{CF}$ model.

Apart from exciton pairing between CFs, Figs. 17(a) and 17(c) reveal a range of FQH states that are not easily explained by the simple CF model. For example, the robust insulating feature at $1/3 + 1/3$ filling corresponds to three distinct possible candidate ground states as illustrated in Figs. 16(a)–16(c): (i) effective IQHE of ${}^2_0\text{CF}$ with $\nu_i^\dagger = 1$; (ii) a Fermi surface of ${}^2_1\text{CF}$ and (iii) effective IQHE of ${}^2_2\text{CF}$ with $\nu_i^\dagger = -1$. Among these candidates, the most intriguing phenomenon results from pairing instability of the Fermi surface, which could give rise to a host of non-Abelian ground states that are proposed as theoretical possibilities for strongly coupled double-layer structure.[70,139–141] Coulomb drag measurements reported in Ref. 5 showed no Hall drag at layer fillings $1/3 + 1/3$, suggesting that the state can be understood as two decoupled single-component FQH states in their sample. However, Coulomb drag measurements in Ref. 4 demonstrated strong interlayer correlations at this layer filling in the strong coupling limit at small d/ℓ_B. More experimental input is needed to identify the nature of electron correlation in this case.

In addition, FQH-like states are observed along trajectories corresponding to half-filled Λ-level, labeled by red solid lines in Fig. 17(b). As this trajectory is unique to the ${}^2_1\text{CF}$ construction, it rules out the possibility of the ground state being an effective integer quantum Hall effect of other ${}^a_b\text{CFs}$. We note that along the red line trajectories, one of the layers remains at constant half-integer $\nu_i^\dagger$, while the other layer varies over a large range of effective filling. A state persisting along this trajectory, therefore, could indicate pairing between CFs within the half-Λ filled layer only, and with no correlation to the other layer. One possibility is that these states may be of a similar origin to the Pfaffian that is believed to describe the half filling even denominator state in single layer systems.[70,142,143] If true, this would be rather remarkable, since the states at the intersections of the lines, at $(3/7, 3/7)$ and $(4/7, 4/7)$ were clearly demonstrated by the drag experiments to involve pairing between CF electrons and CF holes in opposite layers, as discussed below. A different possibility is suggested by an analysis given in the Supplemental Information of Ref. 5. As noted there, the elementary charged excitations of the

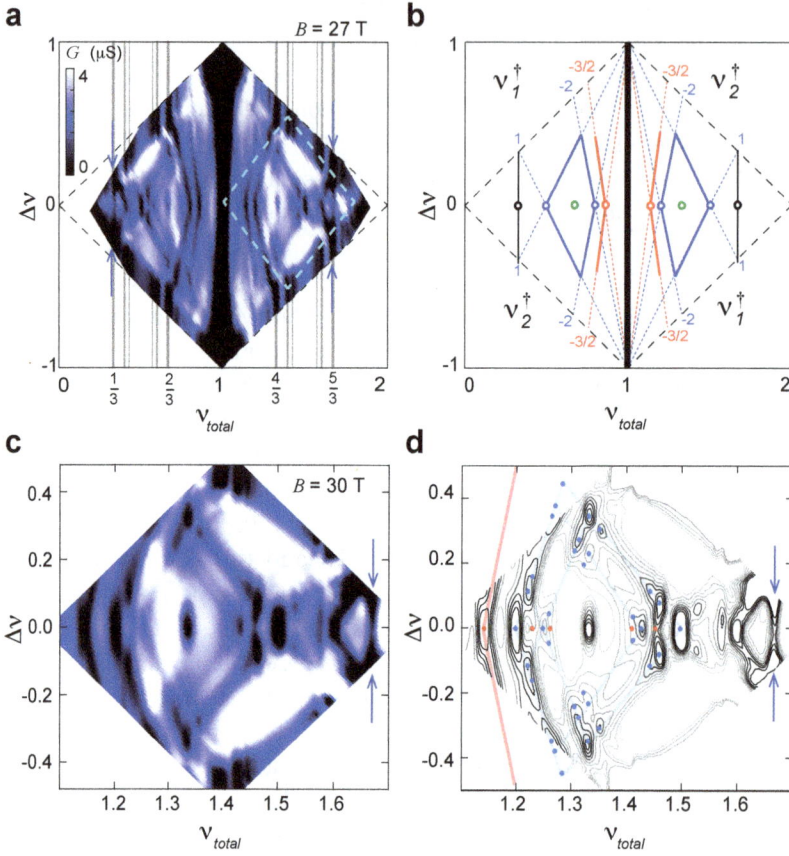

Fig. 17. (a) G versus ν_{total} and $\Delta\nu$ measured at $B = 27$ T. (b) Schematic phase diagram showing constant integer and half-integer CF filling in each graphene layer as dashed blue and red lines, respectively. Features observed in bulk conductance are highlighted as solid blue, red, and black lines. The colored circles correspond to insulating features in Fig. 16(d) under matched density condition. Blue, red and black circles mark integer, even denominator and odd denominator $\nu^{\dagger}$ states, respectively, whereas green circles correspond to $\nu_i = \frac{1}{3}$ and its particle-hole conjugate, where $^2_1$CFs form a Fermi surface phase. (c) High resolution measurement of the region highlighted by a dashed boundary in (a), acquired at $B = 30$ T. (d) Contour plot of bulk conductance data shown in (c). Blue dots correspond to effective integer $^2_1$CF filling in both graphene layers. Red dots mark the effective half filling states. Blue arrows point to the feature corresponding to the (333) state. (From Ref. 4.)

(3/7, 3/7) FQH state should have charge $\pm 3/7$ in one layer and $\mp 4/7$ in the other. Adding quasiparticles, say, of charge $(3/7, -4/7)$, to the parent state would give rise to a semi-quantized state that falls along the red line segment $\nu_1^{\dagger} = -3/2$ on the left side of Fig. 17(b). The other red line segments follow by interchanging layer indices and by particle-hole symmetry within a Landau level. Additional drag measurements should be able to distinguish between possibilities.

The filling $\nu_1 = \nu_2 = 1/6$ corresponds to half-filled Λ-level of $^2_2$CFs. Figures 17(a) and 17(c) demonstrate a robust insulating feature at this filling, which disperses vertically with layer imbalance [solid black lines in Fig. 17(b) and highlighted by blue arrows in Figs. 17(a) and 17(c)]. This robustness against density imbalance is strikingly similar to the bare electron-hole bilayers, where the exciton condensate at half filling of each layer is stable under extreme density imbalance. The filling fraction $\nu = \frac{1}{6}$ ($\nu^\dagger = \frac{1}{3}$), together with the observed insensitivity to layer imbalance is consistent with the exciton condensate described by the Halperin Φ_{333} wavefunction, which has been theoretically predicted to stabilize in double layer systems but not previously observed.[128]

As mentioned above, Coulomb drag measurements in devices with aligned double Hall bar geometry provide crucial additional information for characterizing the FQH states in a double layer system. Examples of such measurements are shown in Fig. 18. Measurements of the Hall voltage and longitudinal voltage drop in the drive and drag layers allow one to define a 4×4 resistivity matrix $\rho_{i\alpha,j\beta}$, with layer indices i, j and cartesian coordinates α, β. According to the layer-coupled CF model, this matrix may be written as $\hat{\rho} = \hat{\rho}^{CF} + \hat{\rho}^{CS}$, where $\hat{\rho}^{CF}$ is the resistivity matrix of the CFs and $\hat{\rho}^{CS}$ is a contribution from the attached Chern–Simons fluxes.[4,5] Specifically, $\hat{\rho}^{CS}$ has zero longitudinal components and has Hall resistivities equal to a or b, in units of h/e^2, for (i, j) in the same or different layers, respectively. For

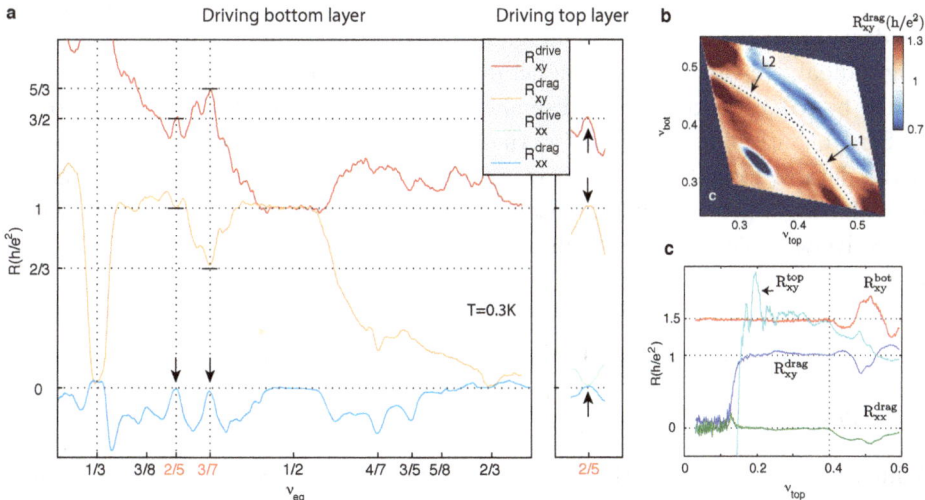

Fig. 18. (a) (Left) Vanishing longitudinal resistance on both graphene layers, (R_{xx}^{drive}, R_{xx}^{drive}), and quantized Hall resistance, (R_{xy}^{drive}, R_{xy}^{drag}) in the drive and drag layers appear at $\nu = \nu_{\text{top}} = \nu_{\text{bot}} = 2/5$ and $3/7$. Short horizontal lines mark the Hall resistance quantization values. (Right) The same measurement as in the left panel, but with the drive and drag layers switched. (b) Hall drag resistance measured around one of the diamond shaped pattern shown in Fig. 17. (c) Line cut through line L2 in (b). It is notable that R_{xy}^{bot} remained constant up to the crossing point between L1 and L2 (vertical dotted line). (From Ref. 5.)

states with integer $\nu_i^\dagger$, where $\hat{\rho}^{CF}$ describes uncoupled fermion species, one finds that the Hall drag is equal to b, while the drive layer Hall resistance for current in layer i is given by $a + (\nu_i^\dagger)^{-1}$. For example, for the FQH state with filling $2/5$ in each layer, the proposed values $a = 2$, $b = 1$, $\nu_i^\dagger = -2$ lead to drive Hall resistance $= 3/2$ and drag Hall resistance $= 1$, in agreement with the experimental results. The longitudinal components of the drive and Hall resistances are found to vanish, as expected for a quantized Hall state.[144]

In the cases where $\nu_i^\dagger$ is half integer, the assumption that the CFs form a Bose condensate of interlayer excitons leads to the prediction that the Hall matrix elements of $\hat{\rho}^{CF}$ have a value $(\nu^\dagger)^{-1}$, independent of the layer indices. Thus, for the state with filling $3/7$ in each layer, the proposed values $a = 2$, $b = 1$, $\nu_i^\dagger = -3/2$ lead to drive Hall resistance $= 5/3$ and drag Hall resistance $= 2/3$, which, again, agrees with the experimental results. By contrast, if instead of excitonic pairing, one had assumed a separate superconducting pairing in each layer, one would have predicted drive Hall resistance $= 4/3$ and drag Hall resistance $= 1$, in disagreement with experiment.

The semi-quantized states, mentioned previously, also give rise to characteristic transport signatures in the double Hall bar geometry. For a semi-quantized state, along a line segment where, say, $\nu_1^\dagger$ is an integer but $\nu_2^\dagger$ is continuously varying, one predicts that the drag Hall resistance will be quantized with a value $= b$, and the drive Hall resistance will be quantized with value $a + (\nu_1^\dagger)^{-1}$ when current is driven in layer 1. On the other hand, when current is driven in layer 2, one predicts that the drive Hall resistance will be unquantized, and there will be a finite value of the longitudinal resistance. The conductance G, measured in the Corbino geometry, should be reduced along a line of semi-quantized states, but not precisely zero.

Semi-quantized states satisfying these predictions may be seen in several places in the data of Fig. 18(b). For example, the blue dotted lines labeled $L1$ and $L2$ are segments where CF filling changes continuously in the top or bottom layer, respectively, while the Λ-level filling is a constant integer value -2 in the other layer.

6. Moiré Superlattices and Fractional Chern Insulators

6.1. *The Hofstadter butterfly*

A final unusual feature of the fractional quantum Hall effect in graphene systems concerns the influence of long wavelength moiré superlattices. Moiré patterns arise at any clean interface between graphene and another crystalline material when the two materials are either lattice mismatched—as with differing and thus lattice mismatched two-dimensional crystals—or when two graphene layers are rotationally misaligned. Moiré superlattices boast two key features for enabling the observation of new quantum Hall phenomena. First, because they arise from the beating of comparatively stiff crystal lattices, they are highly regular, as evidenced by scanning tunneling microscopy studies of graphene on graphite[145,146] and graphene on hBN.[147,148] Second, the moire wavelength, λ_M, can be tuned to around 10–20 nm, closely matching the magnetic length, ℓ_B at moderate magnetic fields. For closely lattice matched materials, such as graphene/hBN, the period of the moiré lattice is given by

$$\lambda_M = \frac{(1+\delta)a}{\sqrt{2(1+\delta)(1-\cos\theta)+\delta^2}}, \tag{27}$$

where δ is the lattice mismatch, $a = 0.246$ nm is the graphene lattice constant, and θ is the interlayer twist angle. For rotationally aligned graphene on hBN, for example, $\delta = 1.8\%$ while $\theta = 0$, so that $\lambda_M \approx 14$ nm.

The interplay between magnetic field and periodic potential in the motion of a quantum mechanical particle has been a topic of theoretical research dating to the work of Azbel'[149] and Zak[150] over half a century ago. The complexity of the problem arises from the general incommensurability between the magnetic length and lattice constant so that the periodicity of the Hamiltonian becomes a rapidly varying function of the dimensionless parameter Φ/Φ_0, where $\Phi = BA$ is the magnetic flux threading a single unit cell of area A and $\Phi_0 = h/e$ is the quantum of magnetic flux. As described below, in simple cases the Hamiltonian reduces to a Harper equation,[151] which describes an electron hopping along a one-dimensional chain in the presence of a commensurate superlattice potential. In its general form, the Harper Hamiltonian is

$$\hat{H} = |m\rangle\langle m+1| + |m\rangle\langle m-1| + 2\cos(2\pi(\omega+m\alpha))|m\rangle\langle m|, \tag{28}$$

where m is the index of the lattice site and ω and α are control parameters. For rational $\alpha = q/p$ (with p and q are mutually prime) the eigenenergies were famously computed numerically by Hofstadter.[152] In this case, the unit cell size for the Harper chain contains p sites, as the substitution $m \rightarrow m + p$ leaves the Hamiltonian invariant; there are thus p bands for each value of ω.

In graphene superlattices, the Hofstadter problem has received the most experimental attention[1,2,13,153–155] at high magnetic fields. In this limit, the lattice potential is weak compared to the cyclotron energy separating Landau levels, so that the effects of the potential only mix states within a single Landau level. This contrasts with the canonical treatments of the Hofstadter problem,[151,152,156,157] in which the magnetic field is weak compared to gaps between lattice bands and only mixes states within a single Bloch band.

To illustrate the basic nature of the problem, we consider a conventional two-dimensional electron gas described by a scalar wave function, in the presence of an additional square lattice periodic potential, $\hat{V}(x,y) = V_0\left(\cos(Qy)+\cos(Qx)\right)$, with $Q = 2\pi/a_0$ and a_0 the lattice constant. At $V_0 = 0$, the eigenfunctions are the conventional Landau level wave functions. We work in the Landau gauge where the vector potential $A = Bx\hat{y}$, so that the eigenfunctions $|N,k_y\rangle$ are distinguished by the Landau level index N which enters the energy $E_N = \hbar\omega_C(N + 1/2)$ and k_y, which is related to the cyclotron guiding center x_c via $x_c = k_y \ell_B^2.$[9] The effect of $V_0 \neq 0$ is to break the translation symmetry of the Hamiltonian, which lifts the degeneracy of electronic states within a Landau level that are distinguished only by their cyclotron guiding centers in the absence of the periodic potential.

We consider for simplicity the effect of V_0 on the lowest $(N = 0)$ Landau level. The problem is then an exercise in degenerate perturbation theory, with unperturbed wavefunctions $\langle x, y | k_y \rangle = \frac{1}{\sqrt{L \ell_B \sqrt{\pi}}} e^{-ik_y y} e^{(x-k_y \ell_B^2)^2/(2\ell_B^2)}$ and perturbation $\hat{V}(x, y)$, which amounts to diagonalizing the Hamiltonian with matrix elements $H_{k_y,k_y'} = \langle k_y' | \hat{V}(x, y) | k_y \rangle$. These can be computed explicitly using the Landau level wavefunctions, giving

$$H_{k_y,k_y'} = \langle k_y' | V_0 \cos(Qy) | k_y \rangle + \langle k_y' | V_0 \cos(Qx) | k_y \rangle \tag{29}$$

$$= V_0 e^{-\ell_B^2 Q^2/4} \left(\frac{1}{2} \left(\delta_{k_y+Q,k_y'} + \delta_{k_y-Q,k_y'} \right) + \delta_{k_y,k_y'} \cos(Q\ell_B^2 k_y) \right). \tag{30}$$

$H_{k_y,k_y'}$ is one-dimensional (in the variable k_y). The maximum bandwidth E_0 acquired by the LL in the presence of the periodic potential is a product of the periodic potential strength V_0 and the exponential LL form factor, which varies with LL index. After noting that $Q^2 \ell_B^2 = 2\pi \frac{\Phi_0}{\Phi}$, this bandwidth is seen to vanish in the $B \to 0$ limit as $E_0 = \frac{V_0}{2} e^{-\ell_B^2 Q^2/4} = \frac{V_0}{2} e^{-\pi \Phi_0/2\Phi}$ as long as one can neglect mixing between Landau levels. Superlattice physics in graphene is thus expected to manifest most strongly at high magnetic fields. $H_{k_y,k_y'}$ is, of course, manifestly periodic in k_y with periodicity Q, so that we expect a continuous spectrum of eigenvalues for k_y^0 in the interval $0 < k_y^0 < Q$, separated into one or more distinct bands. To determine the number of such bands, we consider the set of k_y linked by integer multiples of the superlattice wavevector, $k_y = k_y^0 + mQ$, $m \in \mathbb{Z}$. Making this substitution,

$$H_{m,m'} = 2E_0 \left(\frac{1}{2} (\delta_{m+1,m'} + \delta_{m-1-1,m}) + \delta_{m,m'} \cos(Q^2 \ell_B^2 (k_y^0/Q + m)) \right) \tag{31}$$

or, equivalently,

$$\hat{H}/E_0 = |m\rangle\langle m+1| + |m\rangle\langle m-1| + 2\cos(2\pi(\frac{k_y^0}{Q}\frac{\Phi_0}{\Phi} + m\frac{\Phi_0}{\Phi})) |m\rangle\langle m|. \tag{32}$$

Intuitively, the first two hopping terms originate from the y-dependent part of the potential, with the $\cos(Qy)$ term generating translations by $\pm Q$ in k_y. The third term, arising from the $\cos(Qx)$ term, depends only on x and thus does not mix states with different k_y.

The connection to Eq. (28) is obtained by identifying $\omega = \frac{k_y^0}{Q}\frac{\Phi_0}{\Phi}$ and $\alpha = \Phi_0/\Phi$. For $\Phi/\Phi_0 = p/q$, there are p atoms in the unit cell, and thus p distinct bands. Figures 19(a)–19(b) illustrate this connection for $\Phi/\Phi_0 = 8/9$. At this magnetic field, the area occupied by nine cyclotron orbits is commensurate with that of 16 unit cells, and in the Harper description, the unit cell contains eight atoms. Figure 19(c) shows the resulting spectrum, plotted as a function of the normalized

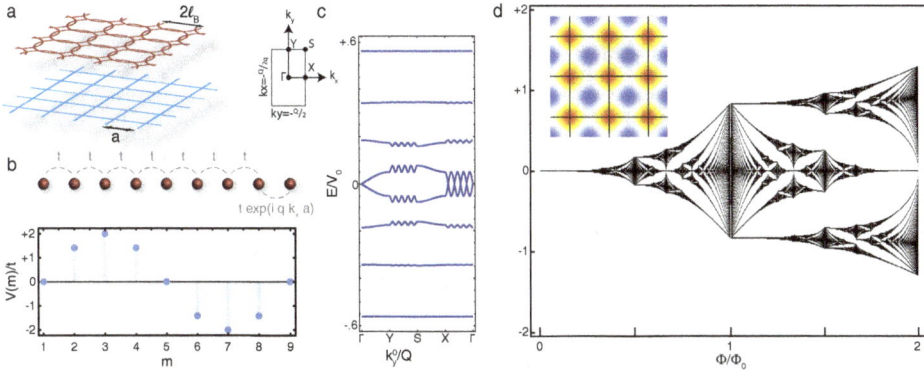

Fig. 19. **Energy spectrum of Hofstadter's butterfly in a square lattice.** (a) Schematic diagram of Hofstadter's butterfly problem for a square lattice with magnetic fields. (b) Equivalent 1D lattice problem with modulated hoping parameters with long commensuration. (c) 1D Energy spectrum computed for (b). (d) Calculated energy spectrum of square Hofstadter's lattice as a function of normalized magnetic field.

momentum k_y^0/Q, with the resulting eight distinct sub-bands. Note that the sum of the sub-bands represents the broadened Landau level, which at this flux acquires a bandwidth comparable to V_0; however, despite this broadening certain individual sub-bands are both well separated in energy from other levels and exceptionally narrow. For a superlattice potential of $V_0 \approx 15$ meV, the flattest sub-bands shown in Fig. 19(c) would have a bandwidth of only several μeV—considerably smaller than the disorder broadening present in even the cleanest two-dimensional systems. From the point of view of electronic correlations, the Hofstadter sub-bands can thus be expected to resemble Landau levels, with intrinsic bandwidth having a negligible effect and the ground state properties entirely a result of the intrinsic Berry curvature and the nature of Coulomb interactions.

Figure 19(d) shows the evolution of the eigenvalues of the square lattice problem described above as a function of the applied flux, calculated for integers $p \in (0, 1000)$ and $q = 500$. The figure, as well as Fig. 20(c), which shows the same calculation for a triangular lattice, illustrates the essential features of the Hofstadter spectrum that arises from solving equations of the Harper type. Most notably, the number of bands is a discontinuous function of the applied flux. Within the apparent bands, moreover, close inspection reveals an infinite fractal hierarchy of gaps. However, while the bands themselves are nowhere continuous, the band gaps cover large, simply connected areas of the Φ–E plane.

It is the Hofstadter mini-gaps that provided the first experimental signatures of the Hofstadter butterfly in experiment.[13,153,154] The experimental system differs from the model discussed above; in particular, the moire pattern generated between graphene and hexagonal boron nitride shares with its parent lattices a triangular symmetry [see Figs. 20(a) and 20(b)]. Figure 20(c) shows the calculated spectrum for a triangular potential for two $N = 0$ Landau levels with a linear energy splitting

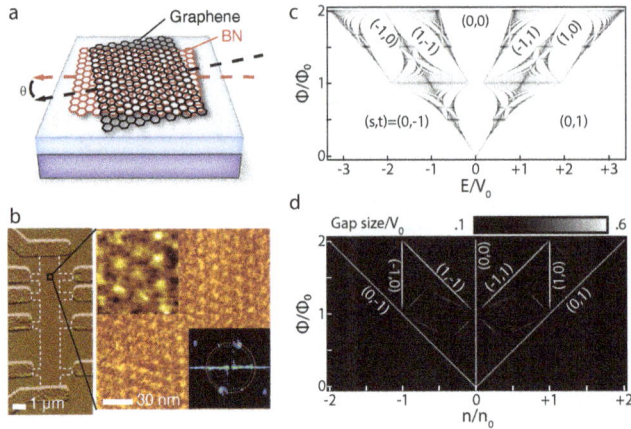

Fig. 20. **Moiré superlattices in graphene-hBN heterostructures.** (a) Schematic of graphene on hexagonal boron nitride showing the resulting triangular Moiré pattern. (b) Atomic force microscopy images showing a graphene Hall bar device and embedded superlattice. Adapted from Ref. 153. (c) Schematic energy level structure for two Landau levels with a triangular lattice potential. A linear-in-B splitting has been introduced phenomenologically. (s, t) quantum numbers are shown for the largest gaps. (d) Simulated Wannier diagram for the spectrum shown in (c). White lines represent energy gaps, shaded according to the gap size so that only the largest gaps are visible.

as might arise from Zeeman or exchange splittings. The basic features of new gaps appearing *within* a Landau level is shared by all Hofstadter systems regardless of the lattice symmetry. Naively, then, the Hofstadter butterfly might manifest through the emergence of new gapped states at partial Landau level filling, which undergoes a highly complex evolution in the magnetic field. However, as first pointed out by Wannier,[158] many of the intricate features of the Hofstadter spectrum are lost upon consideration of the degeneracy of the minibands. The central result is that energy gaps within the Hofstadter spectrum follow a Diophantine equation

$$n/n_0 = t(\Phi/\Phi_0) + s \tag{33}$$

with n the areal carrier density and $n_0 = 1/A$ the density corresponding to one carrier per lattice unit cell. t and s are integer-valued quantum numbers that are unique to each gap. In typical non-spectroscopic experiments, which measure response functions sensitive only to physics near the Fermi level, all gaps (including cyclotron gaps between Landau levels) thus follow simple linear trajectories in the n–Φ plane.

Figure 20(d) shows the 'Wannier plot' equivalent of Fig. 20(c). The figure depicts the n–Φ plane, with energy gaps rendered in white with intensity coded to the gap size. In the absence of a superlattice, the Hofstadter problem reduces to the Landau level problem, with all gaps having $s = 0$. This produces the familiar "Landau fans" typically plotted in density tunable systems, with all gaps extrapolating to $n = 0$ at $\Phi = 0$. The primary signature of the Hofstadter butterfly in the experiment is the appearance of new features associated with energy gaps that do not extrapolate

$n = 0$, or in other words, the appearance of gapped states with $s \neq 0$. Several prominent states of this type are evident in Fig. 20(d) for $\Phi > \Phi_0 = 1$.

Figure 21 shows experimental data from monolayer [Figs. 21(a)–21(c)] and bi-layer graphene [Figs. 21(d) and 21(e)] devices aligned to hexagonal boron nitride, generating a superlattice for which $\Phi = \Phi_0$ at $B \approx 23$ T. Both data sets feature numerous features following linear slopes with $s \neq 0$. As discussed above, in the weak lattice limit, where wave functions in only a single Landau level are mixed by the periodic potential, lattice-associated mini-gaps are negligibly small in the low field limit. For both mono- and bilayer graphene, new gaps associated with the Hofstadter butterfly indeed manifest only at or above $\Phi = \Phi_0$. Note the de-tails of which gaps manifest the difference between the two systems. This is tied to the difference in low-energy Landau level structure. In monolayer graphene, the $N = 0$ LL is well isolated from higher LLs (i.e. the lattice cannot mix states be-tween $N = 0$ and $N \neq 0$ LLs). In this system, in fact, the data show a qualitative match to the theoretical calculation of Fig. 20. In bilayer graphene, in contrast, the near-degeneracy of the $N = 0$ and $N = 1$ lead to strong lattice mixing and a significantly different spectrum.

Like quantum Hall gaps, superlattice induced gaps are characterized by a quan-tized Hall conductivity [Fig. 21(e)], leading to quantized σ_{xy}, vanishing σ_{xx}, and, as in the case of the data of Hunt,[13] quantized two-terminal conductance G. How-ever, unlike conventional quantum Hall gaps, the quantized Hall conductivity is no longer linked to the Landau level filling factor. A very general formula by Streda[159]

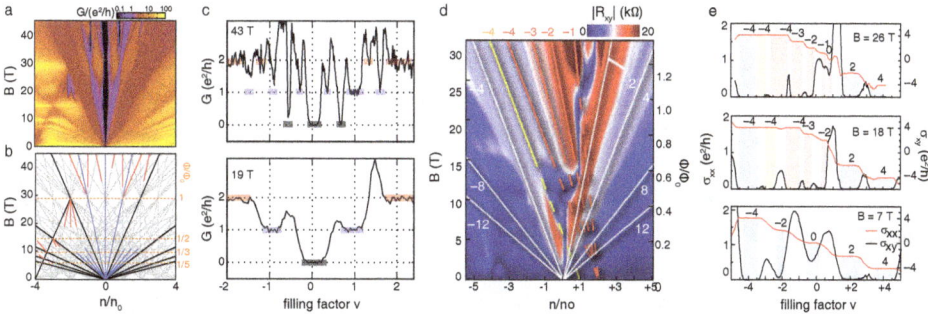

Fig. 21. **Experimental observation of the Hofstadter butterfly in graphene/boron nitride superlattices.** (a) Experimental data measuring G, the two-terminal conductance, for monolayer graphene aligned to hBN. Here, G is effectively equal to R_{xy}^{-1} in the quantum Hall regime. (b) Associated Wannier plot. Lines in red correspond to gapped states arising from Hofstadter minigaps. (c) Line traces above and below $\Phi/\Phi_0 = 1$. At low fields, the total superlattice induced bandwidth is still small, and the quantum sequence is the conventional one for monolayer graphene. Above one flux quantum, the quantization oscillates, as expected from the spectrum shown in Figs. 20(c) and 20(d). Adapted from Ref. 13. (d) Experimental data measuring $|R_{xy}|$, the two-terminal conductance, for bilayer graphene aligned to hBN. Wannier plots are overlaid. (e) Line traces above and below $\Phi = \Phi_0$. Signatures of the butterfly energy at lower magnetic fields in bilayer graphene, owing to the unusual structure of the zero energy Landau level that combined $N = 0$ and $N = 1$ orbital wavefunctions.

posits that for any incompressible gapped state, the Hall conductivity must obey
the relation

$$\sigma_{xy} = \left(e \frac{\partial n_{gap}}{\partial B} \right). \tag{34}$$

In other words, the Hall conductivity is quantized to a value equivalent to the slope
of the line, with t equivalent to the Hall conductivity expressed in natural units,
$t = \frac{h}{e^2} \sigma_{xy}$.

The quantization of t is a result of the general fact that in the absence of time
reversal symmetry, a fully filled electron band is guaranteed to have an integer
quantized Hall conductivity. The quantization can be related to the cumulative
Chern numbers of the occupied bands,[160] or $t = \sum_{occup.} C_i$. The Chern number of
the ith band is defined as the integral of the Berry curvature,

$$C_i = \int_{BZ} \left(\vec{\nabla}_{\vec{k}} \times \vec{a}_{\vec{k},i} \right)_z d^2k, \tag{35}$$

where $a_{\vec{k}}^i = i\langle u_{\vec{k},i}|\nabla_{\vec{k}}|u_{\vec{k},i}\rangle$ is the Berry connection on the Bloch functions. In this
context, the oscillations in t observed as a function of density in Fig. 21(c) result
from the fact that Hofstadter sub-bands, unlike Landau levels with $C = 1$, can carry
Chern numbers that are either positive or negative. Similarly, variations in s reflect
a band property. s corresponds to the change in carrier number that results from
adding a single unit cell to the system at constant magnetic field, or, equivalently,
the change in carrier density that results from changing the unit cell area[161]

$$s = \left(\frac{\partial n}{\partial n_0} \right)_B. \tag{36}$$

In other words, s is simply the Bloch band filling index. Integer quantum Hall states
with translational invariance manifestly have $s = 0$, while Hofstadter butterfly
states contain charge glued to the lattice in their ground state. Changes in s that
appear from gap to gap as sub-bands fill within a lattice-enhanced Landau level,
thus reflect the quantity of charge per unit cell of a particular sub-band.

As a concrete example is the case of $\Phi/\Phi_0 = 3/2$ in monolayer graphene, corresponding to $B = 43$ T in Fig. 21(c). Starting from the $\nu = -1$ Landau level
gap, which has $(s,t) = (0,-1)$, two additional gaps are observed at partial filling
with $(s,t) = (-1,0)$ and $(+1,-1)$, before the $\nu = 0$ gap which has $(s,t) = (0,0)$.
Consequently, the three bands present at this flux must carry $C = 1, -1$, and 1,
respectively, along with charge density of -1, 2, and -1 electron per lattice site;
filling all three leads to $(\Delta s, \Delta t) = (0,1)$ consistent with the unperturbed Landau
level, while filling a subset leads to the correct quantum numbers for a Hofstadter
minigap. The cumulative Chern and lattice quantum numbers of sets of sub-bands
follow naturally by extrapolation of the gaps: for example, at $\Phi/\Phi_0 = 31/20$, there
are 31 bands, clustered into lower and upper groups of 11 and middle grouping of
9. The upper and lower groups are evidently characterized by a cumulative Chern
number of -1 and lattice quantum number of -1, while the middle group has
cumulative Chern number of 1 and lattice quantum number of 2.

6.2. *Electronic interactions in superlattice Chern bands*

While electron interactions have played no role in the theoretical discussions of the Hofstadter problem above, they are of course present in the experiment. The first and simplest effect of interactions is to lift the degeneracy of spin and valley native to graphene via quantum Hall ferromagnetism. Landau level broadening due to the superlattice is, at first glance, inimical to symmetry breaking—bandwidth and exchange interactions directly compete through a Stoner mechanism. Experiments[155] have indeed observed the suppression of quantum Hall ferromagnetism at commensurate fillings, where the bandwidth is largest. In the cleanest samples, and at high magnetic fields, quantum Hall ferromagnetism merely serves to remove the degeneracy completely. This follows from the energy scales in the problem: the superlattice strength of ≈ 15 meV sets the maximum bandwidth of a single Landau level, which is much smaller than the Coulomb scale $e^2/(\varepsilon \ell_B) \approx 11$ meV $\times \sqrt{B[\mathrm{T}]}$ for fields corresponding to $\Phi/\Phi_0 \approx 1$. Quantum Hall ferromagnetism can thus be expected to dominate, splitting the Landau levels into single spin- and valley polarized branches. This is consistent with the observations in monolayer graphene of four copies of the superlattice broadened level at high magnetic fields [Figs. 21(a)–21(c)].

The richest interaction physics, however, occurs within the singly degenerate superlattice enhanced Landau level. Absent interactions, the Landau levels acquire an infinite hierarchy of energy gaps. However, once the energy splittings between Hofstadter sub-bands become sufficiently small compared to the Coulomb scale, interactions can mix different states. In this limit, the hierarchy of Hofstadter gaps terminates, and ground states are obtained from minimizing the Coulomb repulsion energy rather than from single particle band filling. Indeed, experiments have revealed new gapped states not captured within the simple picture that includes only the Hofstadter butterfly spectrum and 'mean-field' like interactions that lift the spin and valley degeneracy. These states manifest as incompressible states in which s and/or t are quantized to *fractional* values.[1,2,162] These states fall into two classes. In the first, s is fractional, while t remains quantized to an integer.[1] Going back to the definition of s in Eq. (36), it follows that these ground states feature a rational fraction of an electron charge glued to each moire lattice site. While this fact could be explained by the existence of fractionally charge excitations, as in the fractional quantum Hall effect, a simpler explanation is that these ground states spontaneously break the superlattice symmetry, enlarging the unit cell. In this picture, fractional s arises from the fact that the actual periodicity of the system is no longer that of the moire lattice. Instead, integer charges are glued to the new, enlarged unit cell. In the case of an $s = m/2$ state, where m is an integer, as appears in Figs. 22(c) and 22(d) and was observed in both monolayer and bilayer moire superlattices,[1,2] this new unit cell is doubled, while for an $s = m/3$ state, it may be trebled. Indeed, numerical simulations using DMRG methods indeed find broken lattice symmetry ground states starting from realistic models of the moire potential, as shown in Fig. 22(f). States of this first class are consequently dubbed symmetry broken Chern insulators (or SBCIs).

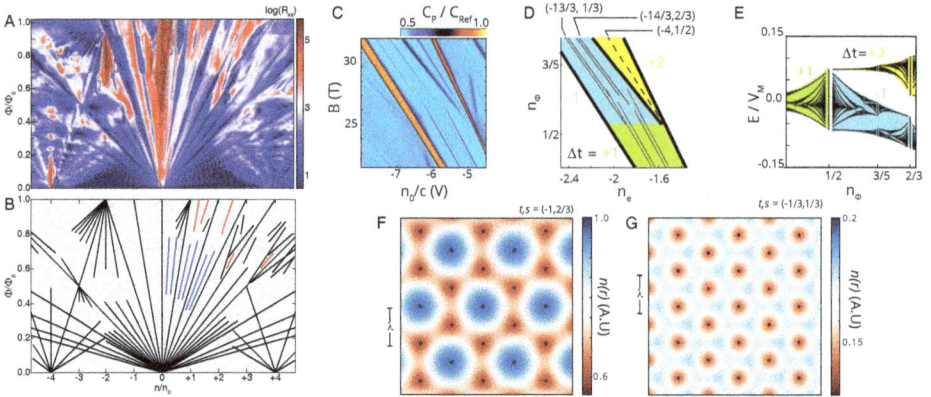

Fig. 22. Correlated states in the Hofstadter butterfly. (a) Experimentally measured R_{xx} in a clean hBN aligned graphene monolayer.[1] **(b)** Wannier plot schematic of (a). Notable in this data are the appearance of several gap features (indicated in red) that have *fractional s* and integer *t*. As discussed in the main text, these states are likely associated with broken *superlattice* symmetry, and have been dubbed Symmetry Broken Chern insulators (SBCI) **(c)** Experimental capacitance data[2] in a clean hBN aligned graphene bilayer. **(d)** Wannier plot schematic and **(e)** corresponding energy band structure, ignoring correlation, for the region shown in (c). Two bands are highlighted—a $C = -1$ band in cyan and a $C = +2$ band in yellow—bracketed in density by gaps with integer *t* and *s*. Within the $C = +2$ band, a gapped state appears at half filling with fractional *s*, again indicating superlattice symmetry breaking; within the $C = -1$ band, states appear at 1/3, 2/5, and 3/5, 2/3 fillings, characterized by both fractional *s* and *t*. These states, known as fractional Chern insulators, are topologically similar to fractional quantum Hall states but bind fractional charges to each superlattice site. **(f)** Real-space density in a symmetry broken Chern insulator state, calculated using density-matrix renormalization group methods.[113] The ground state manifestly breaks translation symmetries of the underlying lattice, indicated by the black dots. **(g)** Real space density for a fractional Chern insulator state. The ground state does not break translation symmetry, and is characterized by fractionally charged quasiparticles in the excitation spectrum.[2]

The second class of states show fractional t in addition to s. By Laughlin's famous argument,[163] originally developed for fractional quantum Hall states within a Landau level, the observation of fractional Hall conductivity implies the existence of a fractionally quantized minimal charge within the excitation spectrum. However, unlike in conventional FQH states where $s = 0$, the finite—and fractional—s implies that fractionally charge anyons are bound to individual lattice sites.[164] DMRG simulations indeed find a ground state that features fractional charges bound to each lattice site, but that does not break the underlying moire lattice symmetry for at least some of the states observed in Ref. 2 [see also Fig. 22(g)]. These states are identical in topology and symmetry to fractional Chern insulators (FCIs) previously proposed as ground states of lattice models in the absence of a magnetic field.[165,166] FCIs can be understood heuristically as a noninteracting Hofstadter butterfly of composite fermions, where the effective magnetic field replaces the physical flux; applying this formulation results in relationships between s and t that align with experiment.[2,162]

References

1. L. Wang, Y. Gao, B. Wen, Z. Han, T. Taniguchi, K. Watanabe, M. Koshino, J. Hone, and C. R. Dean, Evidence for a fractional fractal quantum Hall effect in graphene superlattices, *Science* **350**(6265), 1231–1234 (2015). ISSN 0036-8075, 1095–9203. doi: 10.1126/science.aad2102.
2. E. M. Spanton, A. A. Zibrov, H. Zhou, T. Taniguchi, K. Watanabe, M. P. Zaletel, and A. F. Young, Observation of fractional Chern insulators in a van der Waals heterostructure, *Science* **360**(6384), 62–66 (2018). ISSN 0036-8075, 1095–9203. doi: 10.1126/science.aan8458. URL http://science.sciencemag.org/content/360/6384/62.
3. J. I. A. Li, T. Taniguchi, K. Watanabe, J. Hone, and C. R. Dean, Excitonic superfluid phase in double bilayer graphene, *Nature Phys.* **13**(8), 751–755 (2017). ISSN 1745-2481. doi: 10.1038/nphys4140. URL https://www.nature.com/articles/nphys4140.
4. J. I. A. Li, Q. Shi, Y. Zeng, K. Watanabe, T. Taniguchi, J. Hone, and C. R. Dean, Pairing states of composite fermions in double-layer graphene, *Nature Phys.* **15**(9), 898–903 (2019). ISSN 1745-2481. doi: 10.1038/s41567-019-0547-z. URL https://www.nature.com/articles/s41567-019-0547-z.
5. X. Liu, Z. Hao, K. Watanabe, T. Taniguchi, B. I. Halperin, and P. Kim, Interlayer fractional quantum Hall effect in a coupled graphene double layer, *Nature Phys.* **15**(9), 893–897 (2019). ISSN 1745-2481. doi: 10.1038/s41567-019-0546-0. URL https://www.nature.com/articles/s41567-019-0546-0.
6. J.-H. Chen, C. Jang, S. Adam, M. S. Fuhrer, E. D. Williams, and M. Ishigami, Charged-impurity scattering in graphene, *Nature Phys.* **4**(5), 377–381 (2008). URL http://dx.doi.org/10.1038/nphys935.
7. P. R. Wallace, The band theory of graphite, *Phys. Rev.* **71**(9) (1947). URL http://link.aps.org/doi/10.1103/PhysRev.71.622.
8. P. A. M. Dirac, The quantum theory of the electron, *Proc. Roy. Soc. London. Series A.* **117**(778), 610–624 (1928). URL http://rspa.royalsocietypublishing.org/content/117/778/610.short.
9. L. Landau and E. M. Lifshitz, *Quantum Mechanics (Non-relativistic Theory).* (Butterworth-Heineman, 1977).
10. Y. Zhang, Y.-W. Tan, H. L. Stormer, and P. Kim, Experimental observation of the quantum Hall effect and Berry's phase in graphene, *Nature* **438**(7065), 201–204 (2005). ISSN 0028-0836. doi: 10.1038/nature04235. URL http://www.nature.com/nature/journal/v438/n7065/full/nature04235.html.
11. K. S. Novoselov, A. K. Geim, S. V. Morozov, D. Jiang, M. I. Katsnelson, I. V. Grigorieva, S. V. Dubonos, and A. A. Firsov, Two-dimensional gas of massless Dirac fermions in graphene, *Nature* **438**(7065), 197–200 (2005). ISSN 0028-0836. doi: 10.1038/nature04233. URL http://www.nature.com/nature/journal/v438/n7065/full/nature04233.html.
12. F. Amet, J. R. Williams, K. Watanabe, T. Taniguchi, and D. Goldhaber-Gordon, Insulating behavior at the neutrality point in single-layer graphene, *Phys. Rev. Lett.* **110**(21), 216601 (2013). doi: 10.1103/PhysRevLett.110.216601. URL http://link.aps.org/doi/10.1103/PhysRevLett.110.216601.

13. B. Hunt, J. D. Sanchez-Yamagishi, A. F. Young, M. Yankowitz, B. J. LeRoy, K. Watanabe, T. Taniguchi, P. Moon, M. Koshino, P. Jarillo-Herrero, and R. C. Ashoori, Massive Dirac fermions and Hofstadter butterfly in a van der Waals heterostructure, *Science* **340**(6139), 1427–1430 (2013). ISSN 0036-8075, 1095–9203. doi: 10.1126/science.1237240. URL https://science.sciencemag.org/content/340/6139/1427.

14. C. Toke, P. E. Lammert, V. H. Crespi, and J. K. Jain, Fractional quantum Hall effect in graphene, *Phys. Rev. B.* **74** (2006).

15. C. R. Dean, A. F. Young, P. Cadden-Zimansky, L. Wang, H. Ren, K. Watanabe, T. Taniguchi, P. Kim, J. Hone, and K. L. Shepard, Multicomponent fractional quantum Hall effect in graphene, *Nature Phys.* **7**(9), 693–696 (2011). URL http://dx.doi.org/10.1038/nphys2007.

16. F. Amet, A. J. Bestwick, J. R. Williams, L. Balicas, K. Watanabe, T. Taniguchi, and D. Goldhaber-Gordon, Composite fermions and broken symmetries in graphene, *Nature Commun.* **6**, 5838 (2015). doi: 10.1038/ncomms6838. URL http://www.nature.com/ncomms/2015/150106/ncomms6838/abs/ncomms6838.html.

17. Y. Kim, A. C. Balram, T. Taniguchi, K. Watanabe, J. K. Jain, and J. H. Smet, Even denominator fractional quantum Hall states in higher Landau levels of graphene, *Nature Phys.* **15**(2), 154–158 (2019). ISSN 1745-2473, 1745-2481. doi: 10.1038/s41567-018-0355-x. URL http://www.nature.com/articles/s41567-018-0355-x.

18. S. D. Sarma, S. Adam, E. H. Hwang, and E. Rossi, Electronic transport in two-dimensional graphene, *Rev. Mod. Phys.* **83** (2011).

19. A. H. C. Neto, F. Guinea, N. M. R. Peres, K. S. Novoselov, and A. K. Geim, The electronic properties of graphene, *Rev. Mod. Phys.* **81**(1), (2009). URL http://link.aps.org/abstract/RMP/v81/p109.

20. J. Jung and A. H. MacDonald, Accurate tight-binding models for the pi bands of bilayer graphene, *Phys. Rev. B.* **89**(3), 035405 (2014). doi: 10.1103/PhysRevB.89.035405. URL https://link.aps.org/doi/10.1103/PhysRevB.89.035405.

21. E. McCann and V. I. Fal'ko, Landau-level degeneracy and quantum Hall effect in a graphite bilayer, *Phys. Rev. Lett.* **96**(8) (2006). doi: 10.1103/PhysRevLett.96.086805.

22. A. Varlet, D. Bischoff, P. Simonet, K. Watanabe, T. Taniguchi, T. Ihn, K. Ensslin, M. Mucha-Kruczyski, and V. I. Falko, Anomalous sequence of quantum Hall liquids revealing a tunable Lifshitz transition in bilayer graphene, *Phys. Rev. Lett.* **113**(11), 116602 (2014). doi: 10.1103/PhysRevLett.113.116602. URL https://link.aps.org/doi/10.1103/PhysRevLett.113.116602.

23. K. S. Novoselov, E. McCann, S. V. Morozov, V. I. Falko, M. I. Katsnelson, U. Zeitler, D. Jiang, F. Schedin, and A. K. Geim, Unconventional quantum Hall effect and Berry's phase of 2 in bilayer graphene, *Nature Phys.* **2**(3), 177–180 (2006). ISSN 1745-2481. doi: 10.1038/nphys245. URL https://www.nature.com/articles/nphys245.

24. H. Polshyn, H. Zhou, E. M. Spanton, T. Taniguchi, K. Watanabe, and A. F. Young, Quantitative transport measurements of fractional quantum Hall energy gaps in edgeless graphene devices, *Phys. Rev. Lett.* **121**(22), 226801 (2018). doi: 10.1103/PhysRevLett.121.226801. URL https://link.aps.org/doi/10.1103/PhysRevLett.121.226801.

25. Y. Zeng, J. Li, S. Dietrich, O. Ghosh, K. Watanabe, T. Taniguchi, J. Hone, and C. Dean, High-quality magnetotransport in graphene using the edge-free Corbino geometry, *Phys. Rev. Lett.* **122**(13), 137701 (2019). doi: 10.1103/PhysRevLett.122.137701. URL https://link.aps.org/doi/10.1103/PhysRevLett.122.137701.

26. Z. Jiang, Y. Zhang, H. L. Stormer, and P. Kim, Quantum Hall states near the charge-neutral Dirac point in graphene, *Phys. Rev. Lett.* **99**(10) (2007). URL http://link.aps.org/doi/10.1103/PhysRevLett.99.106802.

27. J. G. Checkelsky, L. Li, and N. P. Ong, Zero-energy state in graphene in a high magnetic field, *Phys. Rev. Lett.* **100**(20) (2008). URL http://link.aps.org/doi/10.1103/PhysRevLett.100.206801.

28. B. I. Halperin, Theory of the quantized Hall conductance, *Helv. Phys. Acta.* **56**, 75–102 (1983).

29. Z. F. Ezawa, *Quantum Hall Effects: Field Theoretical Approach and Related Topics.* (World Scientific, Singapore, 2000).

30. M. Shayegan, E. P. De Poortere, O. Gunawan, Y. P. Shkolnikov, E. Tutuc, and K. Vakili, Two-dimensional electrons occupying multiple valleys in AlAs, *Phys. Stat. Sol. (b).* **243**(14), 3629–3642 (2006). URL http://dx.doi.org/10.1002/pssb.200642212.

31. M. O. Goerbig, Electronic properties of graphene in a strong magnetic field, *Rev. Mod. Phys.* **83**(4), 1193–1243 (2011). URL http://link.aps.org/doi/10.1103/RevModPhys.83.1193.

32. K. Nomura and A. H. MacDonald, Quantum Hall ferromagnetism in graphene, *Phys. Rev. Lett.* **96**(25) (2006). URL http://link.aps.org/doi/10.1103/PhysRevLett.96.256602.

33. Y. Zhang, Z. Jiang, J. P. Small, M. S. Purewal, Y.-W. Tan, M. Fazlollahi, J. D. Chudow, J. A. Jaszczak, H. L. Stormer, and P. Kim, Landau-level splitting in graphene in high magnetic fields, *Phys. Rev. Lett.* **96**(13) (2006). URL http://link.aps.org/doi/10.1103/PhysRevLett.96.136806.

34. C. R. Dean, A. F. Young, I. Meric, C. Lee, L. Wang, S. Sorgenfrei, K. Watanabe, T. Taniguchi, P. Kim, K. L. Shepard, and J. Hone, Boron nitride substrates for high-quality graphene electronics, *Nature Nanotechnol.* **5**, 722–726 (2010). URL http://dx.doi.org/10.1038/nnano.2010.172.

35. J. G. Checkelsky, L. Li, and N. P. Ong, Divergent resistance at the Dirac point in graphene: Evidence for a transition in a high magnetic field, *Phys. Rev. B.* **79**(11) (2009). URL http://link.aps.org/doi/10.1103/PhysRevB.79.115434.

36. X. Du, I. Skachko, F. Duerr, A. Luican, and E. Y. Andrei, Fractional quantum Hall effect and insulating phase of Dirac electrons in graphene, *Nature* **462**(7270), 192–195 (2009). URL http://dx.doi.org/10.1038/nature08522.

37. K. I. Bolotin, F. Ghahari, M. D. Shulman, H. L. Stormer, and P. Kim, Observation of the fractional quantum Hall effect in graphene, *Nature* **462**(7270), 196–199 (2009). URL http://dx.doi.org/10.1038/nature08582.

38. A. F. Young, C. R. Dean, L. Wang, H. Ren, P. Cadden-Zimansky, K. Watanabe, T. Taniguchi, J. Hone, K. L. Shepard, and P. Kim, Spin and valley quantum Hall ferromagnetism in graphene, *Nature Phys.* **8**(7), 550–556 (2012). ISSN 1745-2473. doi: 10.1038/nphys2307. URL https://www.nature.com/nphys/journal/v8/n7/full/nphys2307.html.

39. D. S. Wei, T. v. d. Sar, S. H. Lee, K. Watanabe, T. Taniguchi, B. I. Halperin, and A. Yacoby, Electrical generation and detection of spin waves in a quantum Hall ferromagnet, *Science* **362**(6411), 229–233 (2018). ISSN 0036-8075, 1095-9203. doi: 10.1126/science.aar4061. URL http://science.sciencemag.org/content/362/6411/229.

40. M. Kharitonov, Phase diagram for the =0 quantum Hall state in monolayer graphene, *Phys. Rev. B.* **85**(15), 155439 (2012). URL http://link.aps.org/doi/10.1103/PhysRevB.85.155439.

41. D. A. Abanin, P. A. Lee, and L. S. Levitov, Spin-filtered edge states and quantum Hall effect in graphene, *Phys. Rev. Lett.* **96**(17) (2006). URL http://link.aps.org/doi/10.1103/PhysRevLett.96.176803.

42. H. A. Fertig and L. Brey, Luttinger liquid at the edge of undoped graphene in a strong magnetic field, *Phys. Rev. Lett.* **97**(11) (2006). URL http://link.aps.org/doi/10.1103/PhysRevLett.97.116805.

43. J. Alicea and M. P. A. Fisher, Graphene integer quantum Hall effect in the ferromagnetic and paramagnetic regimes, *Phys. Rev. B.* **74**(7), 075422 (2006). URL http://link.aps.org/doi/10.1103/PhysRevB.74.075422.

44. J. Alicea and M. P. Fisher, Interplay between lattice-scale physics and the quantum Hall effect in graphene, *Solid State Commun.* **143**(11-12), 504–509 (2007). URL http://www.sciencedirect.com/science/article/B6TVW-4P718NV-2/2/254e8a74f041904d3d452619dc93fde5.

45. J. Jung and A. H. MacDonald, Theory of the magnetic-field-induced insulator in neutral graphene sheets, *Phys. Rev. B.* **80**(23) (2009). URL http://link.aps.org/doi/10.1103/PhysRevB.80.235417.

46. A. F. Young, J. D. Sanchez-Yamagishi, B. Hunt, S. H. Choi, K. Watanabe, T. Taniguchi, R. C. Ashoori, and P. Jarillo-Herrero, Tunable symmetry breaking and helical edge transport in a graphene quantum spin Hall state, *Nature* **505**(7484), 528–532 (2014). URL http://dx.doi.org/10.1038/nature12800.

47. L. Veyrat, C. Dprez, A. Coissard, X. Li, F. Gay, K. Watanabe, T. Taniguchi, Z. V. Han, B. A. Piot, H. Sellier, and B. Sacp, Helical quantum Hall phase in graphene on SrTiO_3, arXiv:1907.02299 [cond-mat] (2019). URL http://arxiv.org/abs/1907.02299. arXiv: 1907.02299.

48. B. E. Feldman, B. Krauss, J. H. Smet, and A. Yacoby, Unconventional sequence of fractional quantum Hall states in suspended graphene, *Science* **337**(6099), 1196–1199 (2012). doi: 10.1126/science.1224784. URL http://www.sciencemag.org/content/337/6099/1196.

49. A. A. Zibrov, E. M. Spanton, H. Zhou, C. Kometter, T. Taniguchi, K. Watanabe, and A. F. Young, Even-denominator fractional quantum Hall states at an isospin transition in monolayer graphene, *Nature Phys.* **14**(9), 930–935 (2018). ISSN 1745-2481. doi: 10.1038/s41567-018-0190-0. URL https://www.nature.com/articles/s41567-018-0190-0.

50. Y. Barlas, R. Ct, K. Nomura, and A. H. MacDonald, Intra-Landau-level cyclotron resonance in bilayer graphene, *Phys. Rev. Lett.* **101**(9), 097601 (2008). doi: 10.1103/PhysRevLett.101.097601. URL https://link.aps.org/doi/10.1103/PhysRevLett.101.097601.

51. D. A. Abanin, S. A. Parameswaran, and S. L. Sondhi, Charge 2e skyrmions in bilayer graphene, *Phys. Rev. Lett.* **103**(7), 076802 (2009). doi: 10.1103/PhysRevLett.103.076802. URL https://link.aps.org/doi/10.1103/PhysRevLett.103.076802.

52. J. Lambert and R. Cote, Quantum Hall ferromagnetic phases in the Landau level N=0 of a graphene bilayer, *Phys. Rev. B.* **87**(11), 115415 (2013). doi: 10.1103/PhysRevB.87.115415. URL http://link.aps.org/doi/10.1103/PhysRevB.87.115415.

53. K. Shizuya, Structure and the Lamb-shift-like quantum splitting of the pseudo-zero-mode Landau levels in bilayer graphene, *Phys. Rev. B.* **86**(4), 045431 (2012). doi: 10.1103/PhysRevB.86.045431. URL http://link.aps.org/doi/10.1103/PhysRevB.86.045431.

54. M. Kharitonov, Canted antiferromagnetic phase of the =0 quantum Hall state in bilayer graphene, *Phys. Rev. Lett.* **109**(4), 046803 (2012). doi: 10.1103/PhysRevLett. 109.046803. URL http://link.aps.org/doi/10.1103/PhysRevLett.109.046803.

55. J. Jung, F. Zhang, and A. H. MacDonald, Lattice theory of pseudospin ferromagnetism in bilayer graphene: Competing interaction-induced quantum Hall states, *Phys. Rev. B.* **83**(11), 115408 (2011). doi: 10.1103/PhysRevB.83.115408. URL https://link.aps.org/doi/10.1103/PhysRevB.83.115408.

56. R. Cote, W. Luo, B. Petrov, Y. Barlas, and A. H. MacDonald, Orbital and interlayer skyrmion crystals in bilayer graphene, *Phys. Rev. B.* **82**(24), 245307 (2010). doi: 10.1103/PhysRevB.82.245307. URL https://link.aps.org/doi/10. 1103/PhysRevB.82.245307.

57. B. E. Feldman, J. Martin, and A. Yacoby, Broken-symmetry states and divergent resistance in suspended bilayer graphene, *Nature Phys.* **5**(12), 889–893 (2009). URL http://dx.doi.org/10.1038/nphys1406.

58. R. T. Weitz, M. T. Allen, B. E. Feldman, J. Martin, and A. Yacoby, Broken-symmetry states in doubly gated suspended bilayer graphene, *Science* **330**(6005), 812–816 (2010). doi: 10.1126/science.1194988. URL http://www.sciencemag.org/content/330/6005/812.abstract.

59. Y. Zhao, P. Cadden-Zimansky, Z. Jiang, and P. Kim, Symmetry breaking in the zero-energy Landau level in bilayer graphene, *Phys. Rev. Lett.* **104**(6), 066801 (2010). doi: 10.1103/PhysRevLett.104.066801. URL https://link.aps.org/doi/10.1103/PhysRevLett.104.066801.

60. P. Maher, C. R. Dean, A. F. Young, T. Taniguchi, K. Watanabe, K. L. Shepard, J. Hone, and P. Kim, Evidence for a spin phase transition at charge neutrality in bilayer graphene, *Nature Phys.* **9**(3), 154–158 (2013). ISSN 1745-2473. doi: 10.1038/nphys2528. URL http://www.nature.com/nphys/journal/v9/n3/full/nphys2528.html.

61. J. Velasco Jr, Y. Lee, F. Zhang, K. Myhro, D. Tran, M. Deo, D. Smirnov, A. H. MacDonald, and C. N. Lau, Competing ordered states with filling factor two in bilayer graphene, *Nature Commun.* **5**, 4550 (2014). doi: 10.1038/ncomms5550. URL http://www.nature.com/ncomms/2014/140731/ncomms5550/full/ncomms5550.html.

62. A. Kou, B. E. Feldman, A. J. Levin, B. I. Halperin, K. Watanabe, T. Taniguchi, and A. Yacoby, Electron-hole asymmetric integer and fractional quantum Hall effect in bilayer graphene, *Science* **345**(6192), 55–57 (2014). doi: 10.1126/science.1250270. URL http://www.sciencemag.org/content/345/6192/55.abstract.

63. P. Maher, L. Wang, Y. Gao, C. Forsythe, T. Taniguchi, K. Watanabe, D. Abanin, Z. Papic, P. Cadden-Zimansky, J. Hone, P. Kim, and C. R. Dean, Tunable fractional quantum Hall phases in bilayer graphene, *Science* **345**(6192), 61–64 (2014). doi: 10.1126/science.1252875. URL http://www.sciencemag.org/content/345/6192/61.abstract.

64. K. Lee, B. Fallahazad, J. Xue, D. C. Dillen, K. Kim, T. Taniguchi, K. Watanabe, and E. Tutuc, Chemical potential and quantum Hall ferromagnetism in bilayer graphene, *Science* **345**(6192), 58–61 (2014). doi: 10.1126/science.1251003. URL http://www.sciencemag.org/content/345/6192/58.abstract.

65. Y. Shi, Y. Lee, S. Che, Z. Pi, T. Espiritu, P. Stepanov, D. Smirnov, C. N. Lau, and F. Zhang, Energy gaps and layer polarization of integer and fractional quantum Hall states in bilayer graphene, *Phys. Rev. Lett.* **116**(5), 056601 (2016). doi: 10.1103/PhysRevLett.116.056601. URL http://link.aps.org/doi/10.1103/PhysRevLett.116.056601.

66. A. F. Young and L. S. Levitov, Capacitance of graphene bilayer as a probe of layer-specific properties, *Phys. Rev. B.* **84**(8), 085441 (2011). doi: 10.1103/PhysRevB.84.085441. URL https://link.aps.org/doi/10.1103/PhysRevB.84.085441.

67. B. M. Hunt, J. I. A. Li, A. A. Zibrov, L. Wang, T. Taniguchi, K. Watanabe, J. Hone, C. R. Dean, M. Zaletel, R. C. Ashoori, and A. F. Young, Direct measurement of discrete valley and orbital quantum numbers in bilayer graphene, *Nature Commun.* **8**(1), 948 (2017). ISSN 2041-1723. doi: 10.1038/s41467-017-00824-w. URL https://www.nature.com/articles/s41467-017-00824-w.

68. J. K. Jain, Composite-fermion approach for the fractional quantum Hall effect, *Phys. Rev. Lett.* **63**(2), 199 (1989). URL http://link.aps.org/doi/10.1103/PhysRevLett.63.199.

69. B. I. Halperin, P. A. Lee, and N. Read, Theory of the half-filled Landau level, *Phys. Rev. B.* **47**(12), 7312–7343 (1993). URL http://link.aps.org/doi/10.1103/PhysRevB.47.7312.

70. V. W. Scarola and J. K. Jain, Phase diagram of bilayer composite fermion states, *Phys. Rev. B.* **64**(8), 085313 (2001). doi: 10.1103/PhysRevB.64.085313. URL https://link.aps.org/doi/10.1103/PhysRevB.64.085313.

71. M. Padmanabhan, T. Gokmen, and M. Shayegan, Ferromagnetic fractional quantum Hall states in a valley-degenerate two-dimensional electron system, *Phys. Rev. Lett.* **104**(1), 016805 (2010). doi: 10.1103/PhysRevLett.104.016805. URL http://link.aps.org/doi/10.1103/PhysRevLett.104.016805.

72. B. E. Feldman, A. J. Levin, B. Krauss, D. A. Abanin, B. I. Halperin, J. H. Smet, and A. Yacoby, Fractional quantum Hall phase transitions and four-flux states in graphene, *Phys. Rev. Lett.* **111**(7), 076802 (2013). doi: 10.1103/PhysRevLett.111.076802. URL http://link.aps.org/doi/10.1103/PhysRevLett.111.076802.

73. K. Lai, W. Pan, D. C. Tsui, and Y.-H. Xie, Fractional quantum Hall effect at $=2/3$ and $4/3$ in strained Si quantum wells, *Phys. Rev. B.* **69** (2004).

74. K. Yang, S. Das Sarma, and A. H. MacDonald, Collective modes and skyrmion excitations in graphene SU(4) quantum Hall ferromagnets, *Phys. Rev. B.* **74**(7) (2006). URL http://link.aps.org/doi/10.1103/PhysRevB.74.075423.

75. C. Toke and J. K. Jain, SU(4) composite fermions in graphene: Fractional quantum Hall states without analog in GaAs, *Phys. Rev. B.* **75**(24) (2007).

76. N. Shibata and K. Nomura, Coupled charge and valley excitations in graphene quantum Hall ferromagnets, *Phys. Rev. B.* **77**(23) (2008). URL http://link.aps.org/doi/10.1103/PhysRevB.77.235426.

77. V. M. Apalkov and T. Chakraborty, Fractional quantum Hall states of Dirac electrons in graphene, *Phys. Rev. Lett.* **97**(12), 126801 (2006). doi: 10.1103/PhysRevLett.97.126801. URL https://link.aps.org/doi/10.1103/PhysRevLett.97.126801.

78. N. Shibata and K. Nomura, Fractional quantum Hall effects in graphene and its bilayer, *J. Phys. Soc. Japan* **78**(10) (2009). doi: 10.1143/JPSJ.78.104708. URL http://jpsj.ipap.jp/link?JPSJ/78/104708/.

79. Z. Papic, M. O. Goerbig, and N. Regnault, Atypical fractional quantum Hall effect in graphene at filling factor 1/3, *Phys. Rev. Lett.* **105**(17) (2010). URL http://link.aps.org/doi/10.1103/PhysRevLett.105.176802.

80. F. Ghahari, Y. Zhao, P. Cadden-Zimansky, K. Bolotin, and P. Kim, Measurement of the $=1/3$ fractional quantum Hall energy gap in suspended graphene, *Phys. Rev. Lett.* **106**(4) (2011). URL http://link.aps.org/doi/10.1103/PhysRevLett.106.046801.

81. R. Ribeiro-Palau, S. Chen, Y. Zeng, K. Watanabe, T. Taniguchi, J. Hone, and C. R. Dean, High-quality electrostatically defined Hall bars in monolayer graphene, *Nano Lett.* **19**(4), 2583–2587 (2019). ISSN 1530-6984. doi: 10.1021/acs.nanolett. 9b00351. URL https://doi.org/10.1021/acs.nanolett.9b00351.

82. T. Ando, Screening effect and impurity scattering in monolayer graphene, *J. Phys. Soc. Jpn.* **75** (2006).

83. R. H. Morf, N. dAmbrumenil, and S. Das Sarma, Excitation gaps in fractional quantum Hall states: An exact diagonalization study, *Phys. Rev. B.* **66**(7), 075408 (2002). doi: 10.1103/PhysRevB.66.075408. URL https://link.aps.org/doi/10. 1103/PhysRevB.66.075408.

84. A. C. Balram, U. Wurstbauer, A. Wjs, A. Pinczuk, and J. K. Jain, Fractionally charged skyrmions in fractional quantum Hall effect, *Nature Commun.* **6**, 8981 (2015). ISSN 2041-1723. doi: 10.1038/ncomms9981. URL https://www.nature.com/ articles/ncomms9981.

85. A. F. Dethlefsen, E. Mariani, H.-P. Tranitz, W. Wegscheider, and R. J. Haug, Signatures of spin in the =1/3 fractional quantum Hall effect, *Phys. Rev. B.* **74**(16), 165325 (2006). URL http://link.aps.org/doi/10.1103/PhysRevB.74.165325.

86. F. Schulze-Wischeler, E. Mariani, F. Hohls, and R. J. Haug, Direct measurement of the g factor of composite fermions, *Phys. Rev. Lett.* **92**(15), 156401 (2004). doi: 10.1103/PhysRevLett.92.156401. URL https://link.aps.org/doi/10.1103/ PhysRevLett.92.156401.

87. A. C. Balram, C. Tke, A. Wjs, and J. K. Jain, Fractional quantum Hall effect in graphene: Quantitative comparison between theory and experiment, *Phys. Rev. B.* **92**(7), 075410 (2015). doi: 10.1103/PhysRevB.92.075410. URL https://link.aps. org/doi/10.1103/PhysRevB.92.075410.

88. N. R. Finney, M. Yankowitz, L. Muraleetharan, K. Watanabe, T. Taniguchi, C. R. Dean, and J. Hone, Tunable crystal symmetry in graphene boron nitride heterostructures with coexisting Moiré superlattices, *Nature Nanotechnol.* **14**(11), 1029–1034 (2019). ISSN 1748-3395. doi: 10.1038/s41565-019-0547-2. URL https: //www.nature.com/articles/s41565-019-0547-2.

89. S. Das Sarma and E. H. Hwang, Mobility versus quality in two-dimensional semiconductor structures, *Phys. Rev. B.* **90**(3), 035425 (2014). doi: 10.1103/PhysRevB. 90.035425. URL https://link.aps.org/doi/10.1103/PhysRevB.90.035425.

90. L. Li, F. Yang, G. J. Ye, Z. Zhang, Z. Zhu, W. Lou, X. Zhou, L. Li, K. Watanabe, T. Taniguchi, K. Chang, Y. Wang, X. H. Chen, and Y. Zhang, Quantum Hall effect in black phosphorus two-dimensional electron system, *Nature Nanotechnol.* **11**(7), 593–597 (2016). ISSN 1748-3395. doi: 10.1038/nnano.2016.42. URL https://www. nature.com/articles/nnano.2016.42.

91. D. A. Bandurin, A. V. Tyurnina, G. L. Yu, A. Mishchenko, V. Zlyomi, S. V. Morozov, R. K. Kumar, R. V. Gorbachev, Z. R. Kudrynskyi, S. Pezzini, Z. D. Kovalyuk, U. Zeitler, K. S. Novoselov, A. Patan, L. Eaves, I. V. Grigorieva, V. I. Fal'ko, A. K. Geim, and Y. Cao, High electron mobility, quantum Hall effect and anomalous optical response in atomically thin InSe, *Nature Nanotechnol.* **12**(3), 223–227 (2017). ISSN 1748-3395. doi: 10.1038/nnano.2016.242. URL https://www.nature. com/articles/nnano.2016.242.

92. X. Cui, G.-H. Lee, Y. D. Kim, G. Arefe, P. Y. Huang, C.-H. Lee, D. A. Chenet, X. Zhang, L. Wang, F. Ye, F. Pizzocchero, B. S. Jessen, K. Watanabe, T. Taniguchi, D. A. Muller, T. Low, P. Kim, and J. Hone, Multi-terminal transport measurements of MoS2 using a van der Waals heterostructure device platform, *Nature Nanotechnol.* **10**(6), 534–540 (2015). ISSN 1748-3387. doi: 10.1038/nnano.2015.70. URL http://www.nature.com/nnano/journal/v10/n6/abs/nnano.2015.70.html.

93. A. F. Young, C. R. Dean, I. Meric, S. Sorgenfrei, H. Ren, K. Watanabe, T. Taniguchi, J. Hone, K. L. Shepard, and P. Kim, Electronic compressibility of layer-polarized bilayer graphene, *Phys. Rev. B.* **85**(23), 235458 (2012). doi: 10.1103/PhysRevB.85.235458. URL http://link.aps.org/doi/10.1103/PhysRevB.85.235458.

94. L. Wang, I. Meric, P. Y. Huang, Q. Gao, Y. Gao, H. Tran, T. Taniguchi, K. Watanabe, L. M. Campos, D. A. Muller, J. Guo, P. Kim, J. Hone, K. L. Shepard, and C. R. Dean, One-dimensional electrical contact to a two-dimensional material, *Science* **342**(6158), 614–617 (2013). URL http://www.sciencemag.org/content/342/6158/614.

95. A. A. Zibrov, C. Kometter, H. Zhou, E. M. Spanton, T. Taniguchi, K. Watanabe, M. P. Zaletel, and A. F. Young, Tunable interacting composite fermion phases in a half-filled bilayer-graphene Landau level, *Nature* **549**(7672), 360–364 (2017). ISSN 0028-0836. doi: 10.1038/nature23893. URL http://www.nature.com/nature/journal/v549/n7672/full/nature23893.html?foxtrotcallback=true.

96. J. I. A. Li, C. Tan, S. Chen, Y. Zeng, T. Taniguchi, K. Watanabe, J. Hone, and C. R. Dean, Even denominator fractional quantum Hall states in bilayer graphene, *Science* p. eaao2521 (2017). ISSN 0036-8075, 1095-9203. doi: 10.1126/science.aao2521. URL http://science.sciencemag.org/content/early/2017/10/04/science.aao2521.

97. B. A. Schmidt, K. Bennaceur, S. Bilodeau, G. Gervais, L. N. Pfeiffer, and K. W. West, Second Landau level fractional quantum Hall effects in the Corbino geometry, *Solid State Commun.* **217**, 1–5 (2015). ISSN 0038-1098. doi: 10.1016/j.ssc.2015.05.005. URL http://www.sciencedirect.com/science/article/pii/S0038109815001660.

98. J. Yan and M. S. Fuhrer, Charge transport in dual gated bilayer graphene with Corbino geometry, *Nano Lett.* **10**(11), 4521–4525 (2010). doi: 10.1021/nl102459t. URL http://dx.doi.org/10.1021/nl102459t.

99. Y. Zhao, P. Cadden-Zimansky, F. Ghahari, and P. Kim, Magnetoresistance measurements of graphene at the charge neutrality point, *Phys. Rev. Lett.* **108**(10), 106804 (2012). doi: 10.1103/PhysRevLett.108.106804. URL https://link.aps.org/doi/10.1103/PhysRevLett.108.106804.

100. E. C. Peters, A. J. M. Giesbers, M. Burghard, and K. Kern, Scaling in the quantum Hall regime of graphene Corbino devices, *Appl. Phys. Lett.* **104**(20), 203109 (2014). ISSN 0003-6951. doi: 10.1063/1.4878396. URL https://aip.scitation.org/doi/full/10.1063/1.4878396.

101. M. J. Zhu, A. V. Kretinin, M. D. Thompson, D. A. Bandurin, S. Hu, G. L. Yu, J. Birkbeck, A. Mishchenko, I. J. Vera-Marun, K. Watanabe, T. Taniguchi, M. Polini, J. R. Prance, K. S. Novoselov, A. K. Geim, and M. Ben Shalom, Edge currents shunt the insulating bulk in gapped graphene, *Nature Commun.* **8**, 14552 (2017). ISSN 2041-1723. doi: 10.1038/ncomms14552. URL https://www.nature.com/articles/ncomms14552.

102. M. Kumar, A. Laitinen, and P. Hakonen, Unconventional fractional quantum Hall states and Wigner crystallization in suspended Corbino graphene, *Nature Commun.* **9**(1), 1–8 (2018). ISSN 2041-1723. doi: 10.1038/s41467-018-05094-8. URL https://www.nature.com/articles/s41467-018-05094-8.

103. J. Weis and K. von Klitzing, Metrology and microscopic picture of the integer quantum Hall effect, *Philos. Trans. Roy. Soc. A: Math. Phys. Engin. Sci.* **369**(1953), 3954–3974 (2011). doi: 10.1098/rsta.2011.0198. URL https://royalsocietypublishing.org/doi/10.1098/rsta.2011.0198.

104. Y.-T. Cui, B. Wen, E. Y. Ma, G. Diankov, Z. Han, F. Amet, T. Taniguchi, K. Watanabe, D. Goldhaber-Gordon, C. R. Dean, and Z.-X. Shen, Unconventional correlation between quantum Hall transport quantization and bulk state filling in gated graphene devices, *Phys. Rev. Lett.* **117**(18), 186601 (2016). doi: 10.1103/PhysRevLett.117.186601. URL http://link.aps.org/doi/10.1103/PhysRevLett.117.186601.

105. D. Halbertal, M. B. Shalom, A. Uri, K. Bagani, A. Y. Meltzer, I. Marcus, Y. Myasoedov, J. Birkbeck, L. S. Levitov, A. K. Geim, and E. Zeldov, Imaging resonant dissipation from individual atomic defects in graphene, *Science* **358**(6368), 1303–1306 (2017). ISSN 0036-8075, 1095-9203. doi: 10.1126/science.aan0877. URL http://science.sciencemag.org/content/358/6368/1303.

106. R. Sabo, I. Gurman, A. Rosenblatt, F. Lafont, D. Banitt, J. Park, M. Heiblum, Y. Gefen, V. Umansky, and D. Mahalu, Edge reconstruction in fractional quantum Hall states, *Nature Phys.* **13**(5), 491–496 (2017). ISSN 1745-2481. doi: 10.1038/nphys4010. URL https://www.nature.com/articles/nphys4010.

107. D.-K. Ki, V. I. Falko, D. A. Abanin, and A. F. Morpurgo, Observation of even denominator fractional quantum Hall effect in suspended bilayer graphene, *Nano Lett.* **14**(4), 2135–2139 (2014). ISSN 1530-6984. doi: 10.1021/nl5003922. URL http://dx.doi.org/10.1021/nl5003922.

108. M. Barkeshli, C. Nayak, Z. Papi, A. Young, and M. Zaletel, Topological exciton Fermi surfaces in two-component fractional quantized Hall insulators, *Phys. Rev. Lett.* **121**(2), 026603 (2018). doi: 10.1103/PhysRevLett.121.026603. URL https://link.aps.org/doi/10.1103/PhysRevLett.121.026603.

109. V. M. Apalkov and T. Chakraborty, Stable Pfaffian state in bilayer graphene, *Phys. Rev. Lett.* **107**(18), 186803 (2011). doi: 10.1103/PhysRevLett.107.186803. URL https://link.aps.org/doi/10.1103/PhysRevLett.107.186803.

110. Z. Papic and D. A. Abanin, Topological phases in the zeroth Landau level of bilayer graphene, *Phys. Rev. Lett.* **112**(4), 046602 (2014). doi: 10.1103/PhysRevLett.112.046602. URL http://link.aps.org/doi/10.1103/PhysRevLett.112.046602.

111. M. A. Metlitski, D. F. Mross, S. Sachdev, and T. Senthil, Cooper pairing in non-Fermi liquids, *Phys. Rev. B.* **91**(11), 115111 (2015). doi: 10.1103/PhysRevB.91.115111. URL https://link.aps.org/doi/10.1103/PhysRevB.91.115111.

112. E. H. Rezayi and S. H. Simon, Breaking of particle-hole symmetry by Landau level mixing in the =5/2 quantized Hall state, *Phys. Rev. Lett.* **106**(11), 116801 (2011). doi: 10.1103/PhysRevLett.106.116801. URL http://link.aps.org/doi/10.1103/PhysRevLett.106.116801.

113. M. P. Zaletel, R. S. K. Mong, F. Pollmann, and E. H. Rezayi, Infinite density matrix renormalization group for multicomponent quantum Hall systems, *Phys. Rev. B.* **91**(4), 045115 (2015). doi: 10.1103/PhysRevB.91.045115. URL https://link.aps.org/doi/10.1103/PhysRevB.91.045115.

114. E. H. Rezayi, Landau-level-mixing and the ground state of the 5/2 quantum Hall effect, arXiv:1704.03026 [cond-mat] (2017). URL http://arxiv.org/abs/1704.03026. arXiv: 1704.03026.

115. A. Kumar, G. A. Csathy, M. J. Manfra, L. N. Pfeiffer, and K. W. West, Nonconventional odd-denominator fractional quantum Hall states in the second Landau level, *Phys. Rev. Lett.* **105**(24) (2010). URL http://link.aps.org/doi/10.1103/PhysRevLett.105.246808.

116. M. Levin and B. I. Halperin, Collective states of non-Abelian quasiparticles in a magnetic field, *Phys. Rev. B.* **79**(20), 205301 (2009). doi: 10.1103/PhysRevB.79.205301. URL https://link.aps.org/doi/10.1103/PhysRevB.79.205301.

117. J. P. Eisenstein, H. L. Stormer, L. Pfeiffer, and K. W. West, Evidence for a phase transition in the fractional quantum Hall effect, *Phys. Rev. Lett.* **62**(13) (1989). URL http://link.aps.org/doi/10.1103/PhysRevLett.62.1540.

118. R. H. Morf, Transition from quantum Hall to compressible states in the second Landau level: New light on the =5/2 enigma, *Phys. Rev. Lett.* **80**(7), 1505–1508 (1998). doi: 10.1103/PhysRevLett.80.1505. URL http://link.aps.org/doi/10.1103/PhysRevLett.80.1505.

119. M. R. Peterson, T. Jolicoeur, and S. Das Sarma, Finite-layer thickness stabilizes the Pfaffian state for the 5/2 fractional quantum Hall effect: Wave function overlap and topological degeneracy, *Phys. Rev. Lett.* **101**(1) (2008). URL http://link.aps.org/doi/10.1103/PhysRevLett.101.016807.

120. G. Moore and N. Read, Non-Abelions in the fractional quantum Hall effect, *Nucl. Phys. B.* **360**(2-3), 362–396 (1991). URL http://www.sciencedirect.com/science/article/pii/055032139190407O.

121. M. Greiter, X.-G. Wen, and F. Wilczek, Paired Hall state at half filling, *Phys. Rev. Lett.* **66**(24), 3205–3208 (1991). doi: 10.1103/PhysRevLett.66.3205. URL https://link.aps.org/doi/10.1103/PhysRevLett.66.3205.

122. M. Levin, B. I. Halperin, and B. Rosenow, Particle-hole symmetry and the Pfaffian state, *Phys. Rev. Lett.* **99**(23), 236806 (2007). doi: 10.1103/PhysRevLett.99.236806. URL https://link.aps.org/doi/10.1103/PhysRevLett.99.236806.

123. S.-S. Lee, S. Ryu, C. Nayak, and M. P. A. Fisher, Particle-hole symmetry and the =5/2 quantum Hall state, *Phys. Rev. Lett.* **99**(23), 236807 (2007). doi: 10.1103/PhysRevLett.99.236807. URL https://link.aps.org/doi/10.1103/PhysRevLett.99.236807.

124. C. Nayak, S. H. Simon, A. Stern, M. Freedman, and S. Das Sarma, Non-Abelian anyons and topological quantum computation, *Rev. Mod. Phys.* **80**(3), 1083–1159 (2008). URL http://link.aps.org/doi/10.1103/RevModPhys.80.1083.

125. C. von Keyserlingk, S. Simon, and B. Rosenow, Enhanced bulk-edge Coulomb coupling in fractional Fabry–Perot interferometers, *Phys. Rev. Lett.* **115**(12), 126807 (2015). doi: 10.1103/PhysRevLett.115.126807. URL http://link.aps.org/doi/10.1103/PhysRevLett.115.126807.

126. D. S. Wei, T. van der Sar, J. D. Sanchez-Yamagishi, K. Watanabe, T. Taniguchi, P. Jarillo-Herrero, B. I. Halperin, and A. Yacoby, Mach–Zehnder interferometry using spin- and valley-polarized quantum Hall edge states in graphene, arXiv:1703.00110 [cond-mat] (2017). URL http://arxiv.org/abs/1703.00110. arXiv: 1703.00110.

127. N. R. Cooper and A. Stern, Observable bulk signatures of non-Abelian quantum Hall states, *Phys. Rev. Lett.* **102**(17), 176807 (2009). doi: 10.1103/PhysRevLett.102.176807. URL http://link.aps.org/doi/10.1103/PhysRevLett.102.176807.

128. X.-G. Wen and A. Zee, Neutral superfluid modes and "magnetic" monopoles in multilayered quantum Hall systems, *Phys. Rev. Lett.* **69**(12), 1811–1814 (1992). doi: 10.1103/PhysRevLett.69.1811. URL https://link.aps.org/doi/10.1103/PhysRevLett.69.1811.

129. M. Kellogg, J. P. Eisenstein, L. N. Pfeiffer, and K. W. West, Vanishing Hall resistance at high magnetic field in a double-layer two-dimensional electron system, *Phys. Rev. Lett.* **93**(3), 036801 (2004). doi: 10.1103/PhysRevLett.93.036801. URL https://link.aps.org/doi/10.1103/PhysRevLett.93.036801.

130. E. Tutuc, M. Shayegan, and D. A. Huse, Counterflow measurements in strongly correlated GaAs hole bilayers: Evidence for electron-hole pairing, *Phys. Rev. Lett.* **93**(3), 036802 (2004). doi: 10.1103/PhysRevLett.93.036802. URL https://link.aps.org/doi/10.1103/PhysRevLett.93.036802.

131. J. Eisenstein, Exciton condensation in bilayer quantum Hall systems, *Ann. Rev. Condens. Matt. Phys.* **5**(1), 159–181 (2014). doi: 10.1146/annurev-conmatphys-031113-133832. URL http://dx.doi.org/10.1146/annurev-conmatphys-031113-133832.

132. X. Liu, K. Watanabe, T. Taniguchi, B. I. Halperin, and P. Kim, Quantum Hall drag of exciton condensate in graphene, *Nature Phys.* **13**(8), 746–750 (2017). ISSN 1745-2481. doi: 10.1038/nphys4116. URL https://www.nature.com/articles/nphys4116.

133. A. R. Champagne, A. D. K. Finck, J. P. Eisenstein, L. N. Pfeiffer, and K. W. West, Charge imbalance and bilayer two-dimensional electron systems at $\nu_T = 1$, *Phys. Rev. B.* **78**(20), 205310 (2008). doi: 10.1103/PhysRevB.78.205310. URL https://link.aps.org/doi/10.1103/PhysRevB.78.205310.

134. J. I. A. Li, Q. Shi, Y. Zeng, K. Watanabe, T. Taniguchi, J. Hone, and C. R. Dean, Evidence for pairing states of composite fermions in double-layer graphene, arXiv:1901.03480 [cond-mat] (2019). URL http://arxiv.org/abs/1901.03480. arXiv: 1901.03480.

135. J. Bardeen, L. N. Cooper, and J. R. Schrieffer, Theory of superconductivity, *Phys. Rev.* **108**(5), 1175–1204 (1957). doi: 10.1103/PhysRev.108.1175. URL https://link.aps.org/doi/10.1103/PhysRev.108.1175.

136. M. Randeria and E. Taylor, Crossover from Bardeen–Cooper–Schrieffer to Bose–Einstein condensation and the unitary Fermi gas, *Ann. Rev. Condens. Matt. Phys.* **5**(1), 209–232 (2014). ISSN 1947-5454. doi: 10.1146/annurev-conmatphys-031113-133829. URL https://www.annualreviews.org/doi/10.1146/annurev-conmatphys-031113-133829.

137. A. J. Leggett and S. Zhang. The BEC-BCS crossover: Some history and some general observations. In ed. W. Zwerger, *The BCS-BEC Crossover and the Unitary Fermi Gas*, Lecture Notes in Physics, Springer, Berlin, Heidelberg, (2012), pp. 33–47. ISBN 978-3-642-21978-8. doi: 10.1007/978-3-642-21978-8_2. URL https://doi.org/10.1007/978-3-642-21978-8_2.

138. J. M. Blatt, K. W. Ber, and W. Brandt, Bose–Einstein condensation of excitons, *Phys. Rev.* **126**(5), 1691–1692 (1962). doi: 10.1103/PhysRev.126.1691. URL https://link.aps.org/doi/10.1103/PhysRev.126.1691.

139. E. Ardonne, F. J. M. v. Lankvelt, A. W. W. Ludwig, and K. Schoutens, Separation of spin and charge in paired spin-singlet quantum Hall states, *Phys. Rev. B.* **65**(4), 041305 (2002). doi: 10.1103/PhysRevB.65.041305. URL https://link.aps.org/doi/10.1103/PhysRevB.65.041305.

140. M. Barkeshli and X.-G. Wen, Non-Abelian two-component fractional quantum Hall states, *Phys. Rev. B.* **82**(23), 233301 (2010). doi: 10.1103/PhysRevB.82.233301. URL https://link.aps.org/doi/10.1103/PhysRevB.82.233301.

141. S. Geraedts, M. P. Zaletel, Z. Papi, and R. S. K. Mong, Competing Abelian and non-Abelian topological orders in = 1/3+1/3 quantum Hall bilayers, arXiv:1502.01340 [cond-mat] (2015). URL http://arxiv.org/abs/1502.01340. arXiv: 1502.01340.

142. S. Mukherjee, S. S. Mandal, A. Wjs, and J. K. Jain, Possible anti-Pfaffian pairing of composite fermions at $\nu = 3/8$, *Phys. Rev. Lett.* **109**(25), 256801 (2012). doi: 10.1103/PhysRevLett.109.256801. URL https://link.aps.org/doi/10.1103/PhysRevLett.109.256801.

143. S. Mukherjee, J. K. Jain, and S. S. Mandal, Possible realization of a chiral p-wave paired state in a two-component system, *Phys. Rev. B.* **90**(12), 121305 (2014). doi: 10.1103/PhysRevB.90.121305. URL https://link.aps.org/doi/10.1103/PhysRevB.90.121305.

144. S. R. Renn, Edge excitations and the fractional quantum Hall effect in double quantum wells, *Phys. Rev. Lett.* **68**(5), 658–661 (1992). doi: 10.1103/PhysRevLett.68.658. URL https://link.aps.org/doi/10.1103/PhysRevLett.68.658.

145. G. Li, A. Luican, and E. Y. Andrei, Scanning tunneling spectroscopy of graphene on graphite, *Phys. Rev. Lett.* **102**(17), 176804 (2009). doi: 10.1103/PhysRevLett.102.176804. URL http://link.aps.org/doi/10.1103/PhysRevLett.102.176804.

146. A. Luican, G. Li, A. Reina, J. Kong, R. R. Nair, K. S. Novoselov, A. K. Geim, and E. Y. Andrei, Single-layer behavior and its breakdown in twisted graphene layers, *Phys. Rev. Lett.* **106**(12), 126802 (2011). doi: 10.1103/PhysRevLett.106.126802. URL https://link.aps.org/doi/10.1103/PhysRevLett.106.126802.

147. J. Xue, J. Sanchez-Yamagishi, D. Bulmash, P. Jacquod, A. Deshpande, K. Watanabe, T. Taniguchi, P. Jarillo-Herrero, and B. J. LeRoy, Scanning tunnelling microscopy and spectroscopy of ultra-flat graphene on hexagonal boron nitride, *Nature Materials* **10**(4), 282–285 (2011). URL http://dx.doi.org/10.1038/nmat2968.

148. M. Yankowitz, J. Xue, D. Cormode, J. D. Sanchez-Yamagishi, K. Watanabe, T. Taniguchi, P. Jarillo-Herrero, P. Jacquod, and B. J. LeRoy, Emergence of superlattice Dirac points in graphene on hexagonal boron nitride, *Nature Phys.* **8**(5), 382–386 (2012). URL http://dx.doi.org/10.1038/nphys2272.

149. M. Y. Azbel, Energy spectrum of a conduction electron in a magnetic field, *Sov. Phys. JETP.* **19** (1964).

150. J. Zak, Magnetic translation group, *Phys. Rev.* **134**(6A), A1602–A1606 (1964). doi: 10.1103/PhysRev.134.A1602. URL https://link.aps.org/doi/10.1103/PhysRev.134.A1602.

151. P. G. Harper, The general motion of conduction electrons in a uniform mMagnetic field, with application to the diamagnetism of metals, *Proc. Phys. Soc. Section A.* **68**(10), 879–892 (1955). ISSN 0370-1298. doi: 10.1088/0370-1298/68/10/305. URL https://doi.org/10.1088%2F0370-1298%2F68%2F10%2F305.

152. D. R. Hofstadter, Energy levels and wave functions of Bloch electrons in rational and irrational magnetic fields, *Phys. Rev. B.* **14**(6), 2239–2249 (1976). URL http://link.aps.org/doi/10.1103/PhysRevB.14.2239.

153. C. R. Dean, L. Wang, P. Maher, C. Forsythe, F. Ghahari, Y. Gao, J. Katoch, M. Ishigami, P. Moon, M. Koshino, T. Taniguchi, K. Watanabe, K. L. Shepard, J. Hone, and P. Kim, Hofstadter's butterfly and the fractal quantum Hall effect in Moiré superlattices, *Nature* **497**(7451), 598–602 (2013). URL http://dx.doi.org/10.1038/nature12186.

154. L. A. Ponomarenko, R. V. Gorbachev, G. L. Yu, D. C. Elias, R. Jalil, A. A. Patel, A. Mishchenko, A. S. Mayorov, C. R. Woods, J. R. Wallbank, M. Mucha-Kruczynski, B. A. Piot, M. Potemski, I. V. Grigorieva, K. S. Novoselov, F. Guinea, V. I. Fal'ko, and A. K. Geim, Cloning of Dirac fermions in graphene superlattices, *Nature* **497**(7451), 594–597 (2013). URL http://dx.doi.org/10.1038/nature12187.

155. G. L. Yu, R. V. Gorbachev, J. S. Tu, A. V. Kretinin, Y. Cao, R. Jalil, F. Withers, L. A. Ponomarenko, B. A. Piot, M. Potemski, D. C. Elias, X. Chen, K. Watanabe, T. Taniguchi, I. V. Grigorieva, K. S. Novoselov, V. I. Falko, A. K. Geim, and A. Mishchenko, Hierarchy of Hofstadter states and replica quantum Hall ferromagnetism in graphene superlattices, *Nature Phys.* **10**(7), 525–529 (2014). ISSN 1745-2473. doi: 10.1038/nphys2979. URL http://www.nature.com/nphys/journal/v10/n7/full/nphys2979.html.

156. E. I. Blount, Bloch electrons in a magnetic field, *Phys. Rev.* **126**(5), 1636–1653 (1962). doi: 10.1103/PhysRev.126.1636. URL https://link.aps.org/doi/10.1103/PhysRev.126.1636.

157. G. H. Wannier, Dynamics of band electrons in electric and magnetic fields, *Rev. Mod. Phys.* **34**(4), 645–655 (1962). URL http://link.aps.org/doi/10.1103/RevModPhys.34.645.

158. G. H. Wannier, A result not dependent on rationality for Bloch eElectrons in a magnetic field, *Phys. Stat. Sol. (b).* **88**(2), 757–765 (1978). URL http://dx.doi.org/10.1002/pssb.2220880243.

159. P. Streda, Quantised Hall effect in a two-dimensional periodic potential, *J. Phys. C: Solid State Phys.* **15** (1982).

160. D. J. Thouless, M. Kohmoto, M. P. Nightingale, and M. den Nijs, Quantized Hall conductance in a two-dimensional periodic potential, *Phys. Rev. Lett.* **49**(6) (1982). URL http://link.aps.org/doi/10.1103/PhysRevLett.49.405.

161. A. H. MacDonald, Landau-level subband structure of electrons on a square lattice, *Phys. Rev. B.* **28**(12), 6713–6717 (1983). URL http://link.aps.org/doi/10.1103/PhysRevB.28.6713.

162. B. Cheng, C. Pan, S. Che, P. Wang, Y. Wu, K. Watanabe, T. Taniguchi, S. Ge, R. Lake, D. Smirnov, C. N. Lau, and M. Bockrath, Fractional and symmetry-broken Chern insulators in tunable Moiré superlattices, *Nano Lett.* **19**(7), 4321–4326 (2019). ISSN 1530-6984. doi: 10.1021/acs.nanolett.9b00811. URL https://doi.org/10.1021/acs.nanolett.9b00811.

163. R. B. Laughlin, Anomalous quantum Hall effect: An incompressible quantum fluid with fractionally charged excitations, *Phys. Rev. Lett.* **50**(18), 1395–1398 (1983). URL http://link.aps.org/doi/10.1103/PhysRevLett.50.1395.

164. A. Kol and N. Read, Fractional quantum Hall effect in a periodic potential, *Phys. Rev. B.* **48**(12), 8890–8898 (1993). doi: 10.1103/PhysRevB.48.8890. URL http://link.aps.org/doi/10.1103/PhysRevB.48.8890.

165. E. J. Bergholtz and Z. Liu, Topological flat band models and fractional chern insulators, *Int. J. Mod. Phys. B.* **27**(24), 1330017 (2013). ISSN 0217-9792. doi: 10.1142/S021797921330017X. URL https://www.worldscientific.com/doi/abs/10.1142/S021797921330017X.

166. T. Neupert, C. Chamon, T. Iadecola, L. H. Santos, and C. Mudry, Fractional (Chern and topological) insulators, *Physica Scripta.* **T164**, 014005 (2015). ISSN 1402-4896. doi: 10.1088/0031-8949/2015/T164/014005. URL https://doi.org/10.1088%2F0031-8949%2F2015%2Ft164%2F014005.

Chapter 8

Wavefunctionology: The Special Structure of Certain Fractional Quantum Hall Wavefunctions

Steven H. Simon

Rudolf Peierls Centre for Theoretical Physics,
Clarendon Laboratory, Parks Road,
Oxford, OX1 3PU, United Kingdom
steven.simon@physics.ox.ac.uk

Certain fractional quantum Hall wavefunctions — particularly including the Laughlin, Moore–Read, and Read–Rezayi wavefunctions — have special structure that makes them amenable to analysis using an exeptionally wide range of techniques including conformal field theory (CFT), thin cylinder or torus limit, study of symmetric polynomials and Jack polynomials, and so-called "special" parent Hamiltonians. This review discusses these techniques as well as explaining to what degree some other quantum Hall wavefunctions share this special structure. Along the way we will explore the physics of quantum Hall edges, entanglement spectra, quasiparticles, non-Abelian braiding statistics, and Hall viscosity, among other topics. As compared to a number of other recent reviews, most of this review is written so as to *not* rely on results from conformal field theory — although a short discussion of a few key relations to CFT are included near the end.

Contents

1. Introduction

The study of fractional quantum Hall effect (FQHE) is unusual within condensed matter in that we are often able to write down explicit wavefunctions that are extremely accurate representations of what we believe is realized in experiment. Starting with the seminal work of Laughlin,[1] the field has often progressed by constructing zero-parameter "variational" trial wavefunctions. While these wavefunctions are usually not exact descriptions of any experiment, they are often extremely precise approximations. More importantly these trial wavefunctions are believed to describe the same universal physics as is observed in experiment. A modern approach to this relation is to state that the experimental system and the trial wavefunction may be deformed into each other without closing the excitation gap — thus meaning that they are in the same (topological) phase of matter.

The study of these wavefunctions — their mathematical structure, their excitations, their Hamiltonians, their entanglement properties, the braiding statistics of their quasiparticles, and so forth — has become a huge endeavor. Even though this endeavor has been pursued for almost forty years, even now new ideas are being uncovered about the structure and properties of these wavefunctions. The purpose of this chapter is to review a number of the important advances in understanding that have occurred in the last few decades.

Since the seminal work of Moore and Read,[2] conformal field theory (CFT) has played a central role in the understanding of FQHE. Despite this fact, many of the important results and ideas, although closely related to CFT, can be exposited without detailed discussion of CFT. Particularly since the application of CFT to FQHE has been reviewed elsewhere recently,[3,4] the intent of this review is instead to describes some of the interesting new material with minimal explicit use of CFT. Hence this review should complementary to other already existing reviews. (That said, in the final section of this chapter (Sec. 10), I will turn to briefly discuss a few connections to CFT that are crucial enough, or are modern enough, to warrent additional discussion here.)

Much of this review will focus on the Laughlin states,[1] the Moore–Read state,[5] and the Read–Rezayi series.[6] These wavefunctions are special for several reasons. Perhaps most importantly, they arise as the ground state of simple special parent Hamiltonians such that the ground state as well as the quasihole excitations and edge excitations, have exactly zero interaction energy (see Sec. 3). This property greatly simplifies their understanding. Further, not only the ground state, but also all of the zero interaction energy states (edge excitations and quasiholes) may be explicitly represented as CFT correlators (see Sec. 10).

There will be also be some discussion of other wavefunctions, such as the Gaffnian[7] (and also the Haffnian[8]) which also have very simple special parent Hamiltonians. As we will discuss further these wavefunctions do not correspond to gapped phases of matter, but are nonetheless intseresting. Many other wavefunctions sharing at least some of the nice properties of these wavefunctions can also be constructed using Jack polynomials[9] which we discuss in Sec. 6.

We will also briefly discuss a few aspects of the experimentally important hierarchy[10,11] (or composite fermion[12]) phases of matter although this topic is somewhat de-emphasized given the recent review on hierarchy physics.[4]

We will begin our discussion with some quantum Hall basics in Sec. 2. We introduce the structure of the lowest Landau level in the sphere, cylinder, and disk geometries. We then describe the properties we should demand of our many-body wavefunctions in Sec. 2.2.

In Sec. 3 we discuss parent and "special" Hamiltonians. We begin by discussing in Sec. 3.1 how the Laughlin wavefunction can be constructed as the ground state of a special Hamiltonian. We discuss special parent Hamiltonians in Sec. 3.2 for the Read–Rezayi series with special attention on the Moore–Read state. In Sec. 3.3 the extent to which these ideas can be extended to other wavefunctions is discussed.

Once we have described special parent Hamiltonians for quantum Hall states it is easy to discuss the space of edge excitations as being wavefunctions that do not increase the interaction energy. In Sec. 4 we explore this approach to edge physics in general. We start with the Laughlin edge in Sec. 4.1, and explain how this corresponds to a one-dimensional Bose theory. We also discuss how the inner product between edge state excitations can be described using an effective bosonic edge theory in Sec. 4.1.3. In Sec.4.2 we (briefly) study the edge of hierarchy states. Then in Sec. 4.3 we discuss the more complicated edge of the Moore–Read state (with brief discussion of Read–Rezayi edges).

In Sec. 5 we introduce the idea of a thin-cylinder limit, which simplifies a lot of the intuition about certain quantum Hall states — such as finding the ground state degeneracy on a torus. We show in Sec. 5.1 how the thin limit allows us to determine the edge spectrum for even complicated wavefunctions corresponding to special parent Hamiltonians. In Sec. 6 we explain how the thin-limit is closely related to the Jack polynomial approach to writing wavefunctions for bulk systems.

In Sec. 7 we invoke ideas from quantum information theory and explain the idea of an entanglement spectrum and explain how that is related to the edge spectrum.

In Sec. 8 we turn to describe wavefunctions involving localized quasiholes, beginning with the Laughlin state. First we discuss how the localized quasiholes can be thought of as a superposition of edge excitations. We then move on to the Moore–Read and Read–Rezayi states. The existence of many degenerate quasihole wavefunctions in these cases is a signature of their nonabelian statistics, which we describe briefly in Sec. 8.1. We then describe how nonabelian braiding statistics can be calculated in Sec. 8.1.1 as a combination of monodromy and Berry matrix, and explain the importance of working with an orthonormal holomorphic basis of wavefunctions in Sec. 8.1.2.

In Sec. 9 we discuss Hall viscosity and explain both its relation to Berry phase and also its relation to more conventional fluid dynamical response functions.

In Sec. 10 we turn to give a few comments on conformal field theory and its relationship to the states discussed here. We emphasize the idea of bulk-edge correspondence, and we discuss the importance of conformal blocks and the key issue of their orthonormality. In Sec. 10.2 we discuss what is special about the CFTs that produce well known quantum Hall states.

Finally, in Sec. 11 we offer some brief concluding remarks.

2. Some FQHE Basics

2.1. *Single particle physics*

We begin with some obligatory basics of FQHE, mainly to establish notation and language (see Refs. 12–14 for more details of basics). We consider a two dimensional gas of particles in a high magentic field B perpendicular to the plane of the sample. These particles, with some effective mass m, may be fermions (like electrons, as in actual FQHE experiments to date) or they may be bosons (which may someday[a] be engineered to produce FQHE, see chapter by Nigel Cooper in this volume or Ref. 17). Note that it is a common abuse of nomenclature to call the particles constituting our system "electrons", independent of whether they are bosons or fermions.

The particles interact with each other via some interaction potential (to be specified in more detail below) and we will often confine the particles to a droplet by applying some potential $U(\mathbf{r})$. For simplicity we will ignore any additional degrees of freedom which may exist in certain experiments, such a particle spin (which we can assume is polarized) or valley (as in graphene, see chapter by Dean, Kim, Li, and Young in this volume), or layer index (see Ref. 18). Our Hamiltonian (our "theory of everything") can then be written as

$$H = \left(\sum_i \frac{[\mathbf{p}_i - e\mathbf{A}(\mathbf{r}_i)]^2}{2m^*} \right) + \left(\sum_i U(\mathbf{r_i}) \right) + H_{\text{interaction}}$$

The first term in brackets, the kinetic energy, we assume is the largest scale in the problem. The single particle spectrum of this kinetic energy term breaks into highly degenerate Landau levels with energies $E_n = \hbar\omega_c(n+1/2)$ with $\omega_c = eB/m^*$ being the cyclotron frequency given the effective mass m^*. We assume that a single

---

[a]There is one experimental preprint claiming to have produced FQHE with cold atoms.[15] While this is an extremely interesting work, it has still not been confirmed or followed up. Very recent work[16] has convincingly observed a Laughlin state of two bosons of light interacting via coupling to Rydberg atoms.

Landau level is partially filled, which, without loss of generality we can assume is the lowest Landau level (i.e. $n = 0$).[b]

If we work in symmetric gauge in a planar geometry,[12–14] in the absence of a confining potential U, the single particle eigenstates (of the kinetic term of the Hamiltonian) in the lowest Landau level can be written as

$$\varphi_m(z) \propto z^m \; \mu(z, \bar{z}) \tag{1}$$

with

$$\mu(z, \bar{z}) = e^{-|z|^2/4} \tag{2}$$

where here $z = x + iy$ is the complex representation of the particle position in the plane, and we have set the magnetic length $\ell = \sqrt{\hbar/(eB)}$ to unity. Here, m is the angular momentum of the eigenstate around the origin (the complex phase wrapping m times) and the shape of the wavefunction is roughly an annulus having radius $\sqrt{2m}$ and thickness roughly unity. Thus, the radius is linked to the angular momentum of the eigenstate.

Often we consider fractional quantum Hall systems on surfaces other than a plane. On the sphere,[10] one places a monopole with magentic flux N_ϕ at the center of the sphere, and the lowest Landau level then has exactly $N_\phi + 1$ orbitals. Using[20] stereographic projection from the sphere to the plane, $z = 2R\tan(\theta/2)e^{i\phi}$ with R the radius of the sphere, the wavefunction of these orbitals in an appropriately chosen gauge can be written in form very similar to that of Eq. (1) except that on the sphere

$$\mu(z, \bar{z}) = \left(1 + \frac{|z|^2}{4R^2} \right)^{-1-N_\phi/2} \tag{3}$$

The orbitals z^m again have the shape of an annulus of thickness approximately one, at constant lattitude around the sphere, going from the north pole ($m = 0$) to the south pole ($m = N_\phi$).

On the cylinder geometry of circumference L, we use complex coordinate $w = x + iy$ with the real direction going around the circumference and the imaginary direction going along the length of the cylinder and we define $z = e^{2\pi i w/L}$. The single particle orbitals are of the form[21] of Eq. (1) except on the cylinder

$$\mu(z, \bar{z}) = e^{-|y|^2/2}. \tag{4}$$

The z^m orbitals form rings around the circumference with thickness approximately unity, and whose distance along the cylinder is indexed by m. Note that since the cylinder may be infinitely long in both directions, m can be any integer (including negative). However, we will often consider no orbitals with $m < 0$ to be occupied in order to make more simple analogy with the sphere and plane wavefunctions. One can, of course connect up the ends of the cylinder to form a torus as well.[22]

---

[b]A partially filled higher Landau level can be mapped to a partially filled lowest Landau level at the price of modifying the interaction and confinement terms, so long as we can neglect inter-Landau-level transitions.[13,19]

These three geometries (plane, sphere, cylinder) are all useful. Since the interesting part of the wavefunction are the z^m factors, it is fairly easy to translate physics from one geometry to another. We will typically work in whichever geometry is simplest for elucidating the interesting physics.

2.2. Many body physics

A many-body wavefunction Ψ for N particles is generally written as

$$\Psi(z_1,\ldots,z_N) = \Phi(z_1,\ldots,z_N) \prod_{i=1}^{N} \mu(z_i,\bar{z}_i)$$

where Φ is a polynomial in the $z's$ which is fully symmetric for bosons or fully antisymmetric for fermions. We will typically only write the Φ part of the wavefunction for simplicity.

If the wavefuncton represents an angular momentum eigenstate (as, for example, a ground state would if it is confined in a rotationally invariant potential) then the polynomial Φ should also be homogeneous in degree, since each power of z contributes one quantum of angular momentum to the wavefunction.

The ratio of the number of particles N to the number of available single-particle orbitals N_{orb} in the Landau level is known as the filling fraction

$$\nu = \lim_{N\to\infty} \frac{N}{N_{orb}}$$

To determine the filling fraction of a many-body wavefunction, we need to compare the number of particles to the number of orbitals which are at least partially filled. Indicating orbitals as z^n, let the lowest value of n (the lowest angular momentum orbital) which is at least partially occupied be $n = 0$. On the cylinder this assumption fixes the location of the one of the edges of the quantum Hall droplet, whereas on the sphere or plane this simply states that the quantum Hall droplet covers the north pole or origin of the plane. Then we look for the highest power of z in the wavefunction (which we will call "maxpower") to find the the farthest away orbital which corresponds to the largest radius of the droplet. The total number of orbitals which are at least partially filled is then $N_{orb} = (\text{maxpower} + 1)$. Thus the filling fraction is $\nu = N/(\text{maxpower} + 1)$.

On the sphere we have the more detailed relationship for a quantum Hall state covering the entire sphere

$$N_\phi = \frac{1}{\nu}N - \mathcal{S} \tag{5}$$

where $\mathcal{S}$ is an order one number known as the *shift* of the wavefunction[23] and the flux at the center of the sphere is $N_\phi = \text{maxpower} = N_{orb} - 1$ for the Lowest landau level. This definition of the shift is most cleanly defined on a sphere where one can precisely determine the flux in the sphere and count the number of electrons necessary to fill the sphere (leaving no quasielectrons or quasiholes) for any particular quantum

Hall state. For the disk or cylinder, one defines N_ϕ as $N_\phi = N_{orb} - 1$ where N_{orb} is the number of orbitals at least partially occupied.

Note that even for the integer quantum Hall effect, the shift is an interesting quantity (see for example the discussion in Ref. 12). Filling the lowest Landau level on the sphere requires $N = N_\phi + 1$ electrons, whereas filling the first excited Landau level requires $N = N_\phi + 3$. The shift turns out to be twice the mean orbital spin of an electron in the given state[24,25] — which depends, for example, on which Landau levels we are considering (see also the discussion in Sec. 10 below).

One can multiply a wavefunction by an overall Jastrow factors to produce a new wavefunction

$$\Phi^{\text{new}} = \Phi^{\text{old}} \prod_{i<j} (z_i - z_j)^p \tag{6}$$

$$\nu_{\text{new}} = \frac{1}{\nu_{\text{old}}^{-1} + p}. \tag{7}$$

Indeed, this transformation is the root of the entire composite fermion approach.[12] (See also the chapter by Jainendra Jain in this volume.) Note that if the old wavefunction describes bosons (fermions), the new wavefunction will describe bosons (fermions) for even p and will describe fermions (bosons) for odd p.

There is an immense freedom (i.e. an enormous Hilbert space) for writing down wavefunctions for partially filled Landau levels. For fermions, when there are more orbitals available than there are fermions to fill them, we have an enormous degeneracy of states (N_{orb} choose N) corresponding to deciding where these fermions should go. For bosons since we can multiply occupy single orbitals, a degeneracy ($N_{orb} - 1 + N$ choose N) exists even when there are more bosons than orbitals. This enormous (exponentially large in N for fixed ν) degeneracy is broken by the interactions between the particles and, under appropriate circumstances, a unique FQHE ground state is formed.

We thus turn to consider the interaction term of the Hamiltonian. In real FQHE experiments, the interaction term involves a two-body interaction

$$H_{\text{interaction}} = \sum_{i<j} V(\mathbf{r}_i - \mathbf{r}_j)$$

with V being, for example, the Coulomb interaction. From a theoretical standpoint, however, we will often find it useful to think of more general many-body interactions such as a three-body interaction

$$\sum_{i<j<k} V(\mathbf{r}_i - \mathbf{r}_j, \mathbf{r_i} - \mathbf{r_k})$$

or even interactions involving a larger number of particles. Note that, starting with a two-body interaction, integrating out inter-Landau-level transitions will generate many body interactions.[26–28]

3. FQHE Wavefunctions and Special Parent Hamiltonians

3.1. *Laughlin*

Laughlin's genius was to simply guess the right wavefunction that very accurately described the experimentally observed $\nu = 1/p$ fractional quantum Hall effect:[1]

$$\Phi_{\text{Laughlin}}^{\nu=1/p} = \prod_{i<j}(z_i - z_j)^p \ . \tag{8}$$

Note that for odd p this is an antisymmetric wavefunction appropriate for fermions, whereas for even p it is symmetric and appropriate for bosons. Just to check that this does indeed describe filling fraction $1/p$ we note that the highest power of any z occurring is $z^{p(N-1)}$. Thus, for example, on a sphere, our wavefunction would cover the sphere perfectly if $N_\phi = \text{maxpower} = p(N-1)$. In the large system limit, then we would have a ratio $\nu = N/N_{orb} = N/(N_\phi + 1) \to 1/p$.

Now although this wavefunction is exceedingly accurate for certain real physical systems (say, electrons in high mobility GaAs quantum wells at filling fraction $\nu = 1/3$), for theoretical work it is useful to consider a situation for which this wavefunction will be the *exact* ground state. This will enable us to make precise statements, after which we can think about whether (or to what degree) our statements carry over to the physical experiments. For this reason it is useful to think about so-called *parent Hamiltonians* which give the desired wavefunctions as their unique exact ground states. A Hamiltonian that is "parent" also has the property that the interaction energy is non-negative definite and is exactly zero in the ground state. A further property, which we will call a "special" parent Hamiltonian,[29] is that all low energy edge excitations and quasihole excitations also have zero interaction energy. Much of this review is focused on properties of quantum Hall states that are generated by special parent Hamiltonians.

A simple example is the special parent Hamiltonian[10,30,31] for the bosonic $\nu = 1/2$ Laughlin state (the $p = 2$ case of Eq. (8)) which is obtained by using the interaction

$$H_{\text{interaction}} = V_0 \sum_{i<j} \delta(\mathbf{r}_i - \mathbf{r}_j) \tag{9}$$

with δ being a two-dimensional delta-function and $V_0 > 0$ an interaction energy scale. It is easy to see that the Laughlin $\nu = 1/2$ wavefunction makes the energy of this interaction exactly zero — since in the wavefunction there is zero amplitude for two particles coming to the same position. One should be cautious, however, that the energy would also be zero for any wavefunction that is a fully symmetric polynomial times the Laughlin $\nu = 1/2$ wavefunction (it would have to be a symmetric polynomial to maintain the symmetry of the wavefunction) since the wavefunction would still vanish when any two particles come to the same position. However, multiplying the Laughlin wavefunction by a polynomial will increase the radius of the droplet, since, as discussed above (just below Eq. (1)), increasing the degree of the polynomial increases the radius of the wavefunction. Thus, if we include a weak radially symmetric confining potential $U(|\mathbf{r}|)$, the lowest energy wavefunction, and hence the unique exact ground state, will be the Laughlin wavefunction itself.[c,d] Multiplying by a polynomial generates low energy edge excitations as we will discuss in Sec. 4.

It is obvious that a delta function interaction is ineffective for fermions since two fermions cannot come to the same position anyway. However, we can still write a special parent Hamiltonian for the fermionic $\nu = 1/3$ case by using a Laplacian of a delta function.[10,30,31] In general, special parent Hamiltonians can be written for any Laughlin state $\nu = 1/p$ for arbitrary p by considering high enough derivatives of delta functions.

Perhaps a more transparent way to describe such a parent Hamiltonian interactions is via the idea of pseudopotentials[10,13] often written as V_m. We can write *any* 2-body interaction within a single Landau level as

$$H_{\text{interaction}} = \sum_m \sum_{i<j} V_m P_{ij}^m$$

where P_{ij}^m is a projection operator that projects the wavefunction such that particles i and j have relative angular momentum m, i.e. when[19]

$$\psi_{rel} \sim (z_i - z_j)^m$$

So, for example if we have an interaction with $V_0 > 0$ this gives positive energy to any wavefunction that does not vanish when two particles come to the same position (this is precisely the delta function interaction of Eq. (9)). If we have $V_1 > 0$ this gives energy to a wavefunction that vanishes as $(z_i - z_j)$ when two particles come to the same position (this is the Laplacian of a delta function), and so forth. So for example, if we want to write a special parent Hamiltonian for the Laughlin state of bosons at $\nu = 1/4$ we simply set $V_0 > 0$ and $V_2 > 0$, which gives positive energy unless the wavefunction vanishes as four or more powers when two particles approach each other. By exchange symmetry, a bosonic (fermionic) wavefunction can only vanish as an even (odd) number of powers as two particles approach each other, so that for bosonic (fermionic) systems only the even (odd) pseudopotentials matter.

3.2. *Read–Rezayi and special case of Moore–Read*

The idea of psuedopotentials have been generalized to interactions involving more than two particles.[19,32] A simple application of this idea[33] is to define the Hamiltonian

$$H_k^r = \sum_{i_1 < i_2 < \ldots i_{k+1}} P_{i_1, i_2, \ldots i_{k+1}}^r \tag{10}$$

---

[c]As we will discuss in Sec. 4 below, if the confining potential is quadratic $U(\mathbf{r}) \propto |\mathbf{r}|^2$ there is substantial further simplification and the confinement need not be weak.

[d]It is worth commenting that even for the Laughlin wavefunction with this very simple special parent Hamiltonian, there is no *rigorous* proof that the bulk is gapped in the thermodynamic limit, and this reasonable assumption is justified mainly by numerical work.

where $P^r_{i_1,...i_{k+1}}$ is a projection operator that projects the cluster of $k + 1$ particles to have relative angular momentum greater or equal to r. Thus, H^r_k gives positive energy unless the wavefunction vanishes at least as fast as r powers when $k + 1$ particles come to the same point. For example, the Laughlin $\nu = 1/p$ Hamiltonian is H^p_1.

A crucial example of this construction is given by the Hamiltonian H^2_k which is the special parent Hamiltonian for the bosonic $\mathbb{Z}_k$ Read–Rezayi wavefunction.[6] The explicit real-space form of this Hamiltonian is given by

$$H^2_k = V^{(k)}_0 \sum_{i_1 < i_2 < ... i_{k+1}} \delta(\mathbf{r}_{i_1} - \mathbf{r}_{i_2})\delta(\mathbf{r}_{i_1} - \mathbf{r}_{i_3}) ... \delta(\mathbf{r}_{i_1} - \mathbf{r}_{i_{k+1}}) \tag{11}$$

with $V^{(k)}_0 > 0$. This interaction allows k particles to come to the same point, but gives an energy penalty when $k + 1$ come to the same point. Equivalently H^2_k forbids a cluster of $k + 1$ particles from having zero relative angular momentum, but allows relative angular momentum of 2 hence the superscript on the H^2_k (note that a cluster of bosons cannot[33] have relative angular momentum of 1).

The ground state of the H^2_k Hamiltonian can be written as the following generalization of the Laughlin state (where the number of particles N is a multiple of k), known as the $\mathbb{Z}_k$ Read–Rezayi wavefunction[6,34]

$$\Phi^{\nu=k/2;\text{ bosons}}_{\text{Read-Rezayi}} = \mathbb{S}\left[\left(\prod_{0<i<j\leq N/k}(z_i - z_j)^2\right)\left(\prod_{N/k<i<j\leq 2N/k}(z_i - z_j)^2\right)\cdots\right.$$
$$\left.\cdots\left(\prod_{N(k-1)/k<i<j\leq N}(z_i - z_j)^2\right)\right]. \tag{12}$$

Here, $\mathbb{S}$ means symmetrize over all reordering of the particle numbering. So the scheme here is to divide all the particles into k equal groups and write a Laughlin $\nu = 1/2$ wavefunction for each group, then symmetrize over all different ways you might have chosen these groups. It is easy to see why this is a zero energy wavefunction of the $(k + 1)$-body delta-function: for a cluster of k particles all coming to the same position, the wavefunction need not vanish, since we can put one particle in each of the k groups. However, when the $(k + 1)$st particle comes to that point, then we must have (at least) two particles in (at least) one group and the wavefunction then must vanish. Indeed, while perhaps not entirely obvious, it can be shown that, analogous to the Laughlin case, this is the most dense wavefunction (the lowest degree polynomial) with this property and is hence the ground state. Note that the $k = 1$ case of the Read–Rezayi series is the Laughlin wavefunction and the $k = 2$ case is known as the Moore–Read state[2] which we will discuss in more depth in Sec. 3.2.1.

We can attach Jastrow factors to the the Read–Rezayi series (plugging Eq. (12) in Eq. (6)) to obtain wavefunctions with filling fractions

$$\nu = \frac{k}{2 + kp} \tag{13}$$

with $p \geq 0$ even for bosonic wavefunctions, and $p > 0$ odd for fermionic wavefunctions. Note that $k = 1$ corresponds to the Laughlin series.

It is also straightforward to write special parent Hamiltonians for the Read–Rezayi wavefunctions with Jastrow factors attached. In order to enforce p Jastrow factors in our wavefunction, we must include a two body term (H_1^p) in our Hamiltonian which is precisely the special parent Hamiltonian for Laughlin $\nu = 1/p$. Once we have these p Jastrow factors, bringing $k+1$ particles to the same point will incur $pk(k+1)/2$ powers of z. Thus we can include another term in the Hamiltonian that forbids $k+1$ particles from coming together with relative angular momentum less than or equal to $pk(k+1)/2$. Such a Hamiltonian is exactly $H_k^{pk(k+1)/2+2}$. Thus the total Hamiltonian is $H_k^{pk(k+1))/2} + H_1^p$. In the case where we only want to include a single Jastrow factor for a fermionic wavefunction ($p = 1$), we do not need to include the term H_1^1 since it is trivial for fermions.

By counting powers of z and looking for the highest power that occurs (see the discussion near Eq. (5)), we can find the precise relationship between flux and particle number for the Read–Rezayi wavefunction with p Jastrow factors attached:

$$N_\phi = \left(\frac{2}{k} + p \right) N - (p+2) \tag{14}$$

which clearly gives the filling fraction ν in Eq. (13) in the large system size limit, and gives the shift $\mathcal{S} = p + 2$.

3.2.1. *Special case of Moore–Read*

The $k = 2$ case of the Read–Rezayi series deserves special attention. This is known as the Moore–Read wavefunction,[2] and it can be rewritten in a different form:

$$\Phi_{\text{Moore-Read}}^{\nu=1;\text{bosons}} = \text{Pf}(\mathbf{M}) \prod_{i<j}(z_i - z_j) \tag{15}$$

where here $\mathbf{M}$ is the antisymmetric matrix with components

$$M_{ij} = \frac{1}{z_i - z_j} \tag{16}$$

and zeros along the diagonal. The notation $\text{Pf}(\mathbf{M})$ symbolizes a Pfaffian, which, for an N by N matrix (with N even) is given by

$$\text{Pf}(\mathbf{M}) = \frac{1}{2^N(N!)} \, \mathbb{A} \left[M_{12}M_{34} \ldots M_{(N-1),N} \right]$$

where here $\mathbb{A}$ is the antisymmetrized sum over all permutations of indices.[e] It is interesting to note that all BCS superconducting wavefunctions are essentially Pfaffians.[35] For Cooper pairing of spinless fermions, the BCS wavefunction is precisely a Pfaffian where $M_{ij} = g(\mathbf{r}_i - \mathbf{r}_j)$ with $g(\mathbf{r})$ the (necessarily antisymmetric) wavefunction of a single pair. Thus, the Moore–Read wavefunction is essentially a spinless superconductor.

---

[e]Note the useful identity $[\text{Pf}(\mathbf{M})]^2 = \det(\mathbf{M})$.

3.3. *Other parent cluster Hamiltonians and their wavefunctions*

Parent Hamiltonians have been written down for a number of other wavefunctions, including, in particular, cases including wavefunctions with multiple spin states.[11,20,36–41] Generalizatons of parent Hamiltonians to lattice fractional quantum Hall effect[f] have been pursued by a number of authors.[44–50]

Returning to spinless particles in the continuum, an obvious extension[32,33] of the parent Hamiltonians for Laughlin and Read–Rezayi wavefunctions is to consider Hamiltonians H_k^r that give positive energy to a cluster of $k+1$ particles with relative angular momentum less than some number r (i.e. forces the wavefunction to vanish as at least r powers when $k+1$ particles approach the same point) which we might expect to give a wavefunction at filling fraction $\nu = k/r$ for bosons. An interesting case is that of $k = 2$, $r = 3$, which for bosons gives a $\nu = 2/3$ wavefunction known as the Gaffnian wavefunction.[7] The fermionic version at $\nu = 2/5$ (the wavefunction being obtained by attaching a single Jastrow factor to the bosonic $\nu = 2/3$ wavefunction) is obtained using $k = 2$, $r = 6$. The bosonic Gaffnian wavefunction has extremely high overlap with the Jain $\nu = 2/3$ bosonic quantum Hall state;[7] and similarly the fermionic Gaffnian at $\nu = 2/5$ has extremely high overlap with the Jain $\nu = 2/5$ state. However, where the composite fermion wavefunctions correspond to gapped phases of matter, the Gaffnian, which is related to the nonunitary $M(3,5)$ conformal field theory, is actually gapless,[7,29,51–54] having zero energy neutral excitations in the thermodynamic limit. Indeed, it has been argued that any wavefunction related to a nonunitary conformal field theory must similarly be gapless.[29] A rough argument as to why nonunitarity in a CFT implies gaplessness in the bulk is given by the bulk-edge correspondence which will be discussed in Sec. 10 below.[29] One possible scenario is that the Gaffnian represents a critical point between two gapped quantum Hall phases — although this remains a conjecture.

Another cluster wavefunction worth discussing is known as the Haffnian,[8] corresponding to $k = 2$, $r = 4$ which is a $\nu = 1/2$ wavefunction for bosons, with a fermionic version being $k = 2$, $r = 7$ at $\nu = 1/3$. The Haffnian (related to a unitary, but *non-rational* conformal field theory[55]) is also believed to be gapless and is even more poorly behaved than the Gaffnian — having an extensive ground state degeneracy on a torus in the thermodynamic limit.[32]

Although it may seem that one should be able to generate many new wavefunctions with such clustering Hamiltonians, the situation is not as simple as it seems.[33] For simplicity let us consider the case of boson wavefunctions. For the cases we have discussed thus far (Laughlin, Moore–Read, Read–Rezayi, Gaffnian, Haffnian), we have used the Hamiltonian H_k^r which forbids the wavefunction from vanishing as fewer than r powers when $k+1$ particles approach each other. In each of these cases we get a ground state wavefunction where two special properties

---

[f]Particles hopping on a lattice which exhibit fractional quantum Hall effect are known as Fractional Chern Insulators.[42,43]

hold: (a) the wavefunction does not vanish when k particles approach each other and (b) the wavefunction vanishes as exactly r powers when $k+1$ particles approach each other. As long as these two properties hold we will obtain a wavefunction with filling fraction[33]

$$\nu(k,r) = k/r. \tag{17}$$

However, the ground state of H_k^r need not satisfy these two properties for general k and r. The ground state can vanish when fewer than $k+1$ particles come together, or it can vanish as more than r powers when $k+1$ particles come together.[33] A prime example of this behavior is the Hamiltonian H_2^5 for bosons, which requires that the wavefunction vanish as (at least) $r=5$ powers when $k+1=3$ particles come to the same point. However, the ground state[33] of this Hamiltonian is actually the $\nu=1/2$ Laughlin state, which vanishes when only two particles come to the same point, and vanishes as 6 powers when three come to the same point. It turns out to be possible[56] to analytically construct a special parent Hamiltonian for a wavefunction that vanishes as exactly $r=5$ powers as $k+1=3$ particles come to the same point (related to the nonunitary $M(3,7)$ wavefunction), however it is somewhat more complicated.

In all of the above cases (Laughlin, Moore–Read, Read–Rezayi, Gaffnian, Haffnian, $M(3,7)$), we have specified that the wavefunction must vanish as r powers when $k+1$ particles come to the same point. It has been proposed that specifying these vanishing properties (a so-called "pattern of zeros") can be used to define a wavefunction.[57–59] While some results can be derived from thinking in this language, there is also a limitation. Unfortunately, the simple statement that a wavefunction vanishes as r powers as $k+1$ particles come to the same point is not generally sufficient information to fully define a wavefunction. To clarify this, let us consider a situation where we have only $k+1$ particles. When there is more than one (translationally invariant) polynomial of rth degree in $k+1$ variables, the vanishing properties alone do not define the polynomial.[32,33,56,60] This situation occurs for $k=2, r=6$ and $k=2, r \geq 8$ and all $k>2, r>3$. Instead, in these cases there is a multidimensional space of $(k+1)$-particle wavefunctions which vanish as r powers as all $k+1$ particles come together. This implies that the pattern of zeros approach needs to be supplemented with additional information in order to uniquely define a particular wavefunction.

In these cases where there is a multidimensional space of polynomials available for a $(k+1)$-particle wavefunction, the projection operator in the Hamiltonian of Eq. (10) can be refined and written as a projection onto a subspace of available wavefunctions. We thus choose to give positive energy to some subspace of this while leaving the remaining subspace (possibly a single $(k+1)$-particle wavefunction) at zero energy.[56,61] In this way it is still possible to write parent Hamiltonians for wavefunctions which vanish proportional to a particular polynomial function as some number $(k+1)$ particles come to the same point. In these more complicated cases, special parent Hamiltonians have successfully been constructed for

$k = 2, r = 6$ ($\nu = 1/3$ for bosons as dictated by Eq. (17), which are related to the $N = 1$ superconformal field theories with generic[g] central charge[56,62]). It was also shown[56] that one can construct special parent Hamiltonians corresponding to $M(3, p)$ conformal field theories. These correspond to $k = 2$, $r = p - 2$ and thus to $\nu = 2/(p-2)$ for bosons. Note that except for the simple Moore–Read state (which is $p = 4$) these are all nonunitary conformal field theories and are therefore gapless (see Sec. 10 below).

Other examples of parent Hamiltonians have been explored where the ground state is also picked out by specifying a particular polynomial describing the wavefunction when $k+1$ particles come together. These include the case[61] of $k = 3, r = 4$ ($\nu = 3/4$ for bosons, as dictated by Eq. (17)) related to the S_3 conformal field theories, and also the minimal models of $N = 1$ superconformal field theories $k = 2, r = 6$ ($\nu = 1/3$ for bosons). However, in these cases it has so far not been possible to fully determine how the zero energy excitations match up with the predictions of the CFT. Nonetheless, while it appears possible to construct parent Hamiltonians (but perhaps not special parent Hamiltonians) for wavefunctions which correspond to unitary conformal field theories (and therefore do not run afoul of the rule[29] that nonunitary (or nonrational) CFTs must give gapless wavefunctions) it remains an open question whether these Hamiltonians correspond to gapped phases of matter.

An obvious question is whether one can construct special parent Hamiltonians for the experimentally prominent hierarchy,[10,11] or composite fermion,[12] series of fractional quantum Hall states (see chapter by Jain in this volume). While no exact parent Hamiltonians have been constructed, one approach for studying the composite fermion state at $\nu = 2/3$ for bosons (or correspondingly 2/5 for fermions) uses a delta-function (or derivative of delta function) interaction within two degenerate Landau levels.[63–65] While the resulting wavefunction appears to have many of the properties desired, it is not fully projected to the lowest Landau level, which is problematic. (Generalizing this approach to three degenerate Landau levels[66,67] gives a parent Hamiltonian for a so-called *parton* wavefunction[66,68] related to the $SU(2)_2$ conformal field theory.) In general, determining simple parent Hamiltonians for hierarchy or composite fermion wavefunctions remains an important open problem.[69] It should be noted, however that given any wavefunction corresponding to a gapped phase of matter, one can in principle generate a local parent Hamiltonian, at least numerically.[48,49,70,71] However, the resulting Hamiltonian generically can involve terms with many particle interactions, and will not usually be simple. Further, it is unclear if such a Hamiltonian will have the "special" property that the low energy quasiholes and edge excitations have exactly zero interaction energy.

---

[g]The cases corresponding to the $N = 1$ superconformal minimal series are more complicated.

4. Edge Excitations

We now turn to consider the elementary excitations of the FQHE states described by special parent Hamiltonians. Since the bulk of the system is assumed to be gapped[d], when the confining potential is taken into account, we expect that any low energy excitations will involve excitations of the edge only. The study of quantum Hall edges is a venerable subject,[72,73] and we will only cover some aspects, referring to the literature for a more complete discussion.[74,75]

A physical picture of the edge excitation is given by classical physics. A FQH droplet is assumed to be held in position by some confining potential, which we assume to be rotationally symmetric so that angular momentum remains a good quantum number. This can be thought of as some radial electric field acting on our charged particles and pushing them inwards. On the other hand, we also have a magnetic field perpendicular to the plane of the sample. From basic classical physics, since we have a crossed electric and magnetic field we then obtain a drift velocity for charged particles, pushing them perpendicular to the edge (perpendicular to both electric and magnetic fields). So if a small bump is created at some point along the edge, it will travel around the edge of the droplet at some drift velocity. This is a simple way to picture the edge excitations.

Things simplify quite a bit if we choose to use a parabolic confinement for our quantum Hall droplet on a disk

$$U(\mathbf{r}) = \gamma |\mathbf{r}|^2.$$

In this case, despite this perturbation to the kinetic term of the Hamiltonian, the single particle eigenstates still take the simple form of Eq. (1) (although the magnetic length is altered by the confining potential). With the confining potential, the energy of each single particle eigenstate in the lowest Landau level is now given by

$$E_m^{LLL} = \alpha m + \text{constant} \tag{18}$$

where the prefactor α is set by the strength γ of the confinement.[h] The intuition here is that the eigenstate of angular momemtum m is at radius $r = \sqrt{2m}$ and thus has an energy $\gamma r^2 \propto m$. See the discussion near Eq. (1). Note that in the cylinder geometry one instead obtains an edge excitation of the form of Eq. (18) for a linearly sloped edge potential $U(\mathbf{r}) \sim y$.

The ground state of the special parent Hamiltonian with confinement is the (presumably unique) wavefunction that has zero interaction energy (i.e. satisfies the constraints imposed by the interaction) and also has the minimal total angular momentum, and therefore has the minimal possible confinement energy consistent with zero interaction energy. To look for low energy excitations, we need to find a polynomial wavefunctions of slightly higher degree than the ground state, which still satisfies the zero interaction energy constraint imposed by the special parent Hamiltonian. By being a slightly higher degree polynomial, the droplet will be

---

[h]In the limit of large magentic field, we have $\alpha = 2\gamma$ as expected.

Table 1. Symmetric polynomials listed by degree, and the energy of the eigenstate which results when they multiply the Laughlin ground state.

degree d	symmetric polynomials				$E - E_{\text{ground}}$
1	p_1				α
2	p_2,	$p_1 p_1$			2α
3	p_3,	$p_2 p_1$,	$p_1 p_1 p_1$		3α
4	p_4,	$p_3 p_1$,	$p_2 p_1 p_1$,	$p_2 p_2$, $p_1 p_1 p_1 p_1$	4α

of slightly higher radius, and slightly higher confinement energy. For each added degree of the polynomial, the wavefunction increases its angular momentum by one unit, and the energy increases by one unit of the confinement energy α.

4.1. *Laughlin edge*

For example, consider the special parent Hamiltonian for the Laughlin $\nu = 1/p$ state for a system bosons (in the case of even p) or fermions (for odd p). Here the special parent Hamiltonian interaction is such that we must find a (symmetric for bosons, antisymmetric for fermions) polynomial which vanishes as p or more powers when two particles come to the same position, and we want the polynomial to be of minimal total degree. Thus there should be a factor of $(z_i - z_j)^p$ for each pair of particles. This uniquely picks out the Laughlin $\nu = 1/p$ wavefunction as the ground state as discussed above. To obtain low energy edge excitations, as discussed below Eq. (9), we multiply the Laughlin wavefunction by any symmetric polynomial. The product still satisfies the constraint that the wavefunction vanishes as at least p powers when two particle approach each other and maintains proper bosonic (fermionic) symmetry of the wavefunction for p even (odd). Thus enumerating the possible edge excitations of the Laughlin state amounts to enumerating all possible symmetric polynomials by which we can multiply the ground state wavefunction.[73,76]

There are many ways to achieve the goal of enumerating symmetric polynomials. Perhaps the simplest method is to use the so-called power sum symmetric polynomials.

$$p_m = \sum_{i=1}^{N} z_i^m \tag{19}$$

which form the generators of the *ring* of symmetric polynomials.[i] (We will see other ways to enumerate all symmetric polynomials in Sec. 6 below). We can then make a table (Table 1) of the number of polynomials we can write at degree d, and hence the number of edge eigenstates we can write whose angular momentum is d units more than that of the ground state.

---

[i]A ring is roughly a set of objects that we can add and multiply. The statement that p_m are generators means that sums and products (and multiplying by a constant) of these p_m's form all possible symmetric polynomials.

There is a unique ground state (the Laughlin state). Examining Table 1 we see there is a single eigenstate with energy α greater than the ground state which is given by multiplying the ground state by p_1. The space of eigenstates with energy 2α greater than the ground state is spanned by two wavefunctions, one of which is p_1^2 times the grounds state and the other is p_2 times the ground state (thus the space is two dimensional). For higher angular momentum, eigenstates with angular momentum q units greater than the ground state will have energy $q\alpha$ greater than the ground state, and the number of such eigenstates with will be equal to the number of integer partitions of the integer q. This then gives the well known series 1,2,3,5,7,11, ..., as shown in Table 1.

The wavefunctions built from the products of p_m operators multiplied by the Laughlin wavefunction provide a linearly independent set of states spanning the Hilbert space of excitations at a given angular momentum. However, we are not guaranteed that these wavefunctions are orthogonal to each other (for example, p_1^2 times Laughlin and p_2 times Laughlin span the space of $q = 2$ excitations, but are not strictly orthogonal). However, quite interestingly it turns out that in the thermodynamic limit (large enough droplet and small enough q), these wavefunctions do indeed become orthogonal[77–79] as we will discuss further below in Sec. 4.1.3.

We now ask whether we can write an effective theory of the edge that does not require us to work with wavefunctions for all N particles in the entire FQHE droplet — after all, we are only concerned with (potentially small) deformations of the edge not the entire system. Indeed we can write such a theory, which is known as a chiral Luttinger liquid[73,74,77]

$$H_{\chi LL} = \sum_{n>0}(\alpha n)a_n^\dagger a_n. \tag{20}$$

Here, the operator $a_n^\dagger$ (a_n) is a bosonic operator corresponding to the creation (annihilation) of the nth angular momentum mode of the edge. Since they are bosonic operators they satisfy the usual commutations

$$[a_n, a_m^\dagger] = \delta_{nm}. \tag{21}$$

Here the modes have only positive angular momentum $n > 0$, indicating that modes only travel in one direction (i.e. the excitations are chiral). If we think of the Laughlin ground state as being the "vacuum" $|0\rangle$ of the edge (i.e. the edge with no excitations), we can then make a table of all excitations in order of their angular momentum (and hence of their energy) as shown in Table 2.

We immediately notice that Tables 1 and 2 are essentially identical. With the mapping

$$p_m \leftrightarrow a_m^\dagger$$

there is a precise correspondence between the two descriptions of the edge.[78,79] As noted above, the p_m's give orthogonal states only in the thermodynamic limit (large system and low energy of excitation), whereas the $a_m^\dagger$'s are orthogonal by definition. We discuss this issue in more depth in Sec. 4.1.3.

Table 2. Excitations of the chiral Luttinger liquid (Hamiltonian given in Eq. (20)) and
their corresponding energies. The first column is the angular momentum compared to the
ground state.

$L - L_{ground}$	states				Energy
1	$a_1^\dagger|0\rangle$				α
2	$a_2^\dagger|0\rangle$,	$a_1^\dagger a_1^\dagger|0\rangle$			2α
3	$a_3^\dagger|0\rangle$,	$a_2^\dagger a_1^\dagger|0\rangle$,	$a_1^\dagger a_1^\dagger a_1^\dagger|0\rangle$		3α
4	$a_4^\dagger|0\rangle$,	$a_3^\dagger a_1^\dagger|0\rangle$,	$a_2^\dagger a_1^\dagger a_1^\dagger|0\rangle$,	$a_2^\dagger a_2^\dagger|0\rangle$, $a_1^\dagger a_1^\dagger a_1^\dagger a_1^\dagger|0\rangle$	4α

4.1.1. *Bosonized edge*

Although this material has been discussed many places previously,[4,73–75] it is worth
discussing it again here for completeness. The chiral Luttinger liquid Hamiltonian
density for the $\nu = 1/p$ Laughlin state can be written in terms of a bosonic field
$\phi(x)$ as

$$\mathcal{H} = \frac{v}{4\pi\nu}(\partial_x\phi)^2 \tag{22}$$

where v is the edge velocity, and the commutations of the field ϕ are given by

$$[\phi(x),\phi(x')] = i\nu\pi\,\text{sign}(x - x'), \tag{23}$$

and the charge density along the edge is given by $\rho(x) = \partial_x\phi/(2\pi)$.

To make contact with Eq. (20) we consider the system on a disk as in Eq. (20)
and write $\phi(z)$ with z the complex position around the edge. In this case we can
precisely write

$$\phi(z) = \phi_0 - ia_0 \ln z + i\sum_{n>0}\frac{1}{\sqrt{n}}\left(a_n z^{-n} + a_n^\dagger z^n\right) \tag{24}$$

where the $a_n^\dagger$ are the same bosonic mode creation operators as in Eq. (20) and where
a_0 is the so-called zero-mode operator that counts the total charge in the system
and $[\phi_0, a_0] = i$ defines the conjugate operator ϕ_0 which changes the number of
particles in the system.[55,78,80]

Note that in the language of this bosonic field, we can write a charge bump with
total charge γ at position x along the edge as

$$V_\alpha =: e^{i(\gamma/\nu)\phi(x)} :$$

with colons indicating normal ordering. In particular for $\gamma = 1$ we obtain an electron
creation operator.

4.1.2. *Further Laughlin edge considerations*

Dimensional (RG) arguments tell us that the chiral Luttinger liquid Hamiltonian,
Eq. (20) is the correct low energy long wavelength limit of the Laughlin edge,[73,74]
independent of the Hamiltionian of the underlying problem (i.e. independent of
the details of the interelectron interaction, although the value of α appearing in

the Hamiltonian will depend on details).[j] At higher energy scales or at shorter wavelength the Hamiltonian Eq. 20 (equivalently Eq. (22)) remains exact for the parent Hamiltonian, but for generic inter-electron interactions there will be sub-leading terms in the Hamiltonian which are less relevant in the renormalization group sense. The study of the effect of such nonlinear terms in Luttinger liquids has been a topic of some interest outside of the quantum Hall field,[86] and recently has attracted attention in the quantum Hall context as well.[80,87–91] There was some debate[89,90] over whether, even without long range inter-electron interactions, there could be long-range (so-called "Benjamin–Ono") interaction terms in the effective edge Hamiltonian. However, recent studies[80] support the view that, so long as the inter-electron interaction is not long ranged, additional terms in the effective edge Hamiltonian must also be short ranged. Further, it was found that, for a droplet in a parabolic potential, there are a large number of additional constraints on the form of the edge Hamiltonian. For example, writing down terms of the Hamiltonian in order of decreasing RG relevance, one might expect that the first correction to Eq. (22) would be a term of the form $(\partial \phi)^3$. (Indeed, this is a main correction that is studied in detail in Refs. 86–88.) However with parabolic confinement it was shown that the coefficient of this term must be exactly zero.[80] It remains an outstanding question to determine whether constraints of this type are relevant for experimental systems in non-parabolic confinement.

4.1.3. *Edge state inner products*

As mentioned just before the start of Sec. 4.1.1, the edge state operators $a_m^\dagger$ correspond exactly to the polynomials p_m with which we can multiply the ground state wavefunction — so long as we are in the limit of large system size and low energy. When we are not in this limit the corrections to this orthonogonality are given by an extremely interesting "edge action" construction.[78,79] Defining λ to be a list of nonincreasing integers λ_m, we can write an arbitrary excited edge state as

$$|\lambda\rangle = \prod_m (a_m^\dagger)^{\lambda_m} |0\rangle \tag{25}$$

corresponding to the two dimensional wavefunction

$$\Psi_\lambda = \left(\prod_m p_m^{\lambda_m} \right) \Psi_{Laughlin}^\nu \tag{26}$$

with (now normalized) power sum polynomials (compare Eq. (19))

$$p_m = \frac{1}{\sqrt{\nu}} \sum_{i=1}^{N} \left(\frac{z_i}{R} \right)^m. \tag{27}$$

---

[j]The low energy long wavelength validity of the chiral Luttinger liquid Hamiltonian is true so long as ground state is in the Laughlin phase of matter, and so long as the edge does not "re-construct",[81–85] which is guaranteed when the inter-electron interaction is sufficiently close to the special parent Hamiltonian. In practice there is a very wide range of edge structures and inter-electron interactions for which this remains correct.[81]

with R the radius of the droplet. Note that Eq. (25) is a statement about a state on a one-dimensional edge, whereas Eq. (26) describes the full two-dimensional wavefunction of the droplet. The inner product between two such edge states is generally nontrivial

$$\langle \Psi_\lambda | \Psi_{\lambda'} \rangle = \langle \lambda' | e^{-S} | \lambda \rangle \qquad (28)$$

where S has local interaction terms only, and which operates on the edge space. This operator in the case of a disk geometry can be expanded as

$$S = -\frac{\sqrt{\nu}}{6N}(\partial_z \phi)^3 + \dots$$

where each successive term is smaller by $1/\sqrt{N}$. In the large system size limit, one needs only keep the leading term $e^{-S} \approx 1$ so that states which are orthogonal in the edge space are also orthogonal wavefunctions in the bulk.

It is of course necessary that there is some mapping between states described as edge excitations (Eq. (25)) and states described as bulk wavefunctions (Eq. (26)). What is not obvious is that the form of the mapping should be the exponential of local interaction terms on the edge space and that the operators should be increasingly small as the system gets larger. This form can be justified via intuition from conformal field theory and matrix product states.[78] Although this justification does not constitute a rigorous proof that Eq. (28) is correct, it has also been checked numerically to fairly high precision.[78,79]

4.2. From Laughlin edge to Hierarchy

The Haldane–Halperin hierarchy of quantum Hall states[10,11,92] has recently been reviewed elsewhere[4] so the discussion here will remain brief. As mentioned above, the hierarchy states (which are topologically equivalent to the Jain hierarchy[12,93] and should therefore have similar edge structure), do not have simple parent Hamiltonians. Nonetheless, even without such Hamiltonians that we can solve exactly, we still know the low energy edge structure of these states of matter. Generally the edge will consist of multiple bosonic modes, which we describe by bosonic fields ϕ^a for $a = 1, \dots, n$ for some n, having commutation relations (analogous to Eq. (23))

$$[\phi^a(x), \phi^b(x')] = i\pi K_{ab}^{-1} \text{sign}(x - x')$$

for some symmetric matrix of integers $\mathbf{K}$.

In addition to the $\mathbf{K}$ matrix, a hierarchy quantum Hall states is defined by a vector $\mathbf{t}$ of integers which describe the physical charge of the respective edge modes via $\rho_a(x) = t_a \partial_x \phi^a/(2\pi)$. The filling fraction of the bulk is given by $\nu = \mathbf{t}^T \mathbf{K}^{-1} \mathbf{t}$. For example, in the traditional Haldane–Halperin hierarchy construction $t_1 = 1$ and $t_a = 0$ for $a > 1$, and

$$K_{ab} = p_a \delta_{ab} - \delta_{a,b-1} - \delta_{a,b+1}$$

with (for electrons making up the state) p_1 odd and all other p_a's are even and can be positive or negative but not zero. This generates filling fractions of the continued fraction form

$$\nu = \mathbf{t}^T \mathbf{K}^{-1} \mathbf{t} = \cfrac{1}{p_1 - \cfrac{1}{p_2 - \cfrac{1}{\cdots - \frac{1}{p_n}}}}.$$

Note that one is free to make a basis transformation $\mathbf{K} \to \mathbf{W}^T \mathbf{K} \mathbf{W}$ and $\mathbf{t} \to \mathbf{W}^T \mathbf{t}$ using any integer valued matrix $\mathbf{W}$ of unit determinant.[77,93]

Another useful example is that of the Jain series $\nu = p/(2p+1)$ which has p-dimensional $\mathbf{K}$ matrix and $\mathbf{t}$ vector given by

$$K_{ab} = 2 + \delta_{ab}; \qquad t_a = 1 \tag{29}$$

which can be converted to the hierarchy form via $W_{ab} = \delta_{ab} - \delta_{a-1,b}$.

The Hamiltonian density for the edge is then given (analogous to Eq. (22)) by

$$\mathcal{H} = \frac{1}{4\pi} V_{ab} (\partial_x \phi^a)(\partial_x \phi^b) \tag{30}$$

where V_{ab} is a nonuniversal interaction matrix which determines the n edge mode velocities.

Note that in addition to $\mathbf{K}$ and $\mathbf{t}$, another vector, conventionally called $\mathbf{s}$, representing the spin of the respective edge modes, is required to fully define the topological order of the quantum Hall states.[23,77] This vector also determines the shift of the quantum Hall state (see Eq. (5)) via

$$\mathcal{S} = \frac{2}{\nu} \mathbf{t}^T \mathbf{K}^{-1} \mathbf{s}.$$

For example, for the Jain states described by Eq. (29) the spins are given by $s_a = a + 1/2$. As with the $\mathbf{t}$ vector, the $\mathbf{s}$ vector transforms as $\mathbf{s} \to \mathbf{W}^T \mathbf{s}$ under basis transformation.

To a large extent, one can describe the low energy edge excitations via multiple chiral Luttinger liquid Hamiltonians of the form of Eq. (20) albeit with different velocities for each mode which can be in either the positive or negative directions. One should also be cautious because there generally will also be low energy excitations associated with moving charge from one edge mode to another. Generically at higher energy (or for smaller system sizes) there will be subleading terms of the Hamiltonian Eq. (30) which are irrelevant in the renormalization group sense and are unimportant at low energy and large length scale. The scaling dimension of subleading terms simply counts the total number of derivatives, so for example, $(\partial_x \phi)^3$ and $(\partial_x^2 \phi)^2$ are subleading compared to the leading term $(\partial_x \phi)^2$.

4.3. *Read–Rezayi and (mostly) Moore–Read edges*

Let us now consider the case of a Read–Rezayi quantum Hall droplet. As an example, let us consider the bosonic $\mathbb{Z}_k$ Read–Rezayi state at filling fraction $\nu = k/2$. As mentioned above, the special parent Hamiltonian forces the wavefunction to vanish when $k + 1$ particles come to the same position. The resulting wavefunction is of the form of Eq. (12).

As described above at the beginning of Sec. 4, we consider a droplet with parabolic confinement, so that the edge spectrum is necessarily linear in angular momentum. We must now figure out how many eigenstates there are at each angular momentum. In other words we want to know how many linearly independent homogeneous symmetric polynomials there are at a given degree which satisfy the required condition that they must vanish as $k + 1$ particles come to the same position.[94]

As in the Laughlin case above, it is certainly possible to multiply a Read–Rezayi ground state wavefunction by a symmetric polynomial of qth degree in order to generate a new wavefunction with q units of angular momentum more than the ground state that still satisfies the required vanishing condition. Doing this generates a bosonic edge mode excitation entirely analogous the above described Laughlin case.

However, there are other ways to generate acceptable wavefunctions. In Eq. (12) we could multiply the first group by a symmetric polynomial of q_1^{th} degree in the variables $z_1 \ldots z_{N/k}$, then we multiply the second group by another symmetric polynomial of q_2^{th} degree in the variables $z_{N/k+1}, \ldots z_{2N/k}$ and so forth. Within each group the polynomial still vanishes whenever two particles come to the same point, and since there are k groups, it is necessarily the case that the wavefunction will vanish whenever $k + 1$ particles come to the same point (there must be at least two particles within some group). This strategy gives us vastly more possible ways to generate new polynomials satisfying our clustering rules. Unfortunately, not all such polynomials will be linearly independent from each other — and determining the dimension of the space of eigenstates turns out to be a rather tricky task.[95,96]

For the case of the Moore–Read (the $k = 2$ Read–Rezayi) state, the counting of edge modes was achieved by Milovanovic and Read[21] (along with the counting of edge modes of the 331 state and the Haldane–Rezayi state, which both have multiple spin states). The result of their work is surprisingly simple: assuming an even number of particles in the system, the Moore–Read edge is described by a bosonic mode (as in Laughlin) along with an additional Majorana edge mode. We can thus write an effective theory for the Moore-Read edge as

$$H_{MR-edge} = \sum_{n>0} (an) a_n^\dagger a_n + \sum_{m \geq 0;\, m \text{ odd}} \alpha(m/2)\, \psi_{m/2}^\dagger \psi_{m/2} \tag{31}$$

where a is bosonic (satisfying Eq. (21)) as in the Laughlin case, but ψ is fermionic, satisfying

$$\{\psi_{m/2}^\dagger, \psi_{n/2}\} = \delta_{nm}$$

with the added constraint that the parity of the number of fermions excited on the edge must match the parity of the number of electrons in the system. We can build up a similar table of the possible excitations at small angular momenta

Table 3.   Excitations of the Moore–Read Edge (assuming an even number of particles in the system).

$L - L_{ground}$	states	Energy
1	$a_1^\dagger|0\rangle$	α
2	$a_2^\dagger|0\rangle$, $a_1^\dagger a_1^\dagger|0\rangle$, $\psi_{1/2}^\dagger \psi_{3/2}^\dagger|0\rangle$	2α
3	$a_3^\dagger|0\rangle$, $a_2^\dagger a_1^\dagger|0\rangle$, $a_1^\dagger a_1^\dagger a_1^\dagger|0\rangle$, $a_1^\dagger \psi_{1/2}^\dagger \psi_{3/2}^\dagger|0\rangle$, $\psi_{1/2}^\dagger \psi_{5/2}^\dagger|0\rangle$	3α
4	$a_4^\dagger|0\rangle$, $a_3^\dagger a_1^\dagger|0\rangle$, $a_2^\dagger a_1^\dagger a_1^\dagger|0\rangle$, $a_2^\dagger a_2^\dagger|0\rangle$, $a_1^\dagger a_1^\dagger a_1^\dagger a_1^\dagger|0\rangle$, $a_2^\dagger \psi_{1/2}^\dagger \psi_{3/2}^\dagger|0\rangle$, $a_1^\dagger a_1^\dagger \psi_{1/2}^\dagger \psi_{3/2}^\dagger|0\rangle$, $a_1^\dagger \psi_{1/2}^\dagger \psi_{5/2}^\dagger|0\rangle$, $\psi_{1/2}^\dagger \psi_{7/2}^\dagger|0\rangle$, $\psi_{3/2}^\dagger \psi_{5/2}^\dagger|0\rangle$	4α

which is shown in Table 3. Note that the number of edges states at a given angular momentum is greater than that of the Laughlin case shown in Table 2 (for all angular momentum at least two greater than that of the ground state). This is what we would have suspected from our above argument that we can insert polynomial factors in a greater variety of ways for the Moore–Read and Read–Rezyayi wavefunctions than for the Laughlin state.

As in the case of the Laughlin edge, the effective Hamiltonian Eq. (31) is exact at all energies for the special parent Hamiltonian. Again for generic electron-electron interactions, it is expected that at low energy the form of Eq. (31) remains true[j] if the ground state is in the Moore-Read phase of matter, although the two velocities for the two modes need not be the same (α in the two terms will not be the same).

Analogous to the discussion of Sec. 4.1.3, the different one-dimensional edge excitations (as created by products of $a_m^\dagger$ and ψ_m operators) correspond to orthogonal two-dimensional wavefunctions only in the limit of very large systems. For finite systems, analogous to the Laughlin case, inner products of bulk states can be calculated using the exponential of a local operator S which now operates on a space involving both bosonic and fermionic excitations. See Refs. 78 and 79 for more details.

For the more general Read–Rezayi wavefunctions, the counting of edge excitations was achieved by Refs. 95 and 96, obtaining a spectrum equivalent to a so-called $\hat{su}(2)_k$ current algebra. While this may sound fairly abstract at this point, we will see in Sec. 5 another way to count edge excitations which does not require any knowledge of conformal field theory.

5. Thin Limit

A rather remarkable simplification occurs if one considers quantum Hall systems on a cylindrical or toroidal geometry, and takes the limit that the cylinder radius is very small.[97–104] In this limit the system becomes essentially a one dimensional chain of orbitals (rings around the cylinder indexed by their angular momentum), and the clustering parent Hamiltonians H_k^r discussed above, take a very simple form. For example, the Laughlin $\nu = 1/p$ special parent Hamiltonian H_1^p simply gives positive energy whenever two or more particles occupy p consecutive orbitals. The resulting grounds state wavefunctions (the wavefunctions left with zero energy) in this limit are simple charge density waves.[97] For example the $\nu = 1/3$ Laughlin state on a cylinder is given by the condition that there should be no more than one particle in any three consecutive orbitals. Thus the ground state looks as follows:

$$1\,0\,0\,1\,0\,0\,1\,0\,0\,1\,0\,0\,1$$

where 1 indicates a filled orbital and 0 represents an empty orbital. It is easy to see that for a finite cylinder one obtains maximum density (i.e. the ground state) only when the number of orbitals is $N_{orb} = 3(N-1)+1$ which matches our above counting of powers of z, for example, in Eq. (14) with $q = 1$ and $N_\phi + 1 = N_{orb}$. What is not obvious (and, indeed, is only established by numerical work[98,104,105]) is that this charge density wave is adiabatically connected to the Laughlin ground state. In other words, as the thin cylinder is made thicker, the gapped charge density wave continuously deforms into the gapped Laughlin ground state.

If we instead consider a thin torus rather than a thin cylinder, one discovers there are multiple degenerate ground states. For example, for Laughlin $\nu = 1/3$ we have the three ground states

$$1\,0\,0\,1\,0\,0\,1\,0\,0\,1\,0\,0$$
$$0\,1\,0\,0\,1\,0\,0\,1\,0\,0\,1\,0$$
$$0\,0\,1\,0\,0\,1\,0\,0\,1\,0\,0\,1$$

$$\hookrightarrow \qquad\qquad \hookleftarrow$$

where we should think of the far left as being connected up to the far right (as indicated by the hooked arrows), and here we have $N_{orb} = 3N$. This ground state degeracy persists even when the torus is not thin.[22] The ground state degeneracy on the torus is characteristic of topologically ordered matter[106,107] (indeed, it is often used as a *definition* of topological order). One should be cautious, however, that topological order requires the multiple ground states to be indistinguishable from each other by any local measurement — and this is not the case in the thin-torus limit, where one can measure the position of charges in the charge density wave. However (and this is not an obvious statement!) as the torus is made large in both directions the multiple ground states remain degenerate, and become locally indistinguishable.

Analogous thin torus versions of the general clustering parent Hamiltonians H_k^r follow a very similar rule: Give positive energy to any $k+1$ particles in r consecutive orbitals.

As an example of this let us consider the bosonic Moore–Read state at $\nu = 1$ (this is Read–Rezayi with $k = 2$, having Hamiltonian H_2^2) with the the clustering rule that there should be no more than 2 bosons in 2 consecutive orbitals. The ground state (the highest density state we can make) on a cylinder is given by

$$2\,0\,2\,0\,2\,0\,2\,0\,2$$

where **2** indicates a doubly filled orbital (which is allowed since we have bosons). Again, this gapped charge density wave ground state is continuously connected to the gapped Moore–Read ground state when the cylinder is made thicker. It is easy to count that $N_{orb} = N_\phi + 1 = N - 1$. On a torus, the clustering rule results in three different ground states given by

$$2\,0\,2\,0\,2\,0\,2\,0\,2\,0$$
$$0\,2\,0\,2\,0\,2\,0\,2\,0\,2$$
$$1\,1\,1\,1\,1\,1\,1\,1\,1\,1$$

$$\hookrightarrow \qquad\qquad \hookleftarrow$$

where again we connect up the far left to the far right, and here we have $N_{orb} = N$. As in the Laughlin case, these states remain degenerate even when the torus is not thin. While the three ground states are locally distinguishable in the thin torus limit, they become indistinguishable in the large system (and thick torus) limit, signaling topological order.

For the even more general Read–Rezayi states at $\nu = k/(2 + kp)$ the clustering rule is that there should be (a) no more than one particle in p consecutive orbitals (exactly the rule for Laughlin $\nu = 1/p$) and (b) no more than k particles in $2 + pk$ consecutive orbitals. (Compare this to the cluster Hamiltonian prescription described just after Eq. (13).) As an example, we consider the fermionic $\mathbb{Z}_3$ Read–Rezayi state at filling fraction $\nu = 3/5$ (this is $p = 1$ and $k = 3$ in Eq. (13)). Here the clustering rule is that no more than three particles are allowed in five consecutive orbitals (and no more than one particle in one orbital as required by fermionic statistics). The ground state on the cylinder is given by

$$1\,1\,1\,0\,0\,1\,1\,1\,0\,0\,1\,1\,1\,0\,0\,1\,1\,1$$

with $N_{orb} = N_\phi + 1 = 5N/3 - 2$. On the torus one has ten degenerate ground states given by

$1\,1\,1\,0\,0\,1\,1\,1\,0\,0\,1\,1\,1\,0\,0$	(and four similar translations)
$1\,1\,0\,1\,0\,1\,1\,0\,1\,0\,1\,1\,0\,1\,0$	(and four similar translations)

$$\hookrightarrow \qquad\qquad \hookleftarrow$$

where $N_{orb} = N_\phi + 1 = 5N/3$. Unsurprisingly, the torus ground state degeneracy matches predictions from conformal field theory.[108]

One can construct similar clustering Hamiltonians for other wavefunctions, such as the Gaffnian (the $\nu = 2/3$ bosonic form having the rule of no more than two bosons in three consecutive orbitals) and the Haffnian (the bosonic $\nu = 1/2$ form having the rule of no more than two bosons in four consecutive orbitals. However, here the situation is more complicated. While the Gaffnian and Haffnian hamiltonians may be gapped in the thin cylinder limit, they are not gapped when the cylinder is made thick.[32,109] Nonetheless, the states that are exactly zero energy in the thin torus limit remain exactly zero energy states as the torus is made thicker while additional states come down to zero energy only in the thermodynamic limit.

5.1. *Edge state counting*

The thin cylinder limit for systems with simple clustering rules gives a very clean way of counting the edge modes of a quantum Hall system. Let us begin with the Laughlin $\nu = 1/3$ case where the clustering rule is that there should be no more than one filled orbital in three consecutive orbitals. We will assume a half-infinite quantum Hall system with the quantum Hall state existing on the far left and the vacuum existing on the far right. The ground state is thus

$$\ldots 1\,0\,0\,1\,0\,0\,1\,0\,0\,1\,0\,0\,1\,0\,0\,1 \big| 0\,0\,0\,0\,0\,0\,0\,0\,0\,0 \ldots$$

where the vertical line marks the unexcited edge of the system (the highest angular momentum occupied orbital in the ground state). To create an edge excitation we must promote a fermion to an orbital of higher angular momentum, i.e. we move some occupied orbitals further right. For example, if we want a state with one additional unit of angular momentum we want to promote a fermion to the the right by one more step. There is only one way to do this without violating the clustering rule, given by

$$(\Delta L = 1) \quad \ldots 1\,0\,0\,1\,0\,0\,1\,0\,0\,1\,0\,0\,1\,0\,0\,0 \big| \overset{\frown}{1}\,0\,0\,0\,0\,0\,0\,0\,0\,0 \ldots$$

where we have left the vertical bar in the same place to indicate where the ground state occupation ended. There are then two ways to make an excitation with two units of angular momentum greater than that of the ground state, given by

$$(\Delta L = 2) \quad \ldots 1\,0\,0\,1\,0\,0\,1\,0\,0\,1\,0\,0\,1\,0\,0\,0 \big| 0\,\overset{\frown}{1}\,0\,0\,0\,0\,0\,0\,0\,0 \ldots$$

$$(\Delta L = 2) \quad \ldots 1\,0\,0\,1\,0\,0\,1\,0\,0\,1\,0\,0\,0\,1\,0\,0 \big| 1\,0\,0\,0\,0\,0\,0\,0\,0\,0 \ldots$$

In other words, we may promote the furthest right fermion by two orbitals, or we may promote the two furthest right fermions each by one orbital. For excitations three units of angular momentum greater than that of the ground state, there are three possibilities (promote the furthest right fermion by three orbitals; promote the furthest right fermion by two orbitals and the next furthest right fermion by one orbital; promote the three furthest right fermions each by one orbital). Comparing this counting to that of Tables 2 and 1 we can see that we are describing exactly the same counting (i.e. partitions of integers).

We can also apply the thin cylinder limit to counting edge excitations for more complicated wavefunctions, such as Moore–Read, Read–Rezayi,[42,98–100,103,104,110] and even Gaffnian, Haffnian[32,109] or bilayer wavefunctions.[102] As an example, let us consider the case of the Moore-Read state of bosons at $\nu = 1$. Here, the clustering rule is that there should be no more than 2 bosons in two consecutive orbitals. For a half-infinite cylinder we consider the ground state to be

$$\ldots 2\,0\,2\,0\,2\,0\,2\,0\,2\,0\,2\,0\,2 \big| 0\,0\,0\,0\,0\,0\,0\,0\,0 \ldots$$

where again, we use the vertical line to indicate the position of the furthest right (highest angular momentum) occupied orbital in the ground state. Let us list off the possible edge excitations in order of their angular momentum. There is a single state with angular momentum one unit greater than the ground state

$$(\Delta L = 1) \qquad \ldots 2\,0\,2\,0\,2\,0\,2\,0\,2\,0\,2\,0\,1\,\big|\,1\,0\,0\,0\,0\,0\,0\,0\,0 \ldots$$

There are three wavefunctions with angular momentum two units greater than the ground state

$$(\Delta L = 2) \qquad \ldots 2\,0\,2\,0\,2\,0\,2\,0\,2\,0\,2\,0\,1\,\big|\,0\,1\,0\,0\,0\,0\,0\,0\,0 \ldots$$

$$(\Delta L = 2) \qquad \ldots 2\,0\,2\,0\,2\,0\,2\,0\,2\,0\,2\,0\,0\,\big|\,2\,0\,0\,0\,0\,0\,0\,0\,0 \ldots$$

$$(\Delta L = 2) \qquad \ldots 2\,0\,2\,0\,2\,0\,2\,0\,2\,0\,1\,1\,1\,\big|\,1\,0\,0\,0\,0\,0\,0\,0\,0 \ldots$$

and there are five wavefunctions with angular momentum three units greater than the ground state

$$(\Delta L = 3) \qquad \ldots 2\,0\,2\,0\,2\,0\,2\,0\,2\,0\,2\,0\,1\,\big|\,0\,0\,1\,0\,0\,0\,0\,0\,0 \ldots$$

$$(\Delta L = 3) \qquad \ldots 2\,0\,2\,0\,2\,0\,2\,0\,2\,0\,2\,0\,0\,\big|\,1\,1\,0\,0\,0\,0\,0\,0\,0 \ldots$$

$$(\Delta L = 3) \qquad \ldots 2\,0\,2\,0\,2\,0\,2\,0\,2\,0\,1\,1\,1\,\big|\,0\,1\,0\,0\,0\,0\,0\,0\,0 \ldots$$

$$(\Delta L = 3) \qquad \ldots 2\,0\,2\,0\,2\,0\,2\,0\,2\,0\,1\,1\,0\,\big|\,2\,0\,0\,0\,0\,0\,0\,0\,0 \ldots$$

$$(\Delta L = 3) \qquad \ldots 2\,0\,2\,0\,2\,0\,2\,0\,1\,1\,1\,1\,1\,\big|\,1\,0\,0\,0\,0\,0\,0\,0\,0 \ldots$$

We see that this counting matches up precisely with our description of a bosonic and a Majorana fermionic edge mode as given in Table 3. With a bit of work one can show that the correspondence continues at higher angular momentum as well[96,99,104,110].

This type of thin cylinder state counting is extremely simple. For more general clustering Hamiltonians, including those for the Read–Rezayi series, the Haffnian and the Gaffnian, it can be shown that this type of thin cylinder edge state counting precisely matches the predictions of the corresponding conformal field theory.[110]

6. Squeezing and Jack Polynomials

The thin cylinder limit is advantageous because of the simplicity of the resulting wavefunctions. In the thin limit, each eigenstate is a single Fock state (a single list of occupations) rather than a superposition of many Fock states. When we make the cylinder thick again the eigenstates are not single Fock states but are rather superpositions. Nonetheless some of the structure of the thin cylinder remains.

For simplicity in this section let us assume we are thinking about a system of bosons and we are considering clustering Hamiltonians H_k^r which enforces the constraint that the wavefunction must vanish as at least r powers when $k + 1$

particles come to the same point. For the case of $r = 2$ these produce the $\mathbb{Z}_k$ Read–Rezayi wavefunctions of the form of Eq. (12) (without additional Jastrow factors) as a ground state.

In the limit of a thin cylinder H_k^r imposes a clustering rule that there should be no more than k bosons in r consecutive orbitals. Let us define the Fock states that satisfy this (k, r) clustering condition of the thin cylinder to be called (k, r)-admissable. For thick cylinders (or on the plane or sphere) these (k, r)-admissable Fock states are not eigenstates. However, each admissible Fock state is nonetheless in one-to-one correspondence with a zero energy eigenstates of the corresponding clustering Hamiltonian H_k^r.

To generate a basis of zero energy eigenstates we can use each (k, r)-admissable state as a so-called *root*. Then each root can be superposed with so-called descendant Fock states[111] to create eigenstates. A descendant state is defined from a root state by an operation known as *squeezing* where two particles are moved towards each other by one step in angular momentum. We sometimes say that if one Fock state descends from another we say it is *dominated* by the other state. The descendant states may or may not be admissable. As an example, let us consider one of the root states describing one of the $\Delta L = 3$ edge excitations of the Moore–Read state, and list off some of its descendants. The root state must be (k, r) admissable, which in this case is $(2, 2)$ admissable:

(root-admissable)	$...20202020201 10\vert 2 00000000...$
↓ squeeze	
(admissable)	$...2020202011111\vert 100000000...$
↓ squeeze	
(non-admissable)	$...2020112011120\vert 100000000...$
↓ squeeze	
(non-admissable)	$...2020103011121\vert 000000000...$

To form an eigenstate for a thick cylinder (or sphere or disk) one needs to superpose a root state with its descendants with the proper coefficients. Just knowing that an eigenstate is comprised only of a root and its descendants greatly reduces the size of the Hilbert state that one is considering and this is extremely powerful for numerical work.[111]

The determination of the exact coefficients of the Fock states, and hence the wavefunction itself, appears to be a complicated problem. We must somehow figure out how to superpose the root with all the descendant states in such a way so as to satisfy the clustering Hamiltonian on the non-thin-cylinder geometry. (For the bosonic Read–Rezayi states this would be finding the wavefunction that vanishes as r powers when $k + 1$ particles come to the same point.) Rather incredibly the answer to this problem is given by *Jack Polynomials*[9]

$$\Phi^{\nu=k/r;\text{bosons}} = J_\lambda^\alpha(z_1, \ldots, z_N)$$

where $\alpha = -(k+1)/(r-1)$ and $\boldsymbol{\lambda}$ is a (k,r)-admissable root state (in usual Jack notation this root state is presented as a partition of an integer[9] although in the FQHE world it is more convenient to specify this partition as an admissable Fock state). An enormous amount is known about these polynomials[112,113] and a recent work from the mathematical literature[94] found that for this value of α, at least when k and r are coprime, the resulting polynomials have precisely the desired vanishing properties — that the wavefunction vanishes as r or more powers when $k+1$ particles approach the same position. These Jacks provide explicit wavefunctions for the ground state, corresponding to filling fraction $\nu = k/r$ (as in Eq. (17)), as well as providing explicit wavefunctions that span the space of all the edge excitations (i.e. all lower density wavefunctions that also have zero interaction energy). The Jacks describe the $\mathbb{Z}_k$ Read–Rezayi states $(k,r) = (k,2)$ including the Laughlin $(k=1)$ and Moore–Read $(k=2)$ states, as well as the Gaffnian wavefunction $(k,r) = (2,3)$ and a host of other wavefunctions which have now been identified[114,115] as being described by so-called $WA_{k-1}(k+1, k+r)$ conformal field theories.

For both the bosonic version and the fermionic version (with one Jastrow factor attached) of these wavefunctions, the mathematical structure of the Jack polynomials obey interesting rules whereby wavefunctions on large systems can be constructed efficiently by "sewing" together wavefunctions from small systems.[116,117] These algorithms have allowed numerical work on systems that are far larger than would otherwise be possible. Indeed, the use of Jacks has become an indespensible part of the numerical FQHE toolbox.

The rich mathematical structure of the Jacks has allowed detailed analysis of these (and other) wavefunctions in a host of new ways, both numerical and analytical.[9,114,115,117–124] The Jack approach has been extended, not only to fermionic systems, but also to spin-singlet wavefunction.[116,117,125]

As an interesting aside, we point out that Jack polynomials $J^{\alpha}_{\boldsymbol{\lambda}}$ with different indices $\alpha = 1/p$ turn out to describe the eigenstates of Laughlin $\nu = 1/p$ droplets in an extremely steep confining potential.[126]

7. Entanglement

In the last decade condensed matter theory has increasingly turned to methods of quantum information for the understanding of interacting systems (see for example Ref. 127). A particularly valuable approach is to partition a system into two pieces (a "bipartition") and to examine the quantum entanglement between the pieces.[128] This approach has proven to be extremely valuable in the study of quantum Hall effect as well.

One can generally partition the Hilbert space $\mathcal{H}$ of a system into two pieces

$$\mathcal{H} = \mathcal{H}^A \otimes \mathcal{H}^B$$

This division can be made in a number of different ways. For example, one might put certain orbitals of a system in A and the other orbitals in B (known as an "orbital partition"[129]). Another possibility is that certain particles are in A and the others are in B (a "particle partition"[130]). Yet another possibility is to partition the area of a system into two sub-areas (a and b) and put particles in A if they fall in area a and into B if they fall in area b (a "real-space partition"[131-133]).

A wavefunction $|\Psi\rangle$ for the full system may be written in terms of its partitions using a Schmidt decomposition

$$|\Psi\rangle = \sum_n \lambda_n \, |\psi_n^A\rangle \otimes |\psi_n^B\rangle \tag{32}$$

where the wavefunctions $|\psi_n^A\rangle$ form an orthonormal complete set for the Hilbert space A and the wavefunctions $|\psi_n^B\rangle$ form an orthonormal complete set for the Hilbert space B.

Initial studies[130,134-136] of the entanglement of quantum Hall systems focused on the von Neumann entropy of entanglement given by

$$S_{AB} = -\sum_n \lambda_n \log \lambda_n.$$

In cases where A and B spatially partition a system into two pieces, the entanglement entropy should have the behavior

$$S_{AB} = \alpha L + D + \dots$$

where L is the length of the cut between the two pieces, α is a nonuniversal constant, and D is a subleading piece (known as the total quantum dimension) which encodes information about the topological properties of the phase of matter.[137,138]

In Ref. 129 it was emphasized that the entire spectrum of Schmidt weights λ_n in Eq. (32) contains important information and most of this is thrown away by looking only at the von Neumann entanglement entropy. The full set of Schmidt weights is known as the "entanglement spectrum".

The information contained in the entanglement spectrum is perhaps most clear when we partition the system into pieces A and B so as to conserve certain additional quantum numbers besides particle number N, such as angular momentum L. In this case, the Hilbert space has a "graded" structure

$$\mathcal{H}_{N,L} = \bigoplus_{N_A} \bigoplus_{L_A} \mathcal{H}^A_{L_A, N_A} \otimes \mathcal{H}^B_{L_B, N_B}$$

where $N = N_A + N_B$ and $L = L_A + L_B$. Each term in the Schmidt decomposition must then be similarly labeled with these quantum numbers. For example, we might have a Schmidt weight $\lambda_n(N^A, L^A)$ meaning that in the product $|\psi_n^A\rangle \otimes |\psi_n^B\rangle$, piece A has N^A particles and angular momentum L^A.

Most frequently one considers bipartitioning a sphere in a manner which is symmetric under rotations around the north-south axis (i.e. making a cut along a lattitude) so that the L_z angular momentum is conserved by the cut. A similar possibility is cutting a cylinder around an equator. Let us for now consider either an orbital or real-space partition of this sort which conserves angular momentum.

Fixing N_A, the number of particles that end up on the A side of the cut, the entanglement spectrum is conventionally plotted by showing

$$\xi = -2 \log \lambda$$

as a function of L_A the angular momentum of the particles on the A side of the cut. The values $\xi(L)$ are often referred to as the "entanglement energies" and the reduced density matrix

$$\rho_A = \sum_n \lambda_n |\phi_n^A\rangle\langle\phi_n^A|$$

can be written as $\rho_A = e^{-H_E}$ with H_E being the so-called "entanglement Hamiltonian" whose eigenvectors are $|\phi_n^A\rangle$ and corresponding eigenvalues are ξ_n.

Rather remarkably, the entanglement spectrum that results from this procedure has the same structure as the energy spectrum of the edge of the given quantum Hall state.[129,139] In other words, the entanglement Hamiltonian H_A is similar to the Hamiltonian of a system having N_A particles in a confining potential — having linearly dispersing edge modes of low energy excitations in the long wavelength limit.

For example, for wavefunctions having special parent Hamiltonians, such as the Laughlin wavefunction (or the Read–Rezayi states or any of the other $\nu = k/r$ Jack wavefunctions), the counting of states in the edge spectrum (the number of modes at any given angular momentum L) precisely matches that of the entanglement spectrum. Such countings (so-called "entanglement-spectroscopy") are now commonly used as a way to identify a particular phase of matter from exact diagonalization.[139]

One can also consider wavefunctions from more realistic Hamiltonians. For example, considering a Coulomb interaction between electrons one finds the ground state at $\nu = 1/3$ to be a quantum Hall state, presumably in the Laughlin phase of matter. The corresponding ground state wavefunction of the Coulomb interaction may be extremely similar to the Laughlin wavefunction, but will have some small differences. In the entanglement spectrum, there will be some modes with low entanglement energy which match that of the Laughlin wavefunction, but at high entanglement energy (smaller λ weight in the Schmidt decomposition) there are additional modes indicating some deviation from the ideal wavefunction.[140] A so-called "entanglement gap" separates the low entanglement energy expected modes and the high entanglement energy additional modes.

One can also consider the entanglement spectra of wavefunctions that do not have a simple parent Hamiltonian or simple Jack description. For example, one can consider the entanglement spectrum of a Jain wavefunction.[141–143] As one might expect, one obtains a spectrum corresponding to multiple bosonic modes which matches the presumed edge spectrum, along with extra modes with small weight above an entanglement gap.

There are several discussions of why the edge spectrum and the entanglement spectrum have such similar structure.[78,131,144,145] Although none of these arguments is particularly simple, the approach by Ref. 144 is useful to discuss as it introduces some general ideas (although many of the same ideas reoccur in different form in Ref. 78). One considers splitting the Hamiltonian into the part acting on each part of the partition H_A and H_B as well as a coupling between the two pieces H_{AB}

$$H = H_A + H_B + \lambda H_{AB}.$$

If we consider a case where λ is small (and assuming the system is everywhere gapped) then one can qualitatively understand the structure. If we consider H_A only, acting on a fixed number of particles N_A which are in the A part of the system, we basically have a quantum Hall system with a gapless edge (a quantum Hall droplet). Similarly on the other side of the boundary with H_B we have a system with a gapless edge. The small Hamiltonian λH_{AB} cannot effect the states in the bulk of either the A or B side, since the bulk is gapped. However, λH_{AB} couples (and generically will entangle) the otherwise gapless edge modes of the two sub-systems. This immediately gives us the correct number of states in the entanglement spectrum. That is, the counting of states at each angular momentum matches between the entanglement spectrum and the edge state spectrum.

In some more detail, the two chiral systems with boundary can be thought of as a single achiral system with boundary — and with renormalization group arguments, one can employ ideas of boundary conformal field theory. In particular, at long length scale it is argued that the system must have conformal invariance. It is well known that for rational conformal field theories, the only possible conformally invariant boundary states are so-called Ishibashi states[146] which entangle only states with complementary quantum numbers, in fact giving a maximally entangled state.[78,144] These Ishibashi states have a completely flat entanglement spectrum. This is the fixed point of the renormalization procedure. For a system of finite size, one needs to move back away from the fixed point and examine how the Schmidt weights vary. The leading term is given by a simple linear spectrum in angular momentum

$$H_E = \text{constant} + \kappa L + \dots \tag{33}$$

entirely analogous to a linear spectrum of a confined edge where the constant κ is inversely proportional to the length of the boundary and depends also on the magnitude of the coupling λ between the two sides. Subleading terms are smaller in orders of the length of the boundary (i.e. in powers of $1/\sqrt{N}$).

In Ref. 78 the analysis is taken somewhat further. Focusing on the case where the Hamiltonian is uniform across the entire system, and on cases where there is a special parent Hamiltonian, one can use the ideas of writing the edge state inner products as the exponential of a local operator, as in Sec. 4.1.3, to determine the entanglement Hamiltonian — which analogously can be expanded in terms involving only local operators of an appropriate field. Indeed, the expansion of the subleading terms in the entanglement Hamiltonian is identical to the expansion of the local operator that determines the inner product between bulk wavefunctions as discussed in Sec. 4.1.3 above.

8. Localized Quasiholes

After this extensive discussion of edge excitations of model quantum Hall states, we turn to the question of localized excitations in the bulk. There are naturally two types of low energy "quasiparticle" excitations: The "quasielectron", a minimal localized packet of charge with the same sign as that of the underlying electrons, and the "quasihole", a minimal localized packet of charge with the opposite sign. If we continue to think about the case of simple special parent Hamiltonians, the quasiholes are particularly simple as they have zero interaction energy, like the edge states we have already considered. Indeed, given that our definition of the edge excitations spans the entire space of possible wavefunctions with zero interaction energy, the quasiholes live within the space of what we termed edge excitations. Unless they happen to sit at a particularly symmetric point, like the center of a disk, or a pole of the sphere, the wavefunctions in the presence of quasiholes are not generally angular momentum eigenstates, but are rather superpositions of many different angular momenta. To see how this works, let us consider the case of the well-known Laughlin quasihole

$$\Phi_{qh}(w; z_1, \ldots, z_N) = \left[\prod_{i=1}^{N}(z_i - w)\right] \Phi_{\text{Laughlin}}(z_1, \ldots z_N) \tag{34}$$

The physics here is that the quasihole prefactor pushes charge away from the point w. Indeed, pushing this minimal unit of charge away from the point w is equivalent to Laughlin's adiabatic flux insertion argument.[1] Since insertion of a flux quantum pushes ν units of electron charge away from the flux, we can conclude that the quasihole charge is $-\nu$ times the charge of the underlying electron.

Note that this form of wavefunction is also a polynomial times the Laughlin ground state and is therefore a state with zero interaction energy. However, this polynomial is not homogeneous in degree, and therefore is not an angular momentum eigenstate. Instead, the quasihole factor includes all possible angular momenta in a superposition chosen in such a way that a hole is localized at a single point. To see how this happens, let us expand the quasihole factor into its angular momentum components

$$\prod_i(z_i - w) = (-w)^N + e_1(-w)^{N-1} + e_2(-w)^{N-2} + \ldots e_{N-1})(-w) + e_N$$

where e_j is the symmetric monomoial

$$e_j = \sum_{i_1 < i_2 < \ldots i_j} z_{i_1} z_{i_2} \ldots z_{i_j}.$$

Thus, the quasihole is a "coherent" superposition of many edge state excitations. In this expansion no z_i ever occurs with greater than one power — thus the wavefunction corresponds to increasing N_ϕ by exactly one flux quantum. Note that the component of the wavefunction including e_j has j units of angular momentum (i.e. j powers of z's) added to the Laughlin ground state. Further, the factors of w^p

represent the "wavefunction" of the quasihole. Analogously to having an electron in a z^p orbital, this puts the quasihole in the w^p orbital. Similar coherent state constructions of localized quasiholes have also been achieved for other Jack ground state wavefunctions (including Moore–Read, Read–Rezayi, Gaffnian, etc.) in terms of superpositions of Jack polynomials.[119]

A more direct way to understand the quasiholes of quantum Hall states is via explicit wavefunction expressions. As an example, let us consider the Moore–Read wavefunction written in the form of Eq. (15). Here we can insert two quasiholes, one at position w and another at position w', by rewriting the matrix factor of Eq. (16) representing the pairing wavefunction as[5]

$$M_{ij} = \frac{(z_i - w)(z_j - w')}{z_i - z_j} \tag{35}$$

and we write a wavefunction for all the electrons in the presence of the quasiholes[k] as $\Phi = \mathrm{Pf}(\mathbf{M}) \prod(z_i - z_j)$ as in Eq. (15). In Eq. 35 one particle of the z_i, z_j pair sees a zero of the wavefunction at position w and the other sees a zero of the wavefunction at position w'. Of course in the end the wavefunction is fully (anti)symmetric for (fermions)bosons. It is easy to check that this wavefunction remains a zero energy eigenstate of the three-body special parent Hamiltonian (e.g. for bosons the wavefunction vanishes when three particles come to the same point). Note that if the two quasihole coordinates coincide, $w = w'$ this would be equivalent to multiplying the original Moore–Read wavefunction by a Laughlin quasihole factor similar to Eq. (34), and thus would amount to a compound quasihole of total charge $-\nu$ times the charge of the underlying electron. This charge is split into two when w and w' are moved apart from each other. Thus, each such quasihole has charge $-\nu/2$ times the charge of the constituent "electron". This should fit with our picture where only half of the electrons see a zero of the wavefunction at each quasihole coordinate.

One can add more quasiholes similarly by writing[147]

$$M_{ij} = \frac{\prod_{a=1}^{M}(z_i - w_a)(z_j - w'_a)}{z_i - z_j} \tag{36}$$

which then has $2M$ quasiholes. For some fixed set of $2M$ quasihole coordinates, one then has a choice of which of these coordinates to put in the w group and which to put in the w' group. This gives a large number of different possible quasihole wavefunctions. In fact we have $\frac{1}{2}(2M$ choose $M) = \frac{1}{2}(2M)!/(M!)^2$ different wavefunctions corresponding to the different possible choices. Crucially, only 2^{M-1} of these wavefunctions are linearly independent in agreement with predictions from conformal field theory.[2,3,20,148] A very nontrivial statement is that for a large system with many electrons, and where all of the quasiholes are far apart from each other, there is no local measurement that can distinguish these different wavefunctions from each other. This statement, while not proven, is believed to be true

---

[k]Usually a Pfaffian is defined only for antisymmetric matrices, so we should antisymmetrize M_{ij} before taking the Pfaffian.

and is supported by both numerical[149-152] and theoretical[148,153] work. It is this fact that is the essence of non-Abelian statistics, which we will discuss further in Sec. 8.1 below.

A different form of the Moore–Read wavefunction with quasiholes is given by the generalization of Eq. (12). (Recall that the $k = 2$ case of the Read–Rezayi wavefunction is the Moore-Read wavefunction.) Here we add quasiholes by adding factors similar to that of the Laughlin quasiholes (as in Eq. (34)) to each group of particles in Eq. (12) to obtain[34]

$$\Phi^{\nu=1/2;\text{ bosons}}_{\text{Moore-Read with qh's}} = \mathbb{S}\left[\left(\prod_{a=1}^{M}\prod_{0<s\leq N/2}(z_s - w_a)\prod_{0<i<j\leq N/2}(z_i - z_j)^2\right)\right.$$
$$\left.\left(\prod_{a=1}^{M}\prod_{N/2<s\leq N}(z_s - w'_a)\prod_{N/2<i<j\leq N}(z_i - z_j)^2\right)\right].\text{(37)}$$

Again it is easy to check that this wavefunction is a zero energy state of the special parent Hamiltonian (the wavefunction must vanish when three particles come to the same point, since two of the particles must be in the same group). Similar to the case of Eq. (36) we have a choice of which quasihole coordinates to assign to the w group and which to the w' group. In fact again we have $\frac{1}{2}(2M$ choose $M)$ different possible choices, which span the 2^{M-1}-dimensional space.

This form of wavefunction can easily be generalized to the entire Read–Rezayi series.[34] For the $\mathbb{Z}_k$ Read–Rezayi state with kM quasiholes, we divide the quasihole coordinates into k groups and write

$$\Phi^{\nu=k/2;\text{ bosons}}_{\text{Read-Rezayi with qh's}} = \mathbb{S}\left[\left(\prod_{a=1}^{M}\prod_{0<s\leq N/k}(z_s - w_{a,1})\prod_{0<i<j\leq N/k}(z_i - z_j)^2\right)\right.$$
$$\left(\prod_{a=1}^{M}\prod_{N/k<s\leq 2N/k}(z_s - w_{a,2})\prod_{N/k<i<j\leq 2N/k}(z_i - z_j)^2\right)\cdots$$
$$\cdots\left.\left(\prod_{a=1}^{M}\prod_{N(k-1)/k<s\leq N}(z_s - w_{a,k})\prod_{N(k-1)/k<i<j\leq N}(z_i - z_j)^2\right)\right] \quad\text{(38)}$$

so that in the pth group of particles we have inserted M quasiholes at positions $\{w_{a,p}\}$ for $a = 1\ldots M$. Each $w_{a,p}$ position corresponds to a local depression in density. However, only the z coordinates in the pth group see a zero of the wavefunction at position $w_{a,p}$ for any a. As with the Moore–Read states there is a huge freedom to choose which coordinates to put in which group. And again many of the resulting wavefunctions are not linearly independent. Using a slightly different form of the wavefunction the number of linearly independent wavefunctions is enumerated in Ref. 96. As in the Moore–Read case, when the quasiholes are sufficiently far apart from each other the different linearly independent wavefunctions are presumed to be indistinguishable by any local measurement.

8.1. Non-Abelian statistics in brief

The ideas of non-Abelian statistics have been reviewed elsewhere.[3,154,155] However, for completeness it is worth describing the basic idea here.

Let us consider any of the $\mathbb{Z}_k$ Read–Rezayi states for $k > 1$, and let us suppose we now have some number M of quasiparticles[l] at well separated positions $w_1 \ldots w_M$. There is a multi-dimensional space of degenerate states which can describe quasi-particles at these positions (let us call this dimension D). As we claimed above, the states of this space are indistinguishable by any local measurements. Given any orthonormal basis for this space $|\Psi_n(w_1, \ldots, w_M)\rangle$, a generic wavefunction is some linear combination

$$|\Psi(w_1, \ldots, w_M)\rangle = \sum_{n=1}^{D} A_n |\Psi_n(w_1, \ldots, w_M)\rangle$$

with arbitrary complex coefficients A_n subject to normalization $\sum_n |A_n|^2 = 1$.

Now we consider moving the particles adiabatically. We can imagine that the quasiparticles are attracted to microscopic trapping potentials (small charges) and these traps are slowly moved. Suppose now that at the end of this motion the quasiparticles are again at the same positions $w_1, \ldots w_M$. The wavefunction must again be assembled from the same orthonormal basis, so we write

$$|\Psi'(w_1, \ldots, w_M)\rangle = \sum_{n=1}^{D} A'_n |\Psi_n(w_1, \ldots, w_M)\rangle$$

and we must have

$$A'_m = \sum_n U_{mn} A_n \tag{39}$$

where U_{mn} is a unitary D by D matrix which depends only on the path the particles have taken (assuming the eigenstate energy is always zero[m]). More to the point, up to an Abelian phase the unitary matrix depends only on the topology of the path — i.e. on what knot has been formed by the space-time world lines.

The controlled application of unitary matrices to a given Hilbert space is most of what is required in order to build a quantum computer. All we really need in addition is the ability to initialize and measure the system. The idea of using such braiding to build a quantum computer is known as topological quantum computation.[3,154] The essence of this approach is the following. The wavefunction A_n where we store the quantum information can *only* be altered by braiding quasi-particles around each other. Presuming the temperature is low enough that no stray quasielectron-quasihole pairs are created, the quantum information is thus protected from small amounts of noise. Computations correspond to moving the

---

[l]It is simplest to assume quasiholes with a special parent Hamiltonian so that the energy is simply zero. However, most of the argument is unchanged for quasielectrons and even without a fine-tuned Hamiltonian.

[m]This assumption is appropriate for quasiholes with a special parent Hamiltonian. Without this assumption there is an additional Abelian phase $\exp(i \int E(t)dt)$.

quasiparticles around in particular ways to implement desired unitary transformations.[3,156,157] This motion need not be performed precisely, as any motion that is
topologically equivalent to the desired motion has the same effect (up to an Abelian
phase).

8.1.1. *Calculating the braiding matrix*

An important question is how one actually calculates the matrix U in Eq. (39)
given a particular braid. As above we suppose that for each given set
of positions $\{w_1, \ldots, w_M\}$ there is an orthonormal basis of wavefunctions
$|\Psi_n(w_1, \ldots, w_M; z_1, \ldots, z_N)\rangle$. Here we remind the reader that the z's are physical coordinates of electrons, and the wavefunction must be single valued in these
z's. However, the w's are simply parameters of the wavefunction, and the wavefunction need not be single valued in these coordinates — there may be branch cuts.
If we move the w coordinates around and then bring them back to their original
points, we may return on a different Riemann sheet. We should still have a set
of D wavefunctions spanning the same space. However, if we return on a different
Riemann sheet, we might describe the space in a different basis. A simple example of this would be if we move w_1 all the way around w_2 and we find that Ψ_1
and Ψ_2 have switched with each other due to branch cuts.[n] This effect is known
as the *monodromy* of the wavefunction, and is represented by a matrix $\mathcal{M}$ — this
is the explicit change in the wavefunctions that you get by simply moving the w
coordinates around each other.

However, in addition to the monodromy, there is a second contribution to the
unitary braiding matrix U in Eq. (39). This contribution is a "Berry matrix" term.
Let the positions w_i move adiabatically as a function of some continuous parameter
τ (which may or may not be time) so we have $w_i(\tau)$. The D-dimensional unitary
matrix is then given explicitly by the so-called Berry matrix

$$\mathcal{B} = \mathcal{P} \exp\left[i \int d\tau \left\langle \Psi_n | \frac{\partial}{\partial \tau} | \Psi_m \right\rangle \right] \tag{40}$$

$$\equiv 1 + i \int d\tau \left\langle \Psi_n | \frac{\partial}{\partial \tau} | \Psi_m \right\rangle - \int d\tau \int^\tau d\tau' \sum_q \left\langle \Psi_n | \frac{\partial}{\partial \tau} | \Psi_q \right\rangle \left\langle \Psi_q | \frac{\partial}{\partial \tau'} | \Psi_m \right\rangle + \ldots$$

where $\mathcal{P}$ stands for path ordering of the exponential, the first few terms of which
are expanded out in the second line.

This type of evaluation of the Berry matrix is something one can imagine doing
numerically.[149–152] However, since integrals are often hard to do numerically, a
practical way of evaluating this Berry matrix is given by dividing the path along τ
into many small piece $p = 1, \ldots, \mathcal{N}$ with $\mathcal{N}$ very large (and we have assumed the

---

[n]We could have holomorphic forms like $\Psi_\pm = 1 \pm \sqrt{w_1 - w_2}$ which exchange when w_1 goes around
w_2.

final set of positions w_j is the same as the initial set of positions). We then have°

$$\mathcal{B} = \prod_{p=1}^{N-1} \left\langle \Psi_a(\tau_{p+1}) \middle| \Psi_b(\tau_p) \right\rangle$$

where each term in the product on the right is a matrix with subscript a, b (ranging from 1 to the dimension D), and these matrices must be multiplied together in order to give the final matrix $\mathcal{B}$. Note that although the positions $w_i(\tau)$ at the last step τ_N are the same as the positions at the first step τ_1 the wavefunctions $\Psi_n(\tau_1)$ and $\Psi_n(\tau_N)$ need not be the same as each other due to the monodromy (although the space spanned by all n must be the same space).

The unitary transformation U in Eq. (39) is given by the product of the monodromy and the Berry matrix

$$U = \mathcal{M}\mathcal{B}. \tag{41}$$

The Berry matrix is the result of the adiabatic motion of the particles, whereas the monodromy simply accounts for the fact that when we get the particles back to their original position we may be working in a different basis from when we started.

Note that we are allowed to redefine (rotate) our basis of states arbitrarily at any point along the path (we have only promised that at each point along the path we have an orthonormal basis). Such a redefinition can change both the monodromy and the Berry matrix. However, the final product of the two, the unitary matrix U is independent of such redefinitions.

Numerical calculations of braiding statistics had been done for quasiparticles in abelian quantum Hall states,[158–160] and more recently for non-Abelian quantum Hall states as well.[149–152]

8.1.2. *Holomorphic orthonormal basis*

A great simplification occurs if one works with a basis which is both holomorphic and orthonormal. In the CFT approach it is assumed that the so-called conformal block basis is exactly such a holomorphic, orthonormal basis.[24,153,161] We will return to this assumption, as it is an important one. However, for now, let us simply assume we have a holomorphic orthonormal basis of the following form:

$$\Psi_n(w_1, \ldots, w_M; z_1, \ldots z_N) = \phi_n(w_1, \ldots w_M; z_1, \ldots z_N) \tag{42}$$

$$\times \exp\left[-\frac{1}{4\ell^2} \left(\sum_{n=1}^{N} |z_n|^2 + \frac{e^*}{e} \sum_{n=1}^{M} |w_m|^2 \right) \right]$$

such that

$$\langle \Psi_n | \Psi_m \rangle = \delta_{nm}. \tag{43}$$

where e^* is the charge of the quasiparticle, and we have reinstated the magnetic length $\ell = 1/\sqrt{eB}$ with $\hbar = 1$. The wavefunction ϕ_n is holomorphic in all variables,

---

°This discrete product can in fact be used to define the path-ordered integral in the limit that we have many steps along the path.

and must be single-valued in the z's but may have branch cuts with respect to the w coordinates so there will generally be a nontrivial monodromy from moving the quasiparticles.

Now we calculate the Berry matrix in Eq. (40) explicitly, assuming we have a holomorphic normalized wavefunction. For simplicity, let us assume we are moving only a single quasiparticle at position w while keeping the other quasiparticles fixed. We write

$$\partial_\tau = \frac{\partial w}{\partial \tau} \partial_w + \frac{\partial \bar{w}}{\partial \tau} \partial_{\bar{w}}. \tag{44}$$

Since ϕ_n (in Eq. (42)) is holomorphic, the only contribution to $\partial_{\bar{w}}$ comes from the derivative acting on the gaussian factor and we get

$$\langle \Psi_n | \partial_{\bar{w}} | \Psi_m \rangle = -\delta_{nm} \frac{e^*}{e} \frac{w}{4\ell^2}. \tag{45}$$

In order to evaluate ∂_w, we note that $\partial_w \langle \Psi_n | \Psi_m \rangle = 0$ since the basis is orthonormal for all values of w. We thus have

$$\langle \Psi_n | \partial_w | \Psi_m \rangle = -(\partial_w \langle \Psi_n |) | \Psi_m \rangle = \delta_{nm} \frac{e^*}{e} \frac{\bar{w}}{4\ell^2} \tag{46}$$

where again we have used that the wavefunction ϕ is holomorphic (so ϕ^* is anti-holomorphic). We then have the Berry phase expression (inserting Eq. (44) into Eq. (40))

$$\mathcal{B} = \mathcal{P} \exp \left[i \oint d\tau \left\langle \Psi_n | \frac{\partial}{\partial \tau} | \Psi_m \right\rangle \right]$$

$$= \delta_{nm} \exp \left[i \frac{e^*}{4e\ell^2} \oint (\bar{w} dw - w d\bar{w}) \right] = \delta_{nm} \exp \left[i \frac{Ae^*}{\ell^2 e} \right]$$

where A is the area surrounded by the path of the particle (which we obtained using the complex version of Stokes theorem) and this Berry matrix is now completely diagonal. This phase is precisely the expected Aharonov–Bohm phase we should get for moving a charge e^* around a total magnetic flux AB.

Thus if we work in a holomorphic orthonormal basis, the Berry matrix is simply the Aharonov–Bohm phase. The matrix part of the braiding matrix U in Eq. (39) is entirely from the monodromy contribution in Eq. (41). Thus we can determine the braiding properties of a wavefunction simply by examining the branch cut structure of the holomorphic orthonormal wavefunctions.[24,161]

Thus we see the value of working with a holomorphic orthonormal basis. In the conformal field theory approach, it is assumed that the so-called "conformal blocks" generate just such a basis. Explicit forms of the conformal blocks have been worked out for the case of the Moore–Read state in Refs. 2,147 and 162. We will discuss this issue further in Sec. 10 below. However, we comment here that one can use some of the methods of conformal field theory to try to prove this assumption of orthonormality. While the orthonormality cannot typically be rigorously proven, it can often be reduced to an equivalent statement about screening of some effective plasma which can then be checked by numerics.[3,5,25,40,147,153,161]

One might worry that perturbations to the wavefunction might change the braiding properties. One particular concern that was raised, for example, is that Landau level mixing destroys the nice holomorphic properties of the wavefunctions — and could change the braiding statistics (or ruin their topological nature altogether).[163] However, it has been proven that small perturbations cannot change the non-Abelian part of the braiding — although the Abelian geometric phase *can* in fact be altered.[164]

9. Hall Viscosity

The Berry matrix calculation, as in Eq. (40) describes how the wavefunction varies as some parameter changes adiabatically. In the above case, we are concerned with the change of the wavefunction as quasiparticles are moved around. However, the idea is more general: for any wavefunction which is a function of parameters, a Berry matrix (or Berry phase) can be defined.

A very useful case to study is a Berry phase associated with a change in geometry of a system for the ground state wavefunction. Here we will consider a sheer deformation of the wavefunction on a torus and we will use the Berry connection to calculate a certain non-dissipative sheer viscosity of the system, known as the Hall viscosity.[165,166]

We consider a quantum Hall system on a torus[22] geometry. To describe a torus, we take a complex plane and define a parallelogram via the four points 0, 1, τ and $1+\tau$ where $\tau = \tau_1 + i\tau_2$ with τ_i real. Identifying opposite edges of this parallelogram, we get a torus. We thus have a wavefunction which is a function of two parameters (τ_1, τ_2).

When a normalized wavefunction varies as a function of two parameters, we can define a connection

$$A_\mu = i\langle \Psi | \partial_\mu | \Psi \rangle \tag{47}$$

and then define a curvature scalar[P]

$$F = \epsilon_{\mu\nu} (\partial_\mu A_\nu - \partial_\nu A_\mu).$$

The Hall viscosity is then given by

$$\eta_H = -\frac{4\tau_2^2}{A} F$$

with A the area of the system. This link between a Berry curvature and a viscosity was uncovered in the 1990s in the case of the integer quantum Hall effect,[165] and was extended more recently to the fractional case.[24,25,167,168]

[P]One may be concerned in Eq. (47) that for fractional quantum Hall systems we should be keeping track of the multiple ground states on the torus. However, the inner product in Eq. (47) turns out to be diagonal in this additional index because the different ground states have different quantum numbers in the cases of interest.

The Hall viscosity is a transport coefficient from classical fluid and elasto-dynamics. One defines a stress tensor σ_{ab} via the force f_a

$$\dot{g}_a = f_a = -\partial_b \sigma_{ab}$$

with g_a the momentum density. Then this stress can be expanded

$$\sigma_{ab} = -\lambda_{abcd} u_{cd} - \eta_{abcd} \dot{u}_{cd} + \ldots \tag{48}$$

where u_{cd} is the symmetric local strain tensor

$$u_{cd} = \frac{1}{2}(\partial_c u_d + \partial_d u_c)$$

with u_c the local displacement. In Eq. (48), λ is the elasticity tensor (in an isotropic fluid $\lambda \propto \delta_{ab}\delta_{cd}$), and η is the viscosity. In an isotropic system σ is symmetric, and hence λ and η are symmetric under exchange of their first two indices or under exchange of their last two indices.

Let us now focus on the viscosity term. Note the viscosity gives a response to $\dot{u}_{cd}$, and this can be rewritten as

$$\dot{u}_{cd} = \frac{1}{2}(\partial_c v_d + \partial_d v_c)$$

in terms of the velocity field v (which we would expect to be the important quantity for a fluid). We can then further decompose the viscosity into a symmetric and antisymmetric piece under exchange of first two with last two indices

$$\eta_{abcd} = \eta_{abcd}^S + \eta_{abcd}^A$$

where $\eta_{abcd}^S = \eta_{cdab}^S$ and $\eta_{abcd}^A = -\eta_{cdab}^A$. For an isotropic fluid we expect

$$\eta_{abcd}^A = \eta^H (\delta_{bc}\epsilon_{ad} - \delta_{ad}\epsilon_{bc})$$

where we call η^H the Hall viscosity.[24,25,165,167,168] Like the Hall conductivity, the antisymmetric part here is non-dissipative.

Rather remarkably,[24,25] the Hall viscosity is not only related to the Berry phase but also turns out to be related to the intrinsic orbital angular momentum of the electrons. The orbital angular momentum, in turn, is related to the shift $\mathcal{S}$ of the wavefunction (see Eq. (5)), giving the general result

$$\eta^H = \hbar \frac{\bar{n}\mathcal{S}}{4} \tag{49}$$

where $\bar{n}$ is the average electron density.

The Hall viscosity appears, not only in the fluid dynamics equations, but also as a coefficient of the electromagnetic response of the system in various limits. For example, neglecting Zeeman energy we have the low frequency limit of the Hall conductivity at finite wavevector q given by[167–169]

$$\sigma_{xy}(q) = \frac{e^2}{\hbar}\left(\frac{\nu}{2\pi} + [\eta^H - \hbar\bar{n}](q\ell)^2 + \ldots\right)$$

where $\bar{n}$ is the particle density.

10. Comments on Conformal Field Theory

Although much of the point of this review was to attempt to discuss recent developments without resorting to conformal field theory, it is probably worth making some comments on the CFT mapping. Starting with the work of Moore and Read,[2] this approach to understanding quantum Hall states has been increasingly influential in the field. While certain techniques related to CFT can be applied to almost any FQH state,[3,4] we will see below that the states which are defined by special parent Hamiltonians fit the CFT framework much better than states that do not have such Hamiltonians.

10.1. *The bulk-edge correspondence: In brief*

A conformal field theory is generally defined as a quantum field theory that is invariant under conformal transformations.[55] In 1+1 dimensions (the only case that we will consider) this conformal invariance is such a strong restriction that many properties of such field theories can be obtained exactly. It is crucial to know that $(1+1)$-dimensional CFTs are very closely related to $(2+1)$-dimensional topological quantum field theories[3,55] — a connection that is at the root of the Moore-Read approach.

   The general idea of the FQHE-CFT connection is as follows: as a topologically ordered state of matter, one expects that much of the FQHE physics can be described in terms of a topological quantum field theory — a theory which is invariant under smooth deformations of space and time.[3] To expose the interesting physics that results, we imagine examining a two-dimensional slice of our (2+1)-dimensional space-time manifold. Since the system is topologically invariant we expect that any direction we slice it will be equivalent. What we mean by "equivalent" here is that any type of slice should be described by the same two-dimensional quantum theory. Let us label the three dimensions of our space-time as x, y and t (for time). If we slice the system at a fixed y-coordinate (giving us an x, t plane), the resulting theory is a $(1+1)$-dimensional CFT which describes the dynamics of the gapless FQHE edge. On the hand if we slice the manifold at fixed time (giving us an x, y-plane) the resulting $(2+0)$-dimensional static wavefunction is described by the *same* CFT (we will explain below in more detail how we describe a wavefunction using a CFT). This is an example (indeed, it is perhaps the best explored example) of what is known as "bulk-edge correspondence".

   While CFTs can be either unitary or nonunitary, only the unitary CFTs represent well behaved one dimensional dynamical systems. (Nonunitary CFTs have, for example, scattering matrices that do not conserve amplitude.) Since the one dimensional edge can only be described by a unitary CFT, this implies that either the bulk-edge correspondence somehow fails, or the CFT must be unitary.[29] One way in which this correspondence can indeed fail, is if the bulk is gapless so that edge excitations can leak into the bulk. This gives the rough argument why gapped

FQH states can only be described by unitary CFTs (although a number of other arguments have also been given[24]).

We now explain a bit more detail of the FQHE-CFT correspondence as it is described in the original Moore–Read work.[2] In short, given an appropriate $(1+1)$-dimensional CFT we write a $(2+0)$-dimensional wavefunction as a correlator of the $(1+1)$-dimensional theory

$$\Psi_{2d}(z_1, \ldots, z_N) = \langle \psi_e(z_1) \ldots \psi_e(z_N) \rangle_{1+1\,d}. \tag{50}$$

On the left, $z = x + iy$ represents a position on a two-dimensional plane, whereas on the right $z = x + i\tau$ is a space-time coordinate in a $(1+1)$-dimensional chiral theory (the theory is chiral, depending only on $x + i\tau$, but not $x - i\tau$, which when rotated back to real time gives a theory depending on $x - vt$ but not $x + vt$) with v some velocity. On the right of Eq. (50), ψ_e is an appropriately chosen operator within the $(1+1)$-dimensional theory.[q] The details of this correspondence have been reviewed elsewhere.[3,4] Note that the electron field ψ_e has a (conformal) scaling dimension (or conformal weight) called h_e which describes the long and short distance limits of various correlation functions. For example, we have[r]

$$\langle \psi_e(z) \psi_e^\dagger(z') \rangle = (z - z')^{-2h_e}$$

which (for a two point correlator) is correct at all distance. This scaling dimension turns out to be related to the shift (see Eqs. (5) and (49)) of the quantum hall state (defined in Eq. (50)) via the simple relation $\mathcal{S} = 2h_e$. We thus identify the scaling dimension as being the mean orbital spin of the electron in the state (see the discussion near Eq. (5)).

The correspondence in Eq. (50), admittedly, looks a bit strange. On the left, we have a wavefunction, whereas on the right, we have an expectation. In trying to elucidate the meaning of this correspondence, it is perhaps useful to think about a slightly more general case. Instead of thinking about a wavefunction on a disk geometry, let us think about an annulus (or similarly a cylinder).[78] And instead of thinking about the ground state, let us think about a situation where both inner and outer edges of the annulus are in excited states. Here we can write a more general wavefunction as

$$\Psi_{2d}^{u,v}(z_1, \ldots, z_N) = \langle u | \psi_e(z_1) \ldots \psi_e(z_N) | v \rangle_{1+1\,d} \tag{51}$$

where u and v are excited states of the $(1+1)$-dimensional system. This corresponds to an annulus with excitation $|u\rangle$ on the inner edge and excitation $|v\rangle$ on the outer edge. We can think of these as the in-state and out-states of the $(1+1)$-dimensional CFT.

[q]For the experts we have not written the background charge operator for notational simplicity — it is this background charge operator that gives the ubiquitous gaussian factors of the lowest Landau level wavefunctions, which are the only part of the wavefunction that is not holomorphic. See the discussion in Ref. 4 for example.

[r]Also for the experts, here since we mean the ψ_e field to contain both the charge and neutral sectors, the scaling dimension is the sum of charge and neutral scaling dimensions.

Returning now to a disk geometry (or an annulus where the inner edge remains in the ground state), we can re-examine Eq. (28) and see that this is now rewritten as

$$\langle u|e^{-S}|v\rangle = \int d^2\mathbf{z}_1 d^2\mathbf{z}_2 \ldots d^2\mathbf{z}_N \ \langle u|\psi_e^\dagger(z_N)\ldots\psi_e^\dagger(z_1)|0\rangle \ \langle 0|\psi_e(z_1)\ldots\psi_e(z_N)|v\rangle$$

$$= \left\langle u \left| \int d^2\mathbf{z}_1 d^2\mathbf{z}_2 \ldots d^2\mathbf{z}_N \psi_e^\dagger(z_N)\ldots\psi_e^\dagger(z_1)|0\rangle \ \langle 0|\psi_e(z_1)\ldots\psi_e(z_N) \right| v \right\rangle,$$

the second line making the form of operator e^{-S} obvious. However, the fact that S should be expandable in $1/N$ (and should vanish for large N) is not immediately obvious — nor is the crucial statement that S should be comprised of local operators from the edge CFT.

Note that once we can write all edge excitations as excited states of the CFT, we can form coherent states of these excitations to make localized quasiholes. This, however, can be rewritten using local operators from the CFT

$$\Psi(w_1,\ldots,w_M;z_1,\ldots,z_N) = \langle \psi_{qh}(w_1)\ldots\psi_{qh}(w_M)\psi_e(z_1)\ldots\psi_e(z_N)\rangle \qquad (52)$$

where ψ_{qh} is an appropriately chosen local operator from the $(1+1)$-dimensional CFT. As discussed above in Sec. 8.1.1, we expect that such an expression does not typically represent a single wavefunction but rather represents a vector space of conformal blocks.

One can also now rephrase the orthormality hypothesis in Eq. (43). Written out in terms of the $(1+1)$-dimensional CFT, we are asking that for any fixed w's, we have

$$\delta_{nm} = \int d^2\mathbf{z}_1 \ldots d^2\mathbf{z}_N \ \Psi_n^*(w_1,\ldots,w_M;z_1,\ldots,z_N) \ \Psi_m(w_1,\ldots,w_M;z_1,\ldots,z_N)$$

where here the subscript n and m indicates which conformal block we are considering. It is this statement that has been examined in depth in Refs. 24,149 and 153.

10.2. These few CFTs are apparently special

One can try to build a quantum Hall wavefunction based on many different CFTs[170] — indeed, one only needs to find a CFT with an acceptable operator to act as an electron. While this might sounds simple, in fact, it is not at all clear how many CFTs are actually physically acceptable. With the exception of the Read–Rezayi series (including Laughlin and Moore-Read states) there remains no single-component gapped quantum Hall state generated by a CFT which has all of the nice properties we desire — in particular having a special parent Hamiltonian such that all of the quasiholes of the system have zero interaction energy, and are described by the particle types of the CFT as well.

While construction of such special Hamiltonians corresponding to CFTs has been shown possible for certain gapless states,[7–9] and it looks potentially possible that such a construction might be possible for the (unitary) tricritical Ising model,[56] this has yet to be fully achieved, and it is still not clear that unitarity (and rationality) is sufficient to obtain a gapped quantum Hall state.

Nonetheless, given the close relationship between CFTs and topological quantum field theories in general, it would not be surprising if such a construction could somehow be achieved for any gapped fractional quantum Hall state. Note that substantial progress towards such a construction for composite fermion states has been achieved recently[4] although it falls somewhat short of what has been achieved for the Read–Rezayi series. In particular, there is no special parent Hamiltonian, and the wavefunction is not simply written in the form of Eq. (50), but rather contains a symmetrization or antisymmetrization over a correlator of (at least two) different types of electron field. These shortcomings make further analytic work on these wavefunctions extremely challenging. While better CFT construction of these states may not be possible, it remains tempting to try to find another route. For example, the existence of a special parent Hamiltonian for an unprojected version of a hierarchy state[63–65] seems a possibly promising approach.

10.3. *Further issues with the CFT approach*

In trying to connect more closely to experiment, there are several further concerns related to the CFT approach. For systems with multiple edge modes, a theory can only be conformally invariant if all of the edge modes move at the same velocity.[s] While systems with special parent Hamiltonians and quadratic confinement (see Sec. 4) necessarily have edge modes with the same velocity (which is determined only by the confining potential), more realistic systems generally can have multiple edge velocities. However, as long as all the edge modes are moving in the same direction, this does not typically present a large problem since one can at least imagine deforming the edge potentials locally so as to change the edge velocities arbitrarily. In such cases many of the tools of CFT remain applicable even though the system lacks true conformal invariance.

In more complicated cases edge modes move may in both directions. This occurs, for example, for the experimentally important Jain states with $\nu > 1/2$. In this case we must treat both chiralities of the CFT, the right-movers, which are the holomorphic part and the left-movers, which are the anti-holomorphic part.[t] This then becomes a bit more subtle since the bulk $(2+0)$-dimensional wavefunction remains holomorphic, and the Moore-Read mapping (Eq. (50)) obviously cannot hold precisely, although one might still assume that a topological connection between the edge and bulk theory should still hold.

---

[s]Indeed, with very few conditions added (unitarity, Lorentz invariance, scale invariance) one can assume that any edge which has a single velocity can be described by a CFT.
[t]In some cases one might be able to particle-hole conjugate a system to obtain a different state where all edge modes move in the same direction. This is applicable, for example, for the Jain series at $\nu > 1/2$. However, one can certainly find cases such as $\nu = 2/7$ with edge modes moving in both directions, where this cannot be done.

11. Conclusions

This review has discussed a number of (mostly recent) advances in the understanding of certain fractional quantum Hall wavefunctions and their properties mainly by focusing on the explicit structure of the wavefunctions. To a large extent we focused on wavefunctions that have "special" properties of some sort — usually meaning that they are zero energy states of a special parent Hamiltonian. Such wavefunctions have an enormous amount of mathematical structure that allows understanding at a deep and detailed level which would not be otherwise possible. It is possible, however, that many other fractional quantum Hall states may show very similar properties if examined in the right light. For example, we analytically understand an enormous amount about the Laughlin edge. While we do not have the same analytic handle on general hierarchy or composite fermion states, for example, we expect that many of the same principles apply, and can at least be tested numerically. Even more excitingly, many of the ideas that we can analytically explore in these quantum Hall states may extend to the study of other types of matter altogether. For example, ideas of entanglement spectroscopy, which started in the quantum Hall field,[129] are now very widespread.

The ideas presented in this review show many aspects of the deep mathematical structure present in fractional quantum Hall states. The fact that the study of this structure has continued to develop for well over thirty years is a testament to its richness. Indeed, despite the more than 160 citations included in this review, there are a certainly a huge number of additional closely related works that have not been discussed. One should not interpret this omission as being an opinion that some of these excluded works are somehow less interesting, but rather that they simply did not fit within the particular narrative that I have chosen to follow. Another author certainly could have made other choices and given different emphasis. What is undoubtedly the case, no matter which path one chooses to pursue through the field, is that this topic is full of interesting physics and interesting mathematics — and it is very likely to continue to provide interesting directions for many years to come.

References

1. R. B. Laughlin, Anomalous quantum Hall effect: An incompressible quantum fluid with fractionally charged excitations, *Phys. Rev. Lett.* **50**, 1395–1398 (1983). doi: 10.1103/PhysRevLett.50.1395. URL http://link.aps.org/doi/10.1103/PhysRevLett.50.1395.
2. G. Moore and N. Read, Non-Abelions in the fractional quantum Hall effect, *Nucl. Phys. B* **360**, 362–396 (1991). doi: http://dx.doi.org/10.1016/0550-3213(91)90407-O. URL http://www.sciencedirect.com/science/article/pii/0550321391904070.

3. C. Nayak, S. H. Simon, A. Stern, M. Freedman, and S. Das Sarma, Non-Abelian anyons and topological quantum computation, *Rev. Mod. Phys.* **80**, 1083–1159 (2008). doi: 10.1103/RevModPhys.80.1083. URL http://link.aps.org/doi/10.1103/RevModPhys.80.1083.

4. T. H. Hansson, M. Hermanns, S. H. Simon, and S. F. Viefers, Quantum hall physics: Hierarchies and conformal field theory techniques, *Rev. Mod. Phys.* **89**, 025005 (2017). doi: 10.1103/RevModPhys.89.025005. URL https://link.aps.org/doi/10.1103/RevModPhys.89.025005.

5. G. Moore and N. Seiberg, Classical and quantum conformal field theory, *Comm. Math. Phys.* **123**(2), 177–254 (1989). URL https://projecteuclid.org:443/euclid.cmp/1104178762.

6. N. Read and E. Rezayi, Beyond paired quantum Hall states: Parafermions and incompressible states in the first excited Landau level, *Phys. Rev. B* **59**, 8084–8092 (1999). doi: 10.1103/PhysRevB.59.8084. URL http://link.aps.org/doi/10.1103/PhysRevB.59.8084.

7. S. H. Simon, E. H. Rezayi, N. R. Cooper, and I. Berdnikov, Construction of a paired wave function for spinless electrons at filling fraction $\nu = 2/5$, *Phys. Rev. B* **75**, 075317 (2007). doi: 10.1103/PhysRevB.75.075317. URL http://link.aps.org/doi/10.1103/PhysRevB.75.075317.

8. D. Green. *Strongly Correlated States in Low Dimensions*. PhD thesis, Yale University (2001). URL http://arxiv.org/abs/cond-mat/0202455.

9. B. A. Bernevig and F. D. M. Haldane, Model fractional quantum Hall states and Jack polynomials, *Phys. Rev. Lett.* **100**, 246802 (2008). doi: 10.1103/PhysRevLett.100.246802. URL https://link.aps.org/doi/10.1103/PhysRevLett.100.246802.

10. F. D. M. Haldane, Fractional quantization of the Hall effect: A hierarchy of incompressible quantum fluid states, *Phys. Rev. Lett.* **51**, 605–608 (1983). doi: 10.1103/PhysRevLett.51.605. URL http://link.aps.org/doi/10.1103/PhysRevLett.51.605.

11. B. I. Halperin, Theory of the quantized Hall conductance, *Helv. Phys. Acta.* **56**(1-3), 75 (1983).

12. J. K. Jain, *Composite fermions*. (Cambridge University Press, 2007).

13. R. E. Prange and S. M. Girvin, Eds., *The Quantum Hall Effect, 2ed.* (New York: Springer-Verlag, 1990).

14. T. Chakraborty and P. Pietiläinen, *The Quantum Hall Effects, Fractional and Integer*. (New York: Springer, 1995).

15. N. Gemelke, E. Sarajlic, and S. Chu, Rotating few-body atomic systems in the fractional quantum Hall regime, *arXiv:1007.2677*. (2010). URL https://arxiv.org/abs/1007.2677.

16. L. W. Clark, N. Schine, C. Baum, N. Jia, and J. Simon, Observation of Laughlin states made of light, *arxive:1907.05872*. (2019). URL https://arxiv.org/pdf/1907.05872.pdf.

17. N. R. Cooper, Rapidly rotating atomic gases, *Advances in Physics.* **57**(6), 539–616 (2008). doi: 10.1080/00018730802564122. URL http://dx.doi.org/10.1080/00018730802564122.

18. J. P. Eisenstein. Experimental studies of multicomponent quantum Hall systems. In eds. S. D. Sarma and A. Pinczuk, *Perspectives in Quantum Hall Effects*. Wiley-VCH Verlag (1996).

19. S. H. Simon, E. H. Rezayi, and N. R. Cooper, Pseudopotentials for multiparticle interactions in the quantum Hall regime, *Phys. Rev. B* **75**, 195306 (2007). doi: 10.1103/PhysRevB.75.195306. URL https://link.aps.org/doi/10.1103/PhysRevB.75.195306.

20. N. Read and E. Rezayi, Quasiholes and fermionic zero modes of paired fractional quantum Hall states: The mechanism for non-Abelian statistics, *Phys. Rev. B* **54**, 16864–16887 (1996). doi: 10.1103/PhysRevB.54.16864. URL http://link.aps.org/doi/10.1103/PhysRevB.54.16864.

21. M. Milovanović and N. Read, Edge excitations of paired fractional quantum Hall states, *Phys. Rev. B* **53**, 13559–13582 (1996). doi: 10.1103/PhysRevB.53.13559. URL https://link.aps.org/doi/10.1103/PhysRevB.53.13559.

22. F. D. M. Haldane and E. H. Rezayi, Periodic Laughlin–Jastrow wave functions for the fractional quantized Hall effect, *Phys. Rev. B* **31**, 2529–2531 (1985). doi: 10.1103/PhysRevB.31.2529. URL http://link.aps.org/doi/10.1103/PhysRevB.31.2529.

23. X. G. Wen and A. Zee, Shift and spin vector: New topological quantum numbers for the Hall fluids, *Phys. Rev. Lett.* **69**, 953–956 (1992). doi: 10.1103/PhysRevLett.69.953. URL https://link.aps.org/doi/10.1103/PhysRevLett.69.953.

24. N. Read, Non-Abelian adiabatic statistics and Hall viscosity in quantum Hall states and $p_x + ip_y$ paired superfluids, *Phys. Rev. B* **79**, 045308 (2009). doi: 10.1103/PhysRevB.79.045308. URL http://link.aps.org/doi/10.1103/PhysRevB.79.045308.

25. N. Read and E. H. Rezayi, Hall viscosity, orbital spin, and geometry: Paired superfluids and quantum Hall systems, *Phys. Rev. B* **84**, 085316 (2011). doi: 10.1103/PhysRevB.84.085316. URL http://link.aps.org/doi/10.1103/PhysRevB.84.085316.

26. S. H. Simon and E. H. Rezayi, Landau level mixing in the perturbative limit, *Phys. Rev. B* **87**, 155426 (2013). doi: 10.1103/PhysRevB.87.155426. URL https://link.aps.org/doi/10.1103/PhysRevB.87.155426.

27. I. Sodemann and A. H. MacDonald, Landau level mixing and the fractional quantum Hall effect, *Phys. Rev. B* **87**, 245425 (2013). doi: 10.1103/PhysRevB.87.245425. URL https://link.aps.org/doi/10.1103/PhysRevB.87.245425.

28. M. R. Peterson and C. Nayak, More realistic Hamiltonians for the fractional quantum Hall regime in GaAs and graphene, *Phys. Rev. B* **87**, 245129 (2013). doi: 10.1103/PhysRevB.87.245129. URL https://link.aps.org/doi/10.1103/PhysRevB.87.245129.

29. N. Read, Conformal invariance of chiral edge theories, *Phys. Rev. B* **79**(24), 245304 (2009). doi: 10.1103/physrevb.79.245304. URL https://doi.org/10.1103/physrevb.79.245304.

30. S. A. Trugman and S. Kivelson, Exact results for the fractional quantum Hall effect with general interactions, *Phys. Rev. B* **31**, 280–5284 (1985). doi: 10.1103/PhysRevB.31.5280. URL http://link.aps.org/doi/10.1103/PhysRevB.31.5280.

31. V. L. Pokrovsky and A. L. Talapov, A simple model for fractional Hall effect, *J. Phys. C: Solid State Phys.* **18**(23), L691 (1985). URL http://stacks.iop.org/0022-3719/18/i=23/a=002.

32. C. H. Lee, Z. Papić, and R. Thomale, Geometric construction of quantum Hall clustering hamiltonians, *Phys. Rev. X* **5**, 041003 (2015). doi: 10.1103/PhysRevX.5.041003. URL https://link.aps.org/doi/10.1103/PhysRevX.5.041003.

33. S. H. Simon, E. H. Rezayi, and N. R. Cooper, Generalized quantum Hall projection Hamiltonians, *Phys. Rev. B* **75**, 075318 (2007). doi: 10.1103/PhysRevB.75.075318. URL http://link.aps.org/doi/10.1103/PhysRevB.75.075318.

34. A. Cappelli, L. S. Georgiev, and I. T. Todorov, Parafermion Hall states from coset projections of Abelian conformal theories, *Nucl. Phys. B* **599**(3), 499–530 (2001). ISSN 0550-3213. doi: http://dx.doi.org/10.1016/S0550-3213(00)00774-4. URL `http://www.sciencedirect.com/science/article/pii/S0550321300007744`.

35. P. G. De Gennes, *Superconductivity Of Metals And Alloys (Advanced Books Classics)*. (CRC Press, 2018).

36. E. Ardonne, N. Read, E. Rezayi, and K. Schoutens, Non-Abelian spin-singlet quantum Hall states: Wave functions and quasihole state counting, *Nucl. Phys. B* **607**(3), 549–576 (2001). ISSN 0550-3213. doi: http://dx.doi.org/10.1016/S0550-3213(01)00224-3. URL `http://www.sciencedirect.com/science/article/pii/S0550321301002243`.

37. F. D. M. Haldane and E. H. Rezayi, Spin-singlet wave function for the half-integral quantum Hall effect, *Phys. Rev. Lett.* **60**, 956–959 (1988). doi: 10.1103/PhysRevLett.60.956. URL `https://link.aps.org/doi/10.1103/PhysRevLett.60.956`.

38. J. W. Reijnders, F. J. M. van Lankvelt, K. Schoutens, and N. Read, Quantum Hall states and boson triplet condensate for rotating spin-1 bosons, *Phys. Rev. Lett.* **89**, 120401 (2002). doi: 10.1103/PhysRevLett.89.120401. URL `https://link.aps.org/doi/10.1103/PhysRevLett.89.120401`.

39. J. W. Reijnders, F. J. M. van Lankvelt, K. Schoutens, and N. Read, Rotating spin-1 bosons in the lowest Landau level, *Phys. Rev. A* **69**, 023612 (2004). doi: 10.1103/PhysRevA.69.023612. URL `https://link.aps.org/doi/10.1103/PhysRevA.69.023612`.

40. B. Blok and X. Wen, Many-body systems with non-Abelian statistics, *Nucl. Phys. B* **374**(3), 615–646 (1992). ISSN 0550-3213. doi: https://doi.org/10.1016/0550-3213(92)90402-W. URL `http://www.sciencedirect.com/science/article/pii/055032139290402W`.

41. S. C. Davenport, E. Ardonne, N. Regnault, and S. H. Simon, Spin-singlet Gaffnian wave function for fractional quantum Hall systems, *Phys. Rev. B* **87**, 045310 (2013). doi: 10.1103/PhysRevB.87.045310. URL `https://link.aps.org/doi/10.1103/PhysRevB.87.045310`.

42. E. J. Bergholtz and Z. Liu, Topological flat band models and fractional Chern insulators, *Int. J. Mod. Phys. B* **27**(24), 1330017 (2013). doi: 10.1142/S021797921330017X. URL `http://www.worldscientific.com/doi/abs/10.1142/S021797921330017X`.

43. S. A. Parameswaran, R. Roy, and S. L. Sondhi, Fractional quantum Hall physics in topological flat bands, *Comptes Rendus Physique.* **14**(9), 816–839 (2013). ISSN 1631-0705. doi: https://doi.org/10.1016/j.crhy.2013.04.003. URL `http://www.sciencedirect.com/science/article/pii/S163107051300073X`. Topological insulators / Isolants topologiques.

44. E. Kapit and E. Mueller, Exact parent Hamiltonian for the quantum Hall states in a lattice, *Phys. Rev. Lett.* **105**, 215303 (2010). doi: 10.1103/PhysRevLett.105.215303. URL `https://link.aps.org/doi/10.1103/PhysRevLett.105.215303`.

45. E. Kapit, P. Ginsparg, and E. Mueller, Non-Abelian braiding of lattice bosons, *Phys. Rev. Lett.* **108**, 066802 (2012). doi: 10.1103/PhysRevLett.108.066802. URL `https://link.aps.org/doi/10.1103/PhysRevLett.108.066802`.

46. I. Glasser, J. I. Cirac, G. Sierra, and A. E. B. Nielsen, Exact parent Hamiltonians of bosonic and fermionic Moore–Read states on lattices and local models, *New J. Phys.* **17**(8), 082001 (2015). URL `http://stacks.iop.org/1367-2630/17/i=8/a=082001`.

47. I. Glasser, J. I. Cirac, G. Sierra, and A. E. B. Nielsen, Lattice effects on Laughlin wave functions and parent Hamiltonians, *Phys. Rev. B* **94**, 245104 (2016). doi: 10.1103/PhysRevB.94.245104. URL https://link.aps.org/doi/10.1103/PhysRevB.94.245104.

48. R. Thomale, E. Kapit, D. F. Schroeter, and M. Greiter, Parent Hamiltonian for the chiral spin liquid, *Phys. Rev. B* **80**, 104406 (2009). doi: 10.1103/PhysRevB.80.104406. URL https://link.aps.org/doi/10.1103/PhysRevB.80.104406.

49. M. Greiter, D. F. Schroeter, and R. Thomale, Parent Hamiltonian for the non-Abelian chiral spin liquid, *Phys. Rev. B* **89**, 165125 (2014). doi: 10.1103/PhysRevB.89.165125. URL https://link.aps.org/doi/10.1103/PhysRevB.89.165125.

50. J. Behrmann, Z. Liu, and E. J. Bergholtz, Model fractional Chern insulators, *Phys. Rev. Lett.* **116**, 216802 (2016). doi: 10.1103/PhysRevLett.116.216802. URL https://link.aps.org/doi/10.1103/PhysRevLett.116.216802.

51. T. Jolicoeur, T. Mizusaki, and P. Lecheminant, Absence of a gap in the Gaffnian state, *Phys. Rev. B* **90**, 075116 (2014). doi: 10.1103/PhysRevB.90.075116. URL https://link.aps.org/doi/10.1103/PhysRevB.90.075116.

52. B. A. Bernevig, P. Bonderson, and N. Regnault, Screening behavior and scaling exponents from quantum Hall wavefunctions, *arxiv.* (2012). URL https://arxiv.org/abs/1207.3305.

53. B. Kang and J. E. Moore, Neutral excitations in the Gaffnian state, *Phys. Rev. B* **95**, 245117 (2017). doi: 10.1103/PhysRevB.95.245117. URL https://link.aps.org/doi/10.1103/PhysRevB.95.245117.

54. M. H. Freedman, J. Gukelberger, M. B. Hastings, S. Trebst, M. Troyer, and Z. Wang, Galois conjugates of topological phases, *Phys. Rev. B* **85**, 045414 (2012). doi: 10.1103/PhysRevB.85.045414. URL https://link.aps.org/doi/10.1103/PhysRevB.85.045414.

55. P. Di Francesco, P. Mathieu, and D. Sénéchal, *Conformal Field Theory.* (Springer, New York, 1997).

56. T. S. Jackson, N. Read, and S. H. Simon, Entanglement subspaces, trial wave functions, and special Hamiltonians in the fractional quantum Hall effect, *Phys. Rev. B* **88**, 075313 (2013). doi: 10.1103/PhysRevB.88.075313. URL https://link.aps.org/doi/10.1103/PhysRevB.88.075313.

57. X.-G. Wen and Z. Wang, Topological properties of Abelian and non-Abelian quantum Hall states classified using patterns of zeros, *Phys. Rev. B* **78**, 155109 (2008). doi: 10.1103/PhysRevB.78.155109. URL https://link.aps.org/doi/10.1103/PhysRevB.78.155109.

58. Y.-M. Lu, X.-G. Wen, Z. Wang, and Z. Wang, Non-Abelian quantum Hall states and their quasiparticles: From the pattern of zeros to vertex algebra, *Phys. Rev. B* **81**, 115124 (2010). doi: 10.1103/PhysRevB.81.115124. URL https://link.aps.org/doi/10.1103/PhysRevB.81.115124.

59. X.-G. Wen and Z. Wang. Pattern-of-zeros approach to fractional quantum Hall states and a classification of symmetric polynomial of infinite variables. In eds. C. Bai, J. Fuchs, Y.-Z. Huang, L. Kong, I. Runkel, and C. Schweigert, *Conformal Field Theories and Tensor Categories, Proceedings of a Workshop Held at Beijing International Center for Mathematical Research*, Mathematical Lectures from Peking University. Springer Verlag (2014). URL https://arxiv.org/abs/1203.3268.

60. J. Liptrap. On translation invariant symmetric polynomials and Haldane's conjecture. In eds. K. Lin and W. Wang, Zhenghanand Zhang, *Proceedings of the Nankai International Conference in Memory of Xiao-Song Lin*, Vol. 19, *Nankai Tracts in Mathematics.* World Scientific (2008). URL https://arxiv.org/abs/1004.0364.

61. S. H. Simon, E. H. Rezayi, and N. Regnault, Quantum Hall wave functions based on S_3 conformal field theories, *Phys. Rev. B* **81**, 121301 (2010). doi: 10.1103/PhysRevB.81.121301. URL https://link.aps.org/doi/10.1103/PhysRevB.81.121301.

62. S. H. Simon, Correlators of n = 1 superconformal currents, *J. Phys. A: Math. Theor.* **42**(5), 055402 (2009). URL http://stacks.iop.org/1751-8121/42/i=5/a=055402.

63. J. K. Jain, Theory of the fractional quantum Hall effect, *Phys. Rev. B* **41**, 7653–7665 (1990). doi: 10.1103/PhysRevB.41.7653. URL https://link.aps.org/doi/10.1103/PhysRevB.41.7653.

64. E. H. Rezayi and A. H. MacDonald, Origin of the $\nu = 2/5$ fractional quantum Hall effect, *Phys. Rev. B* **44**, 8395–8398 (1991). doi: 10.1103/PhysRevB.44.8395. URL https://link.aps.org/doi/10.1103/PhysRevB.44.8395.

65. L. Chen, S. Bandyopadhyay, and A. Seidel, Jain-2/5 parent Hamiltonian: Structure of zero modes, dominance patterns, and zero mode generators, *Phys. Rev. B* **95**, 195169 (2017). doi: 10.1103/PhysRevB.95.195169. URL https://link.aps.org/doi/10.1103/PhysRevB.95.195169.

66. X. G. Wen, Non-Abelian statistics in the fractional quantum Hall states, *Phys. Rev. Lett.* **66**, 802–805 (1991). doi: 10.1103/PhysRevLett.66.802. URL https://link.aps.org/doi/10.1103/PhysRevLett.66.802.

67. S. Bandyopadhyay, L. Chen, M. T. Ahari, G. Ortiz, Z. Nussinov, and A. Seidel, Entangled Pauli principles: The DNA of quantum Hall fluids, *Phys. Rev. B* **98**, 161118 (2018). doi: 10.1103/PhysRevB.98.161118. URL https://link.aps.org/doi/10.1103/PhysRevB.98.161118.

68. J. K. Jain, Incompressible quantum Hall states, *Phys. Rev. B* **40**, 8079–8082 (1989). doi: 10.1103/PhysRevB.40.8079. URL https://link.aps.org/doi/10.1103/PhysRevB.40.8079.

69. G. J. Sreejith, M. Fremling, G. S. Jeon, and J. K. Jain, Search for exact local Hamiltonians for general fractional quantum Hall states, *Phys. Rev. B* **98**, 235139 (2018). doi: 10.1103/PhysRevB.98.235139. URL https://link.aps.org/doi/10.1103/PhysRevB.98.235139.

70. E. Chertkov and B. K. Clark, Computational inverse method for constructing spaces of quantum models from wave functions, *Phys. Rev. X* **8**, 031029 (2018). doi: 10.1103/PhysRevX.8.031029. URL https://link.aps.org/doi/10.1103/PhysRevX.8.031029.

71. X.-L. Qi and D. Ranard, Determining a local Hamiltonian from a single eigenstate, *Quantum* **3**, 159 (2019). ISSN 2521-327X. doi: 10.22331/q-2019-07-08-159. URL https://doi.org/10.22331/q-2019-07-08-159.

72. B. I. Halperin, Quantized Hall conductance, current-carrying edge states, and the existence of extended states in a two-dimensional disordered potential, *Phys. Rev. B* **25**, 2185–2190 (1982). doi: 10.1103/PhysRevB.25.2185. URL https://link.aps.org/doi/10.1103/PhysRevB.25.2185.

73. X. G. Wen, Chiral Luttinger liquid and the edge excitations in the fractional quantum Hall states, *Phys. Rev. B* **41**, 12838–12844 (1990). doi: 10.1103/PhysRevB.41.12838. URL https://link.aps.org/doi/10.1103/PhysRevB.41.12838.

74. C. Kane and M. P. A. Fisher. Edge state transport. In eds. S. D. Sarma and A. Pinczuk, *Perspectives in Quantum Hall Effects*. Wiley-VCH Verlag (1996).

75. A. M. Chang, Chiral Luttinger liquids at the fractional quantum Hall edge, *Rev. Mod. Phys.* **75**, 1449–1505 (2003). doi: 10.1103/RevModPhys.75.1449. URL http://link.aps.org/doi/10.1103/RevModPhys.75.1449.

76. M. Stone, Schur functions, chiral bosons, and the quantum-Hall-effect edge states, *Phys. Rev. B* **42**, 8399–8404 (1990). doi: 10.1103/PhysRevB.42.8399. URL https://link.aps.org/doi/10.1103/PhysRevB.42.8399.

77. X.-G. Wen, Topological orders and edge excitations in fractional quantum Hall states, *Adv. Phys.* **44**(5), 405–473 (1995). doi: 10.1080/00018739500101566. URL https://doi.org/10.1080/00018739500101566.

78. J. Dubail, N. Read, and E. H. Rezayi, Edge-state inner products and real-space entanglement spectrum of trial quantum Hall states, *Phys. Rev. B* **86**, 245310 (2012). doi: 10.1103/PhysRevB.86.245310. URL https://link.aps.org/doi/10.1103/PhysRevB.86.245310.

79. R. Fern, R. Bondesan, and S. H. Simon, Structure of edge-state inner products in the fractional quantum Hall effect, *Phys. Rev. B.* **97**(15), 155108 (2018). doi: 10.1103/physrevb.97.155108. URL https://doi.org/10.1103/physrevb.97.155108.

80. R. Fern, R. Bondesan, and S. H. Simon, Effective edge state dynamics in the fractional quantum Hall effect, *Phys. Rev. B* **98**, 155321 (2018). doi: 10.1103/PhysRevB.98.155321. URL https://link.aps.org/doi/10.1103/PhysRevB.98.155321.

81. X. Wan, E. H. Rezayi, and K. Yang, Edge reconstruction in the fractional quantum Hall regime, *Phys. Rev. B* **68**, 125307 (2003). doi: 10.1103/PhysRevB.68.125307. URL https://link.aps.org/doi/10.1103/PhysRevB.68.125307.

82. R. Sabo, I. Gurman, A. Rosenblatt, F. Lafont, D. Banitt, J. Park, M. Heiblum, Y. Gefen, V. Umansky, and D. Mahalu, Edge reconstruction in fractional quantum Hall states, *Nature Phys.* **13**, 491 (2017). URL https://doi.org/10.1038/nphys4010.

83. C. d. C. Chamon and X. G. Wen, Sharp and smooth boundaries of quantum Hall liquids, *Phys. Rev. B* **49**, 8227–8241 (1994). doi: 10.1103/PhysRevB.49.8227. URL http://link.aps.org/doi/10.1103/PhysRevB.49.8227.

84. M. D. Johnson and A. H. MacDonald, Composite edges in the $\nu = 2/3$ fractional quantum Hall effect, *Phys. Rev. Lett.* **67**, 2060–2063 (1991). doi: 10.1103/PhysRevLett.67.2060. URL http://link.aps.org/doi/10.1103/PhysRevLett.67.2060.

85. A. H. MacDonald, Edge states in the fractional-quantum-Hall-effect regime, *Phys. Rev. Lett.* **64**, 220–223 (1990). doi: 10.1103/PhysRevLett.64.220. URL http://link.aps.org/doi/10.1103/PhysRevLett.64.220.

86. A. Imambekov, T. L. Schmidt, and L. I. Glazman, One-dimensional quantum liquids: Beyond the Luttinger liquid paradigm, *Rev. Mod. Phys.* **84**, 1253–1306 (2012). doi: 10.1103/RevModPhys.84.1253. URL https://link.aps.org/doi/10.1103/RevModPhys.84.1253.

87. T. Price and A. Lamacraft, Fine structure of the phonon in one dimension from quantum hydrodynamics, *Phys. Rev. B* **90**, 241415 (2014). doi: 10.1103/PhysRevB.90.241415. URL https://link.aps.org/doi/10.1103/PhysRevB.90.241415.

88. T. Price and A. Lamacraft, Quantum hydrodynamics in one dimension beyond the Luttinger liquid, *arxive/1509.08332*. (2015). URL https://arxiv.org/abs/1509.08332.

89. E. Bettelheim, A. G. Abanov, and P. Wiegmann, Nonlinear quantum shock waves in fractional quantum Hall edge states, *Phys. Rev. Lett.* **97**(24) (2006). doi: 10.1103/physrevlett.97.246401. URL https://doi.org/10.1103/physrevlett.97.246401.

90. P. Wiegmann, Nonlinear hydrodynamics and fractionally quantized solitons at the fractional quantum Hall edge, *Phys. Rev. Lett.* **108**(20) (2012). doi: 10.1103/physrevlett.108.206810. URL https://doi.org/10.1103/physrevlett.108.206810.

91. A. G. Abanov and P. B. Wiegmann, Quantum hydrodynamics, the quantum Benjamin–Ono equation, and the Calogero model, *Phys. Rev. Lett.* **95**(7), 076402 (2005). doi: 10.1103/physrevlett.95.076402. URL https://doi.org/10.1103/physrevlett.95.076402.

92. B. I. Halperin, Statistics of quasiparticles and the hierarchy of fractional quantized Hall states, *Phys. Rev. Lett.* **52**, 1583–1586 (1984). doi: 10.1103/PhysRevLett.52.1583. URL http://link.aps.org/doi/10.1103/PhysRevLett.52.1583.

93. N. Read, Excitation structure of the hierarchy scheme in the fractional quantum Hall effect, *Phys. Rev. Lett.* **65**, 1502–1505 (1990). doi: 10.1103/PhysRevLett.65.1502. URL http://link.aps.org/doi/10.1103/PhysRevLett.65.1502.

94. B. Feigin, M. Jimbo, T. Miwa, and E. Mukhin, A differential ideal of symmetric polynomials spanned by Jack polynomials at $\beta = -(r-1)/(k+1)$, *Int. Math. Res. Not.* **2002**(23), 1223–1237 (2002). doi: 10.1155/S1073792802112050. URL http://dx.doi.org/10.1155/S1073792802112050.

95. E. Ardonne, R. Kedem, and M. Stone, Filling the Bose sea: Symmetric quantum Hall edge states and affine characters, *J. Phys. A: Math. Gen.* **38**(3), 617 (2005). URL http://stacks.iop.org/0305-4470/38/i=3/a=006.

96. N. Read, Wavefunctions and counting formulas for quasiholes of clustered quantum Hall states on a sphere, *Phys. Rev. B* **73**, 245334 (2006). doi: 10.1103/PhysRevB.73.245334. URL https://link.aps.org/doi/10.1103/PhysRevB.73.245334.

97. R. Tao and D. J. Thouless, Fractional quantization of Hall conductance, *Phys. Rev. B* **28**, 1142–1144 (1983). doi: 10.1103/PhysRevB.28.1142. URL http://link.aps.org/doi/10.1103/PhysRevB.28.1142.

98. E. J. Bergholtz and A. Karlhede, Half-filled lowest Landau level on a thin torus, *Phys. Rev. Lett.* **94**, 026802 (2005). doi: 10.1103/PhysRevLett.94.026802. URL http://link.aps.org/doi/10.1103/PhysRevLett.94.026802.

99. E. J. Bergholtz, J. Kailasvuori, E. Wikberg, T. H. Hansson, and A. Karlhede, Pfaffian quantum Hall state made simple: Multiple vacua and domain walls on a thin torus, *Phys. Rev. B* **74**, 081308 (2006). doi: 10.1103/PhysRevB.74.081308. URL http://link.aps.org/doi/10.1103/PhysRevB.74.081308.

100. A. Seidel and D.-H. Lee, Abelian and non-Abelian Hall liquids and charge-density wave: Quantum number fractionalization in one and two dimensions, *Phys. Rev. Lett.* **97**, 056804 (2006). doi: 10.1103/PhysRevLett.97.056804. URL http://link.aps.org/doi/10.1103/PhysRevLett.97.056804.

101. E. J. Bergholtz and A. Karlhede, Quantum Hall system in Tao–Thouless limit, *Phys. Rev. B* **77**, 155308 (2008). doi: 10.1103/PhysRevB.77.155308. URL http://link.aps.org/doi/10.1103/PhysRevB.77.155308.

102. A. Seidel and K. Yang, Halperin (m, m', n) bilayer quantum Hall states on thin cylinders, *Phys. Rev. Lett.* **101**, 036804 (2008). doi: 10.1103/PhysRevLett.101.036804. URL https://link.aps.org/doi/10.1103/PhysRevLett.101.036804.

103. A. Seidel and D.-H. Lee, Domain-wall-type defects as anyons in phase space, *Phys. Rev. B* **76**, 155101 (2007). doi: 10.1103/PhysRevB.76.155101. URL https://link.aps.org/doi/10.1103/PhysRevB.76.155101.

104. A. Seidel, Pfaffian statistics through adiabatic transport in the 1D coherent state representation, *Phys. Rev. Lett.* **101**, 196802 (2008). doi: 10.1103/PhysRevLett.101.196802. URL https://link.aps.org/doi/10.1103/PhysRevLett.101.196802.

105. E. J. Bergholtz and A. Karlhede, A simple view on the quantum Hall system, *arxiv:0611181*. (2006). doi: https://doi.org/10.1007/978-1-4020-8512-3_2. URL https://arxiv.org/abs/cond-mat/0611181.

106. X. G. Wen, Topological order in rigid states, *Int. J. Mod. Phys. B* **04**(02), 239–271 (1990). doi: 10.1142/S0217979290000139. URL https://doi.org/10.1142/S0217979290000139.

107. X. G. Wen and Q. Niu, Ground-state degeneracy of the fractional quantum Hall states in the presence of a random potential and on high-genus Riemann surfaces, *Phys. Rev. B* **41**, 9377–9396 (1990). doi: 10.1103/PhysRevB.41.9377. URL https://link.aps.org/doi/10.1103/PhysRevB.41.9377.

108. E. Ardonne, E. J. Bergholtz, J. Kailasvuori, and E. Wikberg, Degeneracy of non-Abelian quantum Hall states on the torus: Domain walls and conformal field theory, *J. Statistical Mechanics: Theory and Experiment* **2008**(04), P04016 (2008). URL http://stacks.iop.org/1742-5468/2008/i=04/a=P04016.

109. A. Weerasinghe and A. Seidel, Thin torus perturbative analysis of elementary excitations in the Gaffnian and Haldane–Rezayi quantum Hall states, *Phys. Rev. B* **90**, 125146 (2014). doi: 10.1103/PhysRevB.90.125146. URL https://link.aps.org/doi/10.1103/PhysRevB.90.125146.

110. E. Ardonne, Domain walls, fusion rules, and conformal field theory in the quantum Hall regime, *Phys. Rev. Lett.* **102**, 180401 (2009). doi: 10.1103/PhysRevLett.102.180401. URL https://link.aps.org/doi/10.1103/PhysRevLett.102.180401.

111. F. D. M. Haldane, Pauli-like principle for Abelian and non-Abelian FQHE quasiparticles, *Bull. Amer. Phys. Soc.* (2006). URL http://meetings.aps.org/link/BAPS.2006.MAR.P46.7.

112. H. Jack, I.—a class of symmetric polynomials with a parameter, *Proc. Roy. Soc. Edinburgh. Section A. Math. Phys. Sci.* **69**(01), 1–18 (1970). doi: 10.1017/s0080454100008517. URL https://doi.org/10.1017/s0080454100008517.

113. R. P. Stanley, Some combinatorial properties of Jack symmetric functions, *Adv. Math.* **77**(1), 76–115 (1989). doi: 10.1016/0001-8708(89)90015-7. URL https://doi.org/10.1016/0001-8708(89)90015-7.

114. B. A. Bernevig, V. Gurarie, and S. H. Simon, Central charge and quasihole scaling dimensions from model wavefunctions: Toward relating Jack wavefunctions to $\mathcal{W}$-algebras, *J. Phys. A: Math. Theor.* **42**(24), 245206 (2009). doi: 10.1088/1751-8113/42/24/245206. URL https://doi.org/10.1088/1751-8113/42/24/245206.

115. B. Estienne and R. Santachiara, Relating Jack wavefunctions to wa_{k-1} theories, *J. Phys. A: Math. Theor.* **42**(44), 445209 (2009). doi: 10.1088/1751-8113/42/44/445209. URL https://doi.org/10.1088/1751-8113/42/44/445209.

116. B. A. Bernevig and N. Regnault, Anatomy of Abelian and non-Abelian fractional quantum Hall states, *Phys. Rev. Lett.* **103**(20), 206801 (2009). doi: 10.1103/physrevlett.103.206801. URL https://doi.org/10.1103/physrevlett.103.206801.

117. R. Thomale, B. Estienne, N. Regnault, and B. A. Bernevig, Decomposition of fractional quantum Hall model states: Product rule symmetries and approximations, *Phys. Rev. B* **84**, 045127 (2011). doi: 10.1103/PhysRevB.84.045127. URL https://link.aps.org/doi/10.1103/PhysRevB.84.045127.

118. B. A. Bernevig and F. D. M. Haldane, Generalized clustering conditions of Jack polynomials at negative Jack parameter α, *Phys. Rev. B* **77**, 184502 (2008). doi: 10.1103/PhysRevB.77.184502. URL https://link.aps.org/doi/10.1103/PhysRevB.77.184502.

119. B. A. Bernevig and F. D. M. Haldane, Properties of non-Abelian fractional quantum Hall states at filling $\nu = k/r$, *Phys. Rev. Lett.* **101**, 246806 (2008). doi: 10.1103/PhysRevLett.101.246806. URL https://link.aps.org/doi/10.1103/PhysRevLett.101.246806.

120. B. A. Bernevig and F. D. M. Haldane, Clustering properties and model wave functions for non-Abelian fractional quantum Hall quasielectrons, *Phys. Rev. Lett.* **102**, 066802 (2009). doi: 10.1103/PhysRevLett.102.066802. URL https://link.aps.org/doi/10.1103/PhysRevLett.102.066802.

121. B. Estienne, N. Regnault, and R. Santachiara, Clustering properties, Jack polynomials and unitary conformal field theories, *Nucl. Phys. B* **824**(3), 539–562 (2010). doi: 10.1016/j.nuclphysb.2009.09.002. URL https://doi.org/10.1016/j.nuclphysb.2009.09.002.

122. B. Estienne, B. A. Bernevig, and R. Santachiara, Electron-quasihole duality and second-order differential equation for Read–Rezayi and Jack wave functions, *Phys. Rev. B* **82**, 205307 (2010). doi: 10.1103/PhysRevB.82.205307. URL https://link.aps.org/doi/10.1103/PhysRevB.82.205307.

123. K. H. Lee, Z.-X. Hu, and X. Wan, Construction of edge states in fractional quantum Hall systems by Jack polynomials, *Phys. Rev. B* **89**, 165124 (2014). doi: 10.1103/PhysRevB.89.165124. URL https://link.aps.org/doi/10.1103/PhysRevB.89.165124.

124. B. Kuśmierz and A. Wójs, Emergence of Jack ground states from two-body pseudopotentials in fractional quantum Hall systems, *Phys. Rev. B* **97**, 245125 (2018). doi: 10.1103/PhysRevB.97.245125. URL https://link.aps.org/doi/10.1103/PhysRevB.97.245125.

125. B. Estienne and B. A. Bernevig, Spin-singlet quantum Hall states and Jack polynomials with a prescribed symmetry, *Nucl. Phys. B* **857**(2), 185–206 (2012). doi: 10.1016/j.nuclphysb.2011.12.007. URL https://doi.org/10.1016/j.nuclphysb.2011.12.007.

126. R. Fern and S. H. Simon, Quantum Hall edges with hard confinement: Exact solution beyond Luttinger liquid, *Phys. Rev. B* **95**(20), 201108(R) (2017). doi: 10.1103/physrevb.95.201108. URL https://doi.org/10.1103/physrevb.95.201108.

127. B. Zeng, X. Chen, D.-L. Zhou, and X.-G. Wen, *Quantum Information Meets Quantum Matter — From Quantum Entanglement to Topological Phase in Many-Body Systems.* (published online, 2015). URL https://arxiv.org/abs/1508.02595.

128. J. Eisert, M. Cramer, and M. B. Plenio, Colloquium: Area laws for the entanglement entropy, *Rev. Mod. Phys.* **82**, 277–306 (2010). doi: 10.1103/RevModPhys.82.277. URL https://link.aps.org/doi/10.1103/RevModPhys.82.277.

129. H. Li and F. D. M. Haldane, Entanglement spectrum as a generalization of entanglement entropy: Identification of topological order in non-Abelian fractional quantum Hall effect states, *Phys. Rev. Lett.* **101**, 010504 (2008). doi: 10.1103/PhysRevLett.101.010504. URL https://link.aps.org/doi/10.1103/PhysRevLett.101.010504.

130. O. S. Zozulya, M. Haque, K. Schoutens, and E. H. Rezayi, Bipartite entanglement entropy in fractional quantum Hall states, *Phys. Rev. B* **76**, 125310 (2007). doi: 10.1103/PhysRevB.76.125310. URL https://link.aps.org/doi/10.1103/PhysRevB.76.125310.

131. J. Dubail, N. Read, and E. H. Rezayi, Real-space entanglement spectrum of quantum Hall systems, *Phys. Rev. B* **85**, 115321 (2012). doi: 10.1103/PhysRevB.85.115321. URL http://link.aps.org/doi/10.1103/PhysRevB.85.115321.

132. I. D. Rodríguez, S. H. Simon, and J. K. Slingerland, Evaluation of ranks of real space and particle entanglement spectra for large systems, *Phys. Rev. Lett.* **108**, 256806 (2012). doi: 10.1103/PhysRevLett.108.256806. URL http://link.aps.org/doi/10.1103/PhysRevLett.108.256806.

133. A. Sterdyniak, A. Chandran, N. Regnault, B. A. Bernevig, and P. Bonderson, Real-space entanglement spectrum of quantum Hall states, *Phys. Rev. B* **85**, 125308 (2012). doi: 10.1103/PhysRevB.85.125308. URL http://link.aps.org/doi/10.1103/PhysRevB.85.125308.

134. M. Haque, O. Zozulya, and K. Schoutens, Entanglement entropy in fermionic Laughlin states, *Phys. Rev. Lett.* **98**, 060401 (2007). doi: 10.1103/PhysRevLett.98.060401. URL http://link.aps.org/doi/10.1103/PhysRevLett.98.060401.

135. I. D. Rodríguez and G. Sierra, Entanglement entropy of integer quantum Hall states, *Phys. Rev. B* **80**, 153303 (2009). doi: 10.1103/PhysRevB.80.153303. URL https://link.aps.org/doi/10.1103/PhysRevB.80.153303.

136. S. Dong, E. Fradkin, R. G. Leigh, and S. Nowling, Topological entanglement entropy in Chern–Simons theories and quantum Hall fluids, *J. High Energy Phys.* **2008**(05), 016 (2008). URL http://stacks.iop.org/1126-6708/2008/i=05/a=016.

137. M. Levin and X.-G. Wen, Detecting topological order in a ground state wave function, *Phys. Rev. Lett.* **96**, 110405 (2006). doi: 10.1103/PhysRevLett.96.110405. URL https://link.aps.org/doi/10.1103/PhysRevLett.96.110405.

138. A. Kitaev and J. Preskill, Topological entanglement entropy, *Phys. Rev. Lett.* **96**, 110404 (2006). doi: 10.1103/PhysRevLett.96.110404. URL https://link.aps.org/doi/10.1103/PhysRevLett.96.110404.

139. N. Regnault, Entanglement spectroscopy and its application to the quantum Hall effects, *arXiv:1510.07670* (2015). URL http://arxiv.org/abs/1510.07670.

140. A. Sterdyniak, B. A. Bernevig, N. Regnault, and F. D. M. Haldane, The hierarchical structure in the orbital entanglement spectrum of fractional quantum Hall systems, *New J. Phys.* **13**(10), 105001 (2011). URL http://stacks.iop.org/1367-2630/13/i=10/a=105001.

141. I. D. Rodríguez, S. C. Davenport, S. H. Simon, and J. K. Slingerland, Entanglement spectrum of composite fermion states in real space, *Phys. Rev. B* **88**, 155307 (2013). doi: 10.1103/PhysRevB.88.155307. URL http://link.aps.org/doi/10.1103/PhysRevB.88.155307.

142. N. Regnault, B. A. Bernevig, and F. D. M. Haldane, Topological entanglement and clustering of Jain hierarchy states, *Phys. Rev. Lett.* **103**, 016801 (2009). doi: 10.1103/PhysRevLett.103.016801. URL https://link.aps.org/doi/10.1103/PhysRevLett.103.016801.

143. S. C. Davenport, I. D. Rodríguez, J. K. Slingerland, and S. H. Simon, Composite fermion model for entanglement spectrum of fractional quantum Hall states, *Phys. Rev. B* **92**, 115155 (2015). doi: 10.1103/PhysRevB.92.115155. URL http://link.aps.org/doi/10.1103/PhysRevB.92.115155.

144. X.-L. Qi, H. Katsura, and A. W. W. Ludwig, General relationship between the entanglement spectrum and the edge state spectrum of topological quantum states, *Phys. Rev. Lett.* **108**, 196402 (2012). doi: 10.1103/PhysRevLett.108.196402. URL http://link.aps.org/doi/10.1103/PhysRevLett.108.196402.

145. B. Swingle and T. Senthil, Geometric proof of the equality between entanglement and edge spectra, *Phys. Rev. B* **86**, 045117 (2012). doi: 10.1103/PhysRevB.86.045117. URL http://link.aps.org/doi/10.1103/PhysRevB.86.045117.

146. N. Ishibashi, The boundary and crosscap states in conformal field theories, *Mod. Phys. Lett. A* **04**(03), 251–264 (1989). doi: 10.1142/S0217732389000320. URL https://doi.org/10.1142/S0217732389000320.

147. C. Nayak and F. Wilczek, 2n quasihole states realize 2^{n-1}-dimensional spinor braiding statistics in paired quantum Hall states, *Nucl. Phys. B* **479**, 529–553 (1996). doi: http://dx.doi.org/10.1016/0550-3213(96)00430-0. URL http://www.sciencedirect.com/science/article/pii/0550321396004300.

148. N. Read and D. Green, Paired states of fermions in two dimensions with breaking of parity and time-reversal symmetries and the fractional quantum Hall effect, *Phys. Rev. B* **61**, 10267–10297 (2000). doi: 10.1103/PhysRevB.61.10267. URL https://link.aps.org/doi/10.1103/PhysRevB.61.10267.

149. M. Baraban, *Low Energy Excitations in Quantum Condensates*. PhD thesis, Yale University (2010).

150. Y. Tserkovnyak and S. H. Simon, Monte Carlo evaluation of non-Abelian statistics, *Phys. Rev. Lett.* **90**, 016802 (2003). doi: 10.1103/PhysRevLett.90.016802. URL http://link.aps.org/doi/10.1103/PhysRevLett.90.016802.

151. M. Baraban, G. Zikos, N. Bonesteel, and S. H. Simon, Numerical analysis of quasi-holes of the Moore–Read wave function, *Phys. Rev. Lett.* **103**, 076801 (2009). doi: 10.1103/PhysRevLett.103.076801. URL http://link.aps.org/doi/10.1103/PhysRevLett.103.076801.

152. Y.-L. Wu, B. Estienne, N. Regnault, and B. A. Bernevig, Braiding non-Abelian quasiholes in fractional quantum Hall states, *Phys. Rev. Lett.* **113**, 116801 (2014). doi: 10.1103/PhysRevLett.113.116801. URL http://link.aps.org/doi/10.1103/PhysRevLett.113.116801.

153. P. Bonderson, V. Gurarie, and C. Nayak, Plasma analogy and non-Abelian statistics for Ising-type quantum Hall states, *Phys. Rev. B* **83**, 075303 (2011). doi: 10.1103/PhysRevB.83.075303. URL http://link.aps.org/doi/10.1103/PhysRevB.83.075303.

154. A. Kitaev, Fault-tolerant quantum computation by anyons, *Ann. Phys.* **303**(1), 2–30 (2003). ISSN 0003-4916. doi: http://dx.doi.org/10.1016/S0003-4916(02)00018-0. URL http://www.sciencedirect.com/science/article/pii/S0003491602000180.

155. A. Kitaev, Anyons in an exactly solved model and beyond, *Ann. Phys.* **321**(1), 2–111 (2006). ISSN 0003-4916. doi: http://dx.doi.org/10.1016/j.aop.2005.10.005. URL http://www.sciencedirect.com/science/article/pii/S0003491605002381. January Special Issue.

156. L. Hormozi, G. Zikos, N. E. Bonesteel, and S. H. Simon, Topological quantum compiling, *Phys. Rev. B* **75**, 165310 (2007). doi: 10.1103/PhysRevB.75.165310. URL https://link.aps.org/doi/10.1103/PhysRevB.75.165310.

157. B. Field and T. Simula, Introduction to topological quantum computation with non-Abelian anyons, *Quantum Science and Technology* **3**(4), 045004 (2018). URL http://stacks.iop.org/2058-9565/3/i=4/a=045004.

158. H. Kjønsberg and J. Myrheim, Numerical study of charge and statistics of Laughlin quasiparticles, *Int. J. Mod. Phys. A* **14**(04), 537–557 (1999). doi: 10.1142/S0217751X99000270. URL http://www.worldscientific.com/doi/abs/10.1142/S0217751X99000270.

159. G. S. Jeon, K. L. Graham, and J. K. Jain, Fractional statistics in the fractional quantum Hall effect, *Phys. Rev. Lett.* **91**, 036801 (2003). doi: 10.1103/PhysRevLett.91.036801. URL http://link.aps.org/doi/10.1103/PhysRevLett.91.036801.

160. M. P. Zaletel and R. S. K. Mong, Exact matrix product states for quantum Hall wave functions, *Phys. Rev. B* **86**, 245305 (2012). doi: 10.1103/PhysRevB.86.245305. URL https://link.aps.org/doi/10.1103/PhysRevB.86.245305.

161. V. Gurarie and C. Nayak, A plasma analogy and Berry matrices for non-Abelian quantum Hall states, *Nucl. Phys. B* **506**, 685 (1997). doi: http://dx.doi.org/10.1016/S0550-3213(97)00612-3. URL `http://www.sciencedirect.com/science/article/pii/S0550321397006123`.

162. E. Ardonne and G. Sierra, Chiral correlators of the Ising conformal field theory, *J. Phys. A: Math. Theor.* **43**(50), 505402 (2010). URL `http://stacks.iop.org/1751-8121/43/i=50/a=505402`.

163. S. L. Sondhi and S. A. Kivelson, Long-range interactions and the quantum Hall effect, *Phys. Rev. B* **46**, 13319–13325 (1992). doi: 10.1103/PhysRevB.46.13319. URL `https://link.aps.org/doi/10.1103/PhysRevB.46.13319`.

164. S. H. Simon, Effect of Landau level mixing on braiding statistics, *Phys. Rev. Lett.* **100**, 116803 (2008). doi: 10.1103/PhysRevLett.100.116803. URL `https://link.aps.org/doi/10.1103/PhysRevLett.100.116803`.

165. J. E. Avron, R. Seiler, and P. G. Zograf, Viscosity of quantum Hall fluids, *Phys. Rev. Lett.* **75**, 697–700 (1995). doi: 10.1103/PhysRevLett.75.697. URL `http://link.aps.org/doi/10.1103/PhysRevLett.75.697`.

166. C. Hoyos, Hall viscosity, topological states and effective theories, *Int. J. Mod. Phys. B* **28**(15), 1430007 (2014). doi: 10.1142/S0217979214300072. URL `https://doi.org/10.1142/S0217979214300072`.

167. B. Bradlyn, M. Goldstein, and N. Read, Kubo formulas for viscosity: Hall viscosity, ward identities, and the relation with conductivity, *Phys. Rev. B* **86**, 245309 (2012). doi: 10.1103/PhysRevB.86.245309. URL `https://link.aps.org/doi/10.1103/PhysRevB.86.245309`.

168. C. Hoyos and D. T. Son, Hall viscosity and electromagnetic response, *Phys. Rev. Lett.* **108**, 066805 (2012). doi: 10.1103/PhysRevLett.108.066805. URL `https://link.aps.org/doi/10.1103/PhysRevLett.108.066805`.

169. M. Levin and D. T. Son, Particle-hole symmetry and electromagnetic response of a half-filled Landau level, *Phys. Rev. B* **95**, 125120 (2017). doi: 10.1103/PhysRevB.95.125120. URL `https://link.aps.org/doi/10.1103/PhysRevB.95.125120`.

170. B. Estienne, N. Regnault, and R. Santachiara, Clustering properties, Jack polynomials and unitary conformal field theories, *Nucl. Phys. B* **824**(3), 539–562 (2010). ISSN 0550-3213. doi: https://doi.org/10.1016/j.nuclphysb.2009.09.002. URL `http://www.sciencedirect.com/science/article/pii/S0550321309004659`.

© 2020 World Scientific Publishing Company
https://doi.org/10.1142/9789811217494_0009

Chapter 9

Engineering Non-Abelian Quasi-Particles in Fractional Quantum Hall States — A Pedagogical Introduction

Ady Stern

Department of Condensed Matter Physics,
Weizmann Institute of Science,
Rehovot 76100, Israel
adiel.stern@weizmann.ac.il

Non-Abelian quantum Hall states bring to culmination the unique properties of fractionalized topological states of matter, such as fractional quantum numbers, topological ground state degeneracy and anyonic statistics. Unfortunately, they seem to be realized in rather rare conditions. In this chapter we present a pedagogical introduction to recent theoretical proposals for engineering such states. These are based on hybrids of fractional quantum Hall systems with superconductors, on bilayer quantum Hall systems with carefully designed tunnel couplings between the layers and on Chern bands. We also review the wire construction approach to the analysis of non-Abelian quantum Hall states, and focus on a few special cases where this analysis may be carried out explicitly.

Contents

1. Introduction

This chapter deals with non-Abelian quantum Hall states, and with possible constructive ways to engineer them.

    We assume that the reader is familiar with the basic phenomenology of the fractional quantum Hall effect[1,2] — the quantization of the Hall conductivity to a fraction of e^2/h, the energy gap in the bulk, the chiral gapless edge states, the fractionally charged excitations and the anyonic statistics that they follow.[a] We also assume, and hope, that the reader shares our amazement at the beauty of the effect and its phenomenology. With these assumptions in place, it is natural to ask what does it take for a system to display the amazingly beautiful fractional quantum Hall state as its ground state. Our focus in this chapter are non-Abelian quantum Hall states,[5–7] and therefore our discussion will look at possible ways to engineer — in thought experiments and real experiments — situations in which the ground state is a non-Abelian fractional state. The last hundred years have taught us the answer for similar questions regarding other states of matter, for example insulators and superconductors. Insulators are created when an integer number of Bloch bands are filled, and superconductors are created when electrons in a Fermi liquid interact attractively. We are motivated by a search for a similarly simple answer that applies to non-Abelian fractional quantum Hall states.

    We will not give here a full answer to this question. Rather, we will present initial steps of two types. The first type will explore ways by which an Abelian FQHE state may be turned into a state that is, in some sense, non-Abelian.[8–12] The starting point for this part is the realization that the essential aspect of non-Abelian states, the degeneracy of the ground state in the presence of quasi-particles, has a counterpart in Abelian states, the degeneracy of the ground states when the system is put on a torus or a compact surface of a higher genus. The approach we present here finds ways to introduce defects into a planar geometry that effectively makes it a surface of high genus. The ideas presented are all challenging to realize experimentally, but not impossible.

    The second type will explore ways in which parallel one-dimensional quantum wires may be coupled to create two-dimensional non-Abelian states.[13–16] The starting point for this part will be the realization that the two edges of an annular quantum Hall system are just two halves of a one-dimensional system, separated by a gapped bulk. A non-Abelian state with a desired edge structure could then be constructed by first constructing a one-dimensional system with the same edge

---

[a]There are numerous introductory texts to the Quantum Hall States, and it is impossible to cite them all here. Given the author, this chapter shares some lines of thought with Ref. 3. The physics of non-Abelian states is discussed in details in Ref. 4.

structure, and then "inflating" it to have a gapped bulk that separates its two halves.

The structure of this chapter will be the following. In Sec. 2, we will start with a review of some background material — the essential features of non-Abelian FQHE states, and the method of wire construction of fractional Quantum Hall states, a method that will be useful for the rest of our discussion. Section 3 will discuss topological aspects of fractional quantum Hall states and the engineering of non-Abelian quantum Hall states by the introduction of topological defects known as "parafermions". Section 4 employs the wire construction approach to construct Hamiltonians for several types of non-Abelian states. Section 5 briefly discusses the potential of parafermions for quantum computation. Finally, Sec. 6 gives concluding remarks.

2. Background

2.1. *What are non-Abelian quantum Hall states*

Like all fractional quantum Hall states, non-Abelian quantum Hall states[5–7] are two-dimensional many-body systems in a magnetic field, in which the combination of a magnetic field (or effective magnetic field) and an interaction between the constituent particles creates an energy gap between the ground state and the excited states. The state carries quasi-particles, and these quasi-particles are the place where the non-Abelian nature of the state is most pronounced: in the presence of N_{qp} quasi-particles pinned by impurities to fixed, well-separated, positions, the ground state of the system becomes degenerate. For a large N_{qp} the degeneracy is exponential in N_{qp}, i.e. the number of ground states approaches $\Gamma^{N_{qp}}$. The degeneracy is stable, i.e. it does not get split with small variations in the details of the system (see Fig. 1).

At first glance, the dynamics of a system in which there is a stable degeneracy of the ground state is rather limited, since when put in one of the ground states, it stays

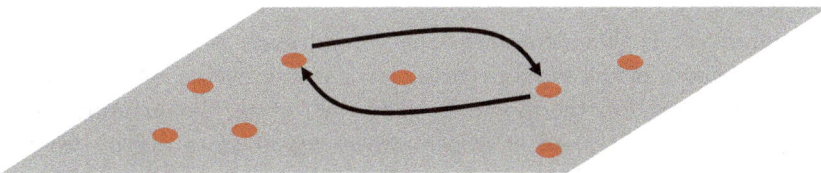

Fig. 1. In a non-Abelian quantum Hall state the presence of localized well-separated quasi-particles and quasi-holes makes the ground state of the system degenerate. In the limit of a large number of quasi-particles N_{qp} the degeneracy approaches $\Gamma^{N_{qp}}$ and Γ is called the quantum dimension of the quasi-particles. An adiabatic interchange of the positions of quasi-particles applies a unitary transformation within the subspace of ground states. The unitary transformations that correspond to different interchanges do not necessarily commute, and hence the name "non-Abelian state".

there forever. However, the degeneracy also introduces a new avenue for dynamics, by carrying out *trajectories in parameter space*.[17] In the simplest example, the parameters are the positions of the quasi-particles. If the system has N electrons (in principle, the constituents particles may also be atoms, or even photons, but we will use electrons for concreteness) and N_{qp} quasi-particles, the subspace of ground states is spanned by a set of many-body wave functions of N arguments $\{r_i\}$ and N_{qp} parameters $\{R_j\}$, to be denoted by $\psi_\ell(\{r_i\}; \{R_j\})$. When the parameters $\{R_j\}$ are varied in time, for example, by varying the positions of the impurities that pin the quasi-particles, the subspace of states changes. When the parameters traverse a cyclic trajectory λ in parameter space, the initial and final subspaces of ground states are identical, and the final state may be compared to the initial one. Then, if the system is initialized in one of the ground states and the parameters are varied adiabatically such that the system stays in the ground state subspace, the final state is related to the initial one by a unitary transformation U acting within this subspace. The transformation U may be expressed in terms of the ground state wavefunctions along the trajectory as

$$U = \hat{P} e^{i \int d\lambda \langle \psi_\ell(\lambda) | \nabla_\lambda | \psi_{\ell'}(\lambda) \rangle} \tag{1}$$

where the integral is taken along the trajectory in parameter space, $\psi_\ell(\lambda)$ are the states spanning the ground state subspace at any point along the trajectory, and $\hat{P}$ is the path-ordering operator.

This transformation is a matrix, and the matrices that correspond to different trajectories generally do not commute, hence the name non-Abelian. As Eq. (1) shows, U is geometric, i.e. independent of the time dependence of the trajectory. Furthermore, apart from an Abelian phase, it is topological: two trajectories that may be deformed to one another without a change in the degeneracy of the ground state lead to the same U. In the simple example where the parameters are the positions of the quasi-particles, two trajectories that may be deformed to one another without crossing a position of a quasi-particle lead to the same U (again, up to an Abelian phase).

The parameter Γ that determines the ground state degeneracy is called the quantum dimension of the quasi-particles.[4] For the simplest cases of non-Abelian quantum Hall states, e.g. the Moore–Read $\nu = 5/2$ state, the quantum dimension is $\Gamma = \sqrt{2}$. Then, every pair of quasi-particles may fuse to one of two possible quasi-particles, which are both Abelian. The Hilbert space of degenerate ground states may be decomposed into product of subspaces, each of which involving only one pair of non-Abelian quasi-particles.

For more sophisticated non-Abelian states the quantum dimension is not of the form $\sqrt{n}$ with integer n. The subspace of degenerate ground states cannot be decomposed into a direct product of local subspaces, and the fusion of any number of non-Abelian quasi-particles ends up with a possible non-Abelian product.

2.2. *Where are non-Abelian quantum Hall states found*

Fractional quantum Hall states are a natural phenomenon, sort of. They emerge in places in nature in which two-dimensional electronic systems occur and happen to be subjected to a strong magnetic field. Truth to be told, they are better thought of as engineered. However, the engineering that we have in mind in this chapter is different from the one that brought von Klitzing *et al.* and Tsui *et al.* to the first quantum Hall observations. While they constructed an interesting system and explored its properties, here we would like to engineer a system for a particular goal — constructing a non-Abelian quantum Hall state.

There are several cases where there are good indications that non-Abelian quantum Hall states should be the ground state of a two-dimensional electronic system in a strong magnetic field for realistic parameters. Most notable are states in the $N = 1$ Landau level, which dominates the physics between $2 < \nu < 4$ for electrons in a single band of quadratic dispersion.

2.3. *How are non-Abelian states identified*

There is an intrinsic conflict when searching for ways to experimentally verify that a quantum Hall state is non-Abelian. The claim to fame of these states is their resilience to external perturbations, while almost any experiment is based on observing the response of a system to a perturbation it is subjected to. Despite this dichotomy, several ways have been proposed to experimentally measure a quantity that identifies a state as non-Abelian. Arguably the most accessible of all is the measurement of the thermal Hall conductivity.[18] The thermal Hall conductivity measures the central charge of the edge, and for most non-Abelian states this central charge is fractional, leading to thermal Hall conductivity that is a fraction of $\pi^2 T/3h$. Indeed, the thermal Hall conductance was recently measured for the $\nu = 5/2$ state, yielding a fractional central charge of $5/2$[19] (see the chapter by Heiblum and Feldman in this book). It should be noted, however, that it is possible for a non-Abelian quantum Hall state to carry an edge state with an integer central charge. We will get back to that. It is also worth mentioning that a thermal Hall conductivity that corresponds to a fractional central charge has recently been measured in a magnetic insulating system, where heat is carried by spin interactions, rather than by charge motion.[20] This observation is mentioned here mostly as a reminder that while this chapter focusses on quantum Hall systems, the concepts we deal with are of broader applicability.

Other ways to ascertain that a quantum Hall state is non-Abelian include a measurement of the ground state degeneracy in the presence of quasi-particles by measuring the entropy that they carry;[21,22] a measurement of the charge associated with the state of quasi-particles when they get close to one another such that the degeneracy is slightly lifted;[23] and a measurement of the interference pattern is in Fabry–Perot and Mach–Zehnder settings.[24–27] Experimental evidence for interference of fractional quasi-particles has been reported in Refs. 28 and 29.

2.4. *Review of the wire construction approach to the quantum Hall effect*

Much of the theoretical research into the fractional quantum Hall effect originates from a frustrating search for a small parameter. The frustration comes from the fact that the obvious small parameter, the ratio of the interaction energy scale $e^2/\epsilon l_B$ to the cyclotron energy $\hbar\omega_c$ (here l_B is the magnetic length, ω_c the cyclotron frequency, and ϵ the dielectric constant), which is reasonably small in experiments, is not useful for a perturbative calculation, due to the immense degeneracy of a ν-filled Landau level, with ν being the fractional filling factor. This may be at the root of the formation of the beautiful world of the FQHE, but it is also a source of difficulty in the theoretical analysis.

Many body theory has already witnessed alternative routes, for cases where the weakness of the interaction cannot serve for a perturbative calculation. Most notably, Fermi liquid theory uses the wave vector, frequency and temperature as the small parameters, when compared to the Fermi wave vector and energy. In the context of the FQHE, we will briefly review here two candidates for small parameters.

The first originates from the Chern–Simons approach, where a flux is attached to the electrons to transform them to composite fermions or composite bosons. The flux constitute of an integer number of flux quanta, which focus into a point flux tube. Perturbative approaches here include smearing of the flux tube onto a large spatial scale, or making the number of flux quanta a small parameter. Both approaches are powerful for analyzing the topological properties of FQHE states, and sometimes (infrequently) have also the power for more quantitaive calculations. In view of the abundance of reviews of this approach in the literature, we will not review it here.

The second approach is the wire construction.[13,14,30,31] This approach starts from the planar system being a set of decoupled wires, and considers the coupling between the wires as the small parameter. At its root lies the vast understanding developed over the years to analyze one-dimensional systems. Following an application of this understanding, the coupling between the wires is chosen to be a relevant perturbation, and its flow under renormalization brings the system into the two-dimensional fractional quantum Hall state. The main power of this approach is in finding the topological properties of the state, and not in quantitative calculations of its properties. By considering coupled-wire models for fractional quantum Hall states, it is possible to analyze properties of phase transitions between different types of such states.

The set-up of the wire construction approach for "simple" FQHE states, such as Laughlin and Jain states, starts from decoupled spin-polarized wires a distance d from one another, with a Fermi momentum k_F, subjected to a perpendicular magnetic field B. Inter-wire coupling comes in the form of many-electron processes that involve electron tunneling between wires and electron back-scattering within wires. Its "goal" is to gap the bulk of the system while leaving chiral modes at the edge.

For this coupling to become relevant it should satisfy the following necessary criteria. First, momentum should be balanced. A charge q that tunnels between two wires a distance nd apart (with n an integer) gets a momentum $qBdn$ from the Lorenz force. The final momentum of the system should equal the sum of the initial momentum and $qBdn$. Second, all electrons involved in the process should originate from, and end at, the Fermi energy. And third, coupling processes that involve different wires should commute with one another.

As examples, let us consider the $\nu = 1$ (see Fig. 2), $\nu = 2$ and $\nu = 1/3$ (see Fig. 3) cases. For $\nu = 1$ we expect single-electron processes to suffice. Indeed, a tunnel process where an electron tunnels from k_F at the wire j to $-k_F$ at the wire $j+1$ gaps the right movers of the j-th wire with the left movers of the $(j+1)$-th wire. When the process is repeated over all $j = 1...N-1$, the bulk becomes gapped, with a single left moving mode at $j = 1$ and a single right-moving mode at $j = N$. For the process to conserve momentum, the condition $eBd = 2\hbar k_F$ should be satisfied. Since the electronic density is $k_F/\pi d$, this condition is nothing but $\nu = 1$.

This example, although the simplest, outlines the general principle: the edges of the $\nu = 1$ IQHE exist already at the single wire level. They are the right- and left-moving modes of a wire of free electrons. The bulk is there just to separate the edge modes from one another such that inter-edge back-scattering becomes exponentially

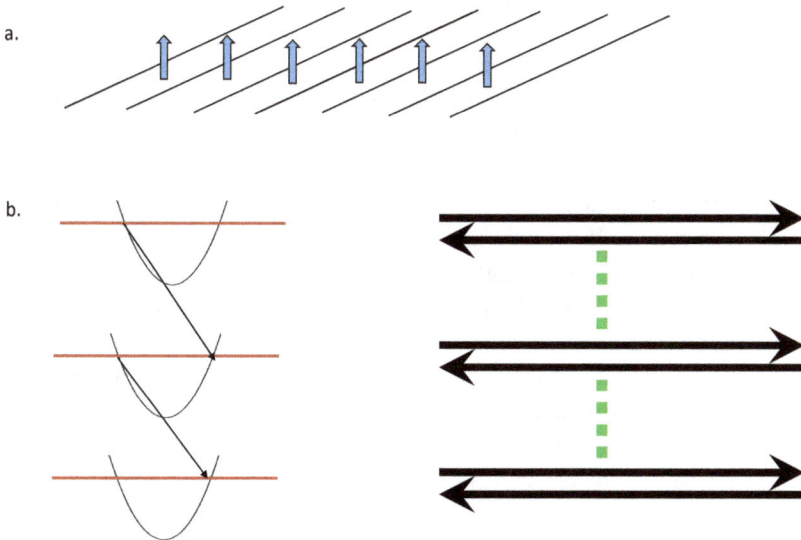

Fig. 2.   (a) The wire construction approach to the quantum Hall effect starts from viewing the plane as made of an array of parallel wires which are subjected to a perpendicular magnetic field, and are coupled by tunneling. (b) The $\nu = 1$ IQHE state is formed when left-moving electrons on one wire back-scatter to become right-moving electrons on the neighboring wire.

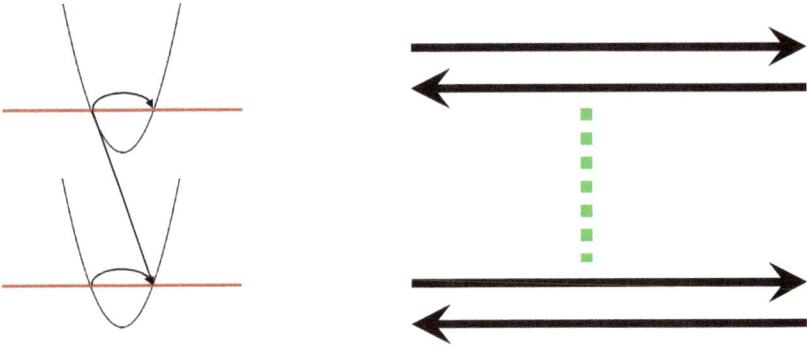

Fig. 3. The $\nu = 1/3$ wire construction involves a many-electron tunneling process, in which one electron tunnels from a right-moving branch in one wire to a left-moving branch in a neighboring wire, while two other electrons back-scatter with no tunneling, one in each of the two wires. Altogether, a momentum of $6k_F$ is being transferred. Indeed, when $6k_F = 2Bd$, the filling factor is $\nu = 1/3$.

suppressed. To that end, the modes are gapped telescopically, right-movers of one wire with the left-movers of the other.

Assuming we keep the magnetic field and the wire separation fixed, the $\nu = 2$ state needs to have a doubled density, which implies a doubled k_F as compared to $\nu = 1$. To keep the momentum balanced, the electrons need to double the momentum they get from the Lorenz force. This is what happens if the electrons are not allowed to tunnel to the nearest neighboring wire, but rather tunnel to the next-nearest-neighbor one. Then, two chiral modes are left ungapped at the edges. That has an interesting consequence: the system effectively breaks into two sub-arrays of wires, which are decoupled from one another — the even numbered wires and the odd numbered ones. This state is analogous to a $\nu = 2$ state constructed from bilayer state in which each layer carries a $\nu = 1$ state, and the two layers are spatially separated. If we now add also tunneling between nearest-neighboring wires, we will make the state analogous to a $\nu = 2$ state in which two Landau levels are filled, and electrons fill the bonding state of the two layers (a symmetric superposition of being in the two layers).

For $\nu = 1/3$ we need multi-electron processes. For the momentum to be balanced at $\nu = 1/3$, we need the momentum balance equation to be $eBd = 6k_F$, rather than $eBd = 2k_F$ which gave us the $\nu = 1$. Thus, on top of the electron that tunnels from k_F at the wire j to $-k_F$ at the wire $j+1$, we have also two electrons back-scattering in the same wire, from k_F to $-k_F$, one at the j-th wire and one at the $(j+1)$-th wire. Altogether, then, this process involves three electrons, and may be obtained from second order virtual transitions of conventional two-electron interaction. Small momentum Luttinger parameters may then be chosen to make this process relevant, such that it leads to gapping of two modes per process. Again, when the process is repeated over all wires, the bulk is gapped, and one chiral edge mode is left ungapped at each edge of the system. The nature of the interaction that gaps the edges is reflected in the current that the edge carries, and leads to a conductivity that is $1/3$ of the conductivity of the $\nu = 1$ edge, as expected from a $\nu = 1/3$ FQH state.

The examples above considered each wire as having electrons of quadratic dispersion. Nothing changes in the analysis if the wires are replaced by a one-dimensional lattice described by a nearest-neighbor tight-binding model. That makes the entire system into a lattice and allows for an extension of the wire approach to discuss Chern bands.

3. Non-Abelian Defects in Abelian Fractional Quantum Hall States

This section will emphasize the concept of topological degeneracy of the ground state.[32] We will first see that the ground state of a fractional quantum Hall state on a torus must be degenerate, and that this degeneracy goes hand-in-hand with the fractionalization of the quasi-particle charge. Then, we will look for ways to engineer further degeneracy of the ground state, by creating fractional quantum Hall states with topological defects known as parafermions (and — in different contexts — also as "genons", "fractional Majorana modes" and "fractional zero modes"). This section will mostly not make use of the wire construction approach, reviewed in the last subsection. This approach will play a key role in the next section.

3.1. *Ground state degeneracy on a torus*

Topological ground state degeneracies are an essential, if not *the* essential, building block for the construction of non-Abelian quasi-particles. Let us then dwell on their origin, in order to understand why a fractionalized state must have a set of degenerate ground states when put on a compact (edge-less) surface with a high genus.

Spectrum degeneracies are part of quantum mechanics since the day the Bohr model for the hydrogen atom was born. However, degeneracies in the spectrum of the hydrogen atom originate from rotational symmetry, and are split when the symmetry is broken. In contrast, the degeneracies we deal with here are topological, i.e. do not originate from a symmetry, and are not split by small changes to the Hamiltonian of the system.

In order to understand the degeneracy of the ground state of a fractionalized state on a torus, we think of the torus as made of two annuli stitched to one another in their edges (see Fig. 4).[33] If both layers are filled with electrons, the construction of a torus would require the annuli to be subjected to opposite magnetic fields, and those would require monopoles for their production. Alternatively, we could have one annulus filled with electrons in a fractional quantum Hall state of ν and the other of holes at the same filling. The total charge on each annulus is an integer. The gap in the bulk allows us to regard the charge as composed of three parts —

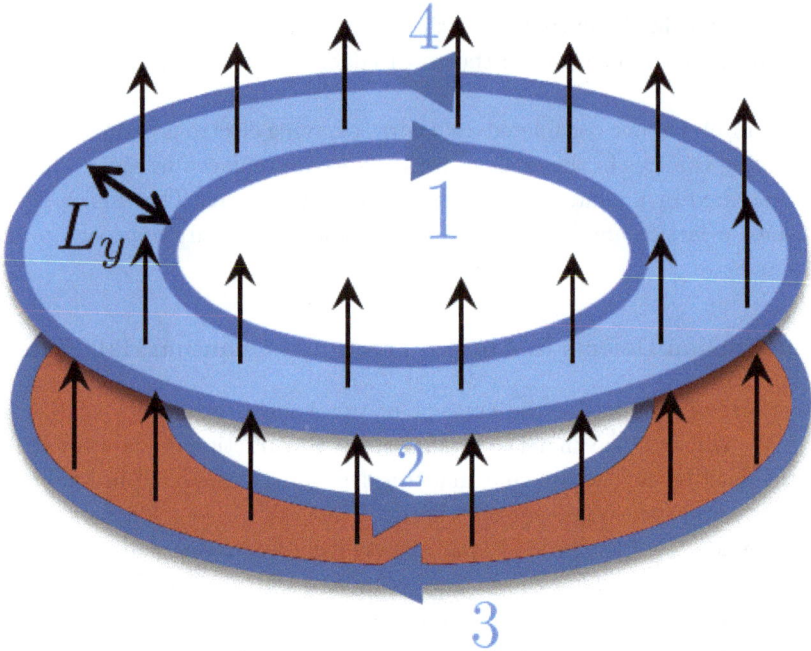

Fig. 4. We analyze a FQHE on a torus by stitching together two annuli in FQHE states, with filling factors $\pm\nu$. The stitching is carried out by coupling together counter-propagating edges on the two annuli, at the interior and at the exterior.

the bulk charge and the charges on the two edges. For simplicity, we think of the bulk as pristine, devoid of any quasi-particles, and we confine ourselves to $\nu = 1/m$. Then, the fractional parts of the charge on the exterior and interior edges satisfy $q_i = -q_e$. These values may be changed only by a transfer of a fractional charge through the bulk, which may be done either by tunnel-coupling the two edges or by inserting a flux to the hole in the annulus.

The two annuli are stitched to form a torus by tunnel-coupling the edge modes of the two exterior edges to one another and the two interior edges to one another. We will describe the stitching in two ways, but we first note that this tunnel-coupling, which couples the edge modes of the two annuli through the vacuum that separates them, is different from the tunnel-coupling mentioned in the previous paragraph, that coupled the edge modes of one annulus through the quantum Hall bulk that separates them. The difference is in the object that is allowed to tunnel: only electrons of integer charge may tunnel through the vacuum, while fractional charges may tunnel across a fractional quantum Hall bulk. Thus, tunneling across the vacuum conserves $q_e^{(j)}$ and $q_i^{(j)}$ (with $j = 1, 2$ enumerating the layers). Tunneling across the annulus does not. Denoting the integer part of the charge on the four edges by $N_e^{(j)}, N_i^{(j)}$, we note that electron tunneling between the edges changes $\Delta N_\alpha = N_\alpha^{(1)} - N_\alpha^{(2)}$ by ± 2 (with $\alpha = i, e$), but does not affect the total charge on the edge $N_\alpha^{tot} = N_\alpha^{(1)} + N_\alpha^{(2)}$. Consequently, an eigenstate of the stitched edges would be characterized by fixed values of N_α^{tot}, q_α^{tot}, Δq_α, but would be a superposition of a broad range of values of ΔN_α.

To find the ground state, we now need to minimize the energy over all these eigenstates. In the absence of disorder, we expect the torus that we form not to have any quasi-particles in its ground state, and therefore we expect $q_\alpha^{tot} = 0$. In contrast, there is no expectation for a particular value of Δq_α in the ground state. Since the charge difference $\Delta N_\alpha + \Delta q_\alpha$ between the two stitched edges has fluctuations that are much larger than one due to the tunneling between the edges, the dependence of the energy on the value of Δq_α becomes exponentially small in the system size. The degenerate ground states are therefore states that are characterized by different values of $\Delta q_\alpha = (-1)^\alpha \ell / m$, with ℓ enumerating the ground states.

The energetics of the ground states may be described also using the field that is conjugate to the charge density, the phase. In this description there are two bosonic fields at the interior and two at the exterior, $\varphi_\alpha^{(i)}(x)$, describing the counter-propagating modes, with $\frac{1}{2\pi}\partial_x \varphi_\alpha^{(i)}(x)$ describing the charge density of the i, α mode. For each α, the total charge density is described by the field $\partial_x \theta(x) = \frac{1}{2}[\partial_x \varphi^{(1)}(x) + \partial_x \varphi^{(2)}(x)]$, while the relative density is described by $\partial_x \phi(x) = \frac{1}{2}[\partial_x \varphi^{(1)}(x) - \partial_x \varphi^{(2)}(x)]$. The commutation relations of the fields are,

$$[\phi(x'), \partial_x \theta(x)] = \frac{i\pi}{m} \delta(x - x') \tag{2}$$

such that the operator $\cos 2m\theta$ transfers an electron between the two counter-propagating modes.

The field theory that describes each of the edges in a Laughlin $\nu = 1/m$ case is the Luttinger liquid theory, in which the Hamiltonian density is

$$\mathcal{H} = \frac{v}{2\pi\nu}\left[K(\partial_x \phi_\alpha)^2 + K^{-1}(\partial_x \theta_\alpha)^2\right] + u_\phi(x)\cos(2m\theta(x)). \tag{3}$$

In Eq. (3) the quadratic terms describe the low energy excitations of the two counter-propagating edges, u is a velocity scale, and the u_ϕ term couples the two counter-propagating edge modes through tunneling. When large enough a cosine term such as appearing here may pin the value of its argument and gap the spectrum of the edge. For simplicity, Eq. (3) neglects small momentum interactions between the edges. The effect of these terms will not be discussed here. More complicated FQH states involve more edges and require more cosine terms for their gapping.

Here, when the tunneling amplitude is large, the degenerate ground states correspond to the pinned values of the cosine term in the Hamiltonian. There are m distinct different minimum points, and hence a ground state degeneracy of m.

It may be illuminating to compare the coupling of the two counter-propagating edge modes to the Josephson tunnel coupling of two superconductors to one another. The superconductors are described by a bosonic field $\phi(x)$, describing the phase difference between the two superconductors, and by its conjugate field $\pi(x)$ describing the difference in charge density. In the absence of tunneling, the relative charge on the two superconductors $\int dx \pi(x)$ is well defined. Tunneling makes this charge fluctuate. The operator that tunnels a Cooper pair across the junction at the point x is $\cos \phi(x)$, and thus the tunneling introduces a potential $\epsilon_J(1 - \cos(\phi))$. This potential makes the phase well defined at its minimum. The new ingredient in our case is the fractional charge on the edge, which stays well defined even in the presence of tunneling, and makes the ground state degenerate. This is manifested in having m minimum points to the cosine term. The m ground states may be spanned by phase states with the pinned value of the phase being $2\pi j/m$, with $j = 1...m$, or by charge states, where the value of the fractional charge is $Q = j/m$ with $j = 1...m$. Each charge state is an equal weight superposition of all phase states, and vice versa. Consequently, a state where the phase is well defined in one of the minimum points may absorb a dipole of fractional charges without being changed.

The ground state degeneracy on the torus depends on the annuli composing the torus being wide enough, that is, on the interior and exterior edges being far enough such that tunneling processes between the interior and the exterior may be neglected. When that is not the case, the degeneracy is lifted by tunneling of a fractional dipole, carrying opposite charges of $\pm 1/m$ between the interior and the exterior. Note that this dipole tunneling takes place between two gapped edges. The gap does not protect the edges against this tunneling due to the available degeneracy.

Much of what comes below is an elaboration on the notion of ground state degeneracy of a fractional quantum Hall state on a topologically non-trivial manifold. Literally speaking, we cannot create fractional quantum Hall state on a torus in a lab, since it would have to enclose sources of magnetic flux, i.e. magnetic monopoles, and such sources have not been found to exist in nature so far. Thus, we have to find ways that allow for the construction of effectively compact, yet topologically non-trivial, manifolds. We already saw how an electron-hole bilayer may mimic a torus. We will see other ways as well. These ways will all require going beyond a simple-minded single-layer fractional quantum Hall device. One way will employ a coupling to a superconductor. Another way will employ well-designed inter-layer couplings. And another will employ lattice defects in Chern insulators. None of these is easy to realize experimentally. Hopefully, none is impossible.

3.2. *Multiple mechanisms to gap counter-propagating edges*

In the previous subsection we saw how two counter-propagating edges may be gapped by tunnel-coupling them to one another, a coupling that back-scatters the direction of motion of each tunneling electron. The next key step in our discussion is the realization that there may be more than one way to gap a set of gapable edges. After describing three examples in which multiple gapping mechanisms exist, we discuss zer-dimensional interfaces between regions of space in which counter-propagating edges are gapped by different mechanisms. We explain that these interfaces host topologically protected zero energy modes, the parafermions, and explore their properties, starting from their enlargement of the degeneracy of the ground state.

3.2.1. *Super-conductivity vs. back-scattering*

The first example is that of two counter-propagating $\nu = \pm 1/m$ edges, as in the bilayer system we considered above. We saw that the edge modes may be gapped by electron tunneling. They may also be gapped by coupling to a superconductor. The physical outcome is different, but the formalism and line of arguments are similar.[9-12] A superconductor exchanges Cooper pairs with the edges, with the two electrons in the Cooper pair emerging from the two edges. This coupling conserves ΔN_α, Δq_α and q_α^{tot}, but makes N_α^{tot} strongly fluctuating. As a consequence, the energy dependence on q_α^{tot} is exponentially small in the size of the system, with the degenerate ground states having $q_\alpha^{tot} = (-1)^\alpha \ell/m$ with ℓ enumerating the m ground states.

Within the bosonized description of the edge, the difference between gapping by back-scattering and gapping by superconductivity is a difference between the arguments of the cosine term. The charge-conserving $\cos m(\phi_1(x) - \phi_2(x))$ is replaced by the pair-creating/annihilating $\cos m(\phi_1(x) + \phi_2(x))$.

3.2.2. *Intra-layer vs. inter-layer back-scattering*

The second example,[8] (see Fig. 5), is that of a bilayer FQHE occupying the x–y plane, where each layer is cut to two half-planes along the y-axis (note that the term bilayer means here two parallel layers that are *not* coupled by tunneling in the bulk. In the graphene literature this system is sometimes referred to as "double-layer"). We would like to consider cases where the two layers are at opposite and at identical filling factors, i.e. $\nu_1 = -\nu_2 = 1/m$ and $\nu_1 = \nu_2 = 1/m$. Along the cut there are four edge modes, two propagating in each direction. The two edge modes within each layer propagate in opposite directions, and the cut may be healed by back-scattering. Intra-layer back-scattering then just eliminates the cut. The relative direction between the edges in different layers depend on the relative sign between ν_1 and ν_2. Since only tunneling between counter-propagating edges leads to gapping, the resulting effective geometry differs between the two cases.

When $\nu_1 = -\nu_2$ inter-layer tunneling mutually gaps counter-propagating edges on the same side of the y-axis ($x = 0$ line). Then, when a region of inter-layer gapping is surrounded by two regions of intra-layer gapping along the y-axis, the effective geometry is that of a hole, very much akin to the hole we described in the stitching of two annuli into a torus, above. The ground state degeneracy introduced by this alternation of gapping mechanisms is the ground state degeneracy of a fractional quantum Hall state on a geometry with a genus. We will come back to situations where gapping mechanisms are alternating below.

(a)

(b)

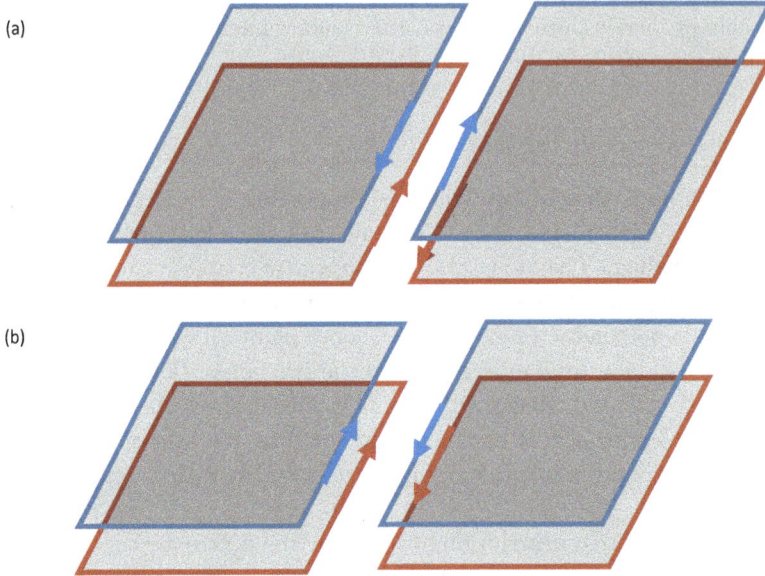

Fig. 5. Multiple gapping mechanisms of counter-propagating edges. A bilayer is cut into two pieces along the y-axis, with two pairs of counter-propagating gapless modes flowing along the cut. There are two ways to gaps these modes in a charge conserving way. One way is by healing the cut, i.e. by back-scattering between edge modes of the same color in the figure. The second way is by backscattering between counter-propagating edge modes in different layers. The precise way depends on the filling factors in the two layers. (a) When the two layers carry opposite filling factors, the edge modes on the same side of the cut mutually gap, effectively creating a quantum Hall state with a hole. (b) When the two layers carry the same filling factors, edge states of opposite side of the gap mutually gap, creating a "crossed" geometry. Notably, this set-up does not involve superconductivity.

When $\nu_1 = \nu_2$, inter-layer back-scattering couples edges on two different sides of the $x = 0$ line. Then, a "crossed" geometry is formed, where the $x < 0$ part of one layer continues to the $x > 0$ part of the other layer.

The bosonized description of the edges of this system involves four bosonic fields $\phi_{i,\alpha}$, where $i = 1, 2$ is the layer and $\alpha = 1, 2$ is the side of the cut. A single cosine term gaps two modes, and therefore we need here two cosine terms. The cuts are healed by the sum of two terms $\sum_{i=1,2} \cos(\phi_{i,1}(x) - \phi_{i,2}(x))$, while inter-layer backscattering involve combinations of fields from the two layers.

Somewhat surprisingly, the resulting ground state degeneracy of the system depends on our assumption of what happens to the edges at $x \to \pm\infty$. To be concrete, let us assume periodic boundary conditions, namely $\phi_i(x \to \infty, y) = \phi_i(x \to -\infty, y)$ and similarly for $y \to \pm\infty$. Then, healing the cut in the first way ends up with a system of two decoupled tori, one for each layer, with a ground state degeneracy of m^2, while the second way ends up with a single torus, with a ground state degeneracy of m.

3.2.3. *Gapping of edge states in Chern insulators*

Chern insulators, exemplified by Haldane's model of electrons on a honeycomb lattice,[34] are quantum Hall states in which the lattice structure is an essential part of the physics. In a tight-binding limit, the lattice Hamiltonian is described in terms of tunneling amplitudes, and the phases of these amplitudes encode the magnetic flux encircled by a closed tunneling trajectory. Since adding a flux quantum to such a trajectory induces an unobservable 2π-phase shift, lattices may break time reversal symmetry even when the average magnetic flux per unit cell vanishes. The Hall conductivity of each of the lattice's Bloch bands is quantized to an integer (in units of e^2/h), and that integer is commonly called the Chern number. With interactions, Abelian and non-Abelian quantum Hall states are known to be formed as well.[35]

In the present context, Chern insulators offer another way of obtaining counter-propagating edge modes that may be gapped in more than one way, as we found above for several bilayer systems.[8] This is a consequence of two observations that we have already reviewed.

The first observation is that we may formulate a lattice model, or a wire con-struction, which is a Chern insulator of Chern number $C = 2$ and is analogous to a bilayer state with a single Landau level in each layer.[8] It was easiest to see it in a wire construction. We simply tune the magnetic field to the value needed for a $\nu = 2$ state, then couple odd-numbered wires to one another to create one layer, and couple even-numbered wires to one another to form the other layer.

The second observation is that in a bilayer fractional quantum Hall state where the two layers have the same filling factor $\nu = 1/m$, a cut along the bilayer creates four gapless edge modes, two on each side of the cut, and allows for two different ways to gap them. Imagine that the cut is being carried out by having the N_0-th and (N_0+1)-th wires disconnected from the four wires to which they were supposed to be coupled, namely, $N_0 - j$, with $j = -2, -1, 2, 3$, and having the right- and left-moving modes of each of the disconnected two wires gap one another. Other than the obvious way to heal the cut, simply by undoing it, the cut may be healed by leaving the N_0-th wire disconnected, and coupling the $(N_0 - 2)$-th wire to the $(N_0 + 1)$-th one. Then, the $(N_0 - 1)$-th couples to the $(N_0 + 2)$-th, and the $(N_0 + 1)$-th closes the disrupted region by coupling to the $(N_0 + 3)$-th wire. In terms of a layer index, this is a "crossed" geometry, in which the two layers are interchanged along the cut. A series of coupling that starts, before the cut, as couplings of even-numbered wires, turns into a coupling of odd-numbered ones.

While so far we have only listed possible gapping mechanisms, we have already hinted that our goal in forming that list is the search for parafermionic zero modes at interface points between regions that are gapped by different gapping mechanisms. Here, such an interface would occur at the ends of dislocations. In situations when a dislocation splits into two disclinations, the parafermionic mode joins one of the disclinations.[36]

The wire construction shows that it is possible to create a $C = 2$ Chern band which easily maps onto a bilayer $\nu = 2$ integer quantum Hall state. In fact, as shown by Barkeshli and Qi,[8] this observation is much more general.[37] The key point in generalizing it is the formulation of a one-to-one mapping between the Hilbert space of a $C = N$ Chern band and an N-layer system in which each layer has a single Landau level. Let us dwell on this mapping. The number of states in a single Landau level is $L_x L_y / 2\pi l_H^2$, with L_x, L_y the dimensions of the system, and $l_H \equiv \sqrt{\hbar c / eB}$ the magnetic length. In the Landau gauge the Landau level states are labeled by a momentum k in the y-direction, and are localized in the x-direction around the point $\bar{x}(k)$. A momentum shift Δk correspond to a position shift $\Delta \bar{x} = \Delta k l_H^2$. The spacing between two consecutive values of k is $\delta k = 2\pi / L_x$, and the range of allowed values for k is of size $L_y / 2\pi l_H^2$, leading altogether to the correct number of states within the Landau level.

The number of states in a single Bloch band on a square monoatomic lattice is the number of sites in the lattice $n_x n_y = \frac{L_x L_y}{a^2}$, where a is the lattice constant. In the Chern band, the role of the Landau states is played by the Wannier states. The latter are not eigenstates of the Hamiltonian. Rather, they are a set of orthogonal states that span the Hilbert space of the band(s) in question. In their original use, Wannier states are meant to provide a basis set of localized states for a given band, which makes them very useful for the study of effects of short-range interactions. For Chern bands it is impossible to define a set of Wannier states that are localized in all directions. Rather, it is useful, for example, for the sake of drawing analogies with Landau levels, to define a basis made of states that are localized in one direction, and extended in the other. Such a basis state is

$$|W(n, k_y)\rangle\rangle = \int dk_x e^{-ik_x n} e^{i\phi(k_x, k_y)} |\psi_{k_x, k_y}\rangle. \tag{4}$$

Here, $|\psi_{k_x,k_y}\rangle$ are the Bloch eigenstates, and the phases $\phi(k_x, k_y)$ are chosen in such a way that the states are maximally localized in the x-direction. The choice that optimizes this localization is

$$\phi(k_x, k_y) = \frac{k_y}{2\pi} \int_0^{2\pi} a_y(0, p_y) dp_y - \int_0^{k_y} a_y(0, p_y) dp_y$$
$$+ \frac{k_x}{2\pi} \int_0^{2\pi} a_x(p_x, k_y) dp_x - \int_0^{k_x} a_x(p_x, k_y) dp_x \tag{5}$$

with $\mathbf{a}(\mathbf{k}) \equiv -i\langle \psi_{k_x, k_y} | \nabla_k | \psi_{k_x, k_y} \rangle$ being the Berry connection of the band.

Note that this basis state is spanned by two quantum numbers, n, k_y, with n taking L_x/a integer values and k_y taking L_y/a values. The states are localized in the x-direction, and the expectation value of the x-coordinate is $x_n(k_y) = n - \frac{1}{2\pi} \int_0^{2\pi} a_x(p_x, k_y) dp_x$. Fixing n, we note that as k_y is varied along its period from 0 to $2\pi/a$, the expectation value $x_n(k_y)$ varies by C unit cells. Thus, the wave functions (4) are not periodic in k_y. When $C = 1$ this definition of $|W(n, k_y)\rangle$ provides a mapping between the single Landau level and the Chern band, where states may be identified by their x-coordinate expectation value. Choosing a convention where the

Landau level state $k = 0$ is localized at $x = 0$, a Landau level state in the Landau gauge whose momentum is k is localized at kl_H^2. The corresponding Wannier state would have n_0, k_{y0}, where $n_0 = [kl_H^2/a]$ and $k_{y0} = \frac{2\pi}{a}(kl_H^2 - n_0 a)$.

When $C = 2$ the number of states in the band does not change, and we expect to map the band to two Landau levels, one in each layer. Thus, the degeneracy of each Landau level should be halved, and a variation of k_y by $2\pi/a$ would shift the x-position by two unit cells. Hence, the boundary conditions with the variation of k_y by 2π now connects all the even-numbered n's to one another, and all the odd-numbered n's to one another, in a way similar to what we saw in the wire construction. The Wannier states defined this way provide a simple way to break the Hilbert space to what is effectively a Hilbert space of two layers. Just as it is for the wires, a dislocation flips between the layers. And just as it is for the bilayer, a cut in the lattice may be healed in two ways, by connecting even-to-even and odd-to-odd or by connecting even-to-odd and odd-to-even, i.e. interchanging the layers. See the two geometries in Fig. 5(b).

All the examples we discussed so far — the two edge modes with their superconductor and backscattering couplings, the four-edge-modes cut in the bilayer of $\nu_1 = -\nu_2$, the cut in the bilayer of $\nu_1 = \nu_2$ and the cut in the $C = 2$ Chern insulator — have a set of edges that may be gapped in two different ways. Each of these ways is described by a set of cosines. In the first example, two edge modes are gapped by one cosine, which may be either charge conserving (back-scattering) or charge non-conserving (coupling to a superconductor). In the other examples, four edge modes are gapped by one of two sets of two cosines. One set corresponds to intra-layer back-scattering, while the other set corresponds to inter-layer back-scattering. In all cases, the gapping is manifested by the argument of the cosine being pinned to a value that minimizes the energy. In all examples the two different gapping mechanisms are "incompatible" with one another. By that, we mean that the arguments of the cosines involved do not commute with one another even at large distances, and hence cannot be simultaneously defined.

At first glance it may seem surprising to see two bosonic fields that do not commute with one another at large mutual distance, at equal time. This becomes more natural, however, when we realize that the fields $\phi(x)$ cannot be translated to electronic operators that act locally. Indeed, it is $\partial_x \phi$, which is related to the electronic density, that is local. Thus, its integral is not.

3.3. *From genus-induced to defect-induced ground state degeneracy*

We now have all the necessary ingredients for introducing parafermionic non-Abelian defects — fractional quantum Hall states, counter-propagating edge modes, and multiple incompatible mechanisms to gap them. We consider a set-up of the type drawn in Fig. 6, where gapless edges are gapped with an alternating set of incompatible gapping mechanisms. The non-Abelian defect is a zero energy mode that occurs at an interface between two incompatible gapping mechanisms. Imagine

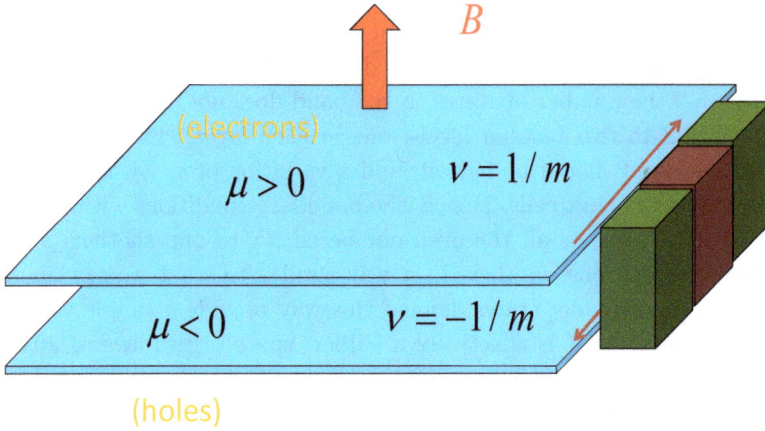

Fig. 6.   A set-up where parafermionic zero modes are created by having two counter-propagating modes gapped alternatingly by two different gapping mechanisms, one by coupling to superconductors and the other coupling the edges by back-scattering.

that a set of gappable edges lies along the y-axis, with the region $y > a$ being gapped by one mechanism, the region $y < -a$ being gapped by the other mechanism, and the region $|y| < a$ being left gapless. Like in any gapless region bounded by gapped regions, a discrete spectrum $\{E_n\}$ of bound states occur. For low energies, we expect a spectrum of excitations proportional to $\pm\epsilon_0(n + \chi)$, with $\epsilon_0 \propto 1/a$ and with integer n and fractional χ. These excitations are created by operators Γ^+ satisfying $[H, \Gamma^+] = E\Gamma^+$. The question of whether there is a state at zero energy, i.e. whether $\chi = 0$, is then determined by the boundary conditions at $x = \pm a$, namely, by the nature of the gapping mechanisms on both sides of the gapless regions. Specifically, when the region is gapped by two incompatible mechanisms, protected zero modes may occur. Those modes then survive even when $a \to 0$.[10]

Let us look at two examples we outlined above. In the case of two counter-propagating $\nu = \pm 1/m$ states, zero modes would occur at the interface between a region gapped by a super-conductor and a region gapped by electron tunneling. For $m = 1$, these are Majorana zero modes. They are unitary operators γ_i (where i enumerates the interface) whose square is a c-number, which may be normalized to one. Consequently, they may be written as hermitian. It may be shown[9,10] that these modes satisfy fermionic commutation relations $\gamma_i\gamma_j = -\gamma_j\gamma_i$ for $i \neq j$.

Being zero modes, the γ_i's commute with the Hamltonian. Being fermionic, they anti-commute with one another. This combination leads to ground state degeneracy. If $|\psi\rangle$ is an eigenstate of H and of γ_1, then $\gamma_2|\psi\rangle$ is an eigenstate of the Hamiltonian with the same energy, yet with a different eigenvalue of γ_1, hence a degenerate ground state. And since $i\gamma_1\gamma_2$ commutes with the Hamiltonian and with all other γ's, each pair of interface modes multiplies the degeneracy of the ground states by two.

For other values of m the unitary operators χ_i satisfy $\chi^{2m} = 1$, and satisfy[9,10] anyonic commutation relations

$$\chi_i \chi_j = e^{i\pi/m} \chi_j \chi_i. \tag{6}$$

These zero modes are commonly referred to as parafermions. It should be noted, however, that the term parafermions carries more than one meaning, and thus may be a source of confusion. Here a parafermion is a localized zero-dimensional zero energy state. It is used also in the context of $1 + 1$ conformal field theories to describe a particular type of a primary field.

Again, having zero modes that do not commute with one another leads to a degeneracy of the ground state. This time, all states $\chi_1^j|\psi\rangle$, with $j = 0...(2m - 1)$ are orthogonal to one another, and the combination $\chi_1^+\chi_2$ commutes with all χ_i ($i \neq 1, 2$). Thus, each pair of interfaces multiplies the degeneracy of the ground state by a factor of $2m$.

The physics of this degeneracy may be understood in the following way. A region that is gapped by coupling to a superconductor (say, the red region in Fig. 6) is surrounded by insulating regions (the green regions in this figure) where the edge is gapped by tunneling and is therefore insulating. Facing the two-dimensional bulk, the red super-conducting region sees quantum Hall states in both layers, which are gapped and have zero longitudinal conductance. Thus, the superconducting region forms a quantum dot surrounded by insulators, except of a coupling to a bulk superconductor. The charge on the quantum dot is quantized in units of the elementary charge $e^* = e/m$, modulo the charge of a Cooper-pair $2e$ which may be exchanged with the bulk superconductor. Thus, the quantum dot has $2m$ charge states. These states are all degenerate in energy, since the coupling to the superconductor screens the charging energy.

The zero mode operator χ_i may be written as a product of two commuting operators $\chi_i = \eta_i \gamma_i$, where γ_i is a Majorana fermion satisfying $\gamma_i^2 = 1$, and η_i is a parafermion satisfying $\eta_i^m = 1$. The former changes the charge by one electron, while the latter changes it by e^* electrons. The decomposition of χ to the product $\eta\gamma$ is the decomposition of the group Z_{2m} to a product $Z_2 \otimes Z_m$, which holds for all odd m's.

To highlight this way of thinking we may define a set of commuting operators $\{e^{i\pi\hat{Q}_i}\}$, one per a superconducting dot, that measures the fractional part of the charge on the dot, taking $2m$ eigenvalues $e^{i\pi j/m}$. This operator is a product of the two zero mode operators that surround the dot, $e^{i\pi\hat{Q}_i} \equiv \chi_{2i-1}^\dagger \chi_{2i}$.

The subspace of degenerate ground states may then be spanned by the eigenstates of the set $\{e^{i\pi\hat{Q}_i}\}$. Note, however, that the product of all eigenvalues is the fractional part of the total charge on the edge. This fractional part must add up to the fractional part of the charge in the bulk to give an integer total charge on the system. For a given bulk charge, then, this requirement imposes a constraint on the product $\prod_i e^{i\pi\hat{Q}_i}$.

The state of such a superconducting quantum dot, that is the fractional charge that it carries, may be measured in two ways. One is by splitting the degeneracy, e.g. by shortening the length of the superconducting region so that interaction between the two ends is no longer negligible, and then using spectroscopic methods. The other is interferometric — by taking a fractional quasi-particle around the superconducting dot.[8–10]

While so far we emphasized the description of the subspace of degenerate ground states in terms of the charge on the quantum dots, this is not the only way to describe it. Alternatively, we could describe the ground states in terms of another set of operators, $\{e^{i\pi\hat{S}_i}\}$, that measures the fractional value of the dipole in the insulating regions. This time $\exp i\pi\hat{S}_i = \chi_{2i}^\dagger\chi_{2i+1}$. The two sets of operators do not commute with one another, but rather satisfy

$$[e^{i\pi\hat{Q}_i}, e^{i\pi\hat{Q}_j}] = [e^{i\pi\hat{S}_i}, e^{i\pi\hat{S}_j}] = 0,$$

$$\left[e^{i\pi\hat{Q}_j}, \prod_{i=1}^{N} e^{i\pi\hat{S}_i}\right] = \left[e^{i\pi\hat{S}_j}, \prod_{i=1}^{N} e^{i\pi\hat{Q}_i}\right] = 0,$$

$$e^{i\pi\hat{Q}_j} e^{i\pi\sum_{k=1}^{l}\hat{S}_k} = e^{i\frac{\pi}{m}\delta_{jl}} e^{i\pi\sum_{k=1}^{l}\hat{S}_k} e^{i\pi\hat{Q}_j}. \tag{7}$$

The subspace of ground states may then be spanned by the set $\{e^{i\pi\hat{Q}_j}\}$ or the set $\{e^{i\pi\hat{S}_j}\}$.

In the case where the system is a bilayer of $\nu_1 = -\nu_2 = 1/m$ the edges are gapped either by intra-layer scattering ("healing of the cut") or by inter-layer scattering. A region where the edges are gapped by inter-layer tunneling surrounded by regions where the cut is healed may be thought of as a tube that connects the two layers to one another. The tube may be regarded as formed by two circular edges (one for each layer) stitched to one another. As we saw in Sec. 3.1 in this case the states of stitched edges may be described by the fractional dipole that they carry, which is a combination of opposite fractional charges in the two layers. This dipole plays the role of $e^{i\pi\hat{Q}}$ in the superconducting regions. However, since there are no superconductors involved, there is only a degeneracy of m per hole.

3.4. *Experimental manifestation of the parafermionic defects*

Experimental proof of braiding of parafermionic defects would be, well, highly desirable. But before aiming so high, there are several experimental set-ups that may provide evidence for the existence of topological ground state degeneracy, and the parafermionic zero energy modes that create it.[10,38] A source of inspiration for these set-ups is the series of experiments conducted in the last few years on one- and two-dimensional systems expected to harbor Majorana zero modes.[39] Examples of the present set-ups are shown in Fig. 7.

A natural signature to look for is the zero bias peak in a conductance spectrum (the differential conductance dI/dV vs V) associated with zero modes, in analogy to that observed in one-dimensional topological superconductors.[39] In Fig. 7(a) current is driven from the source $s1$. A zero bias peak would occur in the differential conductance to $d1$. That peak occurs when the coupling between the gapless

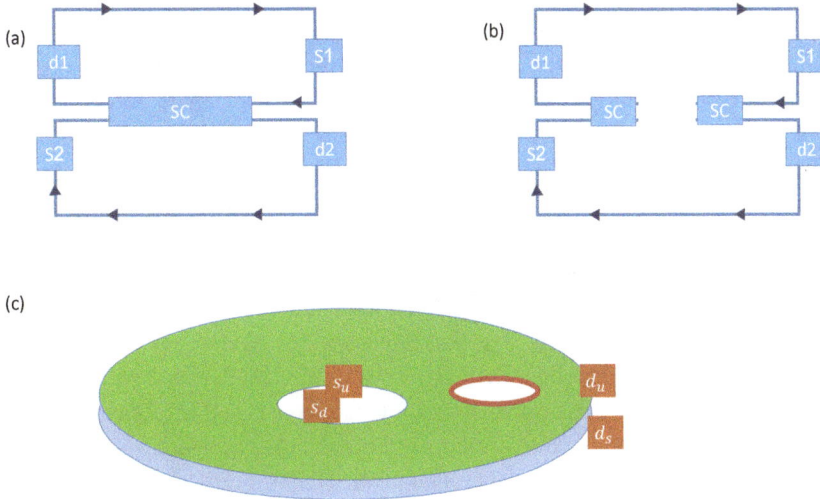

Fig. 7. Three proposed experimental set-ups for searching for signatures of parafermionic defects. The set-ups are described in the text.

edge and the superconductor is through tunneling, and results from the Andreev reflection of the electron that comes from $s1$ to a hole that goes into $d2$. At voltages larger than that of the peak, and smaller than the gap induced by the superconductor in the edge spectrum, electrons from $s1$ would be reflected by normal reflection as electrons to $d2$, such that the current would flow to $d2$. Note that if the superconductor fails to gap the counter-propagating modes, this voltage dependence of the distribution of the current between the two drains will not occur.

A different signature appears in Fig. 7(b), based on the Josephson junction formed by a region gapped by superconductor-backscattering-superconductor segments. In this set-up two counter-propagating edge modes (shown in pale colors) are alternatingly gapped by back-scattering and by super-conductors, in such a way that two isolated superconductors form. The relative phase between the two superconductors is then tuned by a flux loop, and their distance is made small enough to allow for Josephson coupling. The Josephson current depends on the phase difference between the two superconductors, and its periodicity is $4\pi m$,[10] with the $m = 1$ case reproducing the known 4π periodicity for the Josephson effect on the edge of a topological insulator.[40]

A set-up dealing with a Corbino geometry of a bilayer system with $\nu_1 = -\nu_2 = 1/m$ is depicted in Fig. 7(c). This panel shows a top-view of the bilayer system. The two layers are separately contacted by two sources (s_u, s_d in the interior) and two drains (d_u, d_d at the exterior). The separate contacts allow for a measurement of the longitudinal conductance for symmetric and anti-symmetric currents. At the hole the bilayer is depleted, and the two counter-propagating edges of the two layers (depicted in red), are assumed gapped by back-scattering. In the absence of the hole, and under the assumption of no correlation between the layers (a K-matrix of $K = m\sigma_z$), the symmetric and anti-symmetric longitudinal conductances are both very small. The gapped edges surrounding the hole carry a set of m degenerate ground states, which allows for tunneling of $\pm e/m$ dipoles at no energy cost. This degeneracy would manifest itself in a zero bias peak in the asymmetric conductance.

These experiments are all very challenging. By a note of optimism, it is worth mentioning that coupling of integer quantum Hall states to superconductors has already been demonstrated,[41-47] and experiments are underway attempting to couple fractional states to superconductors. The bilayer front is much more advanced, following several decades of research. Most relevant for the present context is to note that bilayers, both of equal sign and opposite sign of charge carriers, have been produced from several materials, with separate contacting to the two layers, in Hall bar and Corbino geometries, and with variable and controlled inter-layer coupling.[48] Of particular importance in this context are bilayers of graphene, where the relative sign of charge carriers is easily controlled by gating.[49] Further on the optimistic note are recent experimental advances in the measurement of entropy[22] and in the measurement of interference in Abelian quantum Hall states.[29]

3.5. Braiding of non-Abelian topological defects

When braiding of non-Abelian quasi-particles is discussed in full-fledged non-Abelian quantum Hall states, the metaphor of braiding has a clear picture behind. The positions $\{R_i(t)\}$ of the quasi-particles are parameters and their world lines are the braids. In the case that we consider, where the defects all lie on one line, the world lines cannot cross one another. The notion of braiding then needs a more careful definition for the present context.

In this section we will discuss a system of zero modes χ_i that lie on one line, at the interfaces between incompatible gapping mechanisms. The system may be initialized with some of these modes being close enough to one another to be coupled. The process of braiding will amount to introducing a set of coupling constants between the zero modes, such that the Hamiltonian of the system will be

$$H = \sum_{i,j} h_{ij}(t)\chi_i^+\chi_j \tag{8}$$

in such a way that *(a)* the instantaneous degeneracy of the Hamiltonian's ground state stays constant at all times; *(b)* the set $\{h_{ij}\}$ comes back to its initial $t = 0$ value after a time T; and *(c)* in a Heisenberg picture, the set of operators $\{\chi_i(T)\}$ is a permutation of the initial set $\{\chi(t = 0)\}$. Similar to the planar case, the unitary operator U that will relate the initial and final zero modes will not vary (up to an Abelian phase) with variations of the trajectory taken in parameter space, so long as the variation does not change the ground state degeneracy. For this unitary operator to be non-trivial, it is required that the trajectory of the coupling constants h_{ij} will be such that it cannot be reduced to the trivial trajectory $h_{ij}(t) = h_{ij}(t = 0)$ without varying this degeneracy.

When compared to the planar braiding process we described above, the difference is that here coupling between zero modes is essential, and that as a consequence, the time-dependent parameters are not the positions of the defects, but their couplings.

Zero modes couple when they get close to one another, and charge tunneling between zero modes take the system from one ground state to another. This is most easily understood when remembering how the space of degenerate ground states is spanned by the charges on the superconductors. We can imagine here several ways to introduce such a coupling. When a back-scattering region shrinks, the two zero modes that border it couple, and the effective Hamiltonian is

$$H_S = \lambda_S \cos\left(\pi \hat{S}_i + \alpha_i\right). \tag{9}$$

This coupling connects different eigenstates of $e^{i\pi \hat{Q}_i}$ and $e^{i\pi \hat{Q}_{i+1}}$, but leaves the product $e^{i\pi(\hat{Q}_i + \hat{Q}_{i+1})}$ intact.

When a superconducting region shrinks, coupling is turned on between the two zero modes that border it, and the effective Hamiltonian is

$$H_Q = \lambda_Q \cos\left(\pi \hat{Q}_i + \alpha_i\right). \tag{10}$$

Both these coupling terms describe the tunneling of a single quasi-particle between two neighboring zero modes. They are not sufficient to carry out a trajectory that interchanges two zero modes. For that, we need more than nearest neighbor coupling. We imagine that the edge on which the zero modes reside is a flexible line that may be deformed to bring zero modes that are not nearest neighbors to proximity. The coupling between the zero modes then depends on the medium that separates the proximate zero modes, through which tunneling takes place. If this medium is the FQHE state, a fractional quasi-particle may tunnel between the zero modes. If it is a vacuum, only electron tunneling may take place. It is the former case that is needed for interchanging zero modes.

Pictorially, the way the interchange is done may be understood by imagining the following situation. You hold two ping-pong balls, one by each hand. You want to interchange them, and you are not allowed to hold two balls by the same hand at any stage of the process. Clearly, your only way to proceed is by finding an empty place, put one of the balls (the right one, say) in that place, transfer the other ball from its original left hand to its destination right hand, and finally pick up the ball that was left aside and hold it in the empty left hand.

To translate this picture to our system, we need to understand how a zero-mode ("ball") is moved to "an empty place", a place where there is no other zero mode. We will be aided by the following observation. Imagine that we have $2N$ interfaces, i.e. $2N$ defects, out of which two are coupled to one another. The $2m$ degeneracy associated with that coupled pair of zero modes is removed by the coupling. Now, if a third defect is coupled to the those two, no further removal of degeneracy takes place. Rather, the coupling of three zero modes to one another always leaves one mode at zero energy. It is not the energy of this zero mode, but rather its wave function, that is affected by the couplings. Thus, imagine that we start with a group of four defects a, b, c, d, where a and b are coupled to one another and therefore move away from zero energy, while c, d are uncoupled and are the zero modes we wish to interchange. First, we copy c to a: we turn on a coupling of b to c, and then turn off the coupling of a to b. The final outcome would be that the zero mode has moved from c ("the full right hand") to a ("the empty place"), without any degeneracy being lifted along the way. Now let us move d ("the left hand") to c ("the now-empty right hand"): we turn on a coupling of b to d, and then turn off the coupling of b to c. Finally, we move a to d: we turn on a coupling of b to a and then turn off the coupling of b to c. At the end of the process, the couplings have gone back to their original values, but c and d were interchanged.

We will not go over the derivation of the unitary transformation associated with parafermionic braiding,[9,10] but rather state the result and comment on it. The unitary transformation associated with interchanging two zero modes that border the same superconducting segment is

$$U = e^{i\alpha(\hat{Q}_j - \frac{k}{m})^2} \tag{11}$$

where $\hat{Q}$ is the charge operator corresponding to the superconducting segment. If the two zero modes border on the same back-scattering segment, the operator $\hat{Q}$ is replaced by $\hat{S}$. The coefficient α depends on the type of tunneling particle involved in the coupling between the zero modes. For electrons $\alpha = m^2/2$. For quasi-particles, it is $m/2$. The value of the integer parameter k depends on the details of the trajectory in the space of coupling constants, but may not be varied without the trajectory crossing a degeneracy point. Note that for electron tunneling, where $\alpha = m^2/2$ the unitary transformation U has the same properties as that of Majorana braiding transformations, namely $U^4 = 1$ and U is periodic in k with a period of two. In contrast, for quasi-particles tunneling it is $U^{2m} = 1$ and U is periodic in k with a period of $2m$. Braiding of zero energy modes that do not border the same segment may be broken into a series of braiding operations of the type already defined.

3.6. Non-Abelian defects vs non-Abelian states

The constructions that we outlined above introduce non-Abelian defects into Abelian quantum Hall states, either by coupling them to superconductors or by manipulating their geometries. It is important to realize the difference between a non-Abelian quantum Hall state and the constructions that we presented so far. In the former, quasi-particles are quantum mechanical degrees of freedom with their own dynamics; the edge states carry fractional central charge; and the ground state degeneracy on a pristine torus, with no quasi-particles in the bulk, reflects the non-Abelian statistics. In the latter, in contrast, none of these requirements is satisfied. The quasi-particles are bound to defects; the edge states carry an integer central charge; and the degeneracy of the ground state on a torus is determined by the Abelian bulk state. Later on we will see the extra steps needed in order to advance from non-Abelian defects to non-Abelian states. While we will carry out this advance on a physical route, a more mathematical route may also be developed.[50]

3.7. *Transition to one dimension*

The constructions that we presented are all based on quantum Hall states, which are inherently two-dimensional. Could we have parafermions of this type in one dimension? After all, Majorana fermions are expected to exist both in 2D systems, e.g. the Moore–Read $\nu = 5/2$ state, and 1D systems, e.g. the Kitaev model of p-wave superconductivity.[51]

General arguments have been presented to show that without the protection of further symmetries, the only non-Abelian defects that may exist in one dimension are Majorana fermions.[52,53] It is constructive to show how this restriction emerges if we try to narrow down the quantum Hall state to a quasi-one dimensional geometry. To that end, we consider an annulus of the bilayer system discussed above, with both edges of the annulus, the interior and the exterior, being gapped by an alternating set of super-conducting regions and back-scattering regions (see Fig. 8). For the simplest illustration, let us consider each edge having two superconducting regions and two back-scattering regions. In principle, we could position the zero energy defects between the regions in such a way that they will keep away from one another even when the annulus is thinned down to a quasi-1D ring. When that is done, there would be no tunnel coupling between the zero modes. Does this imply that the degeneracy of the ground state will stay intact while the annulus is thinned down to become effectively one-dimensional?

The answer is negative,[54] and it has to do with the coupling between two gapped domains of a similar gapping mechanism on the two edges. As we saw in the discussion of a ground state degeneracy on a torus, a region that is gapped by back-scattering may absorb a dipole made of fractional quasi-particle and quasi-hole, without getting out of the subspace of ground states. Similarly, a region that is gapped by a superconductor may absorb a fractional Cooper-pair of charge $2e^*$, without getting out of the subspace of ground states. Consequently, proximity of two regions that are gapped by the same mechanism allow coupling of ground states of both regions, which leads to their energies being split.

For our stitching of the two annuli to form a torus of $\nu = 1/m$, we need to have one gapping mechanism that gaps both the interior and the exterior edges. The resulting ground state degeneracy is m, and when the torus is thinned down to allow for coupling of the two edges the degeneracy is removed. The only way to avoid this removal is by having a symmetry that forbids tunneling between the edges. Translational invariance may play the role of this symmetry.

When the stitching involves two incompatible gapping mechanisms in alternation, the parafermionic interface zero modes increase the degeneracy to $2m$ per

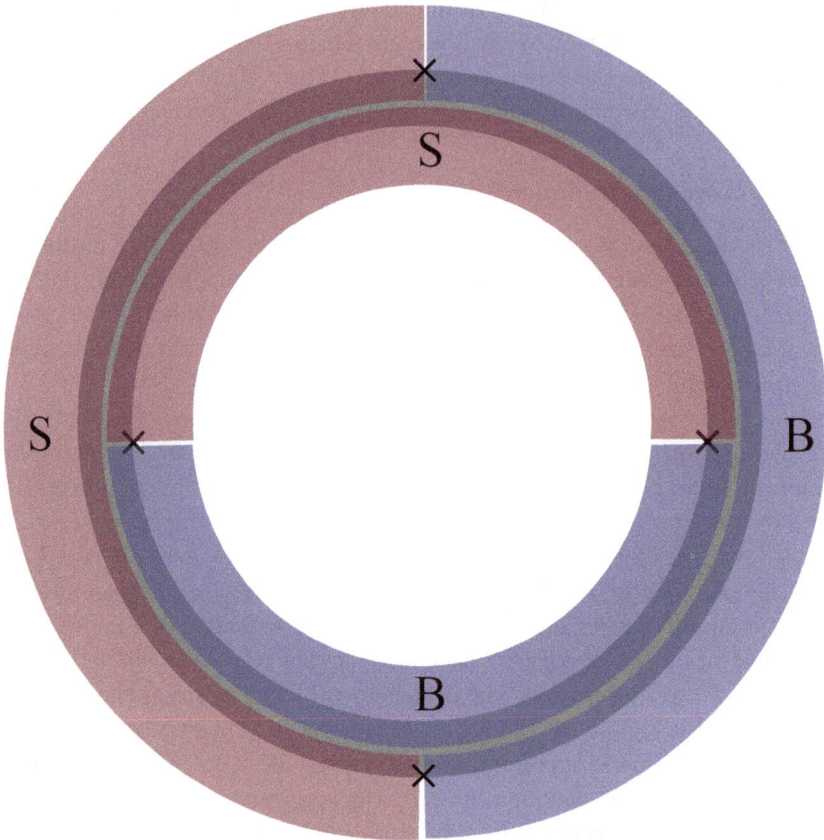

Fig. 8. An annulus of a bilayer $\nu = \pm 1/m$ system has gapless modes at the exterior and the interior, which are gapped by alternating regions of superconductors and back-scattering. Zero energy modes are localized at the interfaces, two at the interior and two at the exterior. The annulus is thinned down such that the interfaces stay far from one another. As explained in the text, tunneling between domains remove all degeneracies except those that are associated with Majorana operators.

region. They do that by allowing also unpaired quasi-particles or quasi-holes to reside within the gapped regions. Indeed, the $2m$ degeneracy is a product of the m degenerate states that we may expect from the analysis of the torus, multiplied by the degeneracy of two that comes together with two Majorana fermions. The coupling between regions of the same mechanism in the exterior and interior removes the degeneracy of m, but leaves the Majorana degeneracy intact. Only coupling between the localized zero modes removes the Majorana-induced degeneracy as well. Thus, in the absence of a symmetry that will forbid tunneling between domains at the interior and the exterior, the thinning down of the annulus results in the restriction of the zero energy defects to Majorana fermions only.

3.8. *One-dimensional phases along fractional edges*

Our discussion so far has concentrated on gapping the edge states of fractionalized phases, and on zero modes that are localized at interface points between different incompatible mechanisms of gapping. This is one corner of a larger issue, which concerns the possible phases that may take place along a one-dimensional edge of a fractionalized phase, but cannot take place in stand-alone one-dimensional systems (at least without special symmetries being imposed).[54–59] Here we would like to briefly discuss several other aspects of this issue.

3.8.1. *Can the edge be gapped?*

The concept of gapping an edge has been central to everything we have done so far. We discussed simple examples where it was rather easy to guess the nature and form of the gapping mechanisms. With more complicated edges, the question is whether the edge may be gapped and will the specification of the gapping mechanisms become more complicated.

For a perturbation that is charge conserving an edge may be gapped only if it does not carry a Hall conductivity. For a perturbation that does not conserve charge, i.e. a coupling to a superconductor, an edge may be gapped only if it carries no thermal Hall conductivity. For Abelian states, where each edge mode carries a central charge of $c = 1$, the latter condition is the requirement that the numbers of right-moving and left-moving edge modes are equal. Those are, however, necessary conditions which are not always sufficient.

A sufficient condition for gappability may be phrased[60] in terms of the properties of the fractional quasi-particles that exist in the bulk whose edge we want to gap. More precisely, it is phrased in terms of the mutual statistics of these quasi-particles. In this language, a gapping mechanism corresponds to a division of the system's quasi-particles into two types, $\mathcal{M}$ and $\mathcal{M}'$. Those quasi-particles that belong to $\mathcal{M}$ have trivial mutual statistics with respect to one another. Those quasi-particles that belong to $\mathcal{M}'$ have non-trivial mutual statistics with respect to at least one member of $\mathcal{M}$. Different gapping mechanisms correspond to different divisions. If no such division exists, the edge may not be gapped.

Let us look at three examples. As a first example, consider the bilayer $\nu = \pm 1/3$ that we considered before. It has nine types of quasi-particles, each defined by two numbers (q_1, q_2), where $q_1, q_2 = 0, \frac{1}{3}, \frac{2}{3}$ denote the fractional charges that are carried by the quasi-particle in each of the layers. The phase accumulated between two such quasi-particles is $e^{i\frac{2\pi}{3}(q_1 q_1' - q_2 q_2')}$. We have two distinct possibilities for $\mathcal{M}$. The first consists of the three quasi-particles (q_1, q_1), and corresponds to gapping with a superconductor. The second consists of the three quasi-particles $(q_1, -q_1)$, and corresponds to gapping with back-scattering.

As the second example, consider the bilayer that we cut to two, having four edges altogether. Here quasi-particles are labeled by four numbers (q_1, q_2, q_3, q_4), taking the same values as before. Now we have four possibilities for $\mathcal{M}$, two of which correspond to charge conserving gapping mechanisms — $\mathcal{M} = (q_1, -q_1, q_2, -q_2)$ and $\mathcal{M} = (q_1, -q_2, q_2, -q_1)$. These two mechanisms correspond to the two ways we discussed by which the edges may be gapped.

Finally, as a third example we consider the edge of the $\nu = 2/3$ state, in which the edge consists of a $\nu = 1$ edge moving in one direction, and a $\nu = -1/3$ edge moving in the other direction. There are three quasi-particles, and there is no way to divide them to the required $\mathcal{M}, \mathcal{M}'$. The edge may not be gapped, even with a perturbation that does not conserve charge.

3.8.2. Dimerizations

Much of our discussion has concentrated on the degeneracy of the ground state in the presence of non-Abelian defects. When these defects are not infinitely far from one another, the degeneracy is split, with the splitting being exponentially small in the ratio of the distance between the zero modes to the localization length of each zero mode. When that ratio is not large enough the degenaracy is well lifted, and interesting phases may be created. Two obvious phases originate from dimerized patterns. In the superconducting–back-scattering example one dimerization pattern couples zero modes which border the same superconducting island. In that pattern, the coupling is of the form $H_c = \lambda \sum_i \cos(\pi Q_i - \alpha)$, i.e. the eigenstates are enumerated by the charge on the superconducting quantum dots. In that pattern the superconducting quantum dots are decoupled from one another, and the edge insulates. A complementary dimerization pattern couples neighboring zero modes that border *neighboring* superconducting islands. In that case, the coupling Hamiltonian tunnels a quasi-particle between two neighboring islands, such that it takes, say, $Q_i \rightarrow Q_i + 1$ and $Q_{i+1} \rightarrow Q_{i+1} - 1$. This coupling term is diagonal in the basis of the $e^{i\pi\hat{S}}$ operator of the insulating region between the superconductors. When this coupling dominates, the edge superconducts.

In the bilayer geometry of $\nu_1 = -\nu_2 = 1/m$ one pattern of dimerization heals the cut in a way that makes the layers effectively planar, while the other pattern of dimerization effectively folds the bilayer system into one-fold single layer at $x < 0$ and another folded single layer at $x > 0$.

More generally, the Hamiltonian of the system when zero modes are coupled to one another may be written in terms of the zero mode operators themselves, χ_i. At the quadratic level, and assuming only coupling between nearest neighbors, the Hamiltonian takes the form

$$H = \sum_j \lambda_j \chi_{j+1}^+ \chi_j + h.c. \tag{12}$$

The two dimerized phases are those in which $\lambda_j^l = \lambda + (-1)^l(-1)^j \delta\lambda$, where the index $l = 1, 2$ enumerates the dimerization type. The extreme dimerization takes place when $\delta\lambda = \lambda$, in which case the spectrum is obviously gapped.

The two dimerization patterns are topologically distinct phases, similar to the two different dimerization patterns of an SSH chain. A transition between them involves a closure of the gap and the formation of a critical phase.

3.8.3. *Critical phases*

When $\delta\lambda = 0$ the period of the discrete translation symmetry of the Hamiltonian changes from two lattice sites to one lattice site, and the phase is gapless. For the case of coupled Majorana zero modes the Hamiltonian is quadratic and its analysis is relatively easy. At the critical phase the edge carries a pair of counter-propagating Majorana modes, carrying a central charge $c = 1/2$.[51] While the dimerized phases are gapped, the critical phase is gapless, yet with some of the degrees of freedom, i.e. part of the central charge, being gapped.

The analysis of the case of coupled parafermion modes is more complicated, and generally requires techniques of conformal field theory.[61] Two cases are of particular interest to us here. For Z_3 parafermions the critical phase carries a central charge of $c = 4/5$. For Z_4 parafermions the critical phase carries a central charge of $c = 1$, but it is an orbifold theory rather than a conformal field theory of a free boson. Both cases will be further elaborated on below. For Z_n parafermions with $n > 4$ and small λ the coupling between the zero modes is an irrelevant perturbation to the $\lambda = 0$ gapless phase of counter-propagating free bosons. Therefore, for weak coupling no degrees of freedom are gapped, and the central charge remains $c = 1$. What happens when the coupling is strong is not fully known.

4. Constructing Non-Abelian States from Wires at Criticality

4.1. *Introduction*

Quantum Hall states, Abelian and non-Abelian alike, must have gapless charged chiral modes at their edges. These modes are one-dimensional Luttinger liquids, and as such each one of them carries a central charge of one. Non-Abelian quantum Hall states carry, on top of the charged chiral modes, also neutral modes that are not Luttinger liquids. These modes are either described by conformal field theories of a fractional central charge, or by orbifold theories with an integer central charge.

The wire construction approach constructs the chiral gapless edge mode by starting with wires that are not chiral and separating their right- and left-moving parts from one another through their coupling to neighboring wires on opposite sides. Its elementary building block is then a "unit cell" (frequently composed of a single wire, but possibly of several wires) that by itself already has two copies of the desired edge modes, one for each chirality. The recipe for this construction then goes along the following steps (a) defining the degrees of freedom within each unit cell, (b) introducing interactions within the unit cell that will bring it to the form of a "thinned down" non-Abelian quantum Hall state. And (c) formulating inter-cell interactions that will "inflate" the quantum Hall state by pushing the gapless modes to the edges of the sample, and separating them by a gapped bulk.

The implementation of this recipe in the Read–Rezayi cases that we will review here is influenced by their physical picture. We therefore start by reviewing this picture.

4.2. *Review of the Read–Rezayi non-Abelian series*

The Read–Rezayi series of non-Abelian states was the first to be introduced as a series of candidate non-Abelian fractional quantum Hall states, and it is also the most thoroughly studied series. Additional types of non-Abelian states were introduced as well, for example, in Refs. 62 and 63. For concreteness, we will mostly focus on the Read–Rezayi series.

Initially, the Read–Rezayi series,[7] which was introduced as a generalization of the non-Abelian $\nu = 5/2$ Moore-Read ("Pfaffian") state,[5] was based on the notion of electron-clustering. The Moore–Read state may be viewed as a state of paired composite fermions that form a p-wave superconductor. The Read–Rezayi states, that correspond to filling factors of $\nu = k/2$ for bosons and $\nu = k/(k+2)$ for fermions, may be viewed as a Bose condensate of clusters of k particles (fermions or bosons) to which ν^{-1} flux quanta are attached to cancel the external magnetic field.

A later way of looking at the Read–Rezayi states starts from k fictitious flavors of bosons. Each of the flavors is at a Laughlin state of filling factor $1/2$, such that the total filling is $k/2$. The symmetrized bosonic wave function is[64]

$$\Psi = \hat{S} \prod_{j=1}^{k} \prod_{l \neq l'}^{N/k} (z_l^j - z_{l'}^j)^2, \tag{13}$$

where N is the number of bosons (assumed divisible by k), the integer indices $l, l' \geq 1$ and $\hat{S}$ is the symmetrizer. This wave function may be shown to be identical to the one obtained initially by Read and Rezayi. Although constructed as a bosonic wave function, Eq. (13) may be easily transformed to be fermionic, by multiplying it by a factor of $\prod(z_l^j - z_{l'}^{j'})$, where the product is over all pairs of electrons, with each pair counted once. We will mostly discuss fermions here.

Following the two ways to construct the same series of wave functions, there are also two ways to construct the series of states from an array of wires. One way starts, effectively, from a unit cell that has k wires, corresponding to the k flavors mentioned above. The total central charge of the unit cell starts as k. Interactions within the cell should then gap part of the cell's degrees of freedom, such that the central charge of the edge will be smaller. The other starts from a unit cell of one wire, but that is a wire in which single electrons coexist with clusters of electrons. Within each wire there are then two pairs of counter-propagating gapless modes, one for single electrons and one for clusters of electrons. Furthermore, each wire should have an interaction term between the two, that breaks a cluster into k electrons and vice versa.

Below, we will review both ways. We will enter some detail regarding $k = 2, 4$ in which calculations may be carried out in ways that make the physics rather transparent, and $k = 3$ which is an interesting representative of all other values. Before delving into the wire construction, we mention that there are other ways to look at non-Abelian states, through the pattern of zeroes in the trial wave functions,[65] and the limit of a thin torus.[66,67] Those are beyond the scope of this chapter.

4.3. *Wire construction for non-Abelian states — the coset construction*

The first wire construction approach for the Read–Rezayi series of states has been initiated by Teo and Kane.[68] It is based on two premises. First, we know what is the edge structure that we need to engineer. In particular, we know the conformal field theory that will describe the neutral mode on the edge. And second, we will use "conventional" one-dimensional quantum wires, each carrying a pair of counter-propagating electronic modes, as the building blocks for the construction of the array.

With these two premises in mind, the Teo–Kane approach is based on the following steps. First, we define a unit cell including n wires with n pairs of counter-propagating edge modes, i.e. a central charge of $c = n$. Second, we break the unit cell into a tensor product of CFTs, in which $c = c_1 + c_2 + c_3$, with c_1, c_2, c_3 being the central charges of three CFTs whose tensor product gives the original CFT. After identifying the three CFTs, gapping terms are introduced to the Hamilotian to gap the bulk. One of the CFTs, say the one with central charge c_3, is gapped within the unit cell it resides in, while the other two CFTs, with central charges c_1, c_2, are gapped telescopically by inter-wire interactions. As a consequence, chiral gapless modes are created at the edge carrying central charges c_1 and c_2. One of the two is charged, with $c = 1$, while the other is neutral, with a fractional central charge. The CFT of the neutral edge is then an $SU(2)_k$, with k being the level.

Since the unit cell from which the planar system will be constructed should include more than one wire, some super-structure, with a periodicity larger than one wire, should be introduced. In the Teo–Kane construction for Read–Rezayi states it is the magnetic field that is assumed to have modulations with a period of several wires.

The same approach may be generalized to more complicated non-Abelian states, based on $SU(N)_k$ CFTs.[69,70] These states involve several electronic flavors (spin states, layer indices, etc.) The application of the wire approach to these states highlights their relation to the Abelian multi-component Halperin states.

The mathematical notion of the breaking of one CFT into a tensor of several CFTs is well developed, but does not easily lend itself to physical intuition. On the other hand, having a starting point where the only microscopic degrees of freedom

are single electrons make this approach a natural one to study situations where a
physical system may go through transitions between different non-Abelian states,
of different clusters.[71-73]

4.4. *Wire construction for the paired states*

The alternative approach, for which the unit cell remains a single wire, is motivated
by the picture of the Read–Rezayi states as based on electron clustering. This
approach regards each wire as composed of two interacting fields, one of single
electrons and the other of clusters of k-electrons. For two particular cases, $k = 2, 4$
this approach allows for a rather detailed analysis of the single wire and the coupled
array of wires.

It is not surprising to find out that the $k = 2$ state of the Read–Rezayi series
is the simplest one to understand. It is based on pairing of fermions into a boson,
which is a notion for which we have a rich set of superconductivity-based theoretical
tools. And its neutral edge mode is a chiral Majorana mode, which is, again, familiar
from the theory of p-wave superconductivity.[6,51]

4.4.1. *A single wire*

For a state based on pairing, it is a reasonable starting point to have wires in which
single electrons and pairs of electrons may coexist. The theory of superconductivity
allows for that, but its use must involve one unusual twist: while the mean field
theory of superconductivity, in the form of the Bogolubov–deGennes Hamiltonian,
gives up charge conservation, the description we construct cannot do that. Thus,
our single wire should have separate degrees of freedom for single electrons and for
pairs of electrons, and an interaction term that converts two electrons into a pair
and vice versa. Writing the electrons in terms of creation and annihilation operators
c^+, c and the pairs in terms of the bosonic creation and annihilation operators $e^{\pm i2\varphi}$,
the Hamiltonian for the single wires is then

$$\mathcal{H} = \mathcal{H}_f^0 + \mathcal{H}_b^0 \tag{14}$$

where

$$\mathcal{H}_b^0 = \frac{v}{2\pi}[K(\partial_x\varphi)^2 + \frac{1}{K}(\partial_x\theta)^2] - 2\mu(\partial_x\theta/2\pi + \bar{\rho}_b) \tag{15}$$

$$H_f^0 = c^\dagger(\epsilon_0 - \partial_x^2/2m - \mu)c + u(c\partial_x c e^{2i\varphi} + h.c.) \tag{16}$$

Here the u-term converts two electrons into a pair and vice versa. It is a p-wave
pairing. The term $\partial_x\theta/2\pi$ describes the fluctuations in the boson density about
an average density $\bar{\rho}_b$, and $[\varphi(x), \theta(x')] = i\pi\Theta(x - x')$. The chemical potential μ
couples to the total charge density $\rho_e = \psi^\dagger\psi + 2\rho_b$.

In the mean field treatment the bosonic operators are assumed to acquire expec-
tation values $u\langle e^{2i\phi}\rangle \equiv \Delta$, and the resulting approximate Hamiltonian loses charge
conservation. In return, it has the advantage of being a quadratic Hamiltonian,

which may be exactly diagonalized. Enforcing charge conservation, we give up the exact solutions.

The mean field wire Hamiltonian has four possible phases. When $\Delta = 0$ the system may be an insulator, if the chemical potential $\mu < 0$ or a Luttinger liquid metal if $\mu > 0$. When $\Delta \neq 0$ the system may be a strong pairing superconductor if $\mu < 0$ or a weak-pairing superconductor if $\mu > 0$. Except the metallic phase, all phases are gapped, and all transitions between phases involve gap closures. Most relevant for the present context, a transition between weak-pairing and strong-pairing superconductivity, i.e. the point where $\mu = 0$ and $\Delta \neq 0$ involves a gap closure and the formation of a pair of counter-propagating Majorana modes. Consequently, when the transition between the phases takes place at a point along the wire, a localized Majorana zero mode occurs at the transition point.

This picture is modified when charge is conserved. Most importantly, when there are charges in the wire there are no gapped phases, since there is always a gapless excitation at which the center of mass is accelerated. This mode carries density and current oscillations, and may be described as a plasmon mode, a phonon mode, or a phase mode, depending on the context. The existence of this gapless mode raises several questions regarding the phases and the phase transitions in the wire. First, at the mean field level the weak superconductor was distinguished from the metallic phase by having an energy gap. With charge conservation there is no such gap. Are these phases still distinct? The answer to this question is negative, as we review below. Second, are the two superconducting phases distinct, and if so what distinguishes them? The answer to this question is affirmative. The two superconducting phases are distinguished by their spectrum of single electron excitation. While both phases are gapless to the introduction of bosonic Cooper pairs, the strong pairing superconductor is gapped to the introduction of single electrons, while the weak pairing phase is gapless. And third, what happens at the phase transition, given that the spectrum is everywhere gapless? The answer to this is in the counting of modes. At both the weak and the strong superconductors the spectrum is gapless, with a pair of counter-propagating free boson mode, of central charge $c = 1$. At the transition another mode becomes gapless. This is a pair of counter-propagating Majorana modes, with central charge $c = 1/2$. Note that in the absence of the pairing interaction the system has two pairs of counter-propagating modes — one of the electrons and one of the pairs — and hence a central charge $c = 2$.

The conclusions outlined in the last paragraph regarding charge-conserving pairing in a single wire may be understood by transforming the electrons/bosons degrees of freedom to charged/neutral ones. This transformation, which is non-local, is carried out by writing the Hamiltonian (14) as $\mathcal{H} \rightarrow U\mathcal{H}U^\dagger$ with

$$U = e^{i\int dx(\psi^\dagger\psi-\bar{\rho}_f)\varphi(x)} \qquad (17)$$

where $\bar{\rho}_f$ is the average fermion density. Denoting the new boson and fermion fields as φ_ρ, θ_ρ and ψ_σ, this has the effect of transforming

$$\psi(x) \to \psi_\sigma(x)e^{-i\varphi_\rho}, \tag{18}$$

$$\varphi(x) \to \varphi_\rho(x), \tag{19}$$

$$\theta(x) \to \theta_\rho(x) + \pi(\bar{\rho}_f x + \int_x^\infty dx \psi_\sigma^\dagger \psi_\sigma). \tag{20}$$

Thus, $\partial_x \theta_\rho/\pi$ now describes the fluctuations in the total electron density about the average value $\bar{\rho}_e = \bar{\rho}_f + 2\bar{\rho}_b$. The fermion field ψ_σ is neutral, but non-local.

The transformed Hamiltonian has three terms,

$$\mathcal{H} = \mathcal{H}_\rho + \mathcal{H}_\sigma + \mathcal{H}_{\text{int}}. \tag{21}$$

The term describing the charged degrees of freedom is

$$\mathcal{H}_\rho = \frac{v}{2\pi}[K_\rho(\partial_x\varphi_\rho)^2 + \frac{1}{K_\rho}(\partial_x\theta_\rho)^2] - \mu(\partial_x\theta_\rho/\pi + \bar{\rho}_e), \tag{22}$$

where K_ρ is the charge Luttinger parameter. The term describing the neutral degrees of freedom is

$$\mathcal{H}_\sigma = \psi_\sigma^\dagger(\epsilon_0 - \partial_x^2/2m)\psi_\sigma + iu(\psi_\sigma\partial_x\psi_\sigma + \psi_\sigma^\dagger\partial_x\psi_\sigma^\dagger) \tag{23}$$

and the interaction term is

$$\mathcal{H}_{\text{int}} = \frac{(\partial_x\varphi_\rho)^2}{2m}\psi_\sigma^\dagger\psi_\sigma - \frac{i\partial_x\varphi_\rho}{m}\psi_\sigma^\dagger\partial_x\psi_\sigma - \frac{v\partial_x\theta_\rho}{K_\rho}(\psi_\sigma^\dagger\psi_\sigma - \bar{\rho}_f). \tag{24}$$

The second term, $\mathcal{H}_\sigma$, can be recognized as a mean field pairing Hamiltonian. It does not conserve the number of fermions, but since the fermions are neutral, this does not imply a violation of charge conservation. This Hamiltonian has a second order topological transition at $\epsilon_0 = 0$. When $\epsilon_0 < 0$ the wire is in the strong superconductivity side, and it is gapped to introducing single fermions. Pairs of fermions may be introduced with no energy gap since they do not operate on $\mathcal{H}_\sigma$. When $\epsilon_0 > 0$ the wire is gapless to single electrons. This may be seen by mapping the neutral fermions onto Ising spins.[15] In electronic terms, the operator that introduces an electron to the wire at no energy cost does so by introducing a 2π twist to the phase of the boson field. At the critical point the wire carries a pair of counter-propagating Majorana modes. It is this critical point which will generate the non-Abelian states in an array of wires.

The interaction term (24) couples the fermions and the bosons, and may be analyzed using renormalization group methods. While the first two terms in (24) are strongly irrelevant, the third is marginal. Its effect was analyzed in Refs. 15,74 and 75, where it was shown that under certain conditions, it may turn the phase transition of a single wire from second to first order. Here we assume that these conditions do not hold for the wires we consider.

4.4.2. *An array of wires*

The last paragraphs carry the seed of the construction of non-Abelian paired states. As we saw, a non-Abelian paired state should have an edge with a charged mode, of integer central charge, and a neutral mode, of a fractional central charge. Here we have them at the single wire level, when the wire is at the transition between weak-pairing and strong-pairing superconducting phases. All we need to do now is to couple wires at this critical point, and gap the modes telescopically between wires. While for a single wire the central charge is $c = 3/2$ only at the critical point, when the wires are coupled, the critical point expands into a phase. Note that the tool we have at our disposal for gapping the modes, inter-wire tunneling, allows only whole electrons to tunnel.

While we expect the non-Abelian states to emerge when the single wires are at their critical points, we may consider the effect of tunneling more generally. When the wires are at the strong pairing phase, they are gapped to tunneling of single electrons, so the inter-wire gapping of their gapless modes would be carried out by tunneling of pairs of electrons. Then, pairs of electrons are the only low energy degrees of freedom, both at the intra- and inter-wire levels. Effectively, then, the wires form bosonic quantum Hall states, where each boson is a pair of electrons, with charge 2. Laughlin states for bosons have filling fractions of $\nu = 1/2m$, with m an integer. With a charge $q = 2e$, the Hall conductivity is $\sigma_{xy} = \frac{q^2}{2mh} = \frac{2e^2}{mh}$. The states are Abelian, and carry a single bosonic edge mode, of central charge $c = 1$.

When the wires are in their weak-pairing phase, they are gapless both to single-electron and to two-electron tunneling. Single electron tunneling allows the construction of the fermionic Laughlin states, with $\nu = 1/(2m+1)$ and $\sigma_{xy} = \frac{e^2}{h(2m+1)}$. Two-electrons tunneling creates the bosonic states described in the previous paragraph. However, here these states are anisotropic, carrying zero energy Majorana modes at wires' ends.

Altogether, then, there are two distinct situations. When σ_{xy} (in units of e^2/h) has an odd denominator, they may be formed in two ways — as bosonic Laughlin states of bosonic filling $1/4(2m+1)$, composed of wires in the stgrong-pairing phase, and as fermionic Laughlin states of fermionic filling $1/(2m+1)$, composed of wires in the weak pairing phase. The wire construction approach allows us to analyze the transition between these two types of states, which is found to be of the Ising type.

When σ_{xy} has an even denominator, the non-Abelian paired quantum Hall state emerges. At these filling fractions, on both sides of the single-wire phase transition the gapped states are created by pair tunneling. At the transition itself, however, single electron tunneling is needed as well, to gap the critical Majorana mode. When both single-electron and two-electron tunneling are allowed, the Majorana modes are gapped telescopically, and a non-Abelian state, with a $c = 1$ charged edge mode and a $c = 1/2$ neutral edge mode, is created. When $\sigma_{xy} = e^2/2h$ this state is the Moore–Read Pfaffian state.

So far, we confined this description to states with a single charge mode and a single neutral mode. A natural generalization direction is that of non-Abelian states with more complicated edges. There are many ways for such generalization, with the simplest one being that of stacking many Pfaffian states together. We focus here on a generalization in which the edge carries one charge mode, of central charge $c = 1$ and any integer number of Majorana modes, with a total central charge $c = n/2$.

The existence of these states may be expected when one regards the Pfaffian state as a p-wave superconductor of composite fermions.[6] The p-wave superconductor belongs to class-D of superconductors that have no time reversal symmetry. This class carries a Z topological classification, where the integer topological invariant n_M counts the number (and direction) of the edge Majorana modes. Since the modes are chiral, interactions would not connect different values of this topological invariant, and — as long as they are weak — would not change the classification.

In the coupled wire construction the creation of a state that carries n_M Majorana modes but only one charged mode requires a tunneling term that tunnels electrons between wires, couples Majorana modes that are n_M wires away from one another, and keeps the momentum balance required for $\nu = 1/2$. This term is a term that tunnels an electron from a wire j to a wire $j + n_M$ and back-scatters a single electron in all the wires in between.[15]

There is an important difference between even and odd values of n_M when it comes to quasi-particles in the bulk. Since the quasi-particles are vortices within the superconcductors of composite fermions, they affect the boundary conditions of the edge Majorana modes. When n_M is odd, an odd number of bulk vortices enforces an odd number of zero energy states on the edge. Since the total number of zero energy states must be even, this enforces each quasi-particle to carry a localized zero energy Majorana mode. Then, braiding quasi-particles affect the zero energy states, and the quasi-particles are non-Abelian. In contrast, when n_M is even the presence of any number of bulk vortices enforces an even number of zero energy states on the edge, and the vortices themselves do not need to carry localized zero energy modes. The states are then Abelian.

The observation that there are infinitely many paired states for $\nu = 1/2$ (or $\nu = 5/2$) puts under the same framework the entire list of states proposed for that filling factor over the years, and suggests that they are part of an infinite set of states. Specifically, the Moore–Read Pfaffian state is, in this language, the $n_M = 1$ state. The Abelian $K = 8$ state is the $n_M = 0$ state. The particle-hole symmetric state, known as PH-Pfaffian, is the $n_M = -1$ state. The $n_M = -2$ state is the so-called (113) state and is Abelian. The $n_M = -3$ state is the particle-hole conjugate of the Pfaffian, known as the anti-Pfaffian. And so the list goes on.

4.5. $k = 4$: Non-Abelian state of electron quartets

A second case for which the wire construction of non-Abelian states may be carried out explicitly to a great detail is that of the Read–Rezayi $k = 4$ state, based on electron quartets. In this case the neutral edge mode carries a central charge $c = 1$, and is described by an orbifold theory.[61,76] This term will be explained below. For now, we just say that the orbifold theory justifies a paraphrase of the old Sondhi–Kivelson statement saying "One is a fraction, too."[b]

Just as in the $k = 2$ case, for $k = 4$ there are infinitely many states, enumerated by two quantum numbers, as was discovered by Barkeshli and Wen.[77] These two numbers will emerge from our analysis below, and will obtain a new physical interpretation.

We explain here the main steps in the analysis of the $k = 4$ non-Abelian quantum Hall states. We skip many of the details, for which we refer the reader to Ref. 16. We follow the same recipe defined in Sec. 4.1 — choosing the right degrees of freedom for each wire; defining the intra-wire interaction; coupling the wires.

4.5.1. *A single wire*

Again, the starting point is a single wire, in which two pairs of counter-propagating modes exist. Here one mode carries single electrons and the other carries quartets of electrons. The total central charge is $c = 2$. Then, there are two interaction terms. The first breaks a quartet into four electrons or glues four electrons into a quartet, while the second back-scatters single electrons from right to left and vice versa. Clearly, the $k = 4$ wire is harder than the $k = 2$ one, since even if we impose an expectation value on the quartet's field, the resulting Hamiltonian is not quadratic in electronic operators. Consequently, the Hamiltonian of the single wire is best written in bosonic terms, using two pairs of fields, one describing quartets of electrons, φ_4, θ_4, and the other describing single electrons, ϕ_{1L}, ϕ_{1R}.

The Hamiltonian has a free boson kinetic term for each of the two pairs, and two interaction terms, one (u) that composes and decomposes clusters and four electrons and one (v) that back-scatters single electrons. Altogether, the Hamiltonian density is,

$$\mathcal{H} = \mathcal{H}_0 + \mathcal{H}_{\text{int}}, \tag{25}$$

where $\mathcal{H}_0$ is the Luttinger liquid Hamiltonian density for the two pairs of fields, and the interaction Hamiltonian density is

$$\mathcal{H}_{int} = u \cos (\varphi_4 - 2\phi_{1L} - 2\phi_{1R}) + v \cos (\phi_{1R} - \phi_{1L}). \tag{26}$$

Just as for the paired case, the non-Abelian state will originate from a critical state between two phases. When the back-scattering v is the dominant interaction term, single electrons are effectively gapped, and quartets are the only degrees of freedom at low energies. The gapping of the electronic mode leaves the system with

---

[b]This has been the title of a paper by Sondhi and Kivelson in 1993, in which they considered the essential effect of electron-electron interaction on the understanding of the low energy excitations of the $\nu = 1$ Integer Quantum Hall state, in contrast to the expectation that interactions are essential only for fractional states. The title was not accepted by the editors of the *Physical Review B* at the time.

a central charge $c = 1$. This phase is analog to the strong pairing phase we had in the previous subsection.

When the quartet-to-electrons u term is the dominant one there is one gapless mode, which is gapless both to single electrons and to quartets. The operator that inserts a single electron with a vanishing energy cost does so with the insertion of a winding to the phase of the bosonic phase field of the clusters.

Remarkably,[61] the critical point may be analyzed exactly due to a hidden $SU(2)$ symmetry that allows us to map the interacting Hamiltonian of the electrons and the quartets onto a quadratic Hamiltonian in terms of composite particles. Technically, we unravel this symmetry by employing one-dimensional bosonization to define new fermions and new bosons. The new particles are frequently not local, but their statistics may be chosen to our convenience. The non-locality requires care, and, in fact, in the present case it is of crucial importance.

The first step of handling the critical state is a change of variables from the set of fields φ_4, θ_4, ϕ_{1L}, ϕ_{1R} to a new set of fields, given by

$$\phi_{\uparrow,R} = \phi_{1R} + \theta_4,$$
$$\phi_{\uparrow,L} = \phi_{1L} - \theta_4, \tag{27}$$
$$\phi_{\downarrow,R} = \frac{1}{2}(-\phi_{1R} - \phi_{1L} + \varphi_4 + 2\theta_4),$$
$$\phi_{\downarrow,L} = \frac{1}{2}(-\phi_{1R} - \phi_{1L} + \varphi_4 - 2\theta_4).$$

The newly defined fields are fermionic. By saying that, we mean that the operators $e^{i\phi_{\alpha,\beta}(x)}$, with α, β being one of the four combinations in (27) are fermionic creation operators. This can be verified by calculating their commutation relations. The fermions carry a chirality symbol L, R and a fictitious spin index $\uparrow, \downarrow$ (we will refer to this fictitious spin as "spin").

The transformation (27) will lead us to an exact solution of the Hamiltonian (25). But before we see that, we must dwell on its two non-trivial aspects, which are eventually what leads to the state being non-Abelian. The first is the $\frac{1}{2}$ factor in the last two lines of (27). Due to this factor, a fermion created by the operator $e^{i\phi_{\downarrow,R}}$ or $e^{i\phi_{\downarrow,L}}$ creates a non-integer number of electrons and quartets. Since the number of electrons on the wire must be an integer, the physical states of the wire are those for which there is an even number of fermions with down spins. Said differently, we are constrained to the subspace $\exp i\pi N_\downarrow = 1$, with $N_\downarrow$ being the number of down spin fermions. If the ϕ's of Eq. (27) are expressed in terms of charge and spin fields $\phi_{\rho,A} \equiv \frac{1}{2}(\phi_\uparrow + \phi_\downarrow)$ and $\phi_{\sigma,A} \equiv \frac{1}{2}(\phi_\uparrow - \phi_\downarrow)$, the constraint corresponds to the identification of ϕ_σ with $\phi_\sigma + \pi$. The second non-trivial aspect of the transformation is that the field $e^{i\theta_4}$, which is a local field in the microscopic Hamiltonian, becomes a non-local field in the transformed basis, and thus forces us to consider some non-local fields in that basis as physical.

Upon this change of variables, the interaction terms should now be expressed in terms of the fermions defined in (27). Obviously, the interaction term is symmetric to interchanging right and left fermions. Less obviously, at the critical point, the interaction term that involved four electrons and one quartet now reduces to a term that is quartic in the spinful fermions. Together with the kinetic term, the Hamiltonian (25) becomes,

$$\mathcal{H}_{\uparrow\downarrow} = -iv_F \sum_{s,A,B} \tau_{AB}^z \psi_{sA}^\dagger \partial_x \psi_{sB} + 2\pi\lambda v_F J_R^x J_L^x, \tag{28}$$

where the sum is over $s = \uparrow, \downarrow$ and $A, B = R, L$,

$$\vec{J}_A = \frac{1}{2} \sum_{r,s=\uparrow,\downarrow} \psi_{rA}^\dagger \vec{\sigma}_{rs} \psi_{sA} \tag{29}$$

and λ may be expressed in terms of the parameters in the original Hamiltonian.

The solvability of this fermionic Hamiltonian is found out when it is bosonized again. The $SU(2)$ current $\vec{J}_A$ in (29) may be bosonized in terms of the spin field $\phi_{\sigma,A}$ to be

$$J_A^x = \frac{1}{2\pi x_c} \cos \phi_{\sigma A}$$

$$J_A^y = \frac{1}{2\pi x_c} \sin \phi_{\sigma A} \tag{30}$$

$$J_A^z = \frac{\partial_x \phi_{\sigma A}}{4\pi}$$

with x_c a short distance cut-off. Then, a rotation of $\pi/2$ around the y-axis turns the interaction term in (28) from a $J_R^x J_L^x$ term, which is a product of two cosines, into a $J_R^z J_L^z$ term, which is a quadratic term, a product of two $\partial_x \phi_\sigma$'s. With that, the entire Hamiltonian becomes quadratic in the bosonic fields, and hence may be diagonalized. No modes are gapped, the central charge remains $c = 2$. The interaction leaves two pairs of counter-propagating modes, but it does affect the eigenstates of the spin mode. Due to the interaction term, proportional to $\lambda \partial_x \phi_{\sigma R} \partial_x \phi_{\sigma L}$, these eigenstates admix the left- and right-moving modes of the fermions ψ_s. The level of admixture depends on λ.

The $\pi/2$ rotation around the y-axis would have been just a renaming of axes, where it not for the constraint, which is also affected. In the rotated frame the constraint requires that any physical operator would be invariant to an interchange of up and down spins. In terms of the field ϕ_σ in the rotated frame, the constraint identifies ϕ_σ with $-\phi_\sigma$, which is the definition of an orbifold.

Altogether, then, the critical state of the single wire differs from that of free fermions by the orbifolding of the free bosonic field theory of the spin sector. This orbifolding does not reduce the central charge, but does constrain the Hilbert space and the allowed operators. The resulting Hamiltonian has then two Luttinger parameters, g_ρ for the charge sector and g_σ for the orbifolded spin sector, and both take continuous values. The parameter g_σ determines the extent to which the eigenstates mix the original left- and right-moving modes. The orbifold theory is familiar from other contexts in physics, such as the Ashkin–Teller model and the four states clock model.[78–81]

4.5.2. An array of wires

The next step is the coupling between wires. As before, we would like to choose tunneling operators that (a) transfer an integer number of electrons and quartets from a wire to its neighbor; (b) keep the balance between the initial momentum, the momentum obtained from the Lorenz force while tunneling, and the final momentum; and (c) are purely chiral, i.e. couple one chirality of the wire of origin to the opposite chirality of the wire of destination. As in all wire constructions, the first two requirements determine a set of allowed filling factors. As for the third requirement, the eigenmodes of the spin sector mix the right- and left-moving fermions of the non-interacting problem to create chiral eigenmodes of the interacting one. There is a set of discrete values of the critical interaction parameter λ for which the eigen-operators of the spin modes create an integer number p fermions of one chirality and $p-1$ fermions of the opposite chirality. This discrete set of λ's are those for which the quantum Hall states are constructed. The most obvious example is the non-interacting case, $\lambda = 0$, for which $p = 1$. In this case the resulting state is Abelian.

The inter-wire tunneling operators need to gap the bulk charge and spin modes. Quartets tunneling may gap the charge modes, but does not couple to the spin modes. Thus, a single electron tunneling is required as well. Generally, the single electron tunneling operator is

$$e^{i\left(\frac{1+m}{2}\phi_{1R}+\frac{1-m}{2}\phi_{1L}+l\theta_k\right)}, \tag{31}$$

with m being an odd integer, and l an integer. This operator becomes relevant at a filling of $\nu = 2/l$. Furthermore, when expressed in terms of the $SU(2)$ fermions, the operators (31) are local only for odd l, imposing the final restriction on our analysis to filling factors of the form $\nu = 2/l$, with odd l.

The momentum balance, and hence the filling factor, do not depend on the value of m in (31). This value is determined by the requirement that the single electron tunneling term gaps the spin mode. Since the eigenvectors of the spin mode couple right- and left-moving electrons in a way that depends on the interaction scale λ, the tunneling operator should depend on the interaction as well. As explained before, for a set of discrete values $\lambda(p)$ there is a single electron operator that couples only to one chirality of the spin mode. For odd p, this happens when $m = (l + p)/2$. Thus, this analysis constructs quantum Hall states at $\nu = \frac{2}{2m-p}$.

A detailed analysis of this array of coupled wires is beyond the scope of the present chapter. As seen in Ref. 16, the combination of the non-locality of some of the operators when transformed to the language of the spinful fermions, with the orbifold constraint on the allowed operators, make the resulting quantum Hall state non-Abelian. Furthermore, different choices of p give rise to different set of non-Abelian quasi-particles. These properties may largely be obtained from the wire construction with no input from the corresponding conformal field theory.

4.6. *Strip-construction of non-Abelian states*

The recipe we used for a wire construction of the $k = 2, 4$ Read–Rezayi states may in principle be used also for $k = 3$, but with one significant difference. In the latter case the analysis of the one-dimensional critical point, from which the non-Abelian state is created, requires conformal field theory, and cannot be carried out using simple-minded bosonization as for $k = 2, 4$. We will not go over this derivation, but rather describe how such a state may be constructed from a "cousin" of the wire construction, the strip construction.

In this approach the system is made of narrow strips of Abelian fractional quantum Hall states, which are parallel to one another (see Fig. 9). The strips are separated by trenches, and are coupled by electron tunneling between their edges, across the vacuum (colored grey in the figure). In this setting, the one-dimensional wires of the previous subsections are replaced by these coupled edge modes, while the vacuum regions of the previous subsections are replaced by the gapped fractional quantum Hall regions (colored orange in the figure).

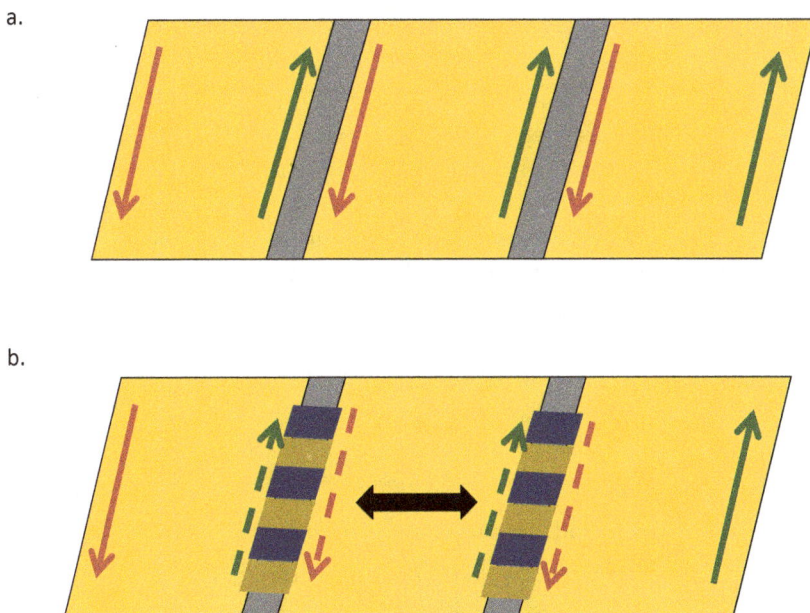

Fig. 9. The strip construction of a non-Abelian fractional quantum Hall state. (a) The starting point — an array of strips, each made of a fractional quantum Hall state, in such a way that pairs of counter-propagating fractional quantum Hall edges are formed (solid arrows). The strips are separated by vacuum trenches. (b) Each pair of edges is coupled by an alternating chain of incompatible gapping mechanisms, forming a chain of coupled parafermions at a critical state. The critical state carries a counter-propagating pair of gapless edge modes, with a reduced central charge (dashed lines). These edge modes are now coupled to one another by tunneling of fractionally-charged quasi-particles across the fractional quantum Hall strips.

The idea behind the present construction is similar to the one behind the wire construction, namely, creating one-dimensional systems with the mode structure we would like the final edge to have, and then gapping these modes telescopically. The one-dimensional systems are created by coupling counter-propagating edges of neighboring strips to a critical state, in the way described in Sec. 3.8.3. For example, a chain of parafermions forms a critical state that carries a pair of counter-propagating edge modes, of a fractional central charge. These phases may be created only on the edge of fractional quantum Hall states, and therefore the strips are made of fractional states. This has the important consequence that the coupling between the neighboring one-dimensional systems is done not through the vacuum, but through a fractionalized medium. Then, there is a richer set of possibilities for tunneling objects — not just electrons, but also fractionally-charged quasi-particles.

The one-dimensional systems described in Sec. 3.8.3 are made of counter-propagating edge modes of fractional quantum Hall states, coupled by incompatible gapping mechanisms in an alternating way, such that a chain of parafermionic zero modes form at the interfaces. These modes are coupled to form a critical phase between two types of dimerizations. For example, if the coupling between the edge modes alternates between superconducting and back-scattering couplings, the critical phase borders between a superconducting phase, in which the parafermions across back-scattering regions dimerize, and a fractional quantum Hall phase, in which the parafermions across the superconducting regions dimerize. The central charge of this critical phase depends on the type of parafermions that are forming it, which in turn depend on the fractional quantum Hall state that the edges border. Furthermore, the quasi-particles that tunnel across the strips to gap the critical one-dimensional modes depend also on the fractional quantum Hall state that forms the strip. Note that since this state may be deconstructed to a set of one-dimensional wires separated by a vacuum, the strip construction is not fundamentally different from the wire construction.

An example for such a construction was worked out in detail in Refs. 82 and 83 where it was shown to create a fractional quantum Hall state with mobile non-Abelian quasi-particles of the Fibonacci-type. The starting point for this construction is an array of strips at the unpolarized $\nu = 2/3$ state. This state is believed to be Abelian, and described by a K-matrix $K = \begin{pmatrix} 1 & 2 \\ 2 & 1 \end{pmatrix}$. It has two counter-propagating modes, as evidenced by the determinant of K being negative. Two neighboring strips then carry two pairs of counter-propagating modes, which may be gapped either by back-scattering (effectively uniting two strips into one) or by superconductivity (here the singlet nature of the unpolarized $\nu = 2/3$ state presumably makes the coupling to the superconductor easier to realize). When both couplings do not involve spin flips, excitations of $e^* = 1/3$ quasi-particles are of high energy, and the low energy subspace involves only quasi-particles of charge 2/3. Two strips that are coupled by an alternating back-scattering and superconductivity segments host a one-dimensional lattice of Z_3 zero modes, and if those are coupled at the critical point, they form a gapless pair of counter-propagating modes, with a central charge of $c = 4/5$.

This pair of modes is then to be telescopically gapped by means of tunneling of quasi-particles across the strips. For that to happen, there must be a tunneling operator that couples counter-propagating modes on the two edges of a single strip, and has matrix elements with the low energy states of these operators. The available tunneling operators tunnel Abelian quasi-particles across the strip. The spinless $2/3$ charged quasi-particles do indeed have matrix elements to the low energy states of the $c = 4/5$ modes, and are therefore able to gap them, leaving at the end of the sample chiral neutral modes of $c = 4/5$, and creating a full-fledged non-Abelian quantum Hall state, with a quasi-particle content similar to that of the Read–Rezayi $k = 3$ state. Due to the coupling to the superconductors, this state does not conserve charge.

This construction is based on a hierarchy of length scales. Largest of all is the bulk energy gap in the strips. Following it is the inter-strip coupling, that reduces the edges from $c = 2$ to $c = 4/5$. And smallest of all is the intra-strip coupling. One can imagine a wire construction that realizes the same structure with the same hierarchy of energy scales, but without the spatial structures of strips and trenches.

The $\nu = 2/3$ is a fruitful arena for searching for different quantum Hall states and their realizations.[73,84,85] It may be a single component state, when it is spin polarized, or a two-component state, when it is unpolarized, or realized as a bilayer state. No less than five different non-Abelian states have been proposed for $\nu = 2/3$, added by several Abelian ones. Numerically, the bilayer state is found to be Abelian, but the single layer $\nu = 2 + \frac{2}{3}$ state has the largest overlaps with the Z_4 Read–Rezayi state.

While the wire construction introduces transitions between different quantum Hall states of identical filling fractions by looking at gap closures in the neutral sectors, the $\nu = 2/3$ introduces another way to describe transitions of this type, through "anyon condenstation".[85] The Z_4 Read–Rezayi state has a non-Abelian quasi-particle that is a boson with respect to itself, call it Φ_b. As such, it may form a condensate in which its number fluctuates macroscopically. Such a condensate is a different phase. Quasi-particles of the Z_4 state that have non-trivial statistics with respect to Φ_b cannot exist as independent quasi-particles in a state in which the number of Φ_b fluctuates. Thus, this state has a reduced set of quasi-particles, which allows for its identification as a bilayer Abelian state.[71]

5. Potential for Computation

The stability associated with topological states of matter makes them appealing for quantum computation. The vision of topological quantum computation based on non-Abelian quasi-particles regards the collection of degenerate ground states as a set of qubits; the braiding of non-Abelian quasi-particles as the means to carry out topologically-protected unitary transformations on these qubits; and the impossibility of distinguishing between ground states through local measurements as the ultimate device against decoherence. As is typically the case, the ideal world differs from the real world in several important respects. First, the ideal model assumes an energy gap between the set of degenerate ground states and the continuum of excited states. In reality, due to disorder there is typically no gap, and at the most there is a mobility gap. Even for a perfectly clean system, the gaps of non-Abelian states are rather small.

Second, the set of unitary transformations that may be carried out by braiding operations is discrete, and therefore the set of topologically protected unitary transformations is discrete as well. For braiding to enable a topologically-protected quantum computation, this set should be dense in the space of all possible unitary transformations, i.e. it should be possible to approximate (with arbitrarily small error) every unitary transformation by means of a series of braiding operations. When that is possible, topological quantum computation is universal. When that is not possible, braiding must be augmented by operations that are not topologically protected.

The simplest non-Abelian type of fractional quantum Hall quasi-particles, those carrying Majorana zero-modes and believed to occur at $\nu = 5/2$, cannot realize universal topological quantum computation. The unitary transformation allowed for Majorana zero modes are given in (11) with $m = 2$ and $k = 0, 1$. For this set to be a universal set, another transformation, $U^{1/2}$, is needed.

Most Read–Rezayi states (all but those with $k = 2, 4$) enable universal quantum computation. Notable among these is the $k = 3$ state, in which the non-Abelian anyons are known as "Fibonacci Anyons", with a quantum dimension given by the golden mean $(1 + \sqrt{5})/2$.

The introduction of the extrinsic topological defects reviewed in Sec. 3.3 naturally raises the question of their computational power. General arguments (see Ref. 86) suggest that anyons whose quantum dimension is a square root of an integer (and thus the Hilbert space of degenerate ground states that they span decomposes into disconnected local subspaces, each spanned by a pair of anyons) cannot realize universal quantum computation. Indeed, this can be shown to be the case for the parafermions we discussed here.[87,88]

Two hopeful directions should however be mentioned. In studies of metaplectic anyons, which are intimately related to the parfermions of Sec. 3.3, it was found that when braiding is supplemented with measurements, a system of metaplectic anyons may realize universal topological quantum computation.[89] Furthermore, when bilayer topological defects ("genons") are implanted in a bi-layer system of the non-Abelian $\nu = 5/2$, their braiding does become capable of universal quantum computation, despite the lack of universality of each of the two layers.[90]

6. Concluding Remarks

The discovery of the fractional quantum Hall effect, about four decades ago, was the first step in the exploration of a fascinating landscape of physical phenomena whose roots lie in topology. In this chapter we reviewed two angles of that landscape — the understanding of non-Abelian quantum Hall states and their engineering. We did that while being aware of the two gaps that confine the field of non-Abelian states: The energy gaps of non-Abelian quantum Hall states are smaller than we would like them to be, while the gap between theoretical concepts and their experimental realization is larger than we would like it to be.

As a way for developing understanding, we reviewed wire constructions of non-Abelian states. These constructions seem to position one-dimensional systems as a very useful starting point for a search for new non-Abelian states. Start the search, they seem to suggest, by finding a one-dimensional system with an interesting gapless spectrum, one that does not conform to the Luttinger liquid paradigm. Critical points at the transition between two phases are good candidates. Once found, identify the right- and left-moving parts of the spectrum, and find perturbations that are able to gap the system by backscattering of electrons. Then, think about an array of such wires, and gap them in a staggered (telescopic) way, the left movers of one wire with the right movers of the next wire. The conditions under which this gapping conserves momentum will determine the filling factor. This is a useful recipe for creating Hamiltonians whose ground states are non-Abelian quantum Hall states. What is missing is a recipe for translating these Hamiltonians into realistic ones, that may be engineered in the lab.

On the engineering side of this chapter we reviewed steps in this direction. We took building blocks that are reasonably well understood, such as single- and double-layer Abelian fractional quantum Hall states, superconductors, ferromagnetic insulators for spin flipping back-scattering etc., and combined them together to create two types of non-Abelian behavior — non-Abelian defects and full-fledged non-Abelian states of matter. The non-Abelian defects were localized at points along gapped abelian fractional edges in which a transition takes place between two incompatible gapping mechanisms. The full fledged non-Abelian states, in which non-Abelian quasi-particles are the low energy degrees of freedom, were created by coupling the defects to one another.

An observation as breathtaking as the quantization of the Hall resistivity to a precision of one part to 10^9 may lead the observer to expect all topological properties of the quantum Hall effect to be as readily observable, with a similar precision. The last decades have taught us to scale down these expectations. Beyond the Hall resistivity, every other topological property of the quantum Hall effect — e.g. fractional charge, thermal Hall conductance, fractional statistics — was found to be a new summit to scale. Based on this short history, it is likely that further experimental and theoretical exploration of non-Abelian physics in the quantum Hall arena will require advances in technological and material aspects, augmented by an increasing level of creativity and ingenuity. This chapter attempted to give motivation for such an exploration.

Acknowledgments

My interest in the subjects discussed in this chapter is about fifteen years old. Throughout the years I have benefitted enormously from many collaborators and colleagues. I am indebted to all of them. I am also grateful to the following sources of funding: Microsoft's Station Q, The European Research Council (Projects MU-NATOP and LEGOTOP), the CRC-183, the Israel Science Foundation and the Israel-US Binational Science Foundation.

References

1. K. V. Klitzing, G. Dorda, and M. Pepper, New method for high-accuracy determination of the fine-structure constant based on quantized Hall resistance, *Phys. Rev. Lett.* **45**, 494–497 (1980). doi: 10.1103/PhysRevLett.45.494. URL https://link.aps.org/doi/10.1103/PhysRevLett.45.494.
2. D. C. Tsui, H. L. Stormer, and A. C. Gossard, Two-dimensional magnetotransport in the extreme quantum limit, *Phys. Rev. Lett.* **48**, 1559–1562 (1982). doi: 10.1103/PhysRevLett.48.1559. URL https://link.aps.org/doi/10.1103/PhysRevLett.48.1559.
3. A. Stern, Anyons and the quantum Hall effecta pedagogical review, *Ann. Phys.* **323**(1), 204–249 (2008). ISSN 0003-4916. doi: https://doi.org/10.1016/j.aop.2007.10.008. URL http://www.sciencedirect.com/science/article/pii/S0003491607001674. January Special Issue 2008.
4. C. Nayak, S. H. Simon, A. Stern, M. Freedman, and S. Das Sarma, Non-Abelian anyons and topological quantum computation, *Rev. Mod. Phys.* **80**, 1083–1159 (2008). doi: 10.1103/RevModPhys.80.1083. URL https://link.aps.org/doi/10.1103/RevModPhys.80.1083.
5. G. Moore and N. Read, Non-Abelions in the fractional quantum Hall effect, *Nucl. Phys. B.* **360**(2), 362–396 (1991). ISSN 0550-3213. doi: https://doi.org/10.1016/0550-3213(91)90407-O. URL http://www.sciencedirect.com/science/article/pii/055032139190407O.
6. N. Read and D. Green, Paired states of fermions in two dimensions with breaking of parity and time-reversal symmetries and the fractional quantum Hall effect, *Phys. Rev. B.* **61**, 10267–10297 (2000). doi: 10.1103/PhysRevB.61.10267. URL https://link.aps.org/doi/10.1103/PhysRevB.61.10267.
7. N. Read and E. Rezayi, Beyond paired quantum Hall states: Parafermions and incompressible states in the first excited Landau level, *Phys. Rev. B.* **59**, 8084–8092 (1999). doi: 10.1103/PhysRevB.59.8084. URL https://link.aps.org/doi/10.1103/PhysRevB.59.8084.
8. M. Barkeshli and X.-L. Qi, Topological nematic states and non-Abelian lattice dislocations, *Phys. Rev. X.* **2**, 031013 (2012). doi: 10.1103/PhysRevX.2.031013. URL https://link.aps.org/doi/10.1103/PhysRevX.2.031013.

9. N. H. Lindner, E. Berg, G. Refael, and A. Stern, Fractionalizing Majorana fermions: Non-Abelian statistics on the edges of Abelian quantum Hall states, *Phys. Rev. X.* **2**, 041002 (2012). doi: 10.1103/PhysRevX.2.041002. URL https://link.aps.org/doi/10.1103/PhysRevX.2.041002.

10. D. J Clarke, J. Alicea, and K. Shtengel, Exotic non-Abelian anyons from conventional fractional quantum Hall states, *Nat. Commun.* **4**, 1348 (2013). doi: 10.1038/ncomms2340.

11. M. Cheng, Superconducting proximity effect on the edge of fractional topological insulators, *Phys. Rev. B.* **86**, 195126 (2012). doi: 10.1103/PhysRevB.86.195126. URL https://link.aps.org/doi/10.1103/PhysRevB.86.195126.

12. A. Vaezi, Fractional topological superconductor with fractionalized Majorana fermions, *Phys. Rev. B.* **87**, 035132 (2013). doi: 10.1103/PhysRevB.87.035132. URL https://link.aps.org/doi/10.1103/PhysRevB.87.035132.

13. C. L. Kane, R. Mukhopadhyay, and T. C. Lubensky, Fractional quantum Hall effect in an array of quantum wires, *Phys. Rev. Lett.* **88**, 036401 (2002). doi: 10.1103/PhysRevLett.88.036401. URL https://link.aps.org/doi/10.1103/PhysRevLett.88.036401.

14. J. C. Y. Teo and C. L. Kane, From Luttinger liquid to non-Abelian quantum Hall states, *Phys. Rev. B.* **89**, 085101 (2014). doi: 10.1103/PhysRevB.89.085101. URL https://link.aps.org/doi/10.1103/PhysRevB.89.085101.

15. C. L. Kane, A. Stern, and B. I. Halperin, Pairing in Luttinger liquids and quantum Hall states, *Phys. Rev. X.* **7**, 031009 (2017). doi: 10.1103/PhysRevX.7.031009. URL https://link.aps.org/doi/10.1103/PhysRevX.7.031009.

16. C. L. Kane and A. Stern, Coupled wire model of Z_4 orbifold quantum Hall states, *Phys. Rev. B.* **98**, 085302 (2018). doi: 10.1103/PhysRevB.98.085302. URL https://link.aps.org/doi/10.1103/PhysRevB.98.085302.

17. C. Nayak and F. Wilczek, 2n-quasihole states realize 2n1-dimensional spinor braiding statistics in paired quantum Hall states, *Nucl. Phys. B.* **479**(3), 529–553 (1996). ISSN 0550-3213. doi: https://doi.org/10.1016/0550-3213(96)00430-0. URL http://www.sciencedirect.com/science/article/pii/0550321396004300.

18. C. L. Kane and M. P. A. Fisher, Quantized thermal transport in the fractional quantum Hall effect, *Phys. Rev. B.* **55**, 15832–15837 (1997). doi: 10.1103/PhysRevB.55.15832. URL https://link.aps.org/doi/10.1103/PhysRevB.55.15832.

19. M. Banerjee, M. Heiblum, V. Umansky, D. E. Feldman, Y. Oreg, and A. Stern, Observation of half-integer thermal Hall conductance, *Nature.* **559**, 205–210 (2018). doi: 10.1038/s41586-018-0184-1.

20. Y. Kasahara, T. Ohnishi, Y. Mizukami, O. Tanaka, S. Ma, K. Sugii, N. Kurita, H. Tanaka, J. Nasu, Y. Motome, T. Shibauchi, and Y. Matsuda, Majorana quantization and half-integer thermal quantum Hall effect in a Kitaev spin liquid, *Nature.* **559**(7713), 227–231 (2018). ISSN 1476-4687. doi: 10.1038/s41586-018-0274-0. URL https://doi.org/10.1038/s41586-018-0274-0.

21. N. R. Cooper and A. Stern, Observable bulk signatures of non-Abelian quantum Hall states, *Phys. Rev. Lett.* **102**, 176807 (2009). doi: 10.1103/PhysRevLett.102.176807. URL https://link.aps.org/doi/10.1103/PhysRevLett.102.176807.

22. N. Hartman, C. Olsen, S. Lscher, M. Samani, S. Fallahi, G. C. Gardner, M. Manfra, and J. Folk, Direct entropy measurement in a mesoscopic quantum system, *Nat. Phys.* **14** (2018). doi: 10.1038/s41567-018-0250-5.

23. G. Ben-Shach, C. Laumann, I. Neder, A. Yacoby, and B. I Halperin, Detecting non-Abelian anyons by charging spectroscopy, *Phys. Rev. Lett.* **110**, 106805 (2013). doi: 10.1103/PhysRevLett.110.106805.

24. A. Stern and B. I. Halperin, Proposed experiments to probe the non-Abelian $\nu = 5/2$ quantum Hall state, *Phys. Rev. Lett.* **96**, 016802 (2006). doi: 10.1103/PhysRevLett. 96.016802. URL https://link.aps.org/doi/10.1103/PhysRevLett.96.016802.

25. P. Bonderson, A. Kitaev, and K. Shtengel, Detecting non-Abelian statistics in the $\nu = 5/2$ fractional quantum Hall state, *Phys. Rev. Lett.* **96**, 016803 (2006). doi: 10.1103/PhysRevLett.96.016803. URL https://link.aps.org/doi/10.1103/PhysRevLett.96.016803.

26. P. Bonderson, K. Shtengel, and J. K. Slingerland, Probing non-Abelian statistics with quasiparticle interferometry, *Phys. Rev. Lett.* **97**, 016401 (2006). doi: 10.1103/PhysRevLett.97.016401. URL https://link.aps.org/doi/10.1103/PhysRevLett.97.016401.

27. D. E. Feldman, Y. Gefen, A. Kitaev, K. T. Law, and A. Stern, Shot noise in an anyonic Mach-Zehnder interferometer, *Phys. Rev. B.* **76**, 085333 (2007). doi: 10.1103/PhysRevB.76.085333. URL https://link.aps.org/doi/10.1103/PhysRevB.76.085333.

28. R. L. Willett, C. Nayak, K. Shtengel, L. N. Pfeiffer, and K. W. West, Magnetic-field-tuned Aharonov-Bohm oscillations and evidence for non-Abelian anyons at $\nu = 5/2$, *Phys. Rev. Lett.* **111**, 186401 (2013). doi: 10.1103/PhysRevLett.111.186401. URL https://link.aps.org/doi/10.1103/PhysRevLett.111.186401.

29. J. Nakamura, S. Fallahi, H. Sahasrabudhe, R. Rahman, S. Liang, G. C. Gardner, and M. J. Manfra, Aharonov-Bohm interference of fractional quantum Hall edge modes, *Nat. Phys.* **15**(6), 563–569 (2019). ISSN 1745-2481. doi: 10.1038/s41567-019-0441-8. URL https://doi.org/10.1038/s41567-019-0441-8.

30. Yakovenko, Victor M. and Goan, Hsi-Sheng, Quantum Hall effect in quasi-one-dimensional conductors: The roles of moving fisdw, finite temperature, and edge states, *J. Phys. I France.* **6**(12), 1917–1937 (1996). doi: 10.1051/jp1:1996197. URL https://doi.org/10.1051/jp1:1996197.

31. S. L. Sondhi and K. Yang, Sliding phases via magnetic fields, *Phys. Rev. B.* **63**, 054430 (2001). doi: 10.1103/PhysRevB.63.054430. URL https://link.aps.org/doi/10.1103/PhysRevB.63.054430.

32. X. Wen, *Quantum Field Theory of Many-Body Systems: From the Origin of Sound to an Origin of Light and Electrons.* Oxford Graduate Texts, (OUP Oxford, 2004). ISBN 9780198530947. URL https://books.google.com/books?id=RYESDAAAQBAJ.

33. E. Sagi, Y. Oreg, A. Stern, and B. I. Halperin, Imprint of topological degeneracy in quasi-one-dimensional fractional quantum Hall states, *Phys. Rev. B.* **91**, 245144 (2015). doi: 10.1103/PhysRevB.91.245144. URL https://link.aps.org/doi/10.1103/PhysRevB.91.245144.

34. F. D. M. Haldane, Model for a quantum Hall effect without landau levels: Condensed-matter realization of the "parity anomaly", *Phys. Rev. Lett.* **61**, 2015–2018 (1988). doi: 10.1103/PhysRevLett.61.2015. URL https://link.aps.org/doi/10.1103/PhysRevLett.61.2015.

35. A. Sterdyniak, C. Repellin, B. A. Bernevig, and N. Regnault, Series of Abelian and non-Abelian states in $c \geq 1$ fractional Chern insulators, *Phys. Rev. B.* **87**, 205137 (2013). doi: 10.1103/PhysRevB.87.205137. URL https://link.aps.org/doi/10.1103/PhysRevB.87.205137.

36. S. Gopalakrishnan, J. C. Y. Teo, and T. L. Hughes, Disclination classes, fractional excitations, and the melting of quantum liquid crystals, *Phys. Rev. Lett.* **111**, 025304 (2013). doi: 10.1103/PhysRevLett.111.025304. URL https://link.aps.org/doi/10.1103/PhysRevLett.111.025304.

37. Z. Liu, G. Möller, and E. J. Bergholtz, Exotic non-Abelian topological defects in lattice fractional quantum Hall states, *Phys. Rev. Lett.* **119**, 106801 (2017). doi: 10.1103/PhysRevLett.119.106801. URL https://link.aps.org/doi/10.1103/PhysRevLett.119.106801.

38. D. J. Clarke, J. Alicea, and K. Shtengel, Exotic circuit elements from zero-modes in hybrid superconductor–quantum-Hall systems, *Nat. Phys.* **10**(11), 877–882 (2014). doi: 10.1038/nphys3114. URL https://doi.org/10.1038/nphys3114.

39. R. M. Lutchyn, E. P. A. M. Bakkers, L. P. Kouwenhoven, P. Krogstrup, C. M. Marcus, and Y. Oreg, Majorana zero modes in superconductor–semiconductor heterostructures, *Nat. Rev. Mater.* **3**(5), 52–68 (2018). doi: 10.1038/s41578-018-0003-1. URL https://doi.org/10.1038/s41578-018-0003-1.

40. L. Fu and C. L. Kane, Josephson current and noise at a superconductor/quantum-spin-Hall-insulator/superconductor junction, *Phys. Rev. B.* **79**, 161408 (2009). doi: 10.1103/PhysRevB.79.161408. URL https://link.aps.org/doi/10.1103/PhysRevB.79.161408.

41. P. Rickhaus, M. Weiss, L. Marot, and C. Schönenberger, Quantum Hall effect in graphene with superconducting electrodes, *Nano Lett.* **12**(4), 1942–1945 (2012). doi: 10.1021/nl204415s. URL https://doi.org/10.1021/nl204415s.

42. Z. Wan, A. Kazakov, M. J. Manfra, L. N. Pfeiffer, K. W. West, and L. P. Rokhinson, Induced superconductivity in high-mobility two-dimensional electron gas in gallium arsenide heterostructures, *Nat. Commun.* **6**(1) (2015). doi: 10.1038/ncomms8426. URL https://doi.org/10.1038/ncomms8426.

43. M. B. Shalom, M. J. Zhu, V. I. Fal'ko, A. Mishchenko, A. V. Kretinin, K. S. Novoselov, C. R. Woods, K. Watanabe, T. Taniguchi, A. K. Geim, and J. R. Prance, Quantum oscillations of the critical current and high-field superconducting proximity in ballistic graphene, *Nat. Phys.* **12**(4), 318–322 (2015). doi: 10.1038/nphys3592. URL https://doi.org/10.1038/nphys3592.

44. F. Amet, C. T. Ke, I. V. Borzenets, J. Wang, K. Watanabe, T. Taniguchi, R. S. Deacon, M. Yamamoto, Y. Bomze, S. Tarucha, and G. Finkelstein, Supercurrent in the quantum Hall regime, *Science* **352**(6288), 966–969 (2016). ISSN 0036-8075. doi: 10.1126/science.aad6203. URL https://science.sciencemag.org/content/352/6288/966.

45. G.-H. Park, M. Kim, K. Watanabe, T. Taniguchi, and H.-J. Lee, Propagation of superconducting coherence via chiral quantum-Hall edge channels, *Scient. Rep.* **7**(1), 10953 (2017). ISSN 2045-2322. doi: 10.1038/s41598-017-11209-w. URL https://doi.org/10.1038/s41598-017-11209-w.

46. S. Matsuo, K. Ueda, S. Baba, H. Kamata, M. Tateno, J. Shabani, C. J. Palmstrm, and S. Tarucha, Equal-spin andreev reflection on junctions of spin-resolved quantum Hall bulk state and spin-singlet superconductor, *Scient. Rep.* **8**(1), 3454 (2018). ISSN 2045-2322. doi: 10.1038/s41598-018-21707-0. URL http://europepmc.org/articles/PMC5823919.

47. M. R. Sahu, X. Liu, A. K. Paul, S. Das, P. Raychaudhuri, J. K. Jain, and A. Das, Inter-Landau-level Andreev reflection at the Dirac point in a graphene quantum Hall state coupled to a nbse$_2$ superconductor, *Phys. Rev. Lett.* **121**, 086809 (2018). doi: 10.1103/PhysRevLett.121.086809. URL https://link.aps.org/doi/10.1103/PhysRevLett.121.086809.

48. J. Eisenstein, Exciton condensation in bilayer quantum Hall systems, *Ann. Rev. Conden. Matt. Phys.* **5**(1), 159–181 (2014). doi: 10.1146/annurev-conmatphys-031113-133832. URL https://doi.org/10.1146/annurev-conmatphys-031113-133832.

49. R. V. Gorbachev, A. K. Geim, M. I. Katsnelson, K. S. Novoselov, T. Tudorovskiy, I. V. Grigorieva, A. H. MacDonald, S. V. Morozov, K. Watanabe, T. Taniguchi, and L. A. Ponomarenko, Strong Coulomb drag and broken symmetry in double-layer graphene, *Nature Phys.* **8**(12), 896–901 (2012). ISSN 1745-2481. doi: 10.1038/nphys2441. URL https://doi.org/10.1038/nphys2441.

50. J. C. Teo, T. L. Hughes, and E. Fradkin, Theory of twist liquids: Gauging an anyonic symmetry, *Ann. Phys.* **360**, 349–445 (2015). doi: 10.1016/j.aop.2015.05.012. URL https://doi.org/10.1016/j.aop.2015.05.012.

51. A. Kitaev, Unpaired Majorana fermions in quantum wires, *Phys. Usp.* **44**(10S), 131 (2001). URL http://iopscience.iop.org/article/10.1070/1063-7869/44/10S/S29/meta.

52. L. Fidkowski and A. Kitaev, Topological phases of fermions in one dimension, *Phys. Rev. B.* **83**, 075103 (2011). doi: 10.1103/PhysRevB.83.075103. URL https://link.aps.org/doi/10.1103/PhysRevB.83.075103.

53. A. M. Turner, F. Pollmann, and E. Berg, Topological phases of one-dimensional fermions: An entanglement point of view, *Phys. Rev. B.* **83**, 075102 (2011). doi: 10.1103/PhysRevB.83.075102. URL https://link.aps.org/doi/10.1103/PhysRevB.83.075102.

54. D. Meidan, E. Berg, and A. Stern, Classification of topological phases of parafermionic chains with symmetries, *Phys. Rev. B.* **95**, 205104 (2017). doi: 10.1103/PhysRevB.95.205104. URL https://link.aps.org/doi/10.1103/PhysRevB.95.205104.

55. Y.-Z. You, C.-M. Jian, and X.-G. Wen, Synthetic non-Abelian statistics by Abelian anyon condensation, *Phys. Rev. B.* **87**, 045106 (2013). doi: 10.1103/PhysRevB.87.045106. URL https://link.aps.org/doi/10.1103/PhysRevB.87.045106.

56. J. Motruk, E. Berg, A. M. Turner, and F. Pollmann, Topological phases in gapped edges of fractionalized systems, *Phys. Rev. B.* **88**, 085115 (2013). doi: 10.1103/PhysRevB.88.085115. URL https://link.aps.org/doi/10.1103/PhysRevB.88.085115.

57. J. Alicea and P. Fendley, Topological phases with parafermions: Theory and blueprints, *Ann. Rev. Conden. Matt. Phys.* **7**(1), 119–139 (2016). doi: 10.1146/annurev-conmatphys-031115-011336. URL https://doi.org/10.1146/annurev-conmatphys-031115-011336.

58. Y. Oreg, E. Sela, and A. Stern, Fractional helical liquids in quantum wires, *Phys. Rev. B.* **89**, 115402 (2014). doi: 10.1103/PhysRevB.89.115402. URL https://link.aps.org/doi/10.1103/PhysRevB.89.115402.

59. J. Klinovaja and D. Loss, Parafermions in an interacting nanowire bundle, *Phys. Rev. Lett.* **112**, 246403 (2014). doi: 10.1103/PhysRevLett.112.246403. URL https://link.aps.org/doi/10.1103/PhysRevLett.112.246403.

60. M. Levin, Protected edge modes without symmetry, *Phys. Rev. X.* **3**, 021009 (2013). doi: 10.1103/PhysRevX.3.021009. URL https://link.aps.org/doi/10.1103/PhysRevX.3.021009.

61. P. Lecheminant, A. O. Gogolin, and A. A. Nersesyan, Criticality in self-dual sine-gordon models, *Nucl. Phys. B.* **639**(3), 502–523 (2002). ISSN 0550-3213. doi: https://doi.org/10.1016/S0550-3213(02)00474-1. URL http://www.sciencedirect.com/science/article/pii/S0550321302004741.

62. P. Bonderson and J. K. Slingerland, Fractional quantum Hall hierarchy and the second Landau level, *Phys. Rev. B.* **78**, 125323 (2008). doi: 10.1103/PhysRevB.78.125323. URL https://link.aps.org/doi/10.1103/PhysRevB.78.125323.

63. E. Ardonne, N. Read, E. Rezayi, and K. Schoutens, Non-Abelian spin-singlet quantum Hall states: Wave functions and quasihole state counting, *Nucl. Phys. B.* **607**(3), 549–576 (2001). ISSN 0550-3213. doi: https://doi.org/10.1016/S0550-3213(01)00224-3. URL http://www.sciencedirect.com/science/article/pii/S0550321301002243.
64. A. Cappelli, L. S. Georgiev, and I. T. Todorov, Parafermion Hall states from coset projections of Abelian conformal theories, *Nucl. Phys. B.* **599**(3), 499–530 (2001). ISSN 0550-3213. doi: https://doi.org/10.1016/S0550-3213(00)00774-4. URL http://www.sciencedirect.com/science/article/pii/S0550321300007744.
65. Y.-M. Lu, X.-G. Wen, Z. Wang, and Z. Wang, Non-Abelian quantum Hall states and their quasiparticles: From the pattern of zeros to vertex algebra, *Phys. Rev. B.* **81**, 115124 (2010). doi: 10.1103/PhysRevB.81.115124. URL https://link.aps.org/doi/10.1103/PhysRevB.81.115124.
66. A. Seidel and D.-H. Lee, Abelian and non-Abelian Hall liquids and charge-density wave: Quantum number fractionalization in one and two dimensions, *Phys. Rev. Lett.* **97**, 056804 (2006). doi: 10.1103/PhysRevLett.97.056804. URL https://link.aps.org/doi/10.1103/PhysRevLett.97.056804.
67. E. J. Bergholtz and A. Karlhede, Half-filled lowest Landau level on a thin torus, *Phys. Rev. Lett.* **94**, 026802 (2005). doi: 10.1103/PhysRevLett.94.026802. URL https://link.aps.org/doi/10.1103/PhysRevLett.94.026802.
68. J. C. Y. Teo and C. L. Kane, From Luttinger liquid to non-Abelian quantum Hall states, *Phys. Rev. B.* **89**, 085101 (2014). doi: 10.1103/PhysRevB.89.085101. URL https://link.aps.org/doi/10.1103/PhysRevB.89.085101.
69. Y. Fuji and P. Lecheminant, Non-Abelian $su(n-1)$-singlet fractional quantum Hall states from coupled wires, *Phys. Rev. B.* **95**, 125130 (2017). doi: 10.1103/PhysRevB.95.125130. URL https://link.aps.org/doi/10.1103/PhysRevB.95.125130.
70. T. Iadecola, T. Neupert, C. Chamon, and C. Mudry, Ground-state degeneracy of non-Abelian topological phases from coupled wires, *Phys. Rev. B.* **99**, 245138 (2019). doi: 10.1103/PhysRevB.99.245138. URL https://link.aps.org/doi/10.1103/PhysRevB.99.245138.
71. M. Barkeshli and X.-G. Wen, Anyon condensation and continuous topological phase transitions in non-Abelian fractional quantum Hall states, *Phys. Rev. Lett.* **105**, 216804 (2010). doi: 10.1103/PhysRevLett.105.216804. URL https://link.aps.org/doi/10.1103/PhysRevLett.105.216804.
72. A. Vaezi and M. Barkeshli, Fibonacci anyons from Abelian bilayer quantum Hall states, *Phys. Rev. Lett.* **113**, 236804 (2014). doi: 10.1103/PhysRevLett.113.236804. URL https://link.aps.org/doi/10.1103/PhysRevLett.113.236804.
73. M. R. Peterson, Y.-L. Wu, M. Cheng, M. Barkeshli, Z. Wang, and S. Das Sarma, Abelian and non-Abelian states in $\nu = 2/3$ bilayer fractional quantum Hall systems, *Phys. Rev. B.* **92**, 035103 (2015). doi: 10.1103/PhysRevB.92.035103. URL https://link.aps.org/doi/10.1103/PhysRevB.92.035103.
74. M. Sitte, A. Rosch, J. S. Meyer, K. A. Matveev, and M. Garst, Emergent Lorentz symmetry with vanishing velocity in a critical two-subband quantum wire, *Phys. Rev. Lett.* **102**, 176404 (2009). doi: 10.1103/PhysRevLett.102.176404. URL http://link.aps.org/doi/10.1103/PhysRevLett.102.176404.
75. O. Alberton, J. Ruhman, E. Berg, and E. Altman, Fate of the Ising quantum critical point coupled to a gapless boson, *arXiv:1609.02599v2.* (2016). URL https://arxiv.org/abs/1609.02599.
76. R. Dijkgraaf, C. Vafa, E. Verlinde, and H. Verlinde, The operator algebra of orbifold models, *Comm. Math. Phys.* **123**(3), 485–526 (1989). URL https://projecteuclid.org:443/euclid.cmp/1104178892.

77. M. Barkeshli and X.-G. Wen, Bilayer quantum Hall phase transitions and the orbifold non-Abelian fractional quantum Hall states, *Phys. Rev. B.* **84**, 115121 (2011). doi: 10.1103/PhysRevB.84.115121. URL https://link.aps.org/doi/10.1103/PhysRevB.84.115121.

78. P. Di Francesco, P. Mathieu, and D. Sénéchal, *Conformal Field Theory*. Graduate texts in Contemporary Physics, (Springer, New York, NY, 1997).

79. P. Ginsparg. Applied conformal field theory. In eds. E. Brézin and J. Zinn-Justin, *Fields, Strings and Critical Phenomena, (Les Houches, Session XLIX)*. (1988). URL https://arxiv.org/abs/hep-th/9108028.

80. E. Fradkin and L. P. Kadanoff, Disorder variables and para-fermions in two-dimensional statistical mechanics, *Nucl. Phys. B.* **170**(1), 1–15 (1980). ISSN 0550-3213. doi: https://doi.org/10.1016/0550-3213(80)90472-1. URL http://www.sciencedirect.com/science/article/pii/0550321380904721.

81. F. Zhang and C. L. Kane, Time-reversal-invariant Z_4 fractional Josephson effect, *Phys. Rev. Lett.* **113**, 036401 (2014). doi: 10.1103/PhysRevLett.113.036401. URL https://link.aps.org/doi/10.1103/PhysRevLett.113.036401.

82. R. S. K. Mong, D. J. Clarke, J. Alicea, N. H. Lindner, P. Fendley, C. Nayak, Y. Oreg, A. Stern, E. Berg, K. Shtengel, and M. P. A. Fisher, Universal topological quantum computation from a superconductor-Abelian quantum Hall heterostructure, *Phys. Rev. X.* **4**, 011036 (2014). doi: 10.1103/PhysRevX.4.011036. URL https://link.aps.org/doi/10.1103/PhysRevX.4.011036.

83. E. M. Stoudenmire, D. J. Clarke, R. S. K. Mong, and J. Alicea, Assembling fibonacci anyons from a F_3 parafermion lattice model, *Phys. Rev. B.* **91**, 235112 (2015). doi: 10.1103/PhysRevB.91.235112. URL https://link.aps.org/doi/10.1103/PhysRevB.91.235112.

84. Y. Alavirad, D. Clarke, A. Nag, and J. D. Sau, F_3 parafermionic zero modes without Andreev backscattering from the 2/3 fractional quantum Hall state, *Phys. Rev. Lett.* **119**, 217701 (2017). doi: 10.1103/PhysRevLett.119.217701. URL https://link.aps.org/doi/10.1103/PhysRevLett.119.217701.

85. A. Vaezi and M. Barkeshli, Fibonacci anyons from Abelian bilayer quantum Hall states, *Phys. Rev. Lett.* **113**, 236804 (2014). doi: 10.1103/PhysRevLett.113.236804. URL https://link.aps.org/doi/10.1103/PhysRevLett.113.236804.

86. E. Rowell, R. Stong, and Z. Wang, On classification of modular tensor categories, *Commun. Math. Phys.* **292**(2), 343–389 (2009). ISSN 1432-0916. doi: 10.1007/s00220-009-0908-z. URL https://doi.org/10.1007/s00220-009-0908-z.

87. M. B. Hastings, C. Nayak, and Z. Wang, Metaplectic anyons, Majorana zero modes, and their computational power, *Phys. Rev. B.* **87**, 165421 (2013). doi: 10.1103/PhysRevB.87.165421. URL https://link.aps.org/doi/10.1103/PhysRevB.87.165421.

88. A. Hutter and D. Loss, Quantum computing with parafermions, *Phys. Rev. B.* **93**, 125105 (2016). doi: 10.1103/PhysRevB.93.125105. URL https://link.aps.org/doi/10.1103/PhysRevB.93.125105.

89. S. X. Cui and Z. Wang, Universal quantum computation with metaplectic anyons, *J. Math. Phys.* **56**(3), 032202 (2015). doi: 10.1063/1.4914941. URL https://doi.org/10.1063/1.4914941.

90. M. Barkeshli, C.-M. Jian, and X.-L. Qi, Twist defects and projective non-Abelian braiding statistics, *Phys. Rev. B.* **87**, 045130 (2013). doi: 10.1103/PhysRevB.87.045130. URL https://link.aps.org/doi/10.1103/PhysRevB.87.045130.

Chapter 10

Fractional Quantum Hall States of Bosons:
Properties and Prospects for Experimental Realization

N. R. Cooper

T.C.M. Group, Cavendish Laboratory, J.J. Thomson Avenue,
University of Cambridge, Cambridge CB3 0HE, United Kingdom
nrc25@cam.ac.uk

An overview is given of experimental settings in which one can expect to observe fractional quantum Hall states of bosons. The focus is placed on ultracold atomic gases, and the regimes most likely to allow the realization of fractional quantum Hall states. The means by which Landau levels, or other topological energy bands, can be generated for cold atoms are summarized. The current theoretical understanding of the likely many-body phases is then presented, focusing on the models that are most readily studied experimentally. The chapter concludes by making contact with other physical platforms where bosonic fractional quantum Hall states are expected to appear: in quantum magnets, engineered qubit arrays and polariton systems.

Contents

1. Introduction

To date all experimental realizations of fractional quantum Hall (FQH) states have been for electrons. The FQH effect was first discovered in modulated-doped gallium arsenide quantum wells,[1] and much of the subsequent exploration of these states has been in similar devices. More recently, studies have progressed to graphene structures, where new features arise, e.g. in superlattice structures.[2]

Despite the experimental prevalence of fermionic FQH states, the theory of the FQH effect can be readily applied to interacting bosonic particles. Notably, the Laughlin wave function[3] describes a FQH state of bosons at filling factor $\nu = 1/p$ when p is an even integer. Similarly, all other model wave functions of fractional quantum Hall states can be converted between fermionic and bosonic variants. It is natural to ask where bosonic FQH states might be found in nature.

In an insightful early contribution, Kalmeyer and Laughlin proposed that the Laughlin state of bosons might describe the ground state of the spin-1/2 Heisenberg antiferromagnet on a triangular lattice.[4] Although subsequent work has shown this not to be the case for that model, the Laughlin state of bosons is believed to describe the ground state of quantum magnets on other frustrated lattices. The relevance of the Laughlin state, and other FQH states, for quantum magnets and related systems will be discussed further below.

The development of the field of ultracold atomic gases has allowed the exploration of quantum many body phases of bosonic and fermionic atoms in a variety of novel settings.[5] These hold the promise of realizing FQH states for both bosonic and fermionic species. In typical settings the bosonic species are more easily cooled to regimes of quantum degeneracy, so the pursuit of bosonic FQH states is a natural goal. Theoretical work has established the conditions under which bosonic FQH states could arise for cold gases of bosons, as well as the forms of these states, in a variety of experimental settings under active investigation. While FQH states take the same qualitative forms for bosons as for fermions, determining the nature of the many body ground state is a delicate issue that depends on a fine balance of energetics related to the specific physical realization. Thus, it is important to study the bosonic models theoretically to assess which, if any, FQH states are stable in realistic physical settings. One notable finding that we highlight below is that the experimentally relevant models for interacting bosons have non-Abelian phases that are more stable than their fermionic counterparts in typical electronic systems.

In Sec. 2 we give an overview of prospects of achieving FQH states for ultracold atomic gases. We focus on bosonic particles, but comment also on fermionic atomic gases in situations for which the cold gas realizations have qualitative differences from conventional electronic systems. For bosonic realizations, we describe theoretical results indicating possible stability of non-Abelian fractional quantum Hall phases, and also for unconventional FQH states on lattices. We conclude in Sec. 3 by describing possibilities for the achievement of FQH states of bosons in other physical settings: in quantum spin systems, synthetic quantum spin systems and photonic materials.

2. Ultracold Atomic Gases

The development of techniques to trap and to cool gases of neutral atoms have made it a routine matter to bring dilute atomic gases into regimes of quantum degeneracy.[5] Although these cold dilute gases exist only as metastable states, the true equilibrium phases being dense crystals, the lifetimes to collapse are very long as a result of the very low densities used, $n_{3D} \sim 10^{14} \mathrm{cm}^{-3}$, which make the three-body collisions required for collapse very rare. Such low densities imply correspondingly low temperatures for the so-called "ultracold" regime of quantum degeneracy. For the thermal de Broglie wavelength, $\lambda_T \sim \hbar/\sqrt{Mk_BT}$ (for atoms of mass M) to be larger than the inter-particle spacing, $n_{3D}^{-1/3}$, requires cooling to below micro-Kelvin temperatures for typical atomic species.

Under these ultracold conditions the thermal wavelength λ_T is also much larger than the range of the inter-particle interaction, so the collisional properties are accurately described by the s-wave scattering length a_s. The s-wave scattering may then be represented by a pairwise contact repulsion $V(\boldsymbol{r}) = g_{3D}\delta^3(\boldsymbol{r})$ with $g_{3D} = 4\pi\hbar^2 a_s/M$, and suitable regularization.[5] For weakly interacting bosons, which form a Bose–Einstein condensate at low temperatures, the strength of interactions is conveniently expressed in terms of the chemical potential $\mu = g_{3D}n_{3D}$.[6]

Throughout this chapter we shall be concerned with quasi-two-dimensional geometries, in which the atoms are restricted to the xy-plane by a tight harmonic confinement in the z-direction. When the energy spacing of the sub-bands of the z confinement, $\hbar\omega_z$, is large compared to the chemical potential, μ, the atoms can be approximated as occupying the lowest sub-band. The effective 2D interaction becomes $V(\boldsymbol{r}) = g_{2D}\delta^2(\boldsymbol{r})$ with $g_{2D} = g_{3D}/(\sqrt{2\pi}a_z)$ and $a_z = \sqrt{\hbar^2/M\omega_z}$. The strength of the interaction between atoms can be widely varied by tuning close to a scattering resonance: either a Feshbach resonance[5] (involving a weakly bound state of the two atoms); or a confinement-induced resonance[7] (involving coupling to transverse sub-band modes). The functional form of the interaction can also be changed, notably by use of atoms (or molecules) with large magnetic (or electric) dipole moments which allow interactions that fall as $1/r^3$ and that can be spatially anisotropic.[8]

2.1. *Landau levels and Chern bands*

In order to search for FQH states of bosonic atoms, stabilized by their inter-particle interactions, one first needs to find ways by which one can form Landau levels — or other suitable flat-band energy spectra — for the individual atoms. To generate Landau levels, one must confront the question of how to cause *neutral* atoms to experience the orbital effects that a uniform magnetic field has on a charged particle. This can be achieved in a variety of ways. We outline some of them here, restricting discussions to the essential physics needed in order to understand the forms of the relevant microscopic models. More comprehensive reviews exist for rotating gases[9,10] and for topological optical lattices.[11]

2.1.1. *Rotation*

An intuitively simple means by which to impose an effective magnetic field on a gas of neutral atoms is to put the gas into rotation. In a frame of reference rotating at angular frequency $\boldsymbol{\Omega}$ an atom of mass M moving at velocity $\boldsymbol{v}$ experiences a Coriolis force $2M\boldsymbol{v} \times \boldsymbol{\Omega}$. This has the same form as a Lorentz force $q\boldsymbol{v} \times \boldsymbol{B}$ on a charge q in magnetic field $\boldsymbol{B}$, provided one makes the association

$$q\boldsymbol{B} = 2M\boldsymbol{\Omega}. \tag{1}$$

This simple classical observation can be made precise, also in its extension to the quantum description, by noting that the transformation to the rotating frame acts to convert the Hamiltonian in the laboratory frame H_0 to

$$H_\Omega = H_0 - \boldsymbol{\Omega} \cdot \boldsymbol{L}, \tag{2}$$

where $\boldsymbol{L} = \boldsymbol{r} \times \boldsymbol{p}$ is the angular momentum operator.[12] For a particle in an isotropic 2D harmonic trap of natural frequency ω_0, the Hamiltonian in a frame rotating at angular velocity $\boldsymbol{\Omega} = \Omega\boldsymbol{e}_z$ is then

$$H_\Omega = \frac{\boldsymbol{p}^2}{2M} + \frac{1}{2}M\omega_0^2 r^2 - \boldsymbol{\Omega} \cdot \boldsymbol{r} \times \boldsymbol{p}. \tag{3}$$

This may be rewritten

$$H_\Omega = \frac{(\boldsymbol{p} - M\boldsymbol{\Omega} \times \boldsymbol{r})^2}{2M} + \frac{1}{2}M(\omega_0^2 - \Omega^2)r^2, \tag{4}$$

showing that it can be viewed as describing a 2D particle of charge q coupled to a vector potential via $q\boldsymbol{A} = M\boldsymbol{\Omega} \times \boldsymbol{r}$. Taking the curl recovers Eq. (1), and gives a flux density

$$n_\phi \equiv \frac{qB}{h} = \frac{2M\Omega}{h}. \tag{5}$$

The harmonic confinement in Eq. (4) is reduced by the centrifugal force, and vanishes at $\Omega = \omega_0$. At this special value, the Hamiltonian describes a free particle in a uniform magnetic field, with cyclotron energy $\hbar\omega_c = 2\hbar\Omega = 2\hbar\omega_0$ and the effective magnetic length $\ell_B = \sqrt{\hbar/qB} = \sqrt{\hbar/2M\omega_0}$. The lowest energy states are the usual lowest Landau level wavefunctions

$$\phi_m(z) \propto z^m \exp(-|z|^2/4), \tag{6}$$

where $z = (x + iy)/\ell_B$, and $m = 0, 1, 2, \ldots$.

Although these wavefunctions have been derived for the fine-tuned case $\Omega = \omega_0$, in fact they are exact energy eigenstates for any rotation rate Ω. This can be understood by noting that the effect of the rotation, Eq. (2), is simply to shift the energies of the 2D harmonic oscillator by an amount that depends on the angular momentum about the z-axis. Since this component of angular momentum is conserved, the energy eigenstates are unaffected by the rotation, Ω. Writing the energy of the 2D oscillator in terms of the radial quantum number $n_r = 0, 1, 2, \ldots$ and angular momentum $m = \ldots, -2, -1, 0, 1, 2, \ldots$, the spectrum is[13]

$$E_\Omega = \hbar\omega_0(2n_r + |m| + 1) - m\hbar\Omega. \tag{7}$$

This spectrum is plotted in Fig. 1 for $\Omega = 0$ and $\Omega = \omega_0$. The special point $\Omega = \omega_0$ recovers the Landau level spectrum, Fig. 1(b). For $\Omega < \omega_0$ the Landau level wavefunctions remain the same, but are shifted in energy by a residual harmonic confinement, leading to the m-dependent term $\hbar(\omega_0 - \Omega)m$ for the orbitals in the lowest Landau level, for which $m > 0$.

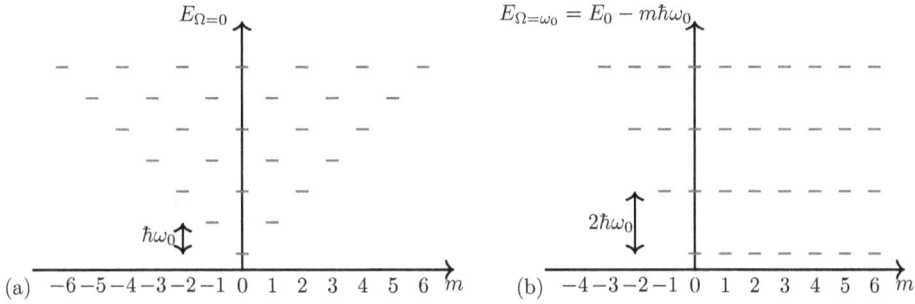

Fig. 1. (a) Spectrum of the isotropic 2D harmonic oscillator, Eq. (7), of natural frequency ω_0 as a function of angular momentum quantum number m. (b) In a frame rotating at angular frequency Ω, the spectrum is $E_\Omega = E_0 - \Omega\hbar m$. For $\Omega = \omega_0$ the spectrum, $E_{\omega_0} = E_0 - m\hbar\omega_0$, is the Landau level spectrum of a charged particle in a uniform magnetic field, with cyclotron energy $2\hbar\omega_0$.

Experiments on atomic Bose–Einstein condensates put into rotation by mechanical stirring have allowed studies of quantum degenerate bosons in regimes in which the chemical potential μ is less than both the sub-band spacing $\hbar\omega_z$ and the cyclotron energy $\hbar\omega_c = 2\hbar\Omega$, that is in the 2D lowest Landau level regime.[14] That said, these studies have so far been restricted to regimes of high particle density where the gases show vortex lattices. The conditions under which FQH states can appear, and the nature of these states, will be discussed in detail below. For now, note that the limitation $\Omega \leq \omega_0$ restricts the flux density (5) to $n_\phi \leq 2M\omega_0/h$. In turn, for a fractional quantum Hall state with 2D particle density $n_{2D} \sim n_\phi$, the interaction energy is of order $\mu \sim g_{2D}n_\phi$, so is restricted to $\mu \lesssim g_{2D}(2M\omega_0/h) \sim \hbar\omega_0(a_s/a_z)$. Typical magnetic traps have a frequency scale $\omega_0 = 2\pi \times (10 - 100)$Hz, and a_s/a_z is a small number, giving an interaction scale μ below the *pico*-Kelvin energy scale. This is an extremely small temperature, much smaller than the temperatures to which it is currently possible to cool atomic gases. The situation can be improved by using for atoms with large a_s (e.g. resonant interactions), or by using very tight traps to increase ω_0, as in Ref. 15 or in quantum gas microscopes.[16]

2.1.2. *Optically dressed states*

The limitation of small flux densities achievable by rotation in typical atomic gases has motivated the development of other ways to generate artificial magnetic fields for cold atoms. One very powerful way to accomplish this, which plays to the strengths of atomic gases, is through optically dressed states.[17]

Optical dressing refers to the use of coherent optical fields to place the atom into a well-defined superposition of internal levels, through the control of the amplitudes and phases of light fields that couple these levels. For a two-level system, coherent driving at frequency ω leads to the coupling

$$\hat{V}(t) = \begin{pmatrix} \epsilon_1 & \Omega\cos(\omega t - \phi) \\ \Omega\cos(\omega t - \phi) & \epsilon_2 \end{pmatrix}, \tag{8}$$

with the matrix expressed in the basis formed by the two internal energy levels. (We use hats to denote operators acting within the space of internal energy levels.) Through a suitable time-dependent gauge transformation one can replace $\cos(\omega t - \phi)$ by $(1/2)[\exp(-i\phi)+\exp(-2i\omega t+i\phi)]$. Then, provided ω is large compared to all other relevant frequency scales, one can make the rotating wave approximation (RWA)[17] by which the remaining oscillating term vanishes, to give a time-independent coupling

$$\hat{V}_{\text{RWA}} = \frac{\epsilon_1 + \epsilon_2}{2}\hat{1} + \frac{\Delta}{2}\hat{\sigma}_z + \frac{\Omega}{2}(\cos\phi\,\hat{\sigma}_x + \sin\phi\,\hat{\sigma}_y), \tag{9}$$

where $\Delta = \epsilon_1 - \epsilon_2 - \hbar\omega$ and $\hat{\sigma}_i$ are the conventional Pauli matrices. Of course, for this simple situation, the phase ϕ could also be removed by a gauge transformation. However, we leave it to emphasize the greater generality of the form of optical coupling: we will be interested in settings where the optical fields lead to *spatially varying* parameters Ω, Δ and ϕ for which one cannot, in general, choose a gauge in which ϕ vanishes everywhere.

The optically dressed states are the eigenstates of this RWA coupling (9). Although presented for a two-level system, these considerations can apply more generally for N_{I} coupled internal states. Consider an atom localized in space at a position $\boldsymbol{r}$, and subjected to local optical fields at this point that couple these N_{I} levels. We denote the dressed states at this location by $|n_{\boldsymbol{r}}\rangle$ with energies $E_n(\boldsymbol{r})$, with $n = 1\ldots N_{\text{I}}$. We use these states as a local basis for a general atomic state

$$|\psi(\boldsymbol{r})\rangle = \sum_n \psi_n(\boldsymbol{r})|n_{\boldsymbol{r}}\rangle. \tag{10}$$

Including the kinetic energy of the atom leads to the full Hamiltonian

$$\hat{H} = \sum_n \left[\frac{\boldsymbol{p}^2}{2M} + E_n(\boldsymbol{r})\right]|n_{\boldsymbol{r}}\rangle\langle n_{\boldsymbol{r}}|. \tag{11}$$

Provided the motion of the particle is slow, in the sense that the typical kinetic energy $\langle\boldsymbol{p}^2\rangle/2M$ is small compared to the energy spacings $E_{n+1} - E_n$, then to a good approximation one can project the Hamiltonian onto states labeled by n alone (ignoring mixing between different dressed states). The effective Hamiltonian for this adiabatic evolution of the nth state, obtained from $H_n\psi_n = \langle n_{\boldsymbol{r}}|H\psi_n|n_{\boldsymbol{r}}\rangle$, is

$$H_n = \frac{(\boldsymbol{p} - q\boldsymbol{A})^2}{2M} + V_n(\boldsymbol{r}), \tag{12}$$

where

$$qA = i\hbar \langle n_{\boldsymbol{r}} | \boldsymbol{\nabla} n_{\boldsymbol{r}} \rangle , \tag{13}$$

$$V_n(\boldsymbol{r}) = E_n(\boldsymbol{r}) + \frac{\hbar^2}{2M} \left(\langle \boldsymbol{\nabla} n_{\boldsymbol{r}} | \boldsymbol{\nabla} n_{\boldsymbol{r}} \rangle - |\langle n_{\boldsymbol{r}} | \boldsymbol{\nabla} n_{\boldsymbol{r}} \rangle|^2 \right) . \tag{14}$$

Note the appearance of a gauge field (13): the Berry connection arising from the (spatial) parallel transport of the local state $|n_{\boldsymbol{r}}\rangle$. The associated Berry phase accumulated as the atom moves in the xy-plane plays the role of the Aharonov–Bohm phase of a charged particle moving in an effective magnetic field. The effective Berry curvature in real space determines the flux density experienced by the particle. This approach was implemented in pioneering experiments reported in Ref. 18, which used three internal states of rubidium-87 to generate an effective magnetic field acting on a Bose–Einstein condensate. The scheme used in this experiment gives a vector potential of order $qA \lesssim \hbar/\lambda$, with λ the optical wavelength. Consequently, the total flux through a region of linear size R is limited to $N_\phi \lesssim \hbar R/\lambda$, and the flux density is limited to $n_\phi \lesssim 1/(R\lambda)$ which is small for typical systems $R \gg \lambda$.

The flux densities achievable using optically dressed states can be vastly increased by forming *optical flux lattices*,[19] in which the optical fields have vortices. The associated phase singularities cause qA to exceed $\hbar/\lambda$, effectively forming Dirac strings, and allowing a magnetic flux density of order $n_\phi \sim N_{\rm I}/\lambda^2$. An example of an optical flux lattice for a two-state system is given by

$$\hat{V}_{\rm RWA}(\boldsymbol{r}) = V_0 \begin{pmatrix} \cos[\boldsymbol{r} \cdot (\boldsymbol{\kappa}_1 + \boldsymbol{\kappa}_2)] & \cos(\boldsymbol{r} \cdot \boldsymbol{\kappa}_1) - i\cos(\boldsymbol{r} \cdot \boldsymbol{\kappa}_2) \\ \cos(\boldsymbol{r} \cdot \boldsymbol{\kappa}_1) + i\cos(\boldsymbol{r} \cdot \boldsymbol{\kappa}_2) & -\cos[\boldsymbol{r} \cdot (\boldsymbol{\kappa}_1 + \boldsymbol{\kappa}_2)] \end{pmatrix} , \tag{15}$$

where $\boldsymbol{\kappa}_{1,2}$ are the two basis vectors of the lattice. Figure 2(c) shows the resulting local flux density for the case where $\boldsymbol{\kappa}_{1,2}$ are of equal length and at 120-degrees to each other, giving a lattice with triangular symmetry.

The local flux density is non-uniform but has a nonzero average. One thus expects the low energy states to mimic those of the continuum Landau level. In this lattice setting, the appropriate way to characterize these 2D energy bands is in terms of their Chern numbers.[20] Indeed, the lowest band of the above optical flux lattice (15) is readily shown to have a Chern number of unit magnitude, $|\mathcal{C}| = 1$, consistent with that of the lowest Landau level. (This result holds true for any finite value of the ratio V_0/E_R of the lattice depth, V_0, to the recoil energy, $E_R = \hbar^2 \kappa^2/2M$, which sets the characteristic kinetic energy.)

The most complete understanding of optical flux lattices is achieved by considering the action of the optical fields in reciprocal space.[21] In essence the optical flux lattice involves a series of momentum exchanges $\boldsymbol{\kappa}_i$ with amplitude and phase defined by the matrix elements $V_{\boldsymbol{\kappa}}^{\alpha'\alpha} = \langle \alpha', \boldsymbol{q} + \boldsymbol{\kappa} | \hat{V}_{\rm RWA} | \alpha, \boldsymbol{q} \rangle$. Here $\alpha, \alpha' = 1, \ldots, N_{\rm I}$ label the undressed internal states. If the optical fields are to define a periodic lattice (as opposed to a quasi-crystal), this set of couplings must form a regular lattice in reciprocal space. For example, Fig. 3(a) shows a generalization of the

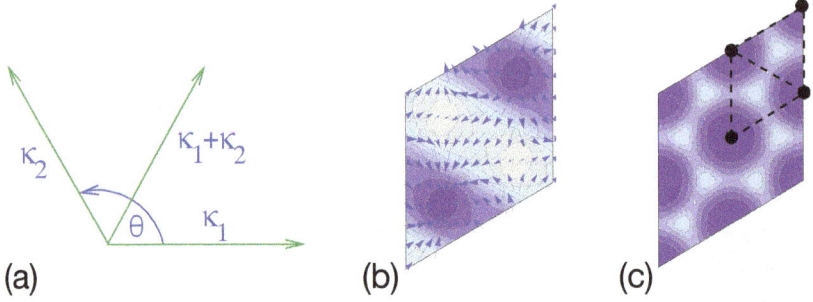

Fig. 2.   Illustration of the optical flux lattice formed by coupling two internal atomic levels in the manner of Eq. (15). (a) The momentum transfers $\kappa_{1,2}$ are depicted at an angle of $\theta = 2\pi/3$. (b) The spatial variation of the lowest energy dressed state n_r in the real space unit cell is illustrated: arrows denote the two components $\langle n_r | \hat{\sigma}_{x,y} | n_r \rangle$; contours and shading denote $\langle n_r | \hat{\sigma}_z | n_r \rangle$. These variations lead to non-trivial Berry connection in real space (13). The local flux density, given by the curl of the real space Berry connection, is spatially varying but with nonzero average. Adapted from Ref. 19.

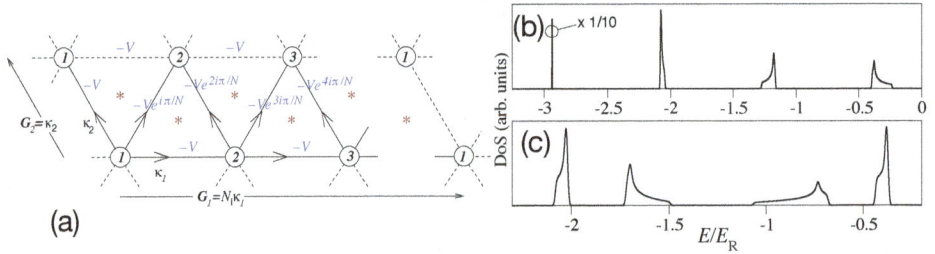

Fig. 3.   Generalization of the triangular optical flux lattice for two internal states, Fig. 2, to situations in which N_I internal states are coupled. (a) The action of the optical fields on the atoms can be represented by a lattice in reciprocal space: the sites are labeled by the undressed internal states $\alpha = 1, 2, \ldots N_I$ and the displacements represent momentum transfers. The (complex) amplitude linking two sites α, α' of this lattice with momentum transfer κ represents the matrix element $V_\kappa^{\alpha'\alpha} = \langle \alpha', \boldsymbol{q} + \boldsymbol{\kappa} | \hat{V}_{\text{RWA}} | \alpha, \boldsymbol{q} \rangle$. Here, the reciprocal space unit cell has an area that scales with the number of internal states N_I. (b) The resulting energy bands can be made to closely approximate Landau levels, appearing as narrow peaks in the density of states (DoS), here shown for $N_I = 4$ internal states. (c) For a related model, in which the phases on all the optical couplings are doubled, the lowest energy band has Chern number $\mathcal{C} = 2$, but still has narrow width in energy. Reproduced from Ref. 21.

triangular optical flux lattice to N_I internal states. Note the expanded BZ, with area that scales with the number of internal states N_I. A similar picture is helpful in understanding strongly correlated phases on topological lattices.[22]

The momentum space picture allows flexible design of optical lattices: (i) with low-energy bands that closely approximate uniform Landau levels, having near uniform Berry curvature in reciprocal space, e.g. Fig. 3(b); (ii) in situations in which there is vanishing net flux through the real space unit cell, yet the lowest energy

band has non-zero Chern number, albeit with lattice effects that cause strong variations in the Berry curvature; (iii) with a lowest energy band that has Chern number
with magnitude larger than unity, Fig. 3(c). Quite generally, one finds that the low
energy bands are topological for a wide range of lattice depth, both for weak lattices
$V_0/E_R \ll 1$ where the bands can be viewed in a "nearly-free electron" approach and
for $V_0/E_R \gg 1$ where a tight-binding model is suitable. The ratio of the band width
to band gap is typically minimized for intermediate lattice depths, $V_0/E_R \sim 1$, and
can be made very small. For example, the lowest band of Fig. 3(b) has a width
$\lesssim 0.005 E_R$ which is about 200 times less than the gap to the next energy band.

The practical implementation of these methods requires careful choice of internal
levels and of the light fields causing the optical coupling.[23] This is particularly
true for cases using $N_{\rm I} > 2$ internal levels, since (electric dipole) selection rules
impose important restrictions on the allowed couplings. For example, a practical
method by which one can form optical flux lattices using the $N_{\rm I} = 3$ spin states
of rubidium-87 involves radio-frequency dressing of the atomic levels into three
non-uniformly spaced states, which are then coupled by two-photon transitions
in a complex light field involving nine frequency components.[24] Quite generally,
the optical flux lattices operate at high flux densities, $n_\phi \sim N_{\rm I}/\lambda^2 \sim N_{\rm I}\kappa^2$. The
interaction energy scale for fractional quantum Hall states with $n_{\rm 2D} \sim n_\phi$ can be
relatively large, i.e. $\mu = g_{\rm 2D} n_{\rm 2D} \sim (a_s/a_z) N_{\rm I} E_R$, with the recoil energy of order
$E_R/h \sim 3 {\rm kHz}$. Moreover, owing to the very narrow energy bands (with width a
small fraction of E_R), this interaction energy can be made large compared to the
bandwidth even for modest values of a_s/a_z, leading to the possibility of strongly
correlated phases.[24]

2.1.3. *Tight-binding lattices*

In the previous sections we have considered atoms that are moving continuously
in real space, and subjected to relatively weak external potentials that are used
to introduce effective magnetic fields. We were driven towards considering atoms
in optical flux lattices, for which the optimal properties arise for relatively shallow
lattices $V_0 \sim E_R$.

Here we consider lattices in the limit in which the potential is sufficiently deep
$V_0 \gtrsim E_R$ that the low-energy bands can be viewed within a tight-binding description. This leads to the connection to tight-binding models that are familiar from
solid state systems. In order to construct topological bands, with non-zero Chern
number, one needs to break time-reversal symmetry in the tight-binding model.
Thus, the goal of the methods described here is to generate inter-site tunnelling
matrix elements with non-zero Peierls phase factors, corresponding to the generation of (local) orbital magnetic fields which break this symmetry. For simplicity
we shall focus on settings in which the potentials are scalar and the atoms have no
internal spin degrees of freedom. However, the ideas presented here can be married
with spin-changing processes.[25,26] Indeed a recent experiment has realized a Chern
band using the techniques of Sec. 2.1.2 within a tight-binding regime.[27]

The methods used to generate topological Chern bands in such tight-binding settings all fall in the class of "Floquet" systems, in which the Hamiltonian is made to vary in time in a periodic matter.[28] When the period T is sufficiently short, such that $\hbar\omega = h/T$ is much larger than other relevant energy scales in the system, the time evolution at long times can be understood in term of the effective Floquet Hamiltonian, H_{eff}. This is defined by the evolution operator over one period, given by the time-ordered integral $\mathcal{T}\exp[-\frac{i}{\hbar}\int_{t_0}^{t_0+T} H(t')\mathrm{d}t'] \equiv \exp(-iH_{\mathrm{eff}}T/\hbar)$. The methods used fall into two broad categories, depending on whether the frequency $\omega \equiv 2\pi/T$ of the time-varying part of the Hamiltonian is resonant or non-resonant with transitions in the spectrum of the static part of the Hamiltonian.

Non-resonant modulation. For non-resonant modulation, and at frequencies that are sufficiently large, the effective Floquet Hamiltonian can be constructed through the application of the Magnus expansion.[28] This expresses the Floquet Hamiltonian as a power law series in $1/\omega$, with a leading term that is the time-averaged Hamiltonian. An important example of this approach is provided by circular shaking of a honeycomb lattice,[29] in which the lattice is displaced by $\mathbf{R}(t) = R_0(\cos\omega t, \sin\omega t)$. In the frame of reference that moves with the lattice, the shaking appears as a force $\mathbf{F}(t) = -M\ddot{\mathbf{R}}(t)$, and the Hamiltonian may be written

$$H(t) = -J_0 \sum_{\langle i,j \rangle} b_i^\dagger b_j - \mathbf{F}(t) \cdot \sum_i \mathbf{r}_i b_i^\dagger b_i , \qquad (16)$$

where the sites i are arranged on a honeycomb lattice at positions $\mathbf{r}_i$, and $\langle i,j \rangle$ denotes the sum over all nearest neighbor pairs. The effective Hamiltonian is equivalent to the tight-binding model of graphene subjected to circularly polarized radiation.[30] Applying the Magnus expansion up to first order in $1/\omega$ leads to an effective Floquet Hamiltonian in which there is a correction to the nearest-neighbor tunnelling amplitude, $J_0 \to J$, and the appearance of a *second*-neighbor tunnelling with an imaginary amplitude[11]

$$H_{\mathrm{eff}} = -J \sum_{\langle i,j \rangle} b_i^\dagger b_j - iJ' \sum_{\langle\langle k,l \rangle\rangle} b_k^\dagger b_l . \qquad (17)$$

This directly realizes the Haldane model,[31] in which time-reversal symmetry is broken by second-neighbor tunnellings (here denoted by double angled brackets). Since the Peierls phase factor generated for second neighbor coupling is $\varphi = \pi/2$ the model is in a regime suitable for the formation of topological bands. This approach was implemented for fermionic atoms in the experiments of Ref. 29, and evidence of the non-zero Berry curvature of the bands was obtained from measurements of the anomalous velocity.

Resonant modulation. A very flexible way in which to imprint tunnelling matrix elements with non-zero Peierls phase factors is to use resonant modulation, or photon-assisted tunnelling.[25] The method is based on taking a static Hamiltonian in which neighboring lattice sites are offset in energy by an amount Δ that is much larger than the inter-site tunnelling matrix element. For example, a strong linear potential gradient in a one-dimensional lattice leads to

$$H_0^{1D} = -J_0 \sum_j \left(b_j^\dagger b_{j+1} + b_{j+1}^\dagger b_j \right) + \sum_j \Delta j b_j^\dagger b_j . \tag{18}$$

For $\Delta \gg J_0$ the energy eigenstates are strongly localized at individual lattice sites. Inter-site motion between two neighboring sites is restored by adding a potential difference between the sites that varies at the resonant frequency $\omega = \Delta/\hbar$. For such resonant coupling, one can apply the RWA, provided the frequency ω is large compared to all other relevant frequency scales, to give a resulting time-independent Hamiltonian. The phase of the resulting tunnelling matrix element can vary in space if there are spatial variations in the relative phases of the modulating potentials. In practice this can be achieved by a two-photon process, with Δ the beat frequency between two light beams, which generate a running wave potential, $H^{1D}(t) = H_0^{1D} + V_0 \sum_j \cos(\Delta t/\hbar + j\varphi)$. The resulting phase of the modulation varies in space in a linear manner

$$H_{RWA}^{1D} = -J \sum_j \left(e^{i\varphi j} b_j^\dagger b_{j+1} + e^{-i\varphi j} b_{j+1}^\dagger b_j \right) , \tag{19}$$

with $J \sim J_0(V_0/\Delta)$.[11] Extending this to a 2D model, in which every row along x is described by the above hopping and in which hopping along y occurs naturally with (real) amplitude J leads to a realization of the (isotropic) Harper–Hofstadter model:[32] hopping on a square lattice with $n_\phi = \varphi/2\pi$ flux quanta in each plaquette. This leads to the spectrum of the Hofstadter butterfly,[33] with a complex set of energy bands which have, in general, non-zero Chern numbers,[20] Fig. 4.

Using this, and related, techniques of resonant modulation, the Harper–Hofstadter model has been implemented in experiments with flux $n_\phi = 1/4$[34,35] and with $n_\phi = 1/2$.[36] Recent experiments have studied the motion of small numbers of particles on a Harper–Hofstadter lattice in a quantum gas microscope with single site resolution.[37] Characteristic energy scales are set by the intersite coupling J, and the onsite interaction energy, U, which can readily be in a strong coupling regime $U \gg J$.

2.2. *Many-body phases*

In the preceding section we have described ways in which cold atoms can be made to experience effective magnetic fields, leading to the formation of Landau levels or other topological energy bands. We now turn to discuss the many-body phases of interacting bosons that occupy these single-particle states. We separate the discussion into two parts: first for interacting bosons in continuum Landau levels associated with a uniform magnetic field, as generated by rotation in a harmonic trap; second for situations in which the atoms experience a periodic lattice potential. For the most part, we shall put the emphasis on the properties of thermodynamically

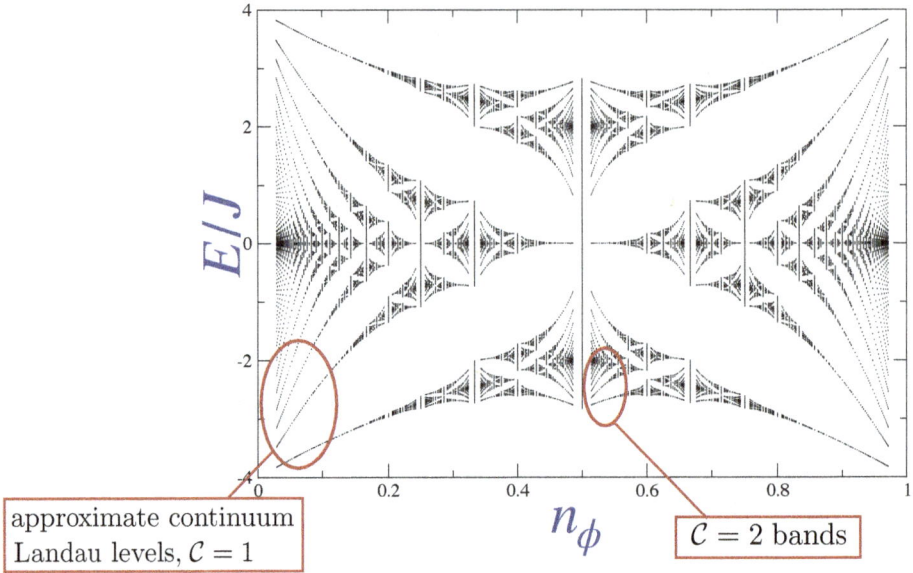

Fig. 4. Hofstadter butterfly energy spectrum of the Harper–Hofstadter model with tunnelling energy J. For flux $n_\phi \ll 1$ the low energy bands recover the continuum Landau levels, each of Chern number $\mathcal{C} = 1$. For n_ϕ close to $1/q$ the low energy bands each have Chern number $\mathcal{C} = q$, indicated here for $\mathcal{C} = 2$ close to $n_\phi = 1/2$. Figure courtesy of G. Möller.

large systems where clear statements can be made regarding the nature of the ground state phase, albeit through the analysis and extrapolation of the results of numerical studies of systems of small numbers of particles.

2.2.1. *Interacting bosons in the lowest Landau level*

Consider a system of N (spinless) bosons in 2D with contact interactions

$$H = \sum_{i=1}^{N} \frac{(\mathbf{p}_i - M\boldsymbol{\Omega} \times \mathbf{r}_i)^2}{2M} + g_{2\mathrm{D}} \sum_{i<j=1}^{N} \delta(\mathbf{r}_i - \mathbf{r}_j). \tag{20}$$

This can be achieved for a rotating gas in a trap when the rotation frequency Ω matches the trap frequency ω_0. For $\Omega < \omega_0$ there is a residual harmonic potential $(1/2)M(\omega_0^2 - \Omega^2)|\mathbf{r}|^2$ which provides an overall residual confinement, Eq. (4). For clarity we omit consideration of this confinement, but note that in general it will lead to an overall non-uniform density distribution.

In the following we describe the nature of the many-body ground states for bosons in the lowest Landau level. Since we are dealing with contact interactions (20), there is only one non-zero Haldane pseudo-potential[38]

$$V_0 = \sqrt{\frac{2}{\pi} \frac{\hbar^2 a_s}{M a_0^2 a_z}} \sim \hbar\omega_0 \frac{a_s}{a_z}, \tag{21}$$

the interaction energy of two particles with zero relative angular momentum. For the most part we shall focus on the ground states of bosons with this simple contact interaction. However, we note that longer-range interactions can be relevant for atoms with dipolar interactions,[8] which contribute to all V_m with even m.

Laughlin State. For a gas of N contact-interacting bosons, rotating at high angular momentum in an isotropic parabolic trap, the $\nu = 1/2$ Laughlin state

$$\Psi_{q=2}^{L}(z_1, z_2, \ldots, z_N) \propto \prod_{i<j=1}^{N} (z_i - z_j)^2 e^{-\sum_i |z_i|^2/4}, \qquad (22)$$

is the exact ground state[13] at total angular momentum $L = N(N-1)$. This result arises from the fact that, for this value of the angular momentum, it is the unique bosonic wave function in the lowest Landau level that vanishes when any pair of particles coincide. Thus, at sufficiently high rotation rate one anticipates that the cold atomic gas will realize this fractional quantum Hall state. This expectation is, of course, borne out in numerical calculations, not only in the disk geometry, Eq. (22), but also in edgeless geometries of the sphere and the torus which are most useful for determining the bulk properties. Figure 5 shows the excitation spectrum for the $\nu = 1/2$ state on a torus. From such studies the bulk gap for creation of a quasi-particle/quasi-hole pair of $\Delta E_{\nu=1/2} \simeq 0.095(5) \times 2\pi V_0$ can be extracted.[10] It is worth noting that the quasi-holes have exactly zero interaction energy for the situation considered, of contact interactions. Thus the quasi-holes behave as

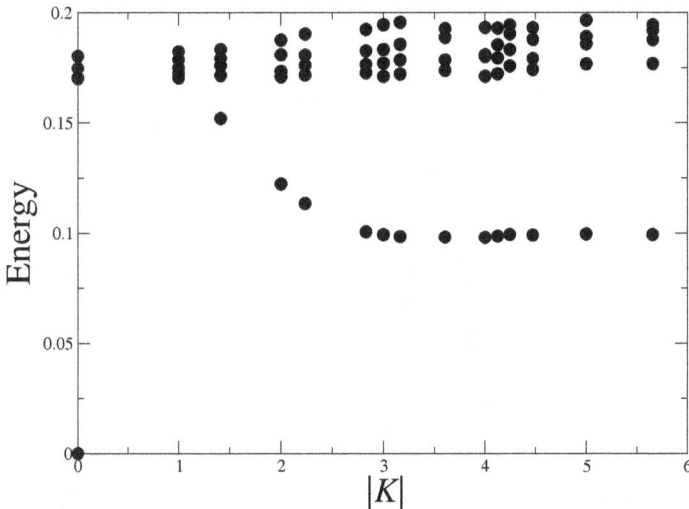

Fig. 5. Excitation spectrum of the $\nu = 1/2$ Laughlin state of bosons with contact repulsion. Computed on a torus for $N = 8$ particles in $N_\phi = 16$ flux quanta. The ground state (two-fold degenerate) is at $E = 0$ and $\boldsymbol{K} = 0$. Energies are measured in units of $2\pi V_0$, with V_0 the Haldane pseudo-potential, Eq. (21), and lengths in units of the magnetic length, ℓ_B.

ideal fractionalized excitations, with vanishing interactions. Indeed, at $\nu < 1/2$, or equivalently $L > N(N-1)$, there are multiple ground states, all with vanishing interaction energy, and with a degeneracy associated with the fractional exclusion statistics of the Laughlin quasi-holes. We shall come back to this in connection with their possible experimental consequences in Sec. 2.3.

That the $\nu = 1/2$ Laughlin state is the exact ground state of a gas of bosons in this experimentally relevant setting — contact interactions, parabolic trap and (potentially) small numbers of particles — is a central result that underpins the expectations that cold atomic gases can be brought to regimes in which fractional quantum Hall physics emerges.

Vortex Lattice. It is helpful to compare the Laughlin state with the vortex lattice phase of a superfluid that is subjected to low, or moderate, rotation rates. Since particle velocity in a superfluid is proportional to the gradient of the phase, the curl of the velocity field vanishes except at singular points — the vortex cores — at which there is a delta-function contribution of strength set by the quantized circulation $\kappa = h/M$. In a uniform setting (e.g. where a potential acts to counterbalance the centrifugal potential) the steady state in the rotating frame consists of a lattice of these quantized vortices, with the density of vortices, n_v, set by Feynman's condition that the mean circulation matches that of rigid body rotation $\boldsymbol{v} = \boldsymbol{\Omega} \times \boldsymbol{r}$, i.e. $n_v(h/M) = \boldsymbol{\nabla} \times (\boldsymbol{\Omega} \times \boldsymbol{r}) = 2\Omega$, that is $n_v = 2\Omega M/h$. Thus, the density of vortices in the vortex lattice is equal to the density of flux quanta (5) of the effective magnetic field (1) associated with the rotation. Indeed this is borne out, even for weakly interacting bosons, in detailed numerical simulations of the vortex distribution.[39] The hydrodynamic behavior of the vortex lattice in the lowest Landau level has been studied in Refs. 40 and 41.

The mean-field vortex lattice phase treats the vortices as classical objects. Going beyond this, to include quantum fluctuations of the vortices, shows that the vortices have a quantum uncertainty in position which is of order the mean inter-particle spacing $\bar{a} = n^{-1/2}$ where n is the 2D density of the bosons.[42] This result expresses the very natural expectation that one cannot locate the core of the vortex to an accuracy better than the inter-particle spacing. Applying a Lindemann criterion for these fluctuations, one asserts that the vortex lattice is stable when the uncertainty in vortex position is smaller than some multiple of the mean vortex spacing, $\bar{n}^{-1/2} < c_L \bar{n}_\phi^{-1/2}$, or equivalently

$$\nu \equiv \frac{n}{n_\phi} > \nu_c. \tag{23}$$

That is, the vortex lattice can be a stable ground state at sufficiently high filling factor.[42] This is consistent with analytic results that apply in the limit $\nu \to \infty$.[43] A simple evaluation of the quantum fluctuations in the vortex lattice leads to $\nu_c \simeq 7$,[42] while an alternative definition, and more detailed calculation, leads to a critical filling factor of $\nu_c = 17$.[40] However, both of these estimates rely on the use of an unknown Lindemann parameter, c_L. Direct calculations of the critical filling factor ν_c have been performed using exact diagonalization for contact interacting bosons on a toroidal geometry. These calculations indicate a transition to a ground state with broken translational invariance, consistent with that of the triangular vortex lattice, at $\nu_c \simeq 6$.[42] Subsequent numerical calculations showed that strong density wave correlations persist down to $\nu = 2$.[44] Thus, the existing numerics suggest a critical filling factor above which the vortex lattice is the ground state of $\nu_c \simeq 2-6$ for contact interactions in the lowest Landau level. These numerics are restricted to very small systems, so are subject to strong finite-size effects.

Composite Fermion States. That the ground state is a vortex lattice at $\nu > \nu_c$ and is the Laughlin liquid at $\nu = 1/2$ leaves open the question of what are the phases in the range $1/2 < \nu < \nu_c$. Early numerical work showed evidence for a successful description of the many-body states in this regime in terms of composite fermions.[45] Here, the composite fermions are formed by binding each boson to a single vortex of the many-body wave function, such that the flux density experienced by the composite fermions is

$$n_\phi^{CF} = n_\phi - n. \tag{24}$$

Treating the composite fermions as non-interacting particles, one expects an incompressible integer quantum Hall state of the composite fermions when p Landau levels are filled, $n/n_\phi^{CF} = \pm p$, leading to the Jain sequence[46]

$$\nu = \frac{p}{p \pm 1}. \tag{25}$$

In addition to the Laughlin state at $\nu = 1/2$, numerical studies on the sphere and the torus have shown evidence for gapped states at $\nu = 2/3$ and $\nu = 3/4$.[42,47] However, there appear to be significant interactions between the composite fermions, so higher members of this sequence do not appear as ground states. Furthermore, as discussed below, the composite fermion states at $\nu = 2, 3/2$ do not appear to describe the ground states of contact interacting bosons at these filling factors.

Moore–Read State. At $\nu = 1$ a model of non-interacting composite fermions would suggest the existence of a *compressible* composite Fermi liquid, analogous to the $\nu = 1/2$ state of fermions.[48] This compressible state is the $p \to \infty$ limit of the sequence (25). This is not found in numerics. Rather the ground state is found to be a robust gapped FQH state. Numerical studies have established this state to be well described by the Moore–Read Pfaffian state,[49] with a particle-hole gap of $\Delta E_{\nu=1} \simeq 0.05 \times 2\pi V_0$.[10,42,47] Note that this is not significantly smaller than the gap for the $\nu = 1/2$ Laughlin state ($\Delta E_{\nu=1/2} \simeq 0.095 \times 2\pi V_0$), indicating that the Moore–Read state is of comparable robustness. Numerical evidence for the Moore–Read state appears from the shift on the sphere, the expected three-fold degeneracy of the ground state on the torus, and adiabatic continuity with the Moore–Read wavefunction which is the exact ground state of a three-body contact interaction. Unlike electrons in the lowest Landau level at $\nu = 1/2$, bosons in the lowest Landau level at $\nu = 1$ do not have an exact particle-hole symmetry. The competition with the possible anti-Pfaffian wavefunction is resolved in favor of the Moore–Read Pfaffian for contact interacting bosons at $\nu = 1$. That this non-Abelian FQH state appears so robustly in the bosonic system is a key motivation for searching for experimental realizations of fractional quantum Hall states of bosons.

Read–Rezayi States. Even more interesting are the ground states that appear at filling factors larger than $\nu = 1$. Numerical studies show that the ground states of contact interacting bosons in the lowest Landau level at $\nu = k/2$, with $k \geq 3$ integer, resemble the Read–Rezayi states.[42] For bosons these can be defined as the states of highest filling factor in the lowest Landau level which are the exact zero energy eigenstate of a $(k + 1)$-body contact interaction.[50] They generalize the Laughlin and Moore–Read states, which are respectively the $k = 1$ and $k = 2$ members of the sequence. For N divisible by k, the states can be written as a symmetrized product over k Laughlin states[42,51]

$$\Psi_k^{\text{RR}}(z_1, z_2, \ldots, z_N) \propto \mathcal{S}\left[\prod_{i<j\in A}^{N/k} (z_i - z_j)^2 \prod_{k<l\in B}^{N/k} (z_k - z_l)^2 \ldots \right] e^{-\Sigma_i |z_i|^2/4}, \quad (26)$$

where $\mathcal{S}$ denotes symmetrization over all ways in which the N particles can be divided into sets $A, B, \ldots$ of N/k particles. The wavefunction (26) vanishes when $k + 1$ particles coincide since at least two of these particles (i and j, say) must be in the same set, and therefore the wavefunction has a factor of $(z_i - z_j)^2$. Like the Moore–Read state the Read–Rezayi states describe non-Abelian phases of matter. However, the relevant anyons step beyond the Ising non-Abelian anyons of the Moore–Read state to allow universal braid statistics.[52]

For contact interactions, the correlation lengths are large, and there are competing phases with broken translational order already for $\nu \geq 2$. (These competing phases include both vortex lattice and stripe/nematic phases.[44]) However, it has been shown that the addition of a small degree of longer-range interaction (i.e. a non-zero Haldane pseudo-potential V_2) can improve stability.[53] This is illustrated for the $k = 3$ Read–Rezayi state in Fig. 6. Recall that there is a competing composite fermion state (25) at $\nu = 3/(3-1) = 3/2$. Nevertheless, the numerical results — in particular the four-fold ground state degeneracy — indicate that this state is not favored compared to the $k = 3$ Read–Rezayi state under the conditions of Fig. 6.

In cold atom experiments, longer-range interactions may arise in atomic species with large dipole moments.[8] These longer-range interactions can also lead to changes in the nature of the vortex lattice phases,[54] and/or the appearance of competing crystalline phases.[55]

Fermions. Although the focus of this chapter is on bosons, we note that there are some aspects of the FQH state of fermionic atoms that are unconventional.

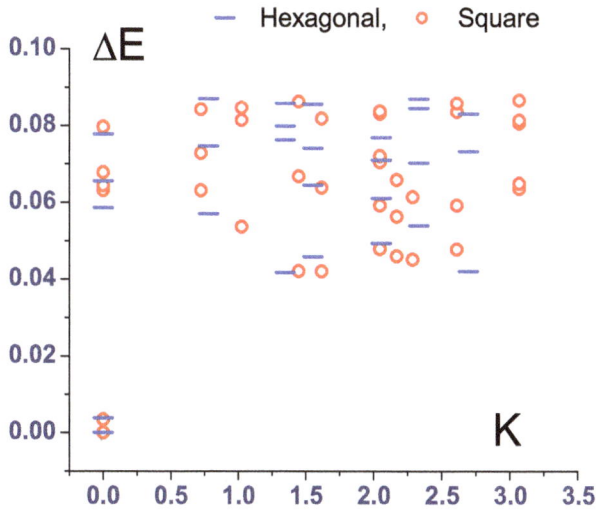

Fig. 6. Excitation spectrum at $\nu = 3/2$ for $N = 13$ bosons on the torus for non-zero range of the interaction giving Haldane pseudo-potential ratio $V_2/V_0 = 0.38$. The near degeneracy of the two levels at $\mathbf{K} = 0$ is as expected for the $k = 3$ Read–Rezayi state. (There is an overall four-fold degeneracy when translational symmetries are taken into account.) The insensitivity to boundary conditions on the torus illustrates convergence. Reproduced from Ref. 53.

Most notable is the availability of gases of two-component fermions for which the superfluid pairing of the fermions can be controlled: spanning the weakly attractive "BCS" regime of Cooper pairs of size large compared to the inter-particle spacing, to the regime of strong binding of pairs into tightly bound bosons of small size which then form a Bose–Einstein condensate (BEC).[5] This leads to interesting possibilities involving the interplay between fermionic pairing and fractional quantum Hall states. For a homogeneous gas without external magnetic field, the transition in the form of ground state is known to be a smooth evolution between BCS and BEC regimes. However, in the presence of a quantizing magnetic field — for which the ground states can have character of the quantum Hall states in 2D, or layered QH states in 3D — one can readily establish that these two limits must be separated by a phase transition.[56] This is most evident for a 2D system of two-component fermions of density n^F at flux n_ϕ^F with filling factor $\nu^F = n^F/n_\phi^F = 2$. For weak attractive interactions between the two components, the ground state is a $\nu^F = 2$ integer quantum Hall state in which the lowest Landau level is filled for both spin components. However, for very strong attractive interactions such that opposite spin fermions pair into bosons with binding energy large compared to the cyclotron energy, the system should be viewed as bosons of density $n^B = n^F/2$ experiencing a flux density $n_\phi^B = 2n_\phi^B$, i.e. at filling factor $\nu^B = \nu^F/4 = 1/2$. Residual contact repulsion between the bosons will stabilize a Laughlin state of these bosons. That there must be a phase transition separating these two regimes is evident from the fact that the edge structure changes: from two modes for the integer quantum Hall state of weakly attractive fermions at $\nu^F = 2$ to the single mode of the Laughlin state of bosons at $\nu^B = 1/2$.[56–58]

2.2.2. Interacting bosons in topological optical lattices

A key feature of optical lattices is that they break continuous translational invariance. Thus, in order that strongly correlated phases akin to FQH states appear in these lattices, one must go beyond the paradigm of continuum Landau levels to allow for this discrete translational invariance. That fractional quantum Hall states can be stable in the presence of periodic density modulations was discussed in early work on the "Hall crystal" phase, in which the density order is spontaneously formed.[59] The consequences of the discrete translational invariance on the form of the Chern–Simons field theoretical description of the Laughlin and Jain states were studied.[60] More recently, this topic has gained attention in connection with the investigation of "fractional Chern insulators"[61–63] — i.e. fractional quantum Hall states formed for particles moving in lattice models which generate bands with non-zero Chern number, with single particle wavefunctions that can differ markedly from those of the continuum Landau level. The optical lattices described in Sec. 2.2.2 provide interesting examples of this physics. Similar effects have been observed for electrons in superlattice structures formed in bilayer-graphene hexagonal boron-nitride devices.[2]

$\mathcal{C} = 1$ bands.

Many of the optical lattices described above have been designed to closely approximate the action of a uniform magnetic field on a charged particle, and thereby to form energy bands that are similar to those of Landau levels. This is the case for the Harper–Hofstadter model at sufficiently small flux per unit cell, n_ϕ, where the magnetic length is much larger than the lattice constant. It is also the case for the optical flux lattices of Sec. 2.1.2. In such cases, one expects that the many-body states for short-range interactions will mirror those described above for contact-interacting bosons in the lowest Landau level — at least in regimes for which the mean interaction strength remains sufficiently small to preclude interband mixing. Although the energy bands can be similar to Landau levels, they are not identical: the bands have some residual dispersion, and the Berry curvature of the energy band is not uniform. Therefore it is important to test the stability of quantum Hall states in the continuum to these settings.

The nature of the ground states of interacting bosons on the Harper–Hofstadter model have been studied using numerical methods, both for hardcore interactions and for contact interactions that do not mix states beyond the lowest energy single-particle band. The results of these studies[64,65] show that the ground state at filling factor $\nu = 1/2$ remains well-described by the Laughlin state for $n_\phi \lesssim 0.3$, as does the composite fermion (25) state at $\nu = 2/3$.[66,67] These states are evidenced by the presence of energy gaps, by the appropriate ground state degeneracies on a

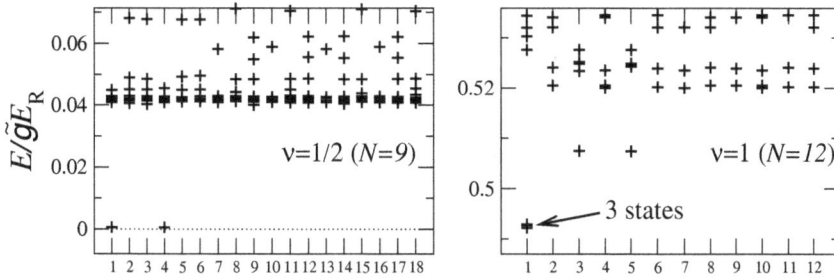

Fig. 7. Many body excitation spectra for bosons in the lowest band of an optical flux lattice designed to mimic the lowest Landau level, using $N_{\rm I} = 3$ hyperfine levels of rubidium-87. (a) At $\nu = 1/2$ ($N = 9$ particles in $N_\phi = N_1 N_2 = 6 \times 3$ states) the ground state is well described by the Laughlin state, with two-fold degeneracy (b) At $\nu = 1$ ($N = 12$ particles in $N_\phi = N_1 N_2 = 6 \times 2$ states) the ground state is well described by the Moore–Read state, with three-fold degeneracy. Energies are measured in units of $\tilde{g} E_R = 3\sqrt{2\pi}(a_s/a_z)(\hbar^2 \kappa^2/m)$ (with κ the characteristic wavevector of the light forming the optical lattice), and the momentum $\boldsymbol{k} = \alpha_1 \boldsymbol{G}_1/N_1 + \alpha_2 \boldsymbol{G}_2/N_2$ is labeled by the index, $i = 1 + \alpha_1 + N_1 \alpha_2$, with $\boldsymbol{G}_{1,2}$ the two smallest reciprocal lattice vectors. Reproduced from Ref. 24.

torus, and by the many-body Chern numbers. The connection with the FQH states of the continuum Landau levels has been explored in Ref. 68, showing how non-uniformities of the geometry of the Bloch wave functions affect the stability of the FQH states.

The optical flux lattices, described in Sec. 2.1.2, allow the formation of bands that closely approximate Landau levels. The Laughlin ($\nu = 1/2$) and Moore–Read ($\nu = 1$) states were studied in detail for the model of Fig. 3(a) for $N_{\rm I} = 3, 4$, and were shown to have similar stability to that found in the continuum Landau levels.[69] Focusing on settings in which the bands closely mimic Landau levels, but with an eye on specific practical implementation for the $N_{\rm I} = 3$ spin states of rubidium-87, a scheme similar to that illustrated in Fig. 3(a) was proposed.[24] This scheme sacrifices uniformity of the Berry curvature of the band to simplify the experimental implementation. The resulting energy bands can still be made very narrow. However, they are of non-zero width and so sufficiently strong interactions are needed to prevent formation of a Bose–Einstein condensate in the lowest energy band minimum (or superposition of degenerate minima). At strong interactions one finds the appearance of a robust Laughlin state at $\nu = 1/2$ and a robust Moore–Read phase at $\nu = 1$, much as in continuum Landau levels. These studies show good prospects for finding the FQH states of bosons of continuum Landau levels within the setting of optical lattices, where densities and interaction energies can be large.

Much is known concerning the interacting ground states in other flat-band models with topological bands.[61,63] Typically such models are formulated as tight-binding models with further neighbor couplings introduced to reduce the bandwidth, and are therefore less readily implemented in cold atom systems. It is interesting to note that, by careful choice of these further neighbor couplings, the wavefunctions can be made identical to those of the continuum Landau level,[70] allowing the development of a model in which a discretized version of the Laughlin wavefunction[71] is the exact many-body ground state.

$\mathcal{C} \neq 1$ **bands.**   The optical lattices described above also provide settings in which the lowest energy band differs qualitatively from a continuum Landau level, by having a Chern number that is not of unit magnitude, $|\mathcal{C}| \neq 1$. For example, as discussed in more detail below, such cases arise for the Harper–Hofstadter model with n_ϕ close to $1/q$, for which the lowest energy band has Chern number $|\mathcal{C}| = q$. For interacting bosons occupying this lowest energy band, one can anticipate strongly correlated phases provided the interaction energy is large compared to the bandwidth. How do these states, formed in Chern bands with $|\mathcal{C}| \neq 1$ relate to fractional quantum Hall states in continuum Landau levels? Remarkably one can gain understanding in this setting by adapting ideas of composite fermion theory to this lattice setting.

The extension of the mean-field composite fermion theory of the continuum Landau levels to the lattice was described in Ref. 60, and explored numerically for contact interacting bosons in Refs. 66, 67 and 72. For bosonic particles, one may form composite fermions by attaching a single vortex. As in the continuum theory, the effective flux density is $n_\phi^{\mathrm{CF}} = n_\phi - n$, Eq. (24). For n and n_ϕ of small magnitude, the relevant single-particle states are approximate continuum Landau levels both for n_ϕ and for n_ϕ^{CF}. The resulting many-body states recover the continuum composite fermion states described above, with bosonic Jain sequence Eq. (25). However, when either n or n_ϕ is of order unity, such that the lattice structure becomes significant, then n_ϕ^{CF} may not be small and the levels that the composite fermions fill can have different character to those of the continuum Landau level.

We illustrate the general approach here by the specific example of the Harper–Hofstadter model at flux n_ϕ close to $1/q$ with q being an integer. In each of these cases, there appears a series of narrow low energy bands in the spectrum which are well separated from each other in energy. For $q = 1$ this condition is equivalent to n_ϕ being close to zero and these narrow bands are the usual Landau levels. These continuum Landau levels and the low energy bands associated with proximity to $n_\phi = 1/2$ are illustrated in Fig. 4.

The low-lying bands of the Harper–Hofstadter model close to $n_\phi = 1/q$ have Chern number $\mathcal{C}$ of magnitude $|q|$. One way to see this is to use a result of Wannier[73] who showed that the number of states per plaquette for each of these bands is $n_s = |qn_\phi - 1|$. By considering one such band to be filled by non-interacting fermions, such that $n = n_s$, and using the Středa formula[74]

$$\sigma_{xy} = e\left(\frac{\partial n}{\partial B}\right)_\mu = \frac{e^2}{h}\left(\frac{\partial n_s}{\partial n_\phi}\right)_\mu = \pm q\frac{e^2}{h} \text{ for } n_\phi \gtrless 1/q \qquad (27)$$

one deduces that the Chern number of the band is $\mathcal{C} = |q|$. Alternatively, one can consider the sequence $n_\phi = \alpha/[|\mathcal{C}|\alpha - \mathrm{sgn}(\mathcal{C})]$ with integer α, which converges to $n_\phi = 1/|\mathcal{C}|$ at large α. Solving the Diophantine equation of Thouless *et al.*[20] at these flux densities to determine the Hall conductance of the low energy bands shows that they each have Chern number $\mathcal{C}$.[72]

Consider the lowest such band occupied with a density n of repulsively interacting bosons. Given that the band consists of n_s states per unit area, it is natural to define the filling per *state* by $\nu_s \equiv n/n_s$. We use the subscript s to emphasize that this is the filling per state rather than per flux quantum, $\nu = n/n_\phi$. (For models that realize bands that closely approximate Landau levels, then $\nu_s = \nu$.) Now consider forming composite fermions by attaching a single vortex, such that a density n of composite fermions experience a resulting flux density $n_\phi^{\mathrm{CF}} = n_\phi - n$. If also $n_\phi^{\mathrm{CF}} \simeq 1/|\mathcal{C}|$, which always applies for low densities $n \ll 1$, these composite fermions will experience low energy $|\mathcal{C}|$ bands of the same form as the original bosons. The number of states per unit area is $n_s^{\mathrm{CF}} = |\mathcal{C} n_\phi^{\mathrm{CF}} - 1| = |\mathcal{C}(n_\phi - n) - 1|$. Treating the composite fermions as non-interacting, one would expect an incompressible state when p bands are filled, $n = \pm p n_s^{\mathrm{CF}}$, leading to the lattice composite fermion sequence[72]

$$\nu_s \equiv \frac{n}{n_s} = \frac{p}{p\mathcal{C} \pm 1}. \tag{28}$$

This generalizes the Jain sequence (25) for conventional Landau levels to the case of topological bands with $|\mathcal{C}| \neq 1$. Numerical exact diagonalizations show evidence for the existence of these states for $\mathcal{C} = 2$[66,72] and $\mathcal{C} = 3$.[72] The same reasoning leads to candidate fermionic states in bands at $\nu_s = \frac{p}{2p\mathcal{C} \pm 1}$;[72] evidence for such a state in a $\mathcal{C} = 2$ band has been found in experimental studies of modulated bilayer graphene.[2]

An interesting case appears for $\mathcal{C} = 2$ and $p = 1$, for which there is a composite fermion state with $\nu_s = p/(2p - 1) = 1$. This state is a bosonic *integer* quantum Hall (BIQH) state. It has no fractionalized excitations (i.e. it has a non-degenerate ground state on a torus) so has short-range entanglement. It is therefore an example of a symmetry-protected topological phase of bosons,[75] the symmetry being the $U(1)$ symmetry associated with conservation of particle number. Consistent with general arguments[76] this state has *even* value of the quantized Hall conductance, $\sigma_{xy} = 2e^2/h$. Numerical results show evidence for the stability of this phase for bosons with contact interactions in the Harper–Hofstadter model.[66,67,72]

Optical flux lattices, described in Sec. 2.1.2, can also be designed for which the real space magnetic field vanishes, yet the Chern number is nonzero, or in which the Chern number has magnitude larger than unity.[21] The strongly correlated phases of models in which the lowest band has $\mathcal{C} = 2$ were studied in Refs. 69 and 77. It was shown that various fractional Chern insulator states can appear. These include so-called "color-entangled" states, at fillings $\nu_s = 1/(1 + \mathcal{C})$ at least for $\mathcal{C} = 2$ and $\mathcal{C} = 3$.[69] These states are related to the fractional quantum Hall states of multi-component quantum Hall systems proposed by Halperin,[78] with $\mathcal{C}$ internal components[79] albeit in a setting without SU($\mathcal{C}$) symmetry. However, they differ in detailed structure. These differences can be exposed in finite-size systems, using an extended zone construction of the form of Fig. 3(a), for which a band of Chern number $\mathcal{C}$ has a number of states that is not divisible by $\mathcal{C}$. Thus, the band cannot be viewed as consisting of $\mathcal{C}$ copies of a band with unit Chern number, so does not have a natural description in terms of $\mathcal{C}$ internal components. Still, the color-entangled fractional Chern insulator states can be well defined even in this case.[22]

2.3. Experimental consequences

There remain experimental challenges in realizing FQH states for atomic gases. However, these appear to be technical, and there are rapid ongoing improvements in capabilities which encourage one to believe that they can be overcome. The theoretical studies provide strong motivations to continue this search, in particular to have experimental access to non-Abelian phases, such as the Moore–Read state at $\nu = 1$ for contact interacting bosons, and to explore FQH states that are stabilized by the lattice including the predicted $\nu_s = 1$ symmetry-protected topological phase of bosons. It is therefore important to consider what are the experimental observables that could be used to probe and to characterize these phases.

The observables that can be accessed in cold gases are rather different to those that are commonly used in electronic systems. In the following we list some of the most natural observables for cold gases.

Equation of State. The confinement of atomic gases in harmonic traps typically leads to inhomogeneous particle density across the sample. This can cause difficulties in the interpretation of measured properties that are averaged across the sample. However, it can be a useful feature for extracting the equation of state of the system if local measurements are made. Specifically, if the trapping potential $V(\boldsymbol{r})$ is sufficiently shallow that the local density varies slowly on the microscopic correlation length, then one may take the mean local particle density at a position $\boldsymbol{r}$ to represent the equilibrium density for a local chemical potential $\mu(\boldsymbol{r}) = \mu - V(\boldsymbol{r})$. Making *in situ* measurements of the expectation value of $n(\boldsymbol{r})$ (over repeated experiments) then allows one to deduce the equation of state $n(\mu)$. This "local density approximation" has been put to use in an accurate determination of the equation of state of strongly interacting atomic gases.[80] The incompressible character of fractional quantum Hall states should appear clearly in such measurements, as a form of "wedding cake" density distribution, with plateaus in the density at quantum Hall states for sufficiently low temperatures.[81]

Local correlations. Rather than constructing the expectation value of the particle density through averaging over repeated images, analysis of each individual image can be made. Each image provides the positions of all of the particles (up to uncertainties from noise and imaging resolution), so contains significant information on the local two-particle (or multi-particle) correlations. For rapidly rotating cases in harmonic traps, the images taken after release of the trap and expansion of the gas give a scaled-up view of the local correlations before expansion.[82] For cold atomic gases formed in "quantum gas microscopes" such images can be taken even *in situ* — with atomic resolution and high fidelity.[16,37] Quantum gas microscopes could also be used to measure correlation functions that can uncover and characterize the gapless edge modes of FQH states.[67] The spatial average of the local two-particle correlation function $\langle n(\boldsymbol{r})n(\boldsymbol{r})\rangle$ can also be obtained by the use of photo-association to convert pairs of atoms in close proximity to molecules.[15]

Transport. Although cold atomic gases are not readily attached to sources and sinks of atomic currents, there are ways in which they can be used to measure transport properties. The use of light beams to shape the cloud into two (large) reservoirs, coupled by a mesoscopic region, has allowed studies of two-point transport properties, including thermo-electrical transport, albeit not yet for systems involving topological bands or phases.[83] The bulk transport of a weakly interacting gas of bosons subjected to the Harper–Hofstadter model at flux $n_\phi = 1/4$ was studied via the motion of the centre of mass of the cloud, allowing a measurement of the Chern number of the lowest band.[84] Using a sum rule, the zero frequency Hall conductivity can be related to the integral over frequencies of the dissipative circular dichroism.[85] This has recently been used to measure the Hall conductivity of non-interacting fermions in the lowest band of the Haldane model.[86]

Spectroscopic probes. Cold atomic gases lend themselves naturally to spectroscopic probes. Bragg spectroscopy, namely a two-photon Bragg scattering process, can be performed with high energy resolution, and with wavevector transfers of order of the inverse optical wavelength, which itself is of order the inverse particle spacing, thereby accessing the full range of relevant momenta.[87] This has been used to measure the (gapless) collective modes of a Bose–Einstein condensate,[88] and the static and dynamic structure factors of strongly interacting phases of fermions.[89] This provides a natural way to probe the (gapped) collective modes of FQH states, such as the roton branch of Fig. 5. Observations of the collective modes of finite-size systems are sensitive to the equation of state of the gas,[90] and could be used to probe the edge structure of FQH states.[91]

Fractionalization. The high degree of control and the potential for precision measurements on cold atomic gases holds promise of finding new ways to probe and detect the particle fractionalization within FQH states of atoms. Precision spectroscopy of rotating atomic clusters has been shown to provide a means to detect Haldane exclusion statistics of quasi-hole excitations.[92] Removing one atom from a finite-size $\nu = 1/2$ Laughlin droplet of bosons leaves the system in an excited state involving two quasi-holes; a count of the number of spectral lines reveals the exclusion statistics of these particles. Although more challenging, it is also possible to envisage means by which to detect braiding statistics.[93,94] An interesting proposal is to introduce additional "impurity" atoms to which anyons of the background FQH state bind, and which may be separately addressed and controlled.[95–98]

Preparation methods. In considering all of these potential experimental observables, an important concern is the means of preparing the system close to its ground state. Typically, *in situ* cooling is unavailable for cold atoms in the complex settings for which the FQH states might arise, so the natural approach is to consider *adiabatic* paths from a more easily prepared many-body ground state to the FQH state. Such trajectories involve a quantum phase transition — from a short-range entangled phase to the FQH state with genuine topological order — so true adiabaticity is only possible in small finite systems.[67,99,100] Theoretical studies have explored the optimal adiabatic paths for small systems[99,101–103] and the use of dissipative methods.[104] If such methods are not available, for large systems one can hope that non-adiabatic effects on crossing the transition into the topologically ordered phase are not too destructive, as in the phase transition between Mott insulator and superfluid phases of strongly interacting lattice bosons.[105]

3. Other Experimental Settings

In Sec. 2 we considered FQH states of bosons formed in ultracold atomic gases. There are a wide variety of other forms of system in which bosonic degrees of freedom can move, without displacing the underlying atoms: a set of localized atoms or molecules (e.g. on a crystalline lattice) can support local spin-flips or other internal excitation which move as bosons through the static lattice; alternatively, the electromagnetic field itself provides a medium through which photons can propagate. In the following we provide an overview of systems in which these bosonic degrees of freedom may form FQH states. For further details, in particular of the topological bands of single-particle excitations in artificial photonic systems, we refer the interested reader to Refs. 106 and 107.

3.1. *Quantum magnets*

A quantum magnet consisting of spin-1/2 degrees of freedom can be readily mapped to a model of hard-core bosons. We take the presence/absence of a boson on a given lattice site r to represent a spin projection $s_r^z = \pm 1/2$, such that the number of bosons on site r is $b_r^\dagger b_r = s_r^z + 1/2$, and define the boson creation and annihilation operators by $b_r^\dagger = s_r^+$, $b_r = s_r^-$. These lead to the commutation relation $[b_r, b_{r'}^\dagger] = \delta_{r,r'}(1 - 2b_r^\dagger b_r)$ which is consistent with the hardcore constraint $(b_r)^2 = (b_r^\dagger)^2 = 0$.

Writing the Heisenberg antiferromagnet in terms of these hardcore bosons

$$H_{\mathrm{AF}} = J \sum_{\langle r, r' \rangle} s_r \cdot s_{r'} = J \sum_{\langle r, r' \rangle} \left[\frac{1}{2}\left(b_r^\dagger b_{r'} + b_{r'}^\dagger b_r \right) + n_r n_{r'} \right] + \text{const.} \qquad (29)$$

shows that it corresponds to both nearest-neighbor tunnelling and nearest-neighbor density interactions. Conservation of the total spin $S^z \equiv \sum_r s_r^z$ is equivalent to conservation of the total number of bosons. Similarly an external field B^z that couples to S^z plays the role of a chemical potential.

Kalmeyer and Laughlin[4] proposed that the groundstate of the spin-1/2 Heisenberg antiferromagnet on the triangular lattice might be a gapped spin-liquid state which is described by the $q = 2$ Laughlin state of bosons (22). That this model is classically frustrated motivates a description in terms of a quantum spin liquid. They investigated this phase of matter using variational studies. Taking N bosons at sites $\{r_1, r_2, \ldots r_N\}$, the spin wave function considered is

$$|\Psi_{\text{KL}}\rangle \propto \sum_{\{r_1, r_2, \ldots, r_N\}} \Psi_{q=2}^{\text{L}}(r_1, r_2, \ldots, r_N) \prod_{n=1}^{N} b_{r_n}^\dagger |\text{vac}\rangle \tag{30}$$

where $\Psi_{q=2}^{\text{L}}$ is the Laughlin wavefunction (22) with $z_i = (x_i + iy_i)/\ell$ and $|\text{vac}\rangle = |\downarrow\rangle_1 \otimes |\downarrow\rangle_2 \ldots |\downarrow\rangle_N$. The wavefunction satisfies the hardcore constraint since $\Psi_2^{\text{L}}(\{r_i\})$ vanishes when any two of its arguments coincide. Choosing N to be one half of the total number of sites of the triangular lattice, corresponding to a state with vanishing total spin component $S^z = 0$, and $4\pi\ell^2 = \sqrt{3}a_0^2$ (with a_0 the nearest neighbor spacing), Kalmeyer and Laughlin determined the variational energy of this state for the Heisenberg Hamiltonian, Eq. (29), and found it to be within 10% of other competing states. They argued for adiabatic continuity between this fractional quantum Hall wavefunction and the true ground state of the Heisenberg antiferromagnet. However, subsequent studies have shown that the ground state of this model is qualitatively distinct: it is magnetically ordered,[108,109] with correlations similar to those of the ground state of the classical antiferromagnet on this lattice.

Nevertheless, the Kalmeyer–Laughlin state remains a viable quantum spin-liquid phase which may appear in other models. Indeed, numerical studies of a frustrated Heisenberg model on the Kagomé lattice, with third-neighbor coupling, show evidence that the ground state spontaneously breaks time-reversal symmetry. It behaves as a gapped chiral spin liquid[110] with qualitative features of topological order (ground state degeneracy on a torus, and entanglement spectrum) that match those of the Kalmeyer–Laughlin state.[4] Such a phase has also been identified, through similar characteristics, for a Kagomé lattice model derived from the low energy limit of the Hubbard model in the presence of an orbital magnetic flux.[111] This flux explicitly breaks time-reversal symmetry and leads to the appearance of a chiral three-spin interaction that stabilizes the Kalmeyer–Laughlin phase.

Certain quantum magnets have been found to exhibit plateaus in the magnetization S_z as a function of external applied field B_z. In the boson picture, this corresponds to the mean boson density having plateaus as a function of chemical potential, i.e. incompressibility. Treating the bosons within a mean field Chern–Simons description, of the same form as the composite fermion construction described above on the Harper–Hofstadter model, has been shown to reproduce the observed magnetization plateaus of the spin-1/2 antiferromagnetic on the Sutherland–Shastry[112] and Kagomé[113] lattices.

These examples draw close connections between standard FQH states of bosons and possible ground states of spin-1/2 quantum magnets. However they constitute a subset of a much broader class of quantum spin liquids[114] that is conceptually linked to fractional quantum Hall states. The deeper connections arise from the existence of topological order and associated fractionalized quasi-particles with anyonic statistics. A notable example is provided by the Kitaev honeycomb model,[115] describing a spin-1/2 lattice model in which the interactions between nearest neighbors are of XX, YY or ZZ type depending on the direction of the bond. This model is exactly solvable in terms of free Majorana fermions. In the presence of a magnetic field, which opens a gap it behaves as a chiral spin liquid with non-Abelian anyons. Recent studies of α-Ru-Cl$_3$ show the appearance of a plateau of the thermal Hall conductivity which is consistent with thermal transport by a Majorana edge mode.[116,117]

3.2. *Engineered qubit arrays*

There are now many experimental systems consisting of arrays of engineered two-level quantum systems which provide a class of many-body systems that are equivalent to spin-1/2 quantum magnets. The only difference from Sec. 3.1 is that the two-level quantum systems are designed and artificially constructed rather than provided by nature.

Atom or molecule arrays. Atoms or molecules subjected to deep optical lattices or strong optical tweezers can be formed into ordered arrays in which the positional degrees of freedom are frozen out. The internal excitations of these frozen atoms/molecules can provide a set of "spin" degrees of freedom. (These involve electronic/spin excitations for atoms and also ro-vibrational excitations for molecules.) The resulting two-level quantum systems can have very large coherence times, and can be coupled by dipolar interactions with timescales large compared to intrinsic decoherence times (e.g. from spontaneous emission). Experiments have shown coherent dynamics of coupled two-level systems, and evidence of strong (hardcore) interactions.[118,119] Theoretical proposals have shown how artificial magnetic fields can be imprinted for the motion of spins in two-dimensional arrays, leading to the possibility of forming FQH states: for rotational states of polar molecules[120] and for internal states of atoms in highly excited "Rydberg" states.[121]

Superconducting qubits. Circuit QED devices, based on superconducting structures, provide one of the leading approaches to generating engineered arrays of two-level quantum systems.[122] The two-level systems have splittings in the microwave frequency regime, so have negligible spontaneous emission, and are readily controlled and coupled by resonant superconducting circuits. An experimental demonstration has been given of the basic elements required to break time-reversal symmetry and to imprint Peierls phase factors on the coupling between the qubits in a single three-site plaquette.[123] This follows the same approach as used in cold atoms,[34,36] described in Sec. 2.1.3, based on site-to-site detuning and re-establishing tunnelling through which Peierls phase factors can be imprinted. It has also been shown theoretically how time-reversal symmetry can be broken in such settings by coupling the qubits by resonator junctions made of passive elements subjected to static electric and magnetic fields.[124] A related, passive, scheme has been proposed based on the use of MW cavities with embedded magnetic materials. Such a MW resonator, designed to be described by the $n_\phi = 1/4$ Harper–Hofstadter model, has been constructed and the (topological) energy bands for photons measured.[125]

Topological Optical Resonators. Approaches similar to those of circuit QED have been pursued for two-level systems that operate in the optical domain. Much progress has been made in understanding how Chern bands can be generated for photonic structures operating in the optical frequency domain,[126] and there has been recent progress in various experimental settings.[107,127,128] Typically, such systems preserve time-reversal symmetry, for example, providing bands with opposite Chern numbers for the two polarizations of the light field. They also typically operate in regimes where interactions between photons are weak. However, strong interactions could, in principle, be introduced, for example, by embedding two-level atoms.[129]

3.3. *Continuum polaritons*

One way to ensure that there are strong interactions between photons is to embed them in a medium in which there is strong coupling such that the appropriate excitation is a polariton: an excitation that has both photon and matter components. Such polaritons can be formed in electronic materials, in which the matter component is an exciton, or in cold atomic gases, in which the matter component is an electronic excitation. The matter components of these polaritons can interact strongly. In both cases, the formation of a coherent polaritonic quasi-particle with long lifetime requires the photon to be held in an optical cavity, reducing its rate of loss.

Analogous to the effect of rotation on an atomic gas, one can engineer an artificial magnetic field for polaritons by rotating the medium through which the light is propagating.[130] This leads to a Landau level spectrum, as in Sec. 2.1.1, and has been shown to allow the formation of a Laughlin state of the polaritons. By coupling to highly-lying "Rydberg" levels of the atoms, the interactions can be made to be very strong and of tunable blockade radius, leading to models that show competing crystalline phases.[55]

Another way in which one can impose rotation on the polariton is to keep the medium stationary and to cause the optical field to rotate, by the use of a cavity that has a twisted character, such that repeated round-trip passage of rays of light within the cavity cause a rotation of the point of intersection of the ray with the medium. This can also be usefully viewed within a Floquet framework, the periodicity set by the time interval between successive transits of the central plane of the cavity.[131] The formation of Landau levels for the cavity modes has been demonstrated in experiment.[132] Theory has shown that, in the presence of a medium in which the light forms polaritons, the associated strong interactions provide ways to prepare the Laughlin state, and to allow measurements of exchange statistics.[133]

4. Concluding Remarks

We have summarized a range of physical settings in which one expects there to appear fractional quantum Hall states of bosons. Although no such state has yet been realized in experiment, this may soon change given the rapid technical advances in the relevant research areas. For bosons within continuum Landau levels, a key result is that the Laughlin state is the exact ground state for the naturally occurring form of two-body contact interactions. Numerical studies indicate stable and robust Moore–Read and Read–Rezayi states, offering the prospect that these systems may allow experimental investigation of non-Abelian phases.

Many of the proposed physical realizations bring in physics that is uncommon in electronic systems. Cold atomic gases could allow for the study of the interplay of strong-pairing superconductivity and FQH physics. Achieving FQH states at high particle density in cold gases naturally leads to lattice-based models, and can allow for novel FQH states that exist only on lattices, notably in bands with Chern number $|\mathcal{C}| > 1$. FQH states of optical photons/polaritons typically involve particle loss, so operation requires pumping and dissipation,[134,135] also bringing novel features compared to the equilibrium situations typically considered for electronic materials.

Theoretical analyses have identified many novel experimental settings in which FQH states of bosons could appear. These theories have been guided by the existing understanding of FQH systems, which has been built on experimental discoveries in electronic systems. It is clear that much remains to be understood about strongly interacting quantum many body systems. The new physical settings for bosonic matter are sufficiently different from electronic systems that existing theories may be inadequate to understand all of their features, and the experimental explorations have the scope to uncover qualitatively new phenomena that were previously unexpected.

Acknowledgments

I have benefitted enormously from discussions and collaborations on these topics with numerous colleagues and co-workers, and I thank them all for these very stimulating interactions. I also gratefully acknowledge research support by EPSRC grant EP/P034616/1 and by a Simons Investigator Award.

References

1. D. C. Tsui, H. L. Stormer, and A. C. Gossard, Two-dimensional magnetotransport in the extreme quantum limit, *Phys. Rev. Lett.* **48**, 1559–1562 (1982).
2. E. M. Spanton, A. A. Zibrov, H. Zhou, T. Taniguchi, K. Watanabe, M. P. Zaletel, and A. F. Young, Observation of fractional Chern insulators in a van der Waals heterostructure, *Science.* **360**(6384), 62–66 (2018).
3. R. B. Laughlin, Anomalous quantum Hall effect: An incompressible quantum fluid with fractionally charged excitations, *Phys. Rev. Lett.* **50**, 1395–1398 (1983).
4. V. Kalmeyer and R. B. Laughlin, Equivalence of the resonating-valence-bond and fractional quantum Hall states, *Phys. Rev. Lett.* **59**, 2095–2098 (1987).
5. I. Bloch, J. Dalibard, and W. Zwerger, Many-body physics with ultracold gases, *Rev. Mod. Phys.* **80**, 885–964 (2008).
6. C. J. Pethick and H. Smith, *Bose–Einstein Condensation in Dilute Gases.* (Cambridge University Press, 2008).
7. M. Olshanii, Atomic scattering in the presence of an external confinement and a gas of impenetrable bosons, *Phys. Rev. Lett.* **81**, 938–941 (1998).
8. M. A. Baranov, M. Dalmonte, G. Pupillo, and P. Zoller, Condensed matter theory of dipolar quantum gases, *Chem. Rev.* **112**(9), 5012–5061 (2012).
9. A. L. Fetter, Rotating trapped Bose–Einstein condensates, *Rev. Mod. Phys.* **81**, 647–691 (2009).
10. N. R. Cooper, Rapidly rotating atomic gases, *Adv. Phys.* **57**(6), 539–616 (2008).
11. N. R. Cooper, J. Dalibard, and I. B. Spielman, Topological bands for ultracold atoms, *Rev. Mod. Phys.* **91**(1), 015005 (2019).
12. L. D. Landau and E. M. Lifshitz, *Statistical Physics Pt 1.* Vol. 5, (Butterworth Heinemann, Oxford, 1981).
13. N. K. Wilkin, J. M. F. Gunn, and R. A. Smith, Do attractive bosons condense?, *Phys. Rev. Lett.* **80**, 2265–2268 (1998).
14. V. Schweikhard, I. Coddington, P. Engels, S. Tung, and E. A. Cornell, Vortex-lattice dynamics in rotating spinor Bose–Einstein condensates, *Phys. Rev. Lett.* **93**(21), 210403 (2004).
15. N. Gemelke, E. Sarajlic, and S. Chu, Rotating few-body atomic systems in the fractional quantum Hall regime, *arXiv e-prints.* art. arXiv:1007.2677 (2010).
16. S. Kuhr, Quantum-gas microscopes: A new tool for cold-atom quantum simulators, *Nat. Sci. Rev.* **3**(2), 170–172 (2016).
17. C. Cohen-Tannoudji, J. Dupont-Roc, and G. Grynberg, *Atom-Photon Interactions.* (Wiley, New York, 1992).
18. Y.-J. Lin, R. L. Compton, A. R. Perry, W. D. Phillips, J. V. Porto, and I. B. Spielman, Bose–Einstein condensate in a uniform light-induced vector potential, *Phys. Rev. Lett.* **102**(13):130401 (2009).
19. N. R. Cooper, Optical flux lattices for ultracold atomic gases, *Phys. Rev. Lett.* **106**(17), 175301 (2011).
20. D. J. Thouless, M. Kohmoto, M. P. Nightingale, and M. den Nijs, Quantized Hall conductance in a two-dimensional periodic potential, *Phys. Rev. Lett.* **49**(6), 405–408 (1982).

21. N. R. Cooper and R. Moessner, Designing topological bands in reciprocal space, *Phys. Rev. Lett.* **109**, 215302 (2012).

22. Y.-L. Wu, N. Regnault, and B. A. Bernevig, Bloch model wave functions and pseudopotentials for all fractional Chern insulators, *Phys. Rev. Lett.* **110**, 106802 (2013).

23. N. R. Cooper and J. Dalibard, Optical flux lattices for two-photon dressed states, *Europhys. Lett.* **95**(6), 66004 (2011).

24. N. R. Cooper and J. Dalibard, Reaching fractional quantum Hall states with optical flux lattices, *Phys. Rev. Lett.* **110**, 185301 (2013).

25. D. Jaksch and P. Zoller, Creation of effective magnetic fields in optical lattices: The Hofstadter butterfly for cold neutral atoms, *New J. Phys.* **5**, 56 (2003).

26. F. Gerbier and J. Dalibard, Gauge fields for ultracold atoms in optical superlattices, *New J. Phys.* **12**(3), 033007 (2010).

27. W. Sun, B.-Z. Wang, X.-T. Xu, C.-R. Yi, L. Zhang, Z. Wu, Y. Deng, X.-J. Liu, S. Chen, and J.-W. Pan, Highly controllable and robust 2d spin-orbit coupling for quantum gases, *Phys. Rev. Lett.* **121**, 150401 (2018).

28. A. Eckardt, Atomic quantum gases in periodically driven optical lattices, *Rev. Mod. Phys.* **89**, 011004 (2017).

29. G. Jotzu, M. Messer, R. Desbuquois, M. Lebrat, T. Uehlinger, D. Greif, and T. Esslinger, Experimental realization of the topological Haldane model with ultracold fermions, *Nature.* **515**(7526), 237–240 (2014).

30. T. Oka and H. Aoki, Photovoltaic Hall effect in graphene, *Phys. Rev. B.* **79**, 081406 (2009).

31. F. D. M. Haldane, Model for a quantum Hall effect without Landau levels: Condensed-matter realization of the "parity anomaly", *Phys. Rev. Lett.* **61**(18), 2015–2018 (1988).

32. P. G. Harper, The general motion of conduction electrons in a uniform magnetic field, with application to the diamagnetism of metals, *Proc. Phys. Soc. A.* **68**(10), 879 (1955).

33. D. R. Hofstadter, Energy levels and wave functions of Bloch electrons in rational and irrational magnetic fields, *Phys. Rev. B.* **14**(6), 2239–2249 (1976).

34. M. Aidelsburger, M. Atala, M. Lohse, J. T. Barreiro, B. Paredes, and I. Bloch, Realization of the Hofstadter Hamiltonian with ultracold atoms in optical lattices, *Phys. Rev. Lett.* **111**, 185301 (2013).

35. M. Aidelsburger, M. Lohse, C. Schweizer, M. Atala, J. T. Barreiro, S. Nascimbène, N. R. Cooper, I. Bloch, and N. Goldman, Measuring the Chern number of Hofstadter bands with ultracold bosonic atoms, *Nat. Phys.* **111**, 162–166 (2015).

36. H. Miyake, G. A. Siviloglou, C. J. Kennedy, W. C. Burton, and W. Ketterle, Realizing the Harper Hamiltonian with laser-assisted tunneling in optical lattices, *Phys. Rev. Lett.* **111**, 185302 (2013).

37. M. E. Tai, A. Lukin, M. Rispoli, R. Schittko, T. Menke, D. Borgnia, P. M. Preiss, F. Grusdt, A. M. Kaufman, and M. Greiner, Microscopy of the interacting Harper–Hofstadter model in the two-body limit, *Nature.* **546**, 519–523 (2017).

38. F. D. M. Haldane, Fractional quantization of the Hall effect: A hierarchy of incompressible quantum fluid states, *Phys. Rev. Lett.* **51**, 605–608 (1983).

39. N. R. Cooper, S. Komineas, and N. Read, Vortex lattices in the lowest Landau level for confined Bose–Einstein condensates, *Phys. Rev. A.* **70**, 033604 (2004).

40. G. Baym, Vortex lattices in rapidly rotating Bose–Einstein condensates: Modes and correlation functions, *Phys. Rev. A.* **69**, 043618 (2004).

41. S. Moroz, C. Hoyos, C. Benzoni, and D. T. Son, Effective field theory of a vortex lattice in a bosonic superfluid, *SciPost Phys.* **5**, 39 (2018).

42. N. R. Cooper, N. K. Wilkin, and J. M. F. Gunn, Quantum phases of vortices in rotating Bose–Einstein condensates, *Phys. Rev. Lett.* **87**, 120405 (2001).
43. E. H. Lieb, R. Seiringer, and J. Yngvason, Yrast line of a rapidly rotating Bose gas: Gross–Pitaevskii regime, *Phys. Rev. A.* **79**, 063626 (2009).
44. N. R. Cooper and E. H. Rezayi, Competing compressible and incompressible phases in rotating atomic Bose gases at filling factor $\nu = 2$, *Phys. Rev. A.* **75**, 013627 (2007).
45. N. R. Cooper and N. K. Wilkin, Composite fermion description of rotating Bose–Einstein condensates, *Phys. Rev. B.* **60**, R16279–R16282 (1999).
46. J. K. Jain, *Composite fermions.* (Cambridge University Press, 2007).
47. N. Regnault and T. Jolicoeur, Quantum Hall fractions in rotating Bose–Einstein condensates, *Phys. Rev. Lett.* **91**, 030402 (2003).
48. B. I. Halperin, P. A. Lee, and N. Read, Theory of the half-filled Landau level, *Phys. Rev. B.* **47**, 7312–7343 (1993).
49. G. Moore and N. Read, Non-Abelions in the fractional quantum Hall effect, *Nucl. Phys. B.* **360**, 362–396 (1991).
50. N. Read and E. Rezayi, Beyond paired quantum Hall states: Parafermions and incompressible states in the first excited Landau level, *Phys. Rev. B.* **59**, 8084–8092 (1999).
51. A. Cappelli, L. S. Georgiev, and I. T. Todorov, Parafermion Hall states from coset projections of Abelian conformal theories, *Nucl. Phys. B.* **599**(3), 499–530 (2001).
52. C. Nayak, S. H. Simon, A. Stern, M. Freedman, and S. D. Sarma, Non-Abelian anyons and topological quantum computation, *Rev. Mod. Phys.* **80**(3):1083 (2008).
53. E. H. Rezayi, N. Read, and N. R. Cooper, Incompressible liquid state of rapidly rotating bosons at filling factor 3/2, *Phys. Rev. Lett.* **95**, 160404 (2005).
54. N. R. Cooper, E. H. Rezayi, and S. H. Simon, Vortex lattices in rotating atomic Bose gases with dipolar interactions, *Phys. Rev. Lett.* **95**, 200402 (2005).
55. F. Grusdt and M. Fleischhauer, Fractional quantum Hall physics with ultracold Rydberg gases in artificial gauge fields, *Phys. Rev. A.* **87**, 043628 (2013).
56. G. Möller and N. R. Cooper, Density waves and supersolidity in rapidly rotating atomic Fermi gases, *Phys. Rev. Lett.* **99**, 190409 (2007).
57. K. Yang and H. Zhai, Quantum Hall transition near a fermion Feshbach resonance in a rotating trap, *Phys. Rev. Lett.* **100**, 030404 (2008).
58. C. Repellin, T. Yefsah, and A. Sterdyniak, Creating a bosonic fractional quantum Hall state by pairing fermions, *Phys. Rev. B.* **96**, 161111 (2017).
59. Z. Tešanović, F. M. C. Axel, and B. I. Halperin, "Hall crystal" versus Wigner crystal, *Phys. Rev. B.* **39**, 8525–8551 (1989).
60. A. Kol and N. Read, Fractional quantum Hall effect in a periodic potential, *Phys. Rev. B.* **48**, 8890–8898 (1993).
61. S. A. Parameswaran, R. Roy, and S. L. Sondhi, Fractional quantum Hall physics in topological flat bands, *Comptes Rendus Physique.* **14**(9), 816–839 (2013).
62. E. J. Bergholtz and Z. Liu, Toplogical flat band models and fractional Chern insulators, *Int. J. Mod. Phys. B.* **27**(24), 1330017 (2013).
63. T. Neupert, C. Chamon, T. Iadecola, L. H. Santos, and C. Mudry, Fractional (Chern and topological) insulators, *Physica Scripta.* **T164**, 014005 (2015).
64. A. S. Sørensen, E. Demler, and M. D. Lukin, Fractional quantum Hall states of atoms in optical lattices, *Phys. Rev. Lett.* **94**(8):086803 (2005).
65. M. Hafezi, A. S. Sørensen, E. Demler, and M. D. Lukin, Fractional quantum Hall effect in optical lattices, *Phys. Rev. A.* **76**(2):023613 (2007).
66. G. Möller and N. R. Cooper, Composite fermion theory for bosonic quantum Hall states on lattices, *Phys. Rev. Lett.* **103**, 105303 (2009).

67. Y.-C. He, F. Grusdt, A. Kaufman, M. Greiner, and A. Vishwanath, Realizing and adiabatically preparing bosonic integer and fractional quantum Hall states in optical lattices, *Phys. Rev. B.* **96**, 201103 (2017).

68. D. Bauer, T. S. Jackson, and R. Roy, Quantum geometry and stability of the fractional quantum Hall effect in the Hofstadter model, *Phys. Rev. B.* **93**, 235133 (2016).

69. A. Sterdyniak, B. A. Bernevig, N. R. Cooper, and N. Regnault, Interacting bosons in topological optical flux lattices, *Phys. Rev. B.* **91**, 035115 (2015).

70. E. Kapit and E. Mueller, Exact parent Hamiltonian for the quantum Hall states in a lattice, *Phys. Rev. Lett.* **105**, 215303 (2010).

71. I. Glasser, J. I. Cirac, G. Sierra, and A. E. B. Nielsen, Lattice effects on Laughlin wave functions and parent Hamiltonians, *Phys. Rev. B.* **94**, 245104 (2016).

72. G. Möller and N. R. Cooper, Fractional Chern insulators in Harper–Hofstadter bands with higher Chern number, *Phys. Rev. Lett.* **115**, 126401 (2015).

73. G. H. Wannier, A result not dependent on rationality for Bloch electrons in a magnetic field, *Physica Status Solidi B.* **88**(2), 757–765 (1978).

74. P. Středa, Quantised Hall effect in a two-dimensional periodic potential, *J. Phys. C: Solid State Physics.* **15**(36), L1299–L1303 (1982).

75. T. Senthil, Symmetry-protected topological phases of quantum matter, *Ann. Rev. Conden. Matt. Phys.* **6**(1), 299–324 (2015).

76. T. Senthil and M. Levin, Integer quantum Hall effect for bosons, *Phys. Rev. Lett.* **110**, 046801 (2013).

77. A. Sterdyniak, N. R. Cooper, and N. Regnault, Bosonic integer quantum Hall effect in optical flux lattices, *Phys. Rev. Lett.* **115**, 116802 (2015).

78. B. I. Halperin, Theory of the quantized Hall resistance, *Helv. Phys. Acta.* **56**, 75 (1983).

79. M. Barkeshli and X.-L. Qi, Topological nematic states and non-Abelian lattice dislocations, *Phys. Rev. X.* **2**, 031013 (2012).

80. M. J. H. Ku, A. T. Sommer, L. W. Cheuk, and M. W. Zwierlein, Revealing the superfluid lambda transition in the universal thermodynamics of a unitary Fermi gas, *Science.* **335**(6068), 563–567 (2012).

81. N. R. Cooper, F. J. M. van Lankvelt, J. W. Reijnders, and K. Schoutens, Quantum Hall states of atomic Bose gases: Density profiles in single-layer and multilayer geometries, *Phys. Rev. A.* **72**, 063622 (2005).

82. N. Read and N. R. Cooper, Free expansion of lowest-Landau-level states of trapped atoms: A wave-function microscope, *Phys. Rev. A.* **68**, 035601 (2003).

83. D. Husmann, M. Lebrat, S. Häusler, J.-P. Brantut, L. Corman, and T. Esslinger, Breakdown of the Wiedemann–Franz law in a unitary Fermi gas, *Proc. Nat. Acad. Sci.* **115**(34), 8563–8568 (2018).

84. M. Aidelsburger, M. Lohse, C. Schweizer, M. Atala, J. T. Barreiro, S. Nascimbène, N. R. Cooper, I. Bloch, and N. Goldman, Measuring the Chern number of Hofstadter bands with ultracold bosonic atoms, *Nat. Phys.* **11**, 162 (2015).

85. D. T. Tran, A. Dauphin, A. G. Grushin, P. Zoller, and N. Goldman, Probing topology by 'heating': Quantized circular dichroism in ultracold atoms, *Sci. Adv.* **3**(8), e1701207 (2017).

86. L. Asteria, D. T. Tran, T. Ozawa, M. Tarnowski, B. S. Rem, N. Fläschner, K. Sengstock, N. Goldman, and C. Weitenberg, Measuring quantized circular dichroism in ultracold topological matter, *Nat. Phys.* **15**, 449 (2019).

87. J. Stenger, S. Inouye, A. P. Chikkatur, D. M. Stamper-Kurn, D. E. Pritchard, and W. Ketterle, Bragg spectroscopy of a Bose–Einstein condensate, *Phys. Rev. Lett.* **82**, 4569–4573 (1999).

88. J. Steinhauer, R. Ozeri, N. Katz, and N. Davidson, Excitation spectrum of a Bose–Einstein condensate, *Phys. Rev. Lett.* **88**, 120407 (2002).
89. S. Hoinka, M. Lingham, K. Fenech, H. Hu, C. J. Vale, J. E. Drut, and S. Gandolfi, Precise determination of the structure factor and contact in a unitary Fermi gas, *Phys. Rev. Lett.* **110**, 055305 (2013).
90. S. Stringari, Collective excitations of a trapped Bose-condensed gas, *Phys. Rev. Lett.* **77**, 2360–2363 (1996).
91. M. A. Cazalilla, N. Barberán, and N. R. Cooper, Edge excitations and topological order in a rotating Bose gas, *Phys. Rev. B.* **71**, 121303 (2005).
92. N. R. Cooper and S. H. Simon, Signatures of fractional exclusion statistics in the spectroscopy of quantum Hall droplets, *Phys. Rev. Lett.* **114**, 106802 (2015).
93. B. Paredes, P. Fedichev, J. I. Cirac, and P. Zoller, 1/2-anyons in small atomic Bose–Einstein condensates, *Phys. Rev. Lett.* **87**, 010402 (2001).
94. E. Kapit, P. Ginsparg, and E. Mueller, Non-Abelian braiding of lattice bosons, *Phys. Rev. Lett.* **108**, 066802 (2012).
95. Y. Zhang, G. J. Sreejith, N. D. Gemelke, and J. K. Jain, Fractional angular momentum in cold-atom systems, *Phys. Rev. Lett.* **113**, 160404 (2014).
96. Y. Zhang, G. J. Sreejith, and J. K. Jain, Creating and manipulating non-Abelian anyons in cold atom systems using auxiliary bosons, *Phys. Rev. B.* **92**, 075116 (2015).
97. D. Lundholm and N. Rougerie, Emergence of fractional statistics for tracer particles in a Laughlin liquid, *Phys. Rev. Lett.* **116**, 170401 (2016).
98. F. Grusdt, N. Y. Yao, D. Abanin, M. Fleischhauer, and E. Demler, Interferometric measurements of many-body topological invariants using mobile impurities, *Nat. Commun.* **7**, 11994 (2016).
99. S. K. Baur, K. R. A. Hazzard, and E. J. Mueller, Stirring trapped atoms into fractional quantum Hall puddles, *Phys. Rev. A.* **78**, 061608 (2008).
100. J. Motruk and F. Pollmann, Phase transitions and adiabatic preparation of a fractional Chern insulator in a boson cold-atom model, *Phys. Rev. B.* **96**, 165107 (2017).
101. M. Popp, B. Paredes, and J. I. Cirac, Adiabatic path to fractional quantum Hall states of a few bosonic atoms, *Phys. Rev. A.* **70**, 053612 (2004).
102. J. Zhang, J. Beugnon, and S. Nascimbene, Creating fractional quantum Hall states with atomic clusters using light-assisted insertion of angular momentum, *Phys. Rev. A.* **94**, 043610 (2016).
103. F. Grusdt, F. Letscher, M. Hafezi, and M. Fleischhauer, Topological growing of Laughlin states in synthetic gauge fields, *Phys. Rev. Lett.* **113**, 155301 (2014).
104. M. Roncaglia, M. Rizzi, and J. I. Cirac, Pfaffian state generation by strong three-body dissipation, *Phys. Rev. Lett.* **104**, 096803 (2010).
105. M. Greiner, O. Mandel, T. Esslinger, T. W. Hänsch, and I. Bloch, Quantum phase transition from a superfluid to a mott insulator in a gas of ultracold atoms, *Nature.* **415**(6867), 39 (2002).
106. M. Aidelsburger, S. Nascimbene, and N. Goldman, Artificial gauge fields in materials and engineered systems, *Comptes Rendus Physique.* **19**(6), 394–432 (2018).
107. T. Ozawa, H. M. Price, A. Amo, N. Goldman, M. Hafezi, L. Lu, M. C. Rechtsman, D. Schuster, J. Simon, O. Zilberberg, and I. Carusotto, Topological photonics, *Rev. Mod. Phys.* **91**, 015006 (2019).
108. L. Capriotti, A. E. Trumper, and S. Sorella, Long-range Néel order in the triangular Heisenberg model, *Phys. Rev. Lett.* **82**, 3899–3902 (1999).
109. S. R. White and A. L. Chernyshev, Néel order in square and triangular lattice Heisenberg models, *Phys. Rev. Lett.* **99**, 127004 (2007).

110. Y.-C. He, D. N. Sheng, and Y. Chen, Chiral spin liquid in a frustrated anisotropic Kagome Heisenberg model, *Phys. Rev. Lett.* **112**, 137202 (2014).

111. B. Bauer, L. Cincio, B. P. Keller, M. Dolfi, G. Vidal, S. Trebst, and A. W. W. Ludwig, Chiral spin liquid and emergent anyons in a Kagome lattice Mott insulator, *Nat. Commun.* **5**, 5137 (2014).

112. G. Misguich, T. Jolicoeur, and S. M. Girvin, Magnetization plateaus of $SrCu_2(BO_3)_2$ from a Chern–Simons theory, *Phys. Rev. Lett.* **87**, 097203 (2001).

113. K. Kumar, K. Sun, and E. Fradkin, Chern–Simons theory of magnetization plateaus of the spin-$\frac{1}{2}$ quantum XXZ Heisenberg model on the Kagome lattice, *Phys. Rev. B.* **90**, 174409 (2014).

114. L. Savary and L. Balents, Quantum spin liquids: A review, *Rep. Prog. Phys.* **80**(1), 016502 (2017).

115. A. Kitaev, Anyons in an exactly solved model and beyond, *Ann. Phys.* **321**(1), 2–111 (2006).

116. Y. Kasahara, K. Sugii, T. Ohnishi, M. Shimozawa, M. Yamashita, N. Kurita, H. Tanaka, J. Nasu, Y. Motome, T. Shibauchi, and Y. Matsuda, Unusual thermal Hall effect in a Kitaev spin liquid candidate α-$RuCl_3$, *Phys. Rev. Lett.* **120**, 217205 (2018).

117. Y. Kasahara, T. Ohnishi, Y. Mizukami, O. Tanaka, S. Ma, K. Sugii, N. Kurita, H. Tanaka, J. Nasu, Y. Motome, T. Shibauchi, and Y. Matsuda, Majorana quantization and half-integer thermal quantum Hall effect in a Kitaev spin liquid, *Nature.* **559**(7713), 227–231 (2018).

118. H. Levine, A. Keesling, A. Omran, H. Bernien, S. Schwartz, A. S. Zibrov, M. Endres, M. Greiner, V. Vuletić, and M. D. Lukin, High-fidelity control and entanglement of Rydberg-atom qubits, *Phys. Rev. Lett.* **121**, 123603 (2018).

119. S. de Léséleuc, V. Lienhard, P. Scholl, D. Barredo, S. Weber, N. Lang, H. P. Büchler, T. Lahaye, and A. Browaeys, Observation of a symmetry-protected topological phase of interacting bosons with Rydberg atoms, *Science* **365**, 775 (2019).

120. N. Y. Yao, A. V. Gorshkov, C. R. Laumann, A. M. Läuchli, J. Ye, and M. D. Lukin, Realizing fractional Chern insulators in dipolar spin systems, *Phys. Rev. Lett.* **110**, 185302 (2013).

121. M. F. Maghrebi, N. Y. Yao, M. Hafezi, T. Pohl, O. Firstenberg, and A. V. Gorshkov, Fractional quantum Hall states of Rydberg polaritons, *Phys. Rev. A.* **91**, 033838 (2015).

122. S. Schmidt and J. Koch, Circuit QED lattices: Towards quantum simulation with superconducting circuits, *Annalen der Physik.* **525**(6), 395–412 (2013).

123. P. Roushan, C. Neill, A. Megrant, Y. Chen, R. Babbush, R. Barends, B. Campbell, Z. Chen, B. Chiaro, A. Dunsworth, A. Fowler, E. Jeffrey, J. Kelly, E. Lucero, J. Mutus, P. J. J. O'Malley, M. Neeley, C. Quintana, D. Sank, A. Vainsencher, J. Wenner, T. White, E. Kapit, H. Neven, and J. Martinis, Chiral ground-state currents of interacting photons in a synthetic magnetic field, *Nat. Phys.* **13**, 146 (2016).

124. J. Koch, A. A. Houck, K. L. Hur, and S. M. Girvin, Time-reversal-symmetry breaking in circuit-QED-based photon lattices, *Phys. Rev. A.* **82**, 043811 (2010).

125. C. Owens, A. LaChapelle, B. Saxberg, B. M. Anderson, R. Ma, J. Simon, and D. I. Schuster, Quarter-flux Hofstadter lattice in a qubit-compatible microwave cavity array, *Phys. Rev. A.* **97**(1), 013818 (2018).

126. S. Raghu and F. D. M. Haldane, Analogs of quantum-Hall-effect edge states in photonic crystals, *Phys. Rev. A.* **78**, 033834 (2008).

127. L. Lu, J. D. Joannopoulos, and M. Soljačić, Topological photonics, *Nat. Phot.* **8**, 821 (2014).

128. S. Mittal, S. Ganeshan, J. Fan, A. Vaezi, and M. Hafezi, Measurement of topological invariants in a 2D photonic system, *Nat. Phot.* **10**, 180 (2016).

129. J. Cho, D. G. Angelakis, and S. Bose, Fractional quantum Hall state in coupled cavities, *Phys. Rev. Lett.* **101**, 246809 (2008).

130. J. Otterbach, J. Ruseckas, R. G. Unanyan, G. Juzeliūnas, and M. Fleischhauer, Effective magnetic fields for stationary light, *Phys. Rev. Lett.* **104**, 033903 (2010).

131. A. Sommer and J. Simon, Engineering photonic Floquet Hamiltonians through Fabry–Pérot resonators, *New J. Phys.* **18**(3), 035008 (2016).

132. N. Schine, A. Ryou, A. Gromov, A. Sommer, and J. Simon, Synthetic Landau levels for photons, *Nature.* **534**, 671 (2016).

133. S. Dutta and E. J. Mueller, Coherent generation of photonic fractional quantum Hall states in a cavity and the search for anyonic quasiparticles, *Phys. Rev. A.* **97**, 033825 (2018).

134. R. O. Umucalilar and I. Carusotto, Fractional quantum Hall states of photons in an array of dissipative coupled cavities, *Phys. Rev. Lett.* **108**, 206809 (2012).

135. I. Carusotto and C. Ciuti, Quantum fluids of light, *Rev. Mod. Phys.* **85**, 299–366 (2013).

Index

www.ingramcontent.com/pod-product-compliance
Lightning Source LLC
Chambersburg PA
CBHW081217220326
41598CB00037B/6802